ALGEBRA AND NUMBER THEORY

ALGEBRA AND NUMBER THEORY

An Integrated Approach

MARTYN R. DIXON
Department of Mathematics
University of Alabama

LEONID A. KURDACHENKO
Department of Algebra
School of Mathematics and Mechanics
National University of Dnepropetrovsk

IGOR YA. SUBBOTIN
Department of Mathematics and Natural Sciences
National University

WILEY

A JOHN WILEY & SONS, INC., PUBLICATION

Published by John Wiley & Sons, Inc., Hoboken, New Jersey.
Published simultaneously in Canada.

For general information on our other products and services or for technical support, please contact our Customer Care Department within the United States at (800) 762-2974, outside the United States at (317) 572-3993 or fax (317) 572-4002.

Wiley also publishes its books in a variety of electronic formats. Some Content that appears in print may not be available in electronic formats. For more information about Wiley products, visit our web site at www.wiley.com.

Library of Congress Cataloging-in-Publication Data:

Dixon, Martyn R. (Martyn Russell), 1955-
Algebra and number theory : an integrated approach / Martyn R. Dixon, Leonid A. Kurdachenko, Igor Ya. Subbotin.
p. cm.
Includes index.
ISBN 978-0-470-49636-7 (cloth)
1. Number theory. 2. Algebra. I. Kurdachenko, L. II. Subbotin, Igor Ya., 1950- III. Title.
QA241.D59 2010
512–dc22

2010003428

Printed in Singapore

10 9 8 7 6 5 4 3 2 1

CONTENTS

PREFACE

Algebra and number theory are two powerful, established branches of modern mathematics at the forefront of current mathematical research which are playing an increasingly significant role in different branches of mathematics (for instance, in geometry, topology, differential equations, mathematical physics, and others) and in many relatively new applications of mathematics such as computing, communications, and cryptography. Algebra also plays a role in many applications of mathematics in diverse areas such as modern physics, crystallography, quantum mechanics, space sciences, and economic sciences.

Preface Historically, algebra and number theory have developed together, enriching each other in the process, and this often makes it difficult to draw a precise boundary separating these subjects. It is perhaps appropriate to say that they actually form one common subject: algebra and number theory. Thus, results in number theory are the basis and "a type of sandbox" for algebraic ideas and, in turn, algebraic tools contribute tremendously to number theory. It is interesting to note that newly developed branches of mathematics such as coding theory heavily use ideas and results from both linear algebra and number theory.

There are three mandatory courses, linear algebra, abstract algebra, and number theory, in all university mathematics programs that every student of mathematics should take. Increasingly, it is also becoming evident that students of computer science and other such disciplines also need a strong background in these three areas. Most of the time, these three disciplines are the subject of different and separate lecture courses that use different books dedicated to each subject individually. In a curriculum that is increasingly stretched by the need to offer traditional favorites, while introducing new applications, we think that it is desirable to introduce a fresh approach to the way these three specific courses are taught. On the basis of the authors' experience, we think that one course, integrating these three disciplines, together with a corresponding book for this integrated course, would be helpful in using class time more efficiently. As an argument supporting this statement, we mention that many theorems in number theory have very simple proofs using algebraic tools. Most importantly, we think the integrated approach will help build a deeper understanding of the subject in the students, as well as improve their retention of knowledge. In this respect, the time-honored European experience of integrated algebra and number theory courses, organically implemented in the university curriculum, would be very efficient.

We have several goals in mind with the writing of this book. One of the most important reasons for writing this book is to give a systematic, integrated, and complete description of the theory of the main number systems that form a basis for the structures that play a central role in various branches of mathematics. Another goal in writing this book was to develop an introductory undergraduate course in number theory and algebra as an integrated discipline. We wanted to write a book that would be appropriate for typical students in computer science or mathematics who possess a certain degree of general mathematical knowledge pertaining to typical students at this stage. Even though it is mathematically quite self-contained, the text will presuppose that the reader is comfortable with mathematical formalism and also has some experience in reading and writing mathematical proofs.

The book consists of 10 chapters. We start our exposition with the elements of set theory (Chapter 1). The next chapter is dedicated to matrices and determinants. This chapter, together with Chapters 4, 5, and 6 covers the main material pertaining to linear algebra. We placed some elements of field theory in Chapter 3 which are needed to describe some of the essential elements of linear algebra (such as vector spaces and bilinear forms) not only over number fields, but over finite fields as well. Chapters 3, 6, 7, and 8 develop the main ideas of algebraic structures, while the final part consisting of Chapters 9 and 10 demonstrates the applications of algebraic ideas to number theory (e.g., Section 9.4). Chapter 10 is dedicated to the development of the rigorous construction of the real number system and its main subsystems. Since the theme of numbers is so very important and plays a key role in the education of prospective mathematicians and mathematics teachers, we have tried to complete this work with all the required but bulky details. We consider this chapter as an important appendix to the main content of the book, and having its own major overlook value.

We would like to extend our sincere appreciation to The University of Alabama (Tuscaloosa, USA), National Dnepropetrovsk University (Ukraine), and National University (California, USA) for their great support of the authors' work.

The authors would also like to thank their wives, Murrie, Tamara, and Milla for all their love and much-needed support while this work was in progress. An endeavor such as this is made lighter by the joy that they bring. The authors would also like to thank their children, Rhiannon, Elena, Daniel, Tat'yana, Victoria, Igor, Janice, and Nicole for providing the necessary distractions when the workday ended.

Finally, we would like to dedicate this book to the memory of two great algebraists. The first of these is Z. I. Borevich who made extremely important contributions in many branches of algebra and number theory. The second is one of the founders of infinite group theory, S. N. Chernikov, who was a teacher and mentor of two of the authors of this book and whose influence has spread far and wide in the world of group theory.

MARTYN R. DIXON
LEONID A. KURDACHENKO
IGOR YA. SUBBOTIN

CHAPTER 1

SETS

1.1 OPERATIONS ON SETS

The concept of a set is one of the very basic concepts of mathematics. In fact, the notion of a set is so fundamental that it is difficult or impossible to use some other more basic notion, which could substitute for it, that is not already synonymous with it, such as family, class, system, collection, assembly, and so on.

Set theory can be used as a faultless material for constructing the rudiments of mathematics only if it is presented axiomatically, a statement that is also true for many other well-developed mathematical theories. This approach to set theory requires a significant amount of time and a relatively high level of audience preparedness. Even though, in this text, we aim to introduce algebraic concepts and facts supporting them using relatively transparent set theoretical and logical justifications, we shall not pursue a really high level of rigor. This is not our goal, and is not realistically possible given that this is a textbook meant primarily for undergraduates. Consequently we shall use a somewhat traditional approach using only what is sometimes termed *naive* set theory.

One of the main founders of set theory was George Cantor (1845–1918), a German mathematician, born in Russia. Set theory is now a ubiquitous part of mathematics, and can be used as a foundation from which much of mathematics can be derived.

A set is a collection of distinct objects which are usually called elements. A set is considered known if a rule is given that allows us to establish whether a particular object is an element of the set or not. The relation of belonging is denoted by the symbol $\in$. So the fact that an element a belongs to a set A will be denoted by $a \in A$. If an object (element) b does not belong to A, we will write $b \notin A$. It is fundamentally important to note that for any object a and for any set A only one of two possible cases can arise, either $a \in A$ or $a \notin A$. This is nothing more than the set theoretical expression of a fundamental law of logic, namely the law of the excluded middle.

If a set A is finite, then A can be defined by *listing all of its elements*. We usually denote such a listing by enclosing the elements of the set within the bracers { and }. Thus the finite set A could be written as follows:

$$A = \{a_1, a_2, \ldots, a_n\}.$$

It is extremely important to understand the notational difference between a and $\{a\}$. The former is usually regarded as denoting an element of a set, whereas the latter indicates the singleton set containing the element a. If A is a set and a is an element of A then we write $a \in A$ to denote that a is an element of A. When $a \in A$ then it is also correct to write $\{a\} \subseteq A$ and conversely if $\{a\} \subseteq A$ then $a \in A$. However, in general, $\{a\} \subseteq A$ does not usually imply that $\{a\} \in A$; it is not usually correct to write $a \in A$ and also $a \subseteq A$, nor are $\{a\} \subseteq A$ and $\{a\} \in A$ usually both correct.

Another way of defining a set is by *assigning a certain property that uniquely characterizes the elements and unifies them in this set*. We can denote such a set A by

$$A = \{x \mid P(x)\},$$

where $P(x)$ denotes this defining property. For example, the set of all real numbers belonging to the segment [2, 5] can be written as $\{x \mid x \in \mathbb{R} \ \& \ 2 \leq x \leq 5\}$. Here $\mathbb{R}$ denotes the set of real numbers. This method of defining a set corresponds to the so-called Cantor Principle of Abstraction that Cantor used as the basis for the definition of a set. According to Cantor, if some property P is given, then one can build a new object—the set of all objects having this property P. The idea of moving from a property P to forming the set of all elements that have this property P is the main essence of Cantor's Principle of Abstraction. For example, the finite set $\{1, 2, 3\}$ could also be defined as $\{x \mid x = 1, \text{ or } x = 2 \text{ or } x = 3\}$.

For some important sets we will use the following conventional notation; we have already seen that $\mathbb{R}$ denotes the set of real numbers, but list this again here.

$\mathbb{N}$ *is the set of all natural numbers;*

$\mathbb{Z}$ *is the set of all integers;*

$\mathbb{Q}$ *is the set of all rational numbers;*

$\mathbb{R}$ *is the set of all real numbers;*

$\mathbb{C}$ *is the set of all complex numbers.*

Note that the *number* 0 *by common agreement is not a natural number*. Therefore, for the set consisting of all natural numbers and the number 0 we will use the notation $\mathbb{N}_0$. This set is the set of (so-called) *whole numbers*.

We shall now introduce various concepts that allow us to talk about sets in a meaningful way.

1.1.1. Definition. *Two sets A and B are called equal, if every element of A is an element of B and conversely, every element of B is an element of A.*

This statement is one of the first set axioms. Together with the principle of abstraction it points out a significant difference between the notions of a property and a set. Indeed, the same set can be determined by different properties. Often, establishing equality between sets determined by different properties leads us to some rather sophisticated mathematical theorems.

A very special important set—*the empty set*—arises naturally, at the outset, in the consideration of sets.

1.1.2. Definition. *A set is said to be empty, if it has no elements.*

Definition 1.1.1 shows that the empty set is unique and we denote it by the symbol $\emptyset$. One can obtain the empty set with the aid of a contradictory property. For example, $\emptyset = \{x \mid x \neq x\}$.

1.1.3. Definition. *A set A is a subset of a set B if every element of A is an element of B. We will denote this by* $A \subseteq B$.

Note that the sets *A and B are equal if and only if* $A \subseteq B$ *and* $B \subseteq A$. From this definition we see that the empty set is a subset of every set. Furthermore, every set A is a subset of itself, so that $A \subseteq A$.

1.1.4. Definition. *A subset A of a set B is called a proper subset of B, if A is a subset of B and* $A \neq B$.

1.1.5. Definition. *Let A be a set. Then the Boolean of the set A, denoted by* $\mathfrak{B}(A)$, *is the set* $\mathfrak{B}(A) = \{X \mid X \subseteq A\}$. *Thus* $\mathfrak{B}(A)$ *denotes the set of all subsets of A.*

We now introduce some operations on sets. Principal among these are the notions of intersection and union.

1.1.6. Definition. *Let A and B be sets. The set* $A \cap B$, *called the intersection of A and B, is the set of elements which belong to both the set A and to the set B. Thus*

$$A \cap B = \big\{x \mid x \in A \text{ and } x \in B\big\}.$$

1.1.7. Definition. *Let A and B be sets. Then the set $A \cup B$, called the union of A and B, is the set of elements which belong to the set A or to the set B. Thus*

$$A \cup B = \{x \mid x \in A \text{ or } x \in B\}.$$

We note that the word "or" used here is used in an inclusive sense.

1.1.8. Definition. *Let A and B be sets. Then the set $A \backslash B$, called the difference of A and B, is the set of elements which belong to the set A but not to the set B. Thus*

$$A \backslash B = \{x \mid x \in A \text{ and } x \notin B\}.$$

If $B \subseteq A$, then $A \backslash B$ is called the complement of the set B in the set A and is often denoted by B^c when A is understood.

1.1.9. Definition. *Let A and B be sets. Then the set $A \triangle B$, called the symmetric difference of A and B, is the set of elements which belong to the set $A \cup B$ but not to the set $A \cap B$. Thus*

$$A \triangle B = (A \cup B) \setminus (A \cap B) = (A \backslash B) \cup (B \backslash A).$$

1.1.10. Theorem. *Let A, B, and C be sets.*

(i) *$A \subseteq B$ if and only if $A \cap B = A$ or $A \cup B = B$.*
In particular, $A \cup A = A = A \cap A$ (the idempotency of intersection and union).

(ii) *$A \cap B = B \cap A$ and $A \cup B = B \cup A$ (the commutative property of intersection and union).*

(iii) *$A \cap (B \cap C) = (A \cap B) \cap C$ and $A \cup (B \cup C) = (A \cup B) \cup C$ (the associative property of intersection and union).*

(iv) *$A \cap (B \cup C) = (A \cap B) \cup (A \cap C)$ and $A \cup (B \cap C) = (A \cup B) \cap (A \cup C)$ (the distributive property).*

(v) *$A \backslash (A \backslash B) = A \cap B$.*

(vi) *$A \backslash (B \cap C) = (A \backslash B) \cup (A \backslash C)$.*

(vii) *$A \backslash (B \cup C) = (A \backslash B) \cap (A \backslash C)$.*

(viii) *$A \triangle B = B \triangle A$.*

(ix) *$A \triangle (B \triangle C) = (A \triangle B) \triangle C$.*

(x) *$A \triangle A = \emptyset$.*

Proof. The proofs of the majority of these assertions are easy to see from the definitions. For example, (viii) follows from the symmetry of the expression $A \triangle B$ in the form $(A \backslash B) \cup (B \backslash A)$. However, to give the reader an indication as to how the proofs of some of the assertions may be written, we give a proof of (iv).

Let $x \in A \cap (B \cup C)$. It follows from the definition that $x \in A$ and $x \in B \cup C$. Since $x \in B \cup C$ either $x \in B$ or $x \in C$ and hence either x is an element of both sets A and B, or x is an element of both sets A and C. Thus $x \in A \cap B$ or $x \in A \cap C$, which is to say that $x \in (A \cap B) \cup (A \cap C)$. This shows that $A \cap (B \cup C) \subseteq (A \cap B) \cup (A \cap C)$.

Conversely, let $x \in (A \cap B) \cup (A \cap C)$. Then $x \in A \cap B$ or $x \in A \cap C$. In each case $x \in A$ and x is an element of one of the sets B or C, that is $x \in A \cap (B \cup C)$. This shows that $(A \cap B) \cup (A \cap C) \subseteq A \cap (B \cup C)$. Note that an alternative method of proof here would be to use the fact that $B \subseteq B \cup C$ and so $A \cap B \subseteq A \cap (B \cup C)$. Likewise $A \cap C \subseteq A \cap (B \cup C)$ and hence $(A \cap B) \cup (A \cap C) \subseteq A \cap (B \cup C)$.

We also note that to prove the important assertion (ix) it is sufficient to show that

$$A \triangle (B \triangle C) = (A \cup B \cup C) \setminus [(A \cap B) \cup (A \cap C) \cup (B \cap C)] = (A \triangle B) \triangle C.$$

We can extend the notions of intersection and union to arbitrary families of sets. Let $\mathfrak{S}$ be a family of sets, so that the elements of $\mathfrak{S}$ are also sets.

1.1.11. Definition. *The intersection of the family $\mathfrak{S}$ is the set of elements which belong to each set S from the family $\mathfrak{S}$ and is denoted by $\cap\mathfrak{S} = \bigcap_{S \in \mathfrak{S}} S$. Thus*

$$\cap\mathfrak{S} = \bigcap_{S \in \mathfrak{S}} S = \{x \mid x \in S \text{ for each set } S \in \mathfrak{S}\}.$$

1.1.12. Definition. *The union of the family $\mathfrak{S}$ is the set of elements which belong to some set S from the family $\mathfrak{S}$ and is denoted by $\cup\mathfrak{S} = \bigcup_{S \in \mathfrak{S}} S$. Thus*

$$\cup\mathfrak{S} = \bigcup_{S \in \mathfrak{S}} S = \{x \mid x \in S \text{ for some set } S \in \mathfrak{S}\}.$$

Next we informally introduce a topic that will be familiar to most readers. Let A and B be sets. A pair of elements (a, b) where $a \in A, b \in B$, that are taken in the given order, is called *an ordered pair*. By definition, $(a, b) = (a_1, b_1)$ if and only if $a = a_1$ and $b = b_1$.

1.1.13. Definition. *Let A and B be sets. Then the set $A \times B$ of all ordered pairs (a, b) where $a \in A, b \in B$ is called the Cartesian product of the sets A and B. If $A = B$, then we call $A \times A$ the Cartesian square of the set A and write $A \times A$ as A^2.*

The real plane $\mathbb{R}^2$ is a natural example of a Cartesian product. The Cartesian product of two segments of the real number line could be interpreted geometrically as a rectangle whose sides are these segments. It is possible to extend the notion of a Cartesian product of two sets to the case of an arbitrary family of sets. First we indicate how to do this for a finite family of sets.

1.1.14. Definition. *Let n be a natural number and let $A_1, \dots, A_n$ be sets. Then the set*

$$A_1 \times \cdots \times A_n = \prod_{1 \le i \le n} A_i$$

of all ordered n-tuples $(a_1, \dots, a_n)$ where $a_j \in A_j$, for $1 \le j \le n$, is called the Cartesian product of the sets $A_1, \dots, A_n$.

Here $(a_1, \dots, a_n) = (b_1, \dots, b_n)$ if and only if $a_1 = b_1, \dots, a_n = b_n$.

The element a_j is called *the jth coordinate or jth component of* $(a_1, \dots, a_n)$.

If $A_1 = \cdots = A_n = A$ we will call $\underbrace{A \times A \times \cdots \times A}_{n}$ *the nth Cartesian power* A^n of the set A.

We shall use the convention that if A is a nonempty set then A^0 will denote a one-element set and we shall denote A^0 by $\{*\}$, where $*$ denotes the unique element of A^0.

Naturally, $A^1 = A$.

It is worth noting that the usual rules for the numerical operations of multiplications and power cannot be extended to the Cartesian product of sets. In particular, the commutative law is not valid in general, which is to say that in general $A \times B \neq B \times A$ if $A \neq B$. The same can also be said for the associative law: it is normally the case that $A \times (B \times C)$, $(A \times B) \times C$ and $A \times B \times C$ are distinct sets.

As mentioned, we can define the Cartesian product of an infinite family of sets. For our purpose it is enough to consider the Cartesian product of a family of sets indexed by the set $\mathbb{N}$. Let $\{A_n \mid n \in \mathbb{N}\}$ be a family of sets, indexed by the natural numbers. We consider infinite ordered tuples $(a_1, \dots, a_n, a_{n+1}, \dots) = (a_n)_{n \in \mathbb{N}}$, where $a_n \in A_n$, for each $n \in \mathbb{N}$. As above, $(a_n)_{n \in \mathbb{N}} = (b_n)_{n \in \mathbb{N}}$ if and only if $a_n = b_n$ for each $n \in \mathbb{N}$. Then the set

$$A_1 \times \cdots \times A_n \times A_{n+1} \times \cdots = \prod_{n \in \mathbb{N}} A_n$$

of all infinite ordered tuples $(a_n)_{n \in \mathbb{N}}$, where $a_n \in A_n$ for each $n \in \mathbb{N}$, is called the Cartesian product of the family of sets $\{A_n \mid n \in \mathbb{N}\}$.

EXERCISE SET 1.1

In each of the following questions explain your reasoning, by giving a proof of your assertion or by using appropriate examples.

1.1.1. Which of the following assertions are valid for all sets A, B, and C?

(i) If $A \notin B$ and $B \notin C$, then $A \notin C$.

(ii) If $A \notin B$ and $B \nsubseteq C$, then $A \notin C$.

(iii) If $A \subseteq B$, $A \neq B$ and $B \subseteq C$, then $C \nsubseteq A$.

(iv) If $A \subseteq B$, $A \neq B$ and $B \in C$, then $A \notin C$.

1.1.2. Give examples of sets A, B, C, D, E satisfying all of the following conditions: $A \subseteq B, A \neq B, B \in C, C \subseteq D, C \neq D, D \subseteq E, D \neq E$.

1.1.3. Give examples of sets A, B, C satisfying all of the following conditions: $A \in B, B \in C$, but $A \notin C$.

1.1.4. Let

$A = \{x \in \mathbb{Z} | x = 2y \text{ for some } y \geq 0\}$;

$B = \{x \in \mathbb{Z} | x = 2y - 1 \text{ for some } y \geq 0\}$;

$C = \{x \in \mathbb{Z} | \ x < 10\}$.

Find $\mathbb{Z} \backslash A$, $\mathbb{Z} \backslash (A \cap B)$, $\mathbb{Z} \backslash C$, $A \backslash (\mathbb{Z} \backslash C)$, $C \backslash (A \cup B)$.

1.1.5. Do there exist nonempty sets A, B, C such that $A \cap B \neq \emptyset$, $A \cap C = \emptyset$, $(A \cap B) \backslash C = \emptyset$?

1.1.6. Let A, B, C be arbitrary sets. Prove that the equation $(A \cap B) \cup C = A \cap (B \cup C)$ is equivalent to $C \subseteq A$.

1.1.7. Let $S_1, \ldots, S_n$ be sets satisfying the following condition: $S_j \subseteq S_{j+1}$ for all $1 \leq i \leq n-1$. Find $S_1 \cap S_2 \cap \cdots \cap S_n$ and $S_1 \cup S_2 \cup \cdots \cup S_n$.

1.1.8. Let $\mathfrak{S} = \{H_n | n \in \mathbb{N}\}$ be a family of sets such that $H_n \subseteq H_{n+1}$ for every $n \in \mathbb{N}$. Let $\mathfrak{R}$ be an infinite subset of $\mathfrak{S}$. Prove that $\cup \mathfrak{S} = \cup \mathfrak{R}$.

1.1.9. Let $\mathfrak{S} = \{H_n | n \in \mathbb{N}\}$ be a family of sets such that $H_n \supseteq H_{n+1}$ for every $n \in \mathbb{N}$. Let $\mathfrak{R}$ be an infinite subset of $\mathfrak{S}$. Prove that $\cap \mathfrak{S} = \cap \mathfrak{R}$.

1.1.10. Let S be the set of all roots of a polynomial $f(X)$. Suppose that $f(X) = g(X)h(X)$. Let S_1 (respectively S_2) be the set of all roots of the polynomial $g(X)$ (respectively $h(X)$). Prove that $S = S_1 \cup S_2$.

1.1.11. Let $g(X)$ and $h(X)$ be polynomials with real coefficients. Let S_1 (respectively S_2) be the set of all real roots of the polynomial $g(X)$ (respectively $h(X)$). Let S be the set of all real roots of the polynomial $f(X) = (g(X))^2 + (h(X))^2$. Prove that $S = S_1 \cap S_2$.

1.1.12. Let A, B, C be sets, suppose that $B \subseteq A$, and that $A \cap C = \emptyset$. Find the solutions X of the following system:

$$\begin{cases} A \backslash X = B \\ X \backslash A = C. \end{cases}$$

1.1.13. Let A, B, C be sets and suppose that $B \subseteq A \subseteq C$. Find the solutions X of the system

$$\begin{cases} A \cap X = B \\ X \cup A = C. \end{cases}$$

1.1.14. Prove that $(a, b) = \{\{a\}, \{a, b\}\} = \{\{c\}, \{c, d\}\} = (c, d)$ if and only if $a = c, b = d$.

1.1.15. Prove that $A \triangle B = C$ is equivalent to $B \triangle C = A$ and $C \triangle A = B$.

1.1.16. Prove that $A \cap (B \triangle C) = (A \cap B) \triangle (A \cap C)$ and $A \triangle (A \triangle B) = B$.

1.1.17. Prove that $\mathfrak{B}(A \cap B) = \mathfrak{B}(A) \cap \mathfrak{B}(B)$ and $\mathfrak{B}(A \cup B) = \{A_1 \cup B_1 \mid A_1 \subseteq A, B_1 \subseteq B\}$.

1.1.18. Prove that the equation $\mathfrak{B}(A) \cup \mathfrak{B}(B) = \mathfrak{B}(A \cup B)$ implies either $A \subseteq B$ or $B \subseteq A$.

1.1.19. Prove that $(A \cap B) \times (C \cap E) = (A \times C) \cap (B \times E)$, $(A \cap B) \times C = (A \times C) \cap (B \times C)$, $A \times (B \cup C) = (A \times B) \cup (A \times C)$, and $(A \cup B) \times (C \cup E) = (A \times C) \cup (B \times C) \cup (A \times E) \cup (B \times E)$.

1.1.20. Let A be a set. A family $\mathfrak{S}$ of subsets of A is called a partition of A if $A = \cup\mathfrak{S}$ and $C \cap D = \emptyset$ whenever C and D are two distinct subsets from $\mathfrak{S}$. Suppose that $\mathfrak{S}$ and $\mathfrak{T}$ are two partitions of A and put $\mathfrak{S} \cap \mathfrak{T} = \{C \cap T \mid C \in \mathfrak{S}, T \in \mathfrak{T}\}$, $\mathfrak{P} = \{X \mid X \in \mathfrak{S} \cap \mathfrak{T} \text{ and } X \neq \emptyset\}$. Is $\mathfrak{P}$ a partition of A?

1.2 SET MAPPINGS

The notion of a mapping (or function) plays a key role in mathematics. At the level of rigor we agreed on, one usually defines the concept of mapping in the following way. We say that we define a mapping from a set A to a set B if for each element of A we (by some rule or law) can associate a uniquely determined element of B. We could stay with this commonly used "definition," that is often used in textbooks ignoring the fact that the term *associate* is still undefined. In this case, it would be enough to define a function as a special kind of correspondence (relation) between two sets. A *correspondence* is simply a set of ordered pairs where the first element of the pair belongs to the first set (the *domain*) and the second element of the pair belongs to the second set (the *range*). For example, the following mapping shows a correspondence from the set A into the set B. The correspondence is defined by the ordered pairs $(1, 8)$, $(1, 2)$, $(2, 9)$, $(7, 6)$, and $(13, 17)$. The domain is the set $\{1, 2, 7, 13\}$. The range is the set $\{2, 8, 6, 9, 17\}$. *A function is a set of ordered pairs in which each element of the domain has only* one *element associated with it in the range*. The correspondence shown above is *not* a function because the element 1 (in the domain) is mapped to two elements in the range, 8 and 2. One correspondence that does define a function is the correspondence $(1, 8)$, $(1, 5)$, $(3, 6)$, $(7, 9)$.

However, we can take another approach and use a more rigorous definition of mapping. This definition is based on the notion of binary correspondence. The reader who does not want to follow this more rigorous approach can simply skip the text up to Definition 1.2.6. A mapping will be rigorously determined if we can list all pairs selected by this correspondence. Taking into account that a set of pairs is connected to the Cartesian product, we make the following definition.

1.2.1. Definition. *Let A and B be sets. A subset Φ of the Cartesian product $A \times B$ is called a correspondence (more precisely, a binary correspondence) between A and B, or a correspondence from A to B. If $A = B$, then the correspondence will be called a* binary relation *on the set A. If $\alpha = (x, y) \in \Phi$, then we say that the elements x and y (in this fixed order) correspond to each other at α.*

Often we will change the notation $(x, y) \in \Phi$ to an equivalent but more natural notation $x\Phi y$. The element x is called *the projection of* α *on* A, and the element y is called *the projection of* α *on* B. We can formalize these expressions by setting $x = \mathbf{pr}_A\alpha$ and $y = \mathbf{pr}_B\alpha$. We also define $\mathbf{pr}_A\Phi = \{\mathbf{pr}_A\alpha \mid \alpha \in \Phi\}$, $\mathbf{pr}_B\Phi = \{\mathbf{pr}_B\alpha \mid \alpha \in \Phi\}$.

For example, let A be the set of all points in the plane and let B be the set of all lines in the plane. We recall that a point P is *incident* with the line λ if P belongs to λ and this *incidence relation* determines a correspondence between the set of points in the plane and the set of lines in this plane. All the lines that go through the point P correspond to P, and all points of the line λ correspond to λ. In this example, an unbounded set of elements of B corresponds to every element of the set A. In general, a fixed element of A can correspond to many, one, or no elements of B.

Here is another example. Let

$$A = \{1, 2, 3, 4, 5\},\ B = \{a, b, c, d\}, \text{ and let}$$
$$\Phi = \{(1, a), (1, c), (2, b), (2, d), (4, a), (5, c)\}.$$

Then we can describe the correspondence Φ with the help of the following simple table:

B/A	1	2	3	4	5
d	·	⊙	·	·	·
c	⊙	·	·	·	⊙
b	·	⊙	·	·	·
a	⊙	·	·	⊙	·

Here all the elements of the set A correspond to the columns, while the elements of B correspond to the rows, the elements of the set $A \times B$ correspond to the points situated on the intersections of columns and rows. The points inside the circles (denoted by ⊙) correspond to the elements of Φ.

When $A = B = \mathbb{R}$, we come to a particularly important case since it arises in numerous applications. In this case, $A \times B = \mathbb{R}^2$ is the real plane. We can think of a binary relation on $\mathbb{R}$ as a set of points in the plane. For example, the relation

$$\Phi = \{(x, y) \mid x^2 + y^2 = 1\}$$

can be illustrated as a circle of radius 1, with center at the origin. Another very useful example is the relation "less than or equal to," denoted as usual by $\leq$, on the set $\mathbb{R}$ of all real numbers.

Next, let n be a positive integer. The relation *congruence modulo* n on $\mathbb{Z}$ is defined as follows. The elements $a, b \in \mathbb{Z}$ are said to be congruent modulo n if $a = b + kn$ for some $k \in \mathbb{Z}$ (thus n divides $a - b$). In this case, we will write $a \equiv b \pmod{n}$. We will consider this relation in detail later.

Very often we can define a correspondence between A and B with the help of some property $P(x, y)$, which connects the element x of A with the element y of B as follows:

$$\Phi = \{(x, y) \mid (x, y) \in A \times B \text{ and } P(x, y) \text{ is valid}\}.$$

This is one of the commonest ways of defining a correspondence.

1.2.2. Definition. *Let A and B be sets and Φ be a correspondence from A to B. Then Φ is said to be a functional correspondence if it satisfies the following conditions:*

- **(F 1)** *for every element $a \in A$ there exists an element $b \in B$ such that $(a, b) \in \Phi$;*
- **(F 2)** *if $(a, b) \in \Phi$ and $(a, c) \in \Phi$, then $b = c$.*

A function or mapping f from a set A to a set B is a triple (A, B, Φ) where Φ is a functional correspondence from A to B. The set A is called the definitional domain or domain of definition of the mapping f; the set B is called the domain of values or value area of the mapping f; a functional correspondence Φ is called the graph of the mapping f. We will write $\Phi = \mathbf{Gr}(f)$. In short, A is called the domain of f and B is called the codomain of f.

If f is a mapping from A to B, then we will denote this symbolically by $f : A \longrightarrow B$.

Condition **(F 1)** implies that $\mathbf{pr}_A \Phi = A$. Thus, together with condition **(F 2)** this means that the mapping f associates a uniquely determined element $b \in B$ with every element $a \in A$.

1.2.3. Definition. *Let $f : A \longrightarrow B$ be a mapping and let $a \in A$. The unique element $b \in B$ such that $(a, b) \in \mathbf{Gr}(f)$ is called the image of a (relative to f) and denoted by $f(a)$.*

In some branches of mathematics, particularly in some algebraic theories, a right-side notation for the image of an element is commonly used; namely, instead of $f(a)$ one uses af. However, in the majority of cases the left-sided notation is generally used and accepted. Taking this into account, we will also employ the left-sided notation for the image of an element.

Every element $a \in A$ has one and only one image relative to f.

1.2.4. Definition. *Let $f : A \longrightarrow B$ be a mapping. If $U \subseteq A$, then put*

$$f(U) = \{f(a) \mid a \in U\}.$$

This set $f(U)$ is called the image of U (relative to f). The image $f(A)$ of the whole set A is called the image of the mapping f and is denoted by $\mathbf{Im}\, f$.

1.2.5. Definition. *Let $f : A \longrightarrow B$ be a mapping. If $b = f(a)$ for some element $a \in A$, then a is called a preimage of b (relative to f). If $V \subseteq B$, then put*

$$f^{-1}(V) = \{a \in A \mid f(a) \in V\}.$$

The set $f^{-1}(V)$ is called the preimage of the set V (relative to f). If $V = \{b\}$ then instead of $f^{-1}(\{b\})$ we will write $f^{-1}(b)$.

Note that in contrast to the image, the element $b \in B$ can have many preimages and may not have any.

We observe that if $A = \emptyset$, there is just one mapping $A \longrightarrow B$, namely, *the empty mapping* in which there is no element to which an image is to be assigned. This might seem strange, but the definition of a mapping justifies this concept. Note that this is true even if B is also empty. By contrast, if $B = \emptyset$ but A is not empty, then there is no mapping from A to B. Generally we will only consider situations when both of the sets A and B are nonempty.

1.2.6. Definition. *The mappings $f : A \longrightarrow B$ and $g : C \longrightarrow D$ are said to be equal if $A = C$, $B = D$ and $f(a) = g(a)$ for each element $a \in A$.*

We emphasize that if the mappings f and g have different codomains, they are not equal even if their domains are equal and $f(a) = g(a)$ for each element $a \in A$.

1.2.7. Definition. *Let $f : A \longrightarrow B$ be a mapping.*

(i) *A mapping f is said to be injective (or one-to-one) if every pair of distinct elements of A have distinct images.*
(ii) *A mapping f is said to be surjective (or onto) if* $\mathbf{Im}\, f = B$.
(iii) *A mapping f is said to be bijective if it is injective and surjective. In this case f is a one-to-one correspondence.*

The following assertion is quite easy to deduce from the definitions and its proof is left to the reader.

1.2.8. Proposition. *Let $f : A \longrightarrow B$ be a mapping. Then*

(i) *f is injective if and only if every element of B has at most one preimage;*

(ii) *f is surjective if and only if every element of B has at least one preimage;*

(iii) *f is bijective if and only if every element of B has exactly one preimage.*

To say that $f : A \longrightarrow B$ is injective means that if $x, y \in A$ and $x \neq y$ then $f(x) \neq f(y)$. Equivalently, to show that f is injective we need to show that if $f(x) = f(y)$ then $x = y$. To show that f is surjective we need to show that if $b \in B$ is arbitrary then there exists $a \in A$ such that $f(a) = b$.

More formally now, we say that a set A *is finite* if there is a positive integer n, for which there exists a bijective mapping $A \longrightarrow \{1, 2, \dots, n\}$. In this case the positive integer n is called the order of the set A and we will write this as $|A| = n$ or **Card** $A = n$. By convention, the empty set is finite and we put $|\emptyset| = 0$. Of course, a set that is not finite is called *infinite*.

1.2.9. Corollary. *Let A and B be finite sets and let $f : A \longrightarrow B$ be a mapping.*

(i) *If f is injective, then $|A| \leq |B|$.*

(ii) *If f is surjective, then $|A| \geq |B|$.*

(iii) *If f is bijective, then $|A| = |B|$.*

These assertions are quite easy to prove and are left to the reader.

1.2.10. Corollary. *Let A be a finite set and let $f : A \longrightarrow A$ be a mapping.*

(i) *If f is injective, then f is bijective.*

(ii) *If f is surjective, then f is bijective.*

Proof.

(i) We first suppose that f is injective and let $A = \{a_1, \dots, a_m\}$. Then $f(a_j) \neq f(a_k)$ whenever $j \neq k$, for $1 \leq j, k \leq m$. It follows that $|\mathbf{Im}\, f| = |\{f(a_1), \dots, f(a_m)\}| = |A|$, and therefore $\mathbf{Im}\, f = A$. Thus f is surjective and an injective, surjective mapping is bijective.

(ii) Next we suppose that f is surjective. Then $f^{-1}(a_j)$ is not empty for $1 \leq j \leq m$. If $f^{-1}(a_j) = f^{-1}(a_k) = x$ then $f(x) = a_j$ and $f(x) = a_k$, so $a_j = a_k$. Thus if $a_j \neq a_k$ then $f(a_j) \neq f(a_k)$, which shows that f is injective. Thus f is bijective.

Let $f : A \longrightarrow B$ be a mapping. Then f induces a mapping from $\mathfrak{B}(A)$ to $\mathfrak{B}(B)$, which associates with each subset U, of A, its image $f(U)$. We will denote this mapping again by f and call it *the extension of the initial function to the Boolean of the set A*.

1.2.11. Theorem. *Let $f : A \longrightarrow B$ be a mapping.*

(i) *If $U \neq \emptyset$, then $f(U) \neq \emptyset$; $f(\emptyset) = \emptyset$.*

(ii) *If* $X \subseteq U \subseteq A$, *then* $f(X) \subseteq f(U)$.
(iii) *If* $X, U \subseteq A$, *then* $f(X) \cup f(U) = f(X \cup U)$.
(iv) *If* $X, U \subseteq A$, *then* $f(X) \cap f(U) \subseteq f(X \cap U)$.

These assertions are very easy to prove and the proofs are omitted, but the method of proof is similar to that given in the proof of Theorem 1.2.12 below. Note that in (iv) the symbol $\subseteq$ cannot be replaced by the symbol $=$, as we see from the following example. Let $f : \mathbb{Z} \longrightarrow \mathbb{Z}$ be the mapping defined as follows. Let $f(x) = x^2$ for each $x \in \mathbb{Z}$, let $X = \{x \in \mathbb{Z} \mid x < 0\}$ and let $U = \{x \in \mathbb{Z} \mid x > 0\}$. Then $X \cap U = \emptyset$ but $f(X) \cap f(U) = U$.

The mapping $f : A \longrightarrow B$ also induces another mapping $g : \mathfrak{B}(B) \longrightarrow \mathfrak{B}(A)$, which associates each subset V of B to its full preimage $f^{-1}(V)$.

1.2.12. Theorem. *Let* $f : A \longrightarrow B$ *be a mapping.*

(i) *If* $Y, V \subseteq B$, *then* $f^{-1}(Y \backslash V) = f^{-1}(Y) \backslash f^{-1}(V)$.
(ii) *If* $Y \subseteq V \subseteq B$, *then* $f^{-1}(Y) \subseteq f^{-1}(V)$.
(iii) *If* $Y, V \subseteq B$, *then* $f^{-1}(Y) \cup f^{-1}(V) = f^{-1}(Y \cup V)$.
(iv) *If* $Y, V \subseteq B$, *then* $f^{-1}(Y) \cap f^{-1}(V) = f^{-1}(Y \cap V)$.

Proof. (i) Let $a \in f^{-1}(Y \backslash V)$. Then $f(a) \in Y \backslash V$, so that $f(a) \in Y$ and $f(a) \notin V$. It follows that $a \in f^{-1}(Y)$ and $a \notin f^{-1}(V)$, so that $a \in f^{-1}(Y) \backslash f^{-1}(V)$. To prove the reverse inclusion we repeat the same arguments in the opposite order.

Assertion (ii) is straightforward to prove.

(iii) We first show that $f^{-1}(Y \cup V) \subseteq f^{-1}(Y) \cup f^{-1}(V)$ and to this end, let $a \in f^{-1}(Y) \cup f^{-1}(V)$. Then $a \in f^{-1}(Y)$ or $a \in f^{-1}(V)$ and hence $f(a) \in Y$ or $f(a) \in V$. Thus, in any case, $f(a) \in Y \cup V$, so that $a \in f^{-1}(Y \cup V)$. We can use similar arguments to obtain the reverse inclusion that $f^{-1}(Y) \cup f^{-1}(V) \subseteq f^{-1}(Y \cup V)$.

Similar arguments can be used to justify (iv).

1.2.13. Definition. *Let* A *be a set. The mapping* $\varepsilon_A : A \longrightarrow A$, *defined by* $\varepsilon_A(a) = a$, *for each* $a \in A$, *is called the identity mapping.*

If C is a subset of A, then the mapping $j_C : C \longrightarrow A$, defined by $j_C(c) = c$ for each element $c \in C$, is called an identical embedding or a canonical injection.

1.2.14. Definition. *Let* $f : A \longrightarrow B$ *and* $g : C \longrightarrow D$ *be mappings. Then we say that* f *is the restriction of* g, *or* g *is an extension of* f, *if* $A \subseteq C$, $B \subseteq D$ *and* $f(a) = g(a)$ *for each element* $a \in A$.

For example, a canonical injection is the restriction of the corresponding identity mapping.

Let A and B be sets. The set of all mappings from A to B is denoted by B^A.

Assume that both sets A and B are finite, say $|A| = k$ and $|B| = n$. We suppose that $A = \{a_1, \ldots, a_k\}$ and that $f : A \longrightarrow B$ is a mapping. If $f \in B^A$ then we define the mapping $\Phi : B^A \longrightarrow B^k$ by $\Phi(f) = (f(a_1), \ldots, f(a_k))$. If $f : A \longrightarrow B$, $g : A \longrightarrow B$ are mappings and $f \neq g$, then there is an element $a_j \in A$ such that $f(a_j) \neq g(a_j)$. It follows that $\Phi(f) \neq \Phi(g)$ and hence Φ is injective. Furthermore, let $(b_1, \ldots, b_k)$ be an arbitrary k-tuple consisting of elements of B. Then we can define a mapping $h : A \longrightarrow B$ by $h(a_1) = b_1, \ldots, h(a_k) = b_k$ and hence $\Phi(h) = (b_1, \ldots, b_k)$. It follows that Φ is surjective and hence bijective. By Corollary 1.2.9, $\left|B^A\right| = \left|B^k\right| = |B|^{|A|}$, a formula which justifies the notation B^A for the set of mappings from A to B. We shall also use this notation for infinite sets.

1.2.15. Definition. *Let A be a set and let B be a subset of A. The mapping $\chi_B : A \longrightarrow \{0, 1\}$ defined by the rule*

$$\chi_B(a) = \begin{cases} 1, & if\ a \in B, \\ 0, & if\ \notin B \end{cases}$$

is called the characteristic function of the subset B.

1.2.16. Theorem. *Let A be a set. Then the mapping $B \longmapsto \chi_B$ is a bijection from the Boolean $\mathfrak{B}(A)$ to the set $\{0, 1\}^A$.*

Proof. By definition, every characteristic function is an element of the set $\{0, 1\}^A$. We show first that the mapping $B \longrightarrow \chi_B$ is surjective by noting that if $f \in \{0, 1\}^A$ and $C = \{c \in A \mid f(c) = 1\}$ then $f = \chi_C$. Next let D and E be two distinct subsets of A. Then, by Definition 1.1.1, either there is an element $d \in D$ such that $d \notin E$, or there is an element $e \in E$ such that $e \notin D$. In the first case, $\chi_D(d) = 1$ and $\chi_E(d) = 0$. In the second case we have $\chi_E(e) = 1$ and $\chi_D(e) = 0$. Hence, in any case, $\chi_D \neq \chi_E$ and it follows that the mapping $B \longrightarrow \chi_B$ is injective and hence bijective.

1.2.17. Corollary. *If A is a finite set then $|\mathfrak{B}(A)| = 2^{|A|}$.*

Indeed, from Theorem 1.2.16 and Corollary 1.2.9 we see that $|\mathfrak{B}(A)| = |\{0, 1\}|^{|A|} = 2^{|A|}$.

For every set A there is an injective mapping from A to $\mathfrak{B}(A)$. For example, $a \longmapsto \{a\}$ for each $a \in A$. However, the following result is also valid and has great implications.

1.2.18. Theorem *(Cantor). Let A be a set. There is no surjective mapping from A onto $\mathfrak{B}(A)$.*

Proof. Suppose, to the contrary, that there is a surjective mapping $f : A \longrightarrow \mathfrak{B}(A)$. Let $B = \{a \in A \mid a \notin f(a)\}$. Since f is surjective there is an element

$b \in A$ such that $B = f(b)$. One of two possibilities occurs, namely, either $b \in B$ or $b \notin B$. If $b \in B$ then $b \in f(b)$. Also however, by the definition of B, $b \notin f(b) = B$, which is a contradiction. Thus $b \notin B$ and, since $B = f(b)$, it follows that $b \notin f(b)$. Again, by the definition of B, we have $b \in B$, which is also a contradiction. Thus in each case we obtain a contradiction, so f cannot be surjective, which proves the result.

1.2.19. Definition. *A set A is called countable if there exists a bijective mapping $f : \mathbb{N} \longrightarrow A$. If an infinite set is not countable then it is said to be uncountable.*

In the case when A is countable we often write $a_n = f(n)$ for each $n \in \mathbb{N}$. Then

$$A = \{a_1, a_2, \ldots, a_n, \ldots\} = \{a_n \mid n \in \mathbb{N}\}.$$

In other words, the elements of a countable set can be indexed (or numbered) by the set of all positive integers. Now we discuss some important properties of countable sets.

1.2.20. Theorem.

(i) *Every infinite set contains a countable subset,*

(ii) *Let A be a countable set and let B be a subset of A. Then either B is finite or B is countable,*

(iii) *The set $\mathbb{N} \times \mathbb{N}$ is countable.*

Proof.

(i) Let A be an infinite set so that, in particular, A is not empty and choose $a_1 \in A$. The subset $A \backslash \{a_1\}$ is also not empty, therefore we can choose an element a_2 in this subset. Since A is infinite, $A \backslash \{a_1, a_2\} \neq \emptyset$, so that we can choose an element a_3 in this subset and so on. This process cannot terminate after a finite number of steps because A is infinite. Hence A contains the infinite subset $\{a_n \mid n \in \mathbb{N}\}$, which is countable.

(ii) Let $A = \{a_n \mid n \in \mathbb{N}\}$. Then there is a least positive integer $k(1)$ such that $a_{k(1)} \in B$ and we put $b_1 = a_{k(1)}$. There is a least positive integer $k(2)$ such that $a_{k(2)} \in B \backslash \{b_1\}$. Put $b_2 = a_{k(2)}$, and so on. If after finitely many steps this process terminates then the subset B is finite. If the process does not terminate then all the elements of B will be indexed by positive whole numbers.

(iii) We can list all the elements of the Cartesian product $\mathbb{N} \times \mathbb{N}$ with the help of the following infinite table.

$$\begin{array}{ccccccc}
(1,1) & (1,2) & (1,3) & \cdots & (1,n) & (1,n+1) & \cdots \\
(2,1) & (2,2) & (2,3) & \cdots & (2,n) & (2,n+1) & \cdots \\
\vdots & \vdots & \vdots & \vdots & \vdots & \vdots & \vdots \\
(k,1) & (k,2) & (k,3) & \ldots & (k,n) & (k,n+1) & \cdots \\
\vdots & \vdots & \vdots & \vdots & \vdots & \vdots & \vdots
\end{array}$$

Notice that if n is a natural number and if the pair (k, t) lies on the nth diagonal (where the diagonal stretches from the bottom left to the top right) then $k + t = n + 1$. Let the set D_n denote the nth diagonal so that

$$D_n = \{(k, t) \mid k + t = n + 1\}.$$

We now list the elements in this table by listing them as they occur on their respective diagonal as follows:

$$\underbrace{(1, 1)}_{D_1}, \underbrace{(1, 2), (2, 1)}_{D_2}, \underbrace{(1, 3), (2, 2), (3, 1)}_{D_3}, \dots, \underbrace{(1, n), (2, n-1), \dots, (n, 1)}_{D_{n+1}}, \dots$$

In this way we obtain a listing of the elements of $\mathbb{N} \times \mathbb{N}$ and it follows that this set is countable.

Assertion (ii) of Theorem 1.2.20 implies that the set of all positive integers in some sense is *the least infinite set*.

1.2.21. Corollary. *Let A and B be sets. If A is countable and there is an injective mapping $f : B \longrightarrow A$, then B is finite or countable.*

Proof. We consider the mapping $f_1 : B \longrightarrow \mathbf{Im}\, f$ defined by $f_1(b) = f(b)$, for each element $b \in B$. By this choice, f_1 is surjective. Since f is injective, f_1 is also injective and hence f_1 is bijective. Finally, Theorem 1.2.20 implies that $\mathbf{Im}\, f$ is finite or countable.

1.2.22. Corollary. *Let $A = \bigcup_{n \in \mathbb{N}} A_n$ where A_n is finite or countable for every $n \in \mathbb{N}$. Then A is finite or countable.*

Proof. The set A_1 is finite or countable, so that $A_1 = \{a_n \mid n \in \Sigma_1\}$, where either $\Sigma_1 = \mathbb{N}$ or $\Sigma_1 = \{1, 2, \dots, k_1\}$ for some $k_1 \in \mathbb{N}$. We introduce a double indexing of the elements of A by setting $b_{1n} = a_n$, for all $n \in \Sigma_1$. Then, by Theorem 1.2.20, $A_2 \backslash A_1$ is finite or countable, and therefore $A_2 \backslash A_1 = \{a_n \mid n \in \Sigma_2\}$ where either $\Sigma_2 = \mathbb{N}$ or $\Sigma_1 = \{1, 2, \dots, k_2\}$, for some $k_2 \in \mathbb{N}$. For this set also we use a double index notation for its elements by setting $b_{2n} = a_n$, for all $n \in \Sigma_2$. Similarly, we index the set $A_3 \backslash (A_2 \cup A_1)$ and so on. Finally, we see that $A = \{b_{ij} \mid (i, j) \in M\}$ where M is a certain subset of the Cartesian product $\mathbb{N} \times \mathbb{N}$.

It is clear, from this construction, that the mapping $b_{ij} \longmapsto (i, j)$ is injective and Corollary 1.2.21 and Theorem 1.2.20 together imply the result.

Since $\mathbb{Z} = \cup_{n \in \mathbb{N}} \{n - 1, -n + 1\}$ we establish the following fact.

1.2.23. Corollary. *The set $\mathbb{Z}$ is countable.*

1.2.24. Corollary. *The set $\mathbb{Q}$ is countable.*

To see this, for each natural number n, put $A_n = \{k/t \mid |k| + |t| = n\}$. Then every subset A_n is finite and $\mathbb{Q} = \bigcup_{n \in \mathbb{N}} A_n$, so that we can apply Corollary 1.2.22.

The implication that we make here is that all countable sets have the same "size" and, in particular, the sets $\mathbb{N}$, $\mathbb{Z}$, and $\mathbb{Q}$ have the same "size."

On the other hand, Theorem 1.2.18 implies that the set $\mathfrak{B}(\mathbb{N})$ is not countable and Theorem 1.2.16 shows that the set $\{0, 1\}^{\mathbb{N}}$ is not countable. Each element of this set can be represented as a countable sequence with terms 0 and 1. With each such sequence $(a_1, a_2, \ldots, a_n, \ldots)$ we can associate the decimal representation of a real number $0 \cdot a_1 a_2 \ldots a_n \ldots$ from the segment $[0, 1]$. It follows from this that *the set* $[0, 1]$, *and hence the set of all real numbers, is uncountable*.

Corollary 1.2.9 shows that to establish the fact that two finite sets have the same "number" of elements there is no need to count these elements. It is sufficient to establish the existence a of bijective mapping between these sets. This idea is the main origin for the abstract notion of a number. Extending this observation to arbitrary sets we arrive at the concept of the *cardinality* of a set.

1.2.25. Definition. *Two sets A and B are called equipollent, if there exists a bijective mapping* $f : A \longrightarrow B$. *We will denote this fact by* $|A| = |B|$, *the cardinal number of A.*

If A and B are finite sets, then the fact that A and B are equipollent means that these sets have the same number of elements. Therefore Cantor introduced the general concept of a cardinal number as a common property of equipollent sets. Two sets A, B have the same cardinality if $|A| = |B|$. On the other hand, every infinite set has the property that it contains a proper subset of the same cardinality. This property is not enjoyed by finite sets. Now we can establish a method of ordering the cardinal numbers.

1.2.26. Definition. *Let A and B be sets. We say that the cardinal number of A is less than or equal to the cardinal number of B (symbolically* $|A| \leq |B|$*), if there is an injective mapping* $f : A \longrightarrow B$.

In this sense, Theorem 1.2.20 shows that the *cardinal number corresponding to a countable set is the smallest one among the infinite cardinal numbers*. Theorem 1.2.18 shows that there are infinitely many different infinite cardinal numbers. We are not going to delve deeply into the arithmetic of cardinal numbers, even though this is a very exciting branch of set theory. However, we present the following important result which, although very natural, is surprisingly difficult to prove. We provide a brief proof, with some details omitted.

1.2.27. Theorem *(Cantor–Bernstein). Let A and B be sets and suppose that* $|A| \leq |B|$ *and* $|B| \leq |A|$. *Then* $|A| = |B|$.

Proof. By writing $A_1 = A \times \{0\}$ and $B_1 = B \times \{1\}$ and noting that $A_1 \cap B_1 = \emptyset$ we may suppose, without loss of generality, that $A \cap B = \emptyset$. Let $f : A \longrightarrow B$ and $g : B \longrightarrow A$ be injective mappings. Consider an arbitrary element $a \in A$. Then either $a \in \mathbf{Im}\, g$ and in this case $a = g(b)$ for some element $b \in B$, or

$a \notin \mathbf{Im}\, g$. Since g is injective, in the first case, the element b satisfying the equation $a = g(b)$ is unique. Similarly, either $b = f(a_1)$ for some unique $a_1 \in A$, or $b \notin \mathbf{Im}\, f$. For an element $x \in A$ we define a sequence $(x_n)_{n \in \mathbb{N}}$ as follows. Put $x_0 = x$, and suppose that, for some $n \geq 0$, we have already defined the element x_n. If n is even, then we define x_{n+1} to be the unique element of the set B such that $x_n = g(x_{n+1})$. If such an element does not exist, the sequence is terminated at x_n. If n is odd, then we define x_{n+1} to be the unique element of the set A such that $x_n = f(x_{n+1})$, with the proviso that the sequence terminates in x_n should such an element x_{n+1} not exist. Only the following two cases are possible:

1. There is an integer n such that the element x_{n+1} does not exist. In this case we call the integer n the *depth* of the element x.

2. The sequence $(x_n)_{n \geq 0}$ is infinite. In this case, it is possible that the set $\{x_n : n \geq 0\}$ is itself finite, as in the case, for example, when $a = g(b)$ and $b = f(a)$. In any case we will say that the element x has infinite depth.

We obtain the following three subsets of A:

the subset A_E consisting of the elements of finite even depth;

the subset A_O consisting of the elements of finite odd depth;

the subset A_∞ consisting of the elements of infinite depth.

We define also similar subsets B_E, B_O, and B_∞ in the set B. From this construction it follows that the restriction of f to A_E is a bijective mapping from A_E to B_O and the restriction of f to A_∞ is a bijective mapping between A_∞ and B_∞. Furthermore, if $x \in A_O$, then there is an element $y \in B_E$ such that $g(y) = x$. Now define a mapping $h : A \longrightarrow B$ by

$$h(x) = \begin{cases} f(x) \text{ if } x \in A_E, \\ f(x) \text{ if } x \in A_\infty, \\ y \text{ if } x \in A_0. \end{cases}$$

It is not hard to prove that h is a bijective mapping and this now proves the theorem.

As an illustration of how the language of sets and mappings may be employed in describing information systems, we consider briefly the concept of *automata*. An automaton is a theoretical device, which is the basic model of a digital computer. It consists of *an input tape*, *an output tape*, and a "*head*," which is able to read symbols on the input tape and print symbols on the output tape. At any instant, the system is in one of a number of states. When the automata reads a symbol on the input tape, it goes to another state and writes a symbol on the output tape.

To make this idea precise we define an automaton A to be a 5-tuple

$$(I, O, S, \nu, \sigma),$$

where I and O are the respective sets of input and output symbols, S is the set of states,

$$\nu : I \times S \longrightarrow O$$

is the output function, and

$$\sigma : I \times S \longrightarrow S$$

is the next state function. The automaton operates in the following manner. If it is in state $s \in S$ and an input symbol $j \in I$ is read, the automaton prints the symbol $\nu(j, s)$ on the output tape and goes to the state $\sigma(j, s)$. Thus the mode of operation is determined by the three sets I, O, S and the two functions ν and σ.

EXERCISE SET 1.2

In each of the following questions explain your reasoning, either by giving a proof of your assertion or a counterexample.

1.2.1. Let $\Phi = \{(x, y) \in \mathbb{N} \times \mathbb{N} \mid 3x = y\}$. Is Φ a functional correspondence?

1.2.2. Let $\Phi = \{(x, y) \in \mathbb{N} \times \mathbb{N} \mid 3x = 5y\}$. Is Φ a functional correspondence?

1.2.3. Let $\Phi = \{(x, y) \in \mathbb{N} \times \mathbb{N} \mid x = 3y\}$. Is Φ a functional correspondence?

1.2.4. Let $\Phi = \{(x, y) \in \mathbb{N} \times \mathbb{N} \mid x^2 = y^2\}$. Is Φ a functional correspondence?

1.2.5. Let $\Phi = \{(x, y) \in \mathbb{N} \times \mathbb{N} \mid x = y^4\}$. Is Φ a functional correspondence?

1.2.6. Let $f : \mathbb{Z} \longrightarrow \mathbb{N}_0$ be the mapping defined by $f(n) = |n|$, where $n \in \mathbb{Z}$. Is f injective? Is f surjective?

1.2.7. Let $f : \mathbb{N} \longrightarrow \{x \in \mathbb{Q} \mid x > 0\}$ be the mapping defined by $f(n) = \frac{n}{n+1}$, where $n \in \mathbb{N}$. Is f injective? Is f surjective?

1.2.8. Let $f : \mathbb{N} \longrightarrow \mathbb{N}$ be the mapping defined by $f(n) = (n + 1)^2$, where $n \in \mathbb{N}$. Is f injective? Is f surjective?

1.2.9. Let $f : \mathbb{N} \longrightarrow \mathbb{N}$ be the mapping defined by $f(n) = \frac{n^2+n}{2}$, where $n \in \mathbb{N}$. Is f injective? Is f surjective?

1.2.10. Let $f : \mathbb{Z} \longrightarrow \mathbb{Z} \times \mathbb{Z}$ be the mapping defined by $f(n) = (n + 1, n)$, where $n \in \mathbb{Z}$. Is f injective? Is f surjective?

1.2.11. Let $f : \mathbb{Z} \longrightarrow \mathbb{Z} \times \mathbb{Z}$ be the mapping defined by $f(n) = (n, n^4)$, where $n \in \mathbb{Z}$. Is f injective? Is f surjective?

1.2.12. Let $f : \mathbb{Z} \longrightarrow \mathbb{N}_0$ be the mapping defined by $f(n) = (n + 1)^2$, where $n \in \mathbb{Z}$. Is f injective? Is f surjective?

1.2.13. Let $f : \mathbb{N}_0 \longrightarrow \mathbb{N}_0$ be the mapping defined by $f(n) = n^2 - 3n$, where $n \in \mathbb{N}_0$. Is f injective? Is f surjective?

1.2.14. Let $f : \mathbb{N}_0 \longrightarrow \mathbb{N}_0$ be the mapping defined by

$$f(n) = \begin{cases} n^2 - 2, & \text{if } n \geq 2 \\ n + 2, & \text{if } n \leq 1 \end{cases}, \qquad n \in \mathbb{N}_0.$$

Is f injective? Is f surjective?

1.2.15. Let $f : \mathbb{N} \longrightarrow \mathfrak{B}(\mathbb{N})$ be the mapping defined by $f(n)$ is the set of all prime divisors of n, where $n \in \mathbb{N}$. Is f injective? Is f surjective?

1.2.16. Construct a bijective mapping from $\mathbb{N}$ to $\mathbb{Z}$.

1.2.17. Let $f_1 : A \longrightarrow B$ and $f_2 : A \longrightarrow B$ be mappings. Prove that the union (respectively the intersection) of $\mathbf{Gr}(f_1)$ and $\mathbf{Gr}(f_2)$ is a graph of some mapping from A to B if and only if $f_1 = f_2$.

1.2.18. Let A and B be finite sets, with $|A| = a, |B| = b$. Find the number of injective mappings from A to B.

1.2.19. Let A be a set. Prove that A is finite if and only if there exists no bijective mapping from A to a proper subset of A.

1.2.20. Let A and B be sets, $U \subseteq A, V \subseteq B$. Prove that $f(U \cap f^{-1}(V)) = f(U) \cap V$.

1.3 PRODUCTS OF MAPPINGS

This section is dedicated to the notion of the product of two mappings. Note, at once, that this product is a partial operation: it is not defined in all cases. If $f : A \longrightarrow B$ and $g : C \longrightarrow D$ are mappings, then the product of g and f is defined only when $B = C$.

1.3.1. Definition. *Let $f : A \longrightarrow B$ and $g : B \longrightarrow C$ be mappings. The mapping $g \circ f$ from A to C, defined by the rule*

$$g \circ f(a) = g(f(a)) \text{ for each } a \in A$$

is called the product or the composite of g and f.

More precisely, first the mapping f acts on the element $a \in A$, and then the mapping g acts on the image, $f(a)$, of a. We agreed to write maps on the left. When we write maps on the right it is logical to write the product in the reverse order. This order is then used to denote the image of an element. Thus, when we write maps on the right, the image of $a \in A$ under the map $f \circ g$ is $a(f \circ g) = (af)g$. It will be convenient to write certain functions known as permutations in this form when we multiply them. Permutations will be introduced later.

We say that the mappings f and g *permute* (or *commute*) if $g \circ f = f \circ g$.

In general, the situation when $g \circ f = f \circ g$ seldom occurs. In fact, let $f : A \longrightarrow B$ and $g : C \longrightarrow D$ be mappings. The product $g \circ f$ is defined if $B = C$ and the product $f \circ g$ is defined if $A = D$. Hence, in order for both products $g \circ f$ and $f \circ g$ to exist it is first necessary that $A = D$ and $B = C$, which means that $f : A \longrightarrow B$ and $g : B \longrightarrow A$. In this case,

$$g \circ f : A \longrightarrow A \text{ and } f \circ g : B \longrightarrow B.$$

In particular, if $A \neq B$ then $g \circ f \neq f \circ g$. However, even when $A = B$, if the set A has at least three elements then there are always mappings $f : A \longrightarrow A$ and $g : A \longrightarrow A$ such that $g \circ f \neq f \circ g$. Indeed, let a_1, a_2, and a_3 be three distinct elements of A. Define the mappings f, g by the rules

$$f(a_1) = a_2, f(a_2) = a_3, f(a_3) = a_1, \text{ and } f(x) = x \text{ whenever } x \notin \{a_1, a_2, a_3\}$$

and

$$g(a_1) = a_3, g(a_2) = a_2, g(a_3) = a_1, \text{ and } g(x) = x \text{ whenever } x \notin \{a_1, a_2, a_3\}.$$

Then we have

$$g \circ f(a_1) = g(f(a_1)) = g(a_2) = a_2 \text{ and } f \circ g(a_1) = f(g(a_1)) = f(a_3) = a_1.$$

It follows that $g \circ f \neq f \circ g$.

However, multiplication of mappings satisfies another important property, namely the associativity property.

1.3.2. Theorem. *Let $f : A \longrightarrow B$, $g : B \longrightarrow C$, and $h : C \longrightarrow D$ be mappings. Then $h \circ (g \circ f) = (h \circ g) \circ f$.*

Proof. We have

$$g \circ f : A \longrightarrow C, h \circ g : B \longrightarrow D \text{ and } h \circ (g \circ f) : A \longrightarrow D, (h \circ g) \circ f : A \longrightarrow D.$$

If a is an arbitrary element of A, then

$$(h \circ (g \circ f))(a) = h((g \circ f)(a)) = h(g(f(a))),$$

whereas

$$((h \circ g) \circ f)(a) = (h \circ g)(f(a)) = h(g(f(a))).$$

Hence $\big(h \circ (g \circ f)\big)(a) = \big((h \circ g) \circ f\big)(a)$ for all $a \in A$ which proves that $(h \circ g) \circ f = h \circ (g \circ f)$.

Let $f : A \longrightarrow B$ be a mapping. It is not hard to see that

$$\varepsilon_B \circ f = f \circ \varepsilon_A = f,$$

so the mappings ε_B and ε_A play the role of "left identity" and "right identity" elements, respectively, for the operation of multiplication of mappings. Also, it should be noted that there is no "universal" identity element for all mappings.

1.3.3. Lemma. *Let $f : A \longrightarrow B$, $g : B \longrightarrow C$ be mappings. If $g \circ f = \varepsilon_A$, then f is an injective mapping and g is a surjective mapping.*

Proof. Suppose that A has elements a and c such that $f(a) = f(c)$. Then

$$a = \varepsilon_A(a) = g \circ f(a) = g(f(a)) = g(f(c)) = g \circ f(c) = \varepsilon_A(c) = c,$$

which shows that f is injective.

Next, let u be an arbitrary element of A. Then

$$u = \varepsilon_A(u) = g \circ f(u) = g(f(u)),$$

and, in particular, $f(u)$ is a preimage of the element u relative to g. It follows that $\mathbf{Im}\ g = A$.

1.3.4. Definition. *Let $f : A \longrightarrow B$ be a mapping. The mapping $g : B \longrightarrow A$ is called a left inverse to f or the retraction associated with f if $g \circ f = \varepsilon_A$.*

The mapping $h : B \longrightarrow A$ is called a right inverse to f or the excision associated with f if $f \circ h = \varepsilon_B$.

Our next theorem gives conditions for the existence of left and right inverses.

1.3.5. Theorem. *Let $f : A \longrightarrow B$ be a mapping. A left inverse to f exists if and only if f is injective. A right inverse to f exists if and only if f is surjective.*

Proof. Suppose first that g is a left inverse of f. Then $g : B \longrightarrow A$ is such that $g \circ f = \varepsilon_A$ and Lemma 1.3.3 shows that f is injective. Conversely, suppose that f is injective. We choose and fix the element u in the set A. If $b \in \mathbf{Im}\ f$, then the element b has a unique preimage a, since f is injective. Put

$$g(b) = \begin{cases} a, & \text{where } f(a) = b \text{ whenever } b \in \mathbf{Im}\ f, \\ u, & \text{if } b \notin \mathbf{Im}\ f. \end{cases}$$

By the definition of g we have, for every element $a \in A$,

$$g \circ f(a) = g(f(a)) = a = \varepsilon_A(a),$$

which shows that g is a left inverse to f.

Now let f be a surjective mapping. Then the preimage of every element $b \in B$ is nonempty. For each element $b \in B$ we choose and fix an element a_b in the set $f^{-1}(b)$. Put $h(b) = a_b$. Then

$$f \circ h(b) = f(h(b)) = f(a_b) = b = \varepsilon_B(b),$$

so that $f \circ h = \varepsilon_B$. This means that h is a right inverse of f. Conversely, if there is a mapping h such that $f \circ h = \varepsilon_B$, then Lemma 1.3.3 implies that the mapping f must be surjective.

The following theorem summarizes some of the main properties of left and right inverses.

1.3.6. Theorem. *Let $f : A \longrightarrow B$, let $f_1 : B \longrightarrow C$ be mappings and let $f_2 = f_1 \circ f$.*

(i) *If f and f_1 are injective, then f_2 is also injective. If g, g_1 are left inverses to f and f_1 respectively, then $g \circ g_1$ is a left inverse to f_2.*

(ii) *If f and f_1 are surjective, then f_2 is also surjective. If g, g_1 are right inverses to f and f_1 respectively, then $g \circ g_1$ is a right inverse to f_2.*

(iii) *If the mapping f_2 is injective, then f is also injective. If g_2 is a left inverse to f_2, then $g_2 \circ f_1$ is a left inverse to f.*

(iv) *If the mapping f_2 is surjective, then f_1 is surjective. If g_2 is a right inverse to f_2, then $f \circ g_2$ is a right inverse to f_1.*

(v) *If f_2 is surjective and f_1 is injective, then f is surjective. If g_2 is a right inverse to f_2, then $g_2 \circ f_1$ is a right inverse to f.*

(vi) *If f_2 is injective and f is surjective, then f_1 is injective. If g_2 is a left inverse to f_2, then $f \circ g_2$ is a left inverse to f_1.*

Proof. (i) Let a_1, a_2 be two distinct elements of the set A. Since the mapping f is injective, we have $f(a_1) = b_1 \neq b_2 = f(a_2)$. The mapping f_1 is also injective, therefore $f_1(b_1) \neq f_1(b_2)$. Thus,

$$\begin{aligned} f_2(a_1) &= f_1 \circ f(a_1) = f_1(f(a_1)) = f_1(b_1) \neq f_1(b_2) = f_1(f(a_2)) \\ &= f_1 \circ f(a_2) = f_2(a_2), \end{aligned}$$

so that f_2 is injective. The equations $g \circ f = \varepsilon_A$, $g_1, \circ f_1 = \varepsilon_B$ imply

$$(g \circ g_1) \circ f_2 = (g \circ g_1) \circ (f_1 \circ f) = g \circ (g_1 \circ f_1) \circ f = g \circ \varepsilon_B \circ f = g \circ f = \varepsilon_A,$$

which shows that $g \circ g_1$ is a left inverse of $f_1 \circ f$.

(ii) Since the mapping f_1 is surjective, $C = \mathbf{Im}\, f_1$ and hence, if $c \in C$, there exists an element $b \in B$ such that $f_1(b) = c$. Since f is also surjective, $b = f(a)$ for some element $a \in A$. Hence,

$$c = f_1(b) = f_1(f(a)) = f_1 \circ f(a) = f_2(a),$$

so that f_2 is also surjective. The equations $f \circ g = \varepsilon_B$ and $f_1 \circ g_1 = \varepsilon_C$ together imply

$$\begin{aligned} f_2 \circ (g \circ g_1) &= (f_1 \circ f) \circ (g \circ g_1) = f_1 \circ (f \circ g) \circ g_1 = f_1 \circ \varepsilon_B \circ g_1 \\ &= f_1 \circ g_1 = \varepsilon_C, \end{aligned}$$

so $g \circ g_1$ is a right inverse of $f_1 \circ f$.

(iii) Suppose that $a_1, a_2 \in A$ are such that $f(a_1) = f(a_2)$. Then $f_1(f(a_1)) = f_1(f(a_2))$ so that

$$f_2(a_1) = f_2(a_2).$$

Since f_2 is injective we deduce that $a_1 = a_2$, which proves that f is injective. Furthermore, $g_2 \circ f_2 = \varepsilon_A$, so we obtain

$$(g_2 \circ f_1) \circ f = g_2 \circ (f_1 \circ f) = g_2 \circ f_2 = \varepsilon_A,$$

and hence $g_2 \circ f_1$ is a left inverse of f.

(iv) Since f_2 is surjective, for every element $c \in C$ there is an element $a \in A$ such that $f_2(a) = c$. Then

$$c = f_2(a) = f_1 \circ f(a) = f_1(f(a)).$$

It follows that the mapping f_1 is also surjective. Furthermore, $f_2 \circ g_2 = \varepsilon_C$ and therefore

$$f_1 \circ (f \circ g_2) = (f_1 \circ f) \circ g_2 = f_2 \circ g_2 = \varepsilon_C,$$

so that $f \circ g_2$ is a right inverse of f_1.

(v) Assertion (iv) implies that the mapping f_1 is surjective and hence, by hypothesis, f_1 is bijective. Let $b \in B$ and $c = f_1(b)$. Since f_2 is surjective, there is an element $a \in A$ such that $f_2(a) = c$. Then

$$c = f_2(a) = f_1 \circ f(a) = f_1(f(a)) \text{ and also } c = f_1(b).$$

Since f_1 is injective it follows that $b = f(a)$, which shows that f is surjective. Furthermore, $f_2 \circ g_2 = \varepsilon_C$. Since f_1 is a bijective mapping, it has a left inverse g_1, so $g_1 \circ f_1 = \varepsilon_B$. Therefore

$$\begin{aligned} f \circ (g_2 \circ f_1) &= (\varepsilon_B \circ f) \circ (g_2 \circ f_1) = (g_1 \circ f_1) \circ f \circ (g_2 \circ f_1) \\ &= g_1 \circ ((f_1 \circ f) \circ g_2) \circ f_1 = g_1 \circ (f_2 \circ g_2) \circ f_1 \\ &= g_1 \circ \varepsilon_C \circ f_1 = g_1 \circ f_1 = \varepsilon_B. \end{aligned}$$

Thus $g_2 \circ f_1$ is a right inverse to f.

(vi) Assertion (iii) implies that the mapping f is injective, and by hypothesis, f is surjective, so f is bijective. Let b_1, b_2 be two distinct elements of B. Since f is bijective, there are distinct elements $a_1, a_2 \in A$ such that $b_1 = f(a_1), b_2 = f(a_2)$. Since the mapping f_2 is injective, $f_2(a_1) \neq f_2(a_2)$. In turn, it follows that

$$f_1(b_1) = f_1(f(a_1)) = f_2(a_1) \neq f_2(a_2) = f_1(f(a_2)) = f_1(b_2),$$

which shows that f_1 is injective.

Furthermore, $g_2 \circ f_2 = \varepsilon_A$. Since f is a bijective mapping, it has a right inverse g, so $f \circ g = \varepsilon_B$. Therefore,

$$\begin{aligned}(f \circ g_2) \circ f_1 &= (f \circ g_2) \circ f_1 \circ \varepsilon_B = (f \circ g_2) \circ f_1 \circ (f \circ g) \\ &= (f \circ g_2) \circ (f_1 \circ f) \circ g = f \circ (g_2 \circ f_2) \circ g \\ &= f \circ \varepsilon_A \circ g = f \circ g = \varepsilon_B.\end{aligned}$$

Thus $f \circ g_2$ is a left inverse to f_1.

We shall use a special notation and terminology for those functions $f : A \longrightarrow A$.

1.3.7. Definition. *Let A be a set. A mapping from A to A is called a transformation of the set A. The set of all transformations of A is denoted by $\mathbf{P}(A)$ or, using our previous notation, A^A.*

We observe that a product of two transformations of A is always defined and is again a transformation. Clearly, multiplication of transformations is associative and in this case there exists an identity element, namely the identity transformation ε_A.

1.3.8. Definition. *Let $f : A \longrightarrow B$ be a mapping. Then the mapping $g : B \longrightarrow A$ is called an inverse of f, if it is simultaneously a left inverse and a right inverse to f, so that $g \circ f = \varepsilon_A$ and $f \circ g = \varepsilon_B$.*

Theorem 1.3.5 shows that a mapping f has an inverse if and only if f is bijective. We note, in this case, that if f has an inverse then it is unique. To show this let $g : B \longrightarrow A$ and $h : B \longrightarrow A$ be mappings satisfying

$$g \circ f = \varepsilon_A,\ f \circ g = \varepsilon_B \text{ and } h \circ f = \varepsilon_A,\ f \circ h = \varepsilon_B.$$

Now consider the product $g \circ f \circ h$. We have

$$h = \varepsilon_A \circ h = (g \circ f) \circ h = g \circ (f \circ h) = g \circ \varepsilon_B = g.$$

Theorem 1.3.5 illustrates how to determine the inverse of a bijective mapping $f : A \longrightarrow B$. If b is an arbitrary element of the set B, then there exists a unique element $a \in A$ such that $f(a) = b$. Put $g(b) = a$. Then

$$g \circ f(a) = g(f(a)) = g(b) = a = \varepsilon_A(a)$$

and

$$f \circ g(b) = f(g(b)) = f(a) = b = \varepsilon_B(b),$$

so that $g \circ f = \varepsilon_A$ and $f \circ g = \varepsilon_B$. Since a bijective mapping f has only one inverse, we use the notation f^{-1} for it and the reader is cautioned not to confuse this with the full preimage and that the notation does not mean $1/f(x)$. We note also that by Lemma 1.3.3 the mapping f^{-1} is also bijective.

Once again, bijective transformations are given special terminology.

1.3.9. Definition. *Let A be a set. A bijective transformation of A is called a permutation of A. The set of all permutations of A is denoted by $\mathbf{S}(A)$. Thus $\phi \in \mathbf{S}(A)$ if and only if $\phi : A \longrightarrow A$ is a bijective mapping.*

If $\varphi, \phi \in S(A)$, then by Theorem 1.3.6 the mapping $\varphi \circ \phi$ is bijective, so that $\varphi \circ \phi \in S(A)$. Multiplication of permutations is associative, and the identity mapping ε_A is a permutation. Finally, if $\varphi \in S(A)$, then φ has an inverse mapping φ^{-1}, which also is a permutation of A.

Later, when we study the main algebraic structures the reader will be introduced to one of the most important and fundamental ideas of modern mathematics—the notion of a group. On the basis of the properties mentioned above, we will conclude that $S(A)$ is a group under the operation of multiplication of permutations and we will discuss some properties of the group of permutations of the set A.

EXERCISE SET 1.3

Be sure to give explanations of your work, either by providing a proof or a counterexample.

1.3.1. Let A be a set and let f be a transformation of A. Suppose that there is a positive integer n such that $f^n = \varepsilon_A$. Prove that f is a permutation of A.

1.3.2. Let A be a finite set and let $|A| = n$. Find $|\mathbf{P}(A)|$.

1.3.3. Let A be a nonempty set. Prove that A is infinite if and only if $\mathbf{S}(A)$ is infinite.

1.3.4. Let A be a nonempty set. Prove that A is infinite if and only if $\mathbf{P}(A)$ is infinite.

1.3.5. Prove that there is a bijective mapping from $A \times B$ to $B \times A$.

1.3.6. Prove that there is a bijective mapping from $A \times (B \times C)$ to $(A \times B) \times C$.

1.3.7. Prove that there is a bijective mapping from $(A \times B)^C$ to $A^C \times B^C$.

1.3.8. Let A be a set consisting of two elements. Is the multiplication on the set $\mathbf{P}(A)$ commutative?

1.3.9. Let $f : \mathbb{N} \longrightarrow \mathbb{Z}$ be the mapping defined by

$$f(n) = \begin{cases} \dfrac{n}{2} - 1 \text{ whenever } n \text{ is even,} \\ -\dfrac{n+1}{2} \text{ whenever } n \text{ is odd.} \end{cases}$$

Is f injective? If yes, find an inverse to f.

1.3.10. Let $f : \mathbb{Q} \longrightarrow \mathbb{Q}$ be the mapping defined by $f(x) = 3x - |x|$, where $x \in \mathbb{Q}$. Is f injective? If yes, find an inverse to f.

1.3.11. Let $f : \mathbb{R} \longrightarrow \mathbb{R}$ be the mapping defined by

$$f(x) = \begin{cases} x^2 \text{ whenever } x \geq 0, \\ x(x-3) \text{ whenever } x < 0. \end{cases}$$

Is f injective? If yes, find an inverse to f.

1.3.12. Let $f : \mathbb{Q} \longrightarrow \mathbb{Q}$ be the mapping defined by

$$f(x) = \begin{cases} \dfrac{x-1}{x+2} \text{ whenever } x \neq -2, \\ 1 \text{ if } x = -2. \end{cases}$$

Is f injective? If yes, find an inverse to f.

1.3.13. Let $f : \mathbb{R} \longrightarrow \mathbb{R}$ be the mapping defined by

$$f(x) = \begin{cases} 1 - x \text{ whenever } x \geq 0, \\ (1-x)^2 \text{ whenever } x < 0. \end{cases}$$

Is f injective? If yes, find an inverse to f.

1.3.14. Let $f : \mathbb{N} \times \mathbb{N} \longrightarrow \mathbb{N}$ be the mapping defined by $f(n, m) = 2^{n-1}(2m - 1)$. Is f injective? If yes, find an inverse to f.

1.3.15. Let $f : \mathbb{Q} \longrightarrow \mathbb{Q}$ be the mapping defined by

$$f(x) = \begin{cases} x + 1 \text{ whenever } x \in \mathbb{Z}, \\ x \text{ whenever } x \notin \mathbb{Z}. \end{cases}$$

Is f injective? If yes, find an inverse to f.

1.3.16. Let f, g, and $h : \mathbb{Q} \longrightarrow \mathbb{Q}$ be mappings defined by $f(x) = \frac{x}{2}$, $g(x) = x + 1$, and $h(x) = x - 1$. Find the products $g \circ f \circ h$, $g \circ h \circ f$, $f \circ g \circ h$, and $h \circ g \circ f$.

1.3.17. Let $f : \mathbb{Q} \longrightarrow \mathbb{Q}$ be the mapping defined by $f(x) = x^2 + 2$, and let $g : \mathbb{Q} \longrightarrow \mathbb{Q}$ be a mapping defined by the rule $g(x) = \frac{x}{2} - 2$. Find the products $g \circ f$, $f \circ g$, $(f \circ g) \circ f$, and $f \circ (g \circ f)$.

1.3.18. Let $f : A \longrightarrow B$ be a mapping and suppose that f has a left inverse (respectively a right inverse). Is this left inverse (respectively right inverse) unique?

1.3.19. Let $f : \mathbb{Q} \longrightarrow \mathbb{Q}$ be the mapping defined by $f(x) = (1 + (1 - x)^{\frac{1}{3}})^{\frac{1}{5}}$. Present f as a product of four mappings.

1.3.20. Let $f : A \longrightarrow B$ be a mapping. Prove the following assertion: f is surjective if and only if for each set U and arbitrary pair of mappings $g : B \longrightarrow U$ and $h : B \longrightarrow U$ the equation $h \circ f = g \circ f$ always implies $h = g$.

1.3.21. Let f, g, and $h : \mathbb{Z} \longrightarrow \mathbb{Z}$ be mappings defined by

$$f(x) = x + 1;$$

$$g(x) = \begin{cases} x \text{ whenever } x \text{ is even,} \\ x + 2 \text{ whenever } x \text{ is odd;} \end{cases}$$

$$h(x) = \begin{cases} x + 2 \text{ whenever } x \text{ is even,} \\ x \text{ whenever } x \text{ is odd.} \end{cases}$$

Prove that f, g, and h are permutations of $\mathbb{Z}$. Find the products $f \circ f$, $g \circ h$, and $h \circ g$.

1.4 SOME PROPERTIES OF INTEGERS

As mentioned already, we will develop the theory of natural numbers and integers later in Chapter 10. For now, we will use the well-known properties of these numbers quite freely. However, there are some important notions and results that are worth recalling now. One of these is The Principal of Mathematical Induction, which we now discuss. The main idea of this principle is the following.

Suppose that for all $n \in \mathbb{N}_0$ we have some assertion $P(n)$. Let us suppose that $P(n)$ is valid for all natural numbers n, where $0 \le n \le k$, and k is a fixed natural number. Such an assumption is commonly called the *induction hypothesis*. Thus, the assertions $P(0), P(1), \ldots, P(k)$ are all valid. If we can show that $P(k + 1)$ is also valid, and this will almost certainly rely on the validity of at least $P(k)$, then we can assert that in this case $P(n)$ is valid for all $n \in \mathbb{N}_0$. This idea can be made more precise using Axiom **(P 4)** of Definition 10.1.1 that occurs later.

We focus our attention on the following important issue. A very key step in a proof by mathematical induction is to check the fact that $P(n)$ takes place for an appropriate small value of n (the so-called basis of the induction, and n

need not be 0 or 1). Neglecting the verification of this step can lead to incorrect assertions. For example, the statement "All people have the same eye color" could be "proven" in the following way. Let $P(n)$ denote the statement that in every set of n people all n people have the same eye color. Indeed, for $n = 1$ the statement $P(1)$ is true. Suppose that $P(k)$ is true and consider a set S consisting of $k + 1$ persons. Then $S = M \cup \{x\}$ where M consists of k people. By the induction hypothesis, all people in the set M have the same eye color. Let $y \in M$ and $T = S \backslash \{y\}$. Then all people from T have the same eye color, by the induction hypothesis and, in particular, x and any person y of M have the same eye color. This means that every person from M has the same eye color as x so that all people in S have the same eye color. Hence the assertion $P(k + 1)$ is true.

The problem with this "proof" is that the first meaningful statement for the assertion that "all people have the same eye color" must deal with at least two people, the case $n = 2$. The statement $P(2)$, that all pairs of people have the same eye color is of course false, so we might say that the induction process never starts. Usually, the problem itself will give you a hint concerning where the induction process should start.

We now illustrate the Principle of Mathematical Induction using some examples. First we will prove the following well-known equation:

$$1 + 3 + 5 + \cdots + (2n - 1) = n^2 \text{ for all natural numbers } n. \tag{1.1}$$

It is clear that if $n = 1$ then this equation is true since $1^2 = 1$ and $2 \times 1 - 1 = 1$ also. To proceed to the inductive step we suppose that we have already proved that Equation 1.1 is valid for all natural numbers $n \leq k$. In particular we have,

$$1 + 3 + \cdots + (2k - 1) = k^2. \tag{1.2}$$

We now use this statement to prove that Equation 1.1 holds in the case when $n = k + 1$. We have, using Equation 1.2,

$$\begin{aligned} 1 + 3 + \cdots + (2k - 1) + (2k + 1) &= [1 + 3 + \cdots + (2k - 1)] + (2k + 1) \\ &= k^2 + (2k + 1) = (k + 1)^2 \end{aligned}$$

By the axiom of mathematical induction we conclude that for any $n \in \mathbb{N}$ the equation $1 + 3 + \cdots + (2n - 1) = n^2$ is true for all natural numbers n.

As another example, we will prove that for any natural number n, $n^3 + 2n$ is divisible by 3.

For the basis step note that if $n = 1$, then $n^3 + 2n = 3$, which is divisible by 3.

Next our induction hypothesis is that the assertion is true for all $n \leq k$, for some natural number k. Thus, in particular 3 divides $k^3 + 2k$. We now use this statement to prove that our assertion is true with $n = k + 1$. To this end consider $(k + 1)^3 + 2(k + 1)$. We have

$$\begin{aligned}(k+1)^3 + 2(k+1) &= (k^3 + 3k^2 + 3k + 1) + (2k + 2) \\ &= (k^3 + 2k) + (3k^2 + 3k + 3) \\ &= (k^3 + 2k) + 3(k^2 + k + 1).\end{aligned}$$

Here $k^3 + 2k$ is divisible by 3, by the induction hypothesis and $3(k^2 + k + 1)$ is obviously divisible by 3. Thus 3 divides $(k + 1)^3 + 2(k + 1)$; so our assertion holds for $n = k + 1$ and the Principle of Mathematical Induction implies that the assertion is true for all $n \in \mathbb{N}$.

As a further illustration of the power of the method we shall prove the well-known binomial theorem

$$\begin{aligned}(x+y)^n = x^n &+ C_1^n x^{n-1} y + C_2^n x^{n-2} y^2 \\ &+ \cdots + C_k^n x^{n-k} y^k + \cdots + y^n.\end{aligned} \tag{1.3}$$

Here x, y are arbitrary numbers, n is a natural number, and the C_k^n are the binomial coefficients that we can compute using the formula

$$C_k^n = \frac{n!}{k!(n-k)!} = \frac{n(n-1)\ldots(n-k+1)}{1 \times 2 \ldots (k-1)k},$$

where we interpret C_k^n to be 0 if $k > n$.

We will prove Formula 1.3 by induction on n.

Clearly, the result is valid for $n = 1, 2$, since $C_1^1 = C_0^1 = 1 = C_0^2 = C_2^2$ and $C_1^2 = 2$. Assume the result is valid for all $n \leq m$ and consider $(x + y)^{m+1}$. We have

$$\begin{aligned}(x+y)^{m+1} &= (x+y)^m (x+y) \\ &= (x^m + C_1^m x^{m-1} y + C_2^m x^{m-2} y^2 + \cdots \\ &\quad + C_k^m x^{m-k} y^k + \cdots + y^m)(x+y) \\ &= x^m (x+y) + C_1^m x^{m-1} y(x+y) + \cdots \\ &\quad + C_k^m x^{m-k} y^k (x+y) + \cdots + y^m (x+y) \\ &= x^{m+1} + x^m y + \cdots + C_{k-1}^m x^{m+2-k} y^{k-1} + C_{k-1}^m x^{m+1-k} y^k \\ &\quad + C_k^m x^{m+1-k} y^k + C_k^m x^{m-k} y^k + \cdots + x y^m + y^{m+1}.\end{aligned}$$

Combining like terms we see that the term $x^{m+1-k} y^k$ has the coefficient

$$C_{k-1}^m + C_k^m = \frac{m!}{(k-1)!(m-k+1)!} + \frac{m!}{k!(m-k)!}$$

$$= \frac{m!}{(k-1)!(m-k)!}\left(\frac{1}{m-k+1}+\frac{1}{k}\right)$$
$$= \frac{m!}{(k-1)!(m-k)!}\frac{m+1}{k(m-k+1)} = \frac{(m+1)!}{k!(m+1-k)!} = C_k^{m+1}.$$

It follows that $(x+y)^{m+1}$ has the stated form and the result now follows by the Principle of Mathematical Induction.

Other important useful results are connected with division of integers. We begin with the following basic result. The reader will observe that this is simply the usual process of long division.

1.4.1. Theorem. *Let $a, b \in \mathbb{Z}$ and $b \neq 0$. Then there are integers q, r such that $a = bq + r$ and $0 \leq r < |b|$. The pair (q, r) having this property is unique.*

Proof. Suppose first that $b > 0$. There exists a positive integer n such that $nb > a$, so $nb - a > 0$. Put

$$M = \{x \mid x \in \mathbb{N}_0 \text{ and } x = nb - a \text{ for some } n \in \mathbb{Z}\}.$$

We have proved that M is not empty. If $0 \in M$, then $bt - a = 0$ for some positive integer t, and $a = bt$. In this case, we can put $q = t, r = 0$. Suppose now that $0 \notin M$. Since M is a subset of $\mathbb{N}_0$, M has a least element $x_0 = bk - a$, where $k \in \mathbb{Z}$. By our assumption, $x_0 \neq 0$. If $x_0 > b$ then $x_0 - b \in \mathbb{N}_0$ and $x_0 - b = b(k-1) - a \in M$. We have $x_0 - (x_0 - b) = b > 0$, so $x_0 > (x_0 - b)$ and we obtain a contradiction with the choice of x_0. Thus $0 < bk - a \leq b$, which we can rearrange by subtracting b, to obtain $-b < b(k-1) - a \leq 0$. It follows that $0 \leq a - b(k-1) < b$. Now put $q = k - 1$ and $r = a - bq$. Then $a = bq + r$, and $0 \leq r < b = |b|$.

Suppose now that $b < 0$. Then $-b > 0$ and, applying the argument above to $-b$, we see that there are integers m, r such that $a = (-b)m + r$ where $0 \leq r < -b = |b|$. Put $q = -m$. Then $a = bq + r$.

To prove the uniqueness claim we suppose also that $a = bq_1 + r_1$ where $0 \leq r_1 < |b|$. We have

$$bq + r = bq_1 + r_1 \text{ or } r - r_1 = bq_1 - bq = b(q_1 - q).$$

If $r = r_1$, then $b(q_1 - q) = 0$ and since $b \neq 0$, then $q_1 - q = 0$, so that $q_1 = q$. Therefore assume that $r_1 \neq r$. Then either $r > r_1$ or $r < r_1$. If $r > r_1$, then

$$0 < r - r_1 \leq |b| \text{ so } 0 < |r - r_1| < |b|, \text{ and } q_1 \neq q.$$

The equation $r - r_1 = b(q_1 - q)$ implies that $|r - r_1| = |b|\,|q - q_1|$ and this shows that $|b| \leq |r - r_1|$, since $|q - q_1| \geq 1$, which is a contradiction.

If we suppose that $r < r_1$, then writing $r_1 - r = b(q - q_1)$ and switching the roles of r and r_1 in the previous argument, we again obtain a contradiction. Hence $r = r_1$ and $q = q_1$.

The previous theorem has some very important consequences.

1.4.2. Definition. *Let $a, b \in \mathbb{Z}$ and $b \neq 0$. Then $a = bq + r$ where $q, r \in \mathbb{Z}$ and $0 \leq r < |b|$. The integer r is called a residue. If $r = 0$, that is $a = bq$, then we say that b is a divisor of a, or that b divides a, or that a is divisible by b. We will write this symbolically as $b \mid a$.*

From this definition we see that $a = a1$, so 1 divides every integer. Similarly, $a = (-a)(-1)$ so ± 1 and $\pm a$ are divisors of a. If b is a divisor of a, then $a = bc = (-b)(-c)$ and so $-b$ is also a divisor of a. Notice also that 0 is divisible by all integers.

We now prove the basic well-known properties of divisibility.

1.4.3. Theorem. *Let $a, b, c \in \mathbb{Z}$. Then the following properties hold:*

(i) *if $a \mid b$ and $b \mid c$, then $a \mid c$;*
(ii) *if $a \mid b$, then $a \mid bc$;*
(iii) *if $a \mid b$, then $ac \mid bc$;*
(iv) *if $c \neq 0$ and $ac \mid bc$, then $a \mid b$;*
(v) *if $a \mid b$ and $c \mid d$, then $ac \mid bd$;*
(vi) *if $a \mid b$ and $a \mid c$, then $a \mid (bk + cl)$ for every $k, l \in \mathbb{Z}$.*

Proof. (i) We have $b = ad$ and $c = bu$ for some $d, u \in \mathbb{Z}$. Then $c = bu = (ad)u = a(du)$, so that $a \mid c$, since $du \in \mathbb{Z}$.

(ii) We have $b = ad$ for some $d \in \mathbb{Z}$. Then $bc = (ad)c = a(dc)$, so that $a \mid bc$.

(iii) We have again $b = ad$. Then $bc = (ad)c = a(dc) = a(cd) = (ac)d$, so that $ac \mid bc$.

(iv) We have $bc = (ac)d = a(cd) = a(dc) = (ad)c$ for some $d \in \mathbb{Z}$. Then $0 = bc - (ad)c = (b - ad)c$. Since $c \neq 0$, Theorem 10.1.11 implies that $b - ad = 0$ and hence $b = ad$.

(v) We have $b = au$ and $d = cv$ for some $u, v \in \mathbb{Z}$. Then

$$\begin{aligned} bd &= (au)(cv) = a(u(cv)) = a((uc)v) \\ &= a((cu)v) = a(c(uv)) = (ac)(uv), \end{aligned}$$

so that $ac \mid bd$.

(vi) We have $b = ad$ and $c = au$ for some $d, u \in \mathbb{Z}$. Then

$$bk + cl = (ad)k + (au)l = a(dk) + a(ul) = a(dk + ul),$$

so that $a \mid bk + cl$.

1.4.4. Definition. *Let $a, b \in \mathbb{Z}$. An integer d is called a greatest common divisor of a and b (which we denote by* **GCD**(a, b)*), if it satisfies the conditions*

(GCD 1) *d divides a and b;*

(GCD 2) *if c divides a and c divides b, then c divides d.*

Furthermore, we define **GCD**$(0, 0)$ *to be* 0.

Clearly if d is a greatest common divisor of a and b, then $-d$ is also a greatest common divisor of a and b. Conversely, if g is another greatest common divisor of a and b, then by (**GCD 2**) $d = gu$ and $g = dv$ for some integers u, v. Then

$$d = gu = (dv)u = d(vu) \text{ and } 0 = d - d(vu) = d(1 - vu).$$

It follows that either $d = 0$ or $1 - vu = 0$. If $d = 0$, then by definition $a = b = 0$. Therefore suppose that $d \neq 0$. Then $1 - uv = 0$ and $uv = 1$. In this case, $1 = |u|\,|v|$, and Theorem 10.1.11(viii) implies that $1 = |u| = |v|$. Hence $u = \pm 1$ and $v = \mp 1$. In particular, $g = \pm d$. Thus we often speak of *the* greatest common divisor of two integers to mean the positive integer satisfying both (**GCD 1**) and (**GCD 2**).

The expression "greatest common divisor of a and b" is really quite descriptive in the sense that if $d =$ **GCD**(a, b) then $|d|$ is the greatest of the divisors of both a and b. For example, for 12 and 30 the numbers $-6, -3, -2, -1, 1, 2, 3, 6$ are common divisors, while 6 and -6 are the greatest common divisors. Of course -6 is minimal in value among the divisors of 12 and 30 but $|-6|$ is the greatest of the divisors.

The following natural question must be raised: for what pairs of integers does the greatest common divisor exist? The following theorem answers this question.

1.4.5. Theorem. *Let a, b be arbitrary integers. Then a and b have a greatest common divisor.*

Proof. Clearly, if $a = b = 0$, then 0 is a greatest common divisor of a and b. Furthermore if $a = 0$ and $b \neq 0$ then **GCD**$(a, b) = b$, with a similar observation if $b = 0$ and $a \neq 0$. Therefore we can assume that a and b are both nonzero. Put

$$M = \{ax + by \mid x, y \in \mathbb{Z}\}.$$

We observe that $a = a1 + b0 \in M$, and $b = a0 + b1 \in M$. Thus, in particular, M is not empty. Furthermore, if $ax + by \in M$, but $ax + by \notin \mathbb{N}$, then

$$-(ax + by) = a(-x) + b(-y) \in M \cap \mathbb{N},$$

so that $M \cap \mathbb{N} \neq \emptyset$. Consequently, $M \cap \mathbb{N}$ has a least element d. Let $d = am + bn$. Suppose that d is not a divisor of a. By Theorem 1.4.1, $a = dq + r$ where $0 < r < d$. Then

$$\begin{aligned} r &= a - dq = a - (am + bn)q \\ &= a - amq - bnq = a(1 - mq) + b(-nq) \in M \cap \mathbb{N}, \end{aligned}$$

and we obtain a contradiction with the choice of d. This contradiction shows that d divides a. Similarly, we can prove that d divides b. Hence d is a common divisor of a and b.

Moreover, if c is a common divisor of a and b then $a = ck$ and $b = cl$, for some integers k and l. We have

$$d = am + bn = (ck)m + (cl)n = c(km) + c(ln) = c(km + ln).$$

Thus c is a divisor of d, so that d satisfies the conditions (**GCD 1**) and (**GCD 2**). Hence d is a greatest common divisor of a and b.

The construction of d in the proof of Theorem 1.4.5 shows that the greatest common divisor satisfies a further interesting property, which we state separately.

1.4.6. Corollary. *Let a, b be arbitrary integers and let $d =$ **GCD**(a, b). Then there are integers m, n such that $d = am + bn$.*

We say that the integers a, b are relatively prime, if **GCD**$(a, b) = \pm 1$.

The following consequence has many important applications in algebra and number theory.

1.4.7. Corollary. *Let a, b be integers. Then a and b are relatively prime if and only if there are integers m, n such that $1 = am + bn$.*

Proof. Indeed, if a and b are relatively prime, then **GCD**$(a, b) = 1$, and we can use Corollary 1.4.6. Conversely, suppose that there are integers m, n such that $1 = am + bn$. Let $d =$ **GCD**(a, b). Then $a = da_1, b = db_1$ for some integers a_1, b_1. We have

$$1 = (da_1)m + (db_1)n = d(a_1m + b_1n).$$

Theorem 10.1.11(viii) implies that $1 = |d|$, and hence a and b are relatively prime.

Our next result is clear intuitively; if we "divide out" the greatest common divisor of two integers then the corresponding quotients have nothing left in common.

1.4.8. Corollary. *Let a, b be integers, let $d = \mathbf{GCD}(a, b)$ and let $a = da_1, b = db_1$. Then a_1 and b_1 are relatively prime.*

Proof. If $\mathbf{GCD}(a_1, b_1) = c > 1$ then $a_1 = ca_2$ and $b_1 = cb_2$, where $b_1, b_2 \in \mathbb{Z}$. Then $dc > d$ is a common divisor of a and b, contrary to the definition of d. The result follows.

Next we establish some further facts about relatively prime integers.

1.4.9. Corollary. *Let a, b, c be integers.*

(i) *If a divides bc and a, b are relatively prime, then a divides c.*
(ii) *If a, b are relatively prime and a, c are also relatively prime, then a and bc are relatively prime.*
(iii) *If a, b divide c and a, b are relatively prime, then ab divides c.*

Proof. (i) By Corollary 1.4.7 there are integers m, n such that $am + bn = 1$. Then

$$\begin{aligned} c &= c(am + bn) = c(am) + c(bn) \\ &= (ca)m + (cb)n = (ac)m + (bc)n = a(cm) + (bc)n. \end{aligned}$$

However $a|bc$ also; so Theorem 1.4.3(vi) shows that a divides $a(cm) + (bc)n$ and hence a divides c.

(ii) By Corollary 1.4.7 there are integers m, n, k, t such that $am + bn = 1$ and $ak + ct = 1$. Then

$$1 = (am + bn)(ak + ct) = a(mak + mct + bnk) + (bc)(nt),$$

and, again using Corollary 1.4.7, we see that a and bc are relatively prime.

(iii) By Corollary 1.4.7 there are integers m, n such that $am + bn = 1$. We have $c = au$ and $c = bv$ for some integers u, v. Then

$$\begin{aligned} c &= c(am + bn) = c(am) + c(bn) = (bv)(am) + (au)(bn) \\ &= (ab)(vm) + (ab)(un) = ab(vm + un). \end{aligned}$$

Thus $ab \mid c$.

Before continuing we make the standard definition of a prime number.

1.4.10. Definition. *Let $a \in \mathbb{Z}$. A divisor of a, which does not coincide with ± 1 or $\pm a$ is called a proper divisor of a. The divisors ± 1 and $\pm a$ are called nonproper divisors of a. A nonzero natural number p is called prime if $p \neq \pm 1$ and p has no proper divisors. An integer that is not prime is called composite.*

We note here that it is a very straightforward argument, using the Principle of Mathematical Induction and Corollary 1.4.9, to prove that if a is prime and $a|b_1b_2 \dots b_n$ then $a|b_i$ for some i. We shall use this fact in the proof of our next result, which is of fundamental importance. This result, which is called the *Fundamental Theorem of Arithmetic* shows that prime numbers are certain kinds of bricks from which each integer is built.

1.4.11. Theorem. *Let a be an integer such that $|a| > 1$. Then $|a|$ is a product of positive primes and this decomposition is unique up to the order of the factors. Furthermore, if $a > 0$, then*

$$a = p_1^{k_1} \dots p_m^{k_m},$$

where $k_1, \dots, k_m$ are positive integers, $p_1, \dots, p_m$ are positive primes, and $p_j \neq p_s$ whenever $j \neq s$. If $a < 0$, then

$$a = -p_1^{k_1} \dots p_m^{k_m}.$$

Proof. Without loss of generality, we may suppose that $a > 0$. We proceed to prove that a is a product of primes, by induction on a. If a is a prime (in particular, $a = 2, 3$), then the result clearly holds. Suppose that a is a not prime and, inductively, that every positive integer u such that $1 < u < a$ decomposes as a product of positive primes. Then a has a proper divisor b, that is $a = bc$ for some integer c. We may suppose that b and c are positive. Then $b < a$, since b is a proper divisor and, for the same reason, $c < a$. By our induction hypothesis b and c can be presented as products of positive primes, so that $a = bc$ has a similar decomposition. Thus, by the Principle of Mathematical Induction, every positive integer greater than 1 is a product of primes.

We now prove uniqueness of the decomposition $a = p_1 \dots p_n$, where $p_1, \dots, p_n$ are primes. We shall prove this by induction on n. To this end, suppose also that $a = q_1 \dots q_t$ where $q_1, \dots, q_t$ are primes. Clearly we may also assume that all prime factors are positive. If $n = 1$, then $a = p_1$ is a prime. We have $p_1 = q_1 \dots q_t$. Since p_1 is prime, its only divisors are $\pm p_1$ and ± 1. Also $q_1 \neq \pm 1$ so it follows that $q_2 \dots q_t = \pm 1$, which means that $t = 1$ and $p_1 = q_1$. Suppose now that $n > 1$ and our assertion already holds for natural numbers that are products of fewer than n primes. We have $p_1 \dots p_n = q_1 \dots q_t$. Clearly, p_1 divides $q_1q_2 \dots q_t$; so, by the observation we made before this theorem we see, by renumbering the q_i if necessary, that p_1 divides q_1. Then since q_1 is prime we have $q_1 = p_1$. Cancelling p_1, it follows that $p_2 \dots p_n = q_2 \dots q_t$. For the natural number $p_2 \dots p_n$ we can apply the induction hypothesis. Thus $n - 1 = t - 1$ and, after some renumbering, $q_j = p_j$ for $j = 2, \dots, n$. Consequently $n = t$ and, since $p_1 = q_1$ also, we now have $p_j = q_j$ for $1 \leq j \leq n$. The result follows.

The existence and uniqueness of the decomposition of integers into prime factors had been assumed as an obvious fact up to the end of the eighteenth century, when the first examples of commutative rings were developed. (A ring is an important algebraic structure that generalizes some number systems—for example, $\mathbb{Z}$ is a ring—and we will study such structures later on in this book.) In some of these examples, the decomposition of elements into prime factors was proved, but the uniqueness of such a decomposition appeared to be false, which was a great surprise for mathematicians of that time. The great German mathematician Carl Friedrich Gauss (1777–1855) who contributed significantly to many fields, including number theory, and was known as the *Prince of Mathematicians*, clearly formulated and proved the Fundamental Theorem of Arithmetic in 1801.

The following absolutely brilliant proof showing that the set of all primes is infinite was published in approximately 300 BC by the father of geometry, Euclid, a great Greek mathematician who wrote one of the most influential mathematical texts of all time, The Elements.

1.4.12. Theorem. *The set of all primes is infinite.*

Proof. Assume the contrary and let $\mathcal{P} = \{p_1, \ldots, p_n\}$ denote the set of all primes. Now, consider the natural number

$$a = 1 + p_1 \ldots p_n.$$

By Theorem 1.4.11, a should be decomposed into a product of primes and so there is a prime $p_j \in \mathcal{P}$ such that $p_j | a$. However p_j also divides $p_1 \ldots p_n$; so, by Theorem 1.4.3(vi), it follows that $p_j | 1$, which is a contradiction. This completes the proof.

The Sieve of Eratosthenes is a simple, ancient algorithm for finding all prime numbers up to a specified integer. It was created by Eratosthenes (276–194 BC), an ancient Greek mathematician. This algorithm consists of the following steps:

1. Write a list of all the natural numbers from 2 to some given number n.
2. Delete from this list all multiples of two (4, 6, 8, etc.).
3. Delete from this list all remaining multiples of three (9, 15, etc.).
4. Find in the list the next remaining prime number 5 and delete all numbers that are multiples of 5, and so on.

Note that the primes $2, 3, \ldots$ do not get deleted in this process—in fact the primes are the only numbers remaining at the end of the process. Moreover, at some stage the process can be stopped. Numbers larger than $\sqrt{n}$ that have not yet been deleted must be prime. For if n is a natural number that is not prime then it must have a prime factor less than $\sqrt{n}$, otherwise n would be a product of at least two natural numbers both strictly larger than $\sqrt{n}$ which is impossible. As a consequence, we need to delete only multiples of primes that are less than or equal to $\sqrt{n}$.

In Theorem 1.4.5 we proved the existence of the greatest common divisor for each pair of integers. However, we did not answer the question of how to find this greatest common divisor. One method of finding the greatest common divisor of the integers a and b would be to find the prime factorizations of a and b and then work as follows. Let $a = p_1^{r_1} \dots p_k^{r_k}$ and $b = p_1^{s_1} \dots p_k^{s_k}$, where $r_j, s_j \geq 0$ for each j. Then it is quite easy to see that $\mathbf{GCD}(a, b) = p_1^{t_1} \dots p_k^{t_k}$, where t_j is the minimum value of r_j and s_j, for each j. The main disadvantage of this method of course is that finding the prime factors of a and b can be difficult. A more practical approaches utilizes a commonly used procedure known as the Euclidean Algorithm which we now describe.

First we note the following statements:

If $b \mid a$ then $\mathbf{GCD}(a, b) = b$.

If $a = bt + r$, for integers t and r, then $\mathbf{GCD}(a, b) = \mathbf{GCD}(b, r)$.

Note that every common divisor of a and b also divides r. So $\mathbf{GCD}(a, b)$ divides r. Since $\mathbf{GCD}(a, b) \mid b$, it is a common divisor of b and r and hence $\mathbf{GCD}(a, b) \leq \mathbf{GCD}(b, r)$. Conversely, since every divisor of b and r also divides a, the reverse is also true. We illustrate this idea with the following example.

Example.

Let $a = 234$, $b = 54$.

$234 = 54 \times 4 + 18$. So $\mathbf{GCD}(234, 54) = \mathbf{GCD}(54, 18)$.

Next, $54 = 18 \times 3$, and $\mathbf{GCD}(54, 18) = 18$.

Therefore, $\mathbf{GCD}(234, 54) = 18$.

Let's describe some details of the Euclidean Algorithm. Suppose that a, b are integers. Since $\mathbf{GCD}(0, x) = x$, for any integer x we can assume that a and b are nonzero. By Theorem 1.4.1, $a = bq_1 + r_1$ where $0 \leq r_1 < |b|$. Again by Theorem 1.4.1, $b = bq_2 + r_2$ where $0 \leq r_2 < r_1$. Next, if $r_2 \neq 0$ then $r_1 = r_2 q_3 + r_3$, where $0 \leq r_3 < r_2$. We continue this process; thus if r_{j-1} and r_j have been obtained with $r_j \neq 0$ then there are integers q_{j+1}, r_{j+1} such that $r_{j-1} = r_j q_{j+1} + r_{j+1}$ with $0 \leq r_{j+1} < r_j$. Since, at each step, $r_{j-1} < r_j$, then this process will terminate in a finite number of steps so that at some step $r_k = 0$. We obtain the following chain of equalities:

$$\begin{aligned} &a = bq_1 + r_1, b = r_1 q_2 + r_2, \\ &r_1 = r_2 q_3 + r_3, \dots, r_{k-3} = r_{k-2} q_{k-1} + r_{k-1}, \\ &r_{k-2} = r_{k-1} q_k + r_k, r_{k-1} = r_k q_{k+1}. \end{aligned} \tag{1.4}$$

It is now possible to prove, using the Principle of Mathematical Induction, that r_k is a common divisor of a and b. Here we just indicate how such a proof would go:

We have

$$r_{k-2} = r_{k-1}q_k + r_k = r_k q_{k+1} q_k + r_k = r_k(q_{k+1}q_k + 1),$$

so that r_k divides r_{k-2}. Further,

$$\begin{aligned} r_{k-3} &= r_{k-2}q_{k-1} + r_{k-1} = r_k(q_{k+1}q_k + 1)q_{k-1} + r_k q_{k+1} \\ &= r_k(q_{k+1}q_k q_{k-1} + q_{k-1} + q_{k+1}), \end{aligned}$$

so that r_k divides r_{k-3}. Moving back up the chain of equalities in Equation 1.4 we finally see that r_k divides a and b.

Now let u be an arbitrary common divisor of a and b. The equation $r_1 = a - bq_1$ shows that u divides r_1. The next equation $r_2 = b - r_1 q_2$ shows that u divides r_2. Continuing to move directly through the chain of equalities in Equation 1.4 we finally see that u divides r_k. This means that r_k is a greatest common divisor of a and b. Again this claim can be proved more formally using the Principle of Mathematical Induction.

Corollary 1.4.6 proves that there are integers x, y such that $r_k = ax + by$. It is important to note that the Euclidian Algorithm allows us to find these numbers x and y. Indeed, we have

$$r_k = r_{k-2} - r_{k-1}q_k.$$

Thus

$$r_{k-1} = r_{k-3} - r_{k-2}q_{k-1},$$

so that

$$\begin{aligned} r_k &= r_{k-2} - (r_{k-3} - r_{k-2}q_{k-1})q_k = r_{k-2} - r_{k-3}q_k + r_{k-2}q_{k-1}q_k \\ &= r_{k-2}(1 + q_{k-1}q_k) - r_{k-3}q_k = r_{k-2}y_1 - r_{k-3}x_1, \text{ say.} \end{aligned}$$

Using the further equation $r_{k-2} = r_{k-4} - r_{k-3}q_{k-2}$, we can prove that $r_k = r_{k-3}y_2 - r_{k-4}x_2$. Continuing in this way and moving back along the chain in Equation 1.4, we finally obtain the equation $r_k = ax + by$. The values of x and y will then be evident.

Of course many standard computer programs will compute the greatest common divisor of two integers in an instant.

EXERCISE SET 1.4

1.4.1. Prove that 3 divides $n^3 - n$ for each positive integer n.

1.4.2. Prove that $n^2 + n$ is even for each positive integer n.

1.4.3. Prove that 8 divides $n^2 - 1$ for each odd positive integer n.

1.4.4. Prove that $1^2 + 2^2 + 3^2 + \cdots + n^2 = \frac{1}{6}n(n+1)(2n+1)$ for each positive integer n.

1.4.5. Prove that $1^3 + 2^3 + 3^3 + \cdots + n^3 = \left(\frac{n(n+1)}{2}\right)^2$ for each positive integer n.

1.4.6. Prove that $1 \times 1! + 2 \times 2! + 3 \times 3! + \cdots + n \times n! = (n+1)! - 1$ for each positive integer n.

1.4.7. Prove that 133 divides $11^{n+2} + 122^{n+1}$ for each integer $n \geq 0$.

1.4.8. Prove that n different lines lying on the same plane and intersecting in the same point divide this plane into $2n$ parts.

1.4.9. Find all positive integers x such that $x^2 + 2x - 3$ is a prime.

1.4.10. Find all positive integers n, k such that $n + k = 221$ and $\mathbf{GCD}(n, k) = 612$.

1.4.11. Suppose n is a two-digit number with the property that if we divide n by the sum of its digits we obtain a quotient of 4 and a remainder of 3. Suppose also that the quotient of n by the product of its digits is 3 and the remainder is 5. What is n?

1.4.12. Find all positive integers x such that 9 divides $x^2 + 2x - 3$.

1.4.13. Prove that $2^n > 2n + 1$ for each integer $n \geq 3$.

1.4.14. Prove that 24 divides $n^4 + 6n^3 + 11n^2 + 6n$ for each positive integer n.

1.4.15. Suppose that $2^n + 1$ is a prime where n is a positive integer. Prove that $n = 2^k$ for some positive integer k.

1.4.16. Let n, k be positive integers. Let $n = kq + r$ where $0 \leq r < k$. Prove that $\mathbf{GCD}(n, k) = \mathbf{GCD}(k, k - r)$.

1.4.17. Find a positive integer k such that $1 + 2 + \cdots + k$ is a three-digit number with all digits equal.

1.4.18. The coefficient of x in the third member of the decomposition of the binomial $(1 + 2x)^n$ is 264. Find the member of this decomposition with the largest coefficient.

1.4.19. Prove that $(k!)^2 \geq k^k$ for all positive integer k.

1.4.20. Prove that 10 divides $k^{73} - k^{37}$ for all positive integer k.

CHAPTER 2

MATRICES AND DETERMINANTS

2.1 OPERATIONS ON MATRICES

Matrices are one of the most useful and prevalent objects in mathematics and its applications. The language of matrices is very convenient and efficient, so scientists use it everywhere. Matrices are particularly useful as a concise means for storing large amounts of information. The idea of a matrix is also a central concept in linear algebra. In this chapter, we begin our study of the main foundations of matrix theory.

We can think of a matrix as a rectangular table of elements, which may be numbers or, more generally, any abstract quantities that can be added and multiplied. The choice of these elements depends on the branch of the particular science and on the specific problem. These elements (the entries of the matrix) could be numbers, or polynomials, or functions, or elements of some abstract algebraic system. We will denote these matrix entries using lower case letters with two indices—the coordinates of this element in the matrix. The first index shows the number of the row in which the element is situated, while the second index is the number of the place of the element in this row, or, which is the same, the number of the column in which this element lies. If the matrix has k rows and n columns then we say that the matrix has dimension $k \times n$ and we can write this matrix in the form

$$\begin{pmatrix} a_{11} & a_{12} & a_{13} & \cdots & a_{1,n-1} & a_{1n} \\ a_{21} & a_{22} & a_{23} & \cdots & a_{2,n-1} & a_{2n} \\ \vdots & \vdots & \vdots & \vdots & \vdots & \vdots \\ a_{k1} & a_{k2} & a_{k3} & \cdots & a_{k,n-1} & a_{kn} \end{pmatrix} \text{ or } \begin{bmatrix} a_{11} & a_{12} & \cdots & a_{1,n-1} & a_{1n} \\ a_{21} & a_{22} & \cdots & a_{2,n-1} & a_{2n} \\ \vdots & \vdots & \vdots & \vdots & \vdots \\ a_{k1} & a_{k2} & \cdots & a_{k,n-1} & a_{kn} \end{bmatrix}.$$

We shall also use the following brief form of matrix notation

$$[a_{ij}]_{1 \le i \le k, 1 \le j \le n} \text{ or } [a_{ij}],$$

when the dimension is reasonably clear. The set of matrices of dimension $k \times n$ whose entries belong to a set S will be denoted by $\mathbf{M}_{k\times n}(S)$. In this chapter, we will mostly think of S as being a subset of the set, $\mathbb{R}$, of real numbers. In this case, we shall say that we are dealing with numerical matrices.

A submatrix is a matrix formed by certain rows and columns from the original matrix.

Thus, if in the matrix $[a_{ij}]$ we choose rows numbered $i(1), i(2), \ldots, i(t)$ and columns numbered $j(1), j(2), \ldots, j(m)$, where $1 \le t \le k$, $1 \le m \le n$, then we will have the following submatrix:

$$\begin{pmatrix} a_{i(1),j(1)} & a_{i(1),j(2)} & \cdots & a_{i(1),j(m)} \\ a_{i(2),j(1)} & a_{i(2),j(2)} & \cdots & a_{i(2),j(m)} \\ \vdots & \vdots & \ddots & \vdots \\ a_{i(t),j(1)} & a_{i(t),j(2)} & \cdots & a_{i(t),j(m)} \end{pmatrix}.$$

In particular, all rows and all columns are submatrices of the original matrix. For example, for the matrix

$$\begin{pmatrix} a_{11} & a_{12} & a_{13} & a_{14} & a_{15} \\ a_{21} & a_{22} & a_{23} & a_{24} & a_{25} \\ a_{31} & a_{32} & a_{33} & a_{34} & a_{35} \end{pmatrix},$$

the following matrices could serve as examples of submatrices:

$$\begin{pmatrix} a_{11} & a_{12} & a_{14} \\ a_{31} & a_{32} & a_{34} \end{pmatrix}, \begin{pmatrix} a_{21} & a_{22} & a_{24} & a_{25} \\ a_{31} & a_{32} & a_{34} & a_{35} \end{pmatrix}, \begin{pmatrix} a_{11} & a_{12} & a_{13} \\ a_{21} & a_{22} & a_{23} \\ a_{31} & a_{32} & a_{33} \end{pmatrix},$$

$$\begin{pmatrix} a_{22} & a_{23} & a_{24} & a_{25} \\ a_{32} & a_{33} & a_{34} & a_{35} \end{pmatrix}, \begin{pmatrix} a_{11} & a_{12} & a_{13} & a_{14} & a_{15} \end{pmatrix}, \begin{pmatrix} a_{13} \\ a_{23} \\ a_{33} \end{pmatrix},$$

$$\begin{pmatrix} a_{11} & a_{12} & a_{13} & a_{14} & a_{15} \\ a_{21} & a_{22} & a_{23} & a_{24} & a_{25} \\ a_{31} & a_{32} & a_{33} & a_{34} & a_{35} \end{pmatrix}, \begin{pmatrix} a_{23} \\ a_{33} \end{pmatrix}, (a_{24}).$$

First we make the following definition of equality of matrices.

2.1.1. Definition. *Two matrices*

$$A = [a_{ij}] \text{ and } B = [b_{ij}]$$

of the set $\mathbf{M}_{k\times n}(S)$ *are said to be equal, if* $a_{ij} = b_{ij}$ *for every pair of indices* (i, j), *where* $1 \leq i \leq k, 1 \leq j \leq n$.

Thus, equal matrices should have the same dimensions and the same elements in the corresponding places.

If in a $k \times n$ matrix the number of rows is equal to the number of columns, so $n = k$, then this matrix is called a *square (or quadratic) matrix*, and the number $n(= k)$ of its rows (or columns) is called *the order of this matrix*.

In particular, a square matrix of order 1 is a 1×1 matrix so is just a single element. The set of all square matrices of order n whose entries belong to S will be denoted by $\mathbf{M}_n(S)$ rather than $\mathbf{M}_{n\times n}(S)$.

Certain special types of matrices crop up on a regular basis. We define some of these next.

2.1.2. Definition. *Let* $A = [a_{ij}]$ *be an* $n \times n$ *numerical matrix.*

(i) *A is called upper triangular, if* $a_{ij} = 0$ *whenever* $i > j$ *and lower triangular if* $a_{ij} = 0$, *whenever* $i < j$.

(ii) *If A is upper or lower triangular then A is called unitriangular, if* $a_{ii} = 1$ *for each* i, $1 \leq i \leq n$.

(iii) *If* $A = [a_{ij}]$ *is triangular then A is called zero triangular if* $a_{ii} = 0$ *for each* i, $1 \leq i \leq n$.

(iv) *A is called diagonal, if* $a_{ij} = 0$ *for every* $i \neq j$.

For example, the matrix

$$\begin{pmatrix} a_{11} & a_{12} & a_{13} \\ 0 & a_{22} & a_{23} \\ 0 & 0 & a_{33} \end{pmatrix}$$

is upper triangular; the matrix

$$\begin{pmatrix} 1 & a_{12} & a_{13} \\ 0 & 1 & a_{23} \\ 0 & 0 & 1 \end{pmatrix}$$

is unitriangular; the matrix

$$\begin{pmatrix} 0 & a_{12} & a_{13} \\ 0 & 0 & a_{23} \\ 0 & 0 & 0 \end{pmatrix}$$

is zero triangular; the matrix

$$\begin{pmatrix} a_{11} & 0 & 0 \\ 0 & a_{22} & 0 \\ 0 & 0 & a_{33} \end{pmatrix}$$

is diagonal.

The power of matrices is perhaps best utilized as a means of storing information. An important part of this is concerned with certain natural operations defined on numerical matrices which we consider next.

2.1.3. Definition. *Let $A = [a_{ij}]$ and $B = [b_{ij}]$ be matrices in the set $\mathbf{M}_{k\times n}(\mathbb{R})$. The sum $A + B$ of these matrices is the matrix $C = [c_{ij}] \in \mathbf{M}_{k\times n}(\mathbb{R})$, whose entries are $c_{ij} = a_{ij} + b_{ij}$ for every pair of indices (i, j), where $1 \le i \le k, 1 \le j \le n$.*

The definition means that we can only add matrices if they have the same dimension and in order to add two matrices of the same dimension we just add the corresponding entries of the two matrices. In this way matrix addition is reduced to the addition of the corresponding entries. Therefore the operation of matrix addition inherits all the properties of number addition.

Thus, addition of matrices is commutative, which means that $A + B = B + A$ for every pair of matrices $A, B \in \mathbf{M}_{k\times n}(\mathbb{R})$. Addition of matrices is associative which means that $(A + B) + C = A + (B + C)$ for every triple of matrices $A, B, C \in \mathbf{M}_{k\times n}(\mathbb{R})$. The set $\mathbf{M}_{k\times n}(\mathbb{R})$ has a zero matrix O each of whose entries is 0. The matrix O is called the (additive) identity element since $A + O = A = O + A$ for each matrix $A \in \mathbf{M}_{k\times n}(\mathbb{R})$. It is not hard to see that for each pair (k, n) there is precisely one $k \times n$ matrix with the property that when it is added to a $k \times n$ matrix A the result is again A. If $A = [a_{ij}] \in \mathbf{M}_{k\times n}(\mathbb{R})$ then the $k \times n$ matrix $-A$ is the matrix whose entries are $-a_{ij}$. It is easy to see from the definition of matrix addition that $A + (-A) = O = -A + A$. This matrix $-A$ is the unique matrix with the property that when it is added to A the result is the matrix O. The matrix $-A$ is called the additive inverse of the matrix A.

Matrix subtraction can be introduced in $\mathbf{M}_{k\times n}(\mathbb{R})$ by using the natural rule that $A - B = A + (-B)$ for every pair of matrices $A, B \in \mathbf{M}_{k\times n}(\mathbb{R})$.

Compared to addition, matrix multiplication looks more sophisticated. We will define it for square matrices first and then will generalize it to rectangular matrices.

2.1.4. Definition. *Let $A = [a_{ij}]$ and $B = [b_{ij}]$ be two matrices in the set $\mathbf{M}_n(\mathbb{R})$. The product AB of these matrices is the matrix $C = [c_{ij}]$, whose elements are*

$$c_{ij} = a_{i1}b_{1j} + a_{i2}b_{2j} + \cdots + a_{in}b_{nj} = \sum_{1\le k\le n} a_{ik}b_{kj}$$

for every pair of indices (i, j), where $1 \le i, j \le n$.

We must observe that matrix multiplication is not commutative as the following example shows. Indeed, let,

$$\begin{pmatrix} 1 & 3 \\ 5 & 2 \end{pmatrix}\begin{pmatrix} 2 & 1 \\ 4 & 3 \end{pmatrix} = \begin{pmatrix} 14 & 10 \\ 18 & 11 \end{pmatrix}, \quad \text{while}$$

$$\begin{pmatrix} 2 & 1 \\ 4 & 3 \end{pmatrix}\begin{pmatrix} 1 & 3 \\ 5 & 2 \end{pmatrix} = \begin{pmatrix} 7 & 8 \\ 19 & 18 \end{pmatrix}.$$

The matrices A, B satisfying $AB = BA$ are called permutable (in other words these matrices commute). However, matrix multiplication does possess another important property—the associative property of matrix multiplication.

2.1.5. Theorem. *For arbitrary matrices $A, B, C \in \mathbf{M}_n(\mathbb{R})$, the following properties hold:*

(i) $(AB)C = A(BC)$.
(ii) $(A + B)C = AC + BC$.
(iii) $A(B + C) = AB + AC$.
(iv) *There exists a matrix $I = I_n \in \mathbf{M}_n(\mathbb{R})$ such that $AI = IA = A$ for each matrix $A \in \mathbf{M}_n(\mathbb{R})$. For a given value of n, I is the unique matrix with this property.*

Proof. (i) We need to show that the corresponding entries of $(AB)C$ and $A(BC)$ are equal. To this end, let

$$A = [a_{ij}],\ B = [b_{ij}],\ \text{and } C = [c_{ij}].$$

Put

$$AB = [d_{ij}],\ BC = [v_{ij}],$$
$$(AB)C = [u_{ij}],\ A(BC) = [w_{ij}].$$

We must show that $u_{ij} = w_{ij}$ for arbitrary (i, j), where $1 \le i, j \le n$. We have

$$u_{ij} = \sum_{1\le k\le n} d_{ik}c_{kj} = \sum_{1\le k\le n}\left(\sum_{1\le m\le n} a_{im}b_{mk}\right)c_{kj} = \sum_{1\le k\le n}\sum_{1\le m\le n}(a_{im}b_{mk})c_{kj}$$

and

$$w_{ij} = \sum_{1\le m\le n} a_{im}v_{mj} = \sum_{1\le m\le n} a_{im}\left(\sum_{1\le k\le n} b_{mk}c_{kj}\right) = \sum_{1\le m\le n}\sum_{1\le k\le n} a_{im}(b_{mk}c_{kj})$$
$$= \sum_{1\le k\le n}\sum_{1\le m\le n} a_{im}(b_{mk}c_{kj}).$$

Since $(a_{im}b_{mk})c_{kj} = a_{im}(b_{mk}c_{kj})$, it follows that $u_{ij} = w_{ij}$ for all pairs (i, j). Hence $(AB)C = A(BC)$.

(ii) We need to show that corresponding entries of $AC + BC$ and $A(B + C)$ are equal. Put

$$AC = [x_{ij}],\ BC = [y_{ij}],\ (A + B)C = [z_{ij}].$$

We shall prove that $z_{ij} = x_{ij} + y_{ij}$ for arbitrary i, j, where $1 \le i, j \le n$. We have

$$z_{ij} = \sum_{1\le k\le n} (a_{ik} + b_{ik})c_{kj} = \sum_{1\le k\le n} a_{ik}c_{kj} + \sum_{1\le k\le n} b_{ik}c_{kj} = x_{ij} + y_{ij}.$$

Thus $(A + B)C = AC + BC$.

The proof of (iii) is similar.

(iv) We define the symbol δ_{ij} (the Kronecker delta) by

$$\delta_{ij} = \begin{cases} 0, & \text{if } i \neq j, \\ 1, & \text{if } i = j. \end{cases}$$

Put

$$I = [\delta_{ij}] = \begin{pmatrix} 1 & 0 & 0 & \cdots & 0 \\ 0 & 1 & 0 & \cdots & 0 \\ \vdots & \vdots & \vdots & \vdots & \vdots \\ 0 & 0 & 0 & \cdots & 1 \end{pmatrix}.$$

Let $AI = [x_{ij}]$ and $IA = [z_{ij}]$. We have

$$x_{ij} = \sum_{1\le k\le n} a_{ik}\delta_{kj} = a_{ij}\delta_{jj} = a_{ij}$$

and

$$z_{ij} = \sum_{1\le k\le n} \delta_{ik}a_{kj} = \delta_{ii}a_{ij} = a_{ij}.$$

It follows that $AI = IA = A$.

In order to prove the uniqueness of I assume that there also exists a matrix U such that $AU = UA = A$ for each matrix $A \in \mathbf{M}_n(\mathbb{R})$. Setting $A = I$, we obtain $IU = I$. Also, though, we know that $IU = U$, from the definition of I, so that $I = U$.

The matrix $I = I_n$ is called the $n \times n$ *identity matrix*.

2.1.6. Definition. *Let $A \in \mathbf{M}_n(\mathbb{R})$. The matrix $U \in \mathbf{M}_n(\mathbb{R})$ is called an inverse matrix or a reciprocal matrix to A if $AU = UA = I$. The matrix A is then said to be invertible or nonsingular.*

It is very straightforward to show that $AO = OA = O$ for all $n \times n$ matrices A, so certainly the zero matrix has no inverse, which is as might be expected. However, many nonzero matrices lack inverses also. The matrix

$$A = \begin{pmatrix} 1 & 0 \\ 0 & 0 \end{pmatrix}$$

is such an example. If A had an inverse

$$\begin{pmatrix} x_{11} & x_{12} \\ x_{21} & x_{22} \end{pmatrix},$$

then we would have

$$\begin{pmatrix} 1 & 0 \\ 0 & 1 \end{pmatrix} = \begin{pmatrix} 1 & 0 \\ 0 & 0 \end{pmatrix} \begin{pmatrix} x_{11} & x_{12} \\ x_{21} & x_{22} \end{pmatrix} = \begin{pmatrix} x_{11} & x_{12} \\ 0 & 0 \end{pmatrix}$$

which is impossible since $0 \neq 1$. This shows that A has no inverse matrix.

If a matrix A has an inverse, then this inverse is unique. Indeed, let U, V be two matrices with the property $AU = UA = I = VA = AV$ and consider the matrix $V(AU)$. We have

$$V(AU) = VI = V \text{ and } V(AU) = (VA)U = IU = U.$$

Thus $V = U$. We will denote the inverse of the matrix A by A^{-1}.

We note that criteria for the existence of an inverse of a given matrix are closely connected to some ideas pertaining to determinants, a topic we shall study in the next section.

Matrix multiplication can be extended to rectangular matrices in general. In the case of two rectangular matrices A and B, their product is defined only if the number of columns of A is equal to the number of rows of B, but the technique of multiplying A and B is the same as described for square matrices. Observe that the number of rows in the product AB is equal to the number of rows in A, while the number of columns of AB is equal to the number of columns in B. Specifically, if $A = [a_{ij}] \in \mathbf{M}_{k\times n}(\mathbb{R})$ and $B = [b_{ij}] \in \mathbf{M}_{n\times t}(\mathbb{R})$, then the product AB of these matrices is the matrix $C = AB = [c_{ij}] \in \mathbf{M}_{k\times t}(\mathbb{R})$, whose elements are

$$c_{ij} = a_{i1}b_{1j} + a_{i2}b_{2j} + \cdots + a_{in}b_{nj} = \sum_{1\leq m\leq n} a_{im}b_{mj}$$

for every pair of indices (i, j), where $1 \leq i \leq k, 1 \leq j \leq t$.

We observe that multiplication of rectangular matrices is associative, when the products are defined. Thus, if $A \in \mathbf{M}_{m\times n}(\mathbb{R})$, $B \in \mathbf{M}_{n\times s}(\mathbb{R})$, and $C \in \mathbf{M}_{s\times t}(\mathbb{R})$, then $A(BC) = (AB)C \in \mathbf{M}_{m\times t}(\mathbb{R})$. The proof of this fact is similar to one that we provided above.

As we mentioned above, matrices form a very useful technical tool which not only provide us with a brief way of writing certain things but also help to clearly show the main ideas. For instance, consider the following system of linear equations:

$$\begin{array}{cccccc} a_{11}x_1 & +a_{12}x_2 & +a_{13}x_3+ & \cdots & +a_{1n}x_n & =b_1 \\ a_{21}x_1 & +a_{22}x_2 & +a_{23}x_3+ & \cdots & +a_{2n}x_n & =b_2 \\ \vdots & \vdots & \vdots & \vdots & \vdots & \vdots \\ a_{k1}x_1 & +a_{k2}x_2 & +a_{k3}x_3+ & \cdots & +a_{kn}x_n & =b_k \end{array}.$$

We can use matrix multiplication to rewrite this system as the following single matrix equation:

$$\begin{pmatrix} a_{11} & a_{12} & a_{13} & \dots & a_{1n} \\ a_{21} & a_{22} & a_{23} & \dots & a_{2n} \\ \vdots & \vdots & \vdots & \vdots & \vdots \\ a_{k1} & a_{k2} & a_{k3} & \dots & a_{kn} \end{pmatrix} \begin{pmatrix} x_1 \\ x_2 \\ \vdots \\ x_k \end{pmatrix} = \begin{pmatrix} b_1 \\ b_2 \\ \vdots \\ b_k \end{pmatrix},$$

which can be written more succinctly as $AX = B$, where $A = (a_{ij})$ is the $k \times n$ matrix of coefficients, $X = (x_j)$ is the $k \times 1$ column vector of unknowns, and $B = (b_j)$ is the $k \times 1$ column vector of constants. This provides us with a very simple way of writing systems of linear equations. We can use the algebra of matrices to help us solve such systems also, as we shall see.

Now we consider multiplication of a matrix by a number, or scalar.

2.1.7. Definition. *Let $A = [a_{ij}]$ be a matrix from the set $\mathbf{M}_{k\times n}(\mathbb{R})$ and let $\alpha \in \mathbb{R}$. The product of the real number α and the matrix A is the matrix $\alpha A = [c_{ij}] \in \mathbf{M}_{k\times n}(\mathbb{R})$, whose entries are defined by $c_{ij} = \alpha a_{ij}$, for every pair of indices (i, j), where $1 \le i \le k$, $1 \le j \le n$.*

Thus, when we multiply a matrix by a real number we multiply each element of the matrix by this number. Here are the main properties of this operation, which can be proved quite easily, in a manner similar to that given in Theorem 2.1.5:

$$\begin{aligned} (\alpha+\beta)A &= \alpha A + \beta A, \\ \alpha(A+B) &= \alpha A + \alpha B, \\ \alpha(\beta A) &= (\alpha\beta)A, \\ 1A &= A, \\ \alpha(AB) &= (\alpha A)B = A(\alpha B). \end{aligned}$$

These equations hold for all real numbers α, β and for all matrices A, B where the multiplication is defined. Note that this operation of multiplying a matrix by a number can be reduced to the multiplication of two matrices since $\alpha A = (\alpha I)A$.

Here is a summary of all the properties we have obtained so far, using our previously established notation.

$$\begin{aligned}
A + B &= B + A, \\
A + (B + C) &= (A + B) + C, \\
A + O &= A, \\
A + (-A) &= O, \\
A(B + C) &= AB + AC, \\
(A + B)C &= AC + BC, \\
A(BC) &= (AB)C, \\
AI &= IA = A, \\
(\alpha + \beta)A &= \alpha A + \beta A, \\
\alpha(A + B) &= \alpha A + \alpha B, \\
\alpha(\beta A) &= (\alpha\beta)A, \\
1A &= A, \\
\alpha(AB) &= (\alpha A)B = A(\alpha B).
\end{aligned}$$

With the aid of these arithmetic operations on matrices, we can define some additional operations on the set $\mathbf{M}_n(\mathbb{R})$. The following two operations are the most useful: the operation of commutation and the operation of transposing.

2.1.8. Definition. *Let $A, B \in \mathbf{M}_n(\mathbb{R})$. The matrix $[A, B] = AB - BA$ is called the commutator of A and B.*

The operation of commutation is anticommutative in the sense that $[A, B] = -[B, A]$. Note also that $[A, A] = O$. This operation is not associative. In fact, by applying Definition 2.1.8, we have

$$[[A, B], C] = [AB - BA, C] = ABC - BAC - CAB + CBA, \text{ whereas}$$
$$[A, [B, C]] = [A, BC - CB] = ABC - ACB - BCA + CBA.$$

Thus, if the associative law were to hold for commutation then it would follow that $BAC + CAB = ACB + BCA$. However, the matrices

$$A = \begin{pmatrix} 0 & 1 \\ 0 & 0 \end{pmatrix}, B = \begin{pmatrix} 1 & 0 \\ 1 & 0 \end{pmatrix}, C = \begin{pmatrix} 0 & 0 \\ 1 & 0 \end{pmatrix}$$

show that this is not true. In fact,

$$\begin{aligned} BAC + CAB &= \begin{pmatrix} 1 & 0 \\ 1 & 0 \end{pmatrix}\begin{pmatrix} 0 & 1 \\ 0 & 0 \end{pmatrix}\begin{pmatrix} 0 & 0 \\ 1 & 0 \end{pmatrix} \\ &\quad + \begin{pmatrix} 0 & 0 \\ 1 & 0 \end{pmatrix}\begin{pmatrix} 0 & 1 \\ 0 & 0 \end{pmatrix}\begin{pmatrix} 1 & 0 \\ 1 & 0 \end{pmatrix} \\ &= \begin{pmatrix} 0 & 1 \\ 0 & 1 \end{pmatrix}\begin{pmatrix} 0 & 0 \\ 1 & 0 \end{pmatrix} + \begin{pmatrix} 0 & 0 \\ 0 & 1 \end{pmatrix}\begin{pmatrix} 1 & 0 \\ 1 & 0 \end{pmatrix} \\ &= \begin{pmatrix} 1 & 0 \\ 1 & 0 \end{pmatrix} + \begin{pmatrix} 0 & 0 \\ 1 & 0 \end{pmatrix} = \begin{pmatrix} 1 & 0 \\ 2 & 0 \end{pmatrix}, \quad \text{whereas} \\ ACB + BCA &= \begin{pmatrix} 0 & 1 \\ 0 & 0 \end{pmatrix}\begin{pmatrix} 0 & 0 \\ 1 & 0 \end{pmatrix}\begin{pmatrix} 1 & 0 \\ 1 & 0 \end{pmatrix} \\ &\quad + \begin{pmatrix} 1 & 0 \\ 1 & 0 \end{pmatrix}\begin{pmatrix} 0 & 0 \\ 1 & 0 \end{pmatrix}\begin{pmatrix} 0 & 1 \\ 0 & 0 \end{pmatrix} \\ &= \begin{pmatrix} 1 & 0 \\ 0 & 0 \end{pmatrix}\begin{pmatrix} 1 & 0 \\ 1 & 0 \end{pmatrix} + \begin{pmatrix} 0 & 0 \\ 0 & 0 \end{pmatrix}\begin{pmatrix} 0 & 1 \\ 0 & 0 \end{pmatrix} \\ &= \begin{pmatrix} 1 & 0 \\ 0 & 0 \end{pmatrix} + \begin{pmatrix} 0 & 0 \\ 0 & 0 \end{pmatrix} = \begin{pmatrix} 1 & 0 \\ 0 & 0 \end{pmatrix}. \end{aligned}$$

Thus, $BAC + CAB \neq ACB + BCA$, so we can see that $[[A, B], C] \neq [A, [B, C]]$.

However, for commutation there is a weakened form of associativity, known as the *Jacobi identity* which, states

$$[[A, B], C] + [[C, A], B] + [[B, C], A] = O.$$

Indeed

$$\begin{aligned} &[[A, B], C] + [[C, A], B] + [[B, C], A] = [AB - BA, C] \\ &\quad + [CA - AC, B] + [BC - CB, A] \\ &= ABC - BAC - CAB + CBA + CAB - ACB - BCA + BAC \\ &\quad + BCA - CBA - ABC + ACB = O. \end{aligned}$$

The second operation of interest is transposition which we now define.

2.1.9. Definition. *Let $A = [a_{ij}]$ be a matrix from the set $\mathbf{M}_{k\times n}(\mathbb{R})$. The transpose of A is the matrix $A^t = [b_{ij}]$ which is the matrix from the set $\mathbf{M}_{n\times k}(\mathbb{R})$ whose entries are $b_{ij} = a_{ji}$. Thus, the rows of A^t are the columns of A, and the columns of A^t are the rows of A. We will say that we obtain A^t by transposition of A.*

Here are the main properties of transposition.

2.1.10. Theorem. *Transposition has the following properties:*

(i) $(A^t)^t = A$, *for all matrices* A.
(ii) $(A + B)^t = A^t + B^t$, *if* $A, B \in \mathbf{M}_{k\times n}(\mathbb{R})$.
(iii) $(AB)^t = B^t A^t$, *if* $A \in \mathbf{M}_{k\times n}(\mathbb{R})$ *and* $B \in \mathbf{M}_{n\times t}(\mathbb{R})$.
(iv) $(A^{-1})^t = (A^t)^{-1}$, *for all invertible square matrices* A. *Thus, if* A^{-1} *exists so does* $(A^t)^{-1}$.
(v) $(\alpha A)^t = \alpha A^t$, *for all matrices* A *and real numbers* α.

Proof. Assertions (i) and (ii) are quite easy to show and the method of proof will be seen in the remaining cases.

(iii) Let $A = [a_{ij}] \in \mathbf{M}_{k\times n}(\mathbb{R})$ and $B = [b_{ij}] \in \mathbf{M}_{n\times t}(\mathbb{R})$. Put

$$AB = C = [c_{ij}] \in \mathbf{M}_{k\times t}(\mathbb{R}),\ A^t = [u_{ij}] \in \mathbf{M}_{n\times k}(\mathbb{R}),$$
$$B^t = [v_{ij}] \in \mathbf{M}_{t\times n}(\mathbb{R}),\ B^t A^t = [w_{ij}] \in \mathbf{M}_{t\times k}(\mathbb{R}).$$

Then

$$c_{ji} = \sum_{1\le m\le n} a_{jm}b_{mi} \text{ and}$$
$$w_{ij} = \sum_{1\le m\le n} v_{im}u_{mj} = \sum_{1\le m\le n} b_{mi}a_{jm} = \sum_{1\le m\le n} a_{jm}b_{mi} = c_{ji}.$$

It follows that $(AB)^t = B^t A^t$.

(iv) If A^{-1} exists then we have $A^{-1}A = AA^{-1} = I$. Using (iii) we obtain

$$I = I^t = (A^{-1}A)^t = A^t(A^{-1})^t \text{ and } I = I^t = (AA^{-1})^t = (A^{-1})^t A^t.$$

Thus, $(A^{-1})^t$ is the inverse of A^t, which is to say that A^t is invertible and $(A^t)^{-1} = (A^{-1})^t$

(v) is also easily shown.

2.1.11. Definition. *The matrix* $A = [a_{ij}] \in \mathbf{M}_n(\mathbb{R})$ *is called symmetric if* $A = A^t$. *In this case* $a_{ij} = a_{ji}$ *for every pair of indices* (i, j), *where* $1 \le i, j \le n$.

The matrix $A = [a_{ij}] \in \mathbf{M}_n(\mathbb{R})$ *is called skew symmetric, if* $A = -A^t$. *Then* $a_{ij} = -a_{ji}$ *for every pair of indices* (i, j), *where* $1 \le i, j \le n$.

We note that if a matrix A has the property that $A = A^t$, then A is necessarily square. Also the elements of the main diagonal of a skew-symmetric matrix must be 0 since, in this case, we have $a_{ii} = -a_{ii}$, or $a_{ii} = 0$ for each i, where $1 \le i \le n$. If A is skew symmetric then $A^t = -A$ clearly.

The following remarkable result illustrates the role that symmetric and skew-symmetric matrices play.

2.1.12. Theorem. *Every square matrix A can be represented in the form $A = S + K$, where S is a symmetric matrix and K is a skew-symmetric matrix. This representation is unique.*

Proof. Let $S = \frac{1}{2}(A + A^t)$ and $K = \frac{1}{2}(A - A^t)$ and note, using Theorem 2.1.10, that

$$S^t = \frac{1}{2}(A + A^t)^t = \frac{1}{2}(A^t + (A^t)^t) = \frac{1}{2}(A^t + A) = S.$$

Thus S is symmetric. Also, again by Theorem 2.1.10,

$$K^t = \frac{1}{2}(A - A^t)^t = \frac{1}{2}(A^t + (-A^t)^t) = \frac{1}{2}(A^t - A) = -\frac{1}{2}(A - A^t) = -K,$$

so that K is skew symmetric. Furthermore, $S + K = \frac{1}{2}(A + A^t) + \frac{1}{2}(A - A^t) = A$. Consequently, it is always possible to write the matrix A as a sum of a symmetric and a skew-symmetric matrix. To show uniqueness, let S_1 be symmetric and let K_1 be skew symmetric such that also $A = S_1 + K_1 = S + K$. Then $-S + S_1 = S_1 - S = K - K_1$. However, $X = S_1 - S$ is symmetric and $X = K - K_1$ is skew symmetric. Then $X = X^t = -X$ and it follows that $X = O$. Therefore, $S = S_1$ and $K = K_1$ and the uniqueness of the expression follows.

EXERCISE SET 2.1

2.1.1. Prove that there are no matrices A and B for which the equation $[A, B] = I$ is valid. **Hint**. Just show that the sum of all elements of the principal diagonal of the matrix $[A, B]$ is equal to 0.

2.1.2. Let A be a diagonal matrix whose diagonal entries are all different. Let B be a matrix such that $AB = BA$. Prove that B is diagonal.

2.1.3. Find all matrices $A \in \mathbf{M}_2(\mathbb{R})$ with the property that $A^2 = O$.

2.1.4. If we interchange rows j and k of a matrix A, what changes does this imply in the matrix AB?

2.1.5. If we interchange columns j and k of a matrix A, what changes does this imply in the matrix AB?

2.1.6. If we add α times row k to row j in the matrix A, what changes does this imply in the matrix AB?

2.1.7. Find A^3 if $A = \begin{pmatrix} 1 & 1 & 1 & \dots & 1 \\ 0 & 1 & 1 & \dots & 1 \\ 0 & 0 & 1 & \dots & 1 \\ \vdots & \vdots & \vdots & \ddots & \vdots \\ 0 & 0 & 0 & \dots & 1 \end{pmatrix}$.

2.1.8. Find $\begin{pmatrix} 1 & 0 & 0 & \dots & 0 \\ 1 & 1 & 0 & \dots & 0 \\ 1 & 1 & 1 & \dots & 0 \\ \vdots & \vdots & \vdots & \ddots & \vdots \\ 1 & 1 & 1 & \dots & 1 \end{pmatrix}^3$.

2.1.9. Find $\begin{pmatrix} 1 & 0 & 0 & 0 & \dots & 0 & 0 \\ 0 & 1 & 0 & 0 & \dots & 0 & 0 \\ 1 & 1 & 1 & 1 & \dots & 1 & 1 \\ 0 & 0 & 0 & 1 & \dots & 0 & 0 \\ \vdots & \vdots & \vdots & \vdots & \ddots & \vdots & \vdots \\ 0 & 0 & 0 & 0 & \dots & 1 & 0 \\ n & 0 & 0 & 0 & \dots & 0 & 1 \end{pmatrix}^3$.

2.1.10. Find $\begin{pmatrix} 1 & 0 & 0 & 0 & \dots & 0 & 0 \\ 2 & 1 & 0 & 0 & \dots & 0 & 0 \\ 3 & 0 & 1 & 0 & \dots & 0 & 0 \\ 4 & 0 & 0 & 1 & \dots & 0 & 0 \\ \vdots & \vdots & \vdots & \vdots & \ddots & \vdots & \vdots \\ n-1 & 0 & 0 & 0 & \dots & 1 & 0 \\ n & 0 & 0 & 0 & \dots & 0 & 1 \end{pmatrix}^3$.

2.1.11. Find $\begin{pmatrix} 1 & 1 & 0 & 0 & 0 & 0 & 0 & 0 & 0 \\ 0 & 1 & 1 & 0 & 0 & 0 & 0 & 0 & 0 \\ 0 & 0 & 1 & 1 & 0 & 0 & 0 & 0 & 0 \\ 0 & 0 & 0 & 1 & 1 & 0 & 0 & 0 & 0 \\ 0 & 0 & 0 & 0 & 1 & 1 & 0 & 0 & 0 \\ 0 & 0 & 0 & 0 & 0 & 1 & 1 & 0 & 0 \\ 0 & 0 & 0 & 0 & 0 & 0 & 1 & 1 & 0 \\ 0 & 0 & 0 & 0 & 0 & 0 & 0 & 1 & 1 \\ 0 & 0 & 0 & 0 & 0 & 0 & 0 & 0 & 1 \end{pmatrix}^8$.

2.1.12. Find the $n \times n$ matrix $\begin{pmatrix} 0 & 0 & 1 & 0 & \dots & 0 & 0 \\ 0 & 0 & 0 & 1 & \dots & 0 & 0 \\ 0 & 0 & 0 & 0 & \dots & 0 & 0 \\ \vdots & \vdots & \vdots & \ddots & \vdots & \vdots & \vdots \\ 0 & 0 & 0 & 0 & \ddots & 0 & 1 \\ 0 & 0 & 0 & 0 & \dots & 0 & 0 \\ n & 0 & 0 & 0 & \dots & 0 & 0 \end{pmatrix}^7$.

2.1.13. Prove that for any square matrix A the product AA^t is a symmetric matrix.

2.1.14. Prove that a product of two symmetric matrices A, B is a symmetric matrix if and only if $AB = BA$.

2.1.15. Prove that a product of two skew-symmetric matrices A, B is a symmetric matrix if and only if $AB = BA$.

2.1.16. Prove that for every pair of symmetric (respectively, skew symmetric) matrices A, B the commutator $[A, B]$ is a skew-symmetric matrix.

2.1.17. Prove that for every pair of symmetric matrices A, B the product $ABAB \ldots ABA$ is also a symmetric matrix.

2.1.18. Let $A \in \mathbf{M}_n(\mathbb{R})$. A matrix A is called nilpotent, if $A^k = O$ for some positive integer k. Let A, B be nilpotent matrices. Prove that $AB = BA$ implies the nilpotency of $A + B$.

2.1.19. Let $A \in \mathbf{M}_n(\mathbb{R})$. A matrix A is called nilpotent, if $A^k = O$ for some positive integer k. The minimal such number k is called the nilpotency class of A. Prove that every zero triangular matrix is nilpotent.

2.1.20. Let A be a nilpotent matrix. Prove that the matrices $I - A$ and $I + A$ are invertible.

2.2 PERMUTATIONS OF FINITE SETS

There is a key numerical characteristic of a matrix called the determinant of the matrix which requires some ideas from permutations of finite sets. The properties of determinants are therefore closely connected with properties of permutations and, for this reason, in this section we shall study some basic properties of permutations. The properties that we discuss now will often be used in the next section when we study determinants.

Let A be a finite set, say $A = \{a_1, a_2, \ldots, a_n\}$. In the case of sets, the order that the elements are written is not important as we saw in Definition 1.1.1. However, there are cases when the order of the elements in a set is important. One such case arose when we considered the Cartesian product. As we saw, the elements of a Cartesian nth power of a set are ordered n-tuples. This means, for example, that the n-tuples $(a_1, a_2, a_3, \ldots, a_n)$ and $(a_2, a_1, a_3, \ldots, a_n)$ are different. An n-tuple consisting of all elements of a finite set $A = \{a_1, a_2, \ldots, a_n\}$ that contains each element from A once and only once is called *a permutation* of the elements $a_1, a_2, \ldots, a_n$. These elements in an n-tuple appear in some order: the tuple has a first element (unless it is empty), a second element (unless its length is less than 2), and so on. For example, if $A = \{1, 2, 3\}$, then (1,2,3) and (3,2,1) are two different ways to list the elements of A in some order; these constitute two different permutations of the numbers 1, 2, 3.

We have already used the term *permutation* to mean a bijective transformation of sets. This term is also widely used in combinatorics but has a different meaning. This often happens in mathematics and can be a cause for confusion, but usually the context should make clear which meaning is in use. In this case, the two concepts are closely related and it should be clear from the context which meaning of permutation is being used.

To justify some of these remarks, let A be a set with n elements, say $A = \{a_1, a_2, \ldots, a_n\}$ and let π denote a permutation of A. For $1 \le j \le n$, let $\pi(a_j) = a_k$, where k is dependent upon j. Then π induces a mapping $\pi_0 : \{1, 2, \ldots, n\} \longrightarrow \{1, 2, \ldots, n\}$ defined by

$$\pi_0(j) = k \text{ whenever } \pi(a_j) = a_k.$$

Thus, $\pi(a_j) = a_{\pi_0(j)}$ for all j such that $1 \le j \le n$. The mapping π_0 is a permutation of $\{1, \ldots, n\}$. To see this, note that if $\pi_0(j) = \pi_0(i)$ then $\pi(a_j) = a_{\pi_0(j)} = a_{\pi_0(i)} = \pi(a_i)$. However, π is a permutation of A so this implies that $a_j = a_i$ and hence $j = i$. Thus, π_0 is injective and hence is bijective by Corollary 1.2.10. Conversely, every permutation σ of $\{1, 2, \ldots, n\}$ gives rise to a permutation ϕ_σ of A. We simply define $\phi_\sigma(a_j) = a_{\sigma(j)}$, for each j such that $1 \le j \le n$. Then, if $\phi_\sigma(a_j) = \phi_\sigma(a_i)$, we have $a_{\sigma(j)} = a_{\sigma(i)}$ and hence $\sigma(j) = \sigma(i)$. Since σ is a permutation of $\{1, 2, \ldots, n\}$ it follows that $j = i$ and hence ϕ_σ is injective. Corollary 1.2.10 further implies that ϕ_σ is bijective and hence a permutation of A. Furthermore, if $\pi \ne \phi$ are two permutations of A then there is an index r such that $\pi(a_r) \ne \phi(a_r)$. It follows that $\pi_0(r) \ne \phi_0(r)$ and therefore $\pi_0 \ne \phi_0$. Consequently, every permutation π of the set A corresponds to precisely one permutation π_0 of $\{1, 2, \ldots, n\}$ and the mapping $\pi \longmapsto \pi_0$ is bijective.

Every algebraic permutation π of the set A is equivalent to a combinatorial permutation, since informally both involve some type of listing of the elements of $\{a_1, a_2, \ldots, a_n\}$. More formally, let σ denote a mapping from $\{1, \ldots, n\}$ to itself and let $(a_{\sigma(1)}, a_{\sigma(2)}, \ldots, a_{\sigma(n)})$ be a combinatorial permutation of the elements $a_1, a_2, \ldots, a_n$. This implies that σ is a bijection and hence a permutation of the set $\{1, \ldots, n\}$. By our analysis above this means that the transformation π of A, defined by the rule $\pi(a_j) = a_{\sigma(j)}$, where $1 \le j \le n$, is a bijection and hence an (algebraic) permutation of A. Thus, every combinatorial permutation of A gives rise to an algebraic permutation of A.

Conversely, let π be a permutation of the set A. Then $\pi(a_j)$ is an element of A and hence $\pi(a_j) = a_{\sigma(j)}$, where $1 \le j \le n$ and σ is a mapping from $\{1, \ldots, n\}$ to itself. Since π is an injective mapping, the elements $a_{\sigma(1)}, a_{\sigma(2)}, \ldots, a_{\sigma(n)}$ are distinct. It follows that $\{a_{\sigma(1)}, a_{\sigma(2)}, \ldots, a_{\sigma(n)}\} = \{a_1, a_2, \ldots, a_n\}$. Hence $(a_{\sigma(1)}, a_{\sigma(2)}, \ldots, a_{\sigma(n)})$ is a combinatorial permutation of the elements $a_1, a_2, \ldots, a_n$. Thus, every algebraic permutation gives rise to a combinatorial permutation.

For our purposes we do not need to focus on the nature of the elements of the given set A. When we study permutations of the elements of A we actually only need to work with their indices, which means that we only work with the set $\{1, 2, \ldots, n\}$.

These arguments show that in order to study permutations of the set $A = \{a_1, a_2, \ldots, a_n\}$, we can study permutations of $\{1, 2, \ldots, n\}$ (notice that the two sets have the same number of elements). Earlier we used the notation $\mathbf{S}(A)$ for the set of permutations of A. However, the notation $\mathbf{S}(\{1, 2, \ldots, n\})$ is cumbersome so we shall instead use the notation $\mathbf{S}_n$ for the set of all permutations of the set $\{1, 2, \ldots, n\}$, which is in accord with standard usage. If $\pi \in \mathbf{S}_n$, then we will say that π is *a permutation of degree n*. Every permutation of degree n can conveniently be written as a matrix consisting of two rows, where the first row has the entries $1, 2, \ldots, n$ and $\pi(m)$ is written in the second row under the entry m in the first row. The permutation π can be written as

$$\begin{pmatrix} 1 & 2 & \ldots & n \\ \pi(1) & \pi(2) & \ldots & \pi(n) \end{pmatrix},$$

which we will call the *tabular form of the permutation*. We note that this is just a notational device; we shall not be adding or multiplying such tabular forms in the manner usually reserved for matrices. Since π is a permutation of the set $\{1, 2, \ldots, n\}$, we see that

$$\{1, 2, \ldots, n\} = \{\pi(1), \pi(2), \ldots, \pi(n)\}.$$

Thus the second row of a tabular form is a permutation of the numbers $1, 2, \ldots, n$. It is not necessary to write all elements of the first row in the natural order from 1 to n, although this is often the way such permutations are written. Sometimes it is convenient to write the first row in a different order. What is most important is that every element of the second row is the image of the corresponding element of the first row situated just above. For example,

$$\begin{pmatrix} 1 & 2 & 3 & 4 & 5 & 6 & 7 & 8 & 9 \\ 4 & 9 & 1 & 7 & 8 & 3 & 5 & 2 & 6 \end{pmatrix} \text{ and } \begin{pmatrix} 2 & 5 & 7 & 1 & 9 & 3 & 6 & 4 & 8 \\ 9 & 8 & 5 & 4 & 6 & 1 & 3 & 7 & 2 \end{pmatrix}$$

are the same permutation. Perhaps, for beginners, in order to better understand permutations, it may be worthwhile to write the permutation with arrows connecting the element of the first row with its image in the second row as in

$$\begin{pmatrix} 1 & 2 & 3 & 4 & 5 & 6 & 7 & 8 & 9 \\ \downarrow & \downarrow & \downarrow & \downarrow & \downarrow & \downarrow & \downarrow & \downarrow & \downarrow \\ 4 & 9 & 1 & 7 & 8 & 3 & 5 & 2 & 6 \end{pmatrix}.$$

This way of writing a permutation will be useful only at the beginning and soon one will feel no need to continue this.

We will multiply permutations by using the general rule of multiplication of mappings, namely composition of functions, introduced in Section 1.3. According to that rule, the product of the two permutations π and σ is the permutation

$$\pi \circ \sigma = \begin{pmatrix} 1 & 2 & \ldots & n \\ \pi(\sigma(1)) & \pi(\sigma(2)) & \ldots & \pi(\sigma(n)) \end{pmatrix}.$$

Thus, to multiply two permutations in tabular form, in the first row of the table corresponding to the permutation σ we choose an arbitrary element i. We locate $\sigma(i)$ in the second row of σ corresponding to i and then find this number $\sigma(i)$ in the first row of the table corresponding to the permutation π. In the second row of the table corresponding to π just under the number $\sigma(i)$, we find the number $\pi(\sigma(i))$. This is the image of i under the product permutation, $\pi \circ \sigma$. A diagram conveniently illustrates this process

$$\begin{array}{cccc} 1 & 2 & \ldots & n \\ \downarrow & \downarrow & & \downarrow \\ \sigma(1) & \sigma(2) & \ldots & \sigma(n) \\ \downarrow & \downarrow & & \downarrow \\ \pi(\sigma(1)) & \pi(\sigma(2)) & \ldots & \pi(\sigma(n)) \end{array}.$$

Given a set A, we next obtain some elements of $\mathbf{S}(A)$.

2.2.1. Lemma. *Let A be a set, let f be a fixed but arbitrary element of $\mathbf{S}(A)$, and let $g \in \mathbf{S}(A)$ be arbitrary. The following mappings are permutations of the set $\mathbf{S}(A)$:*

(i) $\vartheta_1 : g \longrightarrow g^{-1}$;
(ii) $\vartheta_2 : g \longrightarrow f \circ g$;
(iii) $\vartheta_3 : g \longrightarrow g \circ f$.

Proof. (i) Note that if $g \in \mathbf{S}(A)$, then g has an inverse which is also an element of $\mathbf{S}(A)$ so that ϑ_1 is a mapping from $\mathbf{S}(A)$ to itself. We show that ϑ_1 is injective and, to this end, suppose that there are permutations $g_1, g_2 \in \mathbf{S}(A)$ such that $\vartheta_1(g_1) = \vartheta_1(g_2)$. Then $g_1^{-1} = g_2^{-1}$. Since $(g^{-1})^{-1} = g$ it follows that $g_1 = (g_1^{-1})^{-1} = (g_2^{-1})^{-1} = g_2$ and this implies that ϑ_1 is injective. Also if $g \in \mathbf{S}(A)$, then $\vartheta_1(g^{-1}) = (g^{-1})^{-1} = g$ so that ϑ_1 is surjective. Thus, ϑ_1 is bijective.

(ii) Note that when $f, g \in \mathbf{S}(A)$ then $f \circ g \in \mathbf{S}(A)$ so that ϑ_2 is a mapping from $\mathbf{S}(A)$ to itself. To prove that ϑ is injective, let $g_1, g_2 \in \mathbf{S}(A)$ and suppose that $\vartheta_2(g_1) = \vartheta_2(g_2)$. Then we have $f \circ g_1 = f \circ g_2$. Since f^{-1} exists, we may multiply both sides of this equation by f^{-1}. We have

$$\begin{aligned} g_1 &= \varepsilon_A \circ g_1 = (f^{-1} \circ f) \circ g_1 = f^{-1} \circ (f \circ g_1) \\ &= f^{-1} \circ (f \circ g_2) = (f^{-1} \circ f) \circ g_2 = \varepsilon_A \circ g_2 = g_2, \end{aligned}$$

which shows that ϑ_2 is injective. Furthermore, the equation

$$h = \varepsilon_A \circ h = (f \circ f^{-1}) \circ h = f \circ (f^{-1} \circ h) = \vartheta_2(f^{-1} \circ h)$$

implies that ϑ_2 is surjective. Hence ϑ_2 is bijective.

(iii) A similar proof to that in (ii) shows that the mapping $\vartheta_3 : g \longrightarrow g \circ f$ is also bijective.

Permutations interchanging just two integers from the set $\{1, 2, \ldots, n\}$ and leaving all others fixed have special significance.

2.2.2. Definition. *The permutation ι of the set A is called a transposition (more precisely, the transposition of the symbols $k, t \in A$) if $\iota(k) = t, \iota(t) = k$, and $\iota(j) = j$ for all other elements $j \in A$.*

The transposition of k and t will be denoted by ι_{kt} or $(k\ t)$. Thus, a transposition is a permutation that interchanges two selected symbols and leaves all other symbols fixed.

Consider $\iota_{ij}^2 = \iota_{ij} \circ \iota_{ij}$. We have

$$\iota_{ij} \circ \iota_{ij}(i) = \iota_{ij}(\iota_{ij}(i)) = \iota_{ij}(j) = i \text{ and } \iota_{ij} \circ \iota_{ij}(j) = \iota_{ij}(\iota_{ij}(j)) = \iota_{ij}(i) = j.$$

Also, if $k \notin \{i, j\}$, then

$$\iota_{ij} \circ \iota_{ij}(k) = \iota_{ij}(\iota_{ij}(k)) = \iota_{ij}(k) = k.$$

Thus, $\iota_{i,j}(k) = k$ for all $k \in A$ so that $\iota_{ij}^2 = \varepsilon$ is the identity permutation.

We recall that the number of different permutations of elements of the set A consisting of n elements is equal to $n! = 1 \cdot 2 \cdot 3 \cdot \cdots \cdot (n-1) \cdot n$. So we have the following result.

2.2.3. Theorem. $|\mathbf{S}_n| = 1 \cdot 2 \cdot 3 \cdot \cdots \cdot (n-1) \cdot n$.

Proof. The tabular form of the permutation $\pi \in \mathbf{S}_n$ consists of two rows. We can suppose here that the upper row of the tabular form π is $1, 2, \ldots, n$ in this order. The lower row of π is a permutation of the numbers $1, 2, \ldots, n$. Hence, the order of $\mathbf{S}_n$ is equal to the number of different permutations of the numbers $1, 2, \ldots, n$ and this is $n!$.

We now consider all differences $(t-k)$ where $1 \le k < t \le n$, and let $\bigvee_n$ denote the product of such expressions. Then

$$\bigvee_n = \prod_{1 \le k < t \le n} (t-k).$$

If $\pi \in \mathbf{S}_n$, then let

$$\pi\left(\bigvee_n\right) = \prod_{1 \le k < t \le n} (\pi(t) - \pi(k)).$$

For every pair t, k, where $1 \le k < t \le n$, there are natural numbers m, j such that $t = \pi(m)$ and $k = \pi(j)$ so that $t - k = \pi(m) - \pi(j)$. Two cases now occur:

(i) If $m > j$, then $(t-k)$ is a factor in the decomposition of $\pi(\bigvee_n)$.
(ii) If $m < j$, then $\pi(m) = t > k = \pi(j)$.

We say that the natural numbers m, j *form an inversion pair relative to the permutation* π, if $m < j$ but $\pi(m) > \pi(j)$. For example, the permutation

$$\begin{pmatrix} 1 & 2 & 3 & 4 \\ 1 & 4 & 3 & 2 \end{pmatrix}$$

has three inversion pairs namely (2, 3), (2, 4), and (3, 4). Notice that, in case (ii), m, j is an inversion pair. Furthermore, in the second case, the decomposition of $\pi(\bigvee_n)$ has the factor $\pi(j) - \pi(m) = (k - t) = -(t - k)$. Thus, every factor $(t - k)$ of the decomposition $\bigvee_n = \prod_{1 \le k < t \le n}(t - k)$ is also, apart possibly from the sign, a factor in the decomposition $\pi(\bigvee_n) = \prod_{1 \le k < t \le n}(\pi(t) - \pi(k))$. Since the number of factors in $\bigvee_n$ and $\pi(\bigvee_n)$ is the same it follows that

$$\pi\left(\bigvee_n\right) = (-1)^{i(\pi)} \bigvee_n,$$

where $i(\pi)$ denotes the number of inversion pairs, relative to the permutation π. We define $\mathbf{sign}\,\pi = (-1)^{i(\pi)}$ and call $\mathbf{sign}\,\pi$ *the signature of the permutation* π. Consequently, $\pi(\bigvee_n) = \mathbf{sign}\,\pi \cdot \bigvee_n$.

If $\pi, \sigma \in \mathbf{S}_n$ and if $\rho = \pi \circ \sigma$, then

$$\rho\left(\bigvee_n\right) = \prod_{1 \le k < t \le n} (\pi(\sigma(t)) - \pi(\sigma(k))) = \mathbf{sign}\,\rho \cdot \bigvee_n.$$

By using the arguments we saw above, we can prove that

$$(\pi \circ \sigma)\left(\bigvee_n\right) = \mathbf{sign}\,\pi\ \mathbf{sign}\,\sigma \cdot \bigvee_n, \text{ so that } \mathbf{sign}(\pi \circ \sigma) = \mathbf{sign}\,\pi\ \mathbf{sign}\,\sigma.$$

2.2.4. Definition. *The permutation* π *is called even, if* $\mathbf{sign}\,\pi = 1$ *and* π *is called odd, if* $\mathbf{sign}\,\pi = -1$. *Thus,* π *is even precisely when the number of inversion pairs of* π *is even and odd when the number of inversion pairs is odd.*

The equation $\mathbf{sign}(\pi \circ \sigma) = \mathbf{sign}\,\pi\ \mathbf{sign}\,\sigma$ implies that the product of two even permutations is even, the product of two odd permutations is even, and that the product of an even and an odd permutation is odd.

There is a very convenient graphical method for deciding whether a given permutation π is odd or even, based on the following observation. We rewrite the permutation π as two rows of numbers, both in the order $1, 2, \ldots, n$ and then draw a line from each number k to its image $\pi(k)$ in the second row. Let $1 \le j < k \le n$. If (j, k) is not an inversion pair, then the two lines drawn from j to $\pi(j)$ and from k to $\pi(k)$ will not intersect. If the lines do intersect, then this tells us that (j, k) is an inversion pair and the number of such crossovers for all pairs (j, k) determines the number of these. If numbers j and k do not form an inversion pair relative to π, we obtain a picture of the following type:

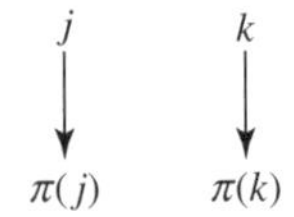

with no crossover of the corresponding lines. If numbers j and k make an inversion pair relative to π, we will have the following picture:

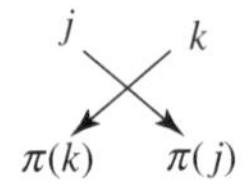

The total number of intersections of these lines is the number of inversion pairs.

We will illustrate this with the following example. Use the same permutation as we already used above:

$$\begin{pmatrix} 1 & 2 & 3 & 4 \\ 1 & 4 & 3 & 2 \end{pmatrix}.$$

The following diagram corresponds to this permutation:

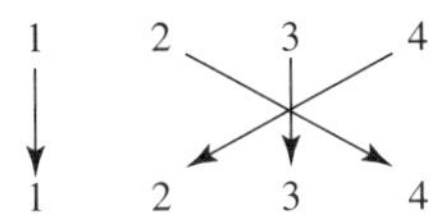

As we can see, there are three intersections corresponding to the three pairs of indices forming inversions (2,3), (2,4), and (3,4). In practice, we often write the permutation π in the usual manner as a matrix, the first row consisting of the elements $\{1, 2, \dots, n\}$ listed in that order. Then, we draw lines from each number in the upper row *to the same number* in the bottom row. This is clearly equivalent to the procedure described above.

We let $\mathbf{A}_n$ denote the subset of $\mathbf{S}_n$ consisting of all even permutations.

2.2.5. Proposition. *Every transposition is an odd permutation.*

Proof. We will find the number of inversion pairs for the transposition ι_{kt} where $k < t$ and both these are fixed. Let i, j be natural numbers such that $1 \leq i < j \leq n$. We wish to determine when this pair forms an inversion pair relative to ι_{kt}. There are several cases to consider. If $\{i, j\} \cap \{k, t\} = \emptyset$, then $\iota_{kt}(i) = i$, $\iota_{kt}(j) = j$, and hence i, j is not an inversion pair since $\iota_{kt}(i) < \iota_{kt}(j)$ in this case. If $j = k$, then $i < j = k < t$ and $\iota_{kt}(i) = i$, $\iota_{kt}(j) = t$. Since $\iota_{kt}(j) = t > k > i = \iota_{kt}(i)$, again the pair i, j is not an inversion pair. In a similar manner, it is possible to show that an inversion pair occurs if $i = k, j = t$ or if $i = k, j < t$ or if $i > k, j = t$. Since k, t are fixed, there is only one inversion pair corresponding to the situation when $i = k, j = t$. When $i = k, j < t$, j could be any of the numbers $k + 1, k + 2, \dots, t - 1$, which gives a total of $t - k - 1$

possible inversion pairs and a similar counting argument shows that there are the same number of inversion pairs when $i > k$, $j = t$. Thus the total number of all such inversion pairs is $2(t - k - 1) + 1$, an odd number. Thus the total number of inversion pairs relative to the transposition ι_{kt} is odd, and hence ι_{kt} is an odd permutation. Since k, t were fixed but arbitrary, the result follows.

Our next result tells us the number of even permutations in the set $\mathbf{S}_n$ and is rather important. We remark that it is evidently the case that $\mathbf{S}_n \backslash \mathbf{A}_n$ is the set of odd permutations in $\mathbf{S}_n$. Of course, no permutation is both even and odd so $\mathbf{S}_n$ is the disjoint union of the sets $\mathbf{A}_n$ and $\mathbf{S}_n \backslash \mathbf{A}_n$.

2.2.6. Proposition. $|\mathbf{A}_n| = \frac{n!}{2}$.

Proof. Let $1 \leq i < j \leq n$ be fixed and consider the mapping $\vartheta : \mathbf{S}_n \longrightarrow \mathbf{S}_n$, which is defined by the rule $\vartheta(\sigma) = \iota_{ij} \circ \sigma$, whenever $\sigma \in \mathbf{S}_n$. By Lemma 2.2.1, the mapping ϑ is a bijection. If π is an even permutation then Proposition 2.2.5 and the remark before Definition 2.2.4 imply that $\iota_{ij} \circ \pi = \vartheta(\pi)$ is an odd permutation. Conversely, if π is an odd permutation, then $\vartheta(\pi)$ is even. Furthermore, in either case, $\vartheta(\iota_{ij} \circ \pi) = \iota_{ij} \circ \iota_{ij} \circ \pi = \pi$. This means that $\vartheta(\mathbf{A}_n) \subseteq \mathbf{S}_n \backslash \mathbf{A}_n \subseteq \vartheta(\mathbf{A}_n)$ and $\vartheta(\mathbf{S}_n \backslash \mathbf{A}_n) \subseteq \mathbf{A}_n \subseteq \vartheta(\mathbf{S}_n \backslash \mathbf{A}_n)$. Since ϑ is bijective,

$$|\mathbf{A}_n| = |\vartheta(\mathbf{A}_n)| = |\mathbf{S}_n \backslash \mathbf{A}_n| .$$

Thus, $|\mathbf{A}_n| = |\mathbf{S}_n \backslash \mathbf{A}_n|$, and our observation before the start of the proof shows that $|\mathbf{A}_n| = \frac{n!}{2}$.

The next theorem shows the key role of transpositions in the theory of permutations.

2.2.7. Theorem. *Every permutation is a product of transpositions.*

Proof. If $\pi \in \mathbf{S}_n$, then put

$$\mathbf{Inv}(\pi) = \{k \mid 1 \leq k \leq n \text{ and } \pi(k) = k\},$$

the set of elements fixed by π. We proceed by induction on the number $r(\pi) = n - |\mathbf{Inv}(\pi)|$. If $r(\pi) = 0$, then $n = |\mathbf{Inv}(\pi)|$. It follows that $\pi = \varepsilon$, the identity permutation. In this case, for example, $\varepsilon = \iota_{ij} \circ \iota_{ij}$ for each transposition ι_{ij}, so certainly the result holds in this case.

Now let $r(\pi) > 0$ and suppose that we have already proved the result for all permutations σ satisfying the condition $r(\sigma) < r(\pi)$.

Since $r(\pi) > 0$, $\mathbf{Inv}(\pi) \neq \{1, 2, \ldots, n\}$. It follows that there is a number k such that $\pi(k) = t \neq k$. Since π is a permutation, it also follows that $\pi(t) \neq t$. Consequently, $k, t \notin \mathbf{Inv}(\pi)$. Now consider the product $\pi_1 = \iota_{kt} \circ \pi$. We have

$$\pi_1(k) = \iota_{kt} \circ \pi(k) = \iota_{kt}(\pi(k)) = \iota_{kt}(t) = k,$$

so that $k \in \mathbf{Inv}(\pi_1)$. If $m \in \mathbf{Inv}(\pi)$, then clearly $m \neq k, m \neq t$, and

$$\pi_1(m) = \iota_{kt} \circ \pi(m) = \iota_{kt}(\pi(m)) = \iota_{kt}(m) = m,$$

so that $m \in \mathbf{Inv}(\pi_1)$. Hence, $\mathbf{Inv}(\pi) \subseteq \mathbf{Inv}(\pi_1)$ and $\mathbf{Inv}(\pi) \neq \mathbf{Inv}(\pi_1)$. This implies that $r(\pi_1) < r(\pi)$. By the induction hypothesis, there exists a decomposition

$$\pi_1 = \upsilon_1 \circ \upsilon_2 \circ \cdots \circ \upsilon_s,$$

where the υ_i are certain transpositions, for $1 \leq i \leq s$. Thus,

$$\iota_{kt} \circ \pi = \upsilon_1 \circ \upsilon_2 \circ \cdots \circ \upsilon_s.$$

Multiplying both sides of this equation, on the left, by the transposition ι_{kt}, and observing that $\varepsilon = \iota_{kt} \circ \iota_{kt}$, we obtain the equation

$$\pi = \iota_{kt} \circ \upsilon_1 \circ \upsilon_2 \circ \cdots \circ \upsilon_s.$$

Thus, π is also a product of transpositions and the induction step is now complete.

We observe that the decomposition of a permutation into a product of transpositions is not unique. One easy example of this is the identity permutation, which can be written as a product of transpositions in many different ways. Less trivially, it is easy to show that the permutation

$$\pi = \begin{pmatrix} 1 & 2 & 3 & 4 & 5 & 6 & 7 \\ 4 & 2 & 1 & 3 & 6 & 7 & 5 \end{pmatrix}$$

has at least two decompositions:

$$\pi = \iota_{57} \circ \iota_{56} \circ \iota_{34} \circ \iota_{13} \text{ and } \pi = \iota_{14} \circ \iota_{34} \circ \iota_{56} \circ \iota_{67}.$$

Let $\pi \in \mathbf{S}_n$. The set

$$\mathbf{Supp}(\pi) = \{1, 2, \ldots, n\} \backslash \mathbf{Inv}(\pi)$$

is called the *support of the permutation* π.

We observe at once that $\mathbf{Supp}(\pi) = \mathbf{Supp}(\pi^{-1})$. To see this, we will prove that $\mathbf{Inv}(\pi) = \mathbf{Inv}(\pi^{-1})$, which immediately implies that $\mathbf{Supp}(\pi) = \mathbf{Supp}(\pi^{-1})$. Let $j \in \mathbf{Inv}(\pi)$ so that $\pi(j) = j$. Applying the permutation π^{-1} to both sides of this equation, we see that $j = \pi^{-1}(j)$ which means that $j \in \mathbf{Inv}(\pi^{-1})$. Thus, $\mathbf{Inv}(\pi) \subseteq \mathbf{Inv}(\pi^{-1})$. Applying the above argument to π^{-1} and remembering that $(\pi^{-1})^{-1} = \pi$, we see that also the inclusion $\mathbf{Inv}(\pi^{-1}) \subseteq \mathbf{Inv}(\pi)$ holds, which shows that $\mathbf{Inv}(\pi) = \mathbf{Inv}(\pi^{-1})$.

2.2.8. Definition. *Let* $1 \leq r \leq n$. *A permutation* π *is called a cycle of length* r *if* $\mathbf{Supp}(\pi) = \{j_1, j_2, \ldots, j_r\}$, *and*

$$\pi(j_1) = j_2, \pi(j_2) = j_3, \ldots, \pi(j_{r-1}) = j_r, \pi(j_r) = j_1.$$

In other words, the permutation π "cycles" the indices $j_1, j_2, \ldots, j_r$ around (thus $j_1 \to j_2 \to j_3 \to \cdots \to j_r \to j_1$) but leaves all other indices fixed. A very convenient shorthand notation is used for cycles. Using the notation introduced above, for the cycle π we write

$$\pi = (j_1\ j_2\ \ldots\ j_r)(j_{r+1})\ldots(j_n),$$

or, more briefly still,

$$\pi = (j_1\ j_2\ \ldots\ j_r).$$

In particular, a transposition is a cycle of length 2. Notice also that, in this notation, it does not matter which j_k is listed first; it is only important that the successor of every index in the cycle is its image. For example,

$$(j_1\ j_2\ \ldots\ j_r) = (j_2\ j_3\ \ldots\ j_r\ j_1) = (j_3\ j_4\ \ldots\ j_r\ j_1\ j_2), \text{ and so on.}$$

Two permutations $\pi, \sigma \in \mathbf{S}_n$ are called *disjoint or independent* if $\mathbf{Supp}(\pi) \cap \mathbf{Supp}(\sigma) = \emptyset$.

We observe the following important property of such permutations, which says that disjoint permutations commute.

2.2.9. Proposition. *Let* $\pi, \sigma \in \mathbf{S}_n$. *If the permutations* π *and* σ *are disjoint, then* $\pi \circ \sigma = \sigma \circ \pi$.

Proof. Let $j \in \{1, 2, \ldots, n\}$. We will prove that $\pi(\sigma(j)) = \sigma(\pi(j))$. If $j \notin \mathbf{Supp}(\pi) \cup \mathbf{Supp}(\sigma)$, then $\pi(\sigma(j)) = j = \sigma(\pi(j))$. Suppose now that $j \in \mathbf{Supp}(\pi)$. Since π, σ are disjoint, it follows that $j \notin \mathbf{Supp}(\sigma)$ so $\sigma(j) = j$ and $\pi(\sigma(j)) = \pi(j)$. If $\pi(j) \in \mathbf{Inv}(\pi)$, then $\pi\,(\pi(j)) = \pi(j)$. Applying π^{-1} to both sides of this equation, we see that $\pi(j) = j$, contrary to the fact that $j \in \mathbf{Supp}(\pi)$. Thus, $\pi(j) \in \mathbf{Supp}(\pi)$ also and the disjointness of σ and π shows that $\sigma(\pi(j)) = \pi(j)$. Thus, $\pi(\sigma(j)) = \sigma(\pi(j))$ whenever $j \in \mathbf{Supp}(\pi)$. A similar argument can be applied when $j \in \mathbf{Supp}(\sigma)$ and this completes the proof.

We end this section with another important theorem.

2.2.10. Theorem. *Every nonidentity permutation is a product of mutually disjoint cycles.*

Proof. Let $\varepsilon \neq \pi \in \mathbf{S}_n$. We will proceed by induction on the number $r(\pi) = |\mathbf{Supp}(\pi)|$. If $r(\pi) = 2$, then, clearly, π is a transposition and hence is a cycle of length 2. Now let $r(\pi) > 2$ and assume that the result is proved for all permutations σ with the property that $r(\sigma) < r(\pi)$. Since $\pi \neq \varepsilon$, there is an index j such that $\pi(j) \neq j$. Consider the numbers

$$j, \pi(j), \pi^2(j) = \pi(\pi(j)), \dots, \pi^m(j)) = \pi\left(\pi^{m-1}(j)\right), \dots$$

These numbers all belong to the finite set $\{1, 2, \dots, n\}$ and hence we can find positive integers t, s such that $t > s$ and $\pi^t(j) = \pi^s(j)$. We now apply the permutation π^{-1} s times to both sides of this equation to deduce that $\pi^{t-s}(j) = j$. Hence, there is a least natural number m such that $\pi^m(j) = j$ and the definition of m implies that the numbers

$$j, \pi(j), \pi^2(j), \dots, \pi^{m-1}(j)$$

are all different. Let

$$\sigma = (j \ \ \pi(j) \ \ \pi^2(j) \ \ \dots \ \ \pi^{m-1}(j)), \text{ and let } \rho = \sigma^{-1} \circ \pi.$$

Clearly, σ is a cycle and $\pi = \sigma \circ \rho$. If $k \in \{j, \pi(j), \pi^2(j), \dots, \pi^{m-1}(j)\}$, then it is easy to see that $\rho(k) = k$. It follows that $\mathbf{Supp}(\sigma) \cap \mathbf{Supp}(\rho) = \emptyset$, and hence $\mathbf{Supp}(\sigma) \subseteq \mathbf{Supp}(\pi)$. Furthermore Proposition 2.2.9 shows that $\sigma \circ \rho = \rho \circ \sigma$.

Now let $k \in \mathbf{Inv}(\pi)$. Since $\mathbf{Supp}(\sigma) \subseteq \mathbf{Supp}(\pi)$, it is clear that $\mathbf{Inv}(\pi) \subseteq \mathbf{Inv}(\sigma)$ and hence $k \in \mathbf{Inv}(\sigma)$. We remarked above that $\mathbf{Inv}(\sigma) = \mathbf{Inv}(\sigma^{-1})$, so we deduce that $k \in \mathbf{Inv}(\rho)$. This shows that $\mathbf{Inv}(\pi) \subseteq \mathbf{Inv}(\rho)$. Furthermore, $j \in \mathbf{Inv}(\rho) \backslash \mathbf{Inv}(\pi)$ so $\mathbf{Inv}(\rho) \neq \mathbf{Inv}(\pi)$. This means that $|\mathbf{Supp}(\sigma)| < |\mathbf{Supp}(\pi)|$. By the induction hypothesis, the permutation ρ is a product of disjoint cycles. Finally $\pi = \sigma \circ \rho$ so π is also a product of cycles, and the equation

$$\mathbf{Supp}(\rho) \cap \mathbf{Supp}(\sigma) = \emptyset$$

shows that these cycles are all disjoint. The result follows.

We make the remark that if $\pi = (j_1 \, j_2 \, \dots \, j_r)$ is a cycle, then $\pi = (j_1 \, j_r) \dots (j_1 \, j_3)(j_1 \, j_2)$ as a product of transpositions. This observation and Theorem 2.2.10 give an alternative proof of Theorem 2.2.7.

EXERCISE SET 2.2

2.2.1. Represent the permutation $\begin{pmatrix} 1 & 2 & 3 & 4 & 5 & 6 & 7 & 8 & 9 & 10 & 11 \\ 3 & 6 & 5 & 11 & 7 & 9 & 8 & 1 & 10 & 2 & 4 \end{pmatrix}$ as a product of transpositions.

2.2.2. Represent the permutation $\begin{pmatrix} 1 & 2 & 3 & 4 & 5 & 6 & 7 & 8 & 9 \\ 6 & 4 & 5 & 7 & 2 & 8 & 3 & 9 & 1 \end{pmatrix}$ as a product of transpositions and find its parity.

2.2.3. Represent the permutation $\begin{pmatrix} 1 & 2 & 3 & 4 & 5 & 6 & 7 & 8 & 9 \\ 8 & 6 & 1 & 7 & 5 & 2 & 9 & 4 & 3 \end{pmatrix}$ as a product of transpositions and find its parity.

2.2.4. Represent the permutation $\begin{pmatrix} 1 & 2 & 3 & 4 & 5 & 6 & 7 & 8 & 9 & 10 \\ 2 & 8 & 3 & 1 & 9 & 6 & 7 & 10 & 5 & 4 \end{pmatrix}$ as a product of transpositions.

2.2.5. Find the parity of the permutation $\begin{pmatrix} 1 & 2 & 3 & 4 & 5 & 6 & 7 \\ 7 & 4 & 3 & 5 & 2 & 1 & 6 \end{pmatrix}$.

2.2.6. Find the parity of the permutation $\begin{pmatrix} 1 & 2 & 3 & 4 & 5 & 6 & 7 \\ 2 & 3 & 1 & 4 & 7 & 5 & 6 \end{pmatrix}$.

2.2.7. Represent the permutation $\begin{pmatrix} 1 & 2 & 3 & 4 & 5 & 6 & 7 & 8 & 9 \\ 5 & 6 & 4 & 7 & 9 & 2 & 8 & 1 & 3 \end{pmatrix}$ as a product of transpositions.

2.2.8. Represent the permutation $\begin{pmatrix} 1 & 2 & 3 & 4 & 5 & 6 & 7 & 8 & 9 \\ 3 & 7 & 4 & 9 & 5 & 2 & 6 & 8 & 1 \end{pmatrix}$ as a product of transpositions.

2.2.9. Represent the permutation $\begin{pmatrix} 1 & 2 & 3 & 4 & 5 & 6 & 7 & 8 & 9 & 10 \\ 2 & 3 & 1 & 8 & 6 & 7 & 5 & 9 & 10 & 4 \end{pmatrix}$ as a product of transpositions and find its parity.

2.2.10. Represent the permutation $\begin{pmatrix} 1 & 2 & 3 & 4 & 5 & 6 & 7 & 8 & 9 \\ 6 & 1 & 5 & 3 & 2 & 4 & 9 & 7 & 8 \end{pmatrix}$ as a product of transpositions and find its parity.

2.2.11. Represent the permutation $\begin{pmatrix} 1 & 2 & 3 & 4 & 5 & 6 & 7 & 8 & 9 \\ 9 & 4 & 2 & 3 & 5 & 7 & 1 & 6 & 8 \end{pmatrix}$ as a product of transpositions and find its parity.

2.2.12. Find the sign of $\pi = \begin{pmatrix} 1 & 2 & 3 & 4 & \dots & 2n-1 & 2n \\ 2 & 1 & 4 & 3 & \dots & 2n & 2n-1 \end{pmatrix}$.

2.2.13. Find the sign of $\pi = \begin{pmatrix} 1 & 2 & 3 & 4 & 5 & 6 & \dots & 3n-2 & 3n-1 & 3n \\ 3 & 2 & 1 & 6 & 5 & 4 & \dots & 3n & 3n-1 & 3n-2 \end{pmatrix}$.

2.2.14. Find the sign of $\pi = \begin{pmatrix} 1 & 2 & 3 & 4 & 5 & 6 & \dots & 2n-3 & 2n-2 & 2n-1 & 2n \\ 3 & 4 & 5 & 6 & 7 & 8 & \dots & 2n-1 & 2n & 1 & 2 \end{pmatrix}$.

2.2.15. Find the sign of $\pi = \begin{pmatrix} 1 & 2 & 3 & 4 & 5 & 6 & \dots & 3n-2 & 3n-1 & 3n \\ 2 & 3 & 1 & 5 & 6 & 4 & \dots & 3n-1 & 3n & 3n-2 \end{pmatrix}$.

2.2.16. Find the sign of $\pi = \begin{pmatrix} 1 & 2 & 3 & 4 & 5 & 6 & \dots & 3n-2 & 3n-1 & 3n \\ 4 & 5 & 6 & 7 & 8 & 9 & \dots & 1 & 2 & 3 \end{pmatrix}$.

2.2.17. Find the permutation π from the equation $\begin{pmatrix} 1 & 2 & 3 & 4 & 5 & 6 \\ 2 & 3 & 5 & 1 & 6 & 4 \end{pmatrix}$

$$\begin{pmatrix} 1 & 2 & 3 & 4 & 5 & 6 \\ \pi(1) & \pi(2) & \pi(3) & \pi(4) & \pi(5) & \pi(6) \end{pmatrix} \begin{pmatrix} 1 & 2 & 3 & 4 & 5 & 6 \\ 3 & 1 & 5 & 2 & 6 & 4 \end{pmatrix}$$
$$= \begin{pmatrix} 1 & 2 & 3 & 4 & 5 & 6 \\ 4 & 3 & 5 & 1 & 6 & 4 \end{pmatrix}.$$

2.2.18. Find the permutation π from the equation $\begin{pmatrix} 1 & 2 & 3 & 4 & 5 & 6 \\ 4 & 3 & 6 & 5 & 2 & 1 \end{pmatrix}$

$$\begin{pmatrix} 1 & 2 & 3 & 4 & 5 & 6 \\ \pi(1) & \pi(2) & \pi(3) & \pi(4) & \pi(5) & \pi(6) \end{pmatrix} \begin{pmatrix} 1 & 2 & 3 & 4 & 5 & 6 \\ 3 & 5 & 2 & 6 & 1 & 4 \end{pmatrix}$$
$$= \begin{pmatrix} 1 & 2 & 3 & 4 & 5 & 6 \\ 6 & 5 & 1 & 2 & 3 & 4 \end{pmatrix}.$$

2.2.19. Find the permutation π from the equation $\begin{pmatrix} 1 & 2 & 3 & 4 & 5 & 6 \\ 5 & 4 & 1 & 2 & 6 & 3 \end{pmatrix}$

$$\begin{pmatrix} 1 & 2 & 3 & 4 & 5 & 6 \\ \pi(1) & \pi(2) & \pi(3) & \pi(4) & \pi(5) & \pi(6) \end{pmatrix} \begin{pmatrix} 1 & 2 & 3 & 4 & 5 & 6 \\ 3 & 6 & 5 & 2 & 1 & 4 \end{pmatrix}$$
$$= \begin{pmatrix} 1 & 2 & 3 & 4 & 5 & 6 \\ 2 & 3 & 5 & 6 & 4 & 1 \end{pmatrix}.$$

2.2.20. Find π^{97} if $\pi = \begin{pmatrix} 1 & 2 & 3 & 4 & 5 & 6 & 7 & 8 & 9 \\ 3 & 6 & 5 & 7 & 4 & 2 & 1 & 9 & 8 \end{pmatrix}$.

2.3 DETERMINANTS OF MATRICES

In this section we introduce a very important numerical characteristic of a square matrix—its determinant. First, we consider some preliminary concepts.

Let $M = \{a_1, a_2, \ldots, a_n\}$ be a finite set of numbers. Sometimes, instead of the commonly used notation $\sum_{i=1}^{n} a_i = a_1 + a_2 + \cdots + a_n$ for the sum of the elements a_i of the set M, we will use the shorter notation $\sum_{a \in M} a$. Note that we can also index a finite set of elements with the help of a segment of the set of natural numbers and also by using other finite sets. For example, if X is a finite set, then we might use X as an index set; using this notation, a set M could be written as $M = \{a_x \mid x \in X\}$. Thus, for the sum of all elements of M, we can also use the notation $\sum_{x \in X} a_x$. If π is a permutation of the finite set X, then

$$\{\pi(x) \mid x \in X\} = X,$$

and hence

$$M = \left\{a_{\pi(x)} \mid x \in X\right\}.$$

It follows that

$$\sum_{x\in X} a_x = \sum_{x\in X} a_{\pi(x)}.$$

We can then use Lemma 2.2.1 to obtain the following result.

2.3.1. Lemma. *Let n be a natural number and let $M = \{a_\pi \mid \pi \in \mathbf{S}_n\}$ be a finite set, indexed by the set of permutations $\mathbf{S}_n$. Then*

(i) $\sum_{\pi\in\mathbf{S}_n} a_\pi = \sum_{\pi\in\mathbf{S}_n} a_{\pi^{-1}}$,
(ii) $\sum_{\pi\in\mathbf{S}_n} a_\pi = \sum_{\pi\in\mathbf{S}_n} a_{\pi\circ\sigma} = \sum_{\pi\in\mathbf{S}_n} a_{\sigma\circ\pi}$, *where σ is a fixed permutation from $\mathbf{S}_n$.*

We will remind the reader of the definition of the determinants of second and third order and then, by analogy, we will introduce the idea of the determinant of a square matrix of arbitrary order. We have

$$\begin{vmatrix} a_{11} & a_{12} \\ a_{21} & a_{22} \end{vmatrix} = a_{11}a_{22} - a_{12}a_{21},$$

$$\begin{vmatrix} a_{11} & a_{12} & a_{13} \\ a_{21} & a_{22} & a_{23} \\ a_{31} & a_{32} & a_{33} \end{vmatrix} = a_{11}\begin{vmatrix} a_{22} & a_{23} \\ a_{32} & a_{33} \end{vmatrix} - a_{12}\begin{vmatrix} a_{21} & a_{23} \\ a_{31} & a_{33} \end{vmatrix} + a_{13}\begin{vmatrix} a_{21} & a_{22} \\ a_{31} & a_{32} \end{vmatrix}$$
$$= a_{11}a_{22}a_{33} + a_{13}a_{21}a_{32} + a_{12}a_{23}a_{31} - a_{13}a_{22}a_{31}$$
$$- a_{12}a_{21}a_{33} - a_{11}a_{23}a_{32}.$$

We wish to discover some common features of these two formulae. First note that in the first case there are two terms in each product whereas in the second there are three terms in each product. In the first case there are 2 terms in total whereas in the second case there are $6 = 3!$ terms. Notice also that in each term the set of second indices occurring is a permutation of the set $\{1, 2\ldots, n\}$, where $n = 2$ or 3. In the first case, therefore, we consider the set $\mathbf{S}_2$ which consists of the two permutations

$$\varepsilon = \begin{pmatrix} 1 & 2 \\ 1 & 2 \end{pmatrix}, \alpha = \begin{pmatrix} 1 & 2 \\ 2 & 1 \end{pmatrix},$$

and we note that $\mathbf{sign}\,\varepsilon = 1$, $\mathbf{sign}\,\alpha = -1$. Consequently, we can write

$$a_{11}a_{22} - a_{12}a_{21} = \mathbf{sign}\,\varepsilon\, a_{1,\varepsilon(1)}a_{2,\varepsilon(2)} + \mathbf{sign}\,\alpha\, a_{1,\alpha(1)}a_{2,\alpha(2)}.$$

Now consider the second case where the determinants are of order 3. In this case, the set $\mathbf{S}_3$ consists of six permutations:

$$\varepsilon = \begin{pmatrix} 1 & 2 & 3 \\ 1 & 2 & 3 \end{pmatrix}, \alpha = \begin{pmatrix} 1 & 2 & 3 \\ 3 & 1 & 2 \end{pmatrix}, \beta = \begin{pmatrix} 1 & 2 & 3 \\ 2 & 3 & 1 \end{pmatrix},$$
$$\gamma = \begin{pmatrix} 1 & 2 & 3 \\ 3 & 2 & 1 \end{pmatrix}, \pi = \begin{pmatrix} 1 & 2 & 3 \\ 1 & 3 & 2 \end{pmatrix}, \sigma = \begin{pmatrix} 1 & 2 & 3 \\ 2 & 1 & 3 \end{pmatrix},$$

where $\mathbf{sign}\,\varepsilon = \mathbf{sign}\,\alpha = \mathbf{sign}\,\beta = 1$ and $\mathbf{sign}\,\gamma = \mathbf{sign}\,\pi = \mathbf{sign}\,\sigma = -1$. Now we have, by inspection,

$$\begin{vmatrix} a_{11} & a_{12} & a_{13} \\ a_{21} & a_{22} & a_{23} \\ a_{31} & a_{32} & a_{33} \end{vmatrix} = \mathbf{sign}\,\varepsilon\; a_{1,\varepsilon(1)}a_{2,\varepsilon(2)}a_{3,\varepsilon(3)} + \mathbf{sign}\,\alpha\; a_{1,\alpha(1)}a_{2,\alpha(2)}a_{3,\alpha(3)}$$
$$+ \mathbf{sign}\,\beta\; a_{1,\beta(1)}a_{2,\beta(2)}a_{3,\beta(3)} + \mathbf{sign}\,\gamma\; a_{1,\gamma(1)}a_{2,\gamma(2)}a_{3,\gamma(3)}$$
$$+ \mathbf{sign}\,\pi\; a_{1,\pi(1)}a_{2,\pi(2)}a_{3,\pi(3)} + \mathbf{sign}\,\sigma\; a_{1,\sigma(1)}a_{2,\sigma(2)}a_{3,\sigma(3)}.$$

For an arbitrary square matrix of dimension n we make the following definition, using the cases when $n = 2, 3$ as our model.

2.3.2. Definition. *Let* $A = [a_{ij}] \in \mathbf{M}_n(\mathbb{R})$. *For each permutation,* $\pi \in \mathbf{S}_n$ *form the product*

$$\mathbf{sign}\,\pi\; a_{1,\pi(1)}a_{2,\pi(2)} \ldots a_{n,\pi(n)}.$$

The sum $\mathbf{det}(A)$ *of all these products is called the determinant of the matrix A. Thus,*

$$\mathbf{det}(A) = \sum_{\pi \in \mathbf{S}_n} \mathbf{sign}\,\pi\; a_{1,\pi(1)}a_{2,\pi(2)} \ldots a_{n,\pi(n)}.$$

We will use the following expanded notation for $\mathbf{det}(A)$:

$$\begin{vmatrix} a_{11} & a_{12} & \ldots & a_{1n} \\ a_{21} & a_{22} & \ldots & a_{2n} \\ \vdots & \vdots & \ddots & \vdots \\ a_{n1} & a_{n2} & \ldots & a_{nn} \end{vmatrix}.$$

Note that the determinant of a matrix of dimension 1 is the number that is in this single-entry matrix.

This definition is difficult to employ even for matrices with relatively small dimensions (for $n = 4$ the sum in the decomposition of the determinant has 24 terms, for $n = 5$ it has 120 terms, and so on). For this reason, we now determine some elementary, but important, properties of determinants that help to evaluate them in a relatively easy way. With the help of these properties we can move from a given matrix to another one with the same determinant, but whose value is more easily calculated.

2.3.3. Proposition. *Let* $A = [a_{ij}] \in \mathbf{M}_n(\mathbb{R})$ *and let* $B = A^t$. *Then* $\mathbf{det}(A) = \mathbf{det}(B)$.

Proof. Let $B = [b_{ij}]$, where $b_{ij} = a_{ji}$, for $1 \leq i, j \leq n$. We have

$$\mathbf{det}(A) = \sum_{\pi \in \mathbf{S}_n} d_\pi, \text{ where } d_\pi = \mathbf{sign}\,\pi\; a_{1,\pi(1)}a_{2,\pi(2)} \ldots a_{n,\pi(n)}.$$

Furthermore,

$$\begin{aligned}\mathbf{det}(B) &= \sum_{\pi \in \mathbf{S}_n} \mathbf{sign}\,\pi\; b_{1,\pi(1)}b_{2,\pi(2)} \ldots b_{n,\pi(n)}\\ &= \sum_{\pi \in \mathbf{S}_n} \mathbf{sign}\,\pi\; a_{\pi(1),1}a_{\pi(2),2} \ldots a_{\pi(n),n}\\ &= \sum_{\pi \in \mathbf{S}_n} \mathbf{sign}\,\pi\; a_{\pi(1),\pi^{-1}\pi(1)}a_{\pi(2),\pi^{-1}\pi(2)} \ldots a_{\pi(n),\pi^{-1}\pi(n)}. \qquad (2.1)\end{aligned}$$

However, π is a permutation so $\{\pi(1), \ldots, \pi(n)\} = \{1, 2, \ldots, n\}$ and also $\mathbf{sign}\,\pi = \mathbf{sign}\,\pi^{-1}$. Thus, $a_{\pi(j),\pi^{-1}\pi(j)} = a_{k,\pi^{-1}(k)}$ and, by rearranging the terms suitably, the last equation of Equation 2.1 becomes

$$\begin{aligned}\mathbf{det}(B) &= \sum_{\pi \in \mathbf{S}_n} \mathbf{sign}\,\pi^{-1}\; a_{1,\pi^{-1}(1)}a_{2,\pi^{-1}(2)} \ldots a_{n,\pi^{-1}(n)}\\ &= \sum_{\pi^{-1} \in \mathbf{S}_n} \mathbf{sign}\,\pi^{-1}\; a_{1,\pi^{-1}(1)}a_{2,\pi^{-1}(2)} \ldots a_{n,\pi^{-1}(n)} = \mathbf{det}(A).\end{aligned}$$

Since the rows of the matrix A^t are the columns of the matrix A, and the columns of A^t are the rows of A, Proposition 2.3.3 shows that the columns and rows of a matrix have the same rights relative to properties of determinants. The proposition implies that, for every assertion concerning the rows of a matrix, we can find a corresponding assertion concerning its columns at least as far as the determinant is concerned. We therefore formulate all determinant properties for rows and remember that corresponding assertions are valid for columns.

Our next proposition presents a rather unusual property of determinants. We note that usually determinants do not exhibit such additive properties.

2.3.4. Proposition. *Let $A = [a_{ij}] \in \mathbf{M}_n(\mathbb{R})$, and suppose that $a_{kj} = b_j + c_j$ for some fixed k, where $1 \le k \le n$. Let $U = [u_{ij}]$ and $V = [v_{ij}]$, where*

$$u_{ij} = \begin{cases} a_{ij}, & \text{if } i \neq k, \\ b_j, & \text{if } i = k. \end{cases} \qquad v_{ij} = \begin{cases} a_{ij}, & \text{if } i \neq k, \\ c_j, & \text{if } i = k. \end{cases}$$

Then $\mathbf{det}(A) = \mathbf{det}(U) + \mathbf{det}(V)$.

Proof. We have

$$\begin{aligned}
\mathbf{det}(A) &= \sum_{\pi \in \mathbf{S}_n} \mathbf{sign}\,\pi\ a_{1,\pi(1)}a_{2,\pi(2)} \dots a_{k-1,\pi(k-1)}a_{k,\pi(k)}a_{k+1,\pi(k+1)} \dots a_{n,\pi(n)} \\
&= \sum_{\pi \in \mathbf{S}_n} \mathbf{sign}\,\pi\ a_{1,\pi(1)}a_{2,\pi(2)} \dots a_{k-1,\pi(k-1)}(b_{\pi(k)} \\
&\quad + c_{\pi(k)})a_{k+1,\pi(k+1)} \dots a_{n,\pi(n)} \\
&= \sum_{\pi \in \mathbf{S}_n} \mathbf{sign}\,\pi\ a_{1,\pi(1)}a_{2,\pi(2)} \dots a_{k-1,\pi(k-1)}b_{\pi(k)}a_{k+1,\pi(k+1)} \dots a_{n,\pi(n)} \\
&\quad + \sum_{\pi \in \mathbf{S}_n} \mathbf{sign}\,\pi\ a_{1,\pi(1)}a_{2,\pi(2)} \dots a_{k-1,\pi(k-1)}c_{\pi(k)}a_{k+1,\pi(k+1)} \dots a_{n,\pi(n)} \\
&= \sum_{\pi \in \mathbf{S}_n} \mathbf{sign}\,\pi\ u_{1,\pi(1)}u_{2,\pi(2)} \dots u_{n,\pi(n)} \\
&\quad + \sum_{\pi \in \mathbf{S}_n} \mathbf{sign}\,\pi\ v_{1,\pi(1)}v_{2,\pi(2)} \dots v_{n,\pi(n)} \\
&= \mathbf{det}(U) + \mathbf{det}(V).
\end{aligned}$$

2.3.5. Proposition. *Let $A = [a_{ij}] \in \mathbf{M}_n(\mathbb{R})$. Suppose that $a_{kj} = \alpha b_j$ for some fixed k, where $1 \le k \le n$ and some fixed $\alpha \in \mathbb{R}$. Let $U = [u_{ij}]$, where*

$$u_{ij} = \begin{cases} a_{ij}, & \text{if } i \neq k, \\ b_j, & \text{if } i = k. \end{cases}$$

Then $\mathbf{det}(A) = \alpha\mathbf{det}(U)$.

Proof. We have

$$\begin{aligned}
\mathbf{det}(A) &= \sum_{\pi \in \mathbf{S}_n} \mathbf{sign}\,\pi\, a_{1,\pi(1)}a_{2,\pi(2)} \dots a_{k-1,\pi(k-1)}a_{k,\pi(k)}a_{k+1,\pi(k+1)} \dots a_{n,\pi(n)} \\
&= \sum_{\pi \in \mathbf{S}_n} \mathbf{sign}\,\pi\, a_{1,\pi(1)}a_{2,\pi(2)} \dots a_{k-1,\pi(k-1)}(\alpha b_{\pi(k)})a_{k+1,\pi(k+1)} \dots a_{n,\pi(n)}
\end{aligned}$$

$$= \alpha \left(\sum_{\pi \in \mathbf{S}_n} \mathbf{sign}\, \pi\, a_{1,\pi(1)} a_{2,\pi(2)} \dots a_{k-1,\pi(k-1)} b_{\pi(k)} a_{k+1,\pi(k+1)} \dots a_{n,\pi(n)} \right)$$

$$= \alpha \left(\sum_{\pi \in \mathbf{S}_n} \mathbf{sign}\, \pi\, u_{1,\pi(1)} u_{2,\pi(2)} \dots u_{n,\pi(n)} \right) = \alpha \mathbf{det}(U).$$

Proposition 2.3.5 shows that if one column (or row) of a matrix A is multiplied by a constant α, then the determinant is also multiplied by α. In particular, by setting $\alpha = 0$ we obtain the following corollary, although this could just be read off from the definition of determinant.

2.3.6. Corollary. *Let $A = [a_{ij}] \in \mathbf{M}_n(\mathbb{R})$ and suppose that $a_{kj} = 0$ for some fixed k, where $1 \leq k \leq n$. Then $\mathbf{det}(A) = 0$.*

Our next result tells us what happens when we interchange two rows of a matrix.

2.3.7. Proposition. *Let $A = [a_{ij}] \in \mathbf{M}_n(\mathbb{R})$, and let k, t be fixed positive integers, where $1 \leq k, t \leq n$. Let $B = [b_{ij}]$, where*

$$b_{ij} = \begin{cases} a_{ij}, & \text{if } i \neq k, t, \\ a_{tj}, & \text{if } i = k, \\ a_{kj}, & \text{if } i = t. \end{cases}$$

Then $\mathbf{det}(B) = -\mathbf{det}(A)$.

Proof. Without loss of generality, we may assume that $k < t$. We have

$$\mathbf{det}(A) = \sum_{\pi \in \mathbf{S}_n} d_\pi,$$

where

$$d_\pi = \mathbf{sign}\, \pi\, a_{1,\pi(1)} a_{2,\pi(2)} \dots a_{n,\pi(n)}.$$

Furthermore,

$$\mathbf{det}(B) = \sum_{\pi \in \mathbf{S}_n} \mathbf{sign}\, \pi\, u_\pi,$$

where

$$\begin{aligned} u_\pi &= b_{1,\pi(1)} b_{2,\pi(2)} \dots b_{k-1,\pi(k-1)} b_{k,\pi(k)} b_{k+1,\pi(k+1)} \dots b_{t-1,\pi(t-1)} \\ &\quad \times b_{t,\pi(t)} b_{t+1,\pi(t+1)} \dots b_{n,\pi(n)} \\ &= a_{1,\pi(1)} a_{2,\pi(2)} \dots a_{k-1,\pi(k-1)} a_{t,\pi(k)} a_{k+1,\pi(k+1)} \dots a_{t-1,\pi(t-1)} \\ &\quad \times a_{k,\pi(t)} a_{t+1,\pi(t+1)} \dots a_{n,\pi(n)}. \end{aligned}$$

Consider the product $\sigma = \pi \circ \iota_{kt}$. Suppose that $i \neq k, t$, then

$$\pi \circ \iota_{kt}(i) = \pi(\iota_{kt}(i)) = \pi(i).$$

Furthermore,

$$\sigma(k) = \pi \circ \iota_{kt}(k) = \pi(\iota_{kt}(k)) = \pi(t) \text{ and } \sigma(t) = \pi \circ \iota_{kt}(t) = \pi(\iota_{kt}(t)) = \pi(k).$$

Since

$$\mathbf{sign}\,\sigma = \mathbf{sign}(\pi \circ \iota_{kt}) = \mathbf{sign}\,\pi\ \mathbf{sign}\,\iota_{kt}$$

and recalling that a transposition is an odd permutation, we obtain

$$\mathbf{sign}\,\sigma = -\mathbf{sign}\,\pi, \text{ or } \mathbf{sign}\,\pi = -\mathbf{sign}\,\sigma.$$

Thus,

$$\mathbf{sign}\,\pi\ b_{1,\pi(1)}b_{2,\pi(2)}\ldots b_{n,\pi(n)} = -\mathbf{sign}\,\sigma\ a_{1,\sigma(1)}a_{2,\sigma(2)}\ldots a_{n,\sigma(n)} = -d_\sigma.$$

We note that as π varies over the elements of $\mathbf{S}_n$, so does σ and hence

$$\mathbf{det}(B) = \sum_{\sigma\in\mathbf{S}_n}(-d_\sigma) = -\left(\sum_{\sigma\in\mathbf{S}_n} d_\sigma\right).$$

Since $\sigma = \pi \circ \iota_{kt}$, Lemma 2.3.1 implies the following equation:

$$\mathbf{det}(B) = \sum_{\sigma\in\mathbf{S}_n}(-d_\sigma) = -\sum_{\pi\in\mathbf{S}_n} d_\sigma = -\mathbf{det}(A).$$

This result tells us that if we interchange two rows (or columns) of a matrix then the sign of the resulting determinant changes.

2.3.8. Corollary. *Let* $A = [a_{ij}] \in \mathbf{M}_n(\mathbb{R})$. *If* A *has two equal columns, then* $\mathbf{det}(A) = 0$.

Proof. Assume that the columns indexed by k and t are equal. If we interchange these columns we obtain a matrix B which is clearly the same as A. However, Proposition 2.3.7 shows that the determinant changes in sign so that $\mathbf{det}(A) = \mathbf{det}(B) = -\mathbf{det}(A)$. It follows easily from this that $\mathbf{det}(A) = 0$.

2.3.9. Corollary. *Let* $A = [a_{ij}] \in \mathbf{M}_n(\mathbb{R})$ *and let* k, t *be fixed positive integers such that* $1 \leq k \neq t \leq n$. *Let* α *be a fixed real number and let* $B = [b_{ij}]$, *where*

$$b_{ij} = \begin{cases} a_{ij}, & \text{if } i \neq k, \\ a_{kj} + \alpha a_{tj}, & \text{if } i = k. \end{cases}$$

Then $\mathbf{det}(B) = \mathbf{det}(A)$.

Proof. Let $U = [u_{ij}]$ where

$$u_{ij} = \begin{cases} a_{ij}, & \text{if } i \neq k, \\ a_{tj}, & \text{if } i = k. \end{cases}$$

By Propositions 2.3.4 and 2.3.5, $\mathbf{det}(B) = \mathbf{det}(A) + \alpha\mathbf{det}(U)$. Corollary 2.3.8 implies that $\mathbf{det}(U) = 0$, since U has two identical columns and hence $\mathbf{det}(B) = \mathbf{det}(A)$.

This result shows that if we add a multiple of one row (or column) of a matrix to another row (respectively, column) of the matrix then the determinant does not change.

The rows of the $m \times n$ matrix A are

$$(a_{11}, a_{12}, \ldots, a_{1n}),$$
$$(a_{21}, a_{22}, \ldots, a_{2n}),$$
$$\vdots,$$
$$(a_{m1}, a_{m2}, \ldots, a_{mn}).$$

Let $\alpha_1, \alpha_2, \ldots, \alpha_m$ be real numbers and, for $1 \leq i \leq n$, let $b_i = \alpha_1 a_{1i} + \alpha_2 a_{2i} + \cdots + \alpha_m a_{mi}$. The n-tuple or row vector $(b_1, b_2, \ldots, b_n)$ is called a *linear combination* of the given rows with coefficients $\alpha_1, \alpha_2, \ldots, \alpha_m$. In this case,

$$(b_1, b_2, \ldots, b_n) = \alpha_1(a_{11}, a_{12}, \ldots, a_{1n}) + \alpha_2(a_{21}, a_{22}, \ldots, a_{2n}) + \cdots$$
$$+ \alpha_m(a_{m1}, a_{m2}, \ldots, a_{mn}).$$

By using Corollary 2.3.9 repeatedly we see that we can keep adding multiples of different rows (or columns) to some row (respectively, column) without changing the determinant.

2.3.10. Corollary. *Let B be a matrix obtained from the matrix A by adding a certain linear combination of certain rows of A to some other row of A. Then* $\mathbf{det}(B) = \mathbf{det}(A)$.

Next, we compute the determinant of an upper triangular matrix. This turns out to be the product of the diagonal elements.

2.3.11. Proposition. *Let $A = [a_{ij}] \in \mathbf{M}_n(\mathbb{R})$ be an upper triangular matrix. Then* $\mathbf{det}(A) = a_{11}a_{22}\ldots a_{nn}$.

Proof. We have

$$\mathbf{det}(A) = \sum_{\pi \in \mathbf{S}_n} \mathbf{sign}\,\pi\; a_{1,\pi(1)}a_{2,\pi(2)}\dots a_{n,\pi(n)}.$$

Let π be a nonidentity permutation. Then, there is a positive integer m such that $\pi(m) \neq m$. Choose the largest positive integer k such that $1 \leq k \leq n$ with the property that $\pi(k) \neq k$. This means that

$$\pi(k+1) = k+1, \dots, \pi(n) = n,$$

so $\pi(k) \notin \{k, k+1, \dots, n\}$, and hence $k > \pi(k)$. Since A is upper triangular it follows that $a_{k,\pi(k)} = 0$. So, in the decomposition of $\mathbf{det}(A)$, only the summand indexed by the identity permutation is nonzero. This means that $\mathbf{det}(A) = a_{11}a_{22}\dots a_{nn}$.

The properties that we have obtained above enable us to use a systematic method for computing the determinant of a matrix of large dimension. This method is the well-known method of Gaussian elimination which is used for solving systems of linear equations and which may be familiar to everybody from a high school algebra course. There it was probably called the method of substitution or elimination. We briefly describe this method here.

Let $A = [a_{ij}] \in \mathbf{M}_n(\mathbb{R})$. To find its determinant we consider the matrix transformations described above that leave the determinant unchanged to within multiplication by -1 and transform the matrix to its upper triangular form. In order to do this we consider the first column of A which consists of the elements a_{i1}, $1 \leq i \leq n$. If all of these elements are zero, then the matrix has a zero column and by Corollary 2.3.6 its determinant is equal to 0. Hence, we may assume that there exists an index k such that $a_{k1} \neq 0$ and without loss of generality we may suppose that $a_{11} \neq 0$. If this is not true we can just interchange the first and kth rows. By Proposition 2.3.7 A and the new matrix have determinants that differ only in sign. Next, we multiply the first row by $\frac{-a_{i1}}{a_{11}}$ and add the result to the ith row, for $2 \leq i \leq n$. By Corollary 2.3.9 such a transformation does not change the determinant. In this way we obtain a matrix all of whose entries in the first column are 0, except for the entry in the first row.

Now consider the second column and apply the same type of transformation as described for the first column. In this case, we only consider the rows 2 through n. Thus, if all entries a_{k2} are 0 for $2 \leq k \leq n$ or if $a_{22} \neq 0$ we do nothing, otherwise we interchange a row below the second with the second row in order to make the entry in the second row and second column of the matrix nonzero. Thus, we may assume $a_{22} \neq 0$ and then we multiply the second row of the matrix by $\frac{-a_{i2}}{a_{22}}$ and add the result to the ith row for $3 \leq i \leq n$ to obtain a matrix whose second column has at most two nonzero elements, which occur in the first and second rows, and whose determinant is equal to the determinant of the original matrix to within multiplication by -1. Performing these operations repeatedly, on succeeding columns, we finally arrive at an upper triangular matrix whose determinant differs by a factor of at most -1 from our original matrix. Since the upper triangular matrix obtained has a very easily computed determinant, by Proposition 2.3.11, this provides a theoretical method for the computation of any determinant.

Now we will apply the properties proved above to compute the determinant of a skew-symmetric matrix A. First, we multiply every row of A by -1 to obtain the matrix $-A$. However, for a skew-symmetric matrix $-A = A^t$, and Propositions 2.3.3 and 2.3.5 can be used repeatedly to obtain

$$\mathbf{det}(A) = \mathbf{det}(A^t) = \mathbf{det}(-A) = (-1)^n \mathbf{det}(A).$$

Thus, if the dimension of a skew-symmetric matrix A is odd then $\mathbf{det}(A) = 0$. For even dimension nothing further will be deduced here.

Here are some more fairly typical problems that deal with the computation of determinants based on the properties proved above.

2.3.12. Example. We will find the determinant of the matrix

$$A = \begin{pmatrix} a_1 & x & \dots & x & x \\ x & a_2 & \dots & x & x \\ \vdots & \vdots & \vdots & \vdots & \vdots \\ x & x & \dots & x & a_n \end{pmatrix}.$$

Subtracting the first row from all the others, we see that, by Corollary 2.3.9, $\mathbf{det}(A) = \mathbf{det}(B)$, where

$$B = \begin{pmatrix} a_1 & x & \dots & x & x \\ x - a_1 & a_2 - x & \dots & 0 & 0 \\ x - a_1 & 0 & \dots & 0 & 0 \\ \vdots & \vdots & \ddots & \vdots & \vdots \\ x - a_1 & 0 & \dots & 0 & a_n - x \end{pmatrix}.$$

Using Proposition 2.3.5 it follows that

$$\mathbf{det}(B) = (a_1 - x)(a_2 - x)\dots(a_n - x)\mathbf{det}(C),$$

where

$$C = \begin{pmatrix} \frac{a_1}{a_1 - x} & \frac{x}{a_2 - x} & \frac{x}{a_3 - x} & \dots & \frac{x}{a_{n-1} - x} & \frac{x}{a_n - x} \\ -1 & 1 & 0 & \dots & 0 & 0 \\ -1 & 0 & 1 & \dots & 0 & 0 \\ \vdots & \vdots & \vdots & \vdots & \vdots & \vdots \\ -1 & 0 & 0 & \dots & 0 & 1 \end{pmatrix}.$$

We have $\frac{a_1}{a_1-x} = 1 + \frac{x}{a_1-x}$. We now add all columns to the first one. By Proposition 2.3.5, $\mathbf{det}(C) = \mathbf{det}(D)$, where

$$D = \begin{pmatrix} b & \dfrac{x}{a_2-x} & \dfrac{x}{a_3-x} & \dots & \dfrac{x}{a_{n-1}-x} & \dfrac{x}{a_n-x} \\ 0 & 1 & 0 & \dots & 0 & 0 \\ 0 & 0 & 1 & \dots & 0 & 0 \\ \vdots & \vdots & \vdots & \vdots & \vdots & \vdots \\ 0 & 0 & 0 & \dots & 0 & 1 \end{pmatrix},$$

and

$$b = 1 + \frac{x}{a_1-x} + \frac{x}{a_2-x} + \dots + \frac{x}{a_n-x}.$$

Applying Propositions 2.3.5 and 2.3.11 we obtain

$$\begin{aligned}\mathbf{det}(A) &= b(a_1-x)\dots(a_n-x) \\ &= x(a_1-x)(a_2-x)\dots(a_n-x)\left(\frac{1}{x} + \frac{1}{a_1-x} + \dots + \frac{1}{a_n-x}\right).\end{aligned}$$

2.3.13. Example. We will find the determinant of the matrix

$$A = \begin{pmatrix} a_1+b_1 & a_1+b_2 & \dots & a_1+b_n \\ a_2+b_1 & a_2+b_2 & \dots & a_2+b_n \\ \vdots & \vdots & \ddots & \vdots \\ a_n+b_1 & a_n+b_2 & \dots & a_n+b_n \end{pmatrix}.$$

The simplest way to proceed here is to subtract row 1 of this matrix from each other row. When we do this Corollary 2.3.9 implies that the determinant does not change and we obtain the new matrix:

$$B = \begin{pmatrix} a_1+b_1 & a_1+b_2 & \dots & a_1+b_n \\ a_2-a_1 & a_2-a_1 & \dots & a_2-a_1 \\ \vdots & \vdots & \ddots & \vdots \\ a_n-a_1 & a_n-a_1 & \dots & a_n-a_1 \end{pmatrix}.$$

There is now a common factor of $a_j - a_1$ in row j for $j \geq 2$. Consequently, if $n \geq 2$ then the matrix B has the same determinant as

$$(a_2-a_1)(a_3-a_1)\dots(a_n-a_1)\begin{pmatrix} a_1+b_1 & a_1+b_2 & \dots & a_1+b_n \\ 1 & 1 & \dots & 1 \\ \vdots & \vdots & \ddots & \vdots \\ 1 & 1 & \dots & 1 \end{pmatrix}.$$

Now if $n \geq 3$, then Corollary 2.3.8 implies that the determinant of B and hence of A is 0. If $n = 2$, then the following easy direct calculation gives the determinant in that case. For,

$$\begin{aligned}\mathbf{det}(A) &= (a_1 + b_1)(a_2 + b_2) - (a_2 + b_1)(a_1 + b_2) \\ &= a_1a_2 + a_1b_2 + b_1a_2 + b_1b_2 - a_2a_1 - a_2b_2 - b_1a_1 - b_1b_2 \\ &= a_1b_2 + b_1a_2 - a_2b_2 - b_1a_1 = (a_2 - a_1)(b_1 - b_2).\end{aligned}$$

EXERCISE SET 2.3

Justify your answers where necessary with a proof or a counterexample.

2.3.1. Using the definition only evaluate $\begin{vmatrix} 0 & 0 & \dots & 0 & 0 & a_{1n} \\ 0 & 0 & \dots & 0 & a_{2n-1} & a_{2n} \\ 0 & 0 & \dots & a_{3n-2} & a_{3n-1} & a_{3n} \\ \vdots & \vdots & \vdots & \vdots & \vdots & \vdots \\ a_{n1} & a_{n2} & \dots & a_{nn-2} & a_{nn-1} & a_{nn} \end{vmatrix}$.

2.3.2. Using the definition only evaluate $\begin{vmatrix} 0 & a_{12} & a_{13} & 0 & 0 \\ a_{21} & a_{22} & a_{23} & a_{24} & a_{25} \\ 0 & a_{32} & a_{33} & 0 & 0 \\ a_{41} & a_{42} & a_{43} & a_{43} & a_{45} \\ 0 & a_{52} & a_{53} & 0 & 0 \end{vmatrix}$.

2.3.3. Evaluate the determinant of the matrix $\begin{vmatrix} a & b & c & 1 \\ b & c & a & 1 \\ c & a & b & 1 \\ \frac{b+c}{2} & \frac{a+c}{2} & \frac{b+a}{2} & 1 \end{vmatrix}$.

2.3.4. Using the definition evaluate $\begin{vmatrix} a_{11} & a_{12} & a_{13} & a_{14} & a_{15} \\ a_{21} & a_{22} & a_{23} & a_{24} & a_{25} \\ a_{31} & a_{32} & 0 & 0 & 0 \\ a_{41} & a_{42} & 0 & 0 & 0 \\ a_{51} & a_{52} & 0 & 0 & 0 \end{vmatrix}$.

2.3.5. Using the definition evaluate $\begin{vmatrix} 1 & 0 & 2 & 0 & 0 \\ 3 & 0 & -1 & 0 & 0 \\ 2 & 0 & 3 & 0 & 0 \\ 1 & 2 & 1 & -1 & 3 \\ -1 & 1 & 7 & 2 & -1 \end{vmatrix}$.

2.3.6. If we rewrite the rows of a matrix in reverse order, then how is the determinant changed?

2.3.7. If we rotate the entries of an $n \times n$ matrix counterclockwise through 90° about the center of the matrix (the point of intersection of the principal and secondary diagonals), then how is the determinant changed? (The principal diagonal is the diagonal from the top left corner of the matrix to the bottom right corner, whereas the secondary diagonal is the diagonal from the bottom left corner to the top right corner of the matrix.)

2.3.8. If we move the first column of an $n \times n$ matrix to the last column and all other columns are moved one column to the left in order, then how is the determinant changed?

2.3.9. Let $A = [a_{ij}] \in \mathbf{M}_5(\mathbb{R})$. Determine the numbers i, j, k such that the product $a_{1i}a_{23}a_{3j}a_{41}a_{5k}$ has a negative sign in its determinant decomposition.

2.3.10. Let $A = [a_{ij}] \in \mathbf{M}_5(\mathbb{R})$. Determine the numbers i, j, k such that the product $a_{12}a_{2i}a_{35}a_{4j}a_{5k}$ has a negative sign in its determinant decomposition.

2.3.11. Let $A = [a_{ij}] \in \mathbf{M}_6(\mathbb{R})$. Determine the numbers i, j, k such that the product $a_{1i}a_{j6}a_{35j}a_{44}a_{51}a_{6k}$ has a negative sign in its determinant decomposition.

2.3.12. Let $A = [a_{ij}] \in \mathbf{M}_n(\mathbb{R})$. What is the sign of $a_{1,n-1}a_{2n}a_{31}a_{42}\ldots a_{n,n-2}$ in its determinant decomposition?

2.3.13. How does the determinant change in an $n \times n$ matrix if the matrix is reflected in its secondary diagonal? (See Problem 2.3.8 for the definition of secondary diagonal.)

2.3.14. Using the properties of determinants prove that $\begin{vmatrix} 1 & 3 & 1 & -1 \\ 1 & 4 & 0 & 2 \\ 0 & 1 & 3 & 5 \\ 6 & -2 & 4 & 4 \end{vmatrix}$ is divisible by 9.

2.3.15. Evaluate the determinant of the matrix $A = [a_{jk}]$ of degree 9, with entries $a_{jk} = \mathbf{min}(j, k)$.

2.3.16. Evaluate $\begin{vmatrix} -5 & -7 & -2 & 2 & -2 & 16 \\ 0 & 0 & 4 & 0 & -5 & 0 \\ 2 & 0 & -2 & 0 & 2 & 0 \\ 6 & 4 & 6 & -1 & 15 & -5 \\ 5 & -4 & 10 & 1 & 14 & 6 \\ 3 & 0 & -2 & 0 & 3 & 0 \end{vmatrix}$.

2.3.17. Evaluate $\begin{vmatrix} 1+x & 1 & 1 & 1 & 1 & 1 \\ 1 & 1+y & 1 & 1 & 1 & 1 \\ 1 & 1 & 1+z & 1 & 1 & 1 \\ 1 & 1 & 1 & 1-z & 1 & 1 \\ 1 & 1 & 1 & 1 & 1-y & 1 \\ 1 & 1 & 1 & 1 & 1 & 1-x \end{vmatrix}$.

2.3.18. Evaluate $\begin{vmatrix} x & y & 0 & 0 & \dots & 0 & 0 \\ 0 & x & y & 0 & \dots & 0 & 0 \\ 0 & 0 & x & y & \dots & 0 & 0 \\ \vdots & \vdots & \vdots & \vdots & \vdots & \vdots & \vdots \\ 0 & 0 & 0 & 0 & \dots & x & y \\ y & 0 & 0 & 0 & \dots & 0 & x \end{vmatrix}$.

2.3.19. Evaluate $\begin{vmatrix} 0 & 1 & 1 & \dots & 1 \\ 1 & 0 & x & \dots & x \\ 1 & x & 0 & \dots & x \\ \vdots & \vdots & \vdots & \ddots & \vdots \\ 1 & x & x & \dots & 0 \end{vmatrix}$.

2.3.20. Evaluate $\begin{vmatrix} 3 & 1 & 0 & 0 & 0 & \dots & 0 & 0 \\ 1 & 3 & 1 & 0 & 0 & \dots & 0 & 0 \\ 0 & 1 & 3 & 1 & 0 & \dots & 0 & 0 \\ 0 & 0 & 1 & 3 & 1 & \dots & 0 & 0 \\ \vdots & \vdots & \vdots & \vdots & \vdots & \vdots & \vdots & \vdots \\ 0 & 0 & 0 & 0 & 0 & \dots & 3 & 1 \\ 0 & 0 & 0 & 0 & 0 & \dots & 1 & 3 \end{vmatrix}$.

2.4 COMPUTING DETERMINANTS

In Section 2.3, we saw one practical method for computing determinants, by means of row reduction. In this section, we offer an alternative approach to this problem which is based on the computation of determinants of submatrices of the given matrix. The ideas we introduce below are essential for this approach.

2.4.1. Definition. *Let $A = [a_{ij}] \in \mathbf{M}_n(\mathbb{R})$ and let $1 \le t \le n$. Select t rows and t columns in the matrix A and form the $t \times t$ submatrix B consisting of the elements situated at the intersections of these chosen rows and columns. Suppose that the selected rows are those numbered $k(1), k(2), \dots, k(t)$ and that the selected columns are those numbered $j(1), j(2), \dots, j(t)$. The determinant of B is called* the minor of degree t *corresponding to rows $k(1), k(2), \dots, k(t)$ and columns $j(1), j(2), \dots, j(t)$, and it will be denoted by*

$$\mathbf{minor}\{\mathbf{k(1)}, \mathbf{k(2)}, \dots, \mathbf{k(t)}; \mathbf{j(1)}, \mathbf{j(2)}, \dots, \mathbf{j(t)}\}.$$

If the chosen rows and columns are deleted from the matrix A, we obtain a submatrix of dimension $n - t$. The determinant of this submatrix is called the complementing minor *to the above constructed minor of degree t and it will be* *denoted by*

$$\mathbf{comp}\{\mathbf{k(1)}, \mathbf{k(2)}, \ldots, \mathbf{k(t)}; \mathbf{j(1)}, \mathbf{j(2)}, \ldots, \mathbf{j(t)}\}.$$

Finally, the cofactor or algebraic complement *to the above constructed minor is the number*

$$A_{\mathbf{k(1)},\mathbf{k(2)},\ldots,\mathbf{k(t)};\mathbf{j(1)},\mathbf{j(2)},\ldots,\mathbf{j(t)}} = (-1)^{k(1)+\cdots+k(t)+j(1)+\cdots+j(t)}\mathbf{comp}\{\mathbf{k(1)}, \ldots, \mathbf{k(t)}; \mathbf{j(1)}, \ldots, \mathbf{j(t)}\}.$$

The following theorem is the basic result for our future considerations.

2.4.2. Theorem. *Let $A = [a_{ij}] \in \mathbf{M}_n(\mathbb{R})$ and let $1 \le t \le n$. Choose t rows and t columns in the matrix A. Then*

(i) *the minor corresponding to these rows and columns is a sum of $t!$ summands, each of which is a product (taken with the sign + or −) of t elements of A;*
(ii) *the algebraic complement to this minor is a sum of $(n-t)!$ summands, each of which is a product (taken with the sign + or −) of $n-t$ elements of A;*
(iii) *the product of the two numbers obtained in (i) and (ii) is a sum of terms each of which comes from the expansion of the determinant of A.*

Proof. Since (i) and (ii) are clear from the definition we simply prove (iii). First we assume that the chosen rows are the first t rows and that the chosen columns are the first t columns. In this case the selected minor, $\Delta = \mathbf{minor}\{\mathbf{1}, \mathbf{2}, \ldots, \mathbf{t}; \mathbf{1}, \mathbf{2}, \ldots, \mathbf{t}\}$, is the determinant of the matrix $B = [b_{ij}] \in \mathbf{M}_t(\mathbb{R})$, where $b_{ij} = a_{ij}$, for $1 \le i, j \le t$ and the algebraic complement to it is $\Gamma = A_{\mathbf{k(1)},\mathbf{k(2)},\ldots,\mathbf{k(t)};\mathbf{j(1)},\mathbf{j(2)},\ldots,\mathbf{j(t)}}$. This coincides with the complementing minor, the determinant of the matrix $C = [c_{ij}] \in \mathbf{M}_{n-t}(\mathbb{R})$, where $c_{ij} = a_{t+i,t+j}$, for $1 \le i, j \le n-t$. Thus, we have

$$\Delta = \mathbf{det}(B) = \sum_{\pi \in \mathbf{S}_t} \mathbf{sign}\,\pi\; b_{1,\pi(1)}b_{2,\pi(2)}\ldots b_{t,\pi(t)} \text{ and}$$

$$\Gamma = \mathbf{det}(C) = \sum_{\sigma \in \mathbf{S}_{n-t}} \mathbf{sign}\,\sigma\; c_{1,\sigma(1)}c_{2,\sigma(2)}\ldots c_{n-t,\sigma(n-t)}.$$

Consider the product

$$\begin{aligned}\Delta\Gamma &= \left(\sum_{\pi \in \mathbf{S}_t} \mathbf{sign}\,\pi\; b_{1,\pi(1)}\ldots b_{t,\pi(t)}\right)\left(\sum_{\sigma \in \mathbf{S}_{n-t}} \mathbf{sign}\,\sigma\; c_{1,\sigma(1)}\ldots c_{n-t,\sigma(n-t)}\right)\\ &= \sum_{\pi \in \mathbf{S}_t}\sum_{\sigma \in \mathbf{S}_{n-t}} \mathbf{sign}\,\pi\;\mathbf{sign}\,\sigma\; b_{1,\pi(1)}\ldots b_{t,\pi(t)}c_{1,\sigma(1)}c_{2,\sigma(2)}\ldots c_{n-t,\sigma(n-t)}\\ &= \sum_{\pi \in \mathbf{S}_t}\sum_{\sigma \in \mathbf{S}_{n-t}} \mathbf{sign}\,\pi\;\mathbf{sign}\,\sigma\; a_{1,\pi(1)}a_{2,\pi(2)}\ldots a_{t,\pi(t)}a_{t+1,t+\sigma(1)}\ldots a_{n,t+\sigma(n-t)}.\end{aligned}$$

Now consider the transformation $\rho = \rho(\pi, \sigma)$ of the set $\{1, 2, \ldots, n\}$ defined as follows. We let

$$\rho(1) = \pi(1), \rho(2) = \pi(2), \ldots, \rho(t) = \pi(t) \text{ and}$$
$$\rho(t+1) = t + \sigma(1), \rho(t+2) = t + \sigma(2), \ldots, \rho(n) = t + \sigma(n-t).$$

We next prove that ρ is a permutation of the set $\{1, 2, \ldots, n\}$, by considering the cases that arise. Let k, m be arbitrary positive integers from the set $\{1, 2, \ldots, n\}$, such that $k < m$. If $m \le t$, then

$$\rho(k) = \pi(k) \ne \pi(m) = \rho(m).$$

When $m > t$ and $k \le t$ we have

$$\rho(k) = \pi(k) \le t < t + \sigma(m-t) = \rho(m).$$

Finally, if $k > t$, then

$$\rho(k) = t + \sigma(k-t) \ne t + \sigma(m-t) = \rho(m).$$

Hence ρ is injective and it follows from Corollary 1.2.1 that ρ is bijective. Hence $\rho \in \mathbf{S}_n$.

We next determine the parity of ρ. Suppose that the pair k, m, where $1 \le k < m \le n$, is an inversion pair relative to ρ. Thus, $\rho(k) > \rho(m)$. We shall again consider the possible cases.

If $m \le t$, then

$$\rho(k) = \pi(k) \ne \pi(m) = \rho(m),$$

and hence k, m is an inversion pair of π.

If $k \le t < m$ we have

$$\rho(k) = \pi(k) \le t < t + \sigma(m-t) = \rho(m),$$

so in this case we do not obtain an inversion pair.

Finally, if $k > t$, then

$$\rho(k) = t + \sigma(k-t) \ne t + \sigma(m-t) = \rho(m),$$

so that $k - t, m - t$ forms an inversion pair relative to σ. We let $i(\rho)$ denote the number of inversion pairs for ρ. The argument above shows that

$$i(\rho) = i(\pi) + i(\sigma),$$

and therefore

$$\mathbf{sign}\,\rho = (-1)^{i(\rho)} = (-1)^{i(\pi)+i(\sigma)} = (-1)^{i(\pi)}(-1)^{i(\sigma)} = (\mathbf{sign}\,\pi)(\mathbf{sign}\,\sigma).$$

Hence, as π runs through all permutations of degree t and σ runs through all permutations of degree $n-t$, the transformation $\rho(\pi,\sigma)$ runs through some subset of $\mathbf{S}_n$. We note that this is a proper subset of $\mathbf{S}_n$ since it is of cardinality $(t!)(n-t)! < n!$, for $0 < t < n$. Going back to our product $\Delta\Gamma$, we can now write

$$\begin{aligned}\Delta\Gamma &= \sum_{\pi\in\mathbf{S}_t}\sum_{\sigma\in\mathbf{S}_{n-t}} \mathbf{sign}\,\pi\ \mathbf{sign}\,\sigma\ a_{1,\pi(1)}\ldots a_{t,\pi(t)}a_{t+1,t+\sigma(1)}\ldots a_{n,t+\sigma(n-t)} \\ &= \sum_{\pi\in\mathbf{S}_t,\sigma\in\mathbf{S}_{n-t}} \mathbf{sign}\,\rho\ a_{1,\rho(1)}a_{2,\rho(2)}\ldots a_{n,\rho(n)} \\ &= \sum_{\substack{\text{some}\\ \text{permutations}\\ \rho\in\mathbf{S}_n}} \mathbf{sign}\,\rho\ a_{1,\rho(1)}a_{2,\rho(2)}\ldots a_{n,\rho(n)}.\end{aligned}$$

So we obtain a part of the sum of the given decomposition of the determinant of the matrix A.

We now consider the general case. We may suppose that

$$k(1) < k(2) < \cdots < k(t) \text{ and } j(1) < j(2) < \cdots < j(t).$$

Recall that, by Proposition 2.3.7, if we interchange two rows or two columns of a matrix A, then the determinant of the new matrix will differ from the determinant of A only by sign. By applying a sequence of such interchanges we shall obtain a new matrix in which the selected rows and columns occur in the upper left corner of the matrix. To show how this is done we first take row $k(1)$ and interchange it with row $k(1)-1$, we then interchange it with row $k(1)-2$, and so on. In this way row $k(1)$ will be moved to the first row and below it, in order will lie rows $1, 2, \ldots, k(1)-1$. We will do $k(1)-1$ interchanges to do this. We then move row $k(2)$ to the second row by using the same procedure. This will take a further $k(2)-2$ interchanges. Continuing this procedure with rows $k(3), \ldots, k(t)$ we shall need a total of

$$(k(1)-1)+\cdots+(k(t)-t) = (k(1)+k(2)+\cdots+k(t)) - (1+2+\cdots+t)$$

row interchanges, a sum we denote by $u(t)$. In a similar manner, we can gather the selected columns $j(1), \ldots, j(t)$ in the left-hand columns of the matrix using a total of

$$(j(1)-1)+\cdots+(j(t)-t) = (j(1)+j(2)+\cdots+j(t)) - (1+2+\cdots+t)$$

column interchanges for this, a sum we denote by $v(t)$. As a result, we obtain a new matrix D in which the selected rows and columns form a submatrix of dimension t, situated in the upper left corner. All other rows and columns will be situated, relative to one another, as they were originally. This means that if we cross out the first t rows and first t columns, we obtain a matrix whose determinant is the complementing minor to the minor of A consisting of the rows numbered $k(1), k(2), \ldots, k(t)$ and columns numbered $j(1), j(2), \ldots, j(t)$. By Proposition 2.3.7

$$\begin{aligned}\mathbf{det}(D) &= (-1)^{u(t)+v(t)}\mathbf{det}(A) \\ &= (-1)^{k(1)+k(2)+\cdots+k(t)+j(1)+j(2)+\cdots+j(t)}\mathbf{det}(A).\end{aligned}$$

It follows that

$$\mathbf{det}(A) = (-1)^{k(1)+k(2)+\cdots+k(t)+j(1)+j(2)+\cdots+j(t)}\mathbf{det}(D).$$

The minor Δ_D of the matrix D corresponding to the rows and columns numbered $1, 2, \ldots, t$ is equal to the minor of A whose rows are numbered $k(1), k(2), \ldots, k(t)$ and whose columns are numbered $j(1), j(2), \ldots, j(t)$, while its complementing minor Γ_D is equal to

$$\mathbf{comp}\{\mathbf{k(1), k(2), \ldots, k(t); j(1), j(2), \ldots, j(t)}\}.$$

We have already proved above that

$$\begin{aligned}\Delta_D\Gamma_D = &\ \mathbf{minor}\{\mathbf{k(1), k(2), \ldots, k(t); j(1), j(2), \ldots, j(t)}\} \\ &\times \mathbf{comp}\{\mathbf{k(1), k(2), \ldots, k(t); j(1), j(2), \ldots, j(t)}\}\end{aligned}$$

is a part of the sum of the decomposition of the determinant of D. Using the equation

$$\mathbf{det}(A) = (-1)^{k(1)+k(2)+\cdots+k(t)+j(1)+j(2)+\cdots+j(t)}\mathbf{det}(D)$$

and the fact that

$$\begin{aligned}A_{\mathbf{k(1),\ldots,k(t);j(1),\ldots,j(t)}} = &\ (-1)^{k(1)+k(2)+\cdots+k(t)+j(1)+j(2)+\cdots+j(t)} \\ &\times \mathbf{comp}\{\mathbf{k(1), \ldots, k(t);\ \ j(1), \ldots, j(t)}\},\end{aligned}$$

we deduce that

$$\mathbf{minor}\{\mathbf{k(1), k(2), \ldots, k(t); j(1), j(2), \ldots, j(t)}\}A_{\mathbf{k(1),k(2),\ldots,k(t);j(1),j(2),\ldots,j(t)}}$$

is a part of the sum which is the decomposition of the matrix A.

Theorem 2.4.2 gives us an alternative theoretical method for computing determinants, as we now show. In the matrix $A = [a_{ij}]$, choose the tth row (or column). Each element a_{tj} (respectively, a_{jt}) of this row (respectively, this column) is considered as a minor of dimension 1 and we multiply it by its cofactor A_{tj} (respectively, A_{jt}).

2.4.3. Theorem *(the decomposition of a determinant by a row or a column). Let* $A = [a_{ij}] \in \mathbf{M}_n(\mathbb{R})$. *Then*

$$\mathbf{det}(A) = \sum_{1 \le j \le n} a_{tj} A_{tj} \text{ (and, respectively, } \mathbf{det}(A) = \sum_{1 \le j \le n} a_{jt} A_{jt}).$$

Proof. By using Proposition 2.3.3, we need to consider only the case when row t is selected in A. Let

$$P_{tj} = \{\pi \in \mathbf{S}_n \mid \pi(t) = j\}.$$

Using arguments similar to those used in the proof of Theorem 2.2.3, it is very easy to see that $|P_{tj}| = (n-1)!$, $P_{tj} \cap P_{tm} = \emptyset$ whenever $j \ne m$, and $\mathbf{S}_n = \bigcup_{1 \le j \le n} P_{tj}$. It follows that

$$\begin{aligned} \mathbf{det}(A) &= \sum_{\pi \in \mathbf{S}_n} \mathbf{sign}\,\pi\ a_{1,\pi(1)} a_{2,\pi(2)} \dots a_{n,\pi(n)} \\ &= \sum_{1 \le j \le n} \left(\sum_{\pi \in P_{tj}} \mathbf{sign}\,\pi\ a_{1,\pi(1)} a_{2,\pi(2)} \dots a_{n,\pi(n)} \right). \end{aligned}$$

Every term of $a_{tj}A_{tj}$ has the factor a_{tj}. If $\mathbf{sign}\,\pi\ a_{1,\pi(1)}a_{2,\pi(2)} \dots a_{n,\pi(n)}$ is such a term, then we have $j = \pi(t)$. A cofactor is distinct from the determinant of a matrix of dimension n by only the factor $(-1)^{t+j}$. Therefore, the decomposition of $a_{tj}A_{tj}$ includes exactly $(n-1)!$ terms. It follows that

$$a_{tj} A_{tj} = \sum_{\pi \in P_{tj}} \mathbf{sign}\,\pi\, a_{1,\pi(1)} a_{2,\pi(2)} \dots a_{n,\pi(n)},$$

and hence

$$\mathbf{det}(A) = \sum_{1 \le j \le n} a_{tj} A_{tj}.$$

We sometimes say that we have expanded the determinant about row t (or column t) when we evaluate the determinant of a matrix using Theorem 2.4.3.

2.4.4. Corollary. *Let* $A = [a_{ij}] \in \mathbf{M}_n(\mathbb{R})$. *Then*

$$\sum_{1 \le j \le n} a_{tj} A_{mj} = \delta_{tm} \mathbf{det}(A) \left(and \sum_{1 \le j \le n} a_{jt} A_{jm} = \delta_{tm} \mathbf{det}(A) \right),$$

for all $1 \le t, m \le n$, *where* δ_{tm} *is the Kronecker symbol.*

Proof. Let $(c_1, c_2, \ldots, c_n)$ be an arbitrary tuple of n real numbers, and replace row t of A by this tuple to obtain a matrix that we denote by B. Thus, if $B = [b_{ij}]$, then

$$b_{ij} = \begin{cases} a_{ij}, & \text{if } i \ne t, \\ c_j, & \text{if } i = t. \end{cases}$$

By Theorem 2.4.3 we have

$$\mathbf{det}(B) = \sum_{1 \le j \le n} b_{jt} B_{tj}.$$

Evidently the cofactor B_{tj} to the element b_{tj} in the matrix B coincides with A_{tj} (in order to obtain it we just cross out the tth row so we eliminate the row that makes the difference between the matrices A and B). By the definition of the elements b_{tj} we have

$$\mathbf{det}(B) = \sum_{1 \le j \le n} c_j A_{tj}.$$

Now let $c_j = a_{mj}$, where $1 \le j \le n$. If $m = t$, then $B = A$, and Theorem 2.4.3 implies that

$$\sum_{1 \le j \le n} a_{tj} A_{tj} = \mathbf{det}(A).$$

On the other hand, if $m \ne t$, the matrix B has two identical rows and Corollary 2.3.8 implies that its determinant is zero. Thus, $\sum_{1 \le j \le n} a_{tj} A_{mj} = 0$. The Kronecker symbol allows us to write the equations we obtained as follows:

$$\sum_{1 \le j \le n} a_{tj} A_{mj} = \delta_{tm} \mathbf{det}(A).$$

The second of our assertions can be obtained in a similar manner.

2.4.5. Example. To demonstrate the application of Theorem 2.4.3, we will find the determinant of the matrix:

$$A_n = \begin{pmatrix} 1 & 1 & 1 & \dots & 1 & 1 \\ a_1 & a_2 & a_3 & \dots & a_{n-1} & a_n \\ a_1^2 & a_2^2 & a_3^2 & \dots & a_{n-1}^2 & a_n^2 \\ \vdots & \vdots & \vdots & \vdots & \vdots & \vdots \\ a_1^{n-1} & a_2^{n-1} & a_3^{n-1} & \dots & a_{n-1}^{n-1} & a_n^{n-1} \end{pmatrix}.$$

This matrix is called *a Vandermonde matrix*.

First, we consider the case when $n = 2$. We have

$$\mathbf{det}(A_2) = a_2 - a_1.$$

We shall prove, by induction, that

$$\mathbf{det}(A_n) = \prod_{1 \le k < t \le n} (a_t - a_k).$$

We suppose that this formula is true for all A_m, where $m < n$. We apply the following transformations to the matrix A. First, we multiply row $(n - 1)$ by $-a_1$ and add the result to row n, then we multiply row $(n - 2)$ by $-a_1$ and add the result to row $(n - 1)$, and so on. Finally, we multiply the first row by $-a_1$ and add the result to the second row. As a result of these transformations, we obtain the following matrix:

$$B_n = \begin{pmatrix} 1 & 1 & \dots & 1 & 1 \\ 0 & a_2 - a_1 & \dots & a_{n-1} - a_1 & a_n - a_1 \\ 0 & a_2^2 - a_1 a_2 & \ddots & a_{n-1}^2 - a_1 a_{n-1} & a_n^2 - a_1 a_n \\ \vdots & \vdots & \vdots & \ddots & \vdots \\ 0 & a_2^{n-1} - a_1 a_2^{n-2} & \dots & a_{n-1}^{n-1} - a_1 a_{n-1}^{n-2} & a_n^{n-1} - a_1 a_n^{n-2} \end{pmatrix}$$

$$= \begin{pmatrix} 1 & 1 & \dots & 1 & 1 \\ 0 & a_2 - a_1 & \dots & a_{n-1} - a_1 & a_n - a_1 \\ 0 & (a_2 - a_1)a_2 & \ddots & (a_{n-1} - a_1)a_{n-1} & (a_n - a_1)a_n \\ \vdots & \vdots & \vdots & \ddots & \vdots \\ 0 & (a_2 - a_1)a_2^{n-2} & \dots & (a_{n-1} - a_1)a_{n-1}^{n-2} & (a_n - a_1)a_n^{n-2} \end{pmatrix}.$$

By Corollary 2.3.9, it follows that $\mathbf{det}(B_n) = \mathbf{det}(A_n)$. Applying Theorem 2.4.3 we expand the determinant of B_n about the first column, and using Proposition 2.3.5 we obtain

$$\mathbf{det}(B_n) = (a_2 - a_1)(a_3 - a_1)\dots(a_n - a_1)\mathbf{det}(A_{n-1}),$$

where we take A_{n-1} to be the matrix A_n with the last row and first column deleted. We now apply our induction hypothesis to A_{n-1}, which gives us the equation

$$\mathbf{det}(A_n) = \mathbf{det}(B_n) = (a_2 - a_1)(a_3 - a_1)\ldots(a_n - a_1)\prod_{2\le k<t\le n}(a_t - a_k)$$

$$= \prod_{1\le k<t\le n}(a_t - a_k).$$

We have now obtained the determinant sought.

2.4.6. Example. The sequence $\{F_n \mid n \in \mathbb{N}\}$ is called the Fibonacci sequence, if $F_1 = F_2 = 1$ and $F_n = F_{n-1} + F_{n-2}$ whenever $n > 2$. A very brief history of this sequence is as follows. The first great mathematician of Medieval Europe, Fibonacci (Leonardo Pisano, 1170 to about 1250) studied the mathematical manuscripts written by the great Arabian and Indian mathematicians. He summarized this knowledge in his famous book *Liber Abaci* (The Book of Counting), published in 1202, and through this book introduced the arabic numerical system to Europeans. Prior to this time, Europeans used Roman numerals that, although useful in some ways, were very inconvenient. This revolutionary move to the arabic system of numeration had profound consequences for European civilization.

Among others, Fibonacci placed the following problem in his book *Liber Abaci*:

How many pairs of rabbits will be produced in a year, beginning with a single pair, if in every month each pair bears a new pair which becomes productive from the second month on?

It is easy to see that one pair will be produced the first month, and one pair also in the second month (since the new pair produced in the first month is not yet mature), and in the third month two pairs will be produced, one by the original pair and one by the pair that was produced in the first month. In the fourth month three pairs will be produced, and in the fifth month five pairs. After this, things expand rapidly (as happens with mice and rabbits), and we get the following sequence of numbers:

$$1, 1, 2, 3, 5, 8, 13, 21, 34, 55, 89, 144, 233, \ldots$$

This is an example of a recursive sequence, obeying the simple rule that to calculate the next term one simply adds the sum of the preceding two:

$$F(1) = 1;\ F(2) = 1;\ F(n) = F(n-1) + F(n-2).$$

As was discovered later, the Fibonacci sequence has a very important place not only in mathematics but also in economics, architecture, the technical and natural sciences, arts and philosophy, medicine, and aesthetics.

Let $d_n = \mathbf{det}(D_n)$, for $n \in \mathbb{N}$, where

$$D_n = \begin{pmatrix} 1 & 1 & 0 & 0 & \dots & 0 & 0 \\ -1 & 1 & 1 & 0 & \dots & 0 & 0 \\ 0 & -1 & 1 & 1 & \dots & 0 & 0 \\ \vdots & \vdots & \vdots & \vdots & \vdots & \vdots & \vdots \\ 0 & 0 & 0 & 0 & \dots & 1 & 1 \end{pmatrix}.$$

When we expand the determinant of this matrix about the first column, we obtain $d_n = d_{n-1} + c_{n-1}$ where $c_{n-1} = \mathbf{det}(C_{n-1})$ and

$$C_{n-1} = \begin{pmatrix} 1 & 0 & 0 & 0 & \dots & 0 & 0 \\ -1 & 1 & 1 & 0 & \dots & 0 & 0 \\ 0 & -1 & 1 & 1 & \dots & 0 & 0 \\ \vdots & \vdots & \vdots & \vdots & \vdots & \vdots & \vdots \\ 0 & 0 & 0 & 0 & \dots & 1 & 1 \end{pmatrix}.$$

Repeating this procedure and expanding the matrix C_{n-1} about the first column, we see that $c_{n-1} = d_{n-2}$. Hence, we have $d_n = d_{n-1} + d_{n-2}$, where $n \in \mathbb{N}$. In particular, we have

$$F_1 = F_2 = \mathbf{det}(D_1),\ F_3 = \mathbf{det}(D_2) \text{ and}$$

$$F_{n+1} = \mathbf{det}(D_n) = \mathbf{det}(D_{n-1}) + \mathbf{det}(D_{n-2}), \quad \text{for } n \in \mathbb{N}.$$

In this way we obtain an interesting symmetrical characteristic of the Fibonacci sequence.

Theorem 2.4.3 can be generalized in the following way.

2.4.7. Theorem *(Pierre–Simon Laplace). Let $A = [a_{ij}] \in \mathbf{M}_n(\mathbb{R})$. In the matrix A choose t rows (respectively, t columns). Multiply every minor of dimension t corresponding to the chosen rows (respectively, columns) by its algebraic complement. The sum of all these products is equal to* $\mathbf{det}(A)$.

Proof. By using Proposition 2.3.3 we need to consider only the case with rows. Let the selected rows be the rows numbered $k(1), k(2), \dots, k(t)$. We recall that

$$\mathbf{det}(A) = \sum_{\pi \in \mathbf{S}_n} \mathbf{sign}\,\pi\ a_{1,\pi(1)}a_{2,\pi(2)}\dots a_{n,\pi(n)}.$$

Consider an arbitrary term $\mathbf{sign}\,\pi\ a_{1,\pi(1)}a_{2,\pi(2)}\dots a_{n,\pi(n)}$ from this sum and within this consider the terms whose first indices belong to the selected rows.

Thus, we consider $\mathbf{sign}\,\pi\ a_{k(1),\pi(k(1))}a_{k(2),\pi(k(2))}\dots a_{k(t),\pi(k(t))}$. This product together with the sign + or − belongs to the decomposition

$$\mathbf{minor}\{\mathbf{k(1)},\mathbf{k(2)},\dots,\mathbf{k(t)};\pi(\mathbf{k(1)}),\pi(\mathbf{k(2)}),\dots,\pi(\mathbf{k(t)})\}.$$

Clearly, the product of all other elements $a_{j,\pi(j)}$, where $j\notin\{k(1),\dots,k(t)\}$ (again with the sign + or −) belongs to the decomposition

$$\mathbf{comp}\{\mathbf{k(1)},\mathbf{k(2)},\dots,\mathbf{k(t)};\pi(\mathbf{k(1)}),\pi(\mathbf{k(2)}),\dots,\pi(\mathbf{k(t)})\}.$$

By Theorem 2.4.2, the term

$$\mathbf{sign}\,\pi\ a_{1,\pi(1)}a_{2,\pi(2)}\dots a_{n,\pi(n)}$$

belongs to the decomposition of the product of

$$\mathbf{minor}\{\mathbf{k(1)},\mathbf{k(2)},\dots,\mathbf{k(t)};\pi(\mathbf{k(1)}),\pi(\mathbf{k(2)}),\dots,\pi(\mathbf{k(t)})\}$$

and

$$\mathbf{A}_{\mathbf{k(1)},\mathbf{k(2)},\dots,\mathbf{k(t)};\pi(\mathbf{k(1)}),\pi(\mathbf{k(2)}),\dots,\pi(\mathbf{k(t)})}.$$

Consequently, every summand of the decomposition of $\mathbf{det}(A)$ involves a product of some minor, corresponding to the chosen rows multiplied by its algebraic complement. The decomposition

$$\mathbf{minor}\{\mathbf{k(1)},\mathbf{k(2)},\dots,\mathbf{k(t)};\pi(\mathbf{k(1)}),\pi(\mathbf{k(2)}),\dots,\pi(\mathbf{k(t)})\}$$

includes $t!$ terms, while the decomposition

$$\mathbf{A}_{\mathbf{k(1)},\mathbf{k(2)},\dots,\mathbf{k(t)};\pi(\mathbf{k(1)}),\pi(\mathbf{k(2)}),\dots,\pi(\mathbf{k(t)})}$$

includes $(n-t)!$ terms. So the decomposition of the product of

$$\mathbf{minor}\{\mathbf{k(1)},\mathbf{k(2)},\dots,\mathbf{k(t)};\pi(\mathbf{k(1)}),\pi(\mathbf{k(2)}),\dots,\pi(\mathbf{k(t)})\}$$

and

$$\mathbf{A}_{\mathbf{k(1)},\mathbf{k(2)},\dots,\mathbf{k(t)};\pi(\mathbf{k(1)}),\pi(\mathbf{k(2)}),\dots,\pi(\mathbf{k(t)})}$$

include $t!(n-t)!$ terms.

Next, we show that the decompositions of the products of two distinct minors corresponding to the chosen rows by their algebraic complements do not include identical terms. Let

$$\{j(1),j(2),\dots,j(t)\}\neq\{s(1),s(2),\dots,s(t)\}$$

and let $\mathbf{sign}\,\pi\ a_{1,\pi(1)}a_{2,\pi(2)}\dots a_{n,\pi(n)}$ belong to a decomposition of the product

$$\mathbf{minor}\{\mathbf{k(1)},\dots,\mathbf{k(t)};\mathbf{j(1)},\dots,\mathbf{j(t)}\}\mathbf{A}_{\mathbf{k(1)},\dots,\mathbf{k(t)};\mathbf{j(1)},\dots,\mathbf{j(t)}};$$

let $\mathbf{sign}\,\sigma\ a_{1,\sigma\pi(1)}a_{2,\sigma(2)}\dots a_{n,\sigma(n)}$ belong to a decomposition of the product

$$\mathbf{minor}\{\mathbf{k(1)},\dots,\mathbf{k(t)};\mathbf{s(1)},\dots,\mathbf{s(t)}\}\mathbf{A}_{\mathbf{k(1)},\dots,\mathbf{k(t)};\mathbf{s(1)},\dots,\mathbf{s(t)}}.$$

This means that

$$\{\pi(k(1)),\pi(k(2)),\dots,\pi(k(t))\}=\{j(1),j(2),\dots,j(t)\}\neq$$
$$\{s(1),s(2),\dots,s(t)\}=\{\sigma(k(1)),\sigma(k(2)),\dots,\sigma(k(t))\}.$$

The total number of minors of dimension t, which corresponds to the selected rows, is equal to the number of combinations $\binom{n}{t}=\frac{n!}{t!(n-t)!}$. Thus the sum of the products of all the minors of dimension t that corresponds to the selected t rows by their algebraic complements gives us $t!(n-t)!\cdot\frac{n!}{t!(n-t)!}=n!$ terms from the decomposition of $\mathbf{det}(A)$. Since the decomposition of $\mathbf{det}(A)$ includes exactly $n!$ terms, we see that the sum of the products of all the minors of dimension t that corresponds to the selected t rows by their algebraic complements is $\mathbf{det}(A)$.

This theorem of Laplace allows us to reduce the computation of determinants of matrices to the computation of determinants of smaller size. However, the number of corresponding smaller matrices could be very large. Therefore, it is most efficient to use this theorem when there are some rows (or columns) with a large number of zeros in the matrix. For example, let

$$A=\begin{pmatrix} a_{11} & a_{12} & \dots & a_{1t} & 0 & \dots & 0\\ a_{21} & a_{22} & \dots & a_{2t} & 0 & \dots & 0\\ \vdots & \vdots & \ddots & \vdots & \vdots & \vdots & \vdots\\ a_{t1} & a_{t2} & \dots & a_{tt} & 0 & \dots & 0\\ a_{t+1,1} & a_{t+1,2} & \dots & a_{t+1,t} & a_{t+1,t+1} & \dots & a_{t+1,n}\\ \vdots & \vdots & \vdots & \vdots & \vdots & \ddots & \vdots\\ a_{n1} & a_{n2} & \dots & a_{nt} & a_{n,t+1} & \dots & a_{nn} \end{pmatrix}.$$

A convenient shorthand is to write this matrix as $\begin{pmatrix} B & O\\ C & D\end{pmatrix}$, where

$$B=\begin{pmatrix} a_{11} & a_{12} & \dots & a_{1t}\\ a_{21} & a_{22} & \dots & a_{2t}\\ \vdots & \vdots & \ddots & \vdots\\ a_{t1} & a_{t2} & \dots & a_{tt}\end{pmatrix},\ O=\begin{pmatrix} 0 & \dots & 0\\ 0 & \dots & 0\\ \vdots & \vdots & \vdots\\ 0 & \dots & 0\end{pmatrix},$$

$$C=\begin{pmatrix} a_{t+1,1} & a_{t+1,2} & \dots & a_{t+1,t}\\ \vdots & \vdots & \ddots & \vdots\\ a_{n1} & a_{n2} & \dots & a_{nt}\end{pmatrix},\ \text{and } D=\begin{pmatrix} a_{t+1,t+1} & \dots & a_{t+1,n}\\ \vdots & \vdots & \vdots\\ a_{n,t+1} & \dots & a_{nn}\end{pmatrix}.$$

Every $\mathbf{minor}\{\mathbf{1}, \mathbf{2}, \ldots, \mathbf{t}; \mathbf{j}(\mathbf{1}), \mathbf{j}(\mathbf{2}), \ldots, \mathbf{j}(\mathbf{t})\}$, where $\{j(1), j(2), \ldots, j(t)\} \neq \{1, 2, \ldots, t\}$, is a determinant of the matrix having a column of zeros and thus is equal to 0 by Corollary 2.3.6. Hence, by Theorem 2.4.7 we obtain the equation $\mathbf{det}(A) = \mathbf{det}(B)\mathbf{det}(D)$.

A similar conclusion can be made for all matrices of the types

$$\begin{pmatrix} B & C \\ O & D \end{pmatrix}, \begin{pmatrix} O & B \\ C & D \end{pmatrix}, \text{ and } \begin{pmatrix} B & C \\ D & O \end{pmatrix}.$$

EXERCISE SET 2.4

Justify your answers where necessary with a proof or a counterexample.

2.4.1. Find all second-order minors of the matrix $\begin{pmatrix} 0 & -1 & 1 \\ 2 & 3 & 0 \\ -2 & 1 & 4 \end{pmatrix}$.

2.4.2. Find the cofactors to all elements of the second row of the matrix $\begin{pmatrix} 4 & -1 & 3 \\ u & v & w \\ -6 & 5 & -2 \end{pmatrix}$.

2.4.3. Find the cofactors to all elements of the third column of the matrix $\begin{pmatrix} 4 & 1 & k & -5 \\ 3 & 2 & m & 1 \\ 2 & -3 & n & 4 \\ -1 & 1 & t & 3 \end{pmatrix}$.

2.4.4. Using Laplace's theorem evaluate the determinant of the matrix $\begin{pmatrix} 5 & 1 & 2 & 7 \\ 3 & 0 & 0 & 2 \\ 1 & 3 & 4 & 5 \\ 2 & 0 & 0 & 3 \end{pmatrix}$.

2.4.5. Using Laplace's theorem evaluate the determinant of the matrix $\begin{pmatrix} 2 & 1 & 4 & 3 & 5 \\ 3 & 4 & 0 & 5 & 0 \\ 3 & 4 & 5 & 2 & 1 \\ 1 & 5 & 2 & 4 & 3 \\ 4 & 6 & 0 & 7 & 0 \end{pmatrix}$.

2.4.6. Using Laplace's theorem evaluate the determinant of the matrix $\begin{pmatrix} 1 & 2 & 3 & 4 & 5 & 3 \\ 6 & 5 & 7 & 8 & 4 & 2 \\ 9 & 8 & 6 & 7 & 0 & 0 \\ 3 & 2 & 4 & 5 & 0 & 0 \\ 3 & 4 & 0 & 0 & 0 & 0 \\ 5 & 6 & 0 & 0 & 0 & 0 \end{pmatrix}$.

2.4.7. Using Laplace's theorem evaluate the determinant of the matrix

$$\begin{pmatrix} 1 & 0 & 2 & 0 & 3 & 0 \\ 5 & 1 & 4 & 2 & 7 & 3 \\ 1 & 0 & 4 & 0 & 9 & 0 \\ 8 & 1 & 5 & 3 & 7 & 6 \\ 1 & 0 & 8 & 0 & 27 & 0 \\ 9 & 1 & 5 & 4 & 3 & 10 \end{pmatrix}.$$

2.4.8. Using Laplace's theorem evaluate the determinant of the matrix

$$\begin{pmatrix} 0 & a & b & c \\ 1 & x & 0 & 0 \\ 1 & 0 & y & 0 \\ 1 & 0 & 0 & z \end{pmatrix}.$$

2.4.9. Using Laplace's theorem evaluate the determinant of the matrix

$$\begin{pmatrix} 1 & x & x & x \\ 1 & a & 0 & 0 \\ 1 & 0 & b & 0 \\ 1 & 0 & 0 & c \end{pmatrix}.$$

2.4.10. Using Laplace's theorem evaluate the determinant of the matrix

$$\begin{pmatrix} 2 & -1 & 3 & 4 & -5 \\ 4 & -2 & 7 & 8 & -7 \\ -6 & 4 & -9 & -2 & 3 \\ 3 & -2 & 4 & 1 & -2 \\ -2 & 6 & 5 & 4 & -3 \end{pmatrix}.$$

2.4.11. Using Laplace's theorem evaluate the determinant of the matrix

$$\begin{pmatrix} 5 & -5 & -3 & 4 & 2 \\ -4 & 4 & 3 & 6 & 3 \\ 3 & -1 & 5 & -9 & -5 \\ -7 & 7 & 6 & 8 & 4 \\ 5 & -3 & 2 & -1 & -2 \end{pmatrix}.$$

2.4.12. Using Laplace's theorem evaluate the determinant of the matrix

$$\begin{pmatrix} 2 & 1 & 2 & 3 & 2 \\ 3 & -2 & 7 & 5 & -1 \\ 3 & -1 & -5 & -3 & -2 \\ 5 & -6 & 4 & 2 & -4 \\ 2 & -3 & 3 & 1 & -2 \end{pmatrix}.$$

2.4.13. Using Laplace's theorem evaluate the determinant of the matrix

$$\begin{pmatrix} 3 & 4 & -3 & -1 & 2 \\ -5 & 6 & 5 & 2 & 3 \\ 4 & -9 & -3 & 7 & -5 \\ -1 & -4 & 1 & 1 & -2 \\ -3 & 7 & 5 & 2 & 3 \end{pmatrix}.$$

2.5 PROPERTIES OF THE PRODUCT OF MATRICES

In this section, we consider some important properties of multiplication of matrices. More precisely, we obtain criteria for a matrix to have an inverse and also establish some techniques for the computation of such an inverse. We introduce the idea of what we call a basic matrix and we study the effects of elementary transformations of matrices. We also introduce some other standard matrices.

The following remarkable theorem will play a key role here.

2.5.1. Theorem. *Let* $A = [a_{ij}], B = [b_{ij}] \in \mathbf{M}_n(\mathbb{R})$.
Then $\mathbf{det}(AB) = \mathbf{det}(A)\mathbf{det}(B)$.

Proof. Let $AB = C = [c_{ij}] \in \mathbf{M}_n(\mathbb{R})$. We consider the following auxiliary matrix $D = [d_{ij}]$ of dimension $2n$:

$$D = \begin{pmatrix} a_{11} & a_{12} & \dots & a_{1,n} & 0 & \dots & 0 \\ a_{21} & a_{22} & \dots & a_{2,n} & 0 & \dots & 0 \\ \vdots & \vdots & \ddots & \vdots & \vdots & \vdots & \vdots \\ a_{n1} & a_{n2} & \dots & a_{n,n} & 0 & \dots & 0 \\ -1 & 0 & \dots & 0 & b_{1,1} & \dots & b_{1,n} \\ \vdots & \vdots & \vdots & \vdots & \vdots & \ddots & \vdots \\ 0 & 0 & \dots & -1 & b_{n,1} & \dots & b_{n,n} \end{pmatrix},$$

which we can write briefly as

$$D = \begin{pmatrix} A & O \\ -I & B \end{pmatrix}.$$

By the arguments we used at the end of Section 2.4, we obtain $\mathbf{det}(D) = \mathbf{det}(A)\mathbf{det}(B)$.

We will transform D using column transformations that do not change the determinant of D but which will allow us to fill the right lower corner of D with zeros. For this, we add the first column of D, multiplied by b_{11} to the $(n+1)$th column. Next, we add the second column, multiplied by b_{21} and add this to the $(n+1)$th column, and so on. Finally, we add the nth column multiplied by $b_{n,1}$ to the $(n+1)$th column. Then we start to repeat this process, first adding the first column of D multiplied by b_{12} to its $(n+2)$th column. Then we add the second column multiplied by b_{22}, and so on. Finally, we add the nth column multiplied by b_{n2}. In general, we add to the $(n+k)$th column of D a linear combination of the first n columns with the elements multiplied, respectively, by the numbers $b_{1k}, b_{2k}, \dots, b_{nk}$. In general, we find, for $i \geq 1$,

$$d_{i,n+k} + b_{1,k}c_1 + b_{2,k}c_2 + \cdots + b_{n,k}c_n,$$

where c_j denotes the jth column of D.

We denote the matrix we obtain by H, so

$$H = \begin{pmatrix} a_{11} & a_{12} & \dots & a_{1,n} & h_{11} & \dots & h_{1,n} \\ a_{21} & a_{22} & \dots & a_{2,n} & h_{21} & \dots & h_{2,n} \\ \vdots & \vdots & \ddots & \vdots & \vdots & \vdots & \vdots \\ a_{n,1} & a_{n,2} & \dots & a_{n,n} & h_{1,n} & \dots & h_{n,n} \\ -1 & 0 & \dots & 0 & 0 & \dots & 0 \\ \vdots & \vdots & \vdots & \vdots & \vdots & \ddots & \vdots \\ 0 & 0 & \dots & -1 & 0 & \dots & 0 \end{pmatrix}.$$

By Corollary 2.3.9, $\mathbf{det}(H) = \mathbf{det}(D)$.

Furthermore, if we recall the sequence of operations that we have performed, it follows that

$$h_{ij} = a_{i1}b_{1j} + a_{i2}b_{2j} + \cdots + a_{in}b_{nj} = c_{ij}, \text{ where } 1 \le i, j \le n.$$

From this it follows also that the matrix AB is the upper right corner of H. Thus

$$H = \begin{pmatrix} A & AB \\ -I & O \end{pmatrix}.$$

Applying the arguments that we developed at the end of Section 2.4, we obtain

$$\mathbf{det}(H) = \mathbf{minor}\{\mathbf{1, 2, \dots, n; n+1, n+2, \dots, 2n}\}\mathbf{A}_{\mathbf{1,2,\dots,n;n+1,n+2,\dots,2n}}.$$

Furthermore,

$$\mathbf{minor}\{\mathbf{1, 2, \dots, n; n+1, n+2, \dots, 2n}\} = \mathbf{det}(AB),$$

and

$$\mathbf{A}_{\mathbf{1,2,\dots,n;n+1,n+2,\dots,2n}} \cdot (-1)^n = (-1)^{1+2+\cdots+n+(n+1)+\cdots+2n} = 1,$$

since the exponent here is $2n^2 + n + n$. Hence, $\mathbf{det}(H) = \mathbf{det}(AB)$ and we have

$$\mathbf{det}(A)\mathbf{det}(B) = \mathbf{det}(D) = \mathbf{det}(H) = \mathbf{det}(AB).$$

We shall apply this theorem to the problem of whether or not a given matrix has an inverse. If the matrix A has an inverse, then by Theorem 2.5.1, we obtain

$$1 = \mathbf{det}(I) = \mathbf{det}(AA^{-1}) = \mathbf{det}(A)\mathbf{det}(A^{-1}),$$

and hence $\mathbf{det}(A) \ne 0$ in this case.

This prompts the following definition.

2.5.2. Definition. *The matrix* $A = [a_{ij}] \in \mathbf{M}_n(\mathbb{R})$ *is called nonsingular if* $\mathbf{det}(A) \neq 0$.

Our short proof above shows that if a matrix has an inverse, then it must be nonsingular and we now show that the converse is also true. We recall that if $A \in \mathbf{M}_n(\mathbb{R})$ is a matrix then A_{ij} is the cofactor of A corresponding to the (i, j) entry of A.

2.5.3. Theorem. *Let* $A = [a_{ij}] \in \mathbf{M}_n(\mathbb{R})$. *Then A has an inverse if and only if it is nonsingular. Moreover, if* $A^{-1} = B = [b_{ij}] \in \mathbf{M}_n(\mathbb{R})$, *then* $b_{ij} = \frac{A_{ji}}{\mathbf{det}(A)}$, *where* $1 \leq i, j \leq n$.

Proof. We observed above that if a matrix has an inverse, then the original matrix is nonsingular. Conversely, suppose that $\mathbf{det}(A) \neq 0$. We shall apply Corollary 2.4.4 to find the inverse matrix.

Put $b_{ij} = \frac{A_{ji}}{\mathbf{det}(A)}$, where $1 \leq i, j \leq n$, and let $B = [b_{ij}]$. Consider the products $AB = [u_{ij}]$ and $BA = [v_{ij}]$. We have

$$u_{ij} = \sum_{1 \leq k \leq n} a_{ik} b_{kj} = \sum_{1 \leq k \leq n} a_{ik} \frac{A_{jk}}{\mathbf{det}(A)} = \frac{1}{\mathbf{det}(A)} \delta_{ij} \mathbf{det}(A) = \delta_{ij} \text{ and}$$

$$v_{ij} = \sum_{1 \leq k \leq n} b_{ik} a_{kj} = \sum_{1 \leq k \leq n} \frac{A_{ki}}{\mathbf{det}(A)} a_{kj} = \frac{1}{\mathbf{det}(A)} \delta_{ji} \mathbf{det}(A) = \delta_{ji} = \delta_{ij},$$

where, as usual, δ_{ij} denotes the Kronecker delta. It follows that $AB = BA = I$, which implies that $B = A^{-1}$.

In particular the inverse of a nonsingular 2×2 matrix $\begin{pmatrix} a & b \\ c & d \end{pmatrix}$ is very easily seen to be

$$\frac{1}{ad - bc} \begin{pmatrix} d & -b \\ -c & a \end{pmatrix}.$$

The set of nonsingular $n \times n$ matrices is very important in mathematics. We let $GL_n(\mathbb{R})$ (respectively, $GL_n(\mathbb{Q})$) denote the subset of $\mathbf{M}_n(\mathbb{R})$ (respectively, $\mathbf{M}_n(\mathbb{Q})$) consisting of all nonsingular matrices.

Next we consider some other properties of matrix multiplication. There are a number of important special matrices and we here define some of these.

Let $E_{km} = [u_{ij}^{(km)}] \in \mathbf{M}_n(\mathbb{R})$ denote the matrix defined by

$$u_{ij}^{(km)} = \begin{cases} 1, & \text{whenever } (i, j) = (k, m), \\ 0, & \text{whenever } (i, j) \neq (k, m). \end{cases}$$

Thus

$$E_{km} = \begin{pmatrix} 0 & 0 & 0 & \dots & 0 & \mathbf{0} & 0 & \dots & 0 & 0 \\ 0 & 0 & 0 & \dots & 0 & \mathbf{0} & 0 & \dots & 0 & 0 \\ \vdots & \vdots & \vdots & \vdots & \vdots & \vdots & \vdots & \vdots & \vdots & \vdots \\ 0 & 0 & 0 & \dots & 0 & \mathbf{0} & 0 & \dots & 0 & 0 \\ \mathbf{0} & \mathbf{0} & \mathbf{0} & \dots & \mathbf{0} & \mathbf{1} & \mathbf{0} & \dots & \mathbf{0} & \mathbf{0} \\ 0 & 0 & 0 & \dots & 0 & \mathbf{0} & 0 & \dots & 0 & 0 \\ \vdots & \vdots & \vdots & \vdots & \vdots & \vdots & \vdots & \vdots & \vdots & \vdots \\ 0 & 0 & 0 & \dots & 0 & \mathbf{0} & 0 & \dots & 0 & 0 \end{pmatrix},$$

where the kth row and the mth column have been typeset in bold.

The matrices E_{km} are called *basic matrices*.

We note that if $A = [a_{ij}] \in \mathbf{M}_n(\mathbb{R})$ then it is clear that

$$A = \sum_{i=1}^{n} \sum_{j=1}^{n} a_{ij} E_{ij}.$$

This means that every matrix is a linear combination of the basic matrices, which can be thought of as some justification for the term "basic."

We shall find the product of two basic matrices. Let $E_{km} E_{rs} = [w_{ij}]$. Then

$$w_{ij} = \sum_{1 \le t \le n} u_{it}^{(km)} u_{tj}^{(rs)}.$$

If $i \neq k$, then $u_{it}^{(km)} = 0$ for each t, so that $w_{ij} = 0$. When $i = k$, $u_{kt}^{(km)} = 0$, if $t \neq m$, and we have $w_{kj} = u_{km}^{(km)} u_{mj}^{(rs)} = u_{mj}^{(rs)}$. Hence, if $m \neq r$ then $w_{kj} = 0$. When $m = r$, $u_{rj}^{rs} = 0$ if $j \neq s$, so $w_{ks} = u_{ms}^{(rs)} = u_{rs}^{(rs)} = 1$ and $w_{kj} = 0$ for $j \neq s$. These calculations establish the following simple rule for multiplying basic matrices:

$$E_{km} E_{rs} = \begin{cases} E_{ks}, & \text{if } m = r, \\ O, & \text{if } m \neq r. \end{cases}$$

We will use this immediately.

2.5.4. Proposition. *Let* $A = [a_{ij}] \in \mathbf{M}_n(\mathbb{R})$. *Then*

$$AE_{km} = \begin{pmatrix} 0 & 0 & \dots & 0 & a_{1k} & 0 & \dots & 0 \\ 0 & 0 & \dots & 0 & a_{2k} & 0 & \dots & 0 \\ \vdots & \vdots & \vdots & \vdots & \vdots & \vdots & \vdots & \vdots \\ 0 & 0 & \dots & 0 & a_{nk} & 0 & \dots & 0 \end{pmatrix},$$

where the column

$$\begin{matrix} a_{1k} \\ a_{2k} \\ \vdots \\ a_{nk} \end{matrix}$$

is the mth column of the matrix AE_{km}. Also

$$E_{km}A = \begin{pmatrix} 0 & 0 & \dots & 0 \\ \vdots & \vdots & \vdots & \vdots \\ 0 & 0 & \dots & 0 \\ a_{m1} & a_{m2} & \dots & a_{mn} \\ 0 & 0 & \dots & 0 \\ \vdots & \vdots & \vdots & \vdots \\ 0 & 0 & \dots & 0 \end{pmatrix},$$

where the row $a_{m1} \quad a_{m2} \quad \dots \quad a_{mn}$ *is the kth row of the matrix $E_{km}A$.*

Proof. We observed above that $A = \sum_{i=1}^{n}\sum_{j=1}^{n} a_{ij}E_{ij}$. Then

$$\begin{aligned} AE_{km} &= \left(\sum_{i=1}^{n}\sum_{j=1}^{n} a_{ij}E_{ij}\right)E_{km} = \sum_{i=1}^{n}\sum_{j=1}^{n} a_{ij}E_{ij}E_{km} \\ &= \sum_{i=1}^{n} a_{ik}E_{ik}E_{km} = \sum_{1\le i\le n} a_{ik}E_{im} \\ &= a_{1k}E_{1m} + a_{2k}E_{2m} + \dots + a_{nk}E_{nm}. \end{aligned}$$

Also

$$\begin{aligned} E_{km}A &= E_{km}\left(\sum_{i=1}^{n}\sum_{j=1}^{n} a_{ij}E_{ij}\right) = \sum_{i=1}^{n}\sum_{j=1}^{n} a_{ij}E_{km}E_{ij} \\ &= \sum_{i=1}^{n}\sum_{j=1}^{n} a_{ij}E_{km}E_{ij} = \sum_{1\le j\le n} a_{mj}E_{km}E_{mj} = \sum_{1\le j\le n} a_{mj}E_{kj} \\ &= a_{m1}E_{k1} + a_{m2}E_{k2} + \dots + a_{mn}E_{kn}. \end{aligned}$$

This proves the required result.

It is useful to verbally describe this multiplication. If we multiply A on the right-hand side by E_{km} (we say that we postmultiply A by E_{km}), then we obtain the matrix whose mth column is the kth column of A, while all other columns are zero. If we multiply A on the left-hand side by E_{km} (we say that we premultiply A by E_{km}), then we obtain the matrix whose kth row is the mth row of A, while all other rows are zero.

As we mentioned already matrix multiplication is not commutative. Indeed, commutativity is a very rare situation even for multiplication of special matrices.

2.5.5. Definition. *Let S be a subset of $\mathbf{M}_n(\mathbb{R})$. The centralizer of the set S is the set all matrices that commute with every matrix from S. Thus the centralizer of S is the set $\{A \in \mathbf{M}_n(\mathbb{R}) : AB = BA$ for all $B \in S\}$.*

The centralizer of the entire set $\mathbf{M}_n(\mathbb{R})$ is called the center of the matrix algebra (we formally define this latter term later). The center will be denoted by $\zeta(\mathbf{M}_n(\mathbb{R}))$.

The identity matrix commutes with every matrix and hence it belongs to the center of $\mathbf{M}_n(\mathbb{R})$. The diagonal matrix αI, where α is any real number, also belongs to this center. Indeed

$$A(\alpha I) = \alpha(AI) = \alpha(IA) = (\alpha I)A = \alpha A.$$

The matrix αI is called a *scalar matrix*. Put

$$\mathbb{R}I = \{\alpha I \mid \alpha \in \mathbb{R}\}$$

and, respectively,

$$\mathbb{Q}I = \{\alpha I \mid \alpha \in \mathbb{Q}\}.$$

Thus the set of all scalar matrices belongs to the center. Every scalar matrix is diagonal. However, not every diagonal matrix belongs to the center, as the following result shows, which gives the effect of premultiplying (or postmultiplying) a matrix A by a diagonal matrix D.

2.5.6. Proposition. *Let $A = [a_{ij}]$, $D = [d_{ij}] \in \mathbf{M}_n(\mathbb{R})$, where D is a diagonal matrix. Left (respectively, right) multiplication of the matrix A by the matrix D is equivalent to multiplying the rows (respectively, columns) of A by the elements $d_{11}, d_{22}, \dots, d_{nn}$.*

Proof. When we premultiply A by D we obtain $DA = C = [c_{ij}]$ and

$$c_{ij} = \sum_{1 \le k \le n} d_{ik}a_{kj} = d_{ii}a_{ij},$$

which proves that the ith row of A is multiplied by d_{ii}. A similar computation shows the result for right multiplication.

Now we can describe the center of the matrix algebra, $\mathbf{M}_n(\mathbb{R})$.

2.5.7. Theorem. $\zeta(\mathbf{M}_n(\mathbb{R})) = \mathbb{R}I$.

Proof. We noted above that the center contains the set of all scalar matrices so $\mathbb{R}I \subseteq \zeta(\mathbf{M}_n(\mathbb{R})$.

To prove the converse, let $A = [a_{ij}] \in \zeta(\mathbf{M}_n(\mathbb{R}))$. The matrix A permutes with every matrix and hence with every basic matrix. Thus

$$AE_{km} = E_{km}A$$

for each pair of indices k, m, where $1 \leq k, m \leq n$.

Put $AE_{km} = [u_{ij}]$ and $E_{km}A = [v_{ij}]$. Suppose that $k \neq m$. Then Proposition 2.5.4 implies that $u_{jm} = a_{jk}$ and $v_{jm} = 0$, whenever $j \neq k$. Thus, if $j \neq k$, then $a_{jk} = 0$, so the matrix A must be diagonal. Applying Proposition 2.5.6 we see that all elements of the main diagonal of A are equal. Consequently, A is a scalar matrix. The theorem follows.

We now consider certain other special matrices related to the basic matrices.

2.5.8. Definition. *Let α be a fixed but arbitrary real number. The matrix $t_{km}(\alpha) = I + \alpha E_{km}$, where $k \neq m$, is called a transvection.*

Clearly, every transvection is a triangular matrix and $\mathbf{det}(t_{km}(\alpha)) = 1$. In particular, Theorem 2.5.3 implies that every transvection has an inverse. To determine such an inverse we consider the product of two transvections. We have

$$t_{km}(\alpha)t_{rs}(\beta) = (I + \alpha E_{km})(I + \beta E_{rs}) = I + \alpha E_{km} + \beta E_{rs} + \alpha\beta E_{km}E_{rs}.$$

From this equation it follows that $t_{km}(\alpha)t_{km}(-\alpha) = I$ and hence $(t_{km}(\alpha))^{-1} = t_{km}(-\alpha)$.

A more intuitive way of seeing this is to observe that $t_{km}(\alpha)$ can be thought of as adding α times row m to row k. With this interpretation the inverse of $t_{km}(\alpha)$ is obtained by just subtracting α times row m from row k, which is to say $t_{km}(\alpha)^{-1} = t_{km}(-\alpha)$. This latter idea—thinking of a transvection as adding a multiple of one row to another—can be seen formally in the next result. It shows that postmultiplication of A by $t_{km}(\alpha)$ is equivalent to adding α times column k to column m and premultiplication of A by $t_{km}(\alpha)$ is equivalent to adding α times row m to row k.

2.5.9. Proposition. *Let* $A = [a_{ij}] \in \mathbf{M}_n(\mathbb{R})$. *Then*

$$At_{km}(\alpha) = \begin{pmatrix} a_{11} & a_{12} & \dots & a_{1,m-1} & a_{1m} + \alpha a_{1k} & a_{1,m+1} & \dots & a_{1n} \\ a_{21} & a_{22} & \dots & a_{2,m-1} & a_{2m} + \alpha a_{2k} & a_{2,m+1} & \dots & a_{2n} \\ \vdots & \vdots & \vdots & \vdots & \vdots & \vdots & \vdots & \vdots \\ a_{n1} & a_{n2} & \dots & a_{n,m-1} & a_{nm} + \alpha a_{nk} & a_{n,m+1} & \dots & a_{nn} \end{pmatrix}$$

and

$$t_{km}(\alpha)A = \begin{pmatrix} a_{11} & a_{12} & \dots & a_{1n} \\ \vdots & \vdots & \vdots & \vdots \\ a_{k-1,1} & a_{k-1,2} & \dots & a_{k-1,n} \\ a_{k1} + \alpha a_{m1} & a_{k2} + \alpha a_{m2} & \dots & a_{kn} + \alpha a_{mn} \\ a_{k+1,1} & a_{k+1,2} & \dots & a_{k+1,n} \\ \vdots & \vdots & \vdots & \vdots \\ a_{n1} & a_{n2} & \dots & a_{nn} \end{pmatrix}.$$

Proof. This assertion follows from the definition of transvection and Proposition 2.5.4.

The next theorem shows the major role that transvections play. Virtually all other matrices can be obtained from them.

2.5.10. Theorem. *Every matrix* $A = [a_{ij}] \in \mathbf{M}_n(\mathbb{R})$ *can be written as a product of certain transvections and a diagonal matrix.*

Proof. Proposition 2.5.9 shows that when we postmultiply A by $t_{km}(\alpha)$ the effect is to add α times column k to column m and the same proposition shows that when we premultiply A by $t_{km}(\alpha)$ the effect is to add α times row m to row k. We now describe the process of decomposing A in the manner sought and note that we have already used some of these same arguments when transforming matrices.

We assume that $A \neq O$. If the first column of A consists entirely of zeros then it is already in a form that will make our final matrix upper triangular and we move to the next column. So we assume that the first column of A is not zero. If $a_{11} = 0$, then, since the first column of A is not zero, we add some row to the first, which is equivalent to premultiplying A by a transvection to obtain a matrix $A_1 = [a_{ij}^{(1)}]$ whose first entry is nonzero. Now we can add multiples of row 1 of the matrix, in turn, to rows $2, 3, \dots, n$ in order to make each entry of the first column of the new matrix, other than the $(1, 1)$ entry, equal to 0. Having done this we now proceed to obtain an upper triangular matrix in the manner suggested in Section2.3, by adding multiples of the succeeding rows to form zeros below the leading diagonal. Each such matrix manipulation can be done by premultiplication by a transvection. Thus, there exist transvections $\mathbf{t}_{k(1),m(1)}(\alpha_1), \dots, \mathbf{t}_{k(t),m(t)}(\alpha_t)$ such that the matrix $B = \mathbf{t}_{k(t),m(t)}(\alpha_t), \dots, \mathbf{t}_{k(1),m(1)}(\alpha_1)A$ is upper triangular.

Next, we reduce the matrix B to diagonal form by postmultiplying it by a sequence of transvections, adding first multiples of column 1 to the columns of B to make all entries in row 1 equal to 0 except possibly for the first. Then we add multiples of column 2 to the columns $3, 4, \ldots, n$ to make all entries in row 2 equal to 0, other than possibly the $(2, 2)$ entry, and proceed in turn with the columns $3, \ldots, n$. This process produces transvections $\mathbf{t}_{q(1),r(1)}(\beta_1), \ldots, \mathbf{t}_{q(s),r(s)}(\beta_s)$ such that

$$\mathbf{t}_{k(t),m(t)}(\alpha_t), \ldots, \mathbf{t}_{k(1),m(1)}(\alpha_1)A\mathbf{t}_{q(s),r(s)}(\beta_s)\ldots\mathbf{t}_{q(1),r(1)}(\beta_1) = D$$

is a diagonal matrix. We have already noted that $(t_{km}(\alpha))^{-1} = t_{km}(-\alpha)$. Therefore, from the above equation, we obtain, premultiplying and postmultiplying by the appropriate inverses,

$$A = \mathbf{t}_{k(1),m(1)}(-\alpha_1), \ldots, \mathbf{t}_{k(t),m(t)}(-\alpha_t)D\mathbf{t}_{q(1),r(1)}(-\beta_1), \ldots, \mathbf{t}_{q(s),r(s)}(-\beta_s).$$

This completes the proof.

We have also used transformations that interchange two rows or columns. Such a transformation is also equivalent to premultiplication or postmultiplication of the given matrix by the following kind of special matrix. Put

$$\mathbf{s}_{km} = \begin{pmatrix} 1 & 0 & \ldots & 0 & \mathbf{0} & 0 & \ldots & 0 & \mathbf{0} & 0 & \ldots & 0 & 0 \\ 0 & 1 & \ldots & 0 & \mathbf{0} & 0 & \ldots & 0 & \mathbf{0} & 0 & \ldots & 0 & 0 \\ \vdots & \vdots & \vdots & \vdots & \vdots & \vdots & \vdots & \vdots & \vdots & \vdots & \vdots & \vdots & \vdots \\ 0 & 0 & \ldots & 1 & \mathbf{0} & 0 & \ldots & 0 & \mathbf{0} & 0 & \ldots & 0 & 0 \\ \mathbf{0} & \mathbf{0} & \ldots & \mathbf{0} & \mathbf{0} & \mathbf{0} & \ldots & \mathbf{0} & \mathbf{1} & \mathbf{0} & \ldots & \mathbf{0} & \mathbf{0} \\ 0 & 0 & \ldots & 0 & \mathbf{0} & 1 & \ldots & 0 & \mathbf{0} & 0 & \ldots & 0 & 0 \\ \vdots & \vdots & \vdots & \vdots & \vdots & \vdots & \vdots & \vdots & \vdots & \vdots & \vdots & \vdots & \vdots \\ 0 & 0 & \ldots & 0 & \mathbf{0} & 0 & \ldots & 1 & \mathbf{0} & 0 & \ldots & 0 & 0 \\ \mathbf{0} & \mathbf{0} & \ldots & \mathbf{0} & \mathbf{1} & \mathbf{0} & \ldots & \mathbf{0} & \mathbf{0} & \mathbf{0} & \ldots & \mathbf{0} & \mathbf{0} \\ 0 & 0 & \ldots & 0 & \mathbf{0} & 0 & \ldots & 0 & \mathbf{0} & 1 & \ldots & 0 & 0 \\ \vdots & \vdots & \vdots & \vdots & \vdots & \vdots & \vdots & \vdots & \vdots & \vdots & \vdots & \vdots & \vdots \\ 0 & 0 & \ldots & 0 & \mathbf{0} & 0 & \ldots & 0 & \mathbf{0} & 0 & \ldots & 1 & 0 \\ 0 & 0 & \ldots & 0 & \mathbf{0} & 0 & \ldots & 0 & \mathbf{0} & 0 & \ldots & 0 & 1 \end{pmatrix}.$$

Here, the first bold row is row k, the second is row m, and the first bold column is column k and the second is column m.

2.5.11. Proposition. *Let* $A = [a_{ij}] \in \mathbf{M}_n(\mathbb{R})$. *Then*

$$A\mathbf{s}_{km} = \begin{pmatrix} a_{11} & a_{12} & \dots & a_{1n} \\ a_{21} & a_{22} & \dots & a_{2n} \\ \vdots & \vdots & \vdots & \vdots \\ a_{k-1,1} & a_{k-1,2} & \dots & a_{k-1,n} \\ a_{m1} & a_{m2} & \dots & a_{mn} \\ a_{k+1,1} & a_{k+1,2} & \dots & a_{k+1,n} \\ \vdots & \vdots & \vdots & \vdots \\ a_{m-1,1} & a_{m-1,2} & \dots & a_{m-1,n} \\ a_{k1} & a_{k2} & \dots & a_{kn} \\ a_{m+1,1} & a_{m+1,2} & \dots & a_{m+1,n} \\ \vdots & \vdots & \vdots & \vdots \\ a_{n1} & a_{m1} & \dots & a_{nn} \end{pmatrix} \quad \textit{and}$$

$$\mathbf{s}_{km}A = \begin{pmatrix} a_{11} & \dots & a_{1,k-1} & a_{1m} & a_{1,k+1} & \dots & a_{1,m-1} & a_{1k} & a_{1,m+1} & \dots & a_{1n} \\ a_{21} & \dots & a_{2,k-1} & a_{2m} & a_{2,k+1} & \dots & a_{2,m-1} & a_{2k} & a_{2,m+1} & \dots & a_{2n} \\ \vdots & \vdots & \vdots & \vdots & \vdots & \vdots & \vdots & \vdots & \vdots & \vdots & \vdots \\ a_{n1} & \dots & a_{n,k-1} & a_{nm} & a_{m,k+1} & \dots & a_{n,m-1} & a_{nk} & a_{n,m+1} & \dots & a_{nn} \end{pmatrix}.$$

Proof. Put $\mathbf{s}_{km} = \left[s_{ij}^{(km)}\right] \in \mathbf{M}_n(\mathbb{R})$, where

$$s_{ij}^{(km)} = \begin{cases} 1, & \text{if } i = j \neq k, m. \\ 1, & \text{if } (i, j) = (k, m) \text{ or } (i, j) = (m, k). \\ 0 & \text{in all other cases.} \end{cases}$$

Also set $A\mathbf{s}_{km} = [v_{ij}]$ and $\mathbf{s}_{km}A = [w_{ij}]$. Then we have

$$v_{ij} = \sum_{1 \leq t \leq n} a_{it} s_{tj}^{(km)}.$$

If $j \neq k, m$, then $s_{tj}^{(km)} = 1$, when $t = j$, and is otherwise 0. Therefore, $v_{ij} = a_{ij} s_{jj}^{(km)} = a_{ij}$ in this case. If $j = k$ then $s_{tk}^{(km)} = 1$, when $t = m$, and is otherwise 0 so $v_{ik} = a_{im} s_{mk}^{(km)} = a_{im}$. If $j = m$ then $s_{tm}^{(km)} = 1$, when $i = k$, and is otherwise 0. Therefore, $v_{im} = a_{ik} s_{km}^{(km)} = a_{ik}$. We can consider the product $\mathbf{s}_{km}A$ in a similar manner and the stated result follows.

EXERCISE SET 2.5

Provide explanations for your work. Justify your answers with a proof or a counterexample where necessary.

2.5.1. Find all matrices that commute with the matrix $\begin{pmatrix} 0 & 1 & 0 & 0 \\ 0 & 0 & 1 & 0 \\ 0 & 0 & 0 & 1 \\ 0 & 0 & 0 & 0 \end{pmatrix}$.

2.5.2. Find all matrices $A \in \mathbf{M}_2(\mathbb{R})$ with the property $A^2 = 0$.

2.5.3. Evaluate A^{-1} where $A = I + E_{12} + 2E_{23}$.

2.5.4. Let $A \in \mathbf{M}_n(\mathbb{R})$ be a nonsingular symmetric matrix. Is A^{-1} also symmetric?

2.5.5. Let $A \in \mathbf{M}_n(\mathbb{R})$ be a nonsingular skew-symmetric matrix. Is A^{-1} skew symmetric?

2.5.6. Obtain the formula for A^{-1} if $A = \begin{pmatrix} a & b \\ c & d \end{pmatrix}$.

2.5.7. Find A^{-1} if $A = \begin{pmatrix} \cos\alpha & -\sin\alpha \\ \sin\alpha & \cos\alpha \end{pmatrix}$.

2.5.8. Find A^{-1} if $A = \begin{pmatrix} 1 & 1 & 1 & 1 \\ 1 & 1 & -1 & -1 \\ 1 & -1 & 1 & -1 \\ 1 & -1 & -1 & 1 \end{pmatrix}$.

2.5.9. Find A^{-1} if $A = \begin{pmatrix} 1 & 2 & 3 & 4 \\ 2 & 3 & 1 & 2 \\ 1 & 1 & 1 & -1 \\ 1 & 0 & -2 & -6 \end{pmatrix}$.

2.5.10. Find A^{-1} if $A = \begin{pmatrix} 1 & 1 & 0 & 0 & 0 & 0 & 0 & 0 & 0 \\ 0 & 1 & 1 & 0 & 0 & 0 & 0 & 0 & 0 \\ 0 & 0 & 1 & 1 & 0 & 0 & 0 & 0 & 0 \\ 0 & 0 & 0 & 1 & 1 & 0 & 0 & 0 & 0 \\ 0 & 0 & 0 & 0 & 1 & 1 & 0 & 0 & 0 \\ 0 & 0 & 0 & 0 & 0 & 1 & 1 & 0 & 0 \\ 0 & 0 & 0 & 0 & 0 & 0 & 1 & 1 & 0 \\ 0 & 0 & 0 & 0 & 0 & 0 & 0 & 1 & 1 \\ 0 & 0 & 0 & 0 & 0 & 0 & 0 & 0 & 1 \end{pmatrix}$.

2.5.11. Find A^{-1} if $A = \begin{pmatrix} 1 & 1 & 1 & \dots & 1 \\ 1 & 0 & 1 & \dots & 1 \\ 1 & 1 & 0 & \dots & 1 \\ \vdots & \vdots & \vdots & \vdots & \vdots \\ 1 & 1 & 1 & \dots & 0 \end{pmatrix}$.

2.5.12. Solve the following matrix equation $\begin{pmatrix} 1 & 2 \\ 3 & 4 \end{pmatrix} X = \begin{pmatrix} 3 & 5 \\ 5 & 9 \end{pmatrix}$.

2.5.13. Solve the following matrix equation $\begin{pmatrix} 3 & -1 \\ 5 & -2 \end{pmatrix} X \begin{pmatrix} 5 & 6 \\ 7 & 8 \end{pmatrix} = \begin{pmatrix} 14 & 16 \\ 9 & 10 \end{pmatrix}$.

2.5.14. Solve the following matrix equation

$$\begin{pmatrix} 1 & 2 & -3 \\ 3 & 2 & -4 \\ 2 & -1 & 0 \end{pmatrix} X = \begin{pmatrix} 1 & -3 & 0 \\ 10 & 2 & 7 \\ 10 & 7 & 8 \end{pmatrix}.$$

2.5.15. Find the centralizer of the set of all matrices of the kind

$$\begin{pmatrix} \alpha & \alpha & \alpha & \dots & \alpha & \alpha \\ 0 & 0 & 0 & \dots & 0 & 0 \\ 0 & 0 & 0 & \dots & 0 & 0 \\ \vdots & \vdots & \vdots & \vdots & \vdots & \vdots \\ 0 & 0 & 0 & \dots & 0 & 0 \\ 0 & 0 & 0 & \dots & 0 & 0 \end{pmatrix}, \text{ where } \alpha \in \mathbb{R}.$$

2.5.16. Represent the matrix

$$\begin{pmatrix} 1 & -1 & 0 \\ 3 & 0 & -2 \\ 4 & 3 & 0 \end{pmatrix}$$

as a product of a transvection and a diagonal matrix.

2.5.17. Represent the matrix

$$\begin{pmatrix} 1 & -7 & 3 \\ 0 & 4 & -1 \\ -1 & 3 & 0 \end{pmatrix}$$

as a product of a transvection and a diagonal matrix.

CHAPTER 3

FIELDS

3.1 BINARY ALGEBRAIC OPERATIONS

In this chapter we consider one of the main concepts in algebra, namely the concept of a field. This chapter is the last chapter of the book that is dedicated to some preliminary concepts. The concept of a field is a key idea for linear algebra. Many books limit consideration of linear algebra to the fields of real and complex numbers. However, vector spaces over finite fields have recently found various important applications in cryptography and other branches of mathematics. For this reason, we think that it is feasible, indeed rather important, to consider vector spaces over arbitrary fields. In turn, in order to introduce fields we need to first consider binary algebraic operations.

The idea of a binary algebraic operation is one of the most fundamental in mathematics. Indeed, we have already informally seen several examples of this concept, so there is a rich variety of concrete algebraic operations whose properties we understand quite well. This gives us a good opportunity to study these examples from a general point of view. The important parts of this concept are generalized in the following definition.

3.1.1. Definition. *Let M be a set. The mapping $\theta : M \times M \longrightarrow M$ from the Cartesian square of M to M is called a binary (algebraic) operation on the set M. Thus, corresponding to every ordered pair (a, b) of elements, where $a, b \in M$, there is a uniquely defined element $\theta(a, b) \in M$. The element $\theta(a, b) \in M$ is called the composition of the elements a and b relative to this operation.*

Notice that there are two important ideas here. One is that $\theta(a, b)$ is an element of M; the other is that $\theta(a, b)$ is uniquely determined by the ordered pair (a, b). Here, we need to say a few words about notation. It is often rather cumbersome to keep referring to the function θ and using the notation $\theta(a, b)$. There are several shorthand symbols that are employed and $\theta(a, b)$ is often written using such special notation. For example, the operation might be denoted by ⧫ and we might then write $\theta(a, b) = a ⧫ b$. We note that, in general, $\theta(a, b)$ will be different from $\theta(b, a)$, which is to say that there is no reason for it to be the case that $a ⧫ b = b ⧫ a$. However, quite often, even the notation $a ⧫ b$ is confusing, and most often we would rather write the operation ⧫ using something more familiar. The most familiar binary operators are $+$ and $\cdot$ and it is these symbols that are most often useful in writing such operations. Thus, instead of writing $a ⧫ b$ we may write $a + b$ or $a \cdot b$. It is important to understand that sometimes these symbols will have familiar meanings, but not always.

In most cases, the operation denoted by the sign $+$ is associated with addition, and the corresponding composition $a + b$ is then called its sum. In this case, we talk about the additive designation of the binary operation.

The operation denoted by the sign $\cdot$ usually is associated with multiplication, and the corresponding composition $a \cdot b$ is called its product. In this case, we talk about the multiplicative designation of the binary operation. Following tradition, we often omit the sign $\cdot$, and denote the product by just ab.

We now consider some examples of binary operations. For the most part, it is quite easy to verify that these are binary operations but they serve to illustrate that binary operations are very familiar to the reader, as we observed earlier.

(i) Addition on any one of the sets $M = \mathbb{N}, \mathbb{Z}, \mathbb{Q}, \mathbb{R}$ is a binary operation. In this case, $\theta(a, b) = a + b$ and this is clearly an element of M. For example, when $M = \mathbb{Z}$ all this says is that the sum of two integers is a uniquely determined third integer.

(ii) Multiplication on any one of the sets $M = \mathbb{N}, \mathbb{Z}, \mathbb{Q}, \mathbb{R}$ is a binary operation. In this case of course, we denote $\theta(a, b)$ by ab.

(iii) Next, let M be an arbitrary set and let $\mathbf{P}(M)$ be the set of all transformations of M. Then, for all $f, g \in \mathbf{P}(M)$, the map θ defined by $\theta(f, g) = f \circ g$ is a binary algebraic operation on the set $\mathbf{P}(M)$.

(iv) Let M be a set and let $\mathfrak{B}(M)$ denote the Boolean of M. For each $X, Y \subseteq M$, each of the mappings

$$(X, Y) \longmapsto X \cup Y, (X, Y) \longmapsto X \cap Y,$$

$$(X, Y) \longmapsto X \backslash Y, (X, Y) \longmapsto X \Delta Y,$$

defines a corresponding binary algebraic operation on the set $\mathfrak{B}(M)$. Note that in the first case, for example, it is more natural to denote the binary operation by $\cup$ here than by $+$ or $\cdot$. We will always use the natural notation when possible.

(v) Addition, multiplication, and commutation of matrices are each binary operations on the set $M_n(\mathbb{R})$.

(vi) Addition and multiplication of real functions (which is to say transformations of the set $\mathbb{R}$) are each binary operations on the set of real functions, $\mathbf{P}(\mathbb{R})$.

(vii) The following mappings

$$(n, k) \longmapsto n^k \text{ and } (n, k) \longmapsto n^k + k^n, \text{ where } n, k \in \mathbb{N},$$

each define binary algebraic operations on the set $\mathbb{N}$.

(viii) The following mappings

$$(n, k) \longmapsto \mathbf{GCD}(n, k) \text{ and } (n, k) \longmapsto \mathbf{LCM}(n, k),$$
$$\text{where } n, k \in \mathbb{Z},$$

define binary algebraic operations on the set $\mathbb{Z}$ (we make the proviso that we take $\mathbf{GCD}(n, k), \mathbf{LCM}(n, k) \geq 0$).

(ix) Addition and taking the cross (or vector) product on the space $\mathbb{R}^3$ are both binary operations.

Certainly, there are also a host of operations that are not binary. For example, if $a, b \in \mathbb{N}$ then the map $\theta(a, b) = a - b$ is not binary since the difference of two natural numbers need not be a natural number.

We next consider some important properties of binary algebraic operations. For the sake of clarity, we will use the multiplicative form of writing a binary operation but will also illustrate the additive form. However, we stress that our binary operations are very much more general than ordinary addition or multiplication.

3.1.2. Definition. *A binary operation on a set M is called commutative if $ab = ba$ for each pair a, b of elements of M.*

For the additive form, commutativity of a and b would be written as

$$a + b = b + a, \text{ where } a, b \in M.$$

Notice that to show that an operation is commutative on M, we have to show that $ab = ba$ is true for all $a, b \in M$. If $ab \neq ba$ for just one pair $a, b \in M$, then the operation is not commutative. Many of the operations listed above are commutative but some are not. The operations of multiplication and addition on the sets of natural numbers, integers, rational, and real numbers are commutative. The operations $\cap$, $\cup$, and Δ on the Boolean $\mathfrak{B}(M)$ are commutative but, since $X \setminus Y \neq Y \setminus X$ in general, $\setminus$ is not a commutative operation unless the underlying set is the empty set. Matrix addition, addition and multiplication of real functions, the operations $\mathbf{GCD}$, $\mathbf{LCM}$ on $\mathbb{Z}$ and vector addition in $\mathbb{R}^3$ are all examples of commutative binary operations. Note that the cross product of two vectors in $\mathbb{R}^3$ is not commutative and, as we have seen above, multiplication of transformations and multiplication of matrices are examples of important binary operations that are not commutative.

If we have three elements $a, b, c \in M$, then we can form the products $a(bc)$ and $(ab)c$ (where we do not change the order in which the elements are written in the product). In general, these two products may be different. For example, it normally matters whether we write $a - (b - c)$ or $(a - b) - c$, when $a, b, c \in \mathbb{Z}$.

3.1.3. Definition. *A binary operation on a set M is called associative if $(ab)c = a(bc)$ for each triple a, b, c of elements of M.*

Written additively, this becomes

$$(a + b) + c = a + (b + c).$$

For four elements a, b, c, d, we can construct a number of different products. For example, we can determine each of the products

$$((ab)c)d, (ab)(cd), (a(bc))d, a(b(cd)), \text{ and } a((bc)d)$$

to name but a few. When the operation is associative, however, all methods of bracketing give the same expression so that there is no need for any complicated bracketing. Thus, for example,

$$((ab)c)d = (ab)(cd).$$

As the next theorem shows, such equations hold in general and the theorem justifies the previous statement. This theorem alone makes associative operations very important.

3.1.4. Theorem. *If the binary operation $\cdot$ defined on the set M is associative and if $a_1, \ldots, a_n$ is any finite subset of M, then the product $a_1 a_2 \ldots a_n$ is unambiguous; any form of bracketing of this product always gives the same element of M.*

Proof. We proceed to prove the result by induction on n, the number of terms in the product. The entries in the product are $a_1, \ldots, a_n$. If $n = 1, 2$ then the result is clear and if $n = 3$, the assertion follows from the associative property.

Suppose now that $n > 3$ and that we have already proved our assertion for all finite products with fewer than n terms in the product. We will show that each product of elements $a_1, a_2, \ldots, a_n$ bracketed in some way and in some fixed order coincides with a fixed product, namely the so-called left-normed order, $(\ldots(((a_1 a_2)a_3)a_4)\ldots)a_{n-1}))a_n$. If the product is of the form La_n where L is the product of the elements $a_1, \ldots, a_{n-1}$ bracketed in some way, then by the induction hypothesis applied to L, we can write L

as $(\ldots(((a_1a_2)a_3)a_4)\ldots)a_{n-2}))a_{n-1}$. Then La_n is the left-normed product $(\ldots(((a_1a_2)a_3)a_4)\ldots)a_{n-1}))a_n$ and the result follows in this case. Otherwise, the product has the form LM where for some natural number t such that $t+1<n$, L is a product of the elements $a_1,\ldots,a_t$ (in that order) and M is a product of the elements $a_{t+1}\ldots a_n$ (in that order) and there is some bracketing among the terms. By the induction hypothesis and associativity, we have

$$\begin{aligned}LM &= (a_1\ldots a_t)(a_{t+1}\ldots a_n) = (a_1\ldots a_t)((a_{t+1}\ldots a_{n-1})a_n)\\ &= ((a_1\ldots a_t)(a_{t+1}\ldots a_{n-1}))a_n = (\ldots(((a_1a_2)a_3)a_4)\ldots)a_{n-1}))a_n.\end{aligned}$$

This proves the result.

By Theorem 3.1.4, the product $a_1\ldots a_n$ is independent of the bracketing that may be assigned, provided the operation is associative. Consequently, there is no ambiguity when we write $a_1\ldots a_n$, and the parentheses will usually not be inserted although, of course, the order in which the elements are written will normally be important. As a convenient shorthand, we write $a_1\ldots a_n$ as $\prod\limits_{1\le i\le n} a_i$.

In the case when $a_1=a_2=\cdots=a_n=a$, we will denote the product $a_1a_2\ldots a_n$ by a^n, and we will call it the nth power of the element a. In this case, the usual “rules of exponents” are a special case of Theorem 3.1.4, which we give as the next corollary.

3.1.5. Corollary. *If a binary operation on a set M is associative, then for each element $a\in M$, and arbitrary $n,m\in\mathbb{N}$,*

$$a^na^m = a^{n+m} \text{ and } (a^n)^m = a^{nm}.$$

When we use additive notation for our binary operation, we use the usual $\sum\limits_{1\le i\le n} a_i$ instead of $\prod\limits_{1\le i\le n} a_i$ and instead of a power of an element, we will use its multiple

$$na = \underbrace{a+\cdots+a}_{n}.$$

In additive notation, Corollary 3.1.5 takes the form

$$na+ma=(n+m)a \text{ and } m(na)=(mn)a.$$

Two elements a and b are called *permutable or commutable* and they are said to commute if

$$ab=ba.$$

For such elements, we have

$$(ab)^n = a^nb^n,$$

for any $n \in \mathbb{N}$, provided the operation is associative. To prove this, we first prove that $ab^n = b^n a$ by induction on n, the case $n = 1$ being the commutativity statement concerning a and b. Taking our induction hypothesis to be that $ab^n = b^n a$ and using Corollary 3.1.5, we have

$$ab^{n+1} = a(b^n b) = (ab^n)b = (b^n a)b = b^n(ab) = b^n(ba) = (b^n b)a = b^{n+1}a,$$

so this result follows by induction. Next, we note that

$$(ab)^2 = abab = a(ba)b = a(ab)b = (aa)(bb) = a^2b^2.$$

Using induction on n and assuming inductively that $(ab)^n = a^n b^n$, we have

$$\begin{aligned}(ab)^{n+1} &= (ab)^n(ab) = (a^n b^n)ab = a^n(b^n a)b = a^n(ab^n(b)) \\ &= (a^n a)(b^n b) = a^{n+1}b^{n+1}.\end{aligned}$$

Using a similar induction argument, we can prove the following generalization.

3.1.6. Proposition. *Let M be a set with an associative binary operation. If $a_1, a_2, \ldots, a_n$ are elements of M such that $a_i a_j = a_j a_i$ for all pairs i, j, where $1 \leq i, j \leq n$, then*

$$(a_1 a_2 \ldots a_n)^m = a_1^m a_2^m \ldots a_n^m,$$

for every $m \in \mathbb{N}$.

In additive notation, this equation takes the form

$$m(a_1 + a_2 + \cdots + a_n) = ma_1 + ma_2 + \cdots + ma_n.$$

Let M be a set with a binary operation. An element $z \in M$ is called *central* if it commutes with every element of M. The set of all central elements of M is called *the center of* M and will be denoted by $\zeta(M)$.

3.1.7. Definition. *A nonempty set S is called a semigroup if S has an associative binary operation defined on it. If this operation is commutative, we will say that S is a commutative semigroup.*

There are many natural examples of semigroups.

(i) The sets of all natural numbers, integers, rational and real numbers form commutative semigroups under the operation of addition. These same sets also form commutative semigroups under the operation of multiplication.

(ii) Next, let M be a fixed but arbitrary set. Theorem 1.3.2 shows that the set $\mathbf{P}(M)$ of all transformations of the set M is a semigroup under the operation

$$(f, g) \longmapsto f \circ g, \text{ where } f, g \in \mathbf{P}(M).$$

As we have already seen, this semigroup is not commutative.

(iii) Let M be a set. Theorem 1.1.10 shows that the Boolean $\mathfrak{B}(M)$ of the set M is a commutative semigroup under each of the operations:

$$(X, Y) \longmapsto X \cap Y, (X, Y) \longmapsto X \cup Y, (X, Y) \longmapsto X \Delta Y,$$

whenever $X, Y \subseteq M$.

(iv) By Theorem 2.15, the set $M_n(\mathbb{R})$ is a commutative semigroup under the operation of matrix addition and a noncommutative semigroup under the operation of matrix multiplication.

(v) The set of all real functions, $f : \mathbb{R} \longrightarrow \mathbb{R}$ is a commutative semigroup under the operations of addition and multiplication.

(vi) The set of all integers is a commutative semigroup under the operations

$$(n, k) \longmapsto \mathbf{GCD}(n, k) \text{ and } (n, k) \longmapsto \mathbf{LCM}(n, k),$$

where $n, k \in \mathbb{Z}$.

(vii) The vector space $\mathbb{R}^3$ is a commutative semigroup under addition of vectors.

Finally, we give one more important example. Let A be a nonempty set, which we will call the alphabet. The elements of A are called the letters of the alphabet. Let $\mathbf{F}_A$ denote the set of all finite tuples of elements of A. Define a (binary) operation on $\mathbf{F}_A$ by the rule

$$(a_1, a_2, \dots, a_n)(b_1, b_2, \dots, b_m) = (a_1, a_2, \dots, a_n, b_1, b_2, \dots, b_m),$$

where $a_i, b_j \in A$, for $1 \leq i \leq n, 1 \leq j \leq m$. This operation is called *concatenation* and it is very easy to see that it is associative. Thus, $\mathbf{F}_A$ is a semigroup under this operation called the free semigroup over the alphabet A.

We can identify a tuple consisting of one element with the element itself. By doing this, we can identify the tuple $(a_1, a_2, \dots, a_n)$ with the formal product $a_1 a_2 \dots a_n$. This allows us to write all the elements w of the free semigroup $\mathbf{F}_A$ in the form $w = a_1 \dots a_n$, which we call a *word* in the alphabet. The number n is called *the length of this word* and two words $a_1 a_2 \dots a_n$ and $b_1 b_2 \dots b_m$, where $a_i, b_j \in A$ are defined to be equal if and only if $n = m$ and $a_t = b_t$ for each $1 \leq t \leq m$.

3.1.8. Definition. *Let M be a set with binary operation. The element $e \in M$ is called a neutral element under this operation if $ae = ea = a$ for each element a of the set M.*

The neutral element is unique whenever it exists. Indeed, if e' is another element with the property $ae' = e'a = a$ for all $a \in M$ then, setting $a = e'$ in the definition of e gives $e'e = ee' = e'$, whereas setting $a = e$ in the definition of e' gives $ee' = e'e = e$ and we obtain $e = e'$.

If the operation on M is written multiplicatively, then the term *identity element* is usually used rather than neutral element and often e is denoted by 1 or 1_m. If we use the additive form, then the neutral element is usually called the *zero element* and is often denoted by 0_M, so that the definition of the zero element is $a + 0_M = 0_M + a = a$ for each element $a \in M$. When the context is clear, we may sometimes omit the subscript here.

(i) The operation of addition on the sets of all natural numbers, integers, rational, and real numbers has a zero element, which is the number 0.

(ii) The operation of multiplication on the sets of all natural numbers, integers, rational, and real numbers has an identity element, which is the number 1.

(iii) Let M be a set and let $\mathbf{P}(M)$ denote the set of all transformations M. We know that $\varepsilon_M \circ f = f \circ \varepsilon_M = f$ for all $f \in \mathbf{P}(M)$. Thus, $\mathbf{P}(M)$ under the operation of composition has an identity element, the permutation ε_M.

(iv) Let M be a set and $\mathfrak{B}(M)$ the Boolean of the set M. The operations

$$(X, Y) \longmapsto X \cap Y, (X, Y) \longmapsto X \cup Y, (X, Y) \longmapsto X \Delta Y,$$
$$\text{where } X, Y \in M,$$

each have neutral elements. For the first operation, this is the set M since $M \cap Y = Y \cap M = Y$ for all $Y \subseteq M$ and, in a similar manner, for the other two operations it is the empty set.

(v) The operation of addition on the set $M_n(\mathbb{R})$ of real matrices has a zero element; namely, the zero matrix O, since $O + A = A + O = A$ for all $A \in M_n(\mathbb{R})$. The operation of multiplication on the set $M_n(\mathbb{R})$ has an identity element, the matrix I, since $AI = IA = A$, for all $A \in M_n(\mathbb{R})$.

(vi) The operation of addition of all real functions has a zero element, the zero function, the function which is 0 for all arguments; the operation of multiplication of all real functions has an identity element, the function which is always equal to 1 for all arguments.

(vii) The operation

$$(n, k) \longmapsto \mathbf{GCD}(n, k), \text{ where } n, k \in \mathbb{Z},$$

has a neutral element, the number 0, since $\mathbf{GCD}(n, 0) = n$.

(viii) The operation of addition on the set $\mathbb{R}^3$ has a zero element, the zero vector.

3.1.9. Definition. *A semigroup that has an identity element is called a semigroup with identity.*

3.1.10. Definition. *Let M be a set with a binary operation. A subset S is called stable under this operation if for each pair of elements $a, b \in S$ the element ab also belongs to S.*

This means that the restriction to S of the binary operation on M is again a binary operation on S. For example, the subset of all even integers is a stable subset of $\mathbb{Z}$ under the operations of addition and multiplication, since the addition and multiplication of two even integers is again even. The proof of the next proposition is quite straightforward, but we illustrate it for the convenience of the reader.

3.1.11. Proposition. *Let M be a set with a binary operation and let $\mathfrak{S}$ be a family of stable subsets of M. Then the intersection $\cap\mathfrak{S} = \cap\{S : S \subset \mathfrak{S}\}$ of all subsets of this family is also stable.*

Proof. If $a, b \in \cap\mathfrak{S}$ then $a, b \in S$ for all $S \in \mathfrak{S}$. Hence $ab \in S$ for all $S \in \mathfrak{S}$, so $ab \in \cap\mathfrak{S}$.

However, we note that a union of even just two stable subsets need not be stable. To see this, consider the subset $2\mathbb{Z}$ of all even integers and the subset $3\mathbb{Z}$ of all integers divisible by 3. Both of these sets are stable under addition, but $2\mathbb{Z} \cup 3\mathbb{Z}$ does not contain $5 = 2 + 3$, and therefore is not stable.

Let M be a set with a binary operation, let C be a subset of M, and let $\mathfrak{S}$ be the family of stable subsets, each of which contains C. Then the intersection $\cap\mathfrak{S}$ is the least stable subset containing C, called the *stable subset generated by C*.

3.1.12. Definition. *Let S be a semigroup. A stable subset R of S is called a subsemigroup if R is a semigroup in its own right under the operation defined on S.*

3.1.13. Definition. *Let M be a set with a binary operation and suppose that there is an identity element e. The element x is called an inverse of the element a if*

$$ax = xa = e.$$

If a has an inverse then we say that a is invertible.

If the operation on M is associative and the element a is invertible, then a has a unique inverse. To see this, let y be an element of M that also satisfies

$$ay = ya = e.$$

Then

$$y = ey = (xa)y = x(ay) = xe = x.$$

We denote the unique inverse of a by a^{-1}. We note that $aa^{-1} = a^{-1}a = e$ and so, evidently,

$$(a^{-1})^{-1} = a.$$

If the operation on M is written additively, then we denote the inverse of a, should it exist, by $-a$, called the negative (or sometimes the opposite) of a. The definition of the negative element takes the form

$$a + (-a) = (-a) + a = 0_M.$$

3.1.14. Proposition. *Let M be a set with an associative binary operation and suppose that M has an identity element e.*

(i) *If the elements $a_1, a_2, \ldots, a_n$ are invertible in M, then the product $a_1 a_2 \ldots a_n$ is also invertible and*

$$(a_1 a_2 \ldots a_n)^{-1} = a_n^{-1} a_{n-1}^{-1} \ldots a_1^{-1}.$$

(ii) *If a is invertible in M then a^n is invertible, for all $n \in \mathbb{N}$, and $(a^n)^{-1} = (a^{-1})^n$.*

Proof. (i) We prove this by induction on n. For the case $n = 2$, we have

$$(a_1 a_2)(a_2^{-1} a_1^{-1}) = a_1(a_2 a_2^{-1})a_1^{-1} = a_1 e a_1^{-1} = a_1 a_1^{-1} = e.$$

Likewise, $(a_2^{-1} a_1^{-1})(a_1 a_2) = e$ so that, by uniqueness of inverses, $(a_1 a_2)^{-1} = a_2^{-1} a_1^{-1}$ and the result holds for $n = 2$. Assuming that the result is true for n, so that $(a_1 \ldots a_n)^{-1} = a_n^{-1} a_{n-1}^{-1} \ldots a_1^{-1}$ we have, using the case $n = 2$ and the induction hypothesis,

$$(a_1 \ldots a_n a_{n+1})^{-1} = [(a_1 \ldots a_n)a_{n+1}]^{-1} = a_{n+1}^{-1}(a_1 \ldots a_n)^{-1} = a_{n+1}^{-1} a_n^{-1} \ldots a_1^{-1}$$

so that the result follows by induction.

(ii) This is clear since $(a^{-1})^n a^n = (a^{-1}a)^n = e = (aa^{-1})^n = (a^n)(a^{-1})^n$, by Proposition 3.1.6, so the result follows by uniqueness of inverses.

We let $\mathbf{U}(P)$ denote the set of invertible elements of the semigroup P. The following can be read off from Proposition 3.1.14.

3.1.15. Corollary. *Let P be a semigroup with identity. Then the subset $\mathbf{U}(P)$ of all invertible elements is stable.*

The existence of an identity element and an inverse to the element a allows us to define all integer powers of a. To do this, we define

$$a^0 = e, \text{ and } a^{-n} = (a^{-1})^n, \text{ whenever } n \in \mathbb{N}.$$

In additive notation, these definitions take the form

$$0a = 0 \text{ and } (-n)a = n(-a).$$

Our next result shows that the usual rules of exponents hold for all integer powers.

3.1.16. Proposition. *Let M be a set together with an associative binary operation and suppose that M has an identity element e. If $a \in M$ is invertible and $m, n \in \mathbb{Z}$ then*

$$a^n a^m = a^{n+m} \text{ and } (a^n)^m = a^{nm}.$$

Proof. If $n, m > 0$, then the assertion follows from Corollary 3.1.5. Furthermore, if one of m or n is 0 then the equalities hold in any case. If $m, n < 0$, then $n = -p, m = -q$, for certain $p, q \in \mathbb{N}$. Then, using the definitions we have,

$$\begin{aligned} a^n a^m &= a^{-p}a^{-q} = (a^{-1})^p (a^{-1})^q = (a^{-1})^{p+q} = a^{-(p+q)} \\ &= a^{-p-q} = a^{n+m} \text{ and} \\ (a^n)^m &= (a^{-p})^{-q} = \left((a^p)^{-1}\right)^{-q} = \left(\left((a^p)^{-1}\right)^{-1}\right)^q = (a^p)^q = a^{pq} = a^{mn}, \end{aligned}$$

using Proposition 3.1.14.

Suppose now that $n > 0$, $-q = m < 0$, and $n > -m = q$. Then

$$a^n a^m = \underbrace{a \dots a}_{n}\underbrace{(a^{-1}) \dots (a^{-1})}_{q} = \underbrace{a \dots a}_{n-q} = a^{n+m}.$$

If $n > 0$, $-q = m < 0$, and $n < -m = q$, then

$$a^n a^m = \underbrace{a \dots a}_{n}\underbrace{(a^{-1}) \dots (a^{-1})}_{q} = \underbrace{a^{-1} \dots a^{-1}}_{q-n} = (a^{-1})^{-(n+m)} = a^{n+m}.$$

For the second equation, if $n > 0$ and $-q = m < 0$, then

$$(a^n)^m = ((a^n)^{-1})^q = ((a^{-1})^n)^q = (a^{-1})^{nq} = (a^{-1})^{-nm} = a^{-(-nm)} = a^{nm}.$$

If $-p = n < 0, m > 0$, then

$$(a^n)^m = ((a^{-1})^p)^m = (a^{-1})^{pm} = (a^{-1})^{-nm} = a^{-(-nm)} = a^{nm}.$$

The result follows.

We next define one of the most important algebraic structures. For now, we shall only give the definition and establish many properties in later chapters.

3.1.17. Definition. *A semigroup G with identity is called a group if every element of G is invertible. Thus, a group is a set G together with a binary algebraic operation $(x, y) \longmapsto xy$, where $x, y \in G$, such that the following conditions (the group axioms) hold:*

(G 1) *The operation is associative so that $x(yz) = (xy)z$ for all $x, y, z \in G$.*

(G 2) *G has an identity element, an element e such that $xe = ex = x$ for all $x \in G$; often 1 or 1_G is used in place of e.*

(G 3) *Every element $x \in G$ has an inverse x^{-1} such that $xx^{-1} = x^{-1}x = e$.*

We note that saying that the operation is binary is really a fourth axiom here, the axiom of closure, which is an alternative way of saying that the operation is binary. If the group operation is commutative, then the group is called abelian (in honor of the great Norwegian mathematician Niels Henrik Abel (1802–1829)).

It is common practice to use addition as the operation when the group is abelian. So, for abelian groups, the group axioms are as follows:

(AG 1) the operation is commutative, so that $x + y = y + x$ for all elements $x, y \in G$;

(AG 2) the operation is associative, so that $x + (y + z) = (x + y) + z$ for all elements $x, y, z \in G$;

(AG 3) G has a zero element, an element 0_G such that $x + 0_G = x$ for all $x \in G$;

(AG 4) every element $x \in G$ has a negative, an element $-x$, such that $x + (-x) = 0_G$.

Let G be an abelian group with additive operation. Then we can define the operation of subtraction by the rule that $x - y = x + (-y)$.

3.1.18. Definition. *Let G be a group. A stable subset H of G is called a subgroup, if H is a group in its own right, under the operation defined on G.*

Certain types of mapping between groups are very important. Indeed we shall use the terminology that we next introduce very often in different contexts.

3.1.19. Definition. *Let M, S be sets with binary operations that we denote by $\star$ and $\diamond$, respectively. Let $f : M \longrightarrow S$ be a mapping. Then f is called a homomorphism, if*

$$f(x \star y) = f(x) \diamond f(y)$$

for arbitrary elements $x, y \in M$.

We say that the mapping f respects the operations. An injective homomorphism is called a monomorphism. *A surjective homomorphism is called an* epimorphism *and a bijective homomorphism is called an isomorphism.*

If $f : M \longrightarrow S$ is an isomorphism, then Theorem 1.3.5 shows that the mapping f has an inverse $f^{-1} : S \longrightarrow M$, which is also bijective. If u, v are arbitrary elements of the set S, then $u = f(x)$ and $v = f(y)$ for certain elements $x, y \in M$, since f is surjective. Furthermore, we now have

$$\begin{aligned} f^{-1}(u \diamond v) &= f^{-1}(f(x) \diamond f(y)) = f^{-1}(f(x \star y)) \\ &= x \star y = f^{-1}(u) \star f^{-1}(v). \end{aligned}$$

This shows that the mapping $f^{-1} : S \longrightarrow M$ is also an isomorphism.

3.1.20. Definition. *Let M, S be sets with binary operations. Then M, S are called* isomorphic *if there exists an isomorphism from M to S and we then write $M \cong S$.*

When two structures M, S are isomorphic in this way, there is no difference between the structures other than the names we give to the elements of the two sets M and S and the names $\star$ and $\diamond$ that we give to the names of the operations. Other than this, the structures of M and S are identical.

If M is a set with a binary operation, then the study of M has two aspects. The first aspect is concerned with the nature of the elements and the structure of M, while the second one concerns properties of the operation. This enables such a study to be conducted from different points of view. We can study the relationship between the elements and the subsets of M and also study individual properties with respect to the given operation. Such an approach is feasible for the study of concrete sets, such as permutations, transformations of the plane and space, symmetries, matrices, and so on. However, we can conduct a study of the properties that does not depend on the nature of the elements and which is completely defined by the operation. This approach is the key approach in algebra and it can be covered very efficiently, thanks to the fundamental notion of isomorphism. Making this more concrete, Gottfried Leibniz (1646–1716) introduced the general notion of an isomorphic relation (which he called a similarity) and pointed out the possibility of the identification of isomorphic operations and relations. He brought attention to a classical example of isomorphism, namely the mapping $x \longmapsto \log x$ from the set of all positive real numbers with the operation of multiplication to the set of all real numbers with the operation of addition. A great French mathematician, Évariste Galois (1811–1832), was also familiar with the idea of isomorphism. He understood that corresponding elements of isomorphic sets M and S have the same properties with respect to the given operation. This notion in its general form was developed in the middle of the nineteenth century. In abstract algebra, we study only such properties that are unchanged by isomorphisms.

EXERCISE SET 3.1

Justify your answers with a proof or a counterexample.

3.1.1. On the set $G = \mathbb{Z} \times \{-1, 1\}$ we define an operation $*$ by the rule $(m, a) * (n, b) = (m + an, ab)$. Is this operation associative or commutative? Has it an identity element? Which elements have inverses?

3.1.2. On a set of four elements define a commutative, associative binary operation having an identity element.

3.1.3. On the set $\mathbb{Z}$ define an operation $\perp$ by the rule $a \perp b = a^2 + b^2$, where $a, b \in \mathbb{Z}$. Is this operation associative or commutative? Has it an identity element?

3.1.4. On the set $\mathbb{Q} \times \mathbb{Q}$ define an operation $\bullet$ by $(a, b) \bullet (c, d) = (ac, b + ad)$. Is this operation associative or commutative? Has it an identity element?

3.1.5. On the set $\mathbb{R}$ define an operation $\bullet$ by the rule $a \bullet b = a + b + ab$. Prove the following:

(i) $a \bullet (b \bullet c) = (a \bullet b) \bullet c$ for all $a, b, c \in \mathbb{R}$.

(ii) $a \bullet b = b \bullet a$ for all $a, b \in \mathbb{R}$.

(iii) If $a \neq -1$, then $a \bullet b = a \bullet c$ if and only if $b = c$.

(iv) Has this operation an identity element?

(v) Which elements have inverses?

3.1.6. On the set $\mathbb{Z}$ define an operation $\perp$ by $a \perp b = 4a + b$, where $a, b \in \mathbb{Z}$. Is this operation associative or commutative? Has it an identity element?

3.1.7. On the set $\mathbb{R} \times \mathbb{R}$ define an operation $\bullet$ by the rule $(a, b) \bullet (c, d) = (ac - bd, bc + ad)$. Is this operation associative or commutative? Has it an identity element?

3.1.8. Let $M = \{u, x, y, z\}$. On M define a binary algebraic operation such that it will be commutative, associative and for which there is an identity element, but M is not a group.

3.1.9. Let $M = \{u, x, y, z\}$. Define a binary algebraic operation on M such that M is a group.

3.1.10. On the set M we define an algebraic binary operation $\clubsuit$ by $a \clubsuit b = a$ for arbitrary $a, b \in M$. Prove that M is a semigroup. Does it have an identity element? If yes, which elements have inverses?

3.1.11. Let M be a set and let $S = \mathfrak{B}(M)$. Is S a semigroup under the operation $\cap$? Is S a semigroup under the operation $\cup$? Are these semigroups isomorphic?

3.1.12. Let $M = \{x, y, z\}$. A binary algebraic operation is defined by the table

	x	y	z
x	x	y	z
y	y	z	x
z	z	x	y

. Is M a group?

3.1.13. Define the binary operation $\blacklozenge$ on $\mathbb{R} \times \mathbb{R}$ by the rule $(a, b) \blacklozenge (c, d) = (ac - 2bd, bc + ad)$. Is the set $\mathbb{R} \times \mathbb{R} \backslash \{(0, 0)\}$ a group under this operation?

3.1.14. Define binary operations $\blacktriangledown$, $\blacktriangle$, and $\blacksquare$ on $\mathbb{Q}$ by the rules $a \blacktriangledown b = a - b + ab$, $a \blacktriangle b = \frac{1}{2}(a + b + ab)$, $a \blacksquare b = \frac{1}{3}(a + b)$.
Of these operations which are associative or commutative? Which have an identity element?

3.1.15. Define a binary operation $\blacktriangledown$ on $\mathbb{R}$ by the rule $a \blacktriangledown b = pa + qb + r$. For which fixed p, q, r, is this operation associative?

3.1.16. Let $\mathbb{Q}^*$ be the set of all nonzero rational numbers. Which of the following properties hold for the operation of division?

(i) $a \div b = b \div a$.

(ii) $(a \div b) \div c = a \div (b \div c)$.

(iii) $((a \div b) \div c) \div d = a \div (b \div (c \div d))$.

(iv) If $a \div b = a \div c$, then $b = c$.

(v) If $b \div a = c \div a$, then $b = c$.

3.2 BASIC PROPERTIES OF FIELDS

Having introduced the concept of a binary algebraic operation in the previous section, we can now introduce further algebraic structures. However, since our very next goal is to introduce the main ideas of linear algebra, including vector spaces over fields, we will focus on the structures (fields) whose properties are needed for this. We will study other algebraic structures later in the book.

3.2.1. Definition. *A set D with two binary algebraic operations, addition and multiplication, is called a division ring if it satisfies the following properties:*

(i) *the addition is commutative, so*

$$x + y = y + x$$

for all elements $x, y \in D$;

(ii) *the addition is associative, so*

$$x + (y + z) = (x + y) + z$$

for all elements $x, y, z \in D$;

(iii) *D has a zero element,* 0_D*, an element with the property that*

$$x + 0_D = 0_D + x = x$$

for all elements $x \in D$*;*

(iv) *each element* $x \in D$ *has an additive inverse (the opposite or negative element),* $-x \in D$*, an element with the property that*

$$x + (-x) = 0_D;$$

(v) *the distributive laws hold in D, so*

$$x(y + z) = xy + xz \text{ and } (x + y)z = xz + yz$$

for all elements $x, y, z \in D$*;*

(vi) *the multiplication is associative, so*

$$x(yz) = (xy)z$$

for all elements $x, y, z \in D$*;*

(vii) *D has a (multiplicative) identity element,* $e \neq 0_D$*, an element with the property that*

$$xe = ex = x$$

for each element $x \in D$*;*

(viii) *each nonzero element* $x \in D$ *has a multiplicative inverse (the reciprocal),* $x^{-1} \in D$*, an element with the property*

$$xx^{-1} = x^{-1}x = e.$$

There are several points that we should mention here. When we think of D, together with the operation of addition only, then we often denote D by D_+ in this case and call D_+ the additive group of the division ring. Likewise, when we only wish to consider the operation of multiplication we talk about the multiplicative group of the division ring D. This is the set of nonzero elements of D, which we denote by $U(D)$ or $D_\times$, under the operation of multiplication.

We also make a remark concerning notation. When $0_D \neq x \in D$ we think of x^{-1} as "x inverse." When $D = \mathbb{Q}$, the set of rational numbers for example, then x^{-1} is the usual reciprocal of x, namely $1/x$. In general, however, in a division ring D, it is usual to write the inverse of x as x^{-1} and not as $1/x$.

The existence of opposite elements allows us to introduce the operation of subtraction into division rings by means of the rule

$$a - b = a + (-b).$$

There are a number of elementary consequences of this definition which, although easy to prove, require rigorous proof, which we now supply.

3.2.2. Proposition. *Let D be a division ring. Then*

(i) $a \cdot 0_D = 0_D \cdot a = 0_D$;
(ii) $a(-b) = (-a)b = -ab$, *for all* $a, b \in D$;
(iii) $a(b - c) = ab - ac$ *and* $(a - b)c = ac - bc$, *for all elements* $a, b, c \in D$.

Proof. In fact, for each $x \in D$ we have $x + 0_D = x$. By the distributivity property we have

$$ax = a(x + 0_D) = ax + a \cdot 0_D.$$

Since the element ax has a negative, $-ax \in D$, adding it to both sides of this equation and using the associative property appropriately, we have

$$0_D = -ax + ax = -ax + ax + a \cdot 0_D = 0_D + a \cdot 0_D = a \cdot 0_D,$$

and, by a similar argument, $0_D \cdot a = 0_D$.

It follows from the definition of the negative and the distributive law that

$$0_D = a \cdot 0_D = a(b + (-b)) = ab + a(-b), \text{ for all } a, b \in D.$$

Thus $a(-b)$ is the negative of ab which means that $a(-b) = -(ab) = -ab$ and we can show similarly that $(-a)b = -ab$.

We can now connect subtraction and multiplication using a variation of the distributive law since

$$a(b - c) = a(b + (-c)) = ab + a(-c) = ab + (-ac) = ab - ac,$$

and similarly

$$(a - b)c = ac - bc.$$

Next we have some standard multiplicative properties.

3.2.3. Proposition. *Let D be a division ring.*

(i) *If* $ab = 0_D$, *then either* $a = 0_D$ *or* $b = 0_D$.
(ii) *If* $ab = ac$ *and* $a \neq 0_D$, *then* $b = c$ *(left cancellation).*
(iii) *If* $ba = ca$ *and* $a \neq 0_D$, *then* $b = c$ *(right cancellation).*

Proof.
(i) Suppose that $a \neq 0_D$. Then a^{-1} exists and we now have

$$0_D = a^{-1}0_D = a^{-1}(ab) = (a^{-1}a)b = eb = b,$$

so that (i) follows.

(ii) If $ab = ac$, then $0_D = ab - ac = a(b - c)$, by Proposition 3.2.2. Since $a \neq 0_D$, it follows from (i) that $b - c = 0_D$, and hence $b = c$.

(iii) This can be proved similarly.

Notice that Proposition 3.2.3(ii) tells us that we can cancel in the division ring D.

3.2.4. Definition. *A division ring D is called a field, if the multiplication of its elements is always commutative. Thus a field has the additional property that $xy = yx$ for all elements $x, y \in D$.*

The sets $\mathbb{Q}$ and $\mathbb{R}$ of rational and real numbers with the natural operations of addition and multiplication are well-known important examples of fields.

Our next example is that of the smallest field. Put $\mathbb{F}_2 = \{0, 1\}$ and define the operation of addition and multiplication by the following rules:

$$\begin{array}{ccc} + & 0 & 1 \\ 0 & 0 & 1 \\ 1 & 1 & 0 \end{array} \quad \text{and} \quad \begin{array}{ccc} \cdot & 0 & 1 \\ 0 & 0 & 0 \\ 1 & 0 & 1 \end{array}.$$

It is easy to prove that $\mathbb{F}_2$ is a field, our first example of a finite field. We can extend this as given below.

Let p be a prime and let $\mathbb{F}_p = \{0, 1, 2, \dots, p - 1\}$. We now define the operations of addition and multiplication (denoted by $\oplus$ and $\otimes$, for now, respectively), by the following rules:

Let $0 \neq k, m \leq p - 1$. If $k + m < p$, then put $k \oplus m = k + m$. Suppose that $k + m \geq p$. By Theorem 1.4.1, there exist positive integers b, r such that $k + m = bp + r$ where $0 \leq r < p$ and moreover, the numbers b, r are uniquely defined. In this case, let $k \oplus m = r$. Similarly, if $km < p$, then let $k \otimes m = km$. Suppose that $km \geq p$. By Theorem 1.4.1, there exist positive integers c, u such that $km = cp + u$ where $0 \leq u < p$, and moreover, the numbers c, u are uniquely defined. In this case, let $k \otimes m = u$.

The zero element here is the number 0. Clearly, also, the negative of k is $p - k$, since $(p - k) + k = p$, which has remainder 0 when divided by p. Furthermore, the equations

$$(k + m) + t = k + (m + t), km = mk, (km)t = k(mt), k(m + t) = km + kt$$

imply

$$(k \oplus m) \oplus t = k \oplus (m \oplus t), k \otimes m = m \otimes k,$$

$$(k \otimes m) \otimes t = k \otimes (m \otimes t) \text{ and } k \otimes (m \oplus t) = (k \otimes m) \oplus (k \otimes t).$$

Clearly, the number 1 is the identity element of $\mathbb{F}_p$. Finally, let $0 < k < p$. Since p is a prime, the integers k, p are relatively prime. Then, by Corollary 1.4.7, there are integers x, y such that $kx + py = 1$. It follows that $kx = 1 + pz$ where $z = -y$. If $x > p$, then $x = pc + a$ where $0 \neq a < p$. Then $kx = kpc + ka$ and $ka = kx - kpc$, so that $ka = 1 + pz - pkc = 1 + p(z - kc)$. It follows that $k \otimes a = 1$ and hence, a is the multiplicative inverse of k.

In this way, all conditions of Definition 3.2.1 hold, which proves that $\mathbb{F}_p$ is a field under the operations $\oplus$ and $\otimes$.

We are always interested in "subobjects" in algebra. Here, we discuss subfields.

3.2.5. Definition. *Let F be a field. A subset H of F is called a subfield if H is stable under both the operation of addition and the operation of multiplication in the field F, and H is itself a field under the same operations.*

3.2.6. Theorem. *Let F be a field. If H is a subfield of F, then H satisfies the following conditions:*

*(**SF 1**) if $x, y \in H$, then $x - y \in H$ and $xy \in H$;*
*(**SF 2**) if $x \in H$, and $x \neq 0_F$, then $x^{-1} \in H$.*

*Conversely, suppose that H has at least two elements. If H satisfies conditions (**SF 1**) and (**SF 2**), then H is a subfield of F.*

Proof. Let H be a subfield of F. In particular, H is a stable subset under the addition and multiplication of the field F. Also H has a zero element 0_H. Thus, $x + 0_H = x$ for each element $x \in H$. By Definition 3.2.1, there is an element $-x \in F$ and we have $-x + x + 0_H = -x + x$. Hence $0_F + 0_H = 0_F$ and it follows that $0_H = 0_F$. By Definition 3.2.1 again, for each element $x \in H$, there is an element $y \in H$ such that $x + y = 0_H$. As we saw above, $0_H = 0_F$, and therefore y is the negative of x in F and hence $y = -x$. In particular, $-x \in H$. Now if x, y are arbitrary elements of H, then $-y \in H$. Since H is a stable subset under addition, $x - y = x + (-y) \in H$. Since H is a stable subset under multiplication, $xy \in H$, so that H satisfies (**SF 1**). By Definition 3.2.1, H has an identity element $e_H \neq 0_H$ and $e_H e_H = e_H$. Since $0_F = 0_H$, $e_H \neq 0_F$, and by Definition 3.2.1, there is an element y such that $ye_H = e_H y = e_F$. Then,

$$e_H = e_F e_H = ye_H e_H = ye_H = e_F.$$

Consequently, $e_F \in H$. By Definition 3.2.1, for each nonzero element $x \in H$ there is an element $z \in H$ such that $xz = e_H$. By what we have proved so far

$e_H = e_F$ and hence, z is the multiplicative inverse of x in F. Hence $z = x^{-1} \in H$. Thus H satisfies (**SF 2**).

Conversely, suppose that H contains at least two elements and satisfies (**SF 1**) and (**SF 2**). Then H contains some nonzero element u. By (**SF 1**) $0_F = u - u \in H$ and by (**SF 2**) $u^{-1} \in H$. Application of (**SF 1**) again implies that $e_F = uu^{-1} \in H$. Next, let x be an arbitrary element of H. By (**SF 1**), $-x = 0_F - x \in H$. Let y be another arbitrary element of H. As above, $-y \in H$. Applying (**SF 1**), we deduce that $x + y = x - (-y) \in H$. Hence H is a stable subset under addition. The condition (**SF 1**) shows that H is a stable subset under multiplication. Thus, the restrictions of addition and multiplication to H are binary operations on H. Conditions (i), (ii), (v), (vi) of Definition 3.2.1 and the commutativity of multiplication hold for H, since these laws are valid for all elements of F. We have proved that $0_F \in H$ and hence 0_F is the zero element of H. We proved above that $-x \in H$ for each element $x \in H$, and also that $e_F \in H$. Thus e_F is the identity element for H. Finally, the condition (viii) of Definition 3.2.1 follows from the condition (**SF 2**).

3.2.7. Corollary. *Let F be a field and $\mathfrak{S}$ be a family of subfields of F. The intersection $\cap\mathfrak{S}$ of all subfields from this family is also a subfield of F.*

Proof. Let $S = \cap\mathfrak{S}$. Since every subfield contains 0_F and e, it follows that $0_F, e \in S$. Next, let $x, y \in S$. If U is an arbitrary element of $\mathfrak{S}$, then $x - y$, $xy \in U$, and therefore $x - y, xy$ lie in the intersection, S, of all elements of $\mathfrak{S}$. Consequently, $x - y, xy \in S$, which shows that S satisfies (**SF 1**). By a similar argument, S satisfies (**SF 2**) and we can now apply Theorem 3.2.6 to deduce the result.

It is worth noting that the union of a collection of fields in general is not a field. However, suppose that M is a set. A family $\mathfrak{L}$ consisting of certain subsets of M is called local, if for each pair of subsets $H, K \in \mathfrak{L}$, there exists a subset $L \in \mathfrak{L}$ such that $H, K \subseteq L$.

A special type of local family is a family that is linearly ordered. A family $\mathfrak{L}$ consisting of subsets of M is called *linearly ordered* if, for each pair of subsets $H, K \in \mathfrak{L}$, either $H \subseteq K$ or $K \subseteq H$.

3.2.8. Corollary. *Let F be a field and let $\mathfrak{L}$ be a local family of subfields of F. Then, the union $\cup\mathfrak{L}$ of all subfields from this family is also a subfield of F.*

Proof. Let $V = \cup\mathfrak{L}$ and let $x, y \in V$. There exist subfields $H, K \in \mathfrak{L}$ such that $x \in H, y \in K$. We choose a subfield $L \in \mathfrak{L}$ that contains both subfields H, K and hence $x, y \in L$. Since L is a subfield, $x - y, xy \in L$ by Theorem 3.2.6. Hence $x - y, xy \in V$. Consequently, V satisfies (**SF 1**). A similar argument enables us to prove that V satisfies (**SF 2**). Now we can apply Theorem 3.2.6 to deduce the result.

3.2.9. Corollary. *Let F be a field and let $\mathfrak{L}$ be a linearly ordered family of subfields of F. Then, the union $\cup\mathfrak{L}$ of all subfields from this family is also a subfield of F.*

3.2.10. Corollary. *Let F be a field and let*

$$H_1 \leq H_2 \leq \cdots \leq H_n \leq \cdots$$

be an ascending chain of subfields of F. Then $\bigcup_{n\in\mathbb{N}} H_n$ is also a subfield of F.

The smallest fields are of some interest to us, as we observe with the following definition.

3.2.11. Definition. *Let F be a field. Then, the intersection F_0 of all subfields of F is called a prime subfield. A field F is called prime, if F coincides with its prime subfield.*

It is then easy to see that if F is a prime field, then F has no proper subfields.

The field $\mathbb{Q}$ of rational numbers is a prime field. To see this, let P be some subfield of $\mathbb{Q}$. From Theorem 3.2.6, it follows that $1 \in P$. By (**SF 1**) we have $2 = 1 + 1 \in P$, $3 = 2 + 1 \in P$, and similarly, for each $n \in \mathbb{N}$ we have $n = n1 \in P$. Again by (**SF 1**) we see that $-n = 0 - n \in P$ for each $n \in \mathbb{N}$, so that $n \in P$ for each $n \in \mathbb{Z}$. Thus $\mathbb{Z} \subseteq P$. If $0 \neq k \in \mathbb{Z}$, then by (**SF 2**) $\frac{1}{k} \in P$. Now, for all $r, k \in \mathbb{Z}$, where $k \neq 0$, we have $\frac{r}{k} = r(\frac{1}{k}) \in P$. Thus $\mathbb{Q} \subseteq P$ so this shows that $P = \mathbb{Q}$.

The field $\mathbb{F}_p$ where p is a prime is a prime field. To see this, let P be some subfield of $\mathbb{F}_p$. Theorem 3.2.6 implies that $1 \in P$. As in the previous paragraph, we see that $2, \ldots, p - 1 \in P$ also and therefore $\mathbb{F}_p \subseteq P$, so $P = \mathbb{F}_p$.

The field $\mathbb{R}$ of real numbers is not prime since it contains $\mathbb{Q}$. Between $\mathbb{Q}$ and $\mathbb{R}$ there are many subfields. Here is a standard method for constructing certain subfields of $\mathbb{R}$ containing $\mathbb{Q}$. However, the reader is cautioned that this by no means exhausts the subfields of $\mathbb{R}$ containing $\mathbb{Q}$.

Let r be a positive integer and suppose that $\sqrt{r} \notin \mathbb{Q}$. Put

$$\mathbb{Q}(\sqrt{r}) = \{a + b\sqrt{r} \mid a, b \in \mathbb{Q}\}.$$

Let α, β be arbitrary elements of $\mathbb{Q}(\sqrt{r})$, say $\alpha = a + b\sqrt{r}$ and $\beta = a_1 + b_1\sqrt{r}$. Easy computations show that

$$\alpha - \beta = (a - a_1) + (b - b_1)\sqrt{r} \text{ and } \alpha\beta = (aa_1 + bb_1r) + (ab_1 + ba_1)\sqrt{r}.$$

It follows that $\alpha - \beta, \alpha\beta \in \mathbb{Q}(\sqrt{r})$. Clearly, $1 \in \mathbb{Q}(\sqrt{r})$. Also if $\alpha \neq 0$, then $a^2 - rb^2 \neq 0$ since $\sqrt{r} \notin \mathbb{Q}$ so

$$\gamma = \frac{a}{a^2 - rb^2} + \left(\frac{-b}{a^2 - rb^2}\right)\sqrt{r} \in \mathbb{Q}(\sqrt{r}).$$

By direct computation it is easy to verify that $\alpha\gamma = \gamma\alpha = 1$, so $\gamma = \alpha^{-1}$. Then Theorem 3.2.6 shows that $\mathbb{Q}(\sqrt{r})$ is a subfield of $\mathbb{R}$ called a *real quadratic field*.

The construction of $\mathbb{Q}(\sqrt{r})$ can be generalized as follows: If F is a subfield of a field K, then we say that K is an extension of F.

When F is a subfield of a field K and α is an element of K, let $\mathfrak{M}$ be the family of subfields of K which contains both F and α. Put $F(\alpha) = \cap\mathfrak{M}$. By Corollary 3.2.7, $F(\alpha)$ is a subfield of K and, by its definition, $F(\alpha)$ is the least subfield containing both F and α. Thus $F(\alpha)$ is an extension of F.

3.2.12. Definition. *Let F be a subfield of a field K and let α be an element of K. The subfield $F(\alpha)$ is called a simple extension of the field F.*

We also say that $F(\alpha)$ is the field obtained from F by adjoining α.

Consequently, $\mathbb{Q}(\sqrt{r})$ is an extension of the prime field $\mathbb{Q}$, by adjoining the element $\sqrt{r}$. It is very easy to generalize this idea. Let F be a subfield of a field K and let M be a subset of K. Let $\mathfrak{M}$ be the family of subfields of K that contains F and M. Put $F(M) = \cap\mathfrak{M}$. By Corollary 3.2.7, $F(M)$ is a subfield of K and, by its definition, $F(M)$ is the smallest subfield which contains both F and M.

3.2.13. Definition. *Let F be a subfield of a field K and M be a subset of K. The subfield $F(M)$ is called the extension of the field F, obtained by adjoining the set M to F.*

We now return to prime subfields and note that the structure of the prime subfield significantly influences the structure of the entire field. To see this, consider the subset $\mathbb{Z}e = \{ne \mid n \in \mathbb{Z}\}$, which we identify with the set of integer multiples of the identity. Two cases arise. If $ne \neq ke$ whenever $n \neq k$, the equation $ne = 0_F$ is possible only when $n = 0$.

The second alternative is that there are integers n, m such that $n \neq m$ but $ne = me$. One of n, m is greater than the other and we may suppose that $n > m$. Then $n - m > 0$ and from the equation $ne = me$ we see that $(n - m)e = 0_F$. Let

$$P = \{k \in \mathbb{N} \mid ke = 0_F\}.$$

The subset P has a least element, t, say, so that t is the least positive integer such that $te = 0_F$. We note that t must be prime. Indeed, if this is not the case, then $t = sr$ where $1 < s < t$ and $1 < r < t$. It follows from the definition of t that $se \neq 0_F$ and $re \neq 0_F$. Then

$$(se)(re) = (sr)(ee) = (sr)e = te = 0_F,$$

which gives a contradiction to Proposition 3.2.3. Consequently, t must be prime. In this case, for each element $a \in F$, we have

$$ta = t(ea) = \underbrace{(ea + \cdots + ea)}_{t} = \underbrace{(e + \cdots + e)}_{t}a = (te)a = 0_F.$$

3.2.14. Definition. *Let F be a field. If $ne \neq ke$ whenever $n \neq k$, then we will say that the field F has characteristic 0 and write* $\mathbf{char}(F) = 0$. *If there is a prime p such that $pe = 0_F$, then we will say that the field F has characteristic p and write* $\mathbf{char}(F) = p$.

Suppose now that $\mathbf{char}(F) = p > 0$. Let n be an arbitrary integer. By Theorem 1.4.1, there are integers q, r such that $n = qp + r$ where $0 \leq r < t$. We have

$$ne = (qp + r)e = qpe + re = q(pe) + re = q0_F + re = 0_F + re = re.$$

It follows that

$$\mathbb{Z}e \subseteq \{0e = 0_F, 1e = e, 2e, \ldots, (p-1)e\}.$$

Using the same argument we see that $0e = 0_F$, $1e = e, 2e, \ldots, (p-1)e$ are all distinct and it follows that

$$\mathbb{Z}e = \{0e = 0_F, 1e = e, 2e, \ldots, (p-1)e\}.$$

Thus, the prime subfield of a field F of characteristic p is $\mathbb{F}_p$.

We next consider certain mappings of fields. It makes sense just to consider mappings of fields that keep the algebraic structure of the field intact. We have seen this kind of idea before.

3.2.15. Definition. *Let F, K be fields. The mapping $f : F \longrightarrow K$ is called a homomorphism if it satisfies the conditions*

$$f(x + y) = f(x) + f(y) \text{ and } f(xy) = f(x)f(y)$$

for all elements $x, y \in F$.

An injective homomorphism is called *a monomorphism* and a surjective homomorphism is called *an epimorphism*. A bijective homomorphism is called an *isomorphism*.

If $f : F \longrightarrow K$ is an isomorphism, then as we noted in Section 3.1 the mapping $f^{-1} : K \longrightarrow F$ is also an isomorphism. The fields F and K are called *isomorphic* if there exists an isomorphism mapping F to K and, in this case, we write $F \cong K$. Clearly, the identity permutation $\varepsilon_F : F \longrightarrow F$ is one example of an isomorphism.

It is very easy to see that if $f : F \longrightarrow K, g : K \longrightarrow L$ are homomorphisms of fields then the product $g \circ f : F \longrightarrow L$ is a homomorphism. If $f : F \longrightarrow K$ is the mapping defined by the rule $f(x) = 0_K$ for each element $x \in F$, then f is a homomorphism called the *zero homomorphism*.

3.2.16. Theorem. *Suppose that F, K are fields and let $f : F \longrightarrow K$ be a homomorphism. The following assertions hold:*

(i) $f(0_F) = 0_K$.

(ii) $f(-x) = -f(x)$ *for all* $x \in F$.

(iii) $f(x - y) = f(x) - f(y)$ *for all* $x, y \in F$.

(iv) *If* f *is a nonzero homomorphism, then* $f(e)$ *is the identity element of the field* K.

(v) *If* f *is a nonzero homomorphism and* x *is a nonzero element of* F*, then* $f(x^{-1}) = (f(x))^{-1}$.

(vi) *Let* H *be a subfield of* F. *If* f *is a nonzero homomorphism, then* $f(H)$ *is a subfield of the field* K. *In particular,* **Im** $f = f(F)$ *is a subfield of* K.

(vii) *If* f *is a nonzero homomorphism, then* f *is a monomorphism. In particular,* $f(F)$ *is isomorphic to some subfield of* K.

Proof.
(i) We have $x + 0_F = x$ for each $x \in f$. Then

$$f(x) + f(0_F) = f(x + 0_F) = f(x).$$

Since $f(x)$ has a negative, $-f(x)$, in K, we add $-f(x)$ to each side of this equation to obtain

$$f(0_F) = 0_K + f(0_F) = -f(x) + f(x) + f(0_F) = -f(x) + f(x) = 0_K,$$

from which we obtain $f(0_F) = 0_K$.

(ii) From the definition of negative element we have $x + (-x) = 0_F$. Thus,

$$0_K = f(0_F) = f(x + (-x)) = f(x) + f(-x) = f(-x) + f(x).$$

This equation shows that the element $f(-x)$ is the negative of $f(x)$, which is to say that $f(-x) = -f(x)$.

(iii) We have

$$\begin{aligned} f(x - y) &= f(x + (-y)) = f(x) + f(-y) \\ &= f(x) + (-f(y)) = f(x) - f(y). \end{aligned}$$

(iv) Suppose that $f(e) = 0_K$. For each element $x \in F$ we have

$$f(x) = f(xe) = f(x)f(e) = f(x)0_K = 0_K,$$

which is a contradiction, since f is not the zero homomorphism. Hence $f(e)$ is a nonzero element of K and thus $f(e)$ has a multiplicative inverse in K. We denote the identity element of K by e_1. From the definition of the identity element, it follows that

$$f(e) = f(ee) = f(e)f(e).$$

Multiplying both sides of this equation by $f(e)^{-1}$, we obtain

$$f(e)^{-1}f(e) = f(e)^{-1}f(e)f(e)$$

and hence conclude that $f(e) = e_1$.

(v) The proof is similar to the proof of (ii).

(vi) Let $u, v \in f(H)$. Then, there are elements $x, y \in H$ such that $u = f(x)$ and $v = f(y)$. Using (iii), we obtain $u - v = f(x) - f(y) = f(x - y)$. Since H is a subfield, $x - y \in H$, so that $f(x - y) = u - v \in f(H)$.

Similarly, $uv = f(x)f(y) = f(xy)$. Since H is a subfield, $xy \in H$, so that $f(xy) = uv \in f(H)$ and $f(H)$ satisfies the condition (**SF 1**).

Suppose now that $u \neq 0_K$. Then (i) shows that $x \neq 0_F$. Since H is a subfield, $x^{-1} \in H$, and by (v), $u^{-1} = f(x)^{-1} = f(x^{-1}) \in f(H)$ and therefore $f(H)$ satisfies the condition (**SF 2**). By Theorem 3.2.6, $f(H)$ is a subfield of K.

(vii) Suppose, for a contradiction, that there are elements $x, y \in F$ such that $x \neq y$ but $f(x) = f(y)$. Then $0_K = f(x) - f(y) = f(x - y)$. Put $z = x - y$. Since $x \neq y$ it follows that $z \neq 0_F$ and hence z has an inverse. By (iv),

$$e_1 = f(e) = f(zz^{-1}) = f(z)f(z^{-1}) = 0_K f(z^{-1}) = 0_K,$$

which gives us the desired contradiction. The result follows.

EXERCISE SET 3.2

3.2.1. Let $P = \{x + y\sqrt{2} \mid x, y \in \mathbb{Q}\}$. Prove directly that P is a subfield of $\mathbb{R}$. Find all isomorphisms $f : P \longrightarrow P$, satisfying the condition $f(x) = x$ for every element $x \in \mathbb{Q}$.

3.2.2. Let $P = \{x + y\sqrt{2} \mid x, y \in \mathbb{Q}\}$. Solve the equation $x^2 - x - 3 = 0$ in P, if possible.

3.2.3. On the set $\mathbb{R} \times \mathbb{R}$ we define operations of addition and multiplication by $(a, b) + (c, d) = (a + c, b + d)$, $(a, b)(c, d) = (ac - 3bd, ad + 2bd + bc)$. Is $\mathbb{R} \times \mathbb{R}$ a field? Find solutions of the equation $X^2 + 1 = 0$ in $\mathbb{R} \times \mathbb{R}$.

3.2.4. Let $\mathbb{F}_5 = \{0, 1, 2, 3, 4\}$ be a field with five elements. Fill out the multiplication and addition tables of its elements:

+	0	1	2	3	4
0	0	1	2	3	4
1	1	2			
2	2				
3	3				
4	4				3

,

×	0	1	2	3	4
0	0	0	0	0	0
1	0	1	2	3	4
2	0	2			3
3	0	3			
4	0	4			

.

3.2.5. Let $\mathbb{F}_4 = \{0, 1, u, v\}$ be a field. Fill out the multiplication and addition tables of its elements:

$+$	0	1	u	v
0	0	1	u	v
1	1	2		
u	u			
v	v			

,

$\times$	0	1	u	v
0				0
1				v
u		u		
0	v	0		u

.

3.2.6. Let F be a field and let $f : F \longrightarrow F$ an isomorphism. Prove that the subset $K = \{x \in F \mid f(x) = x\}$ is a subfield of F.

3.2.7. Let H be the subset of $M_4(\mathbb{R})$ consisting of all matrices of the form $\begin{pmatrix} a & -b & c & d \\ b & a & -d & c \\ -c & d & a & b \\ -d & -c & -b & a \end{pmatrix}$. Prove that H is a division ring.

3.2.8. Let $\mathbb{F}_5 = \{0, 1, 2, 3, 4\}$ be a field. In this field solve the equation $17x = 3$.

3.2.9. Let $P = \{x + y\sqrt{5} \mid x, y \in \mathbb{Q}\}$. Prove directly that P is a subfield of $\mathbb{R}$. Prove that the mapping $f : P \longrightarrow P$, defined by the rule $f(x + y\sqrt{5}) = x - y\sqrt{5}$, is an isomorphism.

3.2.10. Let P be the subset of $M_2(\mathbb{Q})$, consisting of all matrices of the form $\begin{pmatrix} x & y \\ 5y & x \end{pmatrix}$. Prove that P is a field relative to the regular operations of matrix addition and multiplication.

3.2.11. Prove that the fields of the problems 3.2.9 and 3.2.10 are isomorphic.

3.2.12. Let α be a real root of the equation $x^3 = 2$. Put $P = \{x + y\alpha + z\alpha^2 \mid x, y, z \in \mathbb{Q}\}$. Prove directly that P is a subfield of $\mathbb{R}$. Prove that every element of P can be uniquely represented in the given form. Find $(1 - \alpha + \alpha^2)^{-1}$ in the form $x + y\alpha + z\alpha^2$.

3.2.13. Let F be a field of prime characteristic p and let $q = p^k$ where k is a positive integer. Prove that $(x + y)^q = x^q + y^q$ for all $x, y \in F$.

3.2.14. Let F be a finite field and let $p = \mathbf{char}(F)$. Prove that the mapping $x \longrightarrow x^p, x \in F$, is an isomorphism.

3.3 THE FIELD OF COMPLEX NUMBERS

Probably the most important field, having various applications in distinct branches of mathematics, is the field of complex numbers. Complex numbers were invented in connection with the problem of finding roots of polynomials with real coefficients. The polynomial $x^2 + 1$ is the simplest example of a polynomial having no real roots. Finding the roots of this polynomial inevitably leads us to consider the expression $\sqrt{-1}$.

Historically, however, Italian mathematicians flirted with the idea of complex numbers in connection with the formulae they developed expressing the roots of polynomials of degrees 3 and 4 in terms of real coefficients. These investigations were conducted by the great Italian mathematicians Niccolò Fontana Tartaglia (1499–1557), Gerolamo Cardano (1501–1576), and Lodovico Ferrari (1522–1565). They formally worked (under certain restrictions) with expressions containing square roots of negative numbers, but it was a follower of Cardano, Rafael Bombelli (1526–1572), who exhibited complex numbers in a form that is close to the modern one. However, the notion of complex numbers was essentially ignored by most mathematicians of the time and there followed a very long episode of tension between mathematicians who supported the idea of a complex number and opponents who regarded the whole theory as hogwash. The key concept here is the concept of $\sqrt{-1}$ and only Carl Frederic Gauss was able to offer a reasonable and acceptable explanation for it. It was only when developments in mathematics and physics produced a variety of applications of complex numbers that complex numbers came into common use.

The idea of a field extension, introduced in the previous section, allows us to define the complex field in the following natural way.

3.3.1. Definition. *The extension of the field of real numbers obtained by adjoining a root of the polynomial $x^2 + 1$ will be called the field of complex numbers. We denote this field by $\mathbb{C}$ and the root of the polynomial $x^2 + 1$ by i. In this notation, $\mathbb{C} = \mathbb{R}(i)$.*

When we first defined field extensions, we assumed that, to form the extension, we adjoined a set of elements to the original field in such a way that the original field and the set adjoined were subsets of some larger field. In the case of the field of complex numbers, however, we do not know beforehand if there actually exists a field containing the field of real numbers and the number i. Thus, the first question one could address here is the question concerning the existence of the field of complex numbers. For the moment, we assume that this field exists and find the form of its elements. So, let F be a field having a subfield $\mathbb{R}$, and containing i as an element. If K is a subfield of F such that $K \supseteq \mathbb{R}$ and $i \in K$ then, together with each real number x, the field K also contains the element xi. For the same reason, if $x, y \in \mathbb{R}$, then $x + yi \in K$. In this way, the subfield $\mathbb{C}$ which, by our definition, is the intersection of all fields K containing the field $\mathbb{R}$ and the element i, contains the element $x + yi$. Let

$$S = \{x + yi \mid x, y \in \mathbb{R}\}.$$

We shall show that S is a subfield using the criterion we proved in the previous section. If $\alpha = x + yi,\ \beta = u + vi$ then

$$\alpha - \beta = (x-u) + (y-v)i \in S \text{ and}$$
$$\alpha\beta = (x+yi)(u+vi) = (xu - yv) + i(yu + xv) \in S.$$

We note that the latter computation uses the fact that $i^2 = -1$. From the definition of a field, it follows that the identity element (the number 1) of the field $\mathbb{R}$ is the multiplicative identity of the entire field F. However, $1 = 1 + 0i \in S$. Similarly, $i = 0 + 1i \in S$. Finally, let $0 \neq \alpha = x + yi \in S$. Then at least one of x, y is nonzero and hence $x^2 + y^2 \neq 0$. We have

$$(x+yi)(x-yi) = x^2 + y^2,$$

which leads us to the equation

$$(x+yi)^{-1} = \frac{x}{x^2+y^2} + \frac{-y}{x^2+y^2}i \in S.$$

Thus, all conditions of Theorem 3.2.6 are valid for S, and hence S is a subfield of the field F. By the definition of S, we have that $S \supseteq \mathbb{R}$ and $i \in S$ which means that $S \supseteq \mathbb{C} = \mathbb{R}(i)$. On the other hand, we proved that $S \subseteq \mathbb{C}$, which implies that

$$\mathbb{C} = \{x + yi \mid x, y \in \mathbb{R}\}.$$

Suppose that $x + yi = u + vi$ for some real numbers x, y, u, v. Then $x - u = (v - y)i$. Since $i \notin \mathbb{R}$, this equality is possible only in the case when $v - y = 0$ and $x - u = 0$. Therefore $x = u$ and $y = v$. This shows that every complex number can be uniquely represented in the form $x + yi$, where x, y are real numbers. Here x is called *the real part of the complex number* $x + yi$, while yi is called its *imaginary part*.

We shall use the arguments above to prove the existence of the field of complex numbers. We begin with a commonly used representation of complex numbers as points of the coordinate plane; this gives us a geometric model of the field of complex numbers. To do this, we consider the set $\mathbb{R} \times \mathbb{R}$ of all points $\alpha = (x, y)$, where $x, y \in \mathbb{R}$, with operations of addition and multiplication given by the following rules.

If $\alpha = (x, y), \beta = (u, v) \in \mathbb{R} \times \mathbb{R}$, then we define

$$\alpha + \beta = (x+u, y+v) \text{ and } \alpha\beta = (xu - yv, yu + xv).$$

We will call the set $\mathbb{R} \times \mathbb{R}$, together with these operations, *the complex plane*.

We now consider properties of these operations, which are induced by corresponding properties of real numbers.

The operation of addition is commutative since

$$\alpha + \beta = (x + u, y + v) = (u + x, v + y) = \beta + \alpha.$$

The addition is associative. To see this, let $\gamma = (z, w)$. Then

$$\begin{aligned}(\alpha + \beta) + \gamma &= ((x + u) + z, (y + v) + w) \\ &= (x + (u + z), y + (v + w)) = \alpha + (\beta + \gamma),\end{aligned}$$

using the fact that $+$ is an associative operation in $\mathbb{R}$.

There exists a zero element in $\mathbb{R} \times \mathbb{R}$, the pair $(0, 0)$, since

$$(x, y) + (0, 0) = (x + 0, y + 0) = (x, y).$$

For each pair (x, y) there exists an opposite element; this is the pair $(-x, -y)$ since

$$(x, y) + (-x, -y) = (x + (-x), y + (-y)) = (0, 0).$$

The operation of multiplication is commutative since

$$\alpha\beta = (xu - yv, yu + xv) \text{ and } \beta\alpha = (ux - vy, uy + vx),$$

which means that

$$\alpha\beta = \beta\alpha.$$

The multiplication is associative since

$$\begin{aligned}(\alpha\beta)\gamma &= (xu - yv, yu + xv)(z, w) \\ &= ((xu - yv)z - (yu + xv)w, (yu + xv)z + (xu - yv)w) \\ &= (xuz - yvz - yuw - xvw, yuz + xvz + xuw - yvw).\end{aligned}$$

On the other hand,

$$\begin{aligned}\alpha(\beta\gamma) &= (x, y)(uz - vw, vz + uw) \\ &= (x(uz - vw) - y(vz + uw), y(uz - vw) + x(vz + uw)) \\ &= (xuz - xvw - yvz - yuw, yuz - yvw + xvz + xuw).\end{aligned}$$

By comparing the two expressions, it follows that

$$(\alpha\beta)\gamma = \alpha(\beta\gamma).$$

Next, we note that there exists a multiplicative identity element, namely the pair (1, 0) since

$$(x, y)(1, 0) = (x1 - y0, y1 + x0) = (x, y).$$

For each pair $(x, y) \neq (0, 0)$ we note that $x^2 + y^2 \neq 0$ and there exists an inverse element, the pair

$$\left(\frac{x}{x^2 + y^2}, \frac{-y}{x^2 + y^2}\right),$$

since

$$(x, y)\left(\frac{x}{x^2 + y^2}, \frac{-y}{x^2 + y^2}\right) = \left(\frac{x^2 + y^2}{x^2 + y^2}, \frac{yx - xy}{x^2 + y^2}\right) = (1, 0).$$

The distributive property connects addition and multiplication as follows:

$$\begin{aligned}(\alpha + \beta)\gamma &= ((x, y) + (u, v))(z, w) = (x + u, y + v)(z, w)\\ &= ((x + u)z - (y + v)w, (y + v)z + (x + u)w)\\ &= (xz + uz - yw - vw, yz + vz + xw + uw).\end{aligned}$$

Also,

$$\begin{aligned}\alpha\gamma + \beta\gamma &= (x, y)(z, w) + (u, v)(z, w)\\ &= (xz - yw, yz + xw) + (uz - vw, vz + uw)\\ &= (xz - yw + uz - vw, yz + xw + vz + uw).\end{aligned}$$

By comparing these, we see that

$$(\alpha + \beta)\gamma = \alpha\gamma + \beta\gamma.$$

Consequently, $\mathbb{R} \times \mathbb{R}$ is a field with the given operations of addition and multiplication.

The subset $\mathbb{R} \times \{0\} = \{(x, 0) \mid x \in \mathbb{R}\}$ is a subfield of $\mathbb{R} \times \mathbb{R}$. In fact, it is infinite and, using the definitions of addition and multiplication, we have

$$\begin{aligned}&(x, 0) - (y, 0) = (x - y, 0) \in \mathbb{R} \times \{0\} \text{ and}\\ &\quad (x, 0)(y, 0) = (xy - 0 \cdot 0, 0y + x0) = (xy, 0) \in \mathbb{R} \times \{0\}.\end{aligned}$$

Also,

$$\text{if } (x, 0) \neq (0, 0), \text{ then } x \neq 0 \text{ and } (x, 0)^{-1} = (x^{-1}, 0) \in \mathbb{R} \times \{0\}.$$

It follows, using Theorem 3.2.6, that $\mathbb{R} \times \{0\}$ is a subfield of $\mathbb{R} \times \mathbb{R}$.

Now we identify this subfield with the field $\mathbb{R}$. Such an identification of the points on the coordinate axes with the corresponding real numbers is a commonly used procedure, but here we formally justify it.

We define a mapping $f : \mathbb{R} \longrightarrow \mathbb{R} \times \{0\}$ by the rule $f(x) = (x, 0)$. We have

$$f(x + y) = (x + y, 0) = (x, 0) + (y, 0) = f(x) + f(y) \text{ and}$$
$$f(xy) = (xy, 0) = (x, 0)(y, 0) = f(x)f(y).$$

The mapping f is clearly bijective and therefore f is an isomorphism of these fields. Hence, the subfield $\mathbb{R} \times \{0\}$ is isomorphic to the field $\mathbb{R}$ of real numbers and this allows us to identify the pair $(x, 0)$ with the real number x.

We note, using the definitions, that

$$(x, y) = (x, 0) + (0, y) = (x, 0) + (0, 1)(y, 0).$$

Also,

$$(0, 1)^2 = (-1, 0), \text{ so } (0, 1)^2 + (1, 0) = (0, 0).$$

Hence if we put $i = (0, 1)$, then $i^2 = (-1, 0)$. Identifying the pair $(-1, 0)$ with the real number -1, we have

$$(x, y) = (x, 0) + (0, 1)(y, 0) = x + yi.$$

Therefore, our recently constructed field, $\mathbb{R} \times \mathbb{R}$, contains $\mathbb{R}$ (more precisely, an isomorphic copy of $\mathbb{R}$) as a subfield. In addition, this field contains a root of the polynomial $x^2 + 1$. Hence the field $\mathbb{R} \times \mathbb{R}$ contains the subfield $\mathbb{R}(i) = \mathbb{C}$. By our construction, this field coincides with $\mathbb{C}$.

Now we introduce a further model of the complex field, based on matrices, so we call it the matrix model of the field of complex numbers. In order to do this we consider the subset P of the set of matrices, $M_2(\mathbb{R})$, consisting of all matrices of the type

$$\begin{pmatrix} x & y \\ -y & x \end{pmatrix}.$$

The equations

$$\begin{pmatrix} x & y \\ -y & x \end{pmatrix} + \begin{pmatrix} u & v \\ -v & u \end{pmatrix} = \begin{pmatrix} x+u & y+v \\ -y-v & x+u \end{pmatrix} \text{ and}$$
$$\begin{pmatrix} x & y \\ -y & x \end{pmatrix}\begin{pmatrix} u & v \\ -v & u \end{pmatrix} = \begin{pmatrix} xu - yv & xv + yu \\ -xv - yu & xu - yv \end{pmatrix}$$
$$= \begin{pmatrix} xu - yv & xv + yu \\ -(xv + yu) & xu - yv \end{pmatrix}$$

show that the subset P is stable under both addition and multiplication of matrices. Therefore, these operation induce the respective binary operations on P.

Addition of elements of P is commutative and associative, because it is commutative and associative for all elements of $M_2(\mathbb{R})$. The zero matrix belongs to P and hence it is the zero element for P. Furthermore,

$$-\begin{pmatrix} x & y \\ -y & x \end{pmatrix} = \begin{pmatrix} -x & -y \\ y & -x \end{pmatrix} \in P.$$

The operations of addition and multiplication are connected using the distributive property of matrix multiplication over matrix addition. Since multiplication is associative in $M_2(\mathbb{R})$, it is also associative in P. Moreover,

$$\begin{pmatrix} x & y \\ -y & x \end{pmatrix}\begin{pmatrix} u & v \\ -v & u \end{pmatrix} = \begin{pmatrix} xu - yv & xv + yu \\ -xv - yu & xu - yv \end{pmatrix} \text{ and}$$

$$\begin{pmatrix} u & v \\ -v & u \end{pmatrix}\begin{pmatrix} x & y \\ -y & x \end{pmatrix} = \begin{pmatrix} ux - vy & xv + yu \\ -vx - uy & ux - vy \end{pmatrix},$$

so multiplication is commutative in P. Let

$$Y = \begin{pmatrix} x & y \\ -y & x \end{pmatrix}$$

be an arbitrary nonzero element of P. Then, at least one of the numbers x, y is nonzero and hence $\mathbf{det}(Y) = x^2 + y^2 \neq 0$. Thus every nonzero element of P is a nonsingular matrix and therefore, has a multiplicative inverse. Computing this inverse is very easy here, using the algorithm established in Section 2.5, and we see that

$$\begin{pmatrix} x & y \\ -y & x \end{pmatrix}^{-1} = \begin{pmatrix} u & v \\ -v & u \end{pmatrix}$$

where

$$u = \frac{x}{x^2 + y^2}, \ v = \frac{-y}{x^2 + y^2}.$$

This shows that $Y^{-1} \in P$ and hence that P is a field.

We now show that the fields P and $\mathbb{C}$ are isomorphic by defining a mapping $f : \mathbb{C} \longrightarrow P$ using the rule

$$f(x + yi) = \begin{pmatrix} x & y \\ -y & x \end{pmatrix}.$$

This map is easily seen to be bijective. Next, let $\alpha = x + yi$ and $\beta = u + vi$, where $x, y, u, v \in \mathbb{R}$. Then

$$f(\alpha + \beta) = f\big((x + u) + i(y + v)\big) = \begin{pmatrix} x + u & y + v \\ -y - v & x + u \end{pmatrix}, \text{ and}$$

$$f(\alpha) + f(\beta) = f(x + iy) + f(u + iv) = \begin{pmatrix} x & y \\ -y & x \end{pmatrix} + \begin{pmatrix} u & u \\ -v & v \end{pmatrix}.$$

Since the two expressions are the same, we have

$$f(\alpha + \beta) = f(\alpha) + f(\beta).$$

Also,

$$\begin{aligned} f(\alpha\beta) &= f\big((xu - yv) + i(yu + xv)\big) = \begin{pmatrix} xu - yv & xv + yu \\ -xv - yu & xu - yv \end{pmatrix} \\ &= \begin{pmatrix} x & y \\ -y & x \end{pmatrix}\begin{pmatrix} u & u \\ -v & v \end{pmatrix} = f(x + yi)f(u + vi) = f(\alpha)f(\beta). \end{aligned}$$

Thus f is an isomorphism since these equations show that the mapping f respects the operation of addition and multiplication. Consequently, the field $\mathbb{C}$ of complex numbers is isomorphic to the constructed field P. Here $f(\mathbb{R}) = \mathbb{R}I$ is the set of all scalar matrices, which clearly is a subfield of P. Furthermore

$$f(i) = \begin{pmatrix} 0 & 1 \\ -1 & 0 \end{pmatrix} = U,$$

so we can identify the scalar matrix xI with the real number x and we can identify U with i since

$$U^2 = \begin{pmatrix} 0 & 1 \\ -1 & 0 \end{pmatrix}\begin{pmatrix} 0 & 1 \\ -1 & 0 \end{pmatrix} = \begin{pmatrix} -1 & 0 \\ 0 & -1 \end{pmatrix} = -I.$$

In particular, P contains a root of the matrix polynomial $IX^2 + I$. Now we can obtain the usual matrix representation of a complex number:

$$\begin{aligned} \begin{pmatrix} x & y \\ -y & x \end{pmatrix} &= \begin{pmatrix} x & 0 \\ 0 & x \end{pmatrix} + \begin{pmatrix} 0 & y \\ -y & 0 \end{pmatrix} \\ &= \begin{pmatrix} x & 0 \\ 0 & x \end{pmatrix} + \begin{pmatrix} y & 0 \\ 0 & 0 \end{pmatrix}\begin{pmatrix} 0 & 1 \\ -1 & 0 \end{pmatrix} = xI + (yI)U. \end{aligned}$$

Although complex numbers do not admit a natural ordering as the reals do, we can compare complex numbers according to their length.

3.3.2. Definition. *Let $\alpha = x + yi$ be a complex number. The nonnegative real number $\sqrt{x^2 + y^2} = |x + yi|$ is called the modulus or absolute value of the number α.*

We next introduce the notion of complex conjugation.

3.3.3. Theorem. *Let $\Delta\colon \mathbb{C} \longrightarrow \mathbb{C}$ be the mapping defined by $\alpha(x + yi) = x - yi$. Then Δ is an isomorphism. Moreover, $\Delta^2 = \varepsilon_{\mathbb{C}}$ and $\Delta(a) = a$ for each $a \in \mathbb{R}$.*

Proof. Indeed,

$$\begin{aligned}\Delta(x + yi + u + vi) &= \Delta\big((x + u) + (y + v)i\big) = (x + y) - (y + v)i \\ &= (x - yi) + (u - vi) = \Delta(x + yi) + \Delta(u + vi).\end{aligned}$$

$$\begin{aligned}\text{Also } \Delta((x + yi)(u + vi)) &= \Delta\big((xu - yv) + (xv + yu)i\big) \\ &= (xu - yv) - (xv + yu)i \\ &= (x - yi)(u - vi) = \Delta(x + yi)\Delta(u + vi).\end{aligned}$$

The mapping Δ is easily seen to be injective and surjective so Δ is an isomorphism. Finally,

$$\Delta^2(x + yi) = \Delta\big(\Delta(x + yi)\big) = \Delta(x - yi) = x + yi = \varepsilon_{\mathbb{C}}(x + yi),$$

and from the definition of Δ it follows that $\Delta(a) = a$ for every $a \in \mathbb{R}$.

The numbers $x + yi$ and $x - yi$ are called *complex conjugates*. We will denote the complex conjugate of a complex number z by $\overline{z}$. Thus if $z = x + yi$, then $\overline{z} = x - yi$ and

$$\Delta(x + yi) = \overline{x + yi}.$$

Clearly,

$$(x + yi)(x - yi) = x^2 + y^2 = |x + yi|^2.$$

Let $r = |x + yi| = \sqrt{x^2 + y^2} \geq 0$. We can write the complex number $x + yi$ in the form

$$x + yi = r\left(\frac{x}{r} + \frac{y}{r}i\right)$$

and

$$\left|\frac{x}{r} + \frac{y}{r}i\right| = \frac{x^2}{r^2} + \frac{y^2}{r^2} = \frac{x^2 + y^2}{x^2 + y^2} = 1.$$

Therefore, there exists a unique angle ϕ such that $\cos\phi = \frac{x}{r}$ and $\sin\phi = \frac{y}{r}$, where $0 \leq \phi < 2\pi$. The angle ϕ is called the *argument of the complex number* $x + yi$ and is denoted by $\mathbf{arg}(x + yi)$. Thus we have

$$x + yi = r(\cos\phi + i\sin\phi).$$

This form is called *the trigonometric, or modulus-argument, form of the complex number* $x + yi$. The value of r in this expression is uniquely determined by x and y. It is, however, well-known that if ϕ is an angle then the trigonometric functions evaluated at ϕ and $\phi + 2k\pi$ are always the same, for each $k \in \mathbb{Z}$, which is our reason for restricting ϕ to lying in the interval $[0, 2\pi)$.

This form is very convenient in problems connected with multiplication of complex numbers. Indeed, let

$$x + yi = r(\cos\phi + i\sin\phi) \text{ and } u + vi = q(\cos\psi + i\sin\psi).$$

Then

$$\begin{aligned}(x + yi)(u + vi) &= r(\cos\phi + i\sin\phi)q(\cos\psi + i\sin\psi)\\ &= rq\big((\cos\phi\cos\psi - \sin\phi\sin\psi)\\ &\qquad + i(\sin\phi\cos\psi + \cos\phi\sin\psi)\big)\\ &= rq\big(\cos(\phi+\psi) + i\sin(\phi+\psi)\big),\end{aligned}$$

using the well-known trigonometric identities for the angle sum. In this way, we obtain the very convenient formulae

$$\begin{aligned}|(x + yi)(u + vi)| = rq = |x + yi|\,|u + vi| \text{ and}\\ \mathbf{arg}\big((x + yi)(u + vi)\big) = \phi + \psi = \mathbf{arg}(x + yi) + \mathbf{arg}(u + vi).\end{aligned}$$

These formulae can be extended to account for an arbitrary number of factors. In particular, for all $k \in \mathbb{N}$, using induction we have

$$\big(r(\cos\phi + i\sin\phi)\big)^k = r^k(\cos\phi + i\sin\phi)^k = r^k(\cos k\phi + \sin k\phi),$$

which can be written in the form

$$\left|(x + yi)^k\right| = |x + yi|^k \text{ and } \mathbf{arg}\big((x + yi)^k\big) = k\,\mathbf{arg}(x + yi).$$

These formulae constitute what is known as *de Moivre's theorem* and they allow us to take nth roots whenever $n \in \mathbb{N}$.

3.3.4. Theorem. *Let $\alpha = x + yi = r(\cos\phi + i\sin\phi)$ be an arbitrary complex number. Then, there exist exactly k different complex numbers $u + vi = q(\cos\xi + i\sin\xi)$ with the property that $(u + vi)^k = x + yi$ (the complex kth roots of $x + yi$). Furthermore, the values of q and ξ are obtained via the formulae*

$$q = \sqrt[k]{r}, \textit{ the positive kth root of } r,$$

and

$$\xi = \frac{\phi + 2\pi t}{k}, \textit{ where } 0 \le t < k.$$

Proof. By de Moivre's theorem, $\big(q(\cos\xi + i\sin\xi)\big)^k = r(\cos\phi + i\sin\phi)$, so we obtain $q^k = r$ and $k\xi = \phi + 2\pi s$, where $s \in \mathbb{N}$. It follows that $q = \sqrt[k]{r}$ and $\xi = \frac{\phi + 2\pi s}{k}$. By Theorem 1.4.1, we can write $s = km + t$, where $0 \le t \le k - 1$. Therefore

$$\frac{\phi + 2\pi s}{k} = \frac{\phi + 2\pi(km + t)}{k} = \frac{\phi + 2\pi t}{k} + 2\pi m.$$

Then

$$\cos\frac{\phi+2\pi s}{k}=\cos\frac{\phi+2\pi t}{k} \text{ and } \sin\frac{\phi+2\pi s}{k}=\sin\frac{\phi+2\pi t}{k},$$

where $0 \le t \le k-1$. If $\cos\chi = \cos\psi$ and simultaneously $\sin\chi = \sin\psi$, then $\chi - \psi = 2\pi j$ for some integer j. This means that the numbers

$$\sqrt[k]{r}\left(\cos\frac{\phi+2\pi t}{k}+i\sin\frac{\phi+2\pi t}{k}\right), \text{ where } 0 \le t \le k-1,$$

are distinct. Thus we have k different kth roots of $x + yi$.

If $k \in \mathbb{N}$, then the solutions of $z^k = 1$ are called kth roots of unity.

3.3.5. Corollary. *The kth roots of unity are*

$$\xi_j = \cos\frac{2\pi j}{k}+i\sin\frac{2\pi j}{k},$$

where $0 \le j < k$.

Certain kth roots of unity are fundamental, in the sense that they generate the other kth roots of unity.

3.3.6. Definition. *A kth root of unity, ε, is called primitive, if it is not a dth root of unity whenever $d < k$.*

3.3.7. Lemma. *A kth root of unity, ε, is primitive if and only if each kth root of unity, χ, can be written as $\chi = \varepsilon^m$, for some positive integer m.*

Proof. Indeed, let ε be a primitive root and consider the set

$$\{\varepsilon^0 = 1, \varepsilon^1 = \varepsilon, \varepsilon^2, \ldots, \varepsilon^{k-1}\}.$$

Suppose that some elements of this set coincide, so that $\varepsilon^r = \varepsilon^s$, for certain integers r, s such that $0 \le r < s \le k-1$. Then $\varepsilon^{s-r} = 1$ where $0 < s-r \le k-1$, which contradicts the fact that ε is a primitive kth root of unity. Thus the elements $\varepsilon^0 = 1, \varepsilon^1 = \varepsilon, \varepsilon^2, \ldots, \varepsilon^{k-1}$ are all different. Since each of these elements is a kth root of unity and since there are precisely k distinct kth roots of unity it follows that each kth root of unity, χ, is of the form $\chi = \varepsilon^m$ for some positive integer m.

To prove sufficiency we assume that ε is not a primitive kth root of unity so that there exists a least positive integer $n < k$ such that $\varepsilon^n = 1$. Using the arguments above we see that

$$\{\varepsilon^t \mid t \in \mathbb{Z}\} = \{\varepsilon^0, \varepsilon^1, \varepsilon^2, \dots, \varepsilon^{n-1}\},$$

and hence $\left|\left\{\varepsilon^t \mid t \in \mathbb{Z}\right\}\right| = n$. On the other hand, our hypotheses imply that the set $\left\{\varepsilon^t \mid t \in \mathbb{Z}\right\}$ contains all kth roots of unity. However, as we saw above, there are precisely k distinct kth roots of unity so we obtain a contradiction which proves that ε is a primitive kth root of unity.

3.3.8. Theorem. *Let k be a natural number and let ε be a primitive kth root of unity. Then ε^t is a primitive kth root of unity if and only if* $\mathbf{GCD}(k, t) = 1$.

Proof. We suppose first that ε^t is a primitive kth root of unity. If $d = \mathbf{GCD}(k, t) \neq 1$ then $k = dl$ and $t = dr$ for some $r, l \in \mathbb{Z}$. Then $(\varepsilon^t)^l = \varepsilon^{drl} = (\varepsilon^k)^r = 1$, so ε^t is an lth root of unity where $l < k$ which is a contradiction. Thus $\mathbf{GCD}(k, t) = 1$.

Conversely, let $\mathbf{GCD}(k, t) = 1$. By Corollary 1.4.7 there exist integers u, v such that $1 = ku + tv$. We now have, since $\varepsilon^k = 1$,

$$\varepsilon = \varepsilon^1 = \varepsilon^{ku+tv} = \varepsilon^{ku}\varepsilon^{tv} = (1^u)(\varepsilon^t)^v.$$

Hence ε is a power of ε^t. However, every kth root of unity is a power of ε, by Lemma 3.3.7, and hence every kth root of unity is a power of ε^t. It follows, again by Lemma 3.3.7, that ε^t is a primitive kth root of unity.

Finally, we consider an important example of a noncommutative division ring, the *ring of real quaternions*, which was constructed by W. Hamilton in 1843. This concept evolved from the need of mathematicians wanting to develop tools for describing rotations in three dimensional space. Since complex numbers effectively helped to understand rotations in the plane, it was natural to expect that some generalization of complex numbers would be effective in space. This generalization was called the *quaternions* and the theory behind them became important in the development of important concepts such as the vector and scalar product of vectors. The creation of the quaternions and other "hypercomplex systems" inspired mathematicians to work actively in this area.

In order to construct the quaternions, we consider the subset $\mathbb{H}$ of the set $M_2(\mathbb{C})$ consisting of matrices of the type

$$\begin{pmatrix} x & y \\ -\Delta(y) & \Delta(x) \end{pmatrix},$$

where now $x, y \in \mathbb{C}$. Thus if $x = a + bi$ and $y = c + di$, then

$$\begin{pmatrix} x & y \\ -\Delta y & \Delta x \end{pmatrix} = \begin{pmatrix} a + bi & c + di \\ -c + di & a - bi \end{pmatrix}.$$

We show that $\mathbb{H}$ is stable under addition and multiplication. Indeed

$$\begin{pmatrix} x & y \\ -\Delta(y) & \Delta(x) \end{pmatrix} + \begin{pmatrix} u & v \\ -\Delta(v) & \Delta(u) \end{pmatrix}$$

$$= \begin{pmatrix} x+u & y+v \\ -(\Delta(y)+\Delta(v)) & \Delta(x)+\Delta(u) \end{pmatrix} = \begin{pmatrix} x+u & y+v \\ -\Delta(y+v) & \Delta(x+u) \end{pmatrix} \in \mathbb{H}$$

and

$$\begin{pmatrix} x & y \\ -\Delta(y) & \Delta(x) \end{pmatrix} \begin{pmatrix} u & v \\ -\Delta(v) & \Delta(u) \end{pmatrix}$$

$$= \begin{pmatrix} xu - y\Delta(v) & xv + y\Delta(u) \\ -\Delta(y)u - \Delta(x)\Delta(v) & -\Delta(y)v + \Delta(x)\Delta(u) \end{pmatrix}$$

$$= \begin{pmatrix} xu - y\Delta(v) & xv + y\Delta(u) \\ -\Delta(y)\Delta\big(\Delta(u)\big) - \Delta(x)\Delta(v) & \Delta(x)\Delta(u) - \Delta(y)\Delta\big(\Delta(v)\big) \end{pmatrix}$$

$$= \begin{pmatrix} xu - y\Delta(v) & xv + y\Delta(u) \\ -\Delta\big(y\Delta(u) + xv\big) & \Delta\big(xu - y\Delta(v)\big) \end{pmatrix} \in \mathbb{H}.$$

Since, for all matrices of $M_2(\mathbb{C})$, addition is commutative and associative, multiplication is associative and multiplication is distributive over addition, the operations of addition and multiplication in $\mathbb{H}$ possess these same properties. Clearly, the identity matrix I and the zero matrix O belong to $\mathbb{H}$. It easy to see that

$$-\begin{pmatrix} x & y \\ -\Delta(y) & \Delta(x) \end{pmatrix} = \begin{pmatrix} -x & -y \\ \Delta(y) & -\Delta(x) \end{pmatrix} = \begin{pmatrix} -x & -y \\ -\Delta(-y) & \Delta(-x) \end{pmatrix} \in \mathbb{H}.$$

Finally, for an arbitrary nonzero element α of $\mathbb{H}$

$$\alpha = \begin{pmatrix} x & y \\ -\Delta(y) & \Delta(x) \end{pmatrix} \neq \begin{pmatrix} 0 & 0 \\ 0 & 0 \end{pmatrix},$$

and we have

$$\mathbf{det}(\alpha) = x\Delta(x) + y\Delta(y) = |x|^2 + |y|^2 \neq 0.$$

In particular, the matrix α is nonsingular, and has a multiplicative inverse. By Theorem 2.5.3, which holds in the field $\mathbb{C}$,

$$\alpha^{-1} = \begin{pmatrix} x & y \\ -\Delta(y) & \Delta(x) \end{pmatrix}^{-1} = \begin{pmatrix} \dfrac{\Delta x}{\det(\alpha)} & \dfrac{-y}{\det(\alpha)} \\ \dfrac{\Delta(y)}{\det(\alpha)} & \dfrac{x}{\det(\alpha)} \end{pmatrix}$$

$$= \begin{pmatrix} \dfrac{\Delta x}{\det(\alpha)} & \dfrac{-y}{\det(\alpha)} \\ \Delta\left(\dfrac{(y)}{\det(\alpha)}\right) & \Delta\dfrac{(\det(\alpha))}{\Delta x} \end{pmatrix} \in \mathbb{H},$$

and it follows that $\mathbb{H}$ is a division ring.

Next, we will determine the natural form to write the elements of $\mathbb{H}$. Let $x = a + bi$, $y = c + di$. Then

$$\begin{pmatrix} x & y \\ -\Delta(y) & \Delta(x) \end{pmatrix} = \begin{pmatrix} a+bi & c+di \\ -c+di & a-bi \end{pmatrix}$$

$$= \begin{pmatrix} a & 0 \\ 0 & a \end{pmatrix} + \begin{pmatrix} bi & 0 \\ 0 & -bi \end{pmatrix} + \begin{pmatrix} 0 & c \\ -c & 0 \end{pmatrix} + \begin{pmatrix} 0 & di \\ di & 0 \end{pmatrix}$$

$$= a\begin{pmatrix} 1 & 0 \\ 0 & 1 \end{pmatrix} + b\begin{pmatrix} i & 0 \\ 0 & -i \end{pmatrix} + c\begin{pmatrix} 0 & 1 \\ -1 & 0 \end{pmatrix} + d\begin{pmatrix} 0 & i \\ i & 0 \end{pmatrix}.$$

Let

$$\mathbf{I} = \begin{pmatrix} 1 & 0 \\ 0 & 1 \end{pmatrix}, \mathbf{H} = \begin{pmatrix} i & 0 \\ 0 & -i \end{pmatrix}, \mathbf{J} = \begin{pmatrix} 0 & 1 \\ -1 & 0 \end{pmatrix}, \text{ and } \mathbf{K} = \begin{pmatrix} 0 & i \\ i & 0 \end{pmatrix}.$$

By direct calculation we obtain the following multiplication chart for the matrices **I**, **H**, **J**, **K**:

	I	**H**	**J**	**K**
I	**I**	**H**	**J**	**K**
H	**H**	**–I**	**K**	**–J**
J	**J**	**–K**	**–I**	**H**
K	**K**	**J**	**–H**	**–I**

In particular, $\mathbf{HJ} = \mathbf{K}$ and $\mathbf{JH} = -\mathbf{K}$, so that the multiplication in $\mathbb{H}$ is noncommutative. Hence $\mathbb{H}$ is a noncommutative division ring, called the ring of real quaternions and its elements are called the (real) quaternions.

EXERCISE SET 3.3

Give an explanation of your work in the following exercises.

3.3.1. Find necessary and sufficient conditions for the product of two complex numbers z_1z_2 to be a real number.

3.3.2. Find $\mathbf{Re}(\alpha)$ and $\mathbf{Im}(\alpha)$, if $\alpha = (2-3i)^4 + (2+3i)^4$.

3.3.3. Find $\mathbf{Re}(\alpha)$ and $\mathbf{Im}(\alpha)$, if $\alpha = \left(\frac{1-i}{1+i}\right)^{1980}$.

3.3.4. Find $(i+2\alpha^2)(i^3\alpha^2-3)$, if $\alpha = \frac{i-1}{2}$.

3.3.5. Solve the equation $X = -1 - 3X + 2i$.

3.3.6. Solve the equation $X^2 + X = 1$.

3.3.7. Solve the equation $X^2 + 5X + 5 - 3i = 0$.

3.3.8. Solve the equation $X^2 + |X+1| + i = 0$.

3.3.9. Write in the trigonometric form the number $-\frac{1}{2} - \frac{1}{2}i$.

3.3.10. Write in trigonometric form the number $-\cos\sigma + i\sin\sigma$.

3.3.11. Find the sum of all the 15th roots of unity.

3.3.12. Find the sum of the primitive 15th roots of unity.

3.3.13. Given that $\frac{1-i\sqrt{3}}{2}$ is one root of $\sqrt[4]{\alpha}$. Find all other roots of $\sqrt[4]{\alpha}$.

3.3.14. Solve the equation $X^5 + 1 = 0$.

3.3.15. Let α be a complex number with the property $\alpha^k = 1$, where $\alpha \neq 1$. Prove that $1 + \alpha + \alpha^2 + \cdots + \alpha^{k-1} = 0$.

3.3.16. Prove the equation $|\alpha+\beta|^2 = (|\alpha| + |\beta|)^2 + 2(|\alpha\overline{\beta}| + \mathbf{Re}(\alpha\overline{\beta}))$ for all complex numbers α, β.

3.3.17. Let $\phi : \mathbb{C} \longrightarrow \mathbb{C}$ be the mapping defined by the rule $\phi(z) = \frac{az+b}{cz+d}$, $a, b, c, d \in \mathbb{C}$. Find conditions on a, b, c, d for which there are numbers z satisfying the equation $\phi(z) = z$ and find all such z.

3.3.18. Prove that $\mathbb{Q}(i) = \{a + bi \mid a, b \in \mathbb{Q}\}$ is a subfield of $\mathbb{C}$.

3.3.19. Let $x \in \mathbb{C}\backslash\mathbb{R}$; prove that $\mathbb{R}(x) = \mathbb{C}$.

CHAPTER 4

VECTOR SPACES

Linear algebra is one of the oldest branches of mathematics but it is also a branch of mathematics that is still vibrant and very much alive today. Ancient manuscripts include problems that could be solved by doing the basic mathematical operations of addition, subtraction, multiplication, and division step-by-step so that a solution to an equation such as $ax + b = 0$ could be obtained. Of course, the mathematical language used there was very different from the language we use today, but the main ideas were formulated, between 284 and 298 CE, by Diophantus of Alexandria, a great Greek mathematician and he is sometimes called "the father of algebra." Investigation of properties of the linear function $f(x) = ax + b$ and solutions of the equation mentioned above are at the origins of linear algebra. Thus, linear algebra arose as a subject of study because of the practical everyday needs of people. For many years progress in linear algebra was mainly connected with the problems of solving systems of linear equations. Consequently, until the seventeenth century all manuals on algebra were concerned with these themes. Investigations of systems of linear equations in n variables led Leibniz and Cramer to the concept of a determinant.

Linear algebra is also very useful in geometry. Inspired by the ideas of Apollonius, Fermat and, earlier, Descartes, arrived at the idea of analytical geometry and used it to classify plain curves by their degree. They also observed the main principle of representation of straight lines in the plane as linear equations and the conic sections as presentations of equations of the second degree. Fermat's ideas of classification led to the great development of analytical geometry in the eighteenth century because of the works of Clairaut, Euler, P. Cramer, Lagrange, and others. Linear forms were investigated by the great Swiss mathematician and physicist, Euler (1707–1783), who based his classification of plane curves and surfaces on it. Additionally, the theory of differential equations, at the very early stages of its development, requires deep study of systems of linear equations. It was a very logical step to extend these ideas to arbitrary n-dimensional space, which found very important applications in mechanics and physics. The need for an algebraic generalization of the theory of functional equations (differential and integral equations) led to the idea of infinite-dimensional vector spaces. Currently, linear algebra is a foundation of almost all branches of mathematics, theoretical mechanics, physics, mathematical economics, and other sciences.

4.1 VECTOR SPACES

One of the main ideas of linear algebra is the concept of the action of a set on some other set, sometimes called an outer product or scalar multiplication. The origin of this concept is the operation of multiplying a vector by a number.

4.1.1. Definition. *Let M and Ω be sets. We define an action (or outer operation or scalar multiplication) of Ω on M if there is a mapping*

$$\clubsuit : \Omega \times M \longrightarrow M.$$

This means that for every ordered pair (σ, a), where $\sigma \in \Omega$, $a \in M$, there corresponds a uniquely defined element $\clubsuit(\sigma, a)$ of M. The element $\clubsuit(\sigma, a)$ is called the composition of the elements σ and a.

The terminology often used is that $\clubsuit(\sigma, a)$ is a scalar multiple of a. Most of the time, we denote the action of the element σ on the element a using multiplication on the left. Thus, we write $\clubsuit(\sigma, a)$ more simply as $\sigma \cdot a$, or just σa, the dot usually being omitted. We may also sometimes write the action $\clubsuit(\sigma, a)$ as $a \cdot \sigma$, where the multiplication is now done on the right. We may then refer to a left outer multiplication (respectively a right outer multiplication) or a left (right) scalar multiplication. Sometimes an exponential form of writing, a^σ, is used.

Here are some examples of scalar multiplication.

Let G be a group, written multiplicatively. Define the action of $\mathbb{Z}$ on G by

$$(n, g) \longmapsto g^n, \text{ where } n \in \mathbb{Z} \text{ and } g \in G.$$

If we use additive notation for the binary operation on G, then the scalar multiplication would be defined by

$$(n, g) \longmapsto ng, \text{ where } n \in \mathbb{Z} \text{ and } g \in G.$$

Next, let M be a set and let Ω be a subset of $\mathbf{P}(M)$ which, we recall, is the set of transformations of M. Define the action of Ω on M by

$$(f, a) \longmapsto f(a), \text{ where } a \in M \text{ and } f \in \Omega.$$

Finally, let F be a field and let K be a subfield of F (the case $F = K$ is allowed). Define the action of K on F by the rule

$$(b, a) \longmapsto ba, \text{ where } b \in K \text{ and } a \in F.$$

In this case the multiplication defined on F itself becomes the scalar multiplication of K on F. This may seem a little bit confusing initially, but we hope that the context will make our meaning clear.

4.1.2. Definition. *Let M and Ω be sets and suppose that a scalar multiplication of Ω on M is defined. A subset S of M is called stable under this scalar multiplication if $\sigma b \in S$ for each element $b \in S$ and each element $\sigma \in \Omega$.*

This means that the restriction to S of a scalar multiplication is again a scalar multiplication on S. For instance, in the first example every subgroup of G is stable.

One result that the reader will easily see to be true is the following:

4.1.3. Proposition. *Let M and Ω be sets and suppose that a scalar multiplication of Ω on M is defined. If $\mathfrak{S}$ is a family of stable subsets of M, then the intersection $\cap\mathfrak{S}$ of all subsets of this family is also stable.*

The space $\mathbb{R}^3$ from analytic geometry is the first natural example of a vector space. The elements of this space are called vectors (or free vectors) and in calculus we learn to manipulate and picture these. We can multiply them by numbers (scalars) and add them by using the "parallelogram law of vector addition." In our general definition, we will substitute real numbers by elements of an arbitrary field. Of course, in this more general setting, geometric representations may not be as appropriate as they are for $\mathbb{R}^2$ and $\mathbb{R}^3$.

4.1.4. Definition. *Let F be a field and let A be a set. Suppose that an additive binary operation is defined on the set A and that an action of F on A is also defined, which we call (left) scalar multiplication. Then A is a vector space over F or an F-space (or, more precisely, a left vector space), if the following conditions hold:*

(VS 1) *the addition on A is commutative, so*

$$x + y = y + x \text{ for all } x, y \in A;$$

(VS 2) *the addition on A is associative, so*

$$x + (y + z) = (x + y) + z \textit{ for all } x, y, z \in A;$$

(VS 3) *A has a zero element* 0_A*, an element such that*

$$x + 0_A = x \textit{ for all } x \in A;$$

(VS 4) *each element* $x \in A$ *has an additive inverse,* $-x \in A$*, an element satisfying*

$$x + (-x) = 0_A;$$

(VS 5)

$$\alpha(x + y) = \alpha x + \alpha y \textit{ and}$$

$$(\alpha + \beta)x = \alpha x + \beta x \textit{ for all } x, y \in A, \quad \alpha, \beta \in F;$$

(VS 6)

$$\alpha(\beta x) = (\alpha\beta)x \textit{ for all } x \in A, \quad \alpha, \beta \in F;$$

(VS 7) *if e is the identity element of F, then*

$$ex = x \quad for \ all \ x \in A$$

Note that two axioms that do not receive explicit attention as such are the facts that the addition and scalar multiplication are both closed operations. The elements of A are often called vectors, whereas the elements of F are called scalars. Conditions (i)–(iv) show that A is an abelian group under addition. We will say that this is the additive group of the vector space A and denote it by A_+. We note that the existence of additive inverses for each element of A allows us to introduce the operation of subtraction of two elements $a, b \in A$ by the rule

$$a - b = a + (-b).$$

We also note that if, instead of a left scalar multiplication, we use a right scalar multiplication on A, then we use the terminology "right vector space." In this case, conditions **(VS 5)–(VS 7)** are as follows:

(VS 5)

$$(x + y)\alpha = x\alpha + y\alpha \text{ and}$$

$$x(\alpha + \beta) = x\alpha + x\beta \text{ for all } x, y \in A, \quad \alpha, \beta \in F;$$

(VS 6)

$$(x\alpha)\beta = x(\alpha\beta) \text{ for all } x \in A, \quad \alpha, \beta \in F;$$

(VS 7) if e is the identity element of F, then

$$xe = x \text{ for all } x \in A.$$

In algebra, actions are often written on the right whereas in other branches of mathematics left actions are commonly used. For this reason, in this book, we will discuss only left vector spaces but all results proved for left vector spaces have a corresponding right vector space analog.

We next obtain some elementary results, of the type that we have seen before when discussing fields.

4.1.5. Proposition. *Let F be a field and let A be a vector space over F. Then, for all $a, b \in A$ and all $\alpha, \beta \in F$,*

(i) $0_F \cdot a = 0_A$ *and* $\alpha 0_A = 0_A$;
(ii) $\alpha(-a) = (-\alpha)a = -\alpha a$;
(iii) $\alpha(a - b) = \alpha a - \alpha b$ *and* $(\alpha - \beta)a = \alpha a - \beta a$.

Proof. (i) For each $a \in A$, we have $a + 0_A = a$. Then

$$\alpha a = \alpha(a + 0_A) = \alpha a + \alpha 0_A.$$

Since αa has an additive inverse, $-\alpha a$, we have, adding $-\alpha a$ to each side,

$$\begin{aligned} 0_A &= -\alpha a + \alpha a = -\alpha a + (\alpha a + \alpha 0_A) = (-\alpha a + \alpha a) + \alpha 0_A \\ &= 0_A + \alpha 0_A = \alpha 0_A, \end{aligned}$$

and similarly

$$0_F \cdot a = 0_A.$$

(ii) By the definition of additive inverses, we have

$$\begin{aligned} 0_A &= \alpha 0_A = \alpha(a + (-a)) = \alpha a + \alpha(-a) \text{ and} \\ 0_A &= 0_F \cdot a = (\alpha + (-\alpha))a = \alpha a + (-\alpha)a. \end{aligned}$$

These equations show that the products $\alpha(-a)$ and $(-\alpha)a$ are the additive inverse of αa. Since the additive inverse is unique, (ii) follows.

(iii) We have

$$\alpha(a - b) = \alpha(a + (-b)) = \alpha a + \alpha(-b) = \alpha a + (-\alpha b) = \alpha a - \alpha b,$$

and similarly,

$$(\alpha - \beta)a = \alpha a - \beta a.$$

4.1.6. Definition. *Let F be a field and let A be a vector space over F. The subset B of A is called a subspace if B is stable under the operations of addition and scalar multiplication and B is a vector space by restrictions of these operations. In this case, we write $B \leq A$.*

As is often the case, it is much easier to check that a nonempty subset is a subspace than would appear at first sight as we now see.

4.1.7. Theorem. *Let F be a field and let A be a vector space over F. If B is a subspace of A then B satisfies the following conditions:*

(**SS 1**) *if $a, b \in B$ then $a - b \in B$;*
(**SS 2**) *if $\alpha \in F$ and $b \in B$, then $\alpha b \in B$.*

*Conversely, suppose that B is not empty. If B satisfies conditions (**SS 1**) and (**SS 2**), then B is a subspace of A.*

Proof. Let B be a subspace of A. Then B is a stable subset under addition and scalar multiplication. It follows that B has a zero element 0_B. Thus, $x + 0_B = x$ for each element $x \in B$. By Definition 4.1.4, there is an element $-x \in A$ and we have

$$-x + x + 0_B = -x + x \text{ so } 0_A + 0_B = 0_A.$$

It follows that $0_B = 0_A$. Again by Definition 4.1.4 and this observation, we see that for each element $x \in B$ there exists an element $y \in B$ such that $x + y = 0_A$, so y is the negative of x in A. As we know from Section 3.2, the negative element is unique, so that $y = -x$ and hence $-x \in B$. Next, let x, y be arbitrary elements of B. Then, as given earlier, $-y \in B$ and since B is stable under addition,

$$x - y = x + (-y) \in B.$$

Since B is a stable subset under scalar multiplication, $\alpha x \in B$, so that B satisfies (**SS 1**) and (**SS 2**).

Conversely, suppose that $B \neq \emptyset$ and that B satisfies (**SS 1**) and (**SS 2**). If $u \in B$ then, by (**SS 1**), $0_A = u - u \in B$. Further, if $x \in B$ then by (**SS 1**)

$$-x = 0_A - x \in B.$$

Also, if x, y are arbitrary elements of B then $-y \in B$. Using (**SS 1**), we obtain

$$x + y = x - (-y) \in B.$$

Hence B is a stable subset under addition. Condition (**SS 2**) implies that B is a stable subset under scalar multiplication. Thus, the restriction of the addition on B is a binary operation on B and the restriction of the scalar multiplication on B is a scalar multiplication on B. Conditions **(VS 1)**, **(VS 2)**, **(VS 5)**, **(VS 6)**, and **(VS 7)** of Definition 4.1.4 are valid for B, since they are valid for all elements of A. We have already proved that $0_A \in B$, so 0_A is the zero element for B. Also, we proved that $-x \in B$ for each element $x \in B$, so conditions **(VS 3)** and **(VS 4)** of Definition 4.1.4 are satisfied.

4.1.8. Corollary. *Let F be a field and let A be a vector space over F. If $\mathfrak{S}$ is a family of subspaces of A, then the intersection $\cap\mathfrak{S}$ of all subspaces from this family is a subspace of A.*

Proof. Let $S = \cap\mathfrak{S}$. Since every subspace contains 0_A, we have $0_A \in S$, so $S \neq \emptyset$. Let $x, y \in S$. If U is an arbitrary element of $\mathfrak{S}$ then $x - y \in U$, and therefore, $x - y$ lies in the intersection of all elements of $\mathfrak{S}$. But this intersection is S, so that $x - y \in S$. This shows that S satisfies (**SS 1**). Next let $x \in S$ and let $\alpha \in F$. If U is an arbitrary element of $\mathfrak{S}$ then $\alpha x \in U$, and therefore, αx lies in the intersection of all elements of $\mathfrak{S}$. Since this intersection is S, we have $x \in S$. This shows that S satisfies (**SS 2**). Now we apply Theorem 4.1.7 to deduce the result.

It is not hard to see that the union of two subspaces is not always a subspace. However, the following result holds.

4.1.9. Corollary. *Let F be a field and let A be a vector space over F. If $\mathfrak{L}$ is a local family of subspaces of A, then the union $\cup\mathfrak{L}$ of all subspaces from this family is a subspace of A.*

Proof. Let $V = \cup\mathfrak{L}$ and let $x, y \in V$. Then, there exist subspaces $H, K \in \mathfrak{L}$ such that $x \in H$, $y \in K$. Since $\mathfrak{L}$ is a local family, we choose a subspace $L \in \mathfrak{L}$ which contains both subspaces H, K. Then $x, y \in L$ and, since L is a subspace, $x - y \in L$ by Theorem 4.1.7. Hence $x - y \in V$ so V satisfies (**SS 1**). Using a similar argument, we prove that V satisfies (**SS 2**) and we can then apply Theorem 4.1.7.

4.1.10. Corollary. *Let F be a field and let A be a vector space over F. If $\mathfrak{L}$ is a linearly ordered family of subspaces of A, then the union $\cup\mathfrak{L}$ of all subspaces from this family is also a subspace of A.*

4.1.11. Corollary. *Let F be a field and let A be a vector space over F. If*

$$H_1 \leq H_2 \leq \cdots \leq H_n \leq \cdots$$

is an ascending chain of subspaces of A, then $\bigcup_{n \in \mathbb{N}} H_n$ is also a subspace of A.

Now we let A be an F-vector space, let $a_1, \ldots, a_n \in A$ and let $\alpha_1, \ldots, \alpha_n \in F$. Then

$$\alpha_1 a_1 + \cdots + \alpha_n a_n = \sum_{1 \leq j \leq n} \alpha_j a_j$$

is called a linear combination of the elements $a_1, \ldots, a_n$ with coefficients $\alpha_1, \ldots, \alpha_n$. Using mathematical induction, we obtain the following corollary.

4.1.12. Corollary. *Let F be a field and let A be a vector space over F. If B is a subspace of A, then all linear combinations of finitely many elements of B belong to B.*

Note that every vector space A contains two subspaces A and $\{0_A\}$ (which will coincide if $A = \{0_A\}$).

We shall now look at some examples of vector spaces. First, we should mention that $\mathbb{R}^3$ is a vector space over $\mathbb{R}$, but the particular construction appears in a more general context below, so we refrain from a formal proof just yet.

Next let F be a field and let K be a subfield of F. The action of K on F is defined by the rule

$$(b, a) \longmapsto ba, \quad b \in K, \quad a \in F.$$

Since F is a field, F is an abelian group under addition. Condition **(VS 5)** follows from the distributivity of multiplication over addition in F. Condition **(VS 6)** follows from the fact that multiplication in F is associative and condition **(VS 7)** follows from the definition of the identity element. Hence the field F can be considered as a vector space over its subfield K. For example, $\mathbb{R}$ can be considered as a vector space over $\mathbb{Q}$, and $\mathbb{C}$ can be considered as a vector space over $\mathbb{R}$.

For our next example, let $\mathbb{R}^{[a,b]}$ be the set of all real functions defined on the closed interval $[a, b]$. Thus, $f \in \mathbb{R}^{[a,b]}$ if and only if $f : [a, b] \longrightarrow \mathbb{R}$. This set is a vector space over $\mathbb{R}$ using the operations of addition of real functions and multiplication of a real function by a real number. Thus, if $f, g \in \mathbb{R}^{[a,b]}$ then $f + g$, defined by $(f + g)(x) = f(x) + g(x)$, is also an element of $\mathbb{R}^{[a,b]}$. Likewise, if $\alpha \in \mathbb{R}$ then αf is the function defined by $(\alpha f)(x) = \alpha(f(x))$ and $\alpha f \in \mathbb{R}^{[a,b]}$ also. It is then easy to verify that $\mathbb{R}^{[a,b]}$ is a vector space over $\mathbb{R}$. The subset of all continuous functions satisfies conditions **(SS 1)** and **(SS 2)**, precisely because of the well-known facts that a difference of continuous functions is continuous and a scalar multiple of a continuous function is also continuous. Thus, Theorem 4.1.7 shows that this is a subspace of $\mathbb{R}^{[a,b]}$.

The following example is an important one for finite-dimensional vector spaces (where the notion of dimension is defined later).

Let F be a field and let n be a positive integer. Put $A = A_1 \times \cdots \times A_n$ where A_j is a vector space over F, for all j, with $1 \leq j \leq n$. (Sometimes the notation

$A_1 \oplus A_2 \oplus \cdots \oplus A_n$ is used.) We define addition and scalar multiplication on A as follows:

Let $\boldsymbol{a} = (a_1, \ldots, a_n), \boldsymbol{b} = (b_1, \ldots, b_n) \in A$ and let $\alpha \in F$. Then let

$$\boldsymbol{a} + \boldsymbol{b} = (a_1 + b_1, \ldots, a_n + b_n) \text{ and let } \alpha\boldsymbol{a} = (\alpha a_1, \ldots, \alpha a_n).$$

Thus, addition of n-tuples is defined via the addition of the components that are the elements of corresponding vector spaces. Therefore, addition of n-tuples inherits all properties of addition of elements of vector spaces, so addition of n-tuples is commutative, associative, and has a zero element which is the n-tuple $0_A = (0_{A_1}, \ldots, 0_{A_n})$. Also each n-tuple $\boldsymbol{a} = (a_1, \ldots, a_n)$ has an additive inverse, which is the n-tuple $-\boldsymbol{a} = (-a_1, \ldots, -a_n)$.

For the scalar multiplication, we have

$$\begin{aligned}\alpha(\boldsymbol{a} + \boldsymbol{b}) &= \alpha(a_1 + b_1, \ldots, a_n + b_n) = (\alpha(a_1 + b_1), \ldots, \alpha(a_n + b_n)) \\ &= (\alpha a_1 + \alpha b_1, \ldots, \alpha a_n + \alpha b_n) = (\alpha a_1, \ldots, \alpha a_n) + (\alpha b_1, \ldots, \alpha b_n) \\ &= \alpha(a_1, \ldots, a_n) + \alpha(b_1, \ldots, b_n) = \alpha\boldsymbol{a} + \alpha\boldsymbol{b}.\end{aligned}$$

Also,

$$\begin{aligned}(\alpha + \beta)\boldsymbol{a} &= (\alpha + \beta)(a_1, \ldots, a_n) = ((\alpha + \beta)a_1, \ldots, (\alpha + \beta)a_n) \\ &= (\alpha a_1 + \beta a_1, \ldots, \alpha a_n + \beta a_n) = (\alpha a_1, \ldots, \alpha a_n) + (\beta a_1, \ldots, \beta a_n) \\ &= \alpha(a_1, \ldots, a_n) + \beta(a_1, \ldots, a_n) = \alpha\boldsymbol{a} + \beta\boldsymbol{a}.\end{aligned}$$

Furthermore,

$$\begin{aligned}(\alpha(\beta\boldsymbol{a})) &= \alpha(\beta(a_1, \ldots, a_n)) = \alpha(\beta a_1, \ldots, \beta a_n) = (\alpha(\beta a_1), \ldots, \alpha(\beta a_n)) \\ &= ((\alpha\beta)a_1, \ldots, (\alpha\beta)a_n) = \alpha\beta(a_1, \ldots, a_n) = (\alpha\beta)\boldsymbol{a}.\end{aligned}$$

Finally,

$$e\boldsymbol{a} = e(a_1, \ldots, a_n) = (ea_1, \ldots, ea_n) = (a_1, \ldots, a_n) = \boldsymbol{a}.$$

Hence all axioms for a vector space hold.

4.1.13. Definition. *Let F be a field and let $A_1, \ldots, A_n$ be vector spaces over F for all j, where $1 \leq j \leq n$. The vector space $A = A_1 \times \cdots \times A_n$ is called the external direct sum of the vector spaces $A_1, \ldots, A_n$.*

We remarked above that we can consider a field F as a vector space over itself. If $A_1 = \cdots = A_n = F$, then we obtain a vector space A which we denote by F^n. This is the set of all n-tuples with coefficients in F, using componentwise addition and scalar multiplication. We next consider a generalization of this. To this end, let F be a field and let $F^{\mathbb{N}}$ denote the set of all sequences

$$(a_n)_{n\in\mathbb{N}} = (a_1, \dots, a_n, \dots),$$

whose entries belong to the field F.

Two sequences (a_n) and (b_n) are called equal, if $a_n = b_n$ for each $n \in \mathbb{N}$.

The addition and scalar multiplication of sequences are defined using the same model as stated above, namely,

$$(a_n) + (b_n) = (c_n),$$

where $c_n = a_n + b_n$ for all $n \in \mathbb{N}$ and

$$\alpha(a_n) = (\alpha a_n)$$

for each $n \in \mathbb{N}$.

As stated above, it is easy to justify that $F^{\mathbb{N}}$ is a vector space over F. Let

$$F^{(k)} = \{(a_n)_{n\in\mathbb{N}} \mid a_n = 0_F \text{ for all } n > k\}.$$

It is not hard to show that $F^{(k)}$ satisfies conditions (**SS 1**) and (**SS 2**), so Theorem 4.1.7 shows that $F^{(k)}$ is a subspace of $F^{\mathbb{N}}$. It is evident that $F^{(k)}$ appears not to be significantly different from the space F^k defined above. Later, we shall show that these subspaces are isomorphic. Clearly $F^{(k)} \le F^{(k+1)}$ so $F^{(k)} \le F^{(m)}$ whenever $k \le m$.

We write $F^{(\mathbb{N})} = \bigcup_{k\in\mathbb{N}} F^{(k)}$ and note that Corollary 4.1.11 shows that $F^{(\mathbb{N})}$ is a subspace of $F^{\mathbb{N}}$. It is a proper subspace since the sequences in $F^{(\mathbb{N})}$ necessarily terminate, whereas those in $F^{\mathbb{N}}$ need not.

The results of Section 2.1 imply that the set $\mathbf{M}_{k\times n}(\mathbb{R})$ of all $k \times n$ matrices with real coefficients is a vector space over $\mathbb{R}$. We will generalize this case. Let F be a field and let $\mathbf{M}_{k\times n}(F)$ denote the set of all $k \times n$ matrices with coefficients in F. On the set $\mathbf{M}_{k\times n}(F)$, we define the operations of addition and scalar multiplication as we did for numerical matrices. More precisely, let

$$A = [\alpha_{tj}],\ B = [\beta_{tj}] \in \mathbf{M}_{k\times n}(F) \text{ and let } \lambda \in F.$$

Then define

$$A + B = [\gamma_{tj}] \in \mathbf{M}_{k\times n}(F), \text{ where } \gamma_{tj} = \alpha_{tj} + \beta_{tj}, \text{ for } 1 \le t \le k, 1 \le j \le n$$

and

$$\lambda A = [\mu_{tj}], \text{ where } \mu_{tj} = \lambda\alpha_{tj}, \text{ for } 1 \le t \le k, 1 \le j \le n.$$

As for numerical matrices, it is not difficult to prove that the following properties hold:

$$A + B = B + A,$$
$$A + (B + C) = (A + B) + C.$$

These follow because of the commutative and associative properties of addition in F. Next let O be the $k \times n$ matrix all of whose entries are 0_F. This is the additive identity of $\mathbf{M}_{k \times n}(F)$ in the sense that $O + A = A = A + O$ for all $A \in \mathbf{M}_{k \times n}(F)$. Also, for every matrix A there is an additive inverse $-A$, the matrix such that $A + (-A) = O$. If $A = [\alpha_{tj}]$ then $-A = [-\alpha_{tj}]$.

Furthermore, for all $A, B \in \mathbf{M}_{k \times n}(F)$ and all $\lambda, \mu \in F$, we have

$$(\lambda + \mu)A = \lambda A + \mu A, \quad \lambda(A + B) = \lambda A + \lambda B$$
$$\lambda(\mu A) = (\lambda\mu)A \text{ and } eA = A.$$

These follow because of the componentwise definitions of addition and scalar multiplication and because the corresponding properties hold in F. Thus, $\mathbf{M}_{k \times n}(F)$ becomes a vector space over F.

If $k = n$, then we obtain the vector space $\mathbf{M}_n(F)$ of all square matrices of order n, with coefficients in F.

We can also define the concept of the determinant of a matrix $A = [\alpha_{tj}] \in \mathbf{M}_n(F)$, analogous to that done earlier, by

$$\mathbf{det}(A) = \sum_{\pi \in \mathbf{S}_n} \mathbf{sign}\, \pi\ \alpha_{1,\pi(1)}\alpha_{2,\pi(2)} \cdots \alpha_{n,\pi(n)}.$$

When we established the properties of determinants when the coefficients were real, we used the properties of the operations of addition and multiplication such as commutativity and associativity, not the idea that the numbers involved were real numbers. Consequently, the properties of determinants that we established in Sections 2.3 and 2.4 and Theorem 2.5.1 are valid for matrices in $\mathbf{M}_n(F)$.

In a similar fashion, we define the product of matrices $A = [\alpha_{tj}] \in \mathbf{M}_{k \times n}(F)$ and $B = [\beta_{tj}] \in \mathbf{M}_{n \times q}(F)$ by

$$AB = [\gamma_{tj}] \in \mathbf{M}_{k \times q}(F),$$

where

$$\gamma_{tj} = \alpha_{t1}\beta_{1j} + \alpha_{t2}\beta_{2j} + \cdots + \alpha_{tn}\beta_{nj} = \sum_{1 \le l \le n} \alpha_{tl}\beta_{lj}$$

for all pairs t, j, where $1 \le t \le k$ and $1 \le j \le q$.

We note that once again the product is only defined when the number of columns of A is equal to the number of rows of B. Using similar arguments to those used with numerical matrices, it is possible to prove that

$$(AB)C = A(BC),$$
$$(A + B)C = AC + BC, \text{ and}$$
$$A(B + C) = AB + AC.$$

There exists a matrix $I \in \mathbf{M}_n(F)$ such that $AI = IA = A$ for each matrix $A \in \mathbf{M}_n(F)$. Here

$$I = \begin{pmatrix} e & 0 & 0 & 0 & \cdots & 0 & 0 \\ 0 & e & 0 & 0 & \cdots & 0 & 0 \\ \vdots & \vdots & \vdots & \vdots & \vdots & \vdots & \vdots \\ 0 & 0 & 0 & 0 & \cdots & e & 0 \\ 0 & 0 & 0 & 0 & \cdots & 0 & e \end{pmatrix}$$

is the $n \times n$ identity matrix. Let $A \in \mathbf{M}_n(F)$. A matrix U is called the inverse (or reciprocal) of A, if $AU = UA = I$. The matrix A is called nonsingular, if $\mathbf{det}(A) \neq 0_F$. As for numerical matrices, we can prove that a matrix A has an inverse if and only if A is nonsingular. Moreover, in this case, $A^{-1} = [\beta_{tj}] \in \mathbf{M}_n(F)$, where

$$B_{tj} = A_{jt}(\mathbf{det}(A))^{-1}, \text{ for } 1 \leq i, j \leq n$$

and A_{jt} is the cofactor corresponding to a_{jt}. Let E_{km} denote the matrix, whose (k, m) entry is the identity, e, and all others entries are 0_F. As with numerical matrices, we can prove

$$E_{km}E_{rs} = \begin{cases} E_{ks}, & \text{if } m = r, \\ 0, & \text{if m} \neq r. \end{cases}$$

Let $T_n^o(F)$ denote the subset of $\mathbf{M}_n(F)$ consisting of all upper triangular matrices. It is clear that conditions (**SS 1**) and (**SS 2**) are satisfied. So Theorem 4.1.7 shows that $T_n^o(F)$ is a subspace of $\mathbf{M}_n(F)$.

For the same reasons, the sets of all zero-triangular and diagonal matrices are also subspaces of $\mathbf{M}_n(F)$. Put

$$FI = \{\lambda I \mid \lambda \in F\}.$$

As with numerical matrices, λI is called a scalar matrix, so FI is the subset of all scalar matrices.

Next let $A_1, \ldots, A_n$ be subspaces of A and set

$$A_1 + \cdots + A_n = \{a_1 + \cdots + a_n \mid a_j \in A_j, 1 \leq j \leq n\}.$$

The subset $A_1 + \cdots + A_n$ is called the sum of subspaces $A_1, \ldots, A_n$.

4.1.14. Proposition. *Let F be a field and let A be a vector space over F. If $A_1, \ldots, A_n$ are subspaces of A, then their sum $A_1 + \cdots + A_n$ is a subspace of A.*

Proof. Let $x, y \in A_1 + \cdots + A_n$ and let $\alpha \in F$. Then

$$x = a_1 + \cdots + a_n \text{ and } y = b_1 + \cdots + b_n, \text{ where } a_j, b_j \in A_j, \text{ for } 1 \le j \le n.$$

We have

$$x - y = (a_1 + \cdots + a_n) - (b_1 + \cdots + b_n) = (a_1 - b_1) + \cdots + (a_n - b_n).$$

Since A_j is a subspace, Theorem 4.1.7 implies that $(a_j - b_j) \in A_j$, for $1 \le j \le n$. It follows that $x - y \in A_1 + \cdots + A_n$. Furthermore,

$$\alpha x = \alpha(a_1 + \cdots + a_n) = \alpha a_1 + \cdots + \alpha a_n.$$

Since A_j is a subspace, Theorem 4.1.7 again implies that $\alpha a_j \in A_j$, for $1 \le j \le n$. It follows that $\alpha x \in A_1 + \cdots + A_n$ and hence $A_1 + \cdots + A_n$ satisfies conditions (**SS 1**) and (**SS 2**). Once again Theorem 4.1.7 shows that $A_1 + \cdots + A_n$ is a subspace of A.

Certain sums of subspaces play a prominent role in the theory of vector spaces and we define such sums next.

4.1.15. Definition. *Let F be a field, let A be a vector space over F and let $A_1, \ldots, A_n$ be subspaces of A. The subspace $C = A_1 + \cdots + A_n$ is called the internal direct sum of $A_1, \ldots, A_n$, if each element c of C can be represented as $c = a_1 + \cdots + a_n$, where $a_j \in A_j$, for $1 \le j \le n$ and this representation is unique.*

The internal direct sum can be characterized in several different ways as we now show. This allows us to feel free to pick and choose the best method in any given situation of showing that a subspace is a direct sum.

4.1.16. Proposition. *Let F be a field, let A be a vector space over F and let $A_1, \ldots, A_n$ be subspaces of A. Let $C = A_1 + \cdots + A_n$. The following are equivalent:*

(i) *C is the internal direct sum of $A_1, \ldots, A_n$;*
(ii) *$a_1 + \cdots + a_n = 0_A$, where $a_j \in A_j$, for $1 \le j \le n$, if and only if $a_1 = \cdots = a_n = 0_A$;*
(iii) *$A_j \cap \sum_{k \ne j} A_k = \{0_A\}$ for every j, where $1 \le j \le n$.*

Proof. To prove that (i) implies (ii), note that every subspace contains 0_A, so that $0_A \in A_j$, for $1 \le j \le n$ and we obtain

$$a_1 + \cdots + a_n = 0_A + \cdots + 0_A.$$

Then, by (i), the representation of 0_A is unique so

$$a_1 = \cdots = a_n = 0_A.$$

For (ii) implies (iii), we suppose that $x \in A_j \cap \sum_{k \neq j} A_k$. It follows that $x = \sum_{k \neq j} y_k$, where $y_k \in A_k$, for $k \neq j$. Then

$$0_A = y_1 + \cdots + y_{j-1} + (-x) + y_{j+1} + \cdots + y_n.$$

From (ii), we deduce that $y_1 = \cdots = y_{j-1} = -x = y_{j+1} = \cdots = y_n = 0_A$. Hence $A_j \cap \sum_{k \neq j} A_k = \{0_A\}$ for every j, where $1 \leq j \leq n$ and (iii) follows.

Finally, assume that $c = a_1 + \cdots + a_n = b_1 + \cdots + b_n$, where $a_j, b_j \in A_j$, for $1 \leq j \leq n$. It follows that $x = a_j - b_j = \sum_{k \neq j}(b_k - a_k)$. Thus, $x \in A_j \cap \sum_{k \neq j} A_k$, which is 0_A, by hypothesis (iii). So $x = a_j - b_j = 0_A$ and $a_j = b_j$. This is valid for all j, where $1 \leq j \leq n$ and hence (iii) implies (i). The result follows.

EXERCISE SET 4.1

Justify your answers with a proof or a counterexample where appropriate.

4.1.1. Let $A = \mathbb{R}^2$, $M = \mathbb{R}$. Does the mapping $(\alpha, (\beta, \gamma)) \longmapsto (\alpha, \beta, \gamma)$, $\alpha, \beta, \gamma \in \mathbb{R}$ define a scalar multiplication?

4.1.2. Let $A = \mathbb{R}^2, M = \mathbb{R}$. Does the mapping $(\alpha, (\beta, \gamma)) \longmapsto (\alpha + \beta, \gamma)$, $\alpha, \beta, \gamma \in \mathbb{R}$ define a scalar multiplication?

4.1.3. Let $A = \mathbb{R}^2$, $M = \mathbb{R}$. Does the mapping $(\alpha, (\beta, \gamma)) \longmapsto (\alpha\beta, \gamma)$, $\alpha, \beta, \gamma \in \mathbb{R}$ define a scalar multiplication?

4.1.4. Let $A = \mathbb{R}^2, M = \mathbb{R}$. Does the mapping $(\alpha, (\beta, \gamma)) \longmapsto (\alpha, \gamma), \alpha, \beta, \gamma \in \mathbb{R}$ define a scalar multiplication?

4.1.5. Let $B = \{x \in \mathbb{R}^5 \mid x = (\alpha, \beta, 1, 0, 0), \alpha, \beta \in \mathbb{R}\}$. Is B a subspace of $\mathbb{R}^5$?

4.1.6. Let $B = \{x \in \mathbb{R}^5 \mid x = (\alpha, 0, 1, -1, \beta), \alpha, \beta \in \mathbb{R}\}$. Is B a subspace of $\mathbb{R}^5$?

4.1.7. Let $A = \mathbb{R}[X]$ be the vector space of all polynomials with real coefficients, B the subset of polynomials with no real roots. Is B a subspace?

4.1.8. Let A be the vector space of all real functions $f : [0, 2] \longrightarrow \mathbb{R}$, $B = \{f \mid f(1) = 3f(2)\}$. Is B a subspace?

4.1.9. Let A be the vector space of all real functions $f : \mathbb{R} \longrightarrow \mathbb{R}$, $B = \{f \mid f(2x) = (\sin x)f(x)\}$. Is B a subspace?

4.1.10. Let $\mathbb{R}^{\mathbb{N}}$ be the set of all sequences of real numbers, indexed by $\mathbb{N}$. Is the set $\mathbb{R}^{\mathbb{N}}$ a vector space under regular addition of sequences and where scalar multiplication by a real number is done componentwise? Is the subset B of all convergent sequences a subspace of $\mathbb{R}^{\mathbb{N}}$?

4.1.11. Let $\mathbb{R}^{\mathbb{N}}$ be the set of all sequences of real numbers, indexed by $\mathbb{N}$. Let B be the subset of $\mathbb{R}^{\mathbb{N}}$ consisting of all sequences satisfying the Cauchy condition (for every $\varepsilon > 0$ there is a number $n = n(\varepsilon)$ such that $|\alpha_m - \alpha_k| < \varepsilon$ if $m, k > n$). Is the set B of all such sequences a subspace of $\mathbb{R}^{\mathbb{N}}$?

4.1.12. Let $\mathbb{C}^{\mathbb{N}}$ be a set of all sequences of complex numbers, indexed by $\mathbb{N}$. Is this set $\mathbb{C}^{\mathbb{N}}$ a vector space under regular addition of sequences and where scalar multiplication by a complex number is done componentwise? Let B be the subset of $\mathbb{C}^{\mathbb{N}}$ consisting of all sequences (α_n) such that $\sum_{n\in\mathbf{N}} |\alpha_n|$ is convergent. Is the set B of all such sequences a subspace of $\mathbb{C}^{\mathbb{N}}$?

4.1.13. Let $A = \mathbb{R}[X]$ be the vector space of all polynomials with real coefficients and let B be its subspace with the following property: for every $k, 0 \leq k \leq t$, B contains at least one polynomial of degree t. Prove that B coincides with the subspace of all polynomials of degree $\leq t$.

4.1.14. Let B, C, E be subspaces of a vector space A, and let $C \leq B$. Prove that $B \cap (C + E) = C + (B \cap E)$. Is the equation $B \cap (C + E) = (B \cap C) + (B \cap E)$ valid for arbitrary subspaces B, C, E?

4.1.15. Let $A = \mathbb{Q}^{25}$, $B = \{(\alpha_1, \alpha_2, \ldots, \alpha_{25}) \mid \alpha_j = \alpha_{j+1},$ if j is even$\}$. Prove that B is a subspace of A. Find a subspace C that complements B, so $A = B \oplus C$.

4.1.16. Let $A = \mathbb{R}^5, B = \{x \in \mathbb{R}^5 \mid x = (\alpha, 0, \beta, 0, 0), \alpha, \beta \in \mathbb{R}\}$, $C = \{x \in R^5 \mid x = (0, y, 0, 0, y), y \in \mathbb{R}\}$. Find the sum of the subspaces B and C.

4.1.17. Is the set of all nonsingular matrices a subspace of the space $\mathbf{M}_{47}(\mathbb{R})$?

4.1.18. Is the set of all singular matrices a subspace of the space $\mathbf{M}_{47}(\mathbb{R})$?

4.2 DIMENSION

In this section, we consider one of the most important concepts in linear algebra, namely that of a basis and the connected concept of the dimension of a vector space.

Let F be a field, let A be a vector space over F and let M be a subset of A. We consider the family $\mathfrak{S}$ of all subspaces which contain M. By Corollary 4.1.8, $\cap\mathfrak{S}$ is a subspace of the vector space A.

4.2.1. Definition. *Let M be a subset of a vector space A and let $\mathfrak{S}$ be the family of subspaces containing M. The subspace* $\mathbf{Le}(M) = \cap\mathfrak{S}$ *is called the linear envelope of M or the subspace generated by the subset M. We also sometimes say that* $\mathbf{Le}(M)$ *is the subspace spanned by M. The subset M is called a set of generators or a spanning set for* $\mathbf{Le}(M)$. *In particular, if* $\mathbf{Le}(M) = A$, *then we say that*

M generates or spans A. The space A is called finitely generated, if there exists a finite subset M such that $\mathbf{Le}(M) = A$.

If B is a subspace containing M, then B contains $\mathbf{Le}(M)$, by Corollary 4.1.12. Thus $\mathbf{Le}(M)$ is the smallest subspace containing M. It is clear that if M is a subspace of A then $\mathbf{Le}(M) = M$. So we have the following.

4.2.2. Proposition. *Let F be a field and let A be a vector space over F. Suppose that M is a subset of A. The following properties hold:*

(i) $M \subseteq \mathbf{Le}(M)$.
(ii) *If B is a subspace of A and $M \subseteq B$, then* $\mathbf{Le}(M) \leq B$.
(iii) *If B is a subspace of A, then* $\mathbf{Le}(B) = B$. *In particular,* $\mathbf{Le}(\mathbf{Le}(M)) = \mathbf{Le}(M)$ *for every subset M.*
(iv) *If $M \subseteq S$, then* $\mathbf{Le}(M) \leq \mathbf{Le}(S)$.

We find out now what sort of elements belong to the linear envelope of a subset M.

4.2.3. Proposition. *Let F be a field, let A be a vector space over F and let M be a subset of A. Then,* $\mathbf{Le}(M)$ *consists of all linear combinations of all finite subsets of the set M.*

Proof. Let U denote the set of all linear combinations of all finite subsets of M and let $a_1, \dots, a_n$ be arbitrary elements of M. If B is a subspace of A containing M, then by Corollary 4.1.12, every linear combination of elements of M belongs to B. Since this is true for every subspace containing M, every linear combination of the elements $a_1, \dots, a_n$ belongs to $\mathbf{Le}(M)$. Thus $U \leq \mathbf{Le}(M)$.

Now let $x, y \in U$ and let $\gamma \in F$. Then $x = \alpha_1 a_1 + \cdots + \alpha_n a_n$ and $y = \beta_1 b_1 + \cdots + \beta_k b_k$, where $a_1, \dots, a_n, b_1, \dots, b_k \in M$ and $\alpha_1, \dots, \alpha_n, \beta_1, \dots, \beta_k \in F$. We have

$$\begin{aligned} x - y &= (\alpha_1 a_1 + \cdots + \alpha_n a_n) - (\beta_1 b_1 + \cdots + \beta_k b_k) \\ &= \alpha_1 a_1 + \cdots + \alpha_n a_n + (-\beta_1) b_1 + \cdots + (-\beta_k) b_k. \end{aligned}$$

Hence $x - y$ is a linear combination of $a_1, \dots, a_n, b_1, \dots, b_k \in M$, so that $x - y \in U$. Furthermore,

$$\begin{aligned} \gamma x &= \gamma(\alpha_1 a_1 + \cdots + \alpha_n a_n) = \gamma(\alpha_1 a_1) + \cdots + \gamma(\alpha_n a_n) \\ &= (\gamma\alpha_1) a_1 + \cdots + (\gamma\alpha_n) a_n, \end{aligned}$$

so that γx is a linear combination of the elements $a_1, \dots, a_n \in M$ and therefore $\gamma x \in U$. Hence U satisfies conditions (**SS 1**) and (**SS 2**); Theorem 4.1.7 shows that U is a subspace of A. If c is an element of M then $c = ec \in U$ and it follows that $M \subseteq U$. By Proposition 4.2.2, $\mathbf{Le}(M) \subseteq U$ and, since $U \leq \mathbf{Le}(M)$, we have $\mathbf{Le}(M) = U$, which proves the result.

4.2.4. Corollary. *Let F be a field, let A be a vector space over F and let M be a subset of A. If $x \in \mathbf{Le}(M)$, then $x \in \mathbf{Le}(S)$ for some finite subset S of M.*

The next property is very useful.

4.2.5. Lemma. *Let F be a field, let A be a vector space over F and let M be a subset of A. Suppose that x, y are elements with the property that $y \in \mathbf{Le}(M \cup \{x\})$, but $y \notin \mathbf{Le}(M)$. Then $x \in \mathbf{Le}(M \cup \{y\})$.*

Proof. By Corollary 4.2.4, there are elements $a_1, \ldots, a_n \in M$ and $\alpha_1, \ldots, \alpha_n$, $\beta \in F$ such that $y = \alpha_1 a_1 + \cdots + \alpha_n a_n + \beta x$. If $\beta = 0_F$ then $y = \alpha_1 a_1 + \cdots + \alpha_n a_n \in \mathbf{Le}(M)$, contrary to the hypothesis. Hence $\beta \neq 0_F$, so $\beta^{-1} \in F$. We have

$$\beta^{-1} y = (\beta^{-1}\alpha_1)a_1 + \cdots + (\beta^{-1}\alpha_n)a_n + ex$$

so

$$x = \beta^{-1} y - (\beta^{-1}\alpha_1)a_1 - \cdots - (\beta^{-1}\alpha_n)a_n.$$

It follows that $x \in \mathbf{Le}(M \cup \{y\})$.

The following concept is basic to the study of vector spaces.

4.2.6. Definition. *Let F be a field and let A be a vector space over F. A nonempty subset M of A is called free or linearly independent, if $x \notin \mathbf{Le}(M \backslash \{x\})$ for each element $x \in M$.*

Since the linear envelope of every subset is a subspace and therefore contains 0_A, we see that a linearly independent subset cannot contain 0_A. It is also clear that the elements of a linearly independent subset are distinct. A subset that is not linearly independent is called linearly dependent. Part (iii) of the following result is usually the easiest one to check when determining linear independence.

4.2.7. Proposition *(criterion for linear independence). Let F be a field, let A be a vector space over F, and let M be a subset of A.*

(i) *If M is linearly independent then every nonempty subset of M is linearly independent.*
(ii) *An infinite subset M is linearly independent if and only if every finite nonempty subset of M is linearly independent.*
(iii) *The finite subset $S = \{a_1, \ldots, a_n\}$ is linearly independent if and only if the equation $\alpha_1 a_1 + \cdots + \alpha_n a_n = 0_A$ always implies that $\alpha_1 = \cdots = \alpha_n = 0_F$.*

Proof.

(i) Suppose that M is a linearly independent subset and let W be a nonempty subset of M. Suppose, for a contradiction, that W is not linearly independent. Then, by definition, there exists an element $w \in W$ such that $w \in \mathbf{Le}(W\backslash\{w\})$. The inclusion $W \subseteq M$ implies that $W\backslash\{w\} \subseteq M\backslash\{w\}$ and Corollary 4.2.2 shows that $\mathbf{Le}(W\backslash\{w\}) \leq \mathbf{Le}(M\backslash\{w\})$. It follows that $w \in \mathbf{Le}(M\backslash\{w\})$, contradicting the fact that M is linearly independent. Thus, W must also be linearly independent.

(ii) If M is linearly independent, then every finite nonempty subset of M is linearly independent by (i). Conversely, suppose that every nonempty finite subset of M is linearly independent, but that M is not linearly independent. Then there exists an element $x \in M$ such that $x \in \mathbf{Le}(M\backslash\{x\})$. By Corollary 4.2.4, $M\backslash\{x\}$ contains a finite subset T such that $x \in \mathbf{Le}(T)$. Let $Y = T \cup \{x\}$, and note that Y is finite, $x \in Y$ and $x \in \mathbf{Le}(Y \setminus \{x\})$. It follows that Y is linearly dependent and we obtain a contradiction. Therefore M is linearly independent.

(iii) Suppose that S is linearly independent and let $\alpha_1 a_1 + \cdots + \alpha_n a_n = 0_A$. Suppose, for a contradiction, that there is a coefficient α_j such that $\alpha_j \neq 0_F$. Then $\alpha_j a_j = \sum_{k \neq j} \alpha_k a_k$ and, since F is a field, the nonzero element α_j has a multiplicative inverse α_j^{-1}. Therefore, $a_j = \sum_{k \neq j} (\alpha_j^{-1} \alpha_k) a_k$ and it follows that $a_j \in \mathbf{Le}(S\backslash\{a_j\})$, the desired contradiction, since S is linearly independent. Consequently, $\alpha_j = 0_F$ for all j, where $1 \leq j \leq n$.

Conversely, suppose that $\alpha_1 a_1 + \cdots + \alpha_n a_n = 0_A$ always implies that $\alpha_1 = \cdots = \alpha_n = 0_F$. Assume, for a contradiction, that S is not linearly independent. Then there exists an element a_m such that $a_m \in \mathbf{Le}(S\backslash\{a_m\})$. By Proposition 4.2.3, we obtain $a_m = \sum_{k \neq m} \beta_k a_k$ for certain $\beta_k \in F$. It follows that

$$\beta_1 a_1 + \cdots + \beta_{m-1} a_{m-1} + (-e) a_m + \beta_{m+1} a_{m+1} + \cdots + \beta_n a_n = 0_A.$$

Here, the coefficient of a_m is nonzero and the contradiction ensues showing that the subset S is linearly independent.

The following result shows how linearly independent subsets arise.

4.2.8. Lemma. *Let F be a field, let A be a vector space over F and let M be a linearly independent subset of A. If x is an element of A such that $x \notin \mathbf{Le}(M)$, then the subset $M \cup \{x\}$ is linearly independent.*

Proof. By Corollary 4.2.2, $M \subseteq \mathbf{Le}(M)$, so $x \notin M$. Let $S = M \cup \{x\}$ and suppose, for a contradiction, that S is not linearly independent. Then, by definition, there exists an element $y \in S$ such that $y \in \mathbf{Le}(S\backslash\{y\})$. Since $y \in S = M \cup \{x\}$, either $y = x$ or $y \in M$. If we suppose that $y = x$, then $x \in \mathbf{Le}(S\backslash\{x\}) = \mathbf{Le}(M)$, which contradicts the hypothesis that $x \notin \mathbf{Le}(M)$. Thus, we may assume that $y \in M$. Put $T = M\backslash\{y\}$, so $S\backslash\{y\} = T \cup \{x\}$ and then $y \in \mathbf{Le}(T \cup \{x\})$. Since M is a linearly independent subset, $y \notin \mathbf{Le}(T)$ and from Lemma 4.2.5, we deduce that $x \in \mathbf{Le}(T \cup \{y\})$. However, $T \cup \{y\} = M$ and we obtain a contradiction to the hypothesis. This shows that $M \cup \{x\}$ is a linearly independent subset.

As we see in Corollary 4.2.12 whenever $0 \neq x \in A$, the set $\{x\}$ is linearly independent. Then if $y \notin \mathbf{Le}(\{x\})$, we have $\{x, y\}$ is linearly independent and so on. We next define another fundamental notion, that of a basis.

4.2.9. Definition. *Let F be a field and let A be a vector space over F.*

(i) *A nonempty subset M of A is called a basis if it is linearly independent and $\mathbf{Le}(M) = A$.*
(ii) *A subset M of A is called a minimal generating subset for A if $\mathbf{Le}(M) = A$ but $\mathbf{Le}(S) \neq A$ for every proper subset S of M.*
(iii) *A linearly independent subset M of A is called a maximal linearly independent subset if whenever S is a subset of A for which $M \subseteq S$ and $M \neq S$ then S is not linearly independent.*

4.2.10. Theorem. *Let F be a field, let A be a vector space over F and let M be a subset of A. The following are equivalent:*

(i) *M is a basis of A.*
(ii) *M is a maximal linearly independent subset of A.*
(iii) *M is a minimal generating subset for A.*

Proof.
(i) $\Longrightarrow$ (ii) Let M be a basis of A. Then, by definition, M is a linearly independent subset. If S is a subset properly containing M then $S \backslash M$ is nonempty. Let $x \in S \backslash M$. Since M is a basis, $\mathbf{Le}(M) = A$, so $x \in \mathbf{Le}(M)$. Since $x \notin M$, we have $M \subseteq S \backslash \{x\}$. By Corollary 4.2.2, $\mathbf{Le}(M) \leq \mathbf{Le}(S \backslash \{x\})$, so $x \in \mathbf{Le}(S \backslash \{x\})$, from the definition, which shows that S is linearly dependent. Hence M is a maximal linearly independent subset of A.

(ii) $\Longrightarrow$ (i) Let M be a maximal linearly independent subset of A. If we suppose that $\mathbf{Le}(M) \neq A$, then we can choose an element $u \notin \mathbf{Le}(M)$. Lemma 4.2.8 shows that then $M \cup \{u\}$ is a linearly independent subset, contradicting the choice of M. Thus, M is a basis of A.

(i) $\Longrightarrow$ (iii) Let M be a basis of A so that M is a generating set for A. Let T be a proper subset of M. Then $M \backslash T$ is nonempty, and we let $v \in M \backslash T$. Then $T \subseteq M \backslash \{v\}$ and, since M is linearly independent, $v \notin \mathbf{Le}(M \backslash \{v\})$. By Corollary 4.2.4, $\mathbf{Le}(T) \leq \mathbf{Le}(M \backslash \{v\})$, so that $v \notin \mathbf{Le}(T)$. Thus, $\mathbf{Le}(T) \neq A$, so M is a minimal generating subset for A.

(iii) $\Longrightarrow$ (i) Let M be a minimal generating subset for A and suppose that M is not linearly independent. Then, there exists an element $w \in M$ such that $w \in \mathbf{Le}(M \backslash \{w\})$. By Corollary 4.2.4, $M \backslash \{w\} \subseteq \mathbf{Le}(M \backslash \{w\})$, so that $M = (M \backslash \{w\}) \cup \{w\} \subseteq \mathbf{Le}(M \backslash \{w\})$. Again using Corollary 4.2.4, we deduce that $A = \mathbf{Le}(M) \leq \mathbf{Le}(\mathbf{Le}(M \backslash \{w\})) = \mathbf{Le}(M \backslash \{w\})$. Thus, $M \backslash \{w\}$ is a generating set contrary to the definition of M. This contradiction shows that M is linearly independent and hence is a basis for A.

Theorem 4.2.10 gives us several ways to characterize bases but does not answer the question of whether a basis exists or not. The answer to the question of existence is yes, all vector spaces do have a basis, but to prove this is beyond the scope of this book, since it requires a rather advanced axiom of set theory known as Zorn's Lemma.

For finitely generated vector spaces, such deep results are not needed and it is those on which we concentrate.

4.2.11. Theorem. *Let F be a field, let A be a vector space over F and let M be a finite subset of A. Suppose that $A = \mathbf{Le}(M)$. If L is a linearly independent subset of M, then M contains a subset K such that $L \cap K = \emptyset$ and $L \cup K$ is a basis of A.*

Proof. Let

$$\mathfrak{S} = \{X | X \subseteq M, L \cap X = \emptyset \text{ and } L \cup X \text{ is linearly independent}\}.$$

Clearly $\emptyset \in \mathfrak{S}$, so that $\mathfrak{S}$ is not empty. Since M is finite, $\mathfrak{S}$ is also finite and it follows that $\mathfrak{S}$ contains a subset K of M with the largest number of elements. The subset $L \cup K$ is linearly independent, by definition. Suppose, for a contradiction that $\mathbf{Le}(L \cup K) \neq A$. If we suppose that $M \subseteq \mathbf{Le}(L \cup K)$ then, by Corollary 4.2.4,

$$A = \mathbf{Le}(M) \leq \mathbf{Le}(\mathbf{Le}(L \cup K)) = \mathbf{Le}(L \cup K),$$

and we obtain a contradiction with our assumption concerning $L \cup K$. Consequently, M is not a subset of $\mathbf{Le}(L \cup K)$ so we can choose $b \in M$ such that $b \notin \mathbf{Le}(L \cup K)$. Let $T = K \cup \{b\}$. Lemma 4.2.8 proves that $L \cup K \cup \{b\} = L \cup T$ is linearly independent. The inclusion $L \cup K \subseteq \mathbf{Le}(L \cup K)$ shows that $b \notin (L \cup K)$, so that

$$L \cap T = L \cap (K \cup \{b\}) = (L \cap K) \cup (L \cap \{b\}) = \emptyset.$$

It follows that $T \in \mathfrak{S}$. However, $|T| = |K| + 1$, and we obtain a contradiction with the choice of K which shows that $\mathbf{Le}(L \cup K) = A$ and therefore, $L \cup K$ is a basis of A.

4.2.12. Corollary. *Let F be a field, let A be a vector space over F, and let M be a finite subset of A. Suppose that $A = \mathbf{Le}(M)$. Then for each nonzero element a, there exists a finite basis of A containing a.*

Proof. We will show that $\{a\}$ is linearly independent and then deduce the result from Theorem 4.2.11. Indeed, if $\alpha a = 0_A$, where $0 \neq \alpha \in F$, then $\alpha^{-1} \in F$ exists and we have

$$0_A = \alpha^{-1} 0_A = \alpha^{-1}(\alpha a) = (\alpha^{-1}\alpha)a = ea = a.$$

This is a contradiction to the choice of a, which shows that $\alpha = 0_F$. By Proposition 4.2.7, $\{a\}$ is linearly independent and the result follows.

4.2.13. Corollary. *Let F be a field, let A be a vector space over F and let M be a finite subset of A such that $A = \mathbf{Le}(M)$. Then M contains a basis of A.*

Proof. Indeed, M is nonempty and therefore, M contains a nonzero element a. As stated above, the subset $\{a\}$ is linearly independent and Theorem 4.2.11 implies the result.

The following theorem is central in linear algebra. It tells us that the number of elements in a basis of a vector space is an invariant of the space.

4.2.14. Theorem. *Let F be a field and let A be a vector space over F. Suppose that A has a finite basis B. If B_1 is another basis of A, then B_1 is also finite, and moreover, $|B| = |B_1|$.*

Proof. Let $\mathbf{m}(B) = |B| - |B \cap B_1|$. We will use induction on $\mathbf{m}(B)$. If $\mathbf{m}(B) = 0$, then $|B| = |B \cap B_1|$. It follows that $B = B \cap B_1$ so $B \subseteq B_1$. However, by Theorem 4.2.10, every basis is a maximal linearly independent subset, so in this case, $B = B_1$.

Suppose now that $\mathbf{m}(B) = t > 0$ and inductively that, for each basis X of some vector space, with the property that $\mathbf{m}(X) < t$, we have already proved that X is finite and that $|X|$ is invariant. Assume that $B = \{a_1, \dots, a_n\}$ and let $k = n - t$. By renumbering the elements $a_1, \dots, a_n$, if necessary we may suppose that $B \cap B_1 = \{a_1, \dots, a_k\}$. Let $B_2 = B \backslash \{a_{k+1}\}$ and note that, by Theorem 4.2.10, $\mathbf{Le}(B_2) \neq A$. Suppose first that $B_1 \subseteq \mathbf{Le}(B_2)$. Then Corollary 4.2.4 implies that

$$A = \mathbf{Le}(B_1) \leq \mathbf{Le}(\mathbf{Le}(B_2)) = \mathbf{Le}(B_2),$$

a contradiction which shows that $\mathbf{Le}(B_2)$ does not contain B_1. Hence, there exists an element $x \in B_1$ such that $x \notin \mathbf{Le}(B_2)$. Since $B = B_2 \cup \{a_{k+1}\}$, we have $x \in \mathbf{Le}(B_2 \cup \{a_{k+1}\})$, but $x \notin \mathbf{Le}(B_2)$. By Lemma 4.2.5, we deduce that $a_{k+1} \in \mathbf{Le}(B_2 \cup \{x\})$. Now $B_2 \subseteq \mathbf{Le}(B_2 \cup \{x\})$ and $a_{k+1} \in \mathbf{Le}(B_2 \cup \{x\})$ so $B \subseteq \mathbf{Le}(B_2 \cup \{x\})$. Hence

$$A = \mathbf{Le}(B) \leq \mathbf{Le}(\mathbf{Le}(B_2 \cup \{x\})) = \mathbf{Le}(B_2 \cup \{x\}).$$

Since $x \notin \mathbf{Le}(B_2)$, Lemma 4.2.8 shows that $B_2 \cup \{x\}$ is a linearly independent subset. This implies that $B_2 \cup \{x\}$ is a basis for A. Also, $(B_2 \cup \{x\}) \cap B_1 = \{a_1, \dots, a_k, x\}$, so that

$$\mathbf{m}(B_2 \cup \{x\}) = |B_2 \cup \{x\}| - |(B_2 \cup \{x\}) \cap B_1| = n - (k+1) = t - 1.$$

By the induction hypothesis $|B_2 \cup \{x\}| = |B_1|$. However, $|B_2 \cup \{x\}| = |B|$, so that $|B| = |B_1|$ which proves the result.

Corollary 4.2.13 and Theorem 4.2.14 show that if A is a finitely generated vector space, then A has a finite basis B and that each basis of A is finite, of order equal to the order of B. In other words, the number of elements in every basis is an invariant of the vector space A.

4.2.15. Definition. *Let A be a finitely generated nonzero vector space over a field F. The number of elements in an arbitrary basis of A is called the dimension of A and will be denoted by* $\mathbf{dim}_F(A)$. *If A is the zero space, then its dimension is defined to be zero.*

From this moment, instead of the term "finitely generated vector space," we will use the term *finite-dimensional vector space*. Thus, a finite-dimensional vector space is a vector space with a finite basis.

4.2.16. Proposition. *Let A be a finite-dimensional vector space over a field F. Suppose that $\{a_1, \dots, a_n\}$ is a basis of A and let x be an arbitrary element of A. Then*

$$x = \lambda_1 a_1 + \cdots + \lambda_n a_n = \sum_{1 \le j \le n} \lambda_j a_j$$

for certain elements $\lambda_1, \dots, \lambda_n \in F$. Moreover, this representation is unique.

Proof. Since a basis is a subset of generators for A, the latter is the linear envelope of the subset $\{a_1, \dots, a_n\}$. By Proposition 4.2.3, every element of A is a linear combination of the elements $a_1, \dots, a_n$, so that $x = \lambda_1 a_1 + \cdots + \lambda_n a_n$ for certain elements $\lambda_1, \dots, \lambda_n \in F$. Next suppose also that $x = \mu_1 a_1 + \cdots + \mu_n a_n$, where $\mu_1, \dots, \mu_n \in F$. We have

$$\lambda_1 a_1 + \cdots + \lambda_n a_n = x = \mu_1 a_1 + \cdots + \mu_n a_n.$$

It follows that

$$(\lambda_1 - \mu_1)a_1 + \cdots + (\lambda_n - \mu_n)a_n = 0_A.$$

Since a basis is linearly independent, Proposition 4.2.7 implies that

$$\lambda_1 - \mu_1 = \cdots = \lambda_n - \mu_n = 0_F,$$

so

$$\lambda_1 = \mu_1, \quad \lambda_2 = \mu_2, \dots, \lambda_n = \mu_n,$$

which proves the proposition.

We next identify the elements of a vector space with a row vector using the idea of coordinates.

4.2.17. Definition. *Let A be a finite-dimensional vector space over a field F and let $\{a_1, \dots, a_n\}$ be a basis of A. If*

$$x = \lambda_1 a_1 + \cdots + \lambda_n a_n = \sum_{1 \le j \le n} \lambda_j a_j$$

is a representation of the element $x \in A$, then the (unique) elements $\lambda_1, \ldots, \lambda_n \in F$ are called the coordinates of x relative to the basis $\{a_1, \ldots, a_n\}$.

We must stress that usually a vector space can have more than one basis and that indeed, we will regard the basis $\{a_1, \ldots, a_n\}$ and the basis $\{b_1, \ldots, b_n\}$ as different, if the ordered n-tuples $(a_1, \ldots, a_n)$ and $(b_1, \ldots, b_n)$ are different. In this sense, for example, the basis $\{a_1, a_2, \ldots, a_n\}$ and the basis $\{a_2, a_1, a_3, \ldots, a_n\}$ are different.

Let A be a finite-dimensional vector space over a field F and let $\{a_1, \ldots, a_n\}$ and $\{b_1, \ldots, b_n\}$ be bases of A. We write

$$\begin{aligned} b_1 &= \iota_{11} a_1 + \iota_{21} a_2 + \cdots + \iota_{n1} a_n \\ b_2 &= \iota_{12} a_1 + \iota_{22} a_2 + \cdots + \iota_{n2} a_n \\ &\vdots \\ b_n &= \iota_{1n} a_1 + \iota_{2n} a_2 + \cdots + \iota_{nn} a_n. \end{aligned}$$

4.2.18. Definition. *Let A be a finite-dimensional vector space over a field F, and let $\{a_1, \ldots, a_n\}$ and $\{b_1, \ldots, b_n\}$ be two bases of A. Furthermore, let $b_k = \sum_{1 \le j \le n} \iota_{jk} a_j$. The matrix*

$$\begin{pmatrix} \iota_{11} & \iota_{21} & \cdots & \iota_{n1} \\ \iota_{12} & \iota_{22} & \cdots & \iota_{n2} \\ \vdots & \vdots & \ddots & \vdots \\ \iota_{1n} & \iota_{2n} & \cdots & \iota_{nn} \end{pmatrix}$$

is called the transition matrix from the first basis to the second basis.

Thus, in transitioning from the old basis of a_i's to the new basis of b_i's, we write the new basis in terms of the old and use the coefficients obtained to form the rows of the transition matrix.

Let A be a finite-dimensional vector space over F, let $\{a_1, \ldots, a_n\}$, $\{b_1, \ldots, b_n\}$, and $\{c_1, \ldots, c_n\}$ be bases of A. Let $T = [\iota_{jk}] \in \mathbf{M}_n(F)$ be the transition matrix from the first basis to the second basis, let $R = [\rho_{jk}] \in \mathbf{M}_n(F)$ be the transition matrix from the second basis to the third one and let $S = [\sigma_{jk}] \in \mathbf{M}_n(F)$ be the transition matrix from the first basis to the third one. Then, we have $c_k = \sum_{1 \le j \le n} \sigma_{jk} a_j$. On the other hand,

$$c_k = \sum_{1 \le m \le n} \rho_{mk} b_m = \sum_{1 \le m \le n} \rho_{mk} \left(\sum_{1 \le j \le n} \iota_{jm} a_j \right)$$

$$= \sum_{1\le m\le n}\sum_{1\le j\le n} \rho_{mk}\iota_{jm}a_j = \sum_{1\le j\le n}\left(\sum_{1\le m\le n}\iota_{jm}\rho_{mk}\right)a_j.$$

Using Proposition 4.2.16, we deduce that $\sigma_{jk} = \sum_{1\le m\le n}\iota_{jm}\rho_{mk}$, for $1 \le j, k \le n$, which shows that $S = TR$. In particular, the following corollary holds.

4.2.19. Corollary. *The transition matrix from one basis to another is nonsingular.*

Proof. Indeed put $c_j = a_j$ for all j, where $1 \le j \le n$. Then clearly $S = I$ is the identity matrix, and we have $TR = I$ from which it follows that T is nonsingular.

Now, we will find out how the coordinates of an element change during a transition from one basis to another.

Let $\{a_1, \dots, a_n\}$ and $\{b_1, \dots, b_n\}$ be bases of A and let $T = [\iota_{jk}] \in \mathbf{M}_n(F)$ be the transition matrix from the first basis to the second one. For an arbitrary element x of A, we have $x = \sum_{1\le j\le n}\lambda_j a_j$ and $x = \sum_{1\le j\le n}\xi_j b_j$, where $\lambda_1, \dots, \lambda_n$ are the coordinates relative to $\{a_1, \dots, a_n\}$ and $\xi_1, \dots, \xi_n$ are the coordinates relative to $\{b_1, \dots, b_n\}$. We have

$$x = \sum_{1\le k\le n}\xi_k b_k = \sum_{1\le k\le n}\xi_k\left(\sum_{1\le j\le n}\iota_{jk}a_j\right) = \sum_{1\le k\le n}\sum_{1\le j\le n}\xi_k\iota_{jk}a_j$$

$$= \sum_{1\le j\le n}\left(\sum_{1\le k\le n}\iota_{jk}\xi_k\right)a_j.$$

By Proposition 4.2.16, $\lambda_j = \sum_{1\le j\le n}\iota_{jk}\xi_k$, and we arrive at the matrix equation

$$\begin{pmatrix}\lambda_1\\ \lambda_2\\ \cdots\\ \lambda_n\end{pmatrix} = \begin{pmatrix}\iota_{11} & \iota_{12} & \cdots & \iota_{1n}\\ \iota_{21} & \iota_{22} & \cdots & \iota_{2n}\\ \vdots & \vdots & \ddots & \vdots\\ \iota_{n1} & \iota_{n2} & \cdots & \iota_{nn}\end{pmatrix}\begin{pmatrix}\xi_1\\ \xi_2\\ \vdots\\ \xi_n\end{pmatrix}.$$

We note that in this equation, the coefficients are being written as a column vector and in this case, we could express the equation in the matrix form $\mathbf{a} = T^t\mathbf{b}$, where $\mathbf{a}, \mathbf{b}$ represent the column vectors

$$\begin{pmatrix}\lambda_1\\ \lambda_2\\ \vdots\\ \lambda_n\end{pmatrix} \text{ and } \begin{pmatrix}\xi_1\\ \xi_2\\ \vdots\\ \xi_n\end{pmatrix},$$

respectively. Of course, we could also represent this equation using a row vector for the vector of coefficients in which case the equation becomes $\mathbf{a}^t = \mathbf{b}^t T$.

We next consider the subspaces of finite-dimensional vector spaces. As we would intuitively suspect, the dimension of a subspace cannot be larger than the dimension of the original space.

4.2.20. Theorem. *Let A be a finite-dimensional vector space over a field F and let B be a subspace of A. Then B is finite dimensional and $\mathbf{dim}_F(B) \leq \mathbf{dim}_F(A)$. Furthermore, $\mathbf{dim}_F(B) = \mathbf{dim}_F(A)$ if and only if $B = A$.*

Proof. If $B = \{0_A\}$, then $\mathbf{dim}_F(B) = 0 \leq \mathbf{dim}_F(A)$. Therefore, we will assume that B is a nonzero subspace. Suppose that B does not have a finite basis and let $0_A \neq a_1 \in B$. As we have already seen $\{a_1\}$ is linearly independent so, by our assumption, $\mathbf{Le}(\{a_1\}) \neq B$. Therefore, we can choose an element $a_2 \notin \mathbf{Le}(\{a_1\})$. By Lemma 4.2.8, the subset $\{a_1, a_2\}$ is linearly independent, and again, $\mathbf{Le}(\{a_1, a_2\}) \neq B$. In this way, using the same argument, we construct an infinite subset $\{a_n | n \in \mathbb{N}\}$ such that for each n, the subset $\{a_1, \dots, a_n\}$ is linearly independent and $\mathbf{Le}(\{a_1, \dots, a_n\}) \neq B$, for each $n \in \mathbb{N}$. If S is a finite subset of $\{a_n | n \in \mathbb{N}\}$, then there exists a positive integer k such that $S \subseteq \{a_1, \dots, a_k\}$. By Proposition 4.2.7, the subset S is linearly independent and again using Proposition 4.2.7, we see that the set $\{a_n | n \in \mathbb{N}\}$ is also linearly independent. Since the space A is finite dimensional, Theorem 4.2.11 shows that there exists a finite linearly independent subset K such that $\{a_n | n \in \mathbb{N}\} \cup K$ is a basis of A. However, Theorem 4.2.14 shows that each basis of A is finite. This contradiction shows that B has a finite basis $\{b_1, \dots, b_t\}$. Theorem 4.2.11 shows that the subset $\{b_1, \dots, b_t\}$ can be extended to a basis of the entire space A and so it follows that $\mathbf{dim}_F(B) \leq \mathbf{dim}_F(A)$.

Finally, suppose that $\mathbf{dim}_F(B) = \mathbf{dim}_F(A) = n$ and let $\{b_1, \dots, b_n\}$ be a basis of B. If $B \neq A$ then, since $B = \mathbf{Le}(\{b_1, \dots, b_n\})$, Lemma 4.2.8 implies that $\{b_1, \dots, b_n, c\}$ is linearly independent for each element $c \notin B$. Theorem 4.2.11 shows that the subset $\{b_1, \dots, b_n, c\}$ can be extended to a basis U of the entire space A. Consequently, the space A has a basis which contains at least $n + 1$ elements and we obtain a contradiction with Theorem 4.2.14. This proves that $B = A$.

Next we consider the question of the dimension of direct products.

4.2.21. Lemma. *Let A be a vector space over a field F and let C, B be subspaces of A. Suppose that $B \cap C = \{0_A\}$. If M (respectively S) is a linearly independent subset of B (respectively C), then $M \cup S$ is linearly independent.*

Proof. By Proposition 4.2.7, it is sufficient to prove that every finite subset of $M \cup S$ is linearly independent. If K is a finite subset of $M \cup S$, then $K = M_1 \cup S_1$, where M_1 (respectively S_1) is a finite subset of M (respectively S). Therefore, we may assume that the subsets M and S are finite. Let $M = \{a_1, \ldots, a_n\}$ and $S = \{b_1, \ldots, b_k\}$. Choose $\alpha_1, \ldots, \alpha_n, \beta_1, \ldots, \beta_k \in F$ such that

$$\alpha_1 a_1 + \cdots + \alpha_n a_n + \beta_1 b_1 + \cdots + \beta_k b_k = 0_A.$$

Then $\alpha_1 a_1 + \cdots + \alpha_n a_n = (-\beta_1)b_1 + \cdots + (-\beta_k)b_k$, where $\alpha_1 a_1 + \cdots + \alpha_n a_n$ is an element of the subspace B while $(-\beta_1)b_1 + \cdots + (-\beta_k)b_k$ is an element of the subspace C. Since $B \cap C = \{0_A\}$, it follows that

$$\alpha_1 a_1 + \cdots + \alpha_n a_n = (-\beta_1)b_1 + \cdots + (-\beta_k)b_k = 0_A.$$

The subsets M, S are linearly independent, so Proposition 4.2.7 implies that

$$\alpha_1 = \cdots = \alpha_n = \beta_1 = \cdots = \beta_k = 0_F.$$

Again by Proposition 4.2.7, the subset $M \cup S$ is linearly independent.

4.2.22. Proposition. *Let A be a vector space over a field F, let $A_1, \ldots, A_n$ be subspaces of A and let $M_1, \ldots, M_n$ be linearly independent subsets of $A_1, \ldots, A_n$, respectively. If $C = A_1 + \cdots + A_n$ is the internal direct sum of $A_1, \ldots, A_n$, then $M_1 \cup \cdots \cup M_n$ is a linearly independent subset.*

Proof. We use induction on n. If $n = 2$, the assertion follows from Lemma 4.2.21. Suppose inductively that we have already proved that $M_1 \cup \cdots \cup M_{n-1}$ is linearly independent. By Proposition 4.1.14, $(A_1 + \cdots + A_{n-1}) \cap A_n = \{0_A\}$ and Lemma 4.2.21 applies again to give the result.

4.2.23. Corollary. *Let A be a vector space over a field F and let $A_1, \ldots, A_n$ be finite-dimensional subspaces of A. If $C = A_1 + \cdots + A_n$ is the internal direct sum of $A_1, \ldots, A_n$, then $\mathbf{dim}_F(C) = \mathbf{dim}_F(A_1) + \cdots + \mathbf{dim}_F(A_n)$.*

Proof. Let M_j be a basis of A_j, for $1 \leq j \leq n$. By Proposition 4.2.22, $M_1 \cup \cdots \cup M_n$ is linearly independent. Also, if $c \in C$, then $c = c_1 + \cdots + c_n$, where $c_j \in A_j$, for $1 \leq j \leq n$. Since M_j is a basis of A_j, c_j is a linear combination of the elements of M_j, for $1 \leq j \leq n$. It follows that c is a linear combination of the elements from $M_1 \cup \cdots \cup M_n$, so $M_1 \cup \cdots \cup M_n$ is a subset of generators for C. However, $M_1 \cup \cdots \cup M_n$ is linearly independent, so is a basis of the subspace C. The result now follows easily.

4.2.24. Definition. *Let A be a vector space over a field F and let B be a subspace of A. A subspace C is called a complement to B, if $A = B \oplus C$.*

The following assertion is very useful.

4.2.25. Proposition. *Let A be a finite-dimensional vector space over a field F. Then every subspace of A has a complement.*

Proof. Let B be an arbitrary subspace of A. By Theorem 4.2.20, B is finite dimensional so let $\{b_1, \dots, b_k\}$ be a basis for B. Since $M = \{b_1, \dots, b_k\}$ is linearly independent, Theorem 4.2.11 shows that M can be extended to a basis of the entire space A. Thus, there exists a finite subset $S = \{c_1, \dots, c_t\}$ such that $M \cup S$ is a basis of A and we set $C = \mathbf{Le}(S)$. Let a be an arbitrary element of A. By Proposition 4.2.16,

$$a = \beta_1 b_1 + \cdots + \beta_k b_k + \gamma_1 c_1 + \cdots + \gamma_t c_t$$

for certain elements $\beta_1, \dots, \beta_k, \gamma_1, \dots, \gamma_t \in F$. Clearly, $\beta_1 b_1 + \cdots + \beta_k b_k \in B$ and $\gamma_1 c_1 + \cdots + \gamma_t c_t \in C$, so that $a \in B + C$. It follows that $A = B + C$.

Next, let $y \in B \cap C$. Then $y = \lambda_1 b_1 + \cdots + \lambda_k b_k$, where $\lambda_1, \dots, \lambda_k \in F$. On the other hand, $y = \mu_1 c_1 + \cdots + \mu_t c_t$, where $\mu_1, \dots, \mu_t \in F$. We have

$$\lambda_1 b_1 + \cdots + \lambda_k b_k = y = \mu_1 c_1 + \cdots + \mu_t c_t$$

or

$$\lambda_1 b_1 + \cdots + \lambda_k b_k - \mu_1 c_1 - \cdots - \mu_t c_t = 0_A.$$

Since $\{b_1, \dots, b_k, c_1, \dots, c_t\}$ is a basis, Proposition 4.2.7 shows that

$$\lambda_1 = \cdots = \lambda_k = \mu_1 = \cdots = \mu_t = 0_F.$$

Consequently, $y = 0_A$, so that $B \cap C = \{0_A\}$ and it follows that $A = B \oplus C$.

We note that a subspace usually has more than one complement. For example, let A be the vector space, having basis $\{a_1, a_2\}$ and let B be the subspace generated by a_1. We observe that the subset $\{a_1, a_1 + a_2\}$ is also a basis of A. If C (respectively D) is the subspace generated by a_2 (respectively by $a_1 + a_2$), then C, D are complements to the subspace B.

Finally, we would like to consider some important examples of finite dimensional spaces and subspaces. First we consider the most important example for us; namely, the space F^n. Put

$$\begin{aligned}
\mathbf{e}_1 &= (e, 0_F, 0_F, \dots, 0_F, 0_F), \\
\mathbf{e}_2 &= (0_F, e, 0_F, \dots, 0_F, 0_F), \dots, \\
\mathbf{e}_j &= (0_F, 0_F, \dots, 0_F, \underbrace{e}_{j}, 0_F, \dots, 0_F, 0_F), \dots, \\
\mathbf{e}_{n-1} &= (0_F, 0_F, 0_F, \dots, 0_F, e, 0_F), \\
\mathbf{e}_n &= (0_F, 0_F, 0_F, \dots, 0_F, e).
\end{aligned}$$

These elements are linearly independent. For, let $\alpha_1, \dots, \alpha_n$ be elements of F such that $(0_F, 0_F, 0_F, \dots, 0_F, 0_F) = \alpha_1 \mathbf{e}_1 + \alpha_2 \mathbf{e}_2 + \cdots + \alpha_n \mathbf{e}_n$. We have

$$\begin{aligned}\alpha_1\mathbf{e}_1 + \cdots + \alpha_n\mathbf{e}_n &= \alpha_1(e, 0_F, \ldots, 0_F) + \alpha_2(0_F, e, 0_F, \ldots, 0_F) + \cdots \\ &\quad + \alpha_n(0_F, 0_F, \ldots, 0_F, e) \\ &= (\alpha_1, 0_F, \ldots, 0_F) + (0_F, \alpha_2, 0_F, \ldots, 0_F, 0_F) + \cdots \\ &\quad + (0_F, 0_F, 0_F, \ldots, 0_F, \alpha_n) = (\alpha_1, \alpha_2, \ldots, \alpha_n).\end{aligned}$$

It follows that $\alpha_1 = \alpha_2 = \cdots = \alpha_n = 0_F$, and by Proposition 4.2.7, the elements $\mathbf{e}_1, \ldots, \mathbf{e}_n$ are linearly independent. Furthermore, for an arbitrary element $(\gamma_1, \gamma_2, \ldots, \gamma_n)$ of F^n, we have

$$\begin{aligned}(\gamma_1, \ldots, \gamma_n) &= (\gamma_1, 0_F, \ldots, 0_F) + (0_F, \gamma_2, 0_F, \ldots, 0_F) + \cdots \\ &\quad + (0_F, 0_F, \ldots, 0_F, \gamma_n) \\ &= \gamma_1(e, 0_F, \ldots, 0_F) + \gamma_2(0_F, e, 0_F, \ldots, 0_F) + \cdots \\ &\quad + \gamma_n(0_F, \ldots, 0_F, e) \\ &= \gamma_1\mathbf{e}_1 + \gamma_2\mathbf{e}_2 + \cdots + \gamma_n\mathbf{e}_n.\end{aligned}$$

This proves that $\{\mathbf{e}_1, \ldots, \mathbf{e}_n\}$ generates the vector space F^n and is linearly independent. Consequently, this subset is a basis of F^n called the standard or canonical basis of F^n. Thus, the space F^n is finite dimensional and $\mathbf{dim}_F(F^n) = n$.

Now consider the space $F^{\mathbb{N}}$ and its elements

$$\mathbf{e}_j = (\upsilon_{jn})_{n\in\mathbb{N}}, \text{ where } \upsilon_{jj} = e \text{ and } \upsilon_{jn} = 0_F \text{ whenever } j \neq n.$$

As proved above, we can show that the subset $\{\mathbf{e}_j \mid 1 \leq j \leq k\}$ is linearly independent for every $k \in \mathbb{N}$. Proposition 4.2.7 implies that $\{\mathbf{e}_n \mid n \in \mathbb{N}\}$ is linearly independent. Hence, the vector space $F^{\mathbb{N}}$ contains an infinite linearly independent subset and therefore, cannot be finite dimensional. In fact, the arguments stated above allow us to prove that $\{\mathbf{e}_n \mid n \in \mathbb{N}\}$ is a basis for the subspace $F^{(\mathbb{N})}$, but not of $F^{\mathbb{N}}$ which has an uncountable basis.

The vector space $\mathbf{M}_{k\times n}(F)$ is also finite dimensional. Indeed, it is possible using arguments similar to those given above, to show that the subset $\{E_{tj} \mid 1 \leq t \leq k, 1 \leq j \leq n\}$ is a basis of this vector space called the standard or canonical basis of the vector space $\mathbf{M}_{k\times n}(F)$. Hence $\mathbf{dim}_F(\mathbf{M}_{k\times n}(F)) = kn$ and $\mathbf{dim}_F(\mathbf{M}_n(F)) = n^2$, in particular.

EXERCISE SET 4.2

Justify your work, providing a proof or counterexample where necessary.

4.2.1. Prove that the subset $\{(\alpha_{11}, \alpha_{12}, \alpha_{13}, \ldots, \alpha_{1n}), (0, \alpha_{22}, \alpha_{23}, \ldots, \alpha_{2n}), (0, 0, \alpha_{33}, \ldots, \alpha_{3n}), \ldots, (0, 0, 0, \ldots, \alpha_{kk}, \alpha_{k,k+1}, \ldots, \alpha_{kn})\}$ of the vector space $A = \mathbb{Q}^n$ is linearly independent if and only if the numbers α_{11}, $\alpha_{22}, \alpha_{33}, \ldots, \alpha_{kk}$ are nonzero.

4.2.2. Let $\{a_1, a_2, a_3, \ldots, a_m\}$ be a linearly independent subset of a vector space A. Is the subset $\{a_1, a_1 + a_2, a_2 + a_3, \ldots, a_{m-1} + a_m\}$ linearly independent?

4.2.3. Let A be a finite set and let $|A| = n$. On the Boolean $\mathfrak{B}(A)$, we introduce the operation of addition and the operation of scalar multiplication by elements of the field $\mathbb{F}_2 = \{0, 1\}$ using the following rules: $X + Y = (X \cup Y)\backslash(X \cap Y)$, $1X = X, 0X = \emptyset$. Prove that the Boolean $\mathfrak{B}(A)$ is a vector space under these operations. Prove that if $X_1 \subset X_2 \subset \cdots \subset X_n$ and $X_k \neq X_j$ where $k \neq j$, then $X_1, X_2, \ldots, X_n$ are linearly independent. Find a basis and the dimension of $\mathfrak{B}(A)$.

4.2.4. Let B be the subset of the vector space $\mathbf{M}_2(\mathbb{R})$ consisting of all matrices of the form $\begin{pmatrix} \alpha & 0 \\ 2\alpha & 3\alpha \end{pmatrix}$. Is B a subspace? If yes, find $\mathbf{dim}_{\mathbb{R}}(B)$.

4.2.5. Let $A = \mathbb{Q}^{22}$, $B = \{(\alpha_1, \alpha_2, \ldots, \alpha_{22}) \mid \alpha_1 + \alpha_2 + \cdots + \alpha_{22} = 0\}$. Is B a subspace? If yes, find $\mathbf{dim}_{\mathbb{R}}(B)$.

4.2.6. Let $A = \mathbb{R}^{221}$, $B = \{(\alpha_1, \alpha_2, \ldots, \alpha_{221}) \mid 2\alpha_1 = \alpha_{221}\}$. Is B a subspace? If yes, find $\mathbf{dim}_{\mathbb{R}}(B)$.

4.2.7. Let $A = \mathbb{Q}^{23}$, $B = \{(\alpha_1, \alpha_2, \ldots, \alpha_{23}) \mid \alpha_1^2 = \alpha_2, \alpha_2^2 = \alpha_3, \ldots, \alpha_{22}^2 = \alpha_{23}\}$. Is B a subspace? If yes, find $\mathbf{dim}_{\mathbb{R}}(B)$.

4.2.8. Give an example of a nonstandard basis in $\mathbf{M}_3(\mathbb{R})$.

4.2.9. Prove that the subset of all symmetric matrices is a subspace of the vector space $\mathbf{M}_{13}(\mathbb{R})$. Find a basis and the dimension of this subspace.

4.2.10. Is the set of all skew-symmetric matrices a subspace of the vector space $\mathbf{M}_{41}(\mathbb{R})$? If yes, find a basis and the dimension of this subspace.

4.2.11. Let $A = \mathbb{Q}^4$. Do the vectors $(1, 2, 3, 4), (0, 1, -1, 3), (1, 2, 4, 3), (-1, -1, -4, 1)$ form a basis of this space? If yes, find the transition matrix from the standard basis.

4.2.12. Is the subset $\left\{\begin{pmatrix} 1 & -1 \\ 0 & 0 \end{pmatrix}, \begin{pmatrix} 1 & 1 \\ 1 & 0 \end{pmatrix}, \begin{pmatrix} 1 & 1 \\ 0 & 0 \end{pmatrix}, \begin{pmatrix} 1 & 0 \\ 0 & 0 \end{pmatrix}\right\}$ a basis of the vector space $\mathbf{M}_2(\mathbb{Q})$? If yes, find the coordinates of the matrix $\begin{pmatrix} 2 & 5 \\ -2 & 0 \end{pmatrix}$ relative to this basis.

4.2.13. Is the subset $\left\{\begin{pmatrix} 1 & 0 \\ 0 & 0 \end{pmatrix}, \begin{pmatrix} 1 & 1 \\ 1 & 0 \end{pmatrix}, \begin{pmatrix} 1 & 1 \\ 0 & 0 \end{pmatrix}, \begin{pmatrix} 1 & -1 \\ 1 & 1 \end{pmatrix}\right\}$ a basis of the vector space $\mathbf{M}_2(\mathbb{Q})$?

If yes, find the coordinates of the matrix $\begin{pmatrix} 2 & 8 \\ -1 & 3 \end{pmatrix}$ relative to this basis.

4.2.14. In the vector space $\mathbf{M}_2(\mathbb{Q})$, find the transition matrix from the basis $\left\{\begin{pmatrix}1 & -1\\0 & 1\end{pmatrix}, \begin{pmatrix}1 & 1\\1 & 0\end{pmatrix}, \begin{pmatrix}1 & 1\\0 & 0\end{pmatrix}, \begin{pmatrix}1 & 0\\0 & 0\end{pmatrix}\right\}$ to the basis $\left\{\begin{pmatrix}0 & 0\\0 & 1\end{pmatrix}, \begin{pmatrix}0 & 0\\1 & 1\end{pmatrix}, \begin{pmatrix}0 & 1\\1 & 1\end{pmatrix}, \begin{pmatrix}1 & 1\\1 & 1\end{pmatrix}\right\}$.

4.2.15. In the vector space $\mathbb{R}^4$, find a basis containing the vector $(0, 1, 4, 3)$.

4.2.16. In the vector space $\mathbb{R}^4$, find a basis containing the vector $(2, 1, 1, 0)$.

4.2.17. Let $A = \mathbb{Q}^4$. Is the matrix $\begin{pmatrix}1 & 2 & 3 & 4\\0 & 1 & -1 & 3\\1 & 2 & 4 & 3\\-1 & -1 & 4 & 1\end{pmatrix}$ a transition matrix from the standard basis to another one? If yes, find the new basis.

4.2.18. Let $A = \mathbb{Q}^3$, let B be the linear envelope of the subset $\{(1, 2, 1), (1, 1, -1), (1, 3, 3)\}$ and let C be the linear envelope of the subset $\{(2, 3, -1), (1, 1, 2), (1, 1, -3)\}$. Find the bases of the sum and of the intersection of the spaces B and C.

4.2.19. Let $A = \mathbf{NT}_{19}(\mathbb{Q})$ be the subspace of all zero-triangular matrices. Find a basis and the dimension of this subspace.

4.3 THE RANK OF A MATRIX

Matrices are very important tools in linear algebra. In this section, we consider a concept known as the rank of a matrix. This concept is based on the dimension of a space.

4.3.1. Definition. *Let A be a vector space over a field F and let M be a finite subset of A. Then $\mathbf{dim}_F(\mathbf{Le}(M))$ is called the rank of the subset M and is denoted by $\mathbf{rank}(M)$.*

From Corollary 4.2.13, we know that M contains some basis R of the subspace $\mathbf{Le}(M)$. By Theorem 4.2.10, R is a maximal linearly independent subset of $\mathbf{Le}(M)$ and hence R is also a maximal linearly independent subset of M. Thus, we obtain the following characterization of the rank of a subset.

4.3.2. Proposition. *Let F be a field and let A be a vector space over F. Suppose that M is a finite subset of A. Then $\mathbf{rank}(M)$ is equal to the number of elements in every maximal linearly independent subset of M.*

Proof. Let $S = \{a_1, \dots, a_k\}$ be an arbitrary maximal linearly independent subset of M. Clearly, $\mathbf{Le}(M)$ is finite dimensional and we claim that S is indeed a basis of it, from which the result will follow by the definition. If $x \in M \backslash S$, then $S \cup \{x\}$ is linearly dependent so by Proposition 4.2.7, there are scalars $\lambda_1, \dots, \lambda_k, \beta \in F$, not all 0_F

such that

$$\lambda_1 a_1 + \cdots + \lambda_k a_k + \beta x = 0_A. \tag{4.1}$$

If $\beta = 0_F$, then the fact that S is linearly independent and Proposition 4.2.7 imply that $\lambda_1 = \lambda_2 = \cdots = \lambda_k = 0_F$, contrary to the choice of the scalars $\lambda_1, \ldots, \lambda_k, \beta$. Thus, $\beta \neq 0_F$ so, multiplying Equation 4.1 by β^{-1} gives $x = -\beta^{-1}\lambda_1 a_1 - \cdots - \beta^{-1}\lambda_k a_k$. Thus, every element of M is a linear combination of the elements of S. Since every element of $\mathbf{Le}(M)$ is a linear combination of the elements of M, it follows that every element of $\mathbf{Le}(M)$ is a linear combination of the elements of S so that S is indeed a basis of $\mathbf{Le}(M)$ as required.

The following corollary is immediate.

4.3.3. Corollary. *Let F be a field and let A be a vector space over F. Suppose that M is a finite subset of A. Then* $\mathbf{rank}(M) = |M|$ *if and only if M is linearly independent.*

There is a very nice way of applying these results to matrices. Let F be a field and consider the $k \times n$ matrix $A \in M_{k \times n}(F)$, where

$$A = \begin{pmatrix} \alpha_{11} & \alpha_{12} & \alpha_{13} & \ldots & \alpha_{1,n-1} & \alpha_{1n} \\ \alpha_{21} & \alpha_{22} & \alpha_{23} & \ldots & \alpha_{2,n-1} & \alpha_{2n} \\ \vdots & \vdots & \vdots & \vdots & \vdots & \vdots \\ \alpha_{k1} & \alpha_{k2} & \alpha_{k3} & \ldots & \alpha_{k,n-1} & \alpha_{kn} \end{pmatrix}.$$

Every row of this matrix is an n-tuple consisting of elements of F, so, we may consider each row as an element of the vector space F^n. Similarly, every column of this matrix is a k-tuple, with entries in F and so each column can be considered as an element of the vector space F^k.

4.3.4. Definition. *Let F be a field and let $A = [\alpha_{tj}]$ be a $k \times n$ matrix over the field F. Let $\mathbf{R}(A)$ (respectively $\mathbf{C}(A)$) denote the set of all rows (respectively columns) of the matrix A. Then $\mathbf{R}(A)$ (respectively $\mathbf{C}(A)$) is a subset of the vector space F^n (respectively F^k). The numbers* $\mathbf{rank}(\mathbf{R}(A))$ *and* $\mathbf{rank}(C(A))$ *are called the row rank and the column rank of A, respectively.*

We are going to prove that these ranks coincide and exhibit a method for computing them.

4.3.5. Theorem. *Let F be a field and let $A = [\alpha_{tj}]$ be a $k \times n$ matrix over the field F. Suppose that t is a positive integer satisfying the conditions:*

(i) *the matrix A has a nonzero minor of degree t;*
(ii) *each minor of degree $s > t$ is equal to 0_F.*

Then $\mathbf{rank}(\mathbf{C}(A)) = t$.

Proof. We suppose first that $\mathbf{minor}\{1, 2, \ldots, t; 1, 2, \ldots, t\}$ is nonzero and that the corresponding cofactor is denoted by Δ. As we will see later, this will not affect the generality of the result but will significantly simplify the notation. Let $\mathfrak{a}_j$ denote the jth column of the matrix A and consider the matrix

$$B = \begin{pmatrix} \alpha_{11} & \cdots & \alpha_{1t} & \alpha_{1j} \\ \alpha_{21} & \cdots & \alpha_{2t} & \alpha_{2j} \\ \vdots & \vdots & \vdots & \vdots \\ \alpha_{t1} & \cdots & \alpha_{tt} & \alpha_{tj} \\ \alpha_{m1} & \cdots & \alpha_{mt} & \alpha_{mj} \end{pmatrix},$$

where $k+1 \leq j \leq n$ and $1 \leq m \leq k$. If $m \leq t$ then the matrix B has two identical rows, those numbered m and $t+1$, so, by Corollary 2.3.8, $\mathbf{det}(B) = 0_F$. If $m > t$, then

$$\mathbf{det}(B) = \mathbf{minor}\ \{1, 2, \ldots, t, m; 1, 2, \ldots, t, j\}.$$

This minor has degree $t+1$, and by hypothesis (ii), $\mathbf{det}(B) = 0_F$ so, in any case, $\mathbf{det}(B) = 0_F$. Using Theorem 2.4.3, we may expand the determinant of B about the last row. The cofactor corresponding to α_{mj} is $\pm\mathbf{minor}\{1, 2, \ldots, t; 1, 2, \ldots, t\} = \Delta$, whereas the minor corresponding to α_{ms} is the determinant of

$$B_s = \begin{pmatrix} \alpha_{11} & \cdots & \alpha_{1,s-1} & \alpha_{1,s+1} & \cdots & \alpha_{1t} & \alpha_{1j} \\ \alpha_{21} & \cdots & \alpha_{2,s-1} & \alpha_{2,s+1} & \cdots & \alpha_{2t} & \alpha_{2j} \\ \vdots & \vdots & \vdots & \vdots & \vdots & \vdots & \vdots \\ \alpha_{t1} & \cdots & \alpha_{t,s-1} & \alpha_{t,s+1} & \cdots & \alpha_{tt} & \alpha_{tj} \end{pmatrix}.$$

We denote the corresponding cofactor by Δ_s. As we can see, the elements of the row numbered m do not belong to the matrix B_s. Therefore, $\mathbf{det}(B_s)$ and consequently, Δ_s is independent of m. By Theorem 2.4.3,

$$\alpha_{m1}\Delta_1 + \cdots + \alpha_{mt}\Delta_t + \alpha_{mj}\Delta = 0_F.$$

Since Δ is a nonzero element of the field F, Δ has a multiplicative inverse, Δ^{-1}, so we have

$$\alpha_{mj} = \Delta^{-1}(-\Delta_1)\alpha_{m1} + \cdots + \Delta^{-1}(-\Delta_t)\alpha_{mt}.$$

Since this equation is valid for each m, where $1 \leq m \leq k$, we obtain the following linear combination of the columns considered as elements of the vector space F^k:

$$\mathfrak{a}_j = \Delta^{-1}(-\Delta_1)\mathfrak{a}_1 + \cdots + \Delta^{-1}(-\Delta_t)\mathfrak{a}_t, \text{ where } k+1 \leq j \leq n.$$

It follows that $\mathbf{Le}(\mathbf{C}(A))$ is generated by the columns $\mathfrak{a}_1, \ldots, \mathfrak{a}_t$. We next show that the set $\{\mathfrak{a}_1, \ldots, \mathfrak{a}_t\}$ is linearly independent, which implies that it is a basis of $\mathbf{Le}(\mathbf{C}(A))$. Suppose that the contrary is true. Then there exists an index q

such that the column $\mathfrak{a}_q$ is a linear combination of the other columns, say $\mathfrak{a}_q = \sum_{1\leq j\leq t, q\neq j} \lambda_j \mathfrak{a}_j$. Let $\hat{\mathfrak{a}}_j$ denote the column

$$\begin{pmatrix} \alpha_{1j} \\ \alpha_{2j} \\ \vdots \\ \alpha_{tj} \end{pmatrix}.$$

For these "shortened" columns, the same linear combination $\hat{\mathfrak{a}}_q = \sum_{1\leq j\leq t, q\neq j} \lambda_j \hat{\mathfrak{a}}_j$ is true. However, Corollary 2.3.10 shows that in this case, the matrix

$$\begin{pmatrix} \alpha_{11} & \cdots & \alpha_{1t} \\ \alpha_{21} & \cdots & \alpha_{2t} \\ \vdots & \vdots & \vdots \\ \alpha_{t1} & \cdots & \alpha_{tt} \end{pmatrix}$$

has determinant zero, a contradiction which proves that $\{\mathfrak{a}_1, \ldots, \mathfrak{a}_t\}$ is a basis of $\mathbf{Le}(\mathbf{C}(A))$. Thus, $\mathbf{dim}_F \mathbf{Le}(\mathbf{C}(A)) = t$, which proves the result.

Computation of the rank of a matrix appears to require the computation of a possibly very large (but finite!) number of minors of the matrix. However, if we look carefully at the proof of the previous theorem, we see that we did not use the fact that all minors of degree $s > t$ are equal to zero. We actually used only the fact that the minors of degree s *including* the given nonzero minor of degree t are equal to zero. We may infer from this fact that t is the number of columns in a maximal linearly independent subset of the set of all columns. This fact implies that all other minors of degree $s > t$ are equal to zero.

4.3.6. Corollary. *Let F be a field and let $A = [\alpha_{tj}]$ be a $k \times n$ matrix over the field F. Then, the row rank of this matrix coincides with its column rank.*

Proof. Suppose that the column rank of A is denoted by w. By Theorem 4.3.5, there exists a nonzero minor

$$\Delta = \mathbf{minor}\{p(1), p(2), \ldots, p(w); j(1), j(2), \ldots, j(w)\}$$

of degree t. Let $A^t = [\beta_{ij}] \in \mathbf{M}_{n\times k}$ be the transpose of A, so $\beta_{ij} = \alpha_{ji}$. Then $\mathbf{R}(A^t)$ (respectively $\mathbf{C}(A^t)$) is the set of all columns (respectively rows) of the matrix A. Therefore, the column rank of A^t is equal to the row rank of A, and conversely. We will find the column rank of the matrix A^t. By Proposition 2.3.3, the minor of the matrix A^t corresponding to the rows numbered $j(1), j(2), \ldots, j(w)$ and the columns numbered $p(1), p(2), \ldots, p(w)$ is nonzero. Next, choose s arbitrary columns and rows in A^t, where $s > w$. We suppose that the chosen rows are rows $m(1), \ldots, m(s)$ and the chosen columns are $d(1), \ldots, d(s)$. By Proposition 2.3.3, the minor of the matrix A^t corresponding to these rows and columns is equal to the minor of the matrix A consisting of the rows numbered $d(1), \ldots, d(s)$ and columns numbered $m(1), \ldots, m(s)$, and therefore it is equal to zero. Hence, every minor of the matrix A^t of degree $s > w$ is equal to zero. By Theorem 4.3.5, the column rank of A^t is therefore w. Hence the row rank of A is also w, and in particular, the column and the rows ranks of A are equal.

Because of Corollary 4.3.6, we call the common value of the row rank and column rank of a matrix A simply the rank of A and denote it by $\mathbf{rank}(A)$. It will normally be clear what ranks we are talking about.

The rank of a matrix has important applications in solution of systems of linear equations. To see this, we consider the system of linear equations

$$\begin{aligned}
\alpha_{11}x_1 + \alpha_{12}x_2 + \cdots + \alpha_{1n}x_n &= \beta_1 \\
\alpha_{21}x_1 + \alpha_{22}x_2 + \cdots + \alpha_{2n}x_n &= \beta_2 \\
\vdots &= \vdots \\
\alpha_{k-1,1}x_1 + \alpha_{k-1,2}x_2 + \cdots + \alpha_{k-1,n}x_n &= \beta_{k-1} \\
\alpha_{k1}x_1 + \alpha_{k2}x_2 + \cdots + \alpha_{kn}x_n &= \beta_k.
\end{aligned} \tag{4.2}$$

The coefficients α_{tj}, for $1 \le t \le k, 1 \le j \le n$ and elements β_t, for $1 \le t \le k$ belong to F.

4.3.7. Definition. *An n-tuple $(\gamma_1, \ldots, \gamma_n)$ consisting of elements of a field F is called a solution of the system (Eq. 4.2) if every equation from Equation 4.2 becomes an identity after replacing the variables x_j by the corresponding elements γ_j, for $1 \le j \le n$, so $\sum_{1\le j\le n} \alpha_{tj}\gamma_j = \beta_t$ for all t, where $1 \le t \le k$.*

Note that the elements $\gamma_1, \ldots, \gamma_n$ form only one solution $(\gamma_1, \ldots, \gamma_n)$ of the given system, not n solutions. Also a system of linear equations need not have a solution. We next consider the question of the existence of a solution to such a system.

The matrix

$$A = \begin{pmatrix} \alpha_{11} & \alpha_{12} & \cdots & \alpha_{1,n-1} & \alpha_{1n} \\ \alpha_{21} & \alpha_{22} & \cdots & \alpha_{2,n-1} & \alpha_{2n} \\ \vdots & \vdots & \vdots & \vdots & \vdots \\ \alpha_{k1} & \alpha_{k2} & \cdots & \alpha_{k,n-1} & \alpha_{kn} \end{pmatrix}$$

consisting of the coefficients of the variables x_j, where $1 \le j \le n$, is called the coefficient matrix of the system (Eq. 4.2). The matrix

$$A^* = \begin{pmatrix} \alpha_{11} & \alpha_{12} & \cdots & \alpha_{1,n-1} & \alpha_{1n} & \beta_1 \\ \alpha_{21} & \alpha_{22} & \cdots & \alpha_{2,n-1} & \alpha_{2n} & \beta_2 \\ \vdots & \vdots & \vdots & \vdots & \vdots & \vdots \\ \alpha_{k1} & \alpha_{k2} & \cdots & \alpha_{k,n-1} & \alpha_{kn} & \beta_k \end{pmatrix}$$

is called the extended or augmented matrix of the system (Eq. 4.2).

4.3.8. Theorem *(Kronecker–Capelli). The system of linear equations has a solution if and only if the rank of the coefficient matrix of the system is equal to the rank of the augmented matrix of the system.*

Proof. Suppose that the system (Eq. 4.2) has the solution $(\gamma_1, \ldots, \gamma_n)$. We have

$$\begin{aligned}
\alpha_{11}\gamma_1 + \alpha_{12}\gamma_2 + \cdots + \alpha_{1n}\gamma_n &= \beta_1 \\
\alpha_{21}\gamma_1 + \alpha_{22}\gamma_2 + \cdots + \alpha_{2n}\gamma_n &= \beta_2 \\
&\vdots \\
\alpha_{k-1,1}\gamma_1 + \alpha_{k-1,2}\gamma_2 + \cdots + \alpha_{k-1,n}\gamma_n &= \beta_{k-1} \\
\alpha_{k1}\gamma_1 + \alpha_{k2}\gamma_2 + \cdots + \alpha_{kn}\gamma_n &= \beta_k
\end{aligned}$$

Let $\mathfrak{a}_j$ denote the jth column of the matrix A, where $1 \leq j \leq n$, and let $\mathfrak{b}$ denote the column consisting of the elements $\beta_1, \ldots, \beta_k$. Then, this system of equations can be written in the form

$$\gamma_1\mathfrak{a}_1 + \gamma_2\mathfrak{a}_2 + \cdots + \gamma_n\mathfrak{a}_n = \mathfrak{b},$$

as a linear combination of the columns. This equation shows that the column $\mathfrak{b}$ belongs to $\mathbf{Le}(\mathbf{C}(A))$. Thus, $\mathbf{Le}(\mathbf{C}(A)) = \mathbf{Le}(\mathbf{C}(A^*))$ and therefore

$$\mathbf{rank}(A) = \mathbf{dim}_F(\mathbf{Le}(\mathbf{C}(A)) = \mathbf{dim}_F(\mathbf{Le}(\mathbf{C}(A^*)) = \mathbf{rank}(A^*).$$

Conversely, suppose that $\mathbf{rank}(A) = \mathbf{rank}(A^*)$ in which case, we have $\mathbf{dim}_F$ $(\mathbf{Le}(\mathbf{C}(A)) = \mathbf{dim}_F(\mathbf{Le}(\mathbf{C}(A^*)))$. Since $\mathbf{Le}(\mathbf{C}(A))$ is a subspace of $\mathbf{Le}(\mathbf{C}(A^*))$, Theorem 4.2.20 shows that $\mathbf{Le}(\mathbf{C}(A)) = \mathbf{Le}(\mathbf{C}(A^*))$. It follows that the column $\mathfrak{b}$ belongs to $\mathbf{Le}(\mathbf{C}(A))$. By Proposition 4.2.3, every element of $\mathbf{Le}(\mathbf{C}(A))$ is a linear combination of the elements of $\mathbf{C}(A)$, so

$$\gamma_1\mathfrak{a}_1 + \gamma_2\mathfrak{a}_2 + \cdots + \gamma_n\mathfrak{a}_n = \mathfrak{b}.$$

for some elements $\gamma_1, \ldots, \gamma_n \in F$. This leads us to the system of equations,

$$\begin{aligned}
\alpha_{11}\gamma_1 + \alpha_{12}\gamma_2 + \cdots + \alpha_{1n}\gamma_n &= \beta_1 \\
\alpha_{21}\gamma_1 + \alpha_{22}\gamma_2 + \cdots + \alpha_{2n}\gamma_n &= \beta_2 \\
&\vdots \\
\alpha_{k-1,1}\gamma_1 + \alpha_{k-1,2}\gamma_2 + \cdots + \alpha_{k-1,n}\gamma_n &= \beta_{k-1} \\
\alpha_{k1}\gamma_1 + \alpha_{k2}\gamma_2 + \cdots + \alpha_{kn}\gamma_n &= \beta_k,
\end{aligned}$$

which shows that the n-tuple $(\gamma_1, \ldots, \gamma_n)$ is a solution of the system (Eq. 4.2).

Finally we obtain the following result.

4.3.9. Theorem. *Let F be a field, let $A = [\alpha_{tj}] \in \mathbf{M}_{k\times n}(F)$, and let $B = [\beta_{jm}] \in \mathbf{M}_{n\times s}(F)$. Then $\mathbf{rank}(AB) \leq \mathbf{rank}(A)$ and $\mathbf{rank}(AB) \leq \mathbf{rank}(B)$.*

Proof. Let $AB = [\gamma_{tm}] \in \mathbf{M}_{k\times s}(F)$. We consider the mth column of AB which is

$$\alpha_{11}\beta_{1m} + \alpha_{12}\beta_{2m} + \cdots + \alpha_{1n}\beta_{nm} = \gamma_{1m}$$
$$\alpha_{21}\beta_{1m} + \alpha_{22}\beta_{2m} + \cdots + \alpha_{2n}\beta_{nm} = \gamma_{2m}$$
$$\vdots$$
$$\alpha_{k1}\beta_{1m} + \alpha_{k2}\beta_{2m} + \cdots + \alpha_{kn}\beta_{nm} = \gamma_{km}.$$

Let $\mathfrak{a}_j$ denote the jth column of the matrix A, for $1 \leq j \leq n$, and let $\mathfrak{g}_m$ denote the mth column of the matrix AB, for $1 \leq m \leq s$. Then, the system above leads to the following linear combination of columns (as elements of the vector space F^k):

$$\mathfrak{g}_m = \beta_{1m}\mathfrak{a}_1 + \beta_{2m}\mathfrak{a}_2 + \cdots + \beta_{nm}\mathfrak{a}_n.$$

This equation shows that every column of the matrix AB belongs to $\mathbf{Le}(\mathbf{C}(A))$, which means that $\mathbf{Le}(\mathbf{C}(AB)) \leq \mathbf{Le}(\mathbf{C}(A))$ and therefore

$$\mathbf{rank}(AB) = \mathbf{dim}_F(\mathbf{Le}(\mathbf{C}(AB))) \leq \mathbf{dim}_F(\mathbf{Le}(\mathbf{C}(A))) = \mathbf{rank}(A).$$

We note in a similar manner that each arbitrary row of the matrix AB is a linear combination of the rows of the matrix B. Hence every row of the matrix AB belongs to $\mathbf{Le}(\mathbf{R}(B))$. This means that $\mathbf{Le}(\mathbf{R}(AB)) \leq \mathbf{Le}(\mathbf{R}(B))$ and therefore

$$\mathbf{rank}(AB) = \mathbf{dim}_F(\mathbf{Le}(\mathbf{R}(AB))) \leq \mathbf{dim}_F(\mathbf{Le}(\mathbf{R}(B))) = \mathbf{rank}(B).$$

4.3.10. Corollary. *Let F be a field, let $A = [\alpha_{tj}] \in \mathbf{M}_n(F)$ and let $B = [\beta_{jm}] \in \mathbf{M}_{n\times s}(F)$. If A is a nonsingular matrix, then $\mathbf{rank}(AB) = \mathbf{rank}(B)$.*

Proof. Let $C = AB$. By Theorem 4.3.9, $\mathbf{rank}(C) \leq \mathbf{rank}(B)$. Since A is nonsingular, A has a multiplicative inverse, A^{-1} and we have

$$B = IB = (A^{-1}A)B = A^{-1}(AB) = A^{-1}C.$$

Using Theorem 4.3.9, again we deduce that $\mathbf{rank}(B) \leq \mathbf{rank}(C)$, and therefore $\mathbf{rank}(C) = \mathbf{rank}(B)$.

A similar method of proof implies the following corollary also.

4.3.11. Corollary. *Let F be a field, let $A = [\alpha_{tj}] \in \mathbf{M}_{k\times n}$ and let $B = [\beta_{jm}] \in \mathbf{M}_n(F)$. If B is a nonsingular matrix, then $\mathbf{rank}(AB) = \mathbf{rank}(A)$.*

EXERCISE SET 4.3

4.3.1. Let $A = \mathbb{Q}^5$. Find the rank of the subset $M = \{(0, 1, 1, 0, 2), (1, 0, 3, -1, 3), (1, 0, -3, 2, -2), (2, 3, 4, 3, 1), (3, -1, 3, 1, 0)\}$.

4.3.2. Let $A = \mathbb{Q}^3$. Find the rank of the subset $M = \{(1, 1, 1), (\alpha, \beta, \gamma), (\alpha^2, \beta^2, \gamma^2)\}$.

4.3.3. Let $A = \mathbb{R}^5$. Find the rank of the subset $M = \{(0, 0, 1, 1, 2), (-1, 1, 0, 0, 1), (2, -1, 0, 1, -1), (1, 3, -1, 3, 1), (3, 2, 3, 3, 3)\}$.

4.3.4. Find the rank of the matrix $\begin{pmatrix} 2 & 3 & -2 & 0 & 6 \\ 0 & 1 & 5 & 3 & 2 \\ 0 & -1 & 2 & 1 & 0 \\ 1 & 0 & 3 & -1 & 3 \\ 1 & 3 & -3 & 8 & 3 \end{pmatrix}$.

4.3.5. Find the rank of the matrix $\begin{pmatrix} 0 & 3 & -2 & 0 & 6 \\ 2 & 1 & 0 & 3 & 2 \\ 0 & -1 & 2 & 1 & 0 \\ 1 & 0 & 0 & -1 & 3 \\ 1 & 3 & -5 & 8 & 3 \end{pmatrix}$.

4.3.6. Find the rank of the matrix $\begin{pmatrix} 2 & 3 & -2 & 0 & 6 \\ 0 & 1 & 5 & 3 & 2 \\ 0 & -1 & 2 & 1 & 0 \\ 1 & 0 & 3 & -1 & 3 \\ 1 & 3 & -3 & 8 & 3 \end{pmatrix}$.

4.3.7. Find the rank of the matrix $\begin{pmatrix} 0 & 3 & -2 & 0 & 6 \\ 2 & 1 & 0 & 3 & 2 \\ 0 & -1 & 2 & 1 & 0 \\ 1 & 0 & 0 & -1 & 3 \\ 1 & 3 & -5 & 8 & 3 \end{pmatrix}$.

4.3.8. Find the rank of the matrix $\begin{pmatrix} 2 & 3 & 0 & 0 & 6 \\ 0 & 1 & -4 & 3 & 2 \\ 2 & -2 & 2 & 0 & 0 \\ 1 & 0 & 0 & -1 & 3 \\ 1 & 3 & -3 & 8 & 3 \end{pmatrix}$.

4.3.9. Find the rank of the matrix $\begin{pmatrix} 2 & 3 & -2 & 0 & 6 \\ 0 & 1 & -1 & 3 & 2 \\ 0 & -1 & 2 & 1 & 3 \\ 1 & 0 & 0 & -1 & 3 \\ 1 & 3 & -3 & 8 & 3 \end{pmatrix}$.

4.3.10. Find the values of α, such that the matrix

$$\begin{pmatrix} 3 & 1 & 1 & 4 \\ \alpha & 4 & 10 & 1 \\ 1 & 7 & 17 & 3 \\ 2 & 2 & 4 & 3 \end{pmatrix}$$

has the minimal rank and find this rank.

4.4 QUOTIENT SPACES

The planes and straight lines of regular three dimensional space are important subjects of investigation and certain of them are subspaces of this space. Note that Theorem 4.1.7 implies that a subspace of a vector space always contains a zero element, so only such straight lines and planes that pass through the origin are subspaces. We consider how lines and planes not passing through the origin are related to these subspaces in the following example.

Let P be a straight line in $\mathbb{R}^2$ with the equation $y = kx + b$. Then

$$P = \{(x, kx + b) \mid x \in \mathbb{R}\},$$

and we let $P_0 = \{(x, kx) \mid x \in R\}$, which is a subspace of $\mathbb{R}^2$. If (u, v) is an arbitrary element of P, then $(u, v) = (u, ku + b) = (u, ku) + (0, b)$. Thus, to obtain all points of P, we add $(0, b)$ to each of the points of the subspace P_0.

4.4.1. Definition. *Let A be a vector space over a field F and let B be a subspace of A. For each element $x \in A$, let $x + B = \{x + y \mid y \in B\}$. The subset $x + B$ is called an affine subspace of A or a coset of the subspace B, and the element x is called its coset representative.*

We note that $x + B$ is uniquely determined by each of its elements. This means that if $y \in x + B$, then $x + B = y + B$. In fact, $y = x + b_0$ for some element $b_0 \in B$. If $z \in y + B$, then $z = y + b_1$ for some element $b_1 \in B$ so that, by Theorem 4.1.7,

$$z = y + b_1 = (x + b_0) + b_1 = x + (b_0 + b_1) \in x + B.$$

Hence $y + B \subseteq x + B$. On the other hand, $x = y - b_0$, and repeating the same arguments, we obtain the inclusion $x + B \subseteq y + B$, which proves that $y + B = x + B$.

Suppose now that there are two cosets $x + B, y + B$ and that $(y + B) \cap (x + B) \neq \emptyset$. Let $z \in (y + B) \cap (x + B)$. As stated above, the inclusion $z \in y + B$ (respectively, $z \in x + B$) implies that $z + B = y + B$ (respectively $z + B = x + B$). In particular, $x + B = y + B$. The equation $x = x + 0_A$ implies that

$x \in x + B$. This shows that the family of all affine subspaces of A, relative to B, is a partition of A. Indeed, there is an underlying equivalence relation defined on A in which the equivalence classes are precisely the cosets of the subspace B. (See Section 7.2 for details concerning equivalence relations.)

We define an addition and scalar multiplication on the set of all affine subspaces of B using the rules

$$(x + B) + (y + B) = x + y + B$$

and

$$\alpha(x + B) = \alpha x + B,$$

where $x, y \in A, \alpha \in F$. These operations are well defined since if also x_1, y_1 are elements of A such that $x + B = x_1 + B$ and $y + B = y_1 + B$, then we have $x_1 = x + u, y_1 = y + v$ for certain elements of $u, v \in B$. Thus,

$$x_1 + y_1 = (x + u) + (y + v) = (x + y) + (u + v)$$

and

$$\alpha x_1 = \alpha(x + u) = \alpha x + \alpha u.$$

Since B is a subspace, $(u + v), \alpha u \in B$. Therefore

$$x + y + B = x_1 + y_1 + B \text{ and } \alpha x + B = \alpha x_1 + B.$$

Next we show that the set of cosets itself forms a vector space over F with this definition of addition and scalar multiplication. First, we note that if $x, y \in A$, then

$$(x + B) + (y + B) = x + y + B = y + x + B = (y + B) + (x + B);$$

and

$$\begin{aligned}(x + B) + ((y + B) + (z + B)) &= (x + B) + (y + z + B)\\ &= x + (y + z) + B\\ &= ((x + y) + z) + B\\ &= (x + y + B) + (z + B)\\ &= ((x + B) + (y + B)) + (z + B).\end{aligned}$$

Also

$$(x + B) + (0_A + B) = x + 0_A + B = x + B;$$

so that the coset $0_A + B = B$ is the zero element under addition of cosets. Clearly,

$$(x + B) + (-x + B) = (x + (-x)) + B = 0_A + B = B,$$

so that

$$-(x + B) = (-x + B).$$

Also, if $\alpha, \beta \in F$, then

$$\begin{aligned}
\alpha(x + B + y + B) &= \alpha(x + y + B) = \alpha(x + y) + B \\
&= \alpha x + \alpha y + B = \alpha x + B + \alpha y + B \\
&= \alpha(x + B) + \alpha(y + B); \\
(\alpha + \beta)(x + B) &= (\alpha + \beta)x + B = \alpha x + \beta x + B \\
&= (\alpha x + B) + (\beta x + B) = \alpha(x + B) + \beta(x + B); \\
\alpha(\beta(x + B)) &= \alpha(\beta x + B) = \alpha(\beta x) + B = (\alpha\beta)x + B \\
&= (\alpha\beta)(x + B); \text{ and} \\
e(x + B) &= ex + B = x + B.
\end{aligned}$$

This means that the set of all affine subspaces over B is a vector space, using the operations of addition and scalar multiplication defined above.

4.4.2. Definition. *The space of all affine subspaces over the subspace B is called the factor space, or the quotient space, of A over B and is denoted by A/B.*

Note that if $B = \{0_A\}$ is the zero subspace, then $x + B = \{x\}$ and

$$\{x\} + \{y\} = \{x + y\}, \alpha\{x\} = \{\alpha x\},$$

for all $x, y \in A, \alpha \in F$.

This shows that in this case, the quotient space A/B is, to all intents and purposes, the same as A. We will see later that A and A/B are "isomorphic" as vector spaces in this case. If $B = A$, then $x + B = A$ for each $x \in A$. Hence, in this case, $A/B = \{A\}$ is the zero vector space.

4.4.3. Proposition. *Let F be a field, let A be a vector space over F and let B be a subspace of A. Suppose that M is a subset of A with the property that* $\mathbf{Le}(M) = A$. *Then, the subset $\{a + B \mid a \in M\}$ generates the quotient space A/B.*

Proof. Indeed, for each element x of A, we have $x = \lambda_1 a_1 + \cdots + \lambda_k a_k$ for certain elements $\lambda_1, \ldots, \lambda_k \in F$ and $a_1, \ldots, a_k \in M$. Then

$$\begin{aligned}
x + B &= \lambda_1 a_1 + \cdots + \lambda_k a_k + B = (\lambda_1 a_1 + B) + \cdots + (\lambda_k a_k + B) \\
&= \lambda_1(a_1 + B) + \cdots + \lambda_k(a_k + B).
\end{aligned}$$

The result now follows.

For finite-dimensional vector spaces we obtain.

4.4.4. Theorem. *Let F be a field, let A be a finite-dimensional vector space over F and let B be a subspace of A. Then the quotient space A/B is finite dimensional and* $\mathbf{dim}_F(A/B) = \mathbf{dim}_F(A) - \mathbf{dim}_F(B)$.

Proof. By Theorem 4.2.20, B is finite dimensional. Let $M = \{b_1, \ldots, b_k\}$ be a basis of the subspace B. Since M is linearly independent, Theorem 4.2.11 shows that this subset can be extended to a basis of the entire space A. Hence, there exists a finite subset $S = \{c_1, \ldots, c_t\}$ such that $M \cup S$ is a basis of A. By Proposition 4.4.3, the quotient space A/B is generated by $b_1 + B, \ldots, b_k + B, c_1 + B, \ldots, c_t + B$. However, $b_j + B = B$ for all j, where $1 \leq j \leq k$. Hence A/B is generated by $c_1 + B, \ldots, c_t + B$.

Next we show that $c_1 + B, \ldots, c_t + B$ are linearly independent. If $\gamma_1, \ldots, \gamma_t$ are elements of F such that

$$\gamma_1(c_1 + B) + \cdots + \gamma_t(c_t + B) = B,$$

then we have

$$\begin{aligned}\gamma_1(c_1 + B) + \cdots + \gamma_t(c_t + B) &= (\gamma_1 c_1 + B) + \cdots + (\gamma_t c_t + B)\\ &= \gamma_1 c_1 + \cdots + \gamma_t c_t + B.\end{aligned}$$

It follows that $\gamma_1 c_1 + \cdots + \gamma_t c_t \in B$, and since $\{b_1, \ldots, b_k\}$ is a basis of B, there exist elements $\beta_1, \ldots, \beta_k \in F$ such that $\gamma_1 c_1 + \cdots + \gamma_t c_t = \beta_1 b_1 + \cdots + \beta_k b_k$. Then

$$\beta_1 b_1 + \cdots + \beta_k b_k - \gamma_1 c_1 - \cdots - \gamma_t c_t = 0_A.$$

Since $\{b_1, \ldots, b_k, c_1, \ldots, c_t\}$ is a basis of A, Proposition 4.2.7 shows that

$$\beta_1 = \cdots = \beta_k = \gamma_1 = \cdots = \gamma_t = 0_F$$

and hence, again using Proposition 4.2.7, we deduce that $\{c_1 + B, \ldots, c_t + B\}$ is linearly independent. Thus, $\{c_1 + B, \ldots, c_t + B\}$ is a basis of A/B and

$$t = \mathbf{dim}_F(A/B) = (t + k) - t = \mathbf{dim}_F(A) - \mathbf{dim}_F(B),$$

which proves the result.

In the next chapter, we start to discuss mappings between vector spaces. To illustrate what we shall be doing, we consider a straightforward example. Accordingly, let A be a vector space over the field F and let B be a subspace of A. Consider the mapping $\sigma_B : A \longrightarrow A/B$, defined by

$$\sigma_B(x) = x + B, \text{ for } x \in A.$$

Then

$$\sigma_B(x+y) = x+y+B = x+B+y+B = \sigma_B(x)+\sigma_B(y),$$

and

$$\alpha\sigma_B(x) = \alpha(x+B) = \alpha x + B = \sigma_B(\alpha x),$$

whenever $x, y \in A, \alpha \in F$.

The stage is now set for us to discuss the next concept that is important in linear algebra, the idea of mappings that respect the operations of addition and scalar multiplication.

EXERCISE SET 4.4

Justify your work, using a proof or counterexample where appropriate.

4.4.1. Let $A = \mathbb{Q}^{22}$, $B = \{(\alpha_1, \alpha_2, \ldots, \alpha_{22}) \mid \alpha_1 + \alpha_2 + \cdots + \alpha_{22} = 0\}$. Find a basis of A/B.

4.4.2. Let $A = \mathbb{F}_3^{21}$, $B = \{(\alpha_1, \alpha_2, \ldots, \alpha_{21}) \mid 2\alpha_1 = \alpha_{21}\}$. Find the order of A/B.

4.4.3. Let $A = \mathbb{Q}^{221}$, $B = \{(\alpha_1, \alpha_2, \ldots, \alpha_{221}) \mid 2\alpha_1 = \alpha_{221}\}$. Find a basis of A/B.

4.4.4. Let A be the subspace of all symmetric matrices and let B be the subset of all diagonal matrices of the vector space $\mathbf{M}_{11}(\mathbb{R})$. Find the dimension of A/B.

4.4.5. Let $A = \mathbb{Q}^4$ and let B be the linear envelope of the subset $\{(1, 2, 3, 4), (0, 1, -1, 3)\}$. Find a basis of A/B.

4.4.6. Let $A = M_2(\mathbb{Q})$ and B be the linear envelope of the subset consisting of the matrices $\begin{pmatrix} 1 & -1 \\ 0 & 0 \end{pmatrix}, \begin{pmatrix} 1 & 1 \\ 1 & 0 \end{pmatrix}$. Find a basis of A/B.

4.4.7. Let $A = M_2(\mathbb{Q})$ and let B be the linear envelope of the subset consisting of the matrices $\begin{pmatrix} 1 & 1 \\ 0 & 0 \end{pmatrix}, \begin{pmatrix} 1 & 1 \\ 1 & 0 \end{pmatrix}$. Find a basis of A/B.

4.4.8. Let $A = M_2(\mathbb{F}_3)$ and let B be the linear envelope of the subset consisting of the matrices $\begin{pmatrix} 1 & 1 \\ 0 & 0 \end{pmatrix}, \begin{pmatrix} 1 & -1 \\ 1 & 1 \end{pmatrix}$. Find a basis, and the order, of A/B.

CHAPTER 5

LINEAR MAPPINGS

5.1 LINEAR MAPPINGS

For each algebraic structure there is a specific type of mapping which respects that structure. Such a mapping is called a homomorphism of the given structure. For example, in ring theory we deal with ring homomorphisms, in group theory we deal with group homomorphisms, and so on. The term homomorphism is a general term for all areas of algebra but, unfortunately in some respect, the term never caught on in linear algebra, where we use the term *linear mapping*.

5.1.1. Definition. *Let A and V be vector spaces over the same field F. The mapping $f : A \longrightarrow V$ is called a linear mapping, or a homomorphism of vector spaces, if it satisfies the following properties:*

$$f(x + y) = f(x) + f(y) \text{ and } f(\alpha x) = \alpha f(x)$$

for all $x, y \in A, \alpha \in F$. An injective linear mapping is called a monomorphism, a surjective linear mapping is called an epimorphism, and a bijective linear mapping is called an isomorphism.

If $f : A \longrightarrow V$ is an isomorphism, then Theorem 1.3.5 shows that the mapping f has an inverse mapping $f^{-1} : V \longrightarrow A$. If u, v are arbitrary elements of V,

Algebra and Number Theory: An Integrated Approach. By Martyn R. Dixon, Leonid A. Kurdachenko and Igor Ya. Subbotin.

then $u = f(x), v = f(y)$ for certain elements $x, y \in A$. If $\alpha \in F$ we have

$$\begin{aligned} f^{-1}(u+v) &= f^{-1}(f(x)+f(y)) = f^{-1}(f(x+y)) \\ &= x+y = f^{-1}(u)+f^{-1}(v) \text{ and} \\ f^{-1}(\alpha u) &= f^{-1}(\alpha f(x)) = f^{-1}(f(\alpha x)) = \alpha x = \alpha f^{-1}(u). \end{aligned}$$

This implies that the mapping $f^{-1} : V \longrightarrow A$ is also an isomorphism.

5.1.2. Definition. *Let A, V be vector spaces over the field F. Then, A and V are called isomorphic if there exists an isomorphism from one of them to the other and we can write $A \cong_F V$ or $A \cong V$, when the field is understood.*

Clearly, the identity mapping $\varepsilon_A : A \longrightarrow A$ is an isomorphism. Also the mapping $\vartheta : A \longrightarrow V$, the zero mapping, defined by $\vartheta(a) = 0_V$ for all $a \in A$ is linear.

Next, let $\alpha \in F$ be fixed. The mapping $h_\alpha : A \longrightarrow A$ defined by $h_\alpha(x) = \alpha x$ for all $x \in A$ is linear and is called a homothety. Clearly, $h_e = \varepsilon_A$ and $h_0 = \vartheta$.

Furthermore, if $f : A \longrightarrow V$ and $g : V \longrightarrow W$ are linear mappings, then their product $g \circ f$ is also a linear mapping, as can easily be seen.

5.1.3. Proposition. *Let A, V be vector spaces over a field F and let $f : A \longrightarrow V$ be a linear mapping. Then the following properties hold:*

(i) $f(0_A) = 0_V$.

(ii) $f(-x) = -f(x)$ *for all elements* $x \in A$.

(iii) $f(x-y) = f(x) - f(y)$ *for all* $x, y \in A$.

(iv) $f(\alpha_1 x_1 + \cdots + \alpha_n x_n) = \alpha_1 f(x_1) + \cdots + \alpha_n f(x_n)$ *for all* $x_1, \ldots, x_n \in A$ *and* $\alpha_1, \ldots \alpha_n \in F$.

(v) *If B is a subspace of A, then its image $f(B)$ is a subspace of V; in particular, $f(A) = \mathbf{Im} f$ is a subspace of V.*

(vi) *If U is a subspace of V, then its preimage $f^{-1}(U)$ is a subspace of A; in particular,*

$$\mathbf{Ker} f = \{x \in A \mid f(x) = 0_V\} = f^{-1}(\{0_V\})$$

is a subspace of A.

(vii) *If M is a subset of A, then $\mathbf{Le}(f(M)) = f(\mathbf{Le}(M))$.*

Proof.

(i) We have $x + 0_A = x$ for each $x \in A$. Then

$$f(x) + f(0_A) = f(x + 0_A) = f(x).$$

Since $f(x)$ has an additive inverse in V, adding it to both sides of the equation above and using the associative law gives

$$0_V = -f(x) + f(x) = -f(x) + f(x) + f(0_A) = 0_V + f(0_A) = f(0_A).$$

(ii) By definition of additive inverses, $x + (-x) = 0_A$, so that

$$0_V = f(0_A) = f(x + (-x)) = f(x) + f(-x).$$

This shows that $f(-x)$ is the additive inverse of $f(x)$.

(iii) We have

$$f(x - y) = f(x + (-y)) = f(x) + f(-y) = f(x) + (-f(y)) = f(x) - f(y).$$

(iv) We use induction on n. For $n = 2$ we have

$$f(\alpha_1 x_1 + \alpha_2 x_2) = f(\alpha_1 x_1) + f(\alpha_2 x_2) = \alpha_1 f(x_1) + \alpha_2 f(x_2).$$

Suppose, inductively, that we have already proved that

$$f(\alpha_1 x_1 + \cdots + \alpha_{n-1} x_{n-1}) = \alpha_1 f(x_1) + \cdots + \alpha_{n-1} f(x_{n-1}).$$

Then

$$\begin{aligned} f(\alpha_1 x_1 + \cdots + \alpha_n x_n) &= f((\alpha_1 x_1 + \cdots + \alpha_{n-1} x_{n-1}) + \alpha_n x_n) \\ &= f(\alpha_1 x_1 + \cdots + \alpha_{n-1} x_{n-1}) + f(\alpha_n x_n) \\ &= \alpha_1 f(x_1) + \cdots + \alpha_{n-1} f(x_{n-1}) + \alpha_n f(x_n), \end{aligned}$$

by the induction hypothesis.

(v) Let $x, y \in B, \alpha \in F, u = f(x)$ and $v = f(y)$. Then, by Theorem 4.1.7, $x - y, \alpha x \in B$, so that

$$u - v = f(x) - f(y) = f(x - y) \in f(B) \text{ and } \alpha u = \alpha f(x) = f(\alpha x) \in f(B).$$

Thus, by Theorem 4.1.7, $f(B)$ is a subspace of V.

(vi) Let $x, y \in f^{-1}(U)$ and $\alpha \in F$. Then $f(x), f(y) \in U$. Since U is a subspace of V, $f(x) - f(y) = f(x - y) \in U$ and $\alpha f(x) = f(\alpha x) \in U$. This implies that $x - y, \alpha x \in f^{-1}(U)$ and Theorem 4.1.7 implies that $f^{-1}(U)$ is a subspace of A.

(vii) Let $u \in \mathbf{Le}(f(M))$. Then, by Proposition 4.2.3, $u = \alpha_1 w_1 + \cdots + \alpha_n w_n$ for certain elements $w_1, \ldots, w_n \in f(M), \alpha_1, \ldots, \alpha_n \in F$. Hence, there exist $y_1, \ldots, y_n \in M$ such that $w_1 = f(y_1), \ldots, w_n = f(y_n)$. Therefore

$$u = \alpha_1 f(y_1) + \cdots + \alpha_n f(y_n) = f(\alpha_1 y_1 + \cdots + \alpha_n y_n) \in f(\mathbf{Le}(M)).$$

It follows that $\mathbf{Le}(f(M)) \leq f(\mathbf{Le}(M))$. Applying these arguments in reverse order, we obtain the reverse inclusion which proves $\mathbf{Le}(f(M)) = f(\mathbf{Le}(M))$.

The subspace $\mathbf{Ker} f$ is called the kernel of the linear mapping f.

5.1.4. Theorem *(The monomorphism theorem). Let A, V be vector spaces over a field F and let $f : A \longrightarrow V$ be a linear mapping. Then f is a monomorphism if and only if $\mathbf{Ker} f = \{0_A\}$. Furthermore, if f is a monomorphism, then $A \cong \mathbf{Im} f$.*

Proof. Indeed, if f is a monomorphism, then from $x \neq 0_A$ it follows that $f(x) \neq f(0_A) = 0_V$. This means that no nonzero element x of A belongs to $\mathbf{Ker} f$, so $\mathbf{Ker} f = \{0_A\}$.

Conversely, let $\mathbf{Ker} f = \{0_A\}$ and let x, y be elements of A such that $f(x) = f(y)$. Then

$$f(x - y) = f(x) - f(y) = 0_V,$$

so that $x - y \in \mathbf{Ker} f$. It follows that $x - y = 0_A$ so that $x = y$. This proves that f is injective and hence is a monomorphism. Finally, when f is a monomorphism then f also maps A onto $\mathbf{Im}\, f$ so in this case $A \cong \mathbf{Im}\, f$.

5.1.5. Corollary. *Let A, V be vector spaces over a field F and let $f : A \longrightarrow V$ be a monomorphism. If M is a linearly independent subset of A, then $f(M)$ is a linearly independent subset of V.*

Proof. Let S be a finite subset of $f(M)$. Then $S = f(R)$ for some finite subset R of M. To show that $f(M)$ is linearly independent, let $R = \{y_1, \dots, y_n\}$ and let $\alpha_1, \dots, \alpha_n$ be elements of F such that $\alpha_1 f(y_1) + \cdots + \alpha_n f(y_n) = 0_V$. By Proposition 5.1.3,

$$0_V = \alpha_1 f(y_1) + \cdots + \alpha_n f(y_n) = f(\alpha_1 y_1 + \cdots + \alpha_n y_n),$$

which shows that $\alpha_1 y_1 + \cdots + \alpha_n y_n \in \mathbf{Ker} f$. By Theorem 5.1.4, $\mathbf{Ker} f = \{0_A\}$, so that $\alpha_1 y_1 + \cdots + \alpha_n y_n = 0_A$. By Proposition 4.2.7, R is a linearly independent subset and therefore, $\alpha_1 = \cdots = \alpha_n = 0_F$. Using Proposition 4.2.7 again we deduce that S is linearly independent and hence, $f(M)$ is linearly independent.

5.1.6. Corollary. *Let A, V be vector spaces over a field F and let $f : A \longrightarrow V$ be a monomorphism. If A, V have finite dimensions then $\mathbf{dim}_F(A) \leq \mathbf{dim}_F(V)$.*

Proof. Let X be a finite basis of A. By Corollary 5.1.5, $f(X)$ is a linearly independent subset of V. By Proposition 5.1.3, $f(X)$ is a set of generators for $\mathbf{Im} f$, so $f(X)$ is a basis for $\mathbf{Im} f$. By Theorem 4.2.20, $\mathbf{dim}_F(\mathbf{Im} f) \leq \mathbf{dim}_F(V)$ and, since f is injective, $\mathbf{dim}_F(\mathbf{Im} f) = |f(X)| = |X| = \mathbf{dim}_F(A)$.

5.1.7. Theorem *(The epimorphism theorem). Let A, V be vector spaces over a field F and let $f : A \longrightarrow V$ be an epimorphism. Then the space V is isomorphic to the quotient space $A/\mathbf{Ker} f$.*

Proof. Put $B = \mathbf{Ker} f$ and consider the mapping $\psi_f : A/B \longrightarrow V$ defined by $\psi_f(x + B) = f(x)$. We need to be sure that ψ_f is well defined, which means here that it does not depend on the choice of representative of the affine subspace $x + B$. To see this, let y be an element of the space A such that $x + B = y + B$. Then $y \in x + B$, so that $y = x + b$, for some element $b \in B$ and

$$f(y) = f(x + b) = f(x) + f(b) = f(x) + 0_V = f(x),$$

which proves that ψ_f is well defined.

The mapping ψ_f is bijective since, for each element $u \in \mathbf{Im} f$, there exists an element $a \in A$ such that $u = f(a)$ and hence $\psi_f(a + B) = f(a) = u$, which shows that ψ_f is surjective. If $\psi_f(a + B) = \psi_f(c + B)$, then by the definition of ψ_f we have $f(a) = f(c)$, so $0_V = f(a) - f(c) = f(a - c)$ and therefore $a - c \in \mathbf{Ker} f = B$. It follows that $a + B = c + B$, so ψ_f is injective and therefore bijective.

We still need to prove that ψ_f is a homomorphism. However,

$$\begin{aligned}\psi_f((x + B) + (y + B)) &= \psi_f(x + y + B) = f(x + y)\\ &= f(x) + f(y) = \psi_f(x + B) + \psi_f(y + B)\end{aligned}$$

and

$$\psi_f(\alpha(x + B)) = \psi_f(\alpha x + B) = f(\alpha x) = \alpha f(x) = \alpha\psi_f(x + B),$$

for all $x, y \in A, \alpha \in F$.

5.1.8. Corollary. *Let A, V be vector spaces over a field F and let $f : A \longrightarrow V$ be an epimorphism. If A has finite dimension, then V is also finite dimensional and $\mathbf{dim}_F(V) \leq \mathbf{dim}_F(A)$.*

Proof. Let X be a basis of A. By Proposition 5.1.3, $f(X)$ is a set of generators for V. By Corollary 4.2.13, $f(X)$ contains a basis of the space V, so that $\mathbf{dim}_F(V) \leq |f(X)|$. Since $|f(X)| \leq |X| = \mathbf{dim}_F(A)$, the result is proved.

5.1.9. Corollary. *Let A, V be vector spaces over a field F and let $f : A \longrightarrow V$ be an isomorphism. If A has finite dimension and X is a basis of A, then $f(X)$ is a basis of V and $\mathbf{dim}_F(V) = \mathbf{dim}_F(A)$.*

Proof. By Proposition 5.1.3, $f(X)$ is a set of generators for V and, by Corollary 5.1.5, $f(X)$ is a linearly independent subset of V. Hence $f(X)$ is a basis of the space V so that

$$\mathbf{dim}_F(V) = |f(X)| = |X| = \mathbf{dim}_F(A).$$

5.1.10. Theorem *(The first isomorphism theorem). Let A, V be vector spaces over a field F and let $f : A \longrightarrow V$ be a linear mapping. Then $A/\mathbf{Ker} f \cong \mathbf{Im} f \leq V$.*

Proof. Indeed, the restriction of f to the mapping $A \longrightarrow \mathbf{Im}f$ is an epimorphism and from Theorem 5.1.7 we deduce that $\mathbf{Im}\ f \cong A/\mathbf{Ker}f$. Finally, by Proposition 5.1.3, $\mathbf{Im}f$ is a subspace of V.

For finite-dimensional spaces we obtain the following result.

5.1.11. Corollary. *Let A, V be vector spaces over a field F and let $f : A \longrightarrow V$ be a linear mapping. If $\mathbf{dim}_F(A)$ is finite, then $\mathbf{dim}_F(A) = \mathbf{dim}_F(\mathbf{Ker}f) + \mathbf{dim}_F(\mathbf{Im}f)$.*

Proof. By Theorem 5.1.10, $A/\mathbf{Ker}f \cong \mathbf{Im}f$. Corollary 5.1.9 shows that $\mathbf{dim}_F(A/\mathbf{Ker}f) = \mathbf{dim}_F(\mathbf{Im}f)$ and Theorem 4.4.4 implies that $\mathbf{dim}_F(A/\mathbf{Ker}f) = \mathbf{dim}_F(A) - \mathbf{dim}_F(\mathbf{Ker}f)$. Thus,

$$\mathbf{dim}_F(A) = \mathbf{dim}_F(\mathbf{Ker}f) + \mathbf{dim}_F(\mathbf{Im}f).$$

For finite-dimensional vector spaces we obtain the following method of defining linear mappings.

5.1.12. Proposition. *Let A, V be vector spaces over a field F. Suppose that $\mathbf{dim}_F(A)$ is finite and let $\{a_1, \ldots, a_n\}$ be a basis of A. If $\{u_1, \ldots, u_n\}$ are n arbitrary elements of the space V, there exists one and only one linear mapping $f : A \longrightarrow V$ such that $f(a_j) = u_j$, for $1 \leq j \leq n$.*

Proof. Let x be an arbitrary element of A. By Proposition 4.2.16, $x = \sum_{1\leq j\leq n} \xi_j a_j$, where $\xi_j \in F$ for $1 \leq j \leq n$. Define the mapping $f : A \longrightarrow V$ by $f(x) = \sum_{1\leq j\leq n} \xi_j u_j$. If $y \in A$ and $y = \sum_{1\leq j\leq n} \eta_j a_j$, where $\eta_j \in F$ for $1 \leq j \leq n$, then

$$x + y = \sum_{1\leq j\leq n} \xi_j a_j + \sum_{1\leq j\leq n} \eta_j a_j = \sum_{1\leq j\leq n} (\xi_j a_j + \eta_j a_j) = \sum_{1\leq j\leq n} (\xi_j + \eta_j) a_j.$$

By Proposition 4.2.16, this representation is unique. Then

$$\begin{aligned} f(x + y) &= \sum_{1\leq j\leq n} (\xi_j + \eta_j) u_j = \sum_{1\leq j\leq n} (\xi_j u_j + \eta_j u_j) \\ &= \sum_{1\leq j\leq n} \xi_j u_j + \sum_{1\leq j\leq n} \eta_j u_j = f(x) + f(y). \end{aligned}$$

Also, if $\alpha \in F$, then

$$\alpha x = \alpha \left(\sum_{1\leq j\leq n} \xi_j a_j \right) = \sum_{1\leq j\leq n} \alpha(\xi_j a_j) = \sum_{1\leq j\leq n} (\alpha \xi_j) a_j.$$

It follows that

$$f(\alpha x) = \sum_{1 \le j \le n} (\alpha \xi_j) u_j = \sum_{1 \le j \le n} \alpha(\xi_j u_j) = \alpha \left(\sum_{1 \le j \le n} \xi_j u_j \right) = \alpha f(x).$$

This shows that the mapping f is linear. Suppose also that $g : A \longrightarrow V$ is a linear mapping with the property $g(a_j) = u_j$, for $1 \le j \le n$. Then for each element $x \in A$, $x = \sum_{1 \le j \le n} \xi_j a_j$ and we have

$$\begin{aligned} f(x) &= \sum_{1 \le j \le n} \xi_j f(a_j) = \sum_{1 \le j \le n} \xi_j u_j = \sum_{1 \le j \le n} \xi_j g(a_j) \\ &= g\left(\sum_{1 \le j \le n} \xi_j a_j \right) = g(x). \end{aligned}$$

Thus $f = g$, as required.

Consequently, a linear mapping is uniquely determined by assigning images to the basis elements. The following theorem shows that the dimension of a space defines the space up to isomorphism.

5.1.13. Theorem. *Let A, V be vector spaces over a field F. Then A and V are isomorphic if and only if* $\mathbf{dim}_F(A) = \mathbf{dim}_F(V)$.

Proof. If A and V are isomorphic then, by Corollary 5.1.9, $\mathbf{dim}_F(A) = \mathbf{dim}_F(V)$. Conversely, suppose that $\mathbf{dim}_F(A) = \mathbf{dim}_F(V)$. Let $\{a_1, \ldots, a_n\}$ be a basis of A and let $\{v_1, \ldots, v_n\}$ be a basis of V. If x is an arbitrary element of A then, by Proposition 4.2.16, $x = \sum_{1 \le j \le n} \xi_j a_j$, for certain $\xi_j \in F$, where $1 \le j \le n$. Proposition 5.1.12 shows that the mapping $f : A \longrightarrow V$ defined by $f(x) = \sum_{1 \le j \le n} \xi_j v_j$ is linear.

Let u be an arbitrary element of V. Then $u = \sum_{1 \le j \le n} \eta_j v_j$, where $\eta_j \in F$ for $1 \le j \le n$, and we let $y = \sum_{1 \le j \le n} \eta_j a_j$ so that $f(y) = u$. This shows that f is an epimorphism.

To show that f is a monomorphism, let $c \in \mathbf{Ker} f$ and let $c = \sum_{1 \le j \le n} \gamma_j a_j$ be a representation of c relative to the basis $\{a_1, \ldots, a_n\}$, where $\gamma_j \in F$ for $1 \le j \le n$. Then $0_V = f(c) = \sum_{1 \le j \le n} \gamma_j v_j$. Since the subset $\{v_1, \ldots, v_n\}$ is a basis, it is linearly independent and Proposition 4.2.7 implies that $\gamma_1 = \cdots = \gamma_n = 0_F$. Hence $c = 0_A$, and $\mathbf{Ker} f = \{0_A\}$. Theorem 5.1.4 shows that f is a monomorphism and therefore f is an isomorphism.

The following result illustrates the special role of the space F^n.

5.1.14. Corollary. *Let A be a finite-dimensional vector space over the field F. Then $A \cong_F F^n$ where $n =$* $\mathbf{dim}_F(A)$.

Let $\{a_1, \ldots, a_n\}$ be a basis of A and let $x = \sum_{1 \le j \le n} \xi_j a_j$, where $\xi_j \in F$, for $1 \le j \le n$. From Theorem 5.1.13 we see that an isomorphism between A and F^n is defined by $\mathbf{k}(x) = \sum_{1 \le j \le n} \xi_j \mathbf{e}_j$, where $\{\mathbf{e}_1, \ldots, \mathbf{e}_n\}$ is the standard basis of F^n. We remarked in Section 4.2 that $\sum_{1 \le j \le n} \xi_j \mathbf{e}_j = (\xi_1, \ldots, \xi_n)$. Thus, $\mathbf{k}$ maps each element x of the space A to an n-tuple of the coordinates of x in the basis $\{a_1, \ldots, a_n\}$. The mapping $\mathbf{k}$ is called the canonical isomorphism between A and F^n. We should emphasize that the mapping $\mathbf{k}$ depends heavily on the choice of the basis in A.

Let A, V be vector spaces over a field F and let $\mathbf{Hom}_F(A, V)$ denote the set of all linear mappings from A to V. Define addition of linear mappings by if $f, g \in \mathbf{Hom}_F(A, V)$, then $(f + g)(a) = f(a) + g(a)$ for every element $a \in A$. We have

$$\begin{aligned}(f+g)(a+b) &= f(a+b) + g(a+b) = f(a) + f(b) + g(a) + g(b) \\ &= f(a) + g(a) + f(b) + g(b) = (f+g)(a) + (f+g)(b)\end{aligned}$$

and

$$\begin{aligned}(f+g)(\alpha a) &= f(\alpha a) + g(\alpha a) = \alpha f(a) + \alpha g(a) \\ &= \alpha(f(a) + g(a)) = \alpha(f+g)(a),\end{aligned}$$

for all $a, b \in A, \alpha \in F$. It follows that $f + g$ is a linear mapping and that the mapping

$$(f, g) \longmapsto f + g, \text{ where } f, g \in \mathbf{Hom}_F(A, V)$$

defines a binary operation on the set $\mathbf{Hom}_F(A, V)$.

The set $\mathbf{Hom}_F(A, V)$ is an abelian group under this operation since

$$\begin{aligned}(f+g)(a) &= f(a) + g(a) = g(a) + f(a) = (g+f)(a), \\ (f+(g+h))(a) &= f(a) + (g+h)(a) = f(a) + (g(a) + h(a)), \text{ and} \\ ((f+g)+h)(a) &= (f(a) + g(a)) + h(a) = (f(a) + g(a)) + h(a)\end{aligned}$$

for each element $a \in A$. It follows that

$$f + g = g + f \text{ and } f + (g + h) = (f + g) + h$$

for arbitrary $f, g, h \in \mathbf{Hom}_F(A, V)$.

The zero mapping ϑ is the zero element for this operation of addition. Indeed, for each element $a \in A$ we have

$$(f + \vartheta)(a) = f(a) + \vartheta(a) = f(a) + 0_V = f(a),$$

so that $f + \vartheta = f$ for arbitrary $f \in \mathbf{Hom}_F(A, V)$. Finally, let $(-f)(a) = -f(a)$ for each $a \in A$. Clearly $f + (-f) = \vartheta$. So all axioms for an abelian group are valid.

Next, let $f \in \mathbf{Hom}_F(A, V)$ and let $\alpha \in F$. Define the mapping $\alpha f : A \longrightarrow V$ by $(\alpha f)(a) = \alpha f(a)$, for all $a \in A$. If $f, g \in \mathbf{Hom}_F(A, V)$, $\alpha, \beta \in F$, then

$$\begin{aligned}(\alpha(f+g))(a) &= \alpha((f+g)(a)) = \alpha(f(a)+g(a)) = \alpha f(a) + \alpha g(a)\\ &= (\alpha f)(a) + (\alpha g)(a) = (\alpha f + \alpha g)(a),\end{aligned}$$

so that $\alpha(f+g) = \alpha f + \alpha g$.

Furthermore,

$$\begin{aligned}((\alpha+\beta)f)(a) &= (\alpha+\beta)(f(a)) = \alpha f(a) + \beta f(a)\\ &= (\alpha f)(a) + (\beta f)(a) = (\alpha f + \beta f)(a)\end{aligned}$$

and it follows that $(\alpha+\beta)f = \alpha f + \beta f$.

Also,

$$((\alpha\beta)f)(a) = (\alpha\beta)(f(a)) = \alpha(\beta f(a)) = \alpha((\beta f))(a) = (\alpha(\beta f))(a),$$

so that $(\alpha\beta)f = \alpha(\beta f)$. Finally,

$$(ef)(a) = e(f(a)) = f(a),$$

and hence $ef = f$.

Consequently, all conditions of Definition 4.1.4 are valid and the set $\mathbf{Hom}_F(A, V)$ becomes a vector space over the field F. Next, we consider some important special cases.

5.1.15. Definition. *Let A be a vector space over the field F. The linear mapping $f : A \longrightarrow A$ is called a linear transformation of A or a linear operator on A. In this case, we also say that f is an endomorphism of A.*

We write $\mathbf{Hom}_F(A, A) = \mathbf{End}_F(A)$ and, as above, $\mathbf{End}_F(A)$ is a vector space over F. Besides the operations of addition and scalar multiplication, introduced above, the mapping

$$(f, g) \longmapsto f \circ g, \text{ where } f, g \in \mathbf{End}_F(A),$$

introduces a further binary operation on the set $\mathbf{End}_F(A)$. This follows from our remark at the start of this section that a product of two linear mappings is again a linear mapping. Thus, a product of two linear transformations is a linear transformation.

Let $f, g, h \in \mathbf{End}_F(A)$ and let $a \in A$. We have

$$\begin{aligned}(f\circ(g+h))(a) &= f((g+h)(a)) = f(g(a)+h(a)) = f(g(a)) + f(h(a))\\ &= (f\circ g)(a) + (f\circ h)(a) = (f\circ g + f\circ h)(a).\end{aligned}$$

It follows that

$$f \circ (g + h) = f \circ g + f \circ h.$$

In a similar manner, we can also prove that $(g + h) \circ f = g \circ f + h \circ f$. By Theorem 1.3.2, multiplication of mappings is associative and the permutation ε_A, which is a linear transformation of A, is a multiplicative identity. It is also easy to see that

$$\alpha(f \circ g) = (\alpha f) \circ g = f \circ (\alpha g),$$

where $f, g \in \mathbf{End}_F(A), \alpha \in F$.

5.1.16. Definition. *Let A be a vector space over a field F. A linear mapping $f : A \longrightarrow F$ is called a linear functional of A. The vector space $\mathbf{Hom}_F(A, F) = A^*$ is called the dual (or conjugate) space of A.*

5.1.17. Theorem. *Let A be a finite-dimensional vector space over a field F and let $\mathbf{dim}_F(A) = n$. Then $\mathbf{dim}_F(A^*) = n$, and hence A and A^* are isomorphic.*

Proof. Let $\{a_1, \dots, a_n\}$ be a basis of A. If x is an arbitrary element of A then, by Proposition 4.2.16, $x = \sum_{1 \leq j \leq n} \xi_j a_j$, for certain $\xi_j \in F$ with $1 \leq j \leq n$. We define the mapping $\rho_j : A \longrightarrow F$ by $\rho_j(x) = \xi_j$, $1 \leq j \leq n$. Using Proposition 5.1.12 we see that ρ_j is a linear functional.

To show that $\rho_1, \dots, \rho_n$ are linearly independent, let $\gamma_1, \dots, \gamma_n$ be elements of F such that $\sum_{1 \leq j \leq n} \gamma_j \rho_j = \vartheta$. Then

$$0_F = \left(\sum_{1 \leq j \leq n} \gamma_j \rho_j \right) (a_k) = \sum_{1 \leq j \leq n} \gamma_j \rho_j (a_k) = \gamma_k e = \gamma_k, \text{ for } 1 \leq k \leq n$$

and Proposition 4.2.7 shows that $\rho_1, \dots, \rho_n$ are linearly independent.

Next, let f be an arbitrary linear functional. Then

$$f(x) = \sum_{1 \leq j \leq n} \xi_j f\left(a_j\right) = \sum_{1 \leq j \leq n} f\left(a_j\right) \rho_j(x) = \left(\sum_{1 \leq j \leq n} f\left(a_j\right) \rho_j \right) (x),$$

so that $f = \sum_{1 \leq j \leq n} f\left(a_j\right) \rho_j$. Hence, every linear functional is a linear combination of $\rho_1, \dots, \rho_n$, which shows that the subset $\{\rho_1, \dots, \rho_n\}$ is a basis of A^*. Thus, $\mathbf{dim}_F(A^*) = n$ and Theorem 5.1.13 implies that A and A^* are isomorphic.

The basis $\{\rho_1, \dots, \rho_n\}$ constructed above is called the dual of the basis $\{a_1, \dots, a_n\}$ of A. By Theorem 5.1.17, $\mathbf{dim}_F((A^*)^*) = n$ also and therefore the vector spaces A and $(A^*)^* = A^{**}$ are isomorphic. Unlike the isomorphism of A

with A^*, the isomorphism between A and A^{**} does not depend on the basis of A; for this reason, there is a "canonical" isomorphism between A and A^{**} and we now see why this is so.

We define a mapping $\delta : A \longrightarrow A^{**}$ as follows. For every element $x \in A$ we define $\Psi_x : A^* \longrightarrow F$ by $\Psi_x(f) = f(x)$, whenever $f \in A^*$. If $f, g \in A^*$ and $\alpha \in F$, then

$$\Psi_x(f+g) = (f+g)(x) = f(x) + g(x) = \Psi_x(f) + \Psi_x(g) \text{ and}$$
$$\Psi_x(\alpha f) = (\alpha f)(x) = \alpha f(x) = \alpha \Psi_x(f),$$

which shows that Ψ_x is linear. Hence $\Psi_x \in A^{**}$. We now set $\delta(x) = \Psi_x$, where $x \in A$. If $x, y \in A$ then, for each $f \in A^*$, we have

$$\Psi_{x+y}(f) = f(x+y) = f(x) + f(y) = \Psi_x(f) + \Psi_y(f) = (\Psi_x + \Psi_y)(f),$$

which shows that $\Psi_{x+y} = \Psi_x + \Psi_y$. If α is an arbitrary element of F, then

$$\Psi_{\alpha x}(f) = f(\alpha x) = \alpha f(x) = \alpha \Psi_x(f) = (\alpha \Psi_x)(f),$$

which shows that $\Psi_{\alpha x} = \alpha \Psi_x$. Hence we have

$$\delta(x+y) = \Psi_{x+y} = \Psi_x + \Psi_y = \delta(x) + \delta(y) \text{ and}$$
$$\delta(\alpha x) = \Psi_{\alpha x} = \alpha \Psi_x = \alpha \delta(x).$$

These equations show that the mapping δ is linear.

Let $\{a_1, \ldots, a_n\}$ be a basis of A, let $\{\rho_1, \ldots, \rho_n\}$ be a basis of A^* dual to $\{a_1, \ldots, a_n\}$, and let $\{P_1, \ldots, P_n\}$ be a basis of A^{**} dual to $\{\rho_1, \ldots, \rho_n\}$. We have $\delta(a_j) = \Psi_{a_j}$, for $1 \leq j \leq n$. If f is an arbitrary functional then, as proved above, $f = \sum_{1 \leq k \leq n} f(a_k)\, \rho_k$. Hence

$$\Psi_{a_j}(f) = \Psi_{a_j}\left(\sum_{1 \leq k \leq n} f(a_k)\, \rho_k\right) = \sum_{1 \leq k \leq n} f(a_k)\, \Psi_{a_j}(\rho_k)$$
$$= \sum_{1 \leq k \leq n} f(a_k)\, \rho_k(a_j) = f(a_j)e = f(a_j).$$

By definition, $P_j(f) = f(a_j)$, so $\Psi_{a_j}(f) = P_j(f)$, for all $f \in A^*$. Thus, $\Psi_{a_j} = P_j$ and hence $\delta(a_j) = P_j$, for $1 \leq j \leq n$. From the proof of Theorem 5.1.13 we deduce that δ is an isomorphism.

We note that if A is not finite dimensional then the mapping δ is a monomorphism; but it is not always an epimorphism.

We next show that the external direct sum and the internal direct sum can be viewed as being identical. Let $A_1, \ldots, A_n$ be vector spaces over a field F and let D denote the external direct sum of $A_1, \ldots, A_n$. In the vector space D define the subset D_j by

$$a = (a_1, \ldots, a_n) \in D_j \text{ if and only if } a_k = 0_{A_k} \text{ whenever } k \neq j.$$

It is easy to see that D_j is a subspace of D, for $1 \leq j \leq n$. For an arbitrary n-tuple $(a_1, \ldots, a_n)$ we have

$$(a_1, \ldots, a_n) = (a_1, 0_{A_2}, \ldots, 0_{A_n}) + \cdots + (0_{A_1}, \ldots, 0_{A_{n-1}}, a_n),$$

which shows that $D = D_1 + \cdots + D_n$. We next consider the intersection $D_j \cap \sum_{k \neq j} D_k$. If $\mathbf{x} = (x_1, \ldots, x_n) \in D_k$ for $k \neq j$, then $x_j = 0_{A_j}$. Hence, if $\mathbf{y} = (y_1, \ldots, y_n) \in \sum_{k \neq j} D_k$, then $y_j = 0_{A_j}$. If $\mathbf{y} = (y_1, \ldots, y_n) \in D_j \cap \sum_{k \neq j} D_k$, then $y_k = 0_{A_k}$ for $k \neq j$, so that $\mathbf{y} = (0_{A_1}, \ldots, 0_{A_n})$. This shows that D is the internal direct sum of the subspaces $D_1, \ldots, D_n$.

Clearly, the mapping

$$x \longrightarrow (0_{A_1}, \ldots, 0_{A_{j-1}}, x, 0_{A_{j+1}}, \ldots, O_{A_n}), \text{ where } x \in A_j,$$

is an isomorphism of A_j onto D_j, for $1 \leq j \leq n$. Consequently, the external direct sum of the spaces $A_1, \ldots, A_n$ is isomorphic to the internal direct sum of the subspaces $D_1, \ldots, D_n$ where $D_j \cong_F A_j$, for $1 \leq j \leq n$.

Conversely, let A be the internal direct sum of the subspaces $A_1, \ldots, A_n$ and let E denote the external direct sum of $A_1, \ldots, A_n$. Every element $a \in A$ has a representation of the form $a = a_1 + \cdots + a_n$, where $a_j \in A_j$, for $1 \leq j \leq n$ and this representation is unique. The mapping $\upsilon : A \longrightarrow E$ defined by $\upsilon(a) = (a_1, \ldots, a_n)$ is linear. For, if c is another element of A and $c = c_1 + \cdots + c_n$, where $c_j \in A_j$, for $1 \leq j \leq n$, then

$$a + c = a_1 + \cdots + a_n + c_1 + \cdots + c_n = (a_1 + c_1) + \cdots + (a_n + c_n),$$

so that

$$\begin{aligned} \upsilon(a + c) &= (a_1 + c_1, \ldots, a_n + c_n) = (a_1, \ldots, a_n) + (c_1, \ldots, c_n) \\ &= \upsilon(a) + \upsilon(c). \end{aligned}$$

Furthermore, if α is an arbitrary element of F, then

$$\alpha a = \alpha(a_1 + \cdots + a_n) = \alpha a_1 + \cdots + \alpha a_n,$$

so that

$$\upsilon(\alpha a) = (\alpha a_1, \ldots, \alpha a_n) = \alpha(a_1, \ldots, a_n) = \alpha \upsilon(a).$$

It follows that the mapping υ is linear.

Let $(x_1, \ldots, x_n)$ be an arbitrary element of E. Then $x_j \in A_j \leq A$, for $1 \leq j \leq n$, and therefore we can form the sum of the elements $x_1, \ldots, x_n$ in the space A. Let $x = x_1 + \cdots + x_n$, then $\upsilon(x) = (x_1, \ldots, x_n)$ and the mapping υ is injective.

Finally, let $y \in \mathbf{Ker}\,\upsilon$. Then $y = y_1 + \cdots + y_n$, where $y_j \in A_j$, for $1 \leq j \leq n$. The choice of y implies that $\upsilon(y) = (0_A, \ldots, 0_A)$. On the other hand, $\upsilon(y) = (y_1, \ldots, y_n)$ and we deduce that $y_j = 0_A$ for each j, where $1 \leq j \leq n$. Hence $y = 0_A$ and Theorem 5.1.4 shows that υ is a monomorphism and therefore an isomorphism. Consequently, the internal direct sum of the subspaces $A_1, \ldots, A_n$ is isomorphic to the external direct sum. Thus, we no longer need to use the qualifiers "internal" and "external" when discussing direct sums and so refer simply to the "direct sum." It will be clear from the context as to which of the cases, internal or external, is being discussed.

EXERCISE SET 5.1

Justify your work with a proof or a counterexample where necessary.

5.1.1. Let $f : \mathbb{R}^3 \longrightarrow \mathbb{R}^2$ be the mapping defined by $f(\alpha, \beta, \gamma) = (\alpha - \beta, \alpha + \gamma)$. Is this mapping linear?

5.1.2. Let $f : \mathbb{R}^5 \longrightarrow \mathbb{R}^3$ be the mapping defined by $f(\alpha, \beta, \gamma, \lambda, \mu) = (\gamma\lambda, \alpha + \beta, \mu)$. Is this mapping linear?

5.1.3. Let $f : \mathbb{R}^2 \longrightarrow \mathbb{R}^4$ be the mapping defined by $f(\alpha, \beta) = (\alpha, 0, \alpha + \beta, 0)$. Is this mapping linear?

5.1.4. Let $f : \mathbb{R}^4 \longrightarrow \mathbb{R}^3$ be the mapping defined by $f(\alpha, \beta, \gamma, \lambda) = (\alpha, \beta, \gamma + \lambda)$. Is this mapping linear? If yes, find $\mathbf{Im}\,f$ and $\mathbf{Ker}\,f$.

5.1.5. Let $f : \mathbb{R}^5 \longrightarrow \mathbb{R}^3$ be the mapping defined by $f(\alpha, \beta, \gamma, \lambda, \mu) = (0, \alpha + \beta, \mu)$. Is this mapping linear? If yes, find $\mathbf{Im}\,f$ and $\mathbf{Ker}\,f$.

5.1.6. Let $A = \mathbb{R}[X]$ be the vector space of all polynomials with real coefficients and let $f : A \longrightarrow A$ be the mapping defined by $f(g(X)) = g'(X)$ [the derivative of the polynomial $g(X)$]. Prove that f is a linear transformation of the space A. Find $\mathbf{Ker}\,f$ and $\mathbf{Im}\,f$.

5.1.7. Let $A = \mathbb{R}[X]$ be the vector space of all polynomials with real coefficients, let $\phi : A \longrightarrow A$ be the mapping defined by $\varphi(g(X)) = g'(X)$ [the derivative of the polynomial $g(X)$] and let $\psi : A \longrightarrow A$ be the mapping defined by the rule $\varphi(g(X)) = Xg(X)$. Find $\phi \circ \psi^{10} - \psi^{10} \circ \phi$.

5.1.8. Let A be a vector space over a field F and let $f, g \in \mathbf{End}_F(A)$. Prove that $\mathbf{dim}_F(\mathbf{Ker}(f \circ g)) \leq \mathbf{dim}_F(\mathbf{Ker}\,f) + \mathbf{dim}_F(\mathbf{Ker}\,g)$.

5.1.9. Let A be a vector space over a field F and let $f, g \in \mathbf{End}_F(A)$. Prove that $\mathbf{dim}_F(\mathbf{Im}(f + g)) \leq \mathbf{dim}_F(\mathbf{Im}\,f) + \mathbf{dim}_F(\mathbf{Im}\,g)$.

5.1.10. Let A be a vector space over a field F and let $f \in \mathbf{End}_F(A)$. Are the subspaces $\mathbf{Ker}\,f$ and $\mathbf{Im}\,f$ invariant relative to f? [A subspace B is called invariant relative to f if for every $b \in B$ we have $f(b) \in B$].

5.1.11. Let A be a vector space over a field F, let $f, g \in \mathbf{End}_F(A)$ and let B be a subspace of A containing $\mathbf{Im} f$. Is B invariant relative to f?

5.1.12. Let A be a vector space over a field F and let B be a subset of A. Put $B^* = \{f \in A^* \mid f(x) = 0_F \text{ for each } x \in B\}$. Prove that B^* is a subspace of A^*. If B is a subspace of A, then prove that $\mathbf{dim}_F(B) + \mathbf{dim}_F(B^*) = \mathbf{dim}_F(A)$.

5.2 MATRICES OF LINEAR MAPPINGS

In Section 5.1 we began the study of linear mappings of vector spaces. In this section we continue this study in the case of finite-dimensional vector spaces. In this case, matrix techniques can be employed, which prove to be very effective.

Let A, V be finite-dimensional vector spaces over a field F and let $f : A \longrightarrow V$ be a linear mapping. Let $\{a_1, \dots, a_n\}$ be a basis of A and $\{v_1, \dots, v_k\}$ be a basis of V. In Proposition 5.1.12 we proved that a linear mapping f is uniquely defined by assigning the images $f(a_1), \dots, f(a_n)$ to the basis elements $\{a_1, \dots, a_n\}$. By Proposition 4.2.16, each element of V is uniquely defined by its coordinates relative to the basis $\{v_1, \dots, v_k\}$. This holds in particular for $f(a_1), \dots, f(a_n)$ so we have

$$\begin{aligned}
f(a_1) &= \sigma_{11}v_1 + \sigma_{21}v_2 + \cdots + \sigma_{k1}v_k \\
f(a_2) &= \sigma_{12}v_1 + \sigma_{22}v_2 + \cdots + \sigma_{k2}v_k \\
&\vdots \\
f(a_n) &= \sigma_{1n}v_1 + \sigma_{2n}v_2 + \cdots + \sigma_{kn}v_k.
\end{aligned}$$

5.2.1. Definition. *Let A, V be finite-dimensional vector spaces over a field F and let $\{a_1, \dots, a_n\}$, and $\{v_1, \dots, v_k\}$ denote bases of A and V, respectively. Let $f : A \longrightarrow V$ be a linear mapping and suppose that $f(a_m) = \sum_{1 \le j \le k} \sigma_{jm} v_j$, for $1 \le m \le n$. The matrix*

$$\begin{pmatrix}
\sigma_{11} & \sigma_{12} & \dots & \sigma_{1n} \\
\sigma_{21} & \sigma_{22} & \dots & \sigma_{2n} \\
\vdots & \vdots & \ddots & \vdots \\
\sigma_{k1} & \sigma_{k2} & \dots & \sigma_{kn}
\end{pmatrix}$$

is called the matrix of the linear mapping f relative to the bases $\{a_1, \dots, a_n\}$ and $\{v_1, \dots, v_k\}$.

We observe the following important properties of matrices of linear mappings.

5.2.2. Proposition. *Let A, V be finite-dimensional vector spaces over a field F and let $\{a_1, \dots, a_n\}$, and $\{v_1, \dots, v_k\}$ denote bases of A and V, respectively. Let*

$f, g \in \mathbf{Hom}_F(A, V)$ and let $S = [\sigma_{jt}]$, $R = [\rho_{jt}] \in \mathbf{M}_{k\times n}(F)$ denote the matrices of f and g relative to the pair of bases $\{a_1, \ldots, a_n\}$, $\{v_1, \ldots, v_k\}$. Then

(i) *$S + R$ is the matrix of $f + g$ relative to the bases $\{a_1, \ldots, a_n\}$ and $\{v_1, \ldots, v_k\}$;*

(ii) *If $\alpha \in F$, then αS is the matrix of the mapping αf relative to the bases $\{a_1, \ldots, a_n\}$ and $\{v_1, \ldots, v_k\}$.*

Proof.

(i) We have

$$\begin{aligned}(f+g)(a_m) = f(a_m) + g(a_m) &= \sum_{1\le j\le k} \sigma_{jm} v_j + \sum_{1\le j\le k} \rho_{jm} v_j \\ &= \sum_{1\le j\le k} (\sigma_{jm} + \rho_{jm}) v_j\end{aligned}$$

for every m, where $1 \le m \le n$. It follows that $[\sigma_{jt} + \rho_{jt}] = S + R$ is the matrix of the linear mapping $f + g$ relative to the bases $\{a_1, \ldots, a_n\}$ and $\{v_1, \ldots, v_k\}$.

(ii) Likewise, we have

$$(\alpha f)(a_m) = \alpha f(a_m) = \alpha \left(\sum_{1\le j\le k} \sigma_{jm} v_j \right) = \sum_{1\le j\le k} (\alpha \sigma_{jm}) v_j,$$

for every m, where $1 \le m \le n$. It follows that $[\alpha\sigma_{jt}] = \alpha S$ is the matrix of the linear mapping αf relative to the bases $\{a_1, \ldots, a_n\}$ and $\{v_1, \ldots, v_k\}$.

5.2.3. Corollary. *Let A, V be finite-dimensional vector spaces over a field F and suppose that $\mathbf{dim}_F(A) = n$, $\mathbf{dim}_F(V) = k$. Then the vector space $\mathbf{Hom}_F(A, V)$ is isomorphic to $\mathbf{M}_{k\times n}(F)$.*

Proof. Let $\{a_1, \ldots, a_n\}$ and $\{v_1, \ldots, v_k\}$ be bases of A and V respectively. We define the mapping $\Gamma : \mathbf{Hom}_F(A, V) \longrightarrow \mathbf{M}_{k\times n}(F)$ as follows. For each $f \in \mathbf{Hom}_F(A, V)$ let $\Gamma(f) = S$ where S is the matrix of f relative to the bases $\{a_1, \ldots, a_n\}$ and $\{v_1, \ldots, v_k\}$. Proposition 5.2.2 shows that this mapping is linear.

Let $R = [\rho_{jt}] \in \mathbf{M}_{k\times n}(F)$ and define elements $u_1, \ldots, u_n$ of V by $u_m = \sum_{1\le j\le k} \rho_{jm} v_j$, for $1 \le m \le n$. By Proposition 5.1.12, there exists a unique linear mapping $g : A \longrightarrow V$ such that $g(a_m) = u_m = \sum_{1\le j\le k} \rho_{jm} v_j$, for $1 \le m \le n$. Thus, R is the matrix of g relative to the bases $\{a_1, \ldots, a_n\}$ and $\{v_1, \ldots, v_k\}$. Hence the mapping Γ is surjective.

Finally, let $f, g : A \longrightarrow V$ be linear mappings and let $S = [\sigma_{jt}]$, $R = [\rho_{jt}] \in \mathbf{M}_{k\times n}(F)$ be the matrices of f and g relative to the bases $\{a_1, \ldots, a_n\}$ and $\{v_1, \ldots, v_k\}$ respectively. Suppose that $\Gamma(f) = \Gamma(g)$, so that $S = R$. Then

$$g(a_m) = \sum_{1\le j\le k} \rho_{jm} v_j = \sum_{1\le j\le k} \sigma_{jm} v_j = f(a_m), \text{ for } 1 \le m \le n.$$

Let x be an arbitrary element of A. By Proposition 4.2.16, $x = \sum_{1\leq j\leq n} \xi_j a_j$, for certain $\xi_j \in F$. Then

$$f(x) = \sum_{1\leq m\leq n} \xi_m f(a_m) = \sum_{1\leq m\leq n} \xi_m g(a_m) = g(x),$$

which proves that $f = g$. Hence, the mapping Γ is injective and therefore Γ is an isomorphism.

5.2.4. Corollary. *Let A, V be finite-dimensional vector spaces over a field F. Then the vector space $\mathbf{Hom}_F(A, V)$ is finite dimensional and*

$$\mathbf{dim}_F(\mathbf{Hom}_F(A, V)) = \mathbf{dim}_F(A)\mathbf{dim}_F(V).$$

Proof. Let $\mathbf{dim}_F(A) = n$ and $\mathbf{dim}_F(V) = k$. From Corollary 5.2.3 we have the isomorphism $\mathbf{Hom}_F(A, V) \cong_F \mathbf{M}_{k\times n}(F)$ and, by Corollary 5.1.9, $\mathbf{dim}_F(\mathbf{Hom}_F(A, V)) = \mathbf{dim}_F(\mathbf{M}_{k\times n}(F))$. However, $\mathbf{dim}_F(\mathbf{M}_{k\times n}(F)) = nk$ so

$$\mathbf{dim}_F(\mathbf{Hom}_F(A, V)) = nk = \mathbf{dim}_F(A)\mathbf{dim}_F(V).$$

Let $\{a_1, \dots, a_n\}$ and $\{v_1, \dots, v_k\}$ be bases of the spaces A and V respectively. Let $S = [\sigma_{jt}] \in \mathbf{M}_{k\times n}(F)$ denote the matrix of f relative to the bases $\{a_1, \dots, a_n\}$ and $\{v_1, \dots, v_k\}$. Let x be an arbitrary element of A. By Proposition 4.2.16, $x = \sum_{1\leq m\leq n} \xi_m a_m$ and $f(x) = \sum_{1\leq j\leq k} \lambda_j v_j$, for certain $\xi_m, \lambda_j \in F$. Then

$$\begin{aligned} f(x) &= \sum_{1\leq m\leq n} \xi_m f(a_m) = \sum_{1\leq m\leq n} \xi_m \left(\sum_{1\leq j\leq k} \sigma_{jm} v_j \right) \\ &= \sum_{1\leq m\leq n} \sum_{1\leq j\leq k} \xi_m \sigma_{jm} v_j = \sum_{1\leq j\leq k} \left(\sum_{1\leq m\leq n} \sigma_{jm} \xi_m \right) v_j. \end{aligned}$$

By Proposition 4.2.16, $\lambda_j = \sum_{1\leq m\leq n} \sigma_{jm}\xi_m$, for $j = 1, \dots, k$ and we arrive at the matrix equation

$$\begin{pmatrix} \lambda_1 \\ \lambda_2 \\ \vdots \\ \lambda_k \end{pmatrix} = \begin{pmatrix} \sigma_{11} & \sigma_{12} & \dots & \sigma_{1n} \\ \sigma_{21} & \sigma_{22} & \dots & \sigma_{2n} \\ \vdots & \vdots & \ddots & \vdots \\ \sigma_{k1} & \sigma_{k2} & \dots & \sigma_{kn} \end{pmatrix} \begin{pmatrix} \xi_1 \\ \xi_2 \\ \vdots \\ \xi_n \end{pmatrix}.$$

This equation tells us how the coordinates of x relative to $\{a_1, \dots, a_n\}$ are related to the coordinates of $f(x)$ relative to $\{v_1, \dots, v_k\}$ in terms of the matrix of the linear transformation. There is one further characterization of linear mappings that we wish to consider.

5.2.5. Proposition. *Let A, V be finite-dimensional vector spaces over a field F and let $\{a_1, \dots, a_n\}, \{v_1, \dots, v_k\}$ be bases of A and V, respectively. Suppose that $f \in \mathbf{Hom}_F(A, V)$ and that $S = [\sigma_{jt}] \in \mathbf{M}_{k\times n}(F)$ is the matrix of f relative to the bases $\{a_1, \dots, a_n\}$ and $\{v_1, \dots, v_k\}$, respectively. Then $\mathbf{rank}(S) = \mathbf{dim}_F(\mathbf{Im} f)$.*

Proof. Since A is generated by the elements $a_1, \dots, a_n$, $\mathbf{Im} f$ is generated by their images $f(a_1), \dots, f(a_n)$. Let $\kappa : V \longrightarrow F^k$ be the canonical isomorphism. By Corollary 5.1.9, $\mathbf{dim}_F(\mathbf{Im} f) = \mathbf{dim}_F(\kappa(\mathbf{Im} f))$. We have $f(a_m) = \sum_{1\le j\le k} \sigma_{jm} v_j$, so that

$$\kappa(f(a_m)) = (\sigma_{1m}, \dots, \sigma_{km}), \text{ for } 1 \le m \le n.$$

It follows that $\kappa(f(a_m))$ is generated by all the columns of the matrix S. Consequently, $\mathbf{dim}_F(\kappa(\mathbf{Im} f)) = \mathbf{rank}(S)$, as required.

The result proved above shows that the rank of a matrix of a linear mapping f does not change with a basis change and hence is an invariant of the linear mapping, which we denote by $\mathbf{rank}(f)$.

Multiplication of matrices was introduced in Section 2.1 with little justification. The next proposition suggests why we multiply matrices the way we do.

5.2.6. Proposition. *Let A, V, W be finite-dimensional vector spaces over a field F and let $\{a_1, \dots, a_n\}, \{v_1, \dots, v_k\}, \{w_1, \dots, w_t\}$ be bases of A, V, and W, respectively. Suppose that $f : A \longrightarrow V, g : V \longrightarrow W$ are linear mappings. Let $S = [\sigma_{ij}] \in \mathbf{M}_{k\times n}(F)$ be the matrix of the mapping f relative to the bases $\{a_1, \dots, a_n\}$ and $\{v_1, \dots, v_k\}$ and let $R = [\rho_{ij}] \in \mathbf{M}_{t\times k}$ be the matrix of the mapping g relative to the bases $\{v_1, \dots, v_k\}$ and $\{w_1, \dots, w_t\}$. Then RS is the matrix of the mapping $g \circ f$ relative to the bases $\{a_1, \dots, a_n\}$ and $\{w_1, \dots, w_t\}$.*

Proof. Let $M = [\mu_{ij}] \in \mathbf{M}_{t\times n}(F)$ denote the matrix of $g \circ f$ relative to the bases $\{a_1, \dots, a_n\}$ and $\{w_1, \dots, w_t\}$. Then we have

$$(g \circ f)(a_m) = \sum_{1\le s\le t} \mu_{sm} w_s, \text{ for } 1 \le m \le n. \tag{5.1}$$

On the other hand,

$$\begin{aligned}(g \circ f)(a_m) &= g(f(a_m)) = g\left(\sum_{1\le j\le k} \sigma_{jm} v_j\right) = \sum_{1\le j\le k} \sigma_{jm} g(v_j)\\ &= \sum_{1\le j\le k} \sigma_{jm}\left(\sum_{1\le s\le t} \rho_{sj} w_s\right) = \sum_{1\le j\le k}\sum_{1\le s\le t} \sigma_{jm}\rho_{sj} w_s\\ &= \sum_{1\le s\le t}\left(\sum_{1\le j\le k} \rho_{sj}\sigma_{jm}\right) w_s.\end{aligned}$$

By Proposition 4.2.16, the decomposition of $(g \circ f)(a_m)$ given in Equation 5.1 is unique, so that

$$\mu_{sm} = \sum_{1 \le j \le k} \rho_{sj}\sigma_{jm}, \ for\ 1 \le s \le t \text{ and } 1 \le m \le n.$$

It follows that $M = RS$.

The matrix of a linear mapping depends significantly on which bases are being used. If we change the bases chosen in spaces A, V this matrix will be changed also. Our next theorem tells us how this basis change is effected.

5.2.7. Theorem. *Let A, V be finite-dimensional vector spaces over a field F. Let $\{a_1, \ldots, a_n\}, \{b_1, \ldots, b_n\}$ be bases of A and let $\{v_1, \ldots, w_k\}$, $\{w_1, \ldots, w_k\}$ be bases of V. Suppose that $f : A \longrightarrow V$ is a linear mapping. Let $S = [\sigma_{ij}] \in \mathbf{M}_{k\times n}(F)$ be the matrix of f relative to the bases $\{a_1, \ldots, a_n\}$ and $\{v_1, \ldots, v_k\}$ and let $R = [\rho_{ij}] \in \mathbf{M}_{k\times n}(F)$ be the matrix of f relative to the bases $\{b_1, \ldots, b_n\}$ and $\{w_1, \ldots, w_k\}$. Let $T = [\iota_{ij}] \in \mathbf{M}_n(F)$, $Q = [\vartheta_{ij}] \in \mathbf{M}_k(F)$ be the transition matrices from $\{a_1, \ldots, a_n\}$ to $\{b_1, \ldots, b_n\}$ and from $\{v_1, \ldots, v_k\}$ to $\{w_1, \ldots, w_k\}$ respectively. Then $R = Q^{-1}ST$.*

Proof. We have $f(a_m) = \sum_{1\le j\le k} \sigma_{jm} v_j$ and $f(b_m) = \sum_{1\le j\le k} \rho_{jm} w_j$, for $1 \le m \le n$. On the one hand,

$$f(b_m) = \sum_{1\le j\le k} \rho_{jm} w_j = \sum_{1\le j\le k} \rho_{jm}\left(\sum_{1\le s\le k} \vartheta_{sj} v_s\right)$$

$$= \sum_{1\le j\le k}\sum_{1\le s\le k} \rho_{jm}\vartheta_{sj} v_s = \sum_{1\le s\le k}\left(\sum_{1\le j\le k} \vartheta_{sj}\rho_{jm}\right) v_s.$$

On the other hand,

$$f(b_m) = f\left(\sum_{1\le j\le n} \iota_{jm} a_j\right) = \sum_{1\le j\le n} \iota_{jm} f(a_j) = \sum_{1\le j\le n} \iota_{jm}\left(\sum_{1\le s\le k} \sigma_{sj} v_s\right)$$

$$= \sum_{1\le j\le n}\sum_{1\le s\le k} \iota_{jm}\sigma_{sj} v_s = \sum_{1\le s\le k}\left(\sum_{1\le j\le n} \sigma_{sj}\iota_{jm}\right) v_s.$$

By Proposition 4.2.16, the representation of an element as a linear combination of the basis vectors is unique; so

$$\sum_{1\le j\le k} \vartheta_{sj}\rho_{jm} = \sum_{1\le j\le n} \sigma_{sj}\iota_{jm}, \text{ for } 1 \le s \le k, 1 \le m \le n. \tag{5.2}$$

Let

$$QR = [\beta_{sm}] \in \mathbf{M}_{k\times n}(F) \text{ and } ST = [\gamma_{sm}] \in \mathbf{M}_{k\times n}(F).$$

Then, by the definition of multiplication of matrices we obtain

$$\beta_{sm} = \sum_{1\leq j\leq k} \vartheta_{sj}\rho_{jm} \text{ and } \gamma_{sm} = \sum_{1\leq j\leq n} \sigma_{sj}\iota_{jm}.$$

From Equation 5.2 we deduce that $\beta_{sm} = \gamma_{sm}$, for $1 \leq s \leq k$ and $1 \leq m \leq n$, which implies that $QR = ST$. By Corollary 4.2.19, the matrix Q is nonsingular, so Q^{-1} exists, and multiplying both sides of the last equation by Q^{-1}, we obtain $R = Q^{-1}ST$.

We consider next an important special case of linear mappings, namely, the linear transformations.

Let A be the vector space over a field F and let f be a linear transformation of A. Suppose that A is finite dimensional and let $\{a_1, \dots, a_n\}$ be a basis of A. We write each of the elements $f(a_1), \dots, f(a_n)$ in terms of the basis $\{a_1, \dots, a_n\}$, so

$$\begin{aligned}
f(a_1) &= \sigma_{11}a_1 + \sigma_{21}a_2 + \cdots + \sigma_{n1}a_n \\
f(a_2) &= \sigma_{21}a_1 + \sigma_{22}a_2 + \cdots + \sigma_{n2}a_n \\
&\vdots \\
f(a_n) &= \sigma_{1n}a_1 + \sigma_{2n}a_2 + \cdots + \sigma_{nn}a_n.
\end{aligned}$$

5.2.8. Definition. *Let A be a finite-dimensional vector space over a field F and $\{a_1, \dots, a_n\}$ be a basis of A. Let f be a linear transformation of A and let $f(a_m) = \sum_{1\leq j\leq k} \sigma_{jm}v_j$, for $1 \leq m \leq n$. The matrix*

$$\begin{pmatrix}
\sigma_{11} & \sigma_{12} & \dots & \sigma_{1n} \\
\sigma_{21} & \sigma_{22} & \dots & \sigma_{2n} \\
\vdots & \vdots & \ddots & \vdots \\
\sigma_{n1} & \sigma_{n2} & \dots & \sigma_{nn}
\end{pmatrix}$$

is called the matrix of f relative to the basis $\{a_1, \dots, a_n\}$.

From the results proved above for linear mappings we derive the following properties of matrices of linear transformations.

5.2.9. Proposition. *Let A be a finite-dimensional vector space over a field F and let $\{a_1, \dots, a_n\}$ be a basis of A. Let $f, g \in \mathbf{End}_F(A)$ and let $S = [\sigma_{jt}] \in \mathbf{M}_n(F)$, $R = [\rho_{jt}] \in \mathbf{M}_n(F)$ be the matrices of f and g relative to the basis $\{a_1, \dots, a_n\}$. Then*

(i) *$S + R$ is the matrix of $f + g$ relative to $\{a_1, \ldots, a_n\}$.*
(ii) *RS is the matrix of $g \circ f$ relative to $\{a_1, \ldots, a_n\}$.*
(iii) *If $\alpha \in F$ then αS is the matrix of αf relative to the basis $\{a_1, \ldots, a_n\}$.*

This proposition follows from Propositions 5.2.2 and 5.2.6.

5.2.10. Corollary. *Let A be a finite-dimensional vector space over a field F and let $\mathbf{dim}_F(A) = n$. Then there is an isomorphism $\Gamma : \mathbf{End}_F(A) \longrightarrow \mathbf{M}_n(F)$ such that $\Gamma(g \circ f) = \Gamma(g)\Gamma(f)$.*

Proof. Let $\{a_1, \ldots, a_n\}$ be a basis of A. We define the mapping $\Gamma : \mathbf{End}_F(A) \longrightarrow \mathbf{M}_n(F)$ by setting $\Gamma(f) = S$ where f is a linear transformation and S is the matrix of f relative to the basis $\{a_1, \ldots, a_n\}$. Proposition 5.2.9 shows that this mapping respects multiplication and scalar multiplication. As in the proof of Corollary 5.2.3 we can show that Γ is bijective.

5.2.11. Corollary. *Let A be a finite-dimensional vector space over a field F. Then $\mathbf{End}_F(A)$ is finite dimensional and $\mathbf{dim}_F(\mathbf{End}_F(A)) = \mathbf{dim}_F(A)^2$.*

Finally, Theorem 5.2.7 implies the following result.

5.2.12. Corollary. *Let A be a finite-dimensional vector space over a field F and let $\{a_1, \ldots, a_n\}$, $\{b_1, \ldots, b_n\}$ be bases of A. Suppose that f is a linear transformation of A, and let $S = [\sigma_{jt}]$, $R = [\rho_{jt}] \in \mathbf{M}_n(F)$ be the matrices of f relative to $\{a_1, \ldots, a_n\}$ and $\{b_1, \ldots, b_n\}$ respectively. Let $T = [\iota_{jt}] \in \mathbf{M}_n(F)$ be the transition matrix from $\{a_1, \ldots, a_n\}$ to $\{b_1, \ldots, b_n\}$. Then $R = T^{-1}ST$.*

5.2.13. Definition. *Let A be a vector space over a field F. An isomorphism f of A onto itself is called an automorphism of A.*

Thus, an automorphism of A is a bijective linear transformation of A.

5.2.14. Proposition. *Let A be a finite-dimensional vector space over a field F and let f be a linear transformation of A. The following are equivalent:*

(i) *f is a monomorphism.*
(ii) *f is an automorphism.*
(iii) *f is an epimorphism.*

Proof.
(i) $\Longrightarrow$ (ii) Suppose first that f is a monomorphism. By Corollary 5.1.11, $\mathbf{dim}_F(A) = \mathbf{dim}_F(\mathbf{Ker} f) + \mathbf{dim}_F(\mathbf{Im} f)$. Theorem 5.1.4 implies that $\mathbf{Ker} f = \{0_A\}$, so that $\mathbf{dim}_F(A) = \mathbf{dim}_F(\mathbf{Im} f)$. Theorem 4.2.20 shows that in this case, $\mathbf{Im} f = A$. It follows that the mapping f is surjective, so f is bijective and f is an automorphism of A.

The implication (ii) $\Longrightarrow$ (iii) is clear.

(iii) $\Longrightarrow$ (i). We again apply Corollary 5.1.11 and deduce that $\mathbf{dim}_F(A) = \mathbf{dim}_F(\mathbf{Ker} f) + \mathbf{dim}_F(\mathbf{Im} f)$. Since $A = \mathbf{Im} f$, $\mathbf{dim}_F(A) = \mathbf{dim}_F(\mathbf{Im} f)$, so that $\mathbf{dim}_F(\mathbf{Ker} f) = 0$ and hence $\mathbf{Ker} f = \{0_A\}$. Theorem 5.1.4 shows that f is a monomorphism.

5.2.15. Corollary. *Let A be a finite-dimensional vector space over a field F and let f be a linear transformation of A. Then, f is an automorphism of A if and only if* $\mathbf{rank}(f) = \mathbf{dim}_F(A)$.

Thus if f is an automorphism of A, then the matrix of f relative to any basis is nonsingular; and conversely, if the matrix of f relative to some basis is nonsingular, then f is an automorphism.

Corollary 5.2.15 follows from Proposition 5.2.5.

EXERCISE SET 5.2

Justify your work where appropriate using a proof or a counterexample.

5.2.1. Let $f : \mathbb{R}^3 \longrightarrow \mathbb{R}^2$ be the mapping defined by $f(\alpha, \beta, \gamma) = (\alpha - \beta, \alpha + \gamma)$. Find the matrix of f relative to the standard bases over the spaces $\mathbb{R}^3$ and $\mathbb{R}^2$.

5.2.2. Let $f : \mathbb{R}^2 \longrightarrow \mathbb{R}^4$ be the mapping defined by $f(\alpha, \beta) = (\alpha, 0, \alpha + \beta, 0)$. Find its matrix relative to the standard bases over the spaces $\mathbb{R}^2$ and $\mathbb{R}^4$.

5.2.3. Let $f : \mathbb{R}^4 \longrightarrow \mathbb{R}^3$ be the mapping defined by $f(\alpha, \beta, \gamma, \lambda) = (\alpha, \beta, \gamma + \lambda)$. Find the matrix of f relative to the standard bases over the spaces $\mathbb{R}^4$ and $\mathbb{R}^3$.

5.2.4. Let $f : \mathbb{R}^5 \longrightarrow \mathbb{R}^3$ be the mapping defined by $f(\alpha, \beta, \gamma, \lambda, \mu) = (0, \alpha + \beta, \mu)$. Find the matrix of f relative to the standard bases over the spaces $\mathbb{R}^5$ and $\mathbb{R}^3$.

5.2.5. What dimension has the vector space $\mathbf{Hom}_{\mathbb{Q}}(A, U)$ where $A = \mathbf{M}_2(\mathbb{Q})$, $U = \mathbb{Q}^4$.

5.2.6. What dimension has the vector space $\mathbf{Hom}_{\mathbb{Q}}(A, U)$ where $A = \mathbb{Q}^9$, $U = \mathbf{M}_{11}(\mathbb{Q})$.

5.2.7. Prove that the matrices of a linear transformation f relative to two different bases coincide if and only if the transition matrix from the first basis to the second basis permutes with one of the matrices of f relative to one of the given bases.

5.2.8. Let A be a vector space over a field F and let $f, g \in A^*$. Prove that f, g are linearly independent if and only if $\mathbf{Ker} f \cap \mathbf{Ker} g = \{0_A\}$.

5.2.9. Let

$$M = \begin{pmatrix} 1 & 1 & 0 \\ 0 & 0 & 0 \\ 3 & 1 & 0 \\ 0 & 1 & 0 \end{pmatrix}.$$

Is there a linear transformation $f : \mathbb{R}^3 \longrightarrow \mathbb{R}^4$, such that M is its matrix relative to the standard bases of these spaces?

5.2.10. Let f be a linear transformation of the space $A = \mathbb{Q}^3$ having the matrix

$$\begin{pmatrix} 4 & -1 & 0 \\ 2 & 0 & 2 \\ 1 & 1 & 0 \end{pmatrix}$$

relative to the standard basis. Find all f-invariant subspaces. (A subspace B is called invariant relative to f if for every $b \in B$ we have $f(b) \in B$.)

5.2.11. Let f be a linear transformation of the space $A = \mathbb{Q}^3$ having the matrix

$$\begin{pmatrix} 4 & 3 & 3 \\ 2 & 1 & 2 \\ 1 & 0 & 1 \end{pmatrix}$$

relative to the standard basis. Find the matrix of f relative to the bases (1,1,0), (0,1,1), (0,0,1).

5.2.12. Let f be a linear transformation of the space $A = \mathbb{Q}^3$ having the matrix

$$\begin{pmatrix} 4 & 2 & 0 \\ 3 & 1 & 2 \\ 1 & 0 & 1 \end{pmatrix}$$

relative to the standard basis. Find the matrix of f relative to the bases (1,0,0), (1,1,1), (0,1,1).

5.2.13. Let f be a linear transformation of the space $A = \mathbb{F}_3^4$ having the matrix

$$\begin{pmatrix} 1 & 2 & 0 & 1 \\ 0 & 1 & 2 & 0 \\ 1 & 2 & 1 & 0 \\ 2 & 2 & 2 & 1 \end{pmatrix}$$

relative to the standard basis. Find the matrix of f relative to the bases (1,1,1,1), (0,1,1,1), (0,0,1,1), (0,0,0,1).

5.2.14. Let f be a linear transformation of the space $A = \mathbb{Q}^3$ having the matrix

$$\begin{pmatrix} 3 & 2 & 2 & -2 \\ 1 & -1 & 3 & 0 \\ 4 & 1 & 5 & -2 \\ 2 & 3 & -1 & -2 \end{pmatrix}$$

relative to the standard basis. Find $\mathbf{Im} f$ and $\mathbf{Ker} f$.

5.2.15. Let f be a linear transformation of a space A of dimension n. Let M be the matrix of f relative to the basis $a_1, \dots, a_n$. Find the matrix of f relative to the basis $b_1, \dots, b_n$, where $b_1 = a_2, b_2 = a_1$ and $b_j = a_j$ for $j > 2$.

5.2.16. Let f be a linear transformation of a space A of dimension n. Let M be the matrix of f relative to the basis $a_1, \dots, a_n$. Find the matrix of f relative to the basis $b_1, \dots, b_n$, where $b_1 = a_n, b_2 = a_{n-1}, \dots, b_n = a_1$.

5.3 SYSTEMS OF LINEAR EQUATIONS

Systems of linear equations frequently appear in different branches of mathematics and now the subject is very well developed. Several standard methods of solving systems of linear equations have been established. A universal method here is Gaussian elimination and a further well-known technique is Cramer's method, which gives us formulas for the solution of a system of n linear equations in n variables if the corresponding coefficient matrix of the system is nonsingular. In Section 4.3, we began the consideration of systems of linear equations and clarified the question of the existence of a solution of such a system. Now we are in a position to develop a general theory of finding solutions of a system of linear equations.

Let A be a finite-dimensional vector space and let f be a linear transformation of A. Let b be an arbitrary element of A. We consider the question of finding all elements x of the space A, which satisfy the equation $f(x) = b$. Let $\{a_1, \dots, a_n\}$ be a basis of A. By Proposition 4.2.16, every element of A is a linear combination of the elements $a_1, \dots, a_n$ and this representation is unique. So,

$$b = \sum_{1 \le k \le n} \beta_k a_k, \text{ where } \beta_k \in F, \text{ for } 1 \le k \le n.$$

A solution x of the equation $f(x) = b$ can be represented as a linear combination of the basis elements. Since we do not yet know this element, or indeed its expression as a linear combination of the basis vectors, we will write

$$x = \sum_{1 \le j \le n} \xi_j a_j$$

where $\xi_j \in F$ are unknown, for $1 \leq j \leq n$. When we find these coefficients we will find the element x.

Let $S = [\sigma_{jt}] \in \mathbf{M}_n(F)$ denote the matrix of f relative to the basis $\{a_1, \ldots, a_n\}$. We have

$$f(x) = f\left(\sum_{1\leq j\leq n} \xi_j a_j\right) = \sum_{1\leq j\leq n} \xi_j f(a_j) = \sum_{1\leq j\leq n} \xi_j \left(\sum_{1\leq k\leq n} \sigma_{kj} a_k\right)$$

$$= \sum_{1\leq j\leq n}\sum_{1\leq k\leq n} \sigma_{kj}\xi_j a_k = \sum_{1\leq k\leq n}\left(\sum_{1\leq j\leq n} \sigma_{kj}\xi_j\right) a_k.$$

On the other hand,

$$f(x) = b = \sum_{1\leq k\leq n} \beta_k a_k.$$

By Proposition 4.2.16, the representation of an element as a linear combination of the basis vectors is unique, so we obtain

$$\beta_k = \sum_{1\leq j\leq n} \sigma_{kj}\xi_j.$$

This corresponds to the system

$$\begin{aligned}
\sigma_{11}\xi_1 + \sigma_{12}\xi_2 + \cdots + \sigma_{1n}\xi_n &= \beta_1\\
\sigma_{21}\xi_1 + \sigma_{22}\xi_2 + \cdots + \sigma_{2n}\xi_n &= \beta_2\\
&\ \vdots\\
\sigma_{n-1,1}\xi_1 + \sigma_{n-1,2}\xi_2 + \cdots + \sigma_{n-1,n}\xi_n &= \beta_{n-1}\\
\sigma_{n1}\xi_1 + \sigma_{n2}\xi_2 + \cdots + \sigma_{nn}\xi_n &= \beta_n.
\end{aligned}$$

Thus the coordinates of x satisfy the following system of linear equations

$$\begin{aligned}
\sigma_{11}x_1 + \sigma_{12}x_2 + \cdots + \sigma_{1n}x_n &= \beta_1\\
\sigma_{21}x_1 + \sigma_{22}x_2 + \cdots + \sigma_{2n}x_n &= \beta_2\\
&\ \vdots\\
\sigma_{n-1,1}x_1 + \sigma_{n-1,2}x_2 + \cdots + \sigma_{n-1,n}x_n &= \beta_{n-1}\\
\sigma_{n1}x_1 + \sigma_{n2}x_2 + \cdots + \sigma_{nn}x_n &= \beta_n.
\end{aligned} \tag{5.3}$$

Conversely, System 5.3 has coefficient matrix $S = [\sigma_{jt}] \in \mathbf{M}_n(F)$ and, using S, we can define a linear transformation f of A such that the matrix of f relative to the basis $\{a_1, \ldots, a_n\}$ coincides with S. Repeating the arguments above in reverse, we deduce that any solution of this system gives us a solution of $f(x) = b$ that is an n-tuple, formed by the coordinates in the basis $\{a_1, \ldots, a_n\}$.

Thus, we have established a one-to-one relation between equations of the form $f(x) = b$ and systems of linear equations with coefficients from a field F. To solve $f(x) = b$ we need to solve a system of linear equations.

In the case when f is bijective, Proposition 1.2.8 shows that every element of A has one and only one preimage relative to f and it follows that the equation $f(x) = b$ has one and only one solution.

Suppose next that the matrix S of System 5.3 is nonsingular. By Corollary 5.2.15, f is an automorphism of A and, as proved above, the equation $f(x) = b$ has exactly one solution for each element $b \in B$. It follows that System 5.3 has exactly one solution and therefore, we obtain the following statement.

5.3.1. Theorem. *If the matrix of the system of linear equations (Eq. 5.3) is nonsingular, then this system has exactly one solution.*

Now consider the arbitrary system of k linear equations in n unknowns,

$$
\begin{aligned}
\sigma_{11}x_1 + \sigma_{12}x_2 + \cdots + \sigma_{1n}x_n &= \beta_1 \\
\sigma_{21}x_1 + \sigma_{22}x_2 + \cdots + \sigma_{2n}x_n &= \beta_2 \\
&\vdots \\
\sigma_{k-1,1}x_1 + \sigma_{k-1,2}x_2 + \cdots + \sigma_{k-1,k}x_n &= \beta_{k-1} \\
\sigma_{k1}x_1 + \sigma_{k2}x_2 + \cdots + \sigma_{kn}x_n &= \beta_k
\end{aligned} \tag{5.4}
$$

whose coefficients σ_{tj}, where $1 \leq t \leq k$, $1 \leq j \leq n$ and constant terms β_t, $1 \leq t \leq k$, belong to the field F. Theorem 4.3.8 gives us necessary and sufficient conditions for the existence of solutions of this system. Therefore, we will assume that System 5.4 always has solutions.

Many methods of solving systems of linear equations are based on replacing the given system by an equivalent one.

5.3.2. Definition. *Two systems of linear equations in the same number of variables (but possibly a different number of equations) are called equivalent if the sets of their solutions coincide. Thus, every solution of the first system is a solution of the second system, and conversely, every solution of the second system is a solution of the first system.*

We note that when we interchange two equations in System 5.4 we obtain an equivalent system and this corresponds to interchanging two rows of the corresponding augmented matrix. Let $S = [\sigma_{jt}] \in \mathbf{M}_{k\times n}(F)$ be the matrix of System 5.4 and suppose that $\mathbf{rank}(S) = r$. Let

$$
S^* = \begin{pmatrix}
\sigma_{11} & \sigma_{12} & \sigma_{13} & \ldots & \sigma_{1n} & \beta_1 \\
\sigma_{21} & \sigma_{22} & \sigma_{23} & \ldots & \sigma_{2n} & \beta_2 \\
\vdots & \vdots & \vdots & \vdots & \vdots & \vdots \\
\sigma_{k1} & \sigma_{k2} & \sigma_{k3} & \ldots & \sigma_{kn} & \beta_k
\end{pmatrix}
$$

be the augmented matrix of System 5.4. Theorem 4.3.8 shows that $\mathbf{rank}(S) = \mathbf{rank}(S^*) = r$ and, without loss of generality, we can suppose that the first r rows of the matrix S are linearly independent, by interchanging rows if necessary. Then the first r rows of the matrix S^* are also linearly independent, since every linear combination of the first r rows of S^* implies a corresponding linear combination of the first r rows of S. Hence, the first r rows of S^* form a maximal linearly independent subset of the set of all rows of S^*. Thus, every row of S^* is a linear combination of the first r rows. Then, each equation from System 5.4 can be represented as a sum of the first r equations with certain coefficients. This means that every solution of the system

$$\begin{aligned} \sigma_{11}x_1 + \sigma_{12}x_2 + \cdots + \sigma_{1n}x_n &= \beta_1 \\ \sigma_{21}x_1 + \sigma_{22}x_2 + \cdots + \sigma_{2n}x_n &= \beta_2 \\ &\vdots \\ \sigma_{r1}x_1 + \sigma_{r2}x_2 + \cdots + \sigma_{rn}x_n &= \beta_k \end{aligned} \tag{5.5}$$

is a solution of System 5.4 and since every solution of System 5.4 is a solution of System 5.5, the Systems 5.4 and 5.5 are equivalent.

Let S_1 be the matrix of System 5.5 and let S_1^* be the corresponding augmented matrix. Then $\mathbf{rank}(S_1) = \mathbf{rank}(S_1^*) \leq n$. If $r = n$ then, by Theorem 5.3.1, System 5.5 has a unique solution and therefore System 5.4 also does.

Suppose now that $r < n$. Again, by interchanging the order of the equations and by possibly renumbering the variables, we can suppose that **minor** $\{1, 2, \ldots, r; 1, 2, \ldots, r\}$ is nonzero. By transferring the variables $x_{r+1}, \ldots, x_n$ to the right-hand side in all the equations of System 5.5 and by designating values $\gamma_{r+1}, \ldots, \gamma_n \in F$ for these variables, we obtain the system

$$\begin{aligned} \sigma_{11}x_1 + \sigma_{12}x_2 + \cdots + \sigma_{1r}x_r &= \beta_1 - \sigma_{1,r+1}\gamma_{r+1} \cdots - \sigma_{1n}x_n \\ \sigma_{21}x_1 + \sigma_{22}x_2 + \cdots + \sigma_{2r}x_r &= \beta_2 - \sigma_{2,r+1}\gamma_{r+1} \cdots - \sigma_{2n}x_n \\ &\vdots \\ \sigma_{r1}x_1 + \sigma_{r2}x_2 + \cdots + \sigma_{rr}x_r &= \beta_1 - \sigma_{r,r+1}\gamma_{r+1} \cdots - \sigma_{rn}x_n \end{aligned} \tag{5.6}$$

in r variables $x_1, \ldots, x_r$. The coefficient matrix of this system is nonsingular. Therefore, by Theorem 5.3.1, it has a unique solution $\gamma_1, \ldots, \gamma_r$. Clearly, the n-tuple $(\gamma_1, \ldots, \gamma_r, \gamma_{r+1}, \ldots, \gamma_n)$ is a solution of System 5.5. Since the values $\gamma_{r+1}, \ldots, \gamma_n$ for the variables $x_{r+1}, \ldots, x_n$, the so-called free variables, can be chosen in an arbitrary way, we can obtain many solutions of System 5.5.

On the other hand, every solution of System 5.5 can be obtained as follows. Let $(\gamma_1, \ldots, \gamma_r, \gamma_{r+1}, \ldots, \gamma_n)$ be a solution of System 5.5. We choose the values

$\gamma_{r+1}, \ldots, \gamma_n$ for the free variables. Then, the elements $\gamma_1, \ldots, \gamma_r$ will satisfy System 5.6 and will form the unique solution of this system.

So if $r < n$, then the set of solutions can be quite large, particularly if F is infinite (in the cases when $F = \mathbb{R}$ or $F = \mathbb{C}$, for example) when the solution set is also infinite. A natural question arises as to how we can describe the set of solutions. We answer this question next.

We consider System 5.5 again. The number of equations there is less than the number of variables. We add $n - r$ equations with zero coefficients and zero constant terms to this system to obtain a system of n equations in n variables, namely,

$$\begin{aligned} \sigma_{11}x_1 + \sigma_{12}x_2 + \cdots + \sigma_{1n}x_n &= \beta_1 \\ \sigma_{21}x_1 + \sigma_{22}x_2 + \cdots + \sigma_{2n}x_n &= \beta_2 \\ &\vdots \\ \sigma_{n1}x_1 + \sigma_{n2}x_2 + \cdots + \sigma_{nn}x_n &= \beta_n \end{aligned} \tag{5.7}$$

Again, let S denote the coefficient matrix of this system and let S^* denote the corresponding augmented matrix.

Now consider the associated system

$$\begin{aligned} \sigma_{11}x_1 + \sigma_{12}x_2 + \cdots + \sigma_{1n}x_n &= 0_F \\ \sigma_{21}x_1 + \sigma_{22}x_2 + \cdots + \sigma_{2n}x_n &= 0_F \\ &\vdots \\ \sigma_{n1}x_1 + \sigma_{n2}x_2 + \cdots + \sigma_{nn}x_n &= 0_F \end{aligned} \tag{5.8}$$

5.3.3. Definition. *System 5.8 is called the homogeneous system associated with System 5.7.*

Let $(\gamma_1, \ldots, \gamma_n)$ and $(\upsilon_1, \ldots, \upsilon_n)$ be two solutions of System 5.7. Then

$$\sum_{1 \le j \le n} \sigma_{kj}(\gamma_j - \eta_j) = \sum_{1 \le j \le n} \sigma_{kj}\gamma_j - \sum_{1 \le j \le n} \sigma_{kj}\eta_j = \beta_k - \beta_k = 0_F,$$

for $1 \le k \le n$. Hence, the difference of two solutions of System 5.7 is a solution of System 5.8.

Now let $(\gamma_1, \ldots, \gamma_n)$ be a solution of System 5.7 and let $(\eta_1, \ldots, \eta_n)$ be a solution of System 5.8. Then

$$\sum_{1 \le j \le n} \sigma_{kj}(\gamma_j + \eta_j) = \sum_{1 \le j \le n} \sigma_{kj}\gamma_j + \sum_{1 \le j \le n} \sigma_{kj}\eta_j = \beta_k - 0_F = \beta_k,$$

for $1 \leq k \leq n$. Hence the sum of a solution of System 5.7 and a solution of System 5.8 is a solution of System 5.7. From these simple properties we deduce the following result.

5.3.4. Proposition. *Let $(\gamma_1, \dots, \gamma_n)$ be some fixed solution of System 5.7. Then the set of all solutions of System 5.7 coincides with the set of all sums $(\gamma_1, \dots, \gamma_n) + (\eta_1, \dots, \eta_n)$, where $(\eta_1, \dots, \eta_n)$ is an arbitrary solution of System 5.8.*

Thus, in order to obtain all solutions of System 5.7 it is necessary to obtain one solution of System 5.7 and all solutions of System 5.8. We next find all solutions of System 5.8.

Let A be a vector space over a field F, such that $\mathbf{dim}_F(A) = n$. In A, choose a basis $\{a_1, \dots, a_n\}$. As above, we can construct a linear transformation f of A, such that the matrix of f relative to the basis $\{a_1, \dots, a_n\}$ is S. Let $\kappa : V \longrightarrow F^k$ be the canonical isomorphism. If $x = \sum_{1 \leq j \leq n} \xi_j a_j \in \mathbf{Ker} f$, where $\xi_j \in F$ for $1 \leq j \leq n$, then as above $(\xi_1, \dots, \xi_n) = \kappa(x)$ is a solution of System 5.8 and conversely. This implies that $\kappa(\mathbf{Ker} f)$ is the set of all solutions of System 5.8. By Proposition 5.1.3, $\kappa(\mathbf{Ker} f)$ is a subspace of F^n. Thus, the set of all solutions of the homogeneous system (Eq. 5.8) is a subspace of F^n. By Corollary 5.1.11, $\mathbf{dim}_F(\mathbf{Ker} f) = \mathbf{dim}_F(A) - \mathbf{dim}_F(\mathbf{Im} f)$. Proposition 5.2.5 implies that $\mathbf{dim}_F(\mathbf{Im} f) = \mathbf{rank}(S) = r$. Therefore, $\mathbf{dim}_F(\mathbf{Ker} f) = n - r$ and since κ is an isomorphism, Corollary 5.1.9 implies that $\mathbf{dim}_F(\mathbf{Ker} f) = \mathbf{dim}_F(\kappa(\mathbf{Ker} f))$. Hence, the subspace of all solutions of System 5.8 has dimension $n - r$. So we need to find $n - r$ linearly independent solutions of System 5.8 and then all other solutions will be a linear combination of these.

5.3.5. Definition. *A basis of the subspace of all solutions of the homogeneous system (Eq. 5.8) is called a fundamental set of solutions.*

The next question is how to find a fundamental set of solutions of System 5.8. To see how to answer this, let

$$Y = \begin{pmatrix} \eta_{11} & \eta_{12} & \cdots & \eta_{1,n-r} \\ \eta_{21} & \eta_{22} & \cdots & \eta_{2,n-r} \\ \vdots & \vdots & \ddots & \vdots \\ \eta_{n-r,1} & \eta_{n-r,2} & \cdots & \eta_{n-r,n-r} \end{pmatrix}$$

be an arbitrary nonsingular matrix in $\mathbf{M}_{(n-r)\times(n-r)}(F)$. As above, we can suppose that $\mathbf{minor}\{1, 2, \dots, r; 1, 2, \dots, r\}$ is nonzero, for the coefficient matrix of System 5.8. In each of the equations from System 5.8 we move the parts of the equation involving $x_{r+1}, \dots, x_n$ to the right-hand side and assign the coefficients of the jth row of the matrix Y to these variables for their values, where $1 \leq j \leq n-r$. We obtain the following system

$$\begin{aligned}
\sigma_{11}x_1 + \sigma_{12}x_2 + \cdots + \sigma_{1r}x_r &= -\sigma_{1,r+1}\eta_{j1} \cdots - \sigma_{1n}\eta_{j,n-r} \\
\sigma_{21}x_1 + \sigma_{22}x_2 + \cdots + \sigma_{2r}x_r &= -\sigma_{2,r+1}\eta_{j1} \cdots - \sigma_{2n}\eta_{j,n-r} \\
&\vdots \\
\sigma_{r1}x_1 + \sigma_{r2}x_2 + \cdots + \sigma_{rr}x_r &= -\sigma_{r,r+1}\eta_{j1} \cdots - \sigma_{rn}\eta_{j,n-r}
\end{aligned} \tag{5.9}$$

in r variables $x_1, \ldots, x_r$. The coefficient matrix of this system is nonsingular. Therefore, by Theorem 5.3.1, it has a unique solution $\gamma_{j1}, \ldots, \gamma_{jr}$, where $1 \leq j \leq n-r$. We now form the n-tuple $(\gamma_{j1}, \ldots, \gamma_{jr}, \eta_{j1}, \ldots, \eta_{j,n-r})$ which, by construction, is a solution of System 5.8. The system of solutions

$$\begin{gathered}
(\gamma_{11}, \ldots, \gamma_{1r}, \eta_{11}, \ldots, \eta_{1,n-r}), \\
(\gamma_{21}, \ldots, \gamma_{2r}, \eta_{21}, \ldots, \eta_{2,n-r}), \\
\vdots \\
(\gamma_{n-r,1}, \ldots, \gamma_{n-r,r}, \eta_{n-r,1}, \ldots, \eta_{n-r,n-r})
\end{gathered}$$

is a fundamental set of solutions for System 5.8. Indeed, this set is linearly independent, because the matrix consisting of these vectors as rows has a nonzero minor, namely, **minor**$\{1, 2, \ldots, r; r+1, r+2, \ldots, n\}$, of order $n-r$.

EXERCISE SET 5.3

Show your work. Where appropriate, write a proof or give a counterexample.

5.3.1. Find a fundamental system of solutions for the system

$$\begin{aligned}
-4x_1 + (2+2\lambda)x_2 + 2\lambda x_3 + 2\lambda x_4 &= 0 \\
\lambda x_1 + (1+\lambda)x_2 + \lambda x_3 + \lambda x_4 &= 0 \\
\lambda x_1 + (1+\lambda)x_2 - 2x_3 + \lambda x_4 &= 0 \\
-\lambda x_1 + (1+\lambda)x_2 - \lambda x_3 - (2+2\lambda)x_4 &= 0
\end{aligned}$$

5.3.2. Find a fundamental system of solutions for the system

$$\begin{aligned}
3x_1 - 2x_2 + x_3 - x_4 &= 0 \\
-x_1 + x_2 + 4x_3 - 2x_4 &= 0 \\
-2x_1 + 3x_2 - x_3 &= 0
\end{aligned}$$

5.3.3. Find a fundamental system of solutions for the system

$$\begin{aligned} x_1 - x_3 + x_5 &= 0 \\ x_2 - x_4 + x_6 &= 0 \\ x_1 - x_2 + x_5 - x_6 &= 0 \\ x_2 - x_3 + x_6 &= 0 \\ x_1 - x_4 + x_5 &= 0 \end{aligned}$$

5.3.4. For which values of λ does the following system have solutions?

$$\begin{aligned} -x_1 + (1+\lambda)x_2 + (2-\lambda)x_3 + \lambda x_4 &= 3 \\ x_1 - x_2 + (2-\lambda)x_3 + \lambda x_4 &= 2 \\ \lambda x_1 + \lambda x_2 + (2-\lambda)x_3 + \lambda x_4 &= 2 \\ \lambda x_1 + \lambda x2 + (2-\lambda)x_3 - x_4 &= 2 \end{aligned}$$

5.3.5. Find a fundamental system of solutions for the system below, where the coefficients belong to the field $\mathbb{F}_{13} = \mathbb{Z}/13\mathbb{Z}$.

$$\begin{aligned} 4x_1 - 5x_2 - 8x_3 - 9x_4 &= 0 \\ x_1 + 12x_2 - x_3 &= 0 \\ 8x_1 + 11x_2 - x_3 + 7x_4 &= 0 \end{aligned}$$

5.3.6. Find a fundamental system of solutions of the system below, where the coefficients belong to the field $\mathbb{F}_5 = \mathbb{Z}/5\mathbb{Z}$.

$$\begin{aligned} 3x_1 - 2x_2 + x_3 - x_4 &= 0 \\ -x_1 + x_2 + 4x_3 - 2x_4 &= 0 \\ -2x_1 + 3x_2 - x_3 &= 0 \end{aligned}$$

5.3.7. Find a fundamental system of solutions for the system below, where the coefficients belong to the field $\mathbb{F}_3 = \mathbb{Z}/3\mathbb{Z}$.

$$\begin{aligned} 2x_1 - x_2 + 2x_3 - x_4 &= 0 \\ 2x_2 + x_4 &= 0 \\ x_1 - x_2 + 2x_3 - x_4 &= 0 \end{aligned}$$

5.4 EIGENVECTORS AND EIGENVALUES

In Section 5.2 we saw that every linear transformation ϕ corresponds to a matrix relative to some basis. Change of basis naturally implies that this matrix changes and the natural question arises as to which basis gives rise to the matrix of this linear transformation that has the simplest form. We are also interested in determining this basis. The notion of an invariant subspace plays a key role here.

5.4.1. Definition. *Let A be a vector space over a field F and let ϕ be a linear transformation of A. The subspace B of A is called ϕ-invariant, if $\phi(b) \in B$ for each element $b \in B$.*

Trivial examples of ϕ-invariant subspaces include the entire space A and the zero subspace $\{0_A\}$. Other examples of ϕ-invariant subspaces are given in the next proposition.

5.4.2. Proposition. *Let A be a vector space over a field F and let ϕ be a linear transformation of A. Then $\mathbf{Im}\,\phi$ and $\mathbf{Ker}\,\phi$ are ϕ-invariant subspaces of A.*

Proof. If $x \in \mathbf{Im}\,\phi$, then $x = \phi(y)$ for some element $y \in A$ and

$$\phi(x) = \phi(\phi(y)) \in \mathbf{Im}\,\phi$$

also. Hence $\mathbf{Im}\,\phi$ is a ϕ-invariant subspace of A. Next, let $z \in \mathbf{Ker}\,\phi$. Then $\phi(z) = 0_A \in \mathbf{Ker}\,\phi$, so that $\mathbf{Ker}\,\phi$ is also ϕ-invariant.

Let B be a ϕ-invariant subspace of a finite-dimensional space A and let $\{b_1, \dots, b_k\}$ be a basis of B. By Theorem 4.2.11, the linearly independent subset $\{b_1, \dots, b_k\}$ can be extended to a basis of the entire space A, so that there exist elements $b_{k+1}, \dots, b_n$ such that $\{b_1, \dots, b_n\}$ is a basis of A. Let $S = [\sigma_{jt}] \in \mathbf{M}_n(F)$ be the matrix of the linear transformation ϕ relative to the basis $\{b_1, \dots, b_n\}$. Then $\phi(b_j) \in B$, for $1 \le j \le k$, so

$$\phi(b_j) = \sigma_{1j}b_1 + \sigma_{2j}b_2 + \dots + \sigma_{kj}b_k, \text{ for } 1 \le j \le k$$

and $\sigma_{tj} = 0_F$ for $t > k, 1 \le j \le k$. Therefore, the matrix S has the following form:

$$\begin{pmatrix} \sigma_{11} & \sigma_{12} & \dots & \sigma_{1k} & \sigma_{1,k+1} & \dots & \sigma_{1n} \\ \sigma_{21} & \sigma_{22} & \dots & \sigma_{2k} & \sigma_{2,k+1} & \dots & \sigma_{2n} \\ \vdots & \vdots & \ddots & \vdots & \vdots & \vdots & \vdots \\ \sigma_{k1} & \sigma_{k2} & \dots & \sigma_{kk} & \sigma_{k,k+1} & \dots & \sigma_{k1} \\ 0_F & 0_F & \dots & 0_F & \sigma_{k+1,k+1} & \dots & \sigma_{k+1,n} \\ \vdots & \vdots & \vdots & \vdots & \vdots & \ddots & \vdots \\ 0_F & 0_F & \dots & 0_F & \sigma_{n,k+1} & \dots & \sigma_{nn} \end{pmatrix}.$$

Thus, we can see that the existence of a ϕ-invariant subspace significantly simplifies the structure of the matrix S. In particular, if $A = A_1 \oplus \cdots \oplus A_k$ is a direct sum of ϕ-invariant subspaces, then the matrix S has the following form:

$$\begin{pmatrix} S_1 & 0 & \dots & \dots \\ 0 & S_2 & \dots & \dots \\ \vdots & \vdots & \ddots & \vdots \\ 0 & \dots & 0 & S_k \end{pmatrix},$$

where certain square submatrices lie on the main diagonal and all other entries (which are themselves submatrices) are the appropriate size zero matrix. We will say that this matrix has cellwise-diagonal form.

The one-dimensional ϕ-invariant subspaces play a special role. Let B be a ϕ-invariant subspace and suppose that $\mathbf{dim}_F(B) = 1$. If b is a nonzero element of B, then $\{b\}$ is a basis of B. Proposition 4.2.16 shows that every element of B has the form αb for some $\alpha \in F$. Since $\phi(b) \in B, \phi(b) = \gamma b$ for some $\gamma \in F$.

5.4.3. Definition. *Let A be a vector space over a field F and let ϕ be a linear transformation of A. A nonzero element a of A is called an eigenvector of ϕ, if $\phi(a) = \gamma a$ for some $\gamma \in F$. The element γ is called an eigenvalue of the linear transformation ϕ.*

We see that every nonzero element of a ϕ-invariant subspace of dimension one is an eigenvector of ϕ. Every nonzero element c of $\mathbf{Ker}\,\phi$ is also an eigenvector, because $\phi(c) = 0_A = 0_F c$.

Let $\gamma \in F$ and set

$$A(\gamma) = \{x \mid x \in A \text{ and } \phi(x) = \gamma x\}.$$

Of course, $0_A \in A(\gamma)$. The subset $A(\gamma)$ is a subspace of A, called the eigenspace of A corresponding to the eigenvalue γ. Indeed, if $x, y \in A(\gamma)$ and $\alpha \in F$ then

$$\phi(x - y) = \phi(x) - \phi(y) = \gamma x - \gamma y = \gamma(x - y)$$

and

$$\phi(\alpha x) = \alpha\phi(x) = \alpha(\gamma x) = (\alpha\gamma)x = (\gamma\alpha)x = \gamma(\alpha x).$$

It follows from Theorem 4.1.7 that $A(\gamma)$ is a subspace of A. Note that the subspace $A(\gamma)$ could be the zero subspace, but not when γ is an eigenvalue. Moreover, as we will see later, for a finite-dimensional vector space A, there exist only finitely many elements γ such that the subspace $A(\gamma)$ is nonzero.

Now, we will find the elements of $A(\gamma)$. Let $0_A \neq x \in A(\gamma)$ so that $\phi(x) = \gamma x$. On the other hand, $\gamma x = \gamma\varepsilon_A(x)$, so that $\phi(x) = \gamma\varepsilon_A(x)$ and $(\phi - \gamma\varepsilon_A)(x) = 0_A$. Hence the element x belongs to $A(\gamma)$ if and only if $x \in \mathbf{Ker}(\phi - \gamma\varepsilon_A)$. Choose a basis $\{a_1, \dots, a_n\}$ of the space A and let $S = [\sigma_{jt}] \in \mathbf{M}_n(F)$ be the matrix of the linear transformation ϕ relative to the basis $\{a_1, \dots, a_n\}$.

Then, $S - \gamma I$ is the matrix of the linear transformation $\phi - \gamma\varepsilon_A$ relative to this basis. Let $x = \sum_{1 \le m \le n} \xi_m a_m$ be the representation of x in terms of the basis $\{a_1, \dots, a_n\}$. As in Section 5.3, we can show that $x \in \mathbf{Ker}(\phi - \gamma\varepsilon_A)$ if and only if the n-tuple $(\xi_1, \dots, \xi_n)$ is a solution of the system

$$\begin{aligned}
(\sigma_{11} - \gamma)\, x_1 + \sigma_{12}x_2 + \cdots + \sigma_{1n}x_n &= 0_F \\
\sigma_{21}x_1 + (\sigma_{22} - \gamma)\, x_2 + \cdots + \sigma_{2n}x_n &= 0_F \\
&\vdots \\
\sigma_{n1}x_1 + \sigma_{n2}x_2 + \cdots + (\sigma_{nn} - \gamma)\, x_n &= 0_F.
\end{aligned} \tag{5.10}$$

Clearly, the zero n-tuple is always a solution of System 5.10. If we suppose that the matrix $S - \gamma I$ is nonsingular, then by Theorem 5.3.1, System 5.10 has exactly one solution. It follows that the zero n-tuple is the unique solution of System 5.10 and therefore $A(\gamma) = \{0_A\}$, contrary to our assumption, which proves that $\mathbf{det}(S - \gamma I) = 0_F$. Expanding $\mathbf{det}(S - \gamma I)$, we obtain a linear combination of the elements $\gamma^0 = e, \gamma^1 = \gamma, \gamma^2, \dots, \gamma^n$ with coefficients from the field F. Thus, $\mathbf{det}(S - \gamma I)$ is obtained from a polynomial of degree n with coefficients in F, when we substitute γ for X (see Section 7.5).

5.4.4. Definition. *Let F be a field and let $S \in \mathbf{M}_n(F)$. The polynomial $\chi_S(X) = \mathbf{det}(S - XI)$ is called the characteristic polynomial of S.*

Under certain types of transformation $\chi_S(X)$ remains invariant. This is the content of the next proposition.

5.4.5. Proposition. *Let F be a field and let $S \in \mathbf{M}_n(F)$. Suppose that T is a nonsingular matrix of degree n and let $L = T^{-1}ST$. Then $\chi_L(X) = \chi_S(X)$.*

Proof. We have

$$\begin{aligned}
L - XI &= T^{-1}ST - XI = T^{-1}ST - (T^{-1}T)XI = T^{-1}ST - T^{-1}(TXI) \\
&= T^{-1}ST - T^{-1}(XI)T = T^{-1}(S - XI)T.
\end{aligned}$$

It follows that $\chi_L(X) = \mathbf{det}(L - XI) = \mathbf{det}(T^{-1}(S - XI)T)$. Using Theorem 2.5.1 we obtain

$$\begin{aligned}
\chi_L(X) &= \mathbf{det}(T^{-1}(S - XI)T) = \mathbf{det}(T^{-1})\mathbf{det}(S - XE)\mathbf{det}(T) \\
&= \mathbf{det}(S - XE) = \chi_S(X).
\end{aligned}$$

There is a similar definition for the characteristic polynomial of a linear transformation, but we now need Proposition 5.4.5 to make this a well-defined concept.

5.4.6. Definition. *Let A be a finite-dimensional vector space over a field F and let ϕ be a linear transformation of A. If S is the matrix of ϕ relative to some basis* *of A, then $\chi_S(X)$ is called the characteristic polynomial of ϕ and will be denoted by $\chi_\phi(X)$.*

This concept is well defined since it does not depend on the choice of basis of A. In fact, if L is the matrix of ϕ relative to another basis then, by Corollary 5.2.12, $L = T^{-1}ST$ where T is the transition matrix from the first basis to the second one. Proposition 5.4.5 implies that $\chi_L(X) = \chi_S(X)$.

5.4.7. Theorem. *Let A be a finite-dimensional vector space over a field F and let ϕ be a linear transformation of A. The element $\gamma \in F$ is an eigenvalue of ϕ if and only if γ is a root of the characteristic polynomial $\chi_\phi(X)$ of the linear transformation ϕ.*

Proof. If γ is an eigenvalue of the linear transformation ϕ, then there is a nonzero element x of A with the property $\phi(x) = \gamma x$. We have already proved above that in this case, $\mathbf{det}(S - \gamma I) = 0_F$, which means that γ is a root of $\chi_\phi(X)$.

Conversely, suppose that γ is a root of $\chi_\phi(X)$. Then $\chi_\phi(\gamma) = \mathbf{det}(S - \gamma I) = 0_F$, where S is the matrix of ϕ relative to some basis. It follows that $\mathbf{rank}(S - \gamma I) < n$. Using the work of Section 5.3, we see that System 5.10 has more than one solution. In particular, System 5.10 has a nonzero solution $(\xi_1, \ldots, \xi_n)$. Put $x = \sum_{1 \le m \le n} \xi_m a_m$. The results of Section 5.3 imply that $x \in \mathbf{Ker}(\phi - \gamma\varepsilon_A)$, which implies that $\phi(x) = \gamma x$. Thus γ is an eigenvalue of ϕ.

This gives us a method of finding the eigenvalues and eigenvectors of ϕ. By Corollary 7.5.11, the number of roots of the polynomial $\chi_\phi(\gamma)$ is at most $\mathbf{deg}(\chi_\phi(\gamma)) = \mathbf{dim}_F(A)$. However, the weakness of this method lies in finding the roots of $\chi_\phi(\gamma)$, since there are no good general methods of finding roots of polynomials over fields. Therefore, every particular case requires its own consideration. In the case of finite fields it is possible to use computers to find roots, while for the fields $\mathbb{R}$ and $\mathbb{C}$ there are numerous very well-developed approximate methods.

We next note the following important property.

5.4.8. Proposition. *Let A be a finite-dimensional vector space over a field F and let ϕ be a linear transformation of A. Suppose that $\gamma_1, \ldots, \gamma_k$ are the eigenvalues of ϕ, where $\gamma_j \neq \gamma_m$ whenever $j \neq m$. If $a_1, \ldots, a_k$ are nonzero elements of A such that $\phi(a_j) = \gamma_j a_j$, for $1 \le j \le k$, then $\{a_1, \ldots, a_k\}$ is a linearly independent set.*

Proof. We note that a_j is an eigenvector corresponding to the eigenvalue γ_j and use induction on k. If $k = 1$, then the subset $\{a_1\}$ is linearly independent, because a_1 is nonzero.

Suppose that $k > 1$ and that we have already proved that the elements $a_1, \ldots, a_{k-1}$ are linearly independent. Let $\alpha_1, \ldots, \alpha_k$ be elements of F such that $\alpha_1 a_1 + \cdots + \alpha_k a_k = 0_A$. Then

$$\begin{aligned} 0_A &= \gamma_k 0_A = \gamma_k(\alpha_1 a_1 + \cdots + \alpha_{k-1}a_{k-1} + \alpha_k a_k) \\ &= \gamma_k \alpha_1 a_1 + \cdots + \gamma_k \alpha_{k-1} a_{k-1} + \gamma_k \alpha_k a_k. \end{aligned}$$

On the other hand,

$$\begin{aligned} 0_A &= \phi(0_A) = \phi(\alpha_1 a_1 + \cdots + \alpha_{k-1}a_{k-1} + \alpha_k a_k) \\ &= \alpha_1\phi(a_1) + \cdots + \alpha_{k-1}\phi(a_{k-1}) + \alpha_k\phi(a_k) \\ &= \alpha_1\gamma_1 a_1 + \cdots + \alpha_{k-1}\gamma_{k-1}a_{k-1} + \alpha_k\gamma_k a_k \\ &= \gamma_1\alpha_1 a_1 + \cdots + \gamma_{k-1}\alpha_{k-1}a_{k-1} + \gamma_k\alpha_k a_k. \end{aligned}$$

It follows that

$$\begin{aligned} 0_A &= 0_A - 0_A \\ &= (\gamma_k\alpha_1 a_1 + \cdots + \gamma_k\alpha_{k-1}a_{k-1} + \gamma_k\alpha_k a_k) \\ &\quad - (\gamma_1\alpha_1 a_1 + \cdots + \gamma_{k-1}\alpha_{k-1}a_{k-1} + \gamma_k\alpha_k a_k) \\ &= (\gamma_k - \gamma_1)\alpha_1 a_1 + \cdots + (\gamma_k - \gamma_{k-1})\alpha_{k-1}a_{k-1}. \end{aligned}$$

By the induction hypothesis, the elements $a_1, \ldots, a_{k-1}$ are linearly independent and therefore Proposition 4.2.7 implies that

$$0_F = (\gamma_k - \gamma_1)\alpha_1 = \cdots = (\gamma_k - \gamma_{k-1})\alpha_{k-1}.$$

Since $\gamma_k \neq \gamma_j$, for $1 \leq j \leq k-1$, we obtain $\alpha_1 = \cdots = \alpha_{k-1} = 0_F$. Then we have $\alpha_k a_k = 0_A$. Since $a_k \neq 0_A$, we obtain $\alpha_k = 0_F$ and it follows from Proposition 4.2.7 that the subset $\{a_1, \ldots, a_k\}$ is linearly independent.

5.4.9. Theorem. *Let A be a finite-dimensional vector space over a field F, let $\mathbf{dim}_F(A) = n$, and let ϕ be a linear transformation of A. Suppose that in the field F the characteristic polynomial $\chi_\phi(X)$ has the roots $\gamma_1, \ldots, \gamma_n$ such that $\gamma_j \neq \gamma_m$ whenever $j \neq m$. Then A has a basis $\{a_1, \ldots, a_n\}$ such that the matrix of ϕ relative to this basis is diagonal.*

Proof. For each γ_j we find a nonzero eigenvector a_j so that $\phi(a_j) = \gamma_j a_j$, for $1 \leq j \leq n$. By Proposition 5.4.8, the elements $a_1, \ldots, a_n$ are linearly independent and since $\mathbf{dim}_F(A) = n$, the subset $\{a_1, \ldots, a_n\}$ is a basis of A. The equations $\phi(a_j) = \gamma_j a_j$, for $1 \leq j \leq n$, show that the matrix of ϕ relative to this basis is

$$\begin{pmatrix} \gamma_1 & 0 & \ldots & \ldots \\ 0 & \gamma_2 & \ldots & \ldots \\ \vdots & \vdots & \ddots & \vdots \\ 0 & \ldots & 0 & \gamma_k \end{pmatrix}.$$

The result follows.

5.4.10. Corollary. *Let F be a field and let $S \in \mathbf{M}_n(F)$. Suppose that in F the characteristic polynomial $\chi_S(X)$ of the matrix S has the roots $\gamma_1, \ldots, \gamma_n$ such that $\gamma_j \neq \gamma_m$ whenever $j \neq m$. Then, there exists a nonsingular matrix $T \in \mathbf{M}_n(F)$ such that the matrix $T^{-1}ST$ is diagonal.*

Proof. Let A be a vector space over F such that $\mathbf{dim}_F(A) = n$ (we may choose $A = F^n$, for example). Choose some basis $\{c_1, \ldots, c_n\}$. As above, we can define a linear transformation ϕ of A such that the matrix of ϕ relative to the basis $\{c_1, \ldots, c_n\}$ is equal to S. Then $\chi_\phi(X) = \chi_S(X)$. By Theorem 5.4.9, there exists a basis $\{a_1, \ldots, a_n\}$ such that the matrix L of ϕ relative to this basis is diagonal. By Corollary 5.2.12, $L = T^{-1}ST$, where T is the transition matrix from the first basis to the second.

This relatively nice case that we considered does not always apply. If the characteristic polynomial $\chi_\phi(X)$ has no roots in the field F, then ϕ has no eigenvectors. For example, let A be a vector space over $\mathbb{R}$ of dimension 2 and suppose that ϕ is a corresponding linear transformation. Let $S = [\sigma_{jt}] \in \mathbf{M}_2(\mathbb{R})$ be the matrix of ϕ relative to some basis. Then

$$\mathbf{det}(S - XI) = X^2 - (\sigma_{11} + \sigma_{22})X + (\sigma_{11}\sigma_{22} - \sigma_{12}\sigma_{21}).$$

If

$$(\sigma_{11} + \upsilon_{22})^2 - 4(\sigma_{11}\sigma_{22} - \sigma_{12}\sigma_{21}) = (\sigma_{11} - \sigma_{22})^2 + 4\sigma_{12}\sigma_{21} < 0$$

then S has no eigenvalues in $\mathbb{R}$. Therefore, a basis of A for which the matrix corresponding to ϕ is diagonal does not exist. However, over $\mathbb{C}$ it is possible to make such a diagonal matrix, as the reader may verify.

As another example, let A be a vector space over $\mathbb{C}$ of dimension 2 and let ϕ be a linear transformation of A with corresponding matrix $S = [\sigma_{jt}] \in \mathbf{M}_2(\mathbb{C})$ relative to some basis. Then

$$\chi_\phi(X) = X^2 - (\sigma_{11} + \sigma_{22})X + (\sigma_{11}\sigma_{22} - \sigma_{12}\sigma_{21}),$$

and its roots can be found using the quadratic formula to be

$$\gamma_1, \gamma_2 = \left(\frac{\sigma_{11} + \sigma_{22}}{2}\right) \pm \sqrt{\frac{(\sigma_{11} - \sigma_{22})^2}{4} + \sigma_{12}\sigma_{21}}.$$

If $\gamma_1 \neq \gamma_2$, then by Theorem 5.4.9, A has a basis in which the matrix of ϕ is diagonal.

If $\gamma_1 = \gamma_2 = \gamma$ then ϕ may have one or two mutually linearly independent eigenvectors. When there is only one linearly independent eigenvector then there is no basis of A in which the matrix of ϕ is diagonal. A corresponding example is the matrix

$$S = \begin{pmatrix} \gamma & 1 \\ 0 & \gamma \end{pmatrix}.$$

Thus, not all linear transformations have a basis relative to which the corresponding matrix is diagonal. However, something can be salvaged in this situation. It is beyond the scope of this book to give the full details here; so, we briefly indicate what happens.

5.4.11. Definition. *Let F be a field. The matrix $\mathbf{J}_n(\gamma) \in \mathbf{M}_n(F)$ of the form*

$$\begin{pmatrix} \gamma & e & 0_F & \dots & 0_F \\ 0_F & \gamma & e & \dots & 0_F \\ \vdots & \vdots & \vdots & \vdots & e \\ 0_F & 0_F & 0_F & \dots & \gamma \end{pmatrix}$$

is called a Jordan block with eigenvalue γ. A cellwise-diagonal matrix whose blocks are Jordan blocks is called a Jordan matrix.

The following theorem shows the importance of these matrices, where we say nothing more concerning the notion of an algebraically closed field, but the main example is $\mathbb{C}$.

5.4.12. Theorem. *Let A be a finite-dimensional vector space over a field F and let ϕ be a linear transformation of A. Suppose that the field F is algebraically closed. Then A has a basis such that the matrix L of ϕ is a Jordan matrix, and the matrix L is defined with respect to a permutation of its Jordan blocks.*

EXERCISE SET 5.4

5.4.1. Let f be a nonsingular linear transformation of a space A. Prove that f and f^{-1} have the same eigenvectors.

5.4.2. Find the eigenvalues of the linear transformation f of the vector space $A = \mathbb{F}_2^4$ having the matrix

$$\begin{pmatrix} 1 & 0 & 0 & 1 \\ 1 & 0 & 1 & 0 \\ 1 & 1 & 0 & 0 \\ 0 & 0 & 1 & 1 \end{pmatrix}$$

relative to the standard basis.

5.4.3. Find the eigenvalues and eigenvectors of the linear transformation f of the vector space $A = \mathbb{F}_3^3$ having the matrix

$$\begin{pmatrix} 1 & 2 & 0 \\ 2 & 1 & 0 \\ 1 & 0 & 1 \end{pmatrix}$$

relative to the standard basis.

5.4.4. Find the eigenvalues and eigenvectors of the linear transformation f of the vector space $A = \mathbb{F}_3^3$ having the matrix

$$\begin{pmatrix} 0 & 2 & 1 \\ 0 & 1 & 0 \\ 1 & 2 & 0 \end{pmatrix}$$

relative to the standard basis.

5.4.5. Find the eigenvalues and eigenvectors of the linear transformation f of the vector space $A = \mathbb{F}_5^2$ having the matrix

$$\begin{pmatrix} 3 & 4 \\ 1 & 2 \end{pmatrix}$$

relative to the standard basis.

5.4.6. Find the eigenvalues and eigenvectors of the linear transformation f of the vector space $A = \mathbb{R}^3$ whose matrix relative to the standard basis is

$$\begin{pmatrix} 2 & -1 & 2 \\ 5 & -3 & 3 \\ -1 & 0 & -2 \end{pmatrix}$$

5.4.7. Find the eigenvalues and eigenvectors of the linear transformation f of the vector space $A = \mathbb{R}^3$ whose matrix relative to the standard basis is

$$\begin{pmatrix} 0 & 1 & 0 \\ -4 & 4 & 0 \\ -2 & 1 & 2 \end{pmatrix}$$

5.4.8. Find the eigenvalues and eigenvectors of the linear transformation f of the vector space $A = \mathbb{R}^3$ whose matrix relative to the standard basis is

$$\begin{pmatrix} 4 & -5 & 2 \\ 5 & -7 & 3 \\ 6 & -9 & 4 \end{pmatrix}$$

5.4.9. Find the eigenvalues and eigenvectors of the linear transformation f of the vector space $A = \mathbb{R}^3$ whose matrix relative to the standard basis is

$$\begin{pmatrix} 1 & -3 & 3 \\ -2 & -6 & 13 \\ -1 & -4 & 8 \end{pmatrix}$$

5.4.10. Given the eigenvalues of a mapping f, find the eigenvalues of f^2.

5.4.11. Let f, g be a linear transformations of a finite-dimensional vector space A. Suppose that f is nonsingular. Prove that $f \circ g$ and $g \circ f$ have the same characteristics polynomial.

CHAPTER 6

BILINEAR FORMS

6.1 BILINEAR FORMS

In three-dimensional space, $\mathbb{R}^3$, there is a well-known product called the scalar (or dot or inner) product of two vectors. This scalar product is linear in each variable and gives rise to a real number, which enables us to define and compute lengths, angles between vectors, angles between straight lines and planes, and so on. Indeed the notion of a scalar product allows us to define an entire geometry on a space. Bilinear forms, which we consider in this chapter, are a natural generalization of the scalar product idea in arbitrary vector spaces. These forms are very useful not only in linear algebra but also in other different branches of mathematics such as functional analysis, probability theory, quantum mechanics, special relativity theory, and so on.

6.1.1. Definition. *Let A be a vector space over a field F. The mapping $\Phi : A \times A \longrightarrow F$ is called a bilinear form if it is linear in each variable, which means that*

$$\Phi(x + y, z) = \Phi(x, z) + \Phi(y, z) \text{ and } \Phi(\alpha x, y) = \alpha\Phi(x, y),$$

$$\Phi(x, y + z) = \Phi(x, y) + \Phi(x, z) \text{ and } \Phi(x, \alpha y) = \alpha\Phi(x, y)$$

for all $x, y, z \in A, \alpha \in F$.

If a is an element of A, then consider the mappings ${}_a\Phi : A \longrightarrow F$ and $\Phi_a : A \longrightarrow F$ defined by ${}_a\Phi(x) = \Phi(a, x)$ and $\Phi_a(x) = \Phi(x, a)$ for all $x \in A$. In this case, these mappings are both linear functionals and using some well-established techniques and Proposition 5.1.3, we obtain the following result.

6.1.2. Proposition. *Let A be a vector space over a field F and let Φ be a bilinear form on A. Then, the following assertions hold:*

(i) $\Phi(x, 0_A) = \Phi(0_A, x) = 0_F$ *for all* $x \in A$*;*
(ii) $\Phi(-x, y) = \Phi(x, -y) = -\Phi(x, y)$ *for all* $x, y \in A$*;*
(iii) $\Phi(x - y, z) = \Phi(x, z) - \Phi(y, z)$ *for all* $x, y, z \in A$*;*
(iv) $\Phi(x, y - z) = \Phi(x, y) - \Phi(x, z)$ *for all* $x, y, z \in A$.

For the vector space A over the field F, let $\mathbf{Bil}_F(A)$ denote the set of all bilinear forms on A. Define addition of bilinear forms by

$$(\Phi + \Psi)(x, y) = \Phi(x, y) + \Psi(x, y) \text{ for all } x, y \in A,$$

whenever $\Phi, \Psi \in \mathbf{Bil}_F(A)$. As with linear mappings, we can prove that the sum of two bilinear forms is again bilinear. We can also check that addition of forms satisfies the conditions

$$\Phi + \Psi = \Psi + \Phi \text{ and } \Phi + (\Psi + \Gamma) = (\Phi + \Psi) + \Gamma,$$

for all $\Phi, \Psi, \Gamma \in \mathbf{Bil}_F(A)$. Clearly the mapping $\Theta : A \times A \longrightarrow F$ defined by the rule $\Theta(x, y) = 0_F$ for all $x, y \in A$ is bilinear, and from the definition we have

$$\Phi + \Theta = \Phi \text{ for each } \Phi \in \mathbf{Bil}_F(A).$$

Furthermore, put $(-\Phi)(x, y) = -\Phi(x, y)$ for all $x, y \in A$. It is easy to see that $-\Phi$ is a bilinear form and that $\Phi + (-\Phi) = \Theta$. Hence, the set $\mathbf{Bil}_F(A)$ is an abelian group under the operation of addition.

Let $\Phi \in \mathbf{Bil}_F(A)$ and let $\alpha \in F$. Define the mapping $\alpha\Phi : A \times A \longrightarrow F$ by $(\alpha\Phi)(x, y) = \alpha\Phi(x, y)$ for all $x, y \in A$. As for linear mappings, we can show that this scalar multiplication satisfies the conditions

$$\alpha(\Phi + \Psi) = \alpha\Phi + \alpha\Psi,\ (\alpha + \beta)\Phi = \alpha\Phi + \beta\Phi,\ (\alpha\beta)\Phi = \alpha(\beta\Phi), \text{ and } e\Phi = \Phi,$$

for all $\Phi, \Psi \in \mathbf{Bil}_F(A)$, $\alpha, \beta \in F$.

Consequently, all the conditions of Definition 4.1.4 are satisfied and the set $\mathbf{Bil}_F(A)$ becomes a vector space over the field F.

6.1.3. Definition. *Let A be a vector space over a field F and let Φ be a bilinear form on A. Then, Φ is called symmetric, if $\Phi(x, y) = \Phi(y, x)$ for all elements $x, y \in A$. Also Φ is called skew symmetric or symplectic or alternating, if $\Phi(x, y) = -\Phi(y, x)$ for all elements $x, y \in A$.*

If Φ is a symmetric (respectively symplectic) form, then clearly $\alpha\Phi$ is also symmetric (respectively symplectic) for each $\alpha \in F$. We note that if Φ is a symplectic form, then $\Phi(x, x) = -\Phi(x, x)$ so $2\Phi(x, x) = 0_F$. If **char** $F \neq 2$, it follows that $\Phi(x, x) = 0_F$.

Before proceeding, we make some slight notational changes. Suppose first that **char** $F = 0$. Then, $ne \neq 0_F$ for each positive integer n and it follows that ne has a multiplicative inverse $(ne)^{-1}$. For each element $a \in A$ we obtain $(ne)^{-1}(na) = (ne)^{-1}(ne)a = ea = a$. So, we shall write $\frac{1}{n}a$ instead of $(ne)^{-1}a$.

Suppose now that **char** $F = p > 0$ and let n be a positive integer such that **GCD**$(n, p) = 1$. Then $ne \neq 0_F$ and hence, ne has a multiplicative inverse $(ne)^{-1}$. For each element $a \in A$ we again obtain $(ne)^{-1}(na) = (ne)^{-1}(ne)a = ea = a$. Therefore, in this case also, we write $\frac{1}{n}a$ instead of $(ne)^{-1}a$.

The following theorem justifies the importance of symmetric and symplectic forms.

6.1.4. Theorem. *Let A be a vector space over a field F and let Φ be a bilinear form on A. Suppose that* **char** $F \neq 2$. *Then, $\Phi = \Phi_1 + \Phi_2$ where Φ_1 is a symmetric form and Φ_2 is a symplectic form. This representation is unique.*

Proof. Put $\hat{\Phi}(x, y) = \Phi(y, x)$ for all $x, y \in A$ and consider the forms

$$\Phi_3 = \Phi + \hat{\Phi} \text{ and } \Phi_4 = \Phi - \hat{\Phi}.$$

We have

$$\begin{aligned}\Phi_3(y, x) &= \Phi(y, x) + \hat{\Phi}(y, x) = \Phi(y, x) + \Phi(x, y) = \hat{\Phi}(x, y) + \Phi(x, y)\\ &= \Phi_3(x, y)\end{aligned}$$

and

$$\begin{aligned}\Phi_4(y, x) &= \Phi(y, x) - \hat{\Phi}(y, x) = \Phi(y, x) - \Phi(x, y) = -(\Phi(x, y) - \Phi(y, x))\\ &= -\Phi_4(x, y).\end{aligned}$$

Thus Φ_3 is a symmetric bilinear form and Φ_4 is symplectic. Furthermore,

$$\Phi_3(x, y) + \Phi_4(x, y) = \Phi(x, y) + \hat{\Phi}(x, y) + \Phi(x, y) - \hat{\Phi}(x, y) = 2\Phi(x, y).$$

Since **char** $F \neq 2$, it follows that

$$\Phi(x, y) = \frac{1}{2}\Phi_3(x, y) + \frac{1}{2}\Phi_4(x, y).$$

As we remarked above, the form $\frac{1}{2}\Phi_3$ is symmetric and the form $\frac{1}{2}\Phi_4$ is symplectic so let

$$\Phi_1 = \frac{1}{2}\Phi_3 \text{ and } \Phi_2 = \frac{1}{2}\Phi_4.$$

To prove uniqueness, suppose that $\Phi = \Phi_5 + \Phi_6$, where Φ_5 is a symmetric form and Φ_6 is a symplectic form. Then

$$\hat{\Phi}(x, y) = \Phi(y, x) = \Phi_5(y, x) + \Phi_6(y, x) = \Phi_5(x, y) - \Phi_6(x, y).$$

Therefore,

$$\begin{aligned}\Phi_3(x, y) &= \Phi(x, y) + \hat{\Phi}(x, y) = \Phi_5(x, y) + \Phi_6(x, y) + \Phi_5(x, y) - \Phi_6(x, y)\\ &= 2\Phi_5(x, y),\end{aligned}$$

and

$$\begin{aligned}\Phi_4(x, y) &= \Phi(x, y) - \hat{\Phi}(x, y) = \Phi_5(x, y) + \Phi_6(x, y) - \Phi_5(x, y) + \Phi_6(x, y)\\ &= 2\Phi_6(x, y).\end{aligned}$$

It follows that

$$\Phi_5 = \frac{1}{2}\Phi_3 = \Phi_1 \text{ and } \Phi_6 = \frac{1}{2}\Phi_4 = \Phi_2,$$

which proves the uniqueness desired.

Suppose now that the vector space A is finite dimensional. Let $\{a_1, \ldots, a_n\}$ be a basis of A. If x, y are elements of A then, by Proposition 4.2.16, $x = \sum_{1\leq j\leq n} \xi_j a_j$ and $y = \sum_{1\leq t\leq n} \eta_t a_t$, where $\xi_j, \eta_t \in F$, for $1 \leq j, t \leq n$. If Φ is a bilinear form on A, then

$$\begin{aligned}\Phi(x, y) &= \Phi\left(\sum_{1\leq j\leq n} \xi_j a_j, \sum_{1\leq t\leq n} \eta_t a_t\right) = \sum_{1\leq j\leq n} \Phi\left(\xi_j a_j, \sum_{1\leq t\leq n} \eta_t a_t\right)\\ &= \sum_{1\leq j\leq n}\sum_{1\leq t\leq n} \Phi(\xi_j a_j, \eta_t a_t) = \sum_{1\leq j\leq n}\sum_{1\leq t\leq n} \xi_j \Phi(a_j, \eta_t a_t)\\ &= \sum_{1\leq j\leq n}\sum_{1\leq t\leq n} \xi_j \eta_t \Phi(a_j, a_t).\end{aligned}$$

This equation shows that the value of $\Phi(x, y)$ is completely determined by the coordinates of the elements x, y and the elements $\Phi(a_j, a_t)$, for $1 \leq j, t \leq n$. Thus the elements $\Phi(a_j, a_t)$, for $1 \leq j, t \leq n$, determine the form Φ uniquely.

As usual, it is possible to associate a matrix with a bilinear form Φ.

6.1.5. Definition. *Let A be a finite-dimensional vector space over a field F and let $\{a_1, \dots, a_n\}$ be a basis of A. If Φ is a bilinear form on A then put $\sigma_{jt} = \Phi(a_j, a_t)$, for $1 \le j, t \le n$. The matrix*

$$\begin{pmatrix} \sigma_{11} & \sigma_{12} & \dots & \sigma_{1n} \\ \sigma_{21} & \sigma_{22} & \dots & \sigma_{2n} \\ \vdots & \vdots & \ddots & \vdots \\ \sigma_{k1} & \sigma_{k2} & \dots & \sigma_{kn} \end{pmatrix}$$

is called the matrix of Φ relative to the basis $\{a_1, \dots, a_n\}$.

Here are some important properties of the matrix of a bilinear form.

6.1.6. Proposition. *Let A be a finite-dimensional vector space over a field F and let $\{a_1, \dots, a_n\}$ be a basis of A. If $\Phi, \Psi \in \mathbf{Bil}_F(A)$ and $S = [\sigma_{jt}]$, $R = [\rho_{jt}] \in \mathbf{M}_n(F)$ are the matrices of Φ and Ψ relative to the basis $\{a_1, \dots, a_n\}$, then*

(i) *$S + R$ is the matrix of the form $\Phi + \Psi$ relative to the basis $\{a_1, \dots, a_n\}$;*
(ii) *if $\alpha \in F$, then αS is the matrix of the form $\alpha\Phi$ relative to the basis $\{a_1, \dots, a_n\}$.*

Proof.

(i) We have

$$(\Phi + \Psi)(a_j, a_t) = \Phi(a_j, a_t) + \Psi(a_j, a_t) = \sigma_{jt} + \rho_{jt}, \text{ for } 1 \le j, t \le n.$$

It follows that $[\sigma_{jt} + \rho_{jt}] = S + R \in \mathbf{M}_n(F)$ is the matrix of $\Phi + \Psi$ relative to the basis $\{a_1, \dots, a_n\}$.

(ii) We have

$$(\alpha\Phi)(a_j, a_t) = \alpha(\Phi(a_j, a_t)) = \alpha\sigma_{jt}, \text{ for } 1 \le j, t \le n.$$

It follows that $[\alpha\sigma_{jt}] = \alpha S \in \mathbf{M}_n(F)$ is the matrix of $\alpha\Phi$ relative to the basis $\{a_1, \dots, a_n\}$.

6.1.7. Corollary. *Let A be a finite-dimensional vector space over a field F and let $\mathbf{dim}_F(A) = n$. Then, the vector space $\mathbf{Bil}_F(A)$ is isomorphic to $\mathbf{M}_n(F)$.*

Proof. Let $\{a_1, \dots, a_n\}$ be a basis of A. Let $\Phi \in \mathbf{Bil}_F(A)$ and define the mapping $\Gamma : \mathbf{Bil}_F(A) \longrightarrow \mathbf{M}_n(F)$ by $\Gamma(\Phi) = S$, where S is the matrix of Φ relative to the basis $\{a_1, \dots, a_n\}$. Proposition 6.1.6 shows that this mapping is linear.

Let $R = [\rho_{jt}] \in \mathbf{M}_n(F)$. Define the mapping $\Psi : A \times A \longrightarrow F$ as follows: If $x = \sum_{1 \le j \le n} \xi_j a_j$ and $y = \sum_{1 \le t \le n} \eta_t a_t$ are elements of A, then put

$$\Psi(x, y) = \sum_{1 \le j \le n} \sum_{1 \le t \le n} \xi_j \eta_t \rho_{jt}.$$

We will show that Ψ is bilinear. To this end, let z be another element of A, say $z = \sum_{1 \le j \le n} \zeta_j a_j$ and let $\alpha \in F$. Then $x + z = \sum_{1 \le j \le n} (\xi_j + \zeta_j) a_j$, and

$$\begin{aligned}\Psi(x + z, y) &= \sum_{1 \le j \le n} \sum_{1 \le t \le n} (\xi_j + \zeta_j) \eta_t \rho_{jt} \\ &= \sum_{1 \le j \le n} \sum_{1 \le t \le n} \xi_j \eta_t \rho_{jt} + \sum_{1 \le j \le n} \sum_{1 \le t \le n} \zeta_j \eta_t \rho_{jt} = \Psi(x, y) + \Psi(z, y).\end{aligned}$$

Similarly, we can show that $\Psi(x, y + z) = \Psi(x, y) + \Psi(x, z)$. Furthermore, $\alpha x = \sum_{1 \le j \le n} \alpha \xi_j a_j$, and

$$\begin{aligned}\Psi(\alpha x, y) &= \sum_{1 \le j \le n} \sum_{1 \le t \le n} (\alpha \xi_j) \eta_t \rho_{jt} = \sum_{1 \le j \le n} \sum_{1 \le t \le n} \alpha (\xi_j \eta_t \rho_{jt}) \\ &= \alpha \left(\sum_{1 \le j \le n} \sum_{1 \le t \le n} \zeta_j \eta_t \rho_{jt} \right) = \alpha \Psi(x, y).\end{aligned}$$

We can show that $\Psi(x, \alpha y) = \alpha \Psi(x, y)$ in a similar manner.

By definition of Ψ,

$$\Psi(a_m, a_k) = ee\rho_{mk} = \rho_{mk}, \text{ for } 1 \le m, k \le n,$$

which shows that R is the matrix of Ψ relative to the basis $\{a_1, \dots, a_n\}$. Hence the mapping Γ is surjective.

Finally, let Φ, Ψ be bilinear forms on A and let $S = [\sigma_{jt}]$, $R = [\rho_{jt}] \in \mathbf{M}_n(F)$ be the matrices of Φ and Ψ relative to the basis $\{a_1, \dots, a_n\}$. For each pair of elements $x = \sum_{1 \le j \le n} \xi_j a_j$ and $y = \sum_{1 \le t \le n} \eta_t a_t$, we have

$$\Phi(x, y) = \sum_{1 \le j \le n} \sum_{1 \le t \le n} \xi_j \eta_t \sigma_{jt} \text{ and } \Psi(x, y) = \sum_{1 \le j \le n} \sum_{1 \le t \le n} \xi_j \eta_t \rho_{jt}.$$

Suppose that $\Gamma(\Phi) = \Gamma(\Psi)$, so that $S = R$. Then $\sigma_{jt} = \rho_{jt}$ for all j, t, where $1 \le j, t \le n$. It follows that $\Phi(x, y) = \Psi(x, y)$ for all $x, y \in A$, which proves that $\Phi = \Psi$. Hence, the mapping Γ is injective and therefore, Γ is an isomorphism.

The structure of the matrix of a bilinear form is very dependent on the chosen basis. The following theorem describes how a change of basis affects the matrix of the form.

6.1.8. Theorem. *Let A be a finite-dimensional vector space over a field F, and let $\{a_1, \ldots, a_n\}$, $\{b_1, \ldots, b_n\}$ be bases of A. Suppose that Φ is a bilinear form on A and let $S = [\sigma_{jt}]$, $R = [\rho_{jt}] \in \mathbf{M}_n(F)$ denote the matrices of Φ relative to the bases $\{a_1, \ldots, a_n\}$ and $\{b_1, \ldots, b_n\}$, respectively. Let $T = [\iota_{jt}] \in \mathbf{M}_n(F)$ denote the transition matrix from $\{a_1, \ldots, a_n\}$ to $\{b_1, \ldots, b_n\}$. Then $R = T^t ST$.*

Proof. We have

$$\rho_{mk} = \Phi(b_m, b_k) = \Phi\left(\sum_{1\le j\le n} \iota_{jm}a_j, \sum_{1\le t\le n} \iota_{tk}a_t\right) = \sum_{1\le j\le n}\sum_{1\le t\le n} \iota_{jm}\iota_{tk}\Phi(a_j, a_t)$$

$$= \sum_{1\le j\le n}\sum_{1\le t\le n} \iota_{jm}\iota_{tk}\sigma_{jt} = \sum_{1\le j\le n}\sum_{1\le t\le n} \iota_{jm}\sigma_{jt}\iota_{tk}.$$

Let $T^t = [\theta_{jt}] \in \mathbf{M}_n(F)$ so that $\theta_{jt} = \iota_{tj}$, where $1 \le j, t \le n$. We compute the product $T^t ST$. Let $T^t ST = [\gamma_{mk}] \in \mathbf{M}_n(F)$ and let $ST = [\beta_{jt}] \in \mathbf{M}_n(F)$. Then,

$$\gamma_{mk} = \sum_{1\le j\le n} \theta_{mj}\beta_{jk} = \sum_{1\le j\le n} \theta_{mj}\left(\sum_{1\le t\le n} \sigma_{jt}\iota_{tk}\right) = \sum_{1\le j\le n}\sum_{1\le t\le n} \theta_{mj}\sigma_{jt}\iota_{tk}$$

$$= \sum_{1\le j\le n}\sum_{1\le t\le n} \iota_{jm}\sigma_{jt}\iota_{tk}, \text{ for } 1 \le m, k \le n.$$

Comparing this with ρ_{mk} we deduce that $R = T^t ST$, which proves the result.

6.1.9. Corollary. *Let A be finite-dimensional vector space over a field F and let $\{a_1, \ldots, a_n\}$, $\{b_1, \ldots, b_n\}$ be bases of A. Suppose that Φ is a bilinear form on A and that $S = [\sigma_{jt}]$, $R = [\rho_{jt}] \in \mathbf{M}_n(F)$ are the matrices of Φ relative to the bases $\{a_1, \ldots, a_n\}$ and $\{b_1, \ldots, b_n\}$ respectively. Then* **rank**(S) = **rank**(R).

Proof. By Theorem 6.1.8, $R = T^t ST$ where T is the transition matrix from the first basis to the second. By Corollary 4.2.19, the matrix T is nonsingular, so Corollaries 4.3.10 and 4.3.11 together imply that **rank**(R) = **rank**(S).

We introduce the following concept based on these results.

6.1.10. Definition. *Let A be a finite-dimensional vector space over a field F and let $\{a_1, \ldots, a_n\}$ be a basis of A. Suppose Φ is a bilinear form on A and S is the matrix of Φ relative to the basis $\{a_1, \ldots, a_n\}$, then* **rank**(S) *is called the rank of Φ and will be denoted by* **rank**(Φ).

6.1.11. Proposition. *Let A be a finite-dimensional vector space over the field F and let Φ be a bilinear form on A. If Φ is symmetric (respectively skew symmetric), then the matrix of Φ relative to any basis is symmetric (respectively* *skew symmetric). Conversely, suppose that the matrix of Φ relative to some basis is symmetric (respectively skew symmetric). Then Φ is symmetric (respectively skew symmetric).*

Proof. Let $\{a_1, \dots, a_n\}$ be an arbitrary basis of A and let $S = [\sigma_{jt}] \in \mathbf{M}_n(F)$ be the matrix of Φ relative to this basis. If Φ is symmetric (respectively skew symmetric) then $\sigma_{jt} = \Phi(a_j, a_t) = \Phi(a_t, a_j) = \sigma_{tj}$ (respectively $\sigma_{jt} = \Phi(a_j, a_t) = -\Phi(a_t, a_j) = -\sigma_{tj}$), for $1 \leq j, t \leq n$, so S is also symmetric (respectively antisymmetric).

Conversely, let $\{c_1, \dots, c_n\}$ be a basis of A such that the matrix $R = [\rho_{jt}] \in \mathbf{M}_n(F)$ of Φ relative to this basis is symmetric (respectively antisymmetric). For each pair of elements $x = \sum_{1\leq j\leq n} \xi_j c_j$, $y = \sum_{1\leq t\leq njt} \eta_t c_t$, we have

$$\Phi(x, y) = \sum_{1\leq j\leq n} \sum_{1\leq t\leq n} \xi_j \eta_t \rho_{jt} = \sum_{1\leq t\leq n} \sum_{1\leq j\leq n} \eta_t \xi_j \rho_{tj} = \Phi(y, x),$$

and, respectively,

$$\begin{aligned}\Phi(x, y) &= \sum_{1\leq j\leq n} \sum_{1\leq t\leq n} \xi_j \eta_t \rho_{jt} = \sum_{1\leq j\leq n} \sum_{1\leq t\leq n} \eta_t \xi_j (-\rho_{tj}) \\ &= -\left(\sum_{1\leq j\leq n} \sum_{1\leq t\leq n} \eta_t \xi_j \rho_{tj}\right) = -\Phi(y, x).\end{aligned}$$

In the case of symmetric forms, the language of bilinear forms can be translated into the language of quadratic forms.

6.1.12. Definition. *Let A be a finite-dimensional vector space over the field F and let Φ be a bilinear form on A. The mapping $f : A \longrightarrow F$ defined by the rule $f(x) = \Phi(x, x)$ is called the quadratic form associated with the bilinear form Φ.*

By Theorem 6.1.4, $\Phi = \Phi_1 + \Phi_2$ where Φ_1 is a symmetric form and Φ_2 is a symplectic form. Then $\Phi(x, x) = \Phi_1(x, x) + \Phi_2(x, x)$. As we observed above, if $\mathbf{char} F \neq 2$, then $\Phi_2(x, x) = 0_F$. Therefore, it suffices to consider only quadratic forms associated with bilinear symmetric forms.

Let $\{a_1, \dots, a_n\}$ be a basis of the space A and let $S = [\sigma_{jt}] \in \mathbf{M}_n(F)$ be the matrix of Φ relative to this basis. For each element $x = \sum_{1\leq j\leq n} \xi_j a_j$, we have

$$f(x) = \Phi(x, x) = \sum_{1\leq j\leq n} \sum_{1\leq t\leq n} \xi_j \xi_t \sigma_{jt}.$$

The matrix $S = [\sigma_{jt}]$ is called the matrix of the quadratic form f relative to the basis $\{a_1, \dots, a_n\}$. The rule for changing the matrix of a quadratic form at the transition from one basis to another will still be the same as it was for bilinear forms, namely if $\{b_1, \dots, b_n\}$ is another basis and R is the matrix of f relative to this basis, then $R = T^t S T$ where T is the transition matrix from the first basis to the second.

EXERCISE SET 6.1

Write out a proof or give a counterexample. Show your work.

6.1.1. Let $f : \mathbb{R}^3 \longrightarrow \mathbb{R}^3$ be a mapping defined by $f((\alpha, \beta, \gamma), (\lambda, \mu, \nu)) = \alpha\lambda + \beta\mu$. Is this mapping bilinear?

6.1.2. Let $f : \mathbb{R}^3 \longrightarrow \mathbb{R}^3$ be a mapping defined by $f((\alpha, \beta, \gamma), (\lambda, \mu, \nu)) = \alpha\lambda + \gamma\nu$. Is this mapping bilinear?

6.1.3. Relative to the standard basis of the vector space $A = \mathbb{Q}^3$ we define a bilinear form Φ using the matrix

$$\begin{pmatrix} 3 & 1 & -1 \\ 1 & 0 & 2 \\ -1 & 2 & -1 \end{pmatrix}.$$

Find $\Phi(x, y)$ for $x = (1, -2, 0)$, $y = (0, -1, -2)$.

6.1.4. Relative to the standard basis of the vector space $A = \mathbb{Q}^3$ we define a bilinear form Φ using the matrix

$$\begin{pmatrix} 3 & 1 & -6 \\ 1 & 5 & 0 \\ -1 & 2 & -1 \end{pmatrix}.$$

Find $\Phi(x, y)$ for $x = (1, -5, 0)$, $y = (0, -3, -2)$.

6.1.5. Let $A = \mathbb{F}_3^3$ where $\mathbb{F}_3$ is a field of three elements. We define a bilinear form Φ using the matrix

$$\begin{pmatrix} 1 & -1 & -2 \\ 0 & 2 & 0 \\ 1 & 2 & 2 \end{pmatrix}.$$

Find $\Phi(x, y)$ for $x = (1, 1, 0)$, $y = (0, 1, 2)$.

6.1.6. Let $A = \mathbb{F}_5^3$, where $\mathbb{F}_5$ is a field of five elements. We define a bilinear form Φ using the matrix

$$\begin{pmatrix} 1 & 3 & 4 \\ 0 & 2 & 0 \\ 1 & 2 & 3 \end{pmatrix}.$$

Find $\Phi(x, y)$ for $x = (4, 1, 0)$, $y = (3, 0, 2)$.

6.1.7. Relative to the standard basis of the vector space $A = \mathbb{Q}^5$, we define a bilinear form Φ using the matrix

$$\begin{pmatrix} 0 & 0 & 1 & 0 & 0 \\ 1 & 2 & 0 & 1 & 1 \\ 0 & 1 & 1 & 2 & 2 \\ 1 & 0 & 0 & 1 & 2 \\ 2 & 2 & 1 & 0 & 0 \end{pmatrix}.$$

Decompose this form into the sum of a symmetric and an alternating form.

6.1.8. Relative to the standard basis of the vector space $A = \mathbb{Q}^4$, we define a bilinear form Φ using the matrix

$$\begin{pmatrix} 1 & 0 & 1 & 0 \\ 1 & 2 & 0 & 1 \\ 0 & 1 & 1 & 1 \\ 1 & 0 & 0 & 1 \end{pmatrix}.$$

Decompose this form into the sum of a symmetric and an alternating form.

6.1.9. Relative to the standard basis of the vector space $A = \mathbb{F}_5^4$, we define a bilinear form Φ using the matrix

$$\begin{pmatrix} 1 & 0 & 4 & 0 \\ 1 & 2 & 0 & 1 \\ 0 & 1 & 1 & 2 \\ 2 & 0 & 0 & 1 \end{pmatrix}.$$

Decompose this form into the sum of a symmetric and an alternating form.

6.1.10. Let $A = \mathbb{F}_3^3$ where $\mathbb{F}_3$ is the field consisting of three elements. A bilinear form Φ is given relative to the standard basis, as

$$\begin{pmatrix} 1 & -1 & 2 \\ 0 & 2 & 0 \\ 1 & 2 & 2 \end{pmatrix}.$$

Find the matrix of the form relative to the basis $(1,1,0)$, $(0, 1, 1)$, $(0, 0, 1)$.

6.2 CLASSICAL FORMS

In geometry, the important concept of an orthogonal basis has been introduced with the aid of the scalar product. This concept can be extended to a vector space, on which a bilinear form is defined, in the following way.

6.2.1. Definition. *Let A be a vector space over a field F and let Φ be a bilinear form on A. We say that an element $x \in A$ is left orthogonal to $y \in A$ if $\Phi(x, y) = 0_F$. In this case, we say that that y is right orthogonal to x.*

In the general case, left orthogonality does not coincide with right orthogonality. For example, let $\mathbf{dim}_F(A) = 2$ and suppose that A has the basis $\{a_1, a_2\}$ and let the matrix $\begin{pmatrix} \alpha & e \\ 0_F & \gamma \end{pmatrix}$ correspond to the bilinear form Φ. Then $\Phi(a_1, a_2) = e$ and $\Phi(a_2, a_1) = 0_F$, so a_1 is left, but not right, orthogonal to a_2. Of course, if a form Φ is symmetric or symplectic then left orthogonality coincides with right orthogonality.

6.2.2. Definition. *Let A be a vector space over a field F and let Φ be a bilinear form on A. For a subspace M of A we put*

$$^{\perp}M = \{x \mid x \in A \text{ and } \Phi(x, a) = 0_F \text{ for each } a \in M\}$$

and

$$M^{\perp} = \{x \mid x \in A \text{ and } \Phi(a, x) = 0_F \text{ for each } a \in M\}.$$

The subset $^{\perp}M$ (respectively $M^{\perp}$) is called a left (respectively right) orthogonal complement to M. The subset $^{\perp}A$ (respectively $A^{\perp}$) is called a left (respectively right) kernel of Φ.

6.2.3. Proposition. *Let A be a vector space over a field F and let Φ be a bilinear form on A. If M is a subset of A, then $^{\perp}M$ and $M^{\perp}$ are subspaces of A.*

Proof. Clearly $^{\perp}M \neq \emptyset$ since $0_A \in {}^{\perp}M$. Let $x, y \in {}^{\perp}M$, let a be an arbitrary element of M and let $\alpha \in M$. By Proposition 6.1.2,

$$\Phi(x - y, a) = \Phi(x, a) - \Phi(y, a) = 0_F - 0_F = 0_F \text{ and}$$
$$\Phi(\alpha x, a) = \alpha\Phi(x, a) = 0_F.$$

This shows that $x - y, \alpha x \in {}^{\perp}M$. Theorem 4.1.7 implies that $^{\perp}M$ is a subspace and similarly we can deduce also that $M^{\perp}$ is a subspace.

As we have already seen, the subspaces $^{\perp}A$ and $A^{\perp}$ can be different. However, the following result holds.

6.2.4. Theorem. *Let A be a finite-dimensional vector space over a field F and let Φ be a bilinear form on A. Then $\mathbf{dim}_F(^{\perp}A) = \mathbf{dim}_F(A) - \mathbf{rank}(\Phi)$.*

Proof. Let $M = \{a_1, \dots, a_n\}$ be a basis of A and note that it is clearly the case that ${}^{\perp}A \leq {}^{\perp}M$. Let $y \in {}^{\perp}M$ and let $x = \sum_{1 \leq j \leq n} \xi_j a_j$ be an arbitrary element of A, where $\xi_j \in F$, for $1 \leq j \leq n$. Then,

$$\Phi(y, x) = \Phi\left(y, \sum_{1 \leq j \leq n} \xi_j a_j\right) = \sum_{1 \leq j \leq n} \xi_j \Phi(y, a_j) = 0_F.$$

Thus $y \in {}^{\perp}A$, so ${}^{\perp}A = {}^{\perp}M$.

Let $S = [\sigma_{jt}] \in \mathbf{M}_n(F)$ denote the matrix of Φ relative to the basis $\{a_1, \dots, a_n\}$. Let $z = \sum_{1 \leq j \leq n} \zeta_j a_j$ be an arbitrary element of ${}^{\perp}M$, where $\zeta_j \in F$, for $1 \leq j \leq n$. Then

$$\begin{aligned} 0_F &= \Phi(z, a_k) = \Phi\left(\sum_{1 \leq j \leq n} \zeta_j a_j, a_k\right) = \sum_{1 \leq j \leq n} \zeta_j \Phi(a_j, a_k) \\ &= \sum_{1 \leq j \leq n} \zeta_j \sigma_{jk}, \text{ for } 1 \leq k \leq n. \end{aligned}$$

Thus, the n-tuple $(\zeta_1, \dots, \zeta_n)$ is a solution of the system

$$\begin{aligned} \sigma_{11}x_1 + \sigma_{21}x_2 + \cdots + \sigma_{n1}x_n &= 0_F \\ \sigma_{12}x_1 + \sigma_{22}x_2 + \cdots + \sigma_{n2}x_n &= 0_F \\ &\vdots \\ \sigma_{1n}x_1 + \sigma_{2n}x_2 + \cdots + \sigma_{nn}x_n &= 0_F. \end{aligned} \tag{6.1}$$

Conversely, every solution of Equation 6.1 gives the coordinates of some element of ${}^{\perp}M$. We observe that the matrix of the system (Eq. 6.1) is S^t.

Let $\kappa : A \longrightarrow F^n$ be the canonical isomorphism. By the above, $\kappa({}^{\perp}A)$ is the subspace of all solutions of the system (Eq. 6.1). From the results of Section 5.3, we deduce that $\mathbf{dim}_F(\kappa({}^{\perp}A)) = n - \mathbf{rank}(S^t)$ and by Corollary 4.3.6, $\mathbf{rank}(S^t) = \mathbf{rank}(S)$. Therefore, $\mathbf{dim}_F(\kappa({}^{\perp}A)) = n - \mathbf{rank}(S) = n - \mathbf{rank}(\Phi)$. Since κ is an isomorphism, Corollary 5.1.9 implies that $\mathbf{dim}_F(\kappa({}^{\perp}A)) = \mathbf{dim}_F({}^{\perp}A)$, so that $\mathbf{dim}_F({}^{\perp}A) = n - \mathbf{rank}(\Phi) = \mathbf{dim}_F(A) - \mathbf{rank}(\Phi)$.

The proof for the fact about the right kernel is essentially the same as that given above for the left kernel.

6.2.5. Definition. *Let A be a finite-dimensional vector space over a field F and let Φ be a bilinear form on A. The number $\mathbf{dim}_F({}^{\perp}A) = \mathbf{dim}_F(A^{\perp}) = n - \mathbf{rank}(\Phi)$ is called the defect of Φ. A form Φ is called nonsingular, if its defect is 0. When B is a subspace of A, we say that B is nonsingular, if the restriction of Φ to B is a nonsingular form.*

Thus, the form Φ is nonsingular if and only if the matrix of the form is nonsingular. The next result is very easy to prove and is left to the reader.

6.2.6. Proposition. *Let A be a finite-dimensional vector space over a field F and let Φ be a bilinear form on A. Then, the form Φ is nonsingular if and only if for each nonzero element x there is an element y such that $\Phi(x, y) \neq 0_F$.*

6.2.7. Definition. *Let A be a vector space over a field F and let Φ be a bilinear form on A. The form Φ is called classical if the equation $\Phi(x, y) = 0_F$ always implies that $\Phi(y, x) = 0_F$.*

Thus, for a classical form left and right orthogonality coincide.

We saw above that symmetric and symplectic forms are classical. Now we will prove that, in the case when **char** $F \neq 2$, there are no classical forms other than these. We first prove this for nonsingular forms.

6.2.8. Lemma. *Let A be a vector space over a field F and let Φ be a bilinear classical form on A. Suppose that* **char** *$F \neq 2$. If Φ is nonsingular, then Φ is symmetric or symplectic.*

Proof. Let x be an arbitrary nonzero element of A. By Proposition 6.2.6, there exists an element y such that $\Phi(x, y) = \alpha \neq 0_F$. Of course, the element y is also nonzero. Put $u = \alpha^{-1}y$; then

$$\Phi(x, u) = \Phi(x, \alpha^{-1}y) = \alpha^{-1}\Phi(x, y) = \alpha^{-1}\alpha = e.$$

Let z be an arbitrary element of A and let $\Phi(x, z) = \beta$. Then,

$$\Phi(x, z - \beta y) = \Phi(x, z) - \beta\Phi(x, y) = \beta - \beta e = 0_F.$$

It follows, since Φ is classical, that

$$0_F = \Phi(z - \beta y, x) = \Phi(z, x) - \beta\Phi(y, x) \text{ and so}$$

$$\Phi(z, x) = \beta\Phi(y, x) = \Phi(x, z)\Phi(y, x).$$

Let $\Phi(y, x) = \gamma(x)$. Then we have $\Phi(z, x) = \gamma(x)\Phi(x, z)$. The element $\gamma(x)$ does not depend on z and next, we show that it does not depend on x either.

To this end, choose two nonzero elements $x_1, x_2 \in A$. By Proposition 6.2.6, there exist nonzero elements v_1, v_2 such that $\Phi(v_1, x_1) \neq 0_F$ and $\Phi(v_2, x_2) \neq 0_F$. If $\Phi(v_1, x_2) \neq 0_F$, then put $v = v_1$; if $\Phi(v_1, x_2) = 0_F$ but $\Phi(v_2, x_1) = 0_F$, then put $v = v_2$. Suppose that $\Phi(v_1, x_2) = 0_F$ and $\Phi(v_2, x_1) = 0_F$. Set $v = v_1 + v_2$. In this case,

$$\Phi(v, x_1) = \Phi(v_1 + v_2, x_1) = \Phi(v_1, x_1) + \Phi(v_2, x_1) = \Phi(v_1, x_1) \neq 0_F,$$

$$\text{and } \Phi(v, x_2) = \Phi(v_1 + v_2, x_2) = \Phi(v_1, x_2) + \Phi(v_2, x_2) = \Phi(v_2, x_2) \neq 0_F.$$

Hence, there is a v such that $\Phi(v, x_1), \Phi(v, x_2) \neq 0_F$. Now put $\lambda = \Phi(v, x_1)\Phi(v, x_2)^{-1}$, and $w = \lambda x_2$. Then,

$$\Phi(v, w) = \Phi(v, \Phi(v, x_1)\Phi(v, x_2)^{-1}x_2) = \Phi(v, x_1)\Phi(v, x_2)^{-1}\Phi(v, x_2)$$
$$= \Phi(v, x_1) \neq 0_F.$$

It follows that $0_F = \Phi(v, x_1) - \Phi(v, w) = \Phi(v, x_1 - w)$, and therefore, $0_F = \Phi(x_1 - w, v) = \Phi(x_1, v) - \Phi(w, v)$. Hence, $\Phi(x_1, v) = \Phi(w, v) \neq 0$.

Furthermore,

$$\Phi(v, x_1) = \gamma(x_1)\Phi(x_1, v)$$

and

$$\Phi(v, w) = \Phi(v, \lambda x_2) = \lambda\Phi(v, x_2) = \lambda\gamma(x_2)\Phi(x_2, v)$$
$$= \gamma(x_2)\Phi(\lambda x_2, v) = \gamma(x_2)\Phi(w, v).$$

The equation $\Phi(v, x_1) = \Phi(v, w)$ implies that $\gamma(x_1)\Phi(x_1, v) = \gamma(x_2)\Phi(w, v)$ and, since $\Phi(x_1, v) = \Phi(w, v) \neq 0_F$, we see that $\gamma(x_1) = \gamma(x_2)$. Because x_1 and x_2 are arbitrary nonzero elements of A, it follows that there exists a nonzero element $\gamma \in F$ such that $\Phi(z, x) = \gamma\Phi(x, z)$ for all nonzero elements $x, z \in A$.

Finally, let x, z be nonzero elements of A such that $\Phi(z, x) \neq 0_F$. Then

$$\Phi(z, x) = \gamma\Phi(x, z) = \gamma^2\Phi(z, x).$$

It follows that $\gamma^2 - e = 0_F$. Since **char** $F \neq 2$, $e \neq -e$, we have only two possibilities for γ, namely $\gamma = e$ or $\gamma = -e$. In the first case, $\Phi(z, x) = \Phi(x, z)$ for all $x, z \in A$ and in the second case, $\Phi(z, x) = -\Phi(x, z)$ for all $x, z \in A$. Hence, Φ is symmetric or symplectic.

6.2.9. Proposition. *Let A be a finite-dimensional vector space over a field F and let Φ be a classical bilinear form on A. Then $A = A^\perp \oplus B$ and the restriction of Φ to B is a classical nonsingular bilinear form.*

Proof. By Proposition 4.2.25, the subspace $A^\perp$ has a complement B. Suppose that b is an element of B such that $\Phi(b, y) = 0_F$ for all $y \in B$. If x is an arbitrary element of A, then $x = u + v$ where $u \in A^\perp$ and $v \in B$. Then

$$\Phi(b, x) = \Phi(b, u + v) = \Phi(b, u) + \Phi(b, v) = 0_F + 0_F = 0_F.$$

It follows that $b \in A^\perp$, so that $b \in A^\perp \cap B = \{0_A\}$. By Proposition 6.2.6, this means that the restriction of Φ to B is nonsingular and this restriction is clearly a classical form.

We now generalize Lemma 6.2.8 to all cases, not just nonsingular ones.

6.2.10. Theorem. *Let A be a finite-dimensional vector space over a field F and let Φ be a bilinear classical form on A. If* $\mathbf{char} F \neq 2$ *then Φ is symmetric or symplectic.*

Proof. By Proposition 6.2.9, there is a direct decomposition $A = A^{\perp} \oplus B$ such that the restriction of Φ to B is a classical nonsingular form. By Lemma 6.2.8, either $\Phi(u, v) = \Phi(v, u)$ or $\Phi(u, v) = -\Phi(v, u)$ for all $u, v \in B$. Let x, y be arbitrary elements of A. Then $x = a_1 + u$ and $y = a_2 + v$, where $a_1, a_2 \in A^{\perp}$, $u, v \in B$. In the case when Φ restricted to B is symmetric, we have

$$\begin{aligned}\Phi(x, y) &= \Phi(a_1 + u, a_2 + v) = \Phi(a_1, a_2) + \Phi(a_1, v) + \Phi(u, a_2) + \Phi(u, v)\\ &= \Phi(u, v) \text{ and}\\ \Phi(y, x) &= \Phi(a_2 + v, a_1 + u) = \Phi(a_2, a_1) + \Phi(a_2, u) + \Phi(v, a_1) + \Phi(v, u)\\ &= \Phi(v, u).\end{aligned}$$

Therefore, $\Phi(x, y) = \Phi(y, x)$ and it follows that the form Φ is symmetric.

In the case when Φ restricted to B is symplectic, we obtain

$$\Phi(x, y) = \Phi(u, v), \Phi(y, x) = \Phi(v, u) = -\Phi(u, v),$$

so that $\Phi(x, y) = -\Phi(y, x)$, and the form Φ is symplectic on A.

We next consider the structure of those spaces on which a classical bilinear form is given. As in the case of linear transformations, our goal is to choose a basis in which the matrix of the form is as simple as possible, in some sense.

6.2.11. Proposition. *Let A be a finite-dimensional vector space over a field F and let Φ be a bilinear classical form on A. If Φ is nonsingular, then* $\mathbf{dim}_F(B^{\perp}) = \mathbf{dim}_F(A) - \mathbf{dim}_F(B)$ *for each subspace B.*

Proof. If $B = \{0_A\}$, then $B^{\perp} = A$ and $\mathbf{dim}_F(B^{\perp}) = \mathbf{dim}_F(A) - 0$ so we may suppose that B is a nonzero subspace. Let $N = \{a_1, \dots, a_k\}$ be a basis of B. By Theorem 4.2.11, there are elements $\{a_{k+1}, \dots, a_n\}$ of A such that $\{a_1, \dots, a_n\}$ is a basis of A. As in the proof of Theorem 6.2.4, we can show that $B^{\perp} = N^{\perp}$. Let $S = [\sigma_{jt}] \in \mathbf{M}_n(F)$ denote the matrix of Φ relative to $\{a_1, \dots, a_n\}$. Let $x = \sum_{1 \le j \le n} \xi_j a_j$ be an arbitrary element of $N^{\perp}$, where $\xi_j \in F$, for $1 \le j \le n$. Then, for $1 \le m \le k$, we have

$$\begin{aligned}0_F = \Phi(a_m, x) &= \Phi\left(a_m, \sum_{1 \le j \le n} \xi_j a_j\right)\\ &= \sum_{1 \le j \le n} \xi_j \Phi(a_m, a_j) = \sum_{1 \le j \le n} \xi_j \sigma_{mj}.\end{aligned}$$

Therefore, we deduce that the n-tuple $(\xi_1, \ldots, \xi_n)$ is a solution of the system

$$\begin{aligned}
\sigma_{11}x_1 + \sigma_{12}x_2 + \cdots + \sigma_{1n}x_n &= 0_F \\
\sigma_{21}x_1 + \sigma_{22}x_2 + \cdots + \sigma_{2n}x_n &= 0_F \\
&\vdots \\
\sigma_{k1}x_1 + \sigma_{k2}x_2 + \cdots + \sigma_{kn}x_n &= 0_F.
\end{aligned} \tag{6.2}$$

Conversely, every solution of the system (Eq. 6.2) gives the coordinates of some elements of $N^{\perp}$. We remark that the matrix of the system (Eq. 6.2) consists of the first k rows of the matrix S.

Let $\kappa : A \longrightarrow F^n$ be the canonical isomorphism. As above, $\kappa(B^{\perp})$ is the subspace of all solutions of the system (Eq. 6.2). By the results of Section 5.3, $\mathbf{dim}_F(\kappa(A^{\perp})) = n - r$ where r is the rank of the system (Eq. 6.2). By our hypotheses the matrix S is nonsingular, which implies that the set of all rows of S is linearly independent. In particular, the set of the first k rows of S is also linearly independent and it follows that the matrix of the system (Eq. 6.2) has rank k, so $r = k$. Hence $\mathbf{dim}_F(\kappa(B^{\perp})) = n - k$. Since κ is an isomorphism, Corollary 5.1.9 shows that $\mathbf{dim}_F(\kappa(B^{\perp})) = \mathbf{dim}_F(B^{\perp})$, so that $\mathbf{dim}_F(B^{\perp}) = n - k = \mathbf{dim}_F(A) - \mathbf{dim}_F(B)$, as required.

6.2.12. Corollary. *Let A be a finite-dimensional vector space over a field F and let Φ be a classical bilinear form on A. If Φ is nonsingular, then $B = (B^{\perp})^{\perp}$ for each subspace B.*

Proof. For all elements $b \in B$ and $x \in B^{\perp}$, we have $\Phi(b, x) = 0_F$. Since Φ is a classical form, $0_F = \Phi(x, b)$ and therefore, $b \in (B^{\perp})^{\perp}$. Hence, $B \leq (B^{\perp})^{\perp}$. Furthermore, Proposition 6.2.11 implies that

$$\begin{aligned}
\mathbf{dim}_F((B^{\perp})^{\perp}) &= \mathbf{dim}_F(A) - \mathbf{dim}_F(B^{\perp}) \\
&= \mathbf{dim}_F(A) - (\mathbf{dim}_F(A) - \mathbf{dim}_F(B)) = \mathbf{dim}_F(B).
\end{aligned}$$

Now the inclusion $B \leq (B^{\perp})^{\perp}$, together with Theorem 4.2.20, proves that $B = (B^{\perp})^{\perp}$.

6.2.13. Corollary. *Let A be a finite-dimensional vector space over a field F and let Φ be a classical bilinear form on A. If Φ is nonsingular then, for each subspace B, the intersection $B \cap B^{\perp}$ is the kernel of the restriction of Φ to the subspaces B and $B^{\perp}$.*

Proof. Let C denote the kernel of the restriction of Φ to B. If $x \in C$, then $\Phi(b, x) = 0_F$ for all $b \in B$. Thus $x \in B^{\perp}$, so that $x \in B \cap B^{\perp}$. Hence,

$C \leq B \cap B^{\perp}$. Conversely, if $x \in B \cap B^{\perp}$, then x belongs to the kernel of the restriction of Φ on B. It follows that $B^{\perp} \cap (B^{\perp})^{\perp}$ is the kernel of the restriction of Φ on $B^{\perp}$. By Corollary 6.2.12, $(B^{\perp})^{\perp} = B$, which proves that $B \cap B^{\perp} \leq C$. This proves the result.

6.2.14. Theorem. *Let A be a finite-dimensional vector space over a field F and let Φ be a classical bilinear form on A. If Φ is nonsingular, then $A = B \oplus B^{\perp}$ for each nonsingular subspace B. Moreover, $B^{\perp}$ is also nonsingular.*

Proof. Since B is nonsingular, the kernel of the restriction of Φ on B is zero. By Corollary 6.2.13, the kernel coincides with $B \cap B^{\perp}$, so that $B \cap B^{\perp} = \{0_A\}$. By Corollary 4.2.23, $\mathbf{dim}_F(B \oplus B^{\perp}) = \mathbf{dim}_F(B) + \mathbf{dim}_F(B^{\perp})$. By Proposition 6.2.11, $\mathbf{dim}_F(B \oplus B^{\perp}) = \mathbf{dim}_F(B) + \mathbf{dim}_F(A) - \mathbf{dim}_F(B) = \mathbf{dim}_F(A)$ and Theorem 4.2.20 implies that $A = B \oplus B^{\perp}$.

6.2.15. Definition. *Let A be a vector space over a field F and let Φ be a classical bilinear form on A. Suppose that the subspace C is a direct sum of subspaces $A_1, \ldots, A_n$. This direct sum is orthogonal, if $\Phi(x, y) = 0_F$ for all $x \in A_j$ and $y \in A_t$, where $1 \leq j, t \leq n$.*

The following theorem describes the structure of vector spaces on which a classical bilinear form exists.

6.2.16. Theorem. *Let A be a finite-dimensional vector space over a field F and let Φ be a symmetric bilinear form on A. If* **char** *$F \neq 2$, then A is the orthogonal direct sum of the kernel and one-dimensional spaces.*

Proof. First, consider the case when Φ is nonsingular and let a be a nonzero element of A. By Proposition 6.2.6, there exists an element b such that $\Phi(a, b) \neq 0_F$. If $\Phi(a, a) \neq 0_F$, then put $a_1 = a$. If $\Phi(a, a) = 0_F$ but $\Phi(b, b) \neq 0_F$, then put $a_1 = b$. If $\Phi(a, a) = 0_F$ and $\Phi(b, b) = 0_F$ and if $a_1 = a + b$, then

$$\begin{aligned}\Phi(a_1, a_1) &= \Phi(a + b, a + b)\\ &= \Phi(a, a) + \Phi(a, b) + \Phi(b, a) + \Phi(b, b) = 2\Phi(a, b).\end{aligned}$$

Since $\Phi(a, b) \neq 0_F$ and **char** $F \neq 2$, we know $2\Phi(a, b) \neq 0_F$. Hence, in any case, there exists an element a_1 such that $\Phi(a_1, a_1) \neq 0_F$. Let A_1 be the subspace, generated by a_1 so that A_1 is nonsingular. From Theorem 6.2.14 we deduce that $A = A_1 \oplus A_1^{\perp}$ and the subspace $A_1^{\perp}$ is nonsingular. As above, there exists an element $a_2 \in A_1^{\perp}$ such that $\Phi(a_2, a_2) \neq 0_F$ and we let A_2 be the subspace generated by a_2. Clearly, every element of A_1 is orthogonal to every element of A_2. Put $B = A_1 \oplus A_2$. Then, by Proposition 4.2.22, $\{a_1, a_2\}$ is a basis of B. The matrix

$$\begin{pmatrix} \Phi(a_1, a_1) & 0_F \\ 0_F & \Phi(a_2, a_2) \end{pmatrix}$$

is the matrix of the restriction of Φ to B, relative to this basis. Since this matrix is nonsingular, B is nonsingular. Again using Theorem 6.2.14, we obtain the orthogonal direct decomposition $A = B \oplus B^{\perp} = A_1 \oplus A_2 \oplus B^{\perp}$, where the subspace $B^{\perp}$ is nonsingular. The argument above can be repeated as often as we like and, after finitely many steps, we obtain a decomposition of A into an orthogonal direct sum of one-dimensional spaces.

Next we consider the general case. By Proposition 6.2.9, $A = A^{\perp} \oplus C$ where C is a nonsingular subspace. As above, C is an orthogonal direct sum of subspaces of dimension 1, which proves the result.

6.2.17. Definition. *Let A be a vector space over a field F and let Φ be a symmetric bilinear form on A. A subset $\{a_1, \dots, a_m\}$ of A is called orthogonal, if $\Phi(a_j, a_t) = 0_F$ for all j, t where $1 \le j, t \le n$. If an orthogonal subset $\{a_1, \dots, a_n\}$ is a basis of A, then we say that $\{a_1, \dots, a_n\}$ is an orthogonal basis of A.*

6.2.18. Corollary. *Let A be a finite-dimensional vector space over a field F and let Φ be a symmetric bilinear form on A. If* **char** *$F \neq 2$, then A has an orthogonal basis.*

Proof. By Theorem 6.2.16, $A = A^{\perp} \oplus A_1 \oplus \dots \oplus A_k$ where this direct sum is orthogonal and $\mathbf{dim}_F(A_j) = 1$, for $1 \le j \le k$. In each subspace A_j choose a nonzero element a_j, where $1 \le j \le k$, and let $\{a_{k+1}, \dots, a_n\}$ be a basis of $A^{\perp}$. Proposition 4.2.22 shows that $\{a_1, \dots, a_n\}$ is a basis of A. By this choice, this basis is orthogonal.

6.2.19. Corollary. *Let A be a finite-dimensional vector space over a field F and let Φ be a bilinear symmetric form on A. If* **char** *$F \neq 2$, then A has a basis $\{a_1, \dots, a_n\}$, relative to which the matrix of Φ has the form*

$$\begin{pmatrix} \gamma_1 & 0_F & \dots & 0_F & 0_F & \dots & 0_F \\ 0_F & \gamma_2 & \dots & 0_F & 0_F & \dots & 0_F \\ \vdots & \vdots & \ddots & \vdots & \vdots & \vdots & \vdots \\ 0_F & 0_F & \dots & \gamma_r & 0_F & \dots & 0_F \\ 0_F & 0_F & \dots & 0_F & 0_F & \dots & 0_F \\ \vdots & \vdots & \vdots & \vdots & \vdots & \ddots & \vdots \\ 0_F & 0_F & \dots & 0_F & 0_F & \dots & 0_F \end{pmatrix}$$

*where $\gamma_j \neq 0_F$, $1 \le j \le r$, and $r =$ **rank**(Φ).*

6.2.20. Corollary. *Let F be a field and let S be a symmetric matrix of degree n. If* **char** $F \neq 2$, *then there exists a nonsingular matrix $T \in \mathbf{M}_n(F)$ such that the matrix $T^t ST$ is diagonal.*

Proof. Let A be a vector space over a field F such that $\mathbf{dim}_F(A) = n$ (we may choose $A = F^n$, for example). Choose some basis $\{a_1, \ldots, a_n\}$. By Corollary 6.1.7, there exists a bilinear form Φ such that the matrix of Φ relative to the basis $\{a_1, \ldots, a_n\}$ is S. By Proposition 6.1.11, Φ is symmetric and by Corollary 6.2.19, there exists a basis of A such that the matrix L of Φ relative to this basis is diagonal. By Theorem 6.1.8, $L = T^t ST$ where T is the transition matrix from the first basis to the second one.

Next, we describe the spaces with a symplectic bilinear form. We cannot use orthogonal bases here since, as we have seen above, for such forms Φ we have $\Phi(x, x) = 0_F$. However symplectic planes, as defined below, are fundamental to our study.

6.2.21. Definition. *A vector space of dimension 2 over a field F with* **char** $F \neq 2$, *on which a nonsingular symplectic bilinear form is given, is called a symplectic plane.*

Let A be a symplectic plane and let $\{b_1, b_2\}$ be a basis of A. Then $\Phi(b_1, b_2) = \alpha \neq 0_F$. Put $a_1 = b_1, a_2 = \alpha^{-1} b_2$. Then $\{a_1, a_2\}$ is also a basis of A and $\Phi(a_1, a_2) = e$. Hence, the matrix of Φ relative to the basis $\{a_1, a_2\}$ is

$$\begin{pmatrix} 0_F & e \\ -e & 0_F \end{pmatrix}.$$

6.2.22. Theorem. *Let A be a finite-dimensional vector space over a field F and let Φ be a symplectic bilinear form on A. If* **char** $F \neq 2$, *then A is an orthogonal direct sum of the kernel and symplectic planes.*

Proof. First, consider the case when the form Φ is nonsingular. Let a be a nonzero element of A. By Proposition 6.2.6, there exists an element b such that $\Phi(a, b) \neq 0_F$. Let A_1 be the subspace, generated by a, b. Then A_1 is a nonsingular subspace of dimension 2. From Theorem 6.2.14, we deduce that $A = A_1 \oplus A_1^{\perp}$ and that $A_1^{\perp}$ is nonsingular. We apply the same arguments to the subspace $A_1^{\perp}$ and continue in this way. Repeating these arguments, we see that in finitely many steps we obtain a decomposition of A into an orthogonal direct sum of symplectic planes.

For the general case, Proposition 6.2.9 implies that $A = A^{\perp} \oplus C$ where C is a nonsingular subspace. As above, C is an orthogonal direct sum of symplectic planes, which proves the result.

6.2.23. Corollary. *Let A be a finite-dimensional vector space over a field F and let Φ be a symplectic bilinear form on A. If* **char** $F \neq 2$, *then A has a basis*

$\{a_1, \dots, a_n\}$, *relative to which the matrix of* Φ *has the form*

$$\begin{pmatrix}
0_F & e & 0_F & 0_F & 0_F & 0_F & \dots & 0_F & 0_F & 0_F & 0_F & 0_F & 0_F & \dots & 0_F \\
-e & 0_F & 0_F & 0_F & 0_F & 0_F & \dots & 0_F & 0_F & 0_F & 0_F & 0_F & 0_F & \dots & 0_F \\
0_F & 0_F & 0_F & e & 0_F & 0_F & \dots & 0_F & 0_F & 0_F & 0_F & 0_F & 0_F & \dots & 0_F \\
-e & 0_F & -e & 0_F & 0_F & 0_F & \dots & 0_F & 0_F & 0_F & 0_F & 0_F & 0_F & \dots & 0_F \\
\vdots & \vdots & \vdots & \vdots & \vdots & \vdots & \vdots & \vdots & \vdots & \vdots & \vdots & \vdots & \vdots & \vdots & \vdots \\
\mathbf{0}_F & \mathbf{0}_F & \mathbf{0}_F & \mathbf{0}_F & \mathbf{0}_F & \mathbf{0}_F & \dots & \mathbf{0}_F & \mathbf{e} & \mathbf{0}_F & \mathbf{0}_F & \mathbf{0}_F & \mathbf{0}_F & \dots & \mathbf{0}_F \\
\mathbf{0}_F & \mathbf{0}_F & \mathbf{0}_F & \mathbf{0}_F & \mathbf{0}_F & \mathbf{0}_F & \dots & -\mathbf{e} & \mathbf{0}_F & \mathbf{0}_F & \mathbf{0}_F & \mathbf{0}_F & \mathbf{0}_F & \dots & \mathbf{0}_F \\
0_F & 0_F & 0_F & 0_F & 0_F & 0_F & \dots & 0_F & 0_F & 0_F & 0_F & 0_F & 0_F & \dots & 0_F \\
\vdots & \vdots & \vdots & \vdots & \vdots & \vdots & \vdots & \vdots & \vdots & \vdots & \vdots & \vdots & \vdots & \vdots & \vdots \\
0_F & 0_F & 0_F & 0_F & 0_F & 0_F & \dots & 0_F & 0_F & 0_F & 0_F & 0_F & 0_F & \dots & 0_F
\end{pmatrix},$$

where $2r = \mathbf{rank}(\Phi)$ *and the boldface rows are the rows numbered* $2r - 1$ *and* $2r$.

6.2.24. Corollary. *Let* F *be a field and let* S *be a skew-symmetric matrix of order* n. *If* $\mathbf{char}\, F \neq 2$, *then there exists a nonsingular matrix* $T \in \mathbf{M}_n(F)$ *such that the matrix* $T^t ST$ *has the form*

$$\begin{pmatrix}
0_F & e & 0_F & 0_F & 0_F & 0_F & \dots & 0_F & 0_F & 0_F & 0_F & 0_F & 0_F & \dots & 0_F \\
-e & 0_F & 0_F & 0_F & 0_F & 0_F & \dots & 0_F & 0_F & 0_F & 0_F & 0_F & 0_F & \dots & 0_F \\
0_F & 0_F & 0_F & e & 0_F & 0_F & \dots & 0_F & 0_F & 0_F & 0_F & 0_F & 0_F & \dots & 0_F \\
-e & 0_F & -e & 0_F & 0_F & 0_F & \dots & 0_F & 0_F & 0_F & 0_F & 0_F & 0_F & \dots & 0_F \\
\vdots & \vdots & \vdots & \vdots & \vdots & \vdots & \vdots & \vdots & \vdots & \vdots & \vdots & \vdots & \vdots & \vdots & \vdots \\
\mathbf{0}_F & \mathbf{0}_F & \mathbf{0}_F & \mathbf{0}_F & \mathbf{0}_F & \mathbf{0}_F & \dots & \mathbf{0}_F & \mathbf{e} & \mathbf{0}_F & \mathbf{0}_F & \mathbf{0}_F & \mathbf{0}_F & \dots & \mathbf{0}_F \\
\mathbf{0}_F & \mathbf{0}_F & \mathbf{0}_F & \mathbf{0}_F & \mathbf{0}_F & \mathbf{0}_F & \dots & -\mathbf{e} & \mathbf{0}_F & \mathbf{0}_F & \mathbf{0}_F & \mathbf{0}_F & \mathbf{0}_F & \dots & \mathbf{0}_F \\
0_F & 0_F & 0_F & 0_F & 0_F & 0_F & \dots & 0_F & 0_F & 0_F & 0_F & 0_F & 0_F & \dots & 0_F \\
\vdots & \vdots & \vdots & \vdots & \vdots & \vdots & \vdots & \vdots & \vdots & \vdots & \vdots & \vdots & \vdots & \vdots & \vdots \\
0_F & 0_F & 0_F & 0_F & 0_F & 0_F & \dots & 0_F & 0_F & 0_F & 0_F & 0_F & 0_F & \dots & 0_F
\end{pmatrix},$$

where $2r = \mathbf{rank}(S)$ *and the boldface rows are the rows numbered* $2r - 1$ *and* $2r$.

Above, we proved the existence of an orthogonal basis for a space with a given symmetric bilinear form, and the proof is constructive in that it can be used to actually find such a basis. However, the proof suggests that this is a very long and tedious process. Thus, the first step is to find an orthogonal complement for the element a_1 and for this, we need to find a fundamental system of solutions of a system of $n - 1$ equations; at the second step we need to solve a system of $n - 2$ equations, and so on. The language of quadratic forms helps us to develop more efficient methods of finding an orthogonal basis. One of these methods is a classical method of the great French mathematician, Lagrange (1736–1813) involving the reduction of a quadratic form to a diagonal form. If $\{a_1, \ldots, a_n\}$ is an orthogonal basis, and f is a quadratic form associated with a symmetric bilinear form Φ, then $f(x) = \sum_{1\leq j\leq n} \xi_j^2\gamma_j$, where $x = \sum_{1\leq j\leq n} \xi_j a_j$ and $\gamma_j = \Phi(a_j, a_j)$, for $1 \leq j \leq n$. Such a quadratic form is called a diagonal or normal form. In the general case, a quadratic form is a function of the coordinates of an element. The main idea of the Lagrange method is step-by-step transformation of coordinates. As we saw in Section 4.2, transformation of coordinates is realized by the rule

$$\begin{pmatrix}\lambda_1\\ \lambda_2\\ \vdots\\ \lambda_n\end{pmatrix} = \begin{pmatrix} \iota_{11} & \iota_{12} & \ldots & \iota_{1n}\\ \iota_{21} & \iota_{22} & \ldots & \iota_{2n}\\ \vdots & \vdots & \ddots & \vdots\\ \iota_{n1} & \iota_{n2} & \ldots & \iota_{nn}\end{pmatrix}\begin{pmatrix}\xi_1\\ \xi_2\\ \vdots\\ \xi_n\end{pmatrix}.$$

In this way, for each coordinate transformation we can find a corresponding transition matrix from one basis to the other.

For the basis $\{a_1, \ldots, a_n\}$, the quadratic form f takes the form

$$f(x) = \sum_{1\leq j\leq n}\sum_{1\leq t\leq n} x_j x_t \sigma_{jt}, \tag{6.3}$$

where the entries in the n-tuple $(x_1, \ldots, x_n)$ are the coordinates of x relative to $\{a_1, \ldots, a_n\}$. We assume, as usual, that $\mathbf{char}F \neq 2$. Suppose first that there is a positive integer j such that $\sigma_{jj} \neq 0_F$. By relabeling, if necessary, we may suppose that $\sigma_{11} \neq 0_F$. Then, we can write Equation 6.3 as

$$f(x) = \sigma_{11}^{-1}(\sigma_{11}x_1 + \sigma_{12}x_2 + \cdots + \sigma_{1n}x_n)^2 + g(x_2, \ldots, x_n),$$

where $g(x_2, \ldots, x_n)$ is a quadratic form in $x_2, \ldots, x_n$. Put

$$\begin{aligned} y_1 &= \sigma_{11}x_1 + \sigma_{12}x_2 + \cdots + \sigma_{1n}x_n\\ y_2 &= x_2\\ &\vdots\\ y_n &= x_n. \end{aligned}$$

The corresponding transition matrix is

$$\begin{pmatrix} \sigma_{11} & \sigma_{12} & \ldots & \sigma_{1n}\\ 0_F & e & \ldots & 0_F\\ \vdots & \vdots & \ddots & \vdots\\ 0_F & 0_F & \ldots & e\end{pmatrix},$$

which is nonsingular. Thus $f(y_1, \ldots, y_n) = \sigma_{11}^{-1}y_1^2 + g(y_2, \ldots, y_n)$. Suppose now that $\sigma_{jj} = 0_F$ for all j, where $1 \leq j \leq n$. Since the form f is nonsingular, there are indices j, t such that $\sigma_{jt} \neq 0_F$ and, by relabeling the basis vectors if necessary, we may assume that $\sigma_{12} \neq 0_F$. We now consider the following transformation of variables:

$$\begin{aligned} x_1 &= z_1 + z_2 \\ x_2 &= z_1 - z_2 \\ x_3 &= z_3 \\ &\vdots \\ x_n &= z_n. \end{aligned}$$

Then, $2\sigma_{12}x_1x_2 = 2\sigma_{12}z_1^2 - 2\sigma_{12}z_2^2$ and the quadratic form is now of the same type as the one we considered in the previous case. Now apply similar transformations to the quadratic form $g(x_2, \dots, x_n)$ and repeat this process. In a finite number of steps we obtain the expression

$$f = \gamma_1 u_1^2 + \gamma_2 u_2^2 + \cdots + \gamma_r u_r^2,$$

where $r \leq n$. The corresponding basis is orthogonal.

EXERCISE SET 6.2

Show your work, giving a proof or counterexample, where necessary.

6.2.1. Let $f : \mathbb{Q}^2 \times \mathbb{Q}^2 \longrightarrow \mathbb{Q}$ be a bilinear form defined by $f((\alpha, \beta), (\gamma, \lambda)) = \alpha\gamma + 2\alpha\lambda + 3\beta\gamma + 6\beta\lambda$. Find the right kernel.

6.2.2. Let $f : \mathbb{Q}^2 \times \mathbb{Q}^2 \longrightarrow \mathbb{Q}$ be a bilinear form defined by $f((\alpha, \beta), (\gamma, \lambda)) = \alpha\gamma + 2\alpha\lambda + 3\beta\gamma + 6\beta\lambda$. Find the left kernel.

6.2.3. A bilinear form $f : \mathbb{R}^3 \times \mathbb{R}^3 \longrightarrow \mathbb{R}$ is given relative to the standard basis using the matrix

$$\begin{pmatrix} 2 & -1 & 0 \\ 1 & -1 & -2 \\ 1 & 0 & 0 \end{pmatrix}.$$

Find the left kernel.

6.2.4. A bilinear form $f : \mathbb{R}^3 \times \mathbb{R}^3 \longrightarrow \mathbb{R}$ is given relative to the standard basis using the matrix

$$\begin{pmatrix} 2 & -1 & 0 \\ 1 & -1 & -2 \\ 1 & 0 & 0 \end{pmatrix}.$$

Find the right kernel.

6.2.5. Over the space $A = \mathbb{Q}^3$ a bilinear form with the matrix

$$\begin{pmatrix} 2 & 3 & 5 \\ 1 & 0 & 4 \\ 1 & 3 & 1 \end{pmatrix}$$

relative to the standard basis is given. Find the left orthogonal complement to the subspace $\mathbb{Q}a$ where $a = (0, 2, 5)$.

6.2.6. Over the space $A = \mathbb{R}^4$ a bilinear form with the matrix

$$\begin{pmatrix} 1 & 1 & 3 & 4 \\ 2 & 0 & 0 & 8 \\ 3 & 0 & 0 & 2 \\ 3 & 1 & 3 & 1 \end{pmatrix}$$

relative to the standard basis is given. Find the left orthogonal complement to the subspace $\mathbb{Q}a$ where $a = (0, 1, 3, 5)$.

6.2.7. Over the space $A = \mathbb{R}^4$ a bilinear form with the matrix

$$\begin{pmatrix} 1 & 1 & 3 & 4 \\ 2 & 0 & 0 & 8 \\ 3 & 0 & 0 & 2 \\ 3 & 1 & 3 & 1 \end{pmatrix}$$

relative to the standard basis is given. Find the right orthogonal complement to the subspace $\mathbb{Q}a$ where $a = (0, 1, 3, 5)$.

6.2.8. Over the space $A = \mathbb{Q}^3$ a bilinear form with the matrix

$$\begin{pmatrix} 2 & 3 & 5 \\ 1 & -1 & 0 \\ 1 & 0 & 1 \end{pmatrix}$$

relative to the standard basis is given. Find the right orthogonal complement to the subspace $\mathbb{Q}a$ where $a = (0, 2, 5)$.

6.2.9. Over the space $A = \mathbb{Q}^3$ a bilinear form with the matrix

$$\begin{pmatrix} 3 & -1 & 0 \\ 2 & 0 & 1 \\ 2 & 1 & -1 \end{pmatrix}$$

relative to the standard basis is given. Find the right orthogonal complement to the set $\{(1, 1, 0), (0, -1, 1)\}$.

6.2.10. Over the space $A = \mathbb{R}^3$ a bilinear form with the matrix

$$\begin{pmatrix} 9 & 5 & 0 \\ 4 & 9 & 5 \\ 0 & 4 & 9 \end{pmatrix}$$

relative to the standard basis is given. Find the right orthogonal complement to the set $\{(1, 1, 0), (0, 1, 2)\}$.

6.2.11. Over the space $A = \mathbb{R}^3$ a bilinear form with the matrix

$$\begin{pmatrix} 9 & 5 & 0 \\ 4 & 9 & 5 \\ 0 & 4 & 9 \end{pmatrix}$$

relative to the standard basis is given. Find the left orthogonal complement to the set $\{(1, 1, 0), (0, 1, 2)\}$.

6.2.12. Find an orthogonal basis of the space $A = \mathbb{Q}^3$ with a bilinear form given relative to the standard basis by the matrix

$$\begin{pmatrix} 1 & 2 & -1 \\ 2 & -1 & 0 \\ -1 & 0 & 0 \end{pmatrix}.$$

6.2.13. Find an orthogonal basis of the space $A = \mathbb{Q}^3$ with a bilinear form given relative to the standard basis by the matrix

$$\begin{pmatrix} 3 & 1 & -1 \\ 1 & 0 & 2 \\ -1 & 2 & -1 \end{pmatrix}.$$

6.2.14. Over the space $A = \mathbb{R}^3$ a bilinear form with the matrix

$$\begin{pmatrix} 0 & 1 & 3 \\ -1 & 0 & 2 \\ -3 & -2 & -1 \end{pmatrix}$$

relative to the standard basis is given. Do the elements $(1, 1, 1)$ and $(2, 3, 0)$ generate a hyperbolic plane?

6.2.15. Over the space $A = \mathbb{Q}^3$ an alternating form with the matrix

$$\begin{pmatrix} 0 & 1 & 1 \\ -1 & 0 & 3 \\ -1 & -3 & 0 \end{pmatrix}$$

relative to the standard basis is given. Decompose the space A into a direct sum of a hyperbolic plane and the kernel of the form.

6.2.16. Over the space $A = \mathbb{Q}^4$ a bilinear form with the matrix

$$\begin{pmatrix} 1 & 5 & 7 & 0 \\ -5 & -1 & 0 & 3 \\ 7 & 0 & 2 & -1 \\ 0 & 3 & -1 & 6 \end{pmatrix}$$

relative to the standard basis is given. Find an orthogonal basis of the space.

6.2.17. Over the space $A = F^4$, where $F = \mathbb{Z}/3\mathbb{Z}$, a bilinear form with the matrix

$$\begin{pmatrix} 1 & 1 & 0 & 1 \\ 1 & 0 & 1 & 0 \\ 0 & 1 & 1 & 0 \\ 1 & 0 & 0 & 1 \end{pmatrix}$$

relative to the standard basis is given. Find an orthogonal basis of the space.

6.3 SYMMETRIC FORMS OVER $\mathbb{R}$

In this section, we consider symmetric bilinear forms over the field $\mathbb{R}$ of real numbers. Let A be a finite-dimensional vector space over $\mathbb{R}$, on which a symmetric bilinear form Φ is given. As we proved in Corollary 6.2.18, the space A has an orthogonal basis, say $\{a_1, \dots, a_n\}$ and $\Phi(a_j, a_j)$ is a real number for $1 \le j \le n$. Hence, we have the three possibilities; namely, $\Phi(a_j, a_j) > 0$, $\Phi(a_j, a_j) < 0$, or $\Phi(a_j, a_j) = 0$. The number of elements of the basis satisfying this latter equation is clearly equal to the dimension of the kernel of the form, so it is the same for every orthogonal basis. We now show that the number of elements a_j for which $\Phi(a_j, a_j) > 0$ is also an invariant of the space.

6.3.1. Proposition. *Let A be a vector space over $\mathbb{R}$ and let Φ be a nonsingular symmetric bilinear form on A. Let $\{a_1, \dots, a_n\}$ and $\{c_1, \dots, c_n\}$ be two orthogonal bases of A. Let $\rho = \{j \mid 1 \le j \le n$ and $\Phi(a_j, a_j) > 0\}$ and let $\rho_1 = \{j \mid 1 \le j \le n$ and $\Phi(c_j, c_j) > 0\}$. Then $|\rho| = |\rho_1|$.*

Proof. By relabeling the basis vectors, if necessary, we can suppose that $\rho = \{1, \dots, k\}$ and $\rho_1 = \{1, \dots, t\}$. Consider the subset $\{a_1, \dots, a_k, c_{t+1}, \dots, c_n\}$. We shall show that it is linearly independent. To this end, let $\alpha_1, \dots, \alpha_k, \gamma_{t+1}, \dots, \gamma_n$ be real numbers such that $\alpha_1 a_1 + \cdots + \alpha_k a_k + \gamma_{t+1} c_{t+1} + \cdots + \gamma_n c_n = 0$. It follows that $\gamma_{t+1} c_{t+1} + \cdots + \gamma_n c_n = -\alpha_1 a_1 - \cdots - \alpha_k a_k$. Then

$$\begin{aligned} &\Phi(\gamma_{t+1} c_{t+1} + \cdots + \gamma_n c_n, \gamma_{t+1} c_{t+1} + \cdots + \gamma_n c_n) \\ &= \Phi(-\alpha_1 a_1 - \cdots - \alpha_k a_k, -\alpha_1 a_1 - \cdots - \alpha_k a_k). \end{aligned}$$

Since both bases are orthogonal,

$$\begin{aligned}\Phi(\gamma_{t+1}c_{t+1} + \cdots + \gamma_n c_n, \gamma_{t+1}c_{t+1} + \cdots + \gamma_n c_n) \\ = \gamma_{t+1}^2\Phi(c_{t+1}, c_{t+1}) + \cdots + \gamma_n^2\Phi(c_n, c_n)\end{aligned}$$

and

$$\begin{aligned}\Phi(-\alpha_1 a_1 - \cdots - \alpha_k a_k, -\alpha_1 a_1 - \cdots - \alpha_k a_k) \\ = \alpha_1^2\Phi(a_1, a_1) + \cdots + \alpha_k^2\Phi(a_k, a_k),\end{aligned}$$

so that

$$\begin{aligned}\alpha_1^2\Phi(a_1, a_1) + \cdots + \alpha_k^2\Phi(a_k, a_k) \\ = \gamma_{t+1}^2\Phi(c_{t+1}, c_{t+1}) + \cdots + \gamma_n^2\Phi(c_n, c_n).\end{aligned} \tag{6.4}$$

However, $\Phi(a_j, a_j) > 0$, for $1 \leq j \leq k$ and $\Phi(c_j, c_j) < 0$, for $t + 1 \leq j \leq n$. Thus Equation 6.4 shows that

$$\alpha_1^2 = \cdots = \alpha_k^2 = \gamma_{t+1}^2 = \cdots = \gamma_n^2 = 0,$$

which implies that $\alpha_1 = \cdots = \alpha_k = \gamma_{t+1} = \cdots = \gamma_n = 0$. Proposition 4.2.7 shows that the subset $\{a_1, \ldots, a_k, c_{t+1}, \ldots, c_n\}$ is linearly independent and Theorem 4.2.11 implies that this subset can be extended to a basis of the entire space. However, by Theorem 4.2.14, each basis of A has exactly n elements and it follows that $|\{a_1, \ldots, a_k, c_{t+1}, \ldots, c_n\}| = k + n - t \leq n$. Then $k - t \leq 0$ and hence, $k \leq t$.

Next, we consider the subset $\{a_{k+1}, \ldots, a_n, c_1, \ldots, c_t\}$ and proceed as above. Those arguments show that this subset is also linearly independent and that $t \leq k$. Consequently, $k = t$.

6.3.2. Theorem *(Sylvester). Let A be a vector space over $\mathbb{R}$ and let Φ be a symmetric bilinear form on A. Let $\{a_1, \ldots, a_n\}$ and $\{c_1, \ldots, c_n\}$ be two orthogonal bases of A. Let*

$$\begin{aligned}&\rho = \{j \mid 1 \leq j \leq n \text{ and } \Phi(a_j, a_j) > 0\}, \rho_1 = \{j \mid 1 \leq j \leq n \\ &\quad \text{and } \Phi(c_j, c_j) > 0\}; \\ &\nu = \{j \mid 1 \leq j \leq n \text{ and } \Phi(a_j, a_j) < 0\}, \nu_1 = \{j \mid 1 \leq j \leq n \\ &\quad \text{and } \Phi(c_j, c_j) < 0\}; \\ &\xi = \{j \mid 1 \leq j \leq n \text{ and } \Phi(a_j, a_j) = 0\}, \xi_1 = \{j \mid 1 \leq j \leq n \\ &\quad \text{and } \Phi(c_j, c_j) = 0\}.\end{aligned}$$

Then $|\rho| = |\rho_1|$, $|\nu| = |\nu_1|$, $|\xi| = |\xi_1|$.

Proof. By relabeling the basis vectors, if necessary, we may suppose that $\rho = \{1, \ldots, k\}$, $\rho_1 = \{1, \ldots, t\}$, $\nu = \{k+1, \ldots, r\}$, $\nu_1 = \{t+1, \ldots, s\}$, $\xi = \{r+1, \ldots, n\}$, and $\xi_1 = \{s+1, \ldots, n\}$. As we mentioned above, the number of basis elements a_j for which $\Phi(a_j, a_j) = 0$ is equal to the dimension of the kernel of the form, so $n - r = n - s$ and hence $r = s$. The result now follows for nonsingular forms using Proposition 6.3.1.

Let K denote the kernel of the form Φ and consider the quotient space A/K. Define a mapping $\Phi^* : A/K \times A/K \longrightarrow \mathbb{R}$ by $\Phi^*(x+K, y+K) = \Phi(x, y)$. We show first that this mapping is well defined, which here means that it does not depend on the choice of cosets representatives. Let x_1, y_1 be elements such that $x_1 + K = x + K$ and $y_1 + K = y + K$. Then, $x_1 = x + z_1$, $y_1 = y + z_2$ where $z_1, z_2 \in K$ and we have

$$\begin{aligned}\Phi(x_1, y_1) = \Phi(x + z_1, y + z_2) &= \Phi(x, y) + \Phi(x, z_2) + \Phi(z_1, y) + \Phi(z_1, z_2) \\ &= \Phi(x, y).\end{aligned}$$

This shows that Φ^* is a well-defined mapping. We next show that Φ^* is bilinear. We have

$$\begin{aligned}\Phi^*(x + K + u + K, y + K) &= \Phi^*(x + u + K, y + K) = \Phi(x + u, y) \\ &= \Phi(x, y) + \Phi(u, y) \\ &= \Phi^*(x + K, y + K) + \Phi^*(u + K, y + K),\end{aligned}$$

and similarly

$$\Phi^*(x + K, u + K + y + K) = \Phi^*(x + K, u + K) + \Phi^*(x + K, y + K).$$

Furthermore, if $\alpha \in \mathbb{R}$ then

$$\begin{aligned}\Phi^*(\alpha(x + K), y + K) &= \Phi^*(\alpha x + K, y + K) = \Phi(\alpha x, y) = \alpha\Phi(x, y) \\ &= \alpha\Phi^*(x + K, y + K)\end{aligned}$$

and, similarly,

$$\Phi^*(x, \alpha(y + K)) = \alpha\Phi^*(x + K, y + K).$$

This shows that Φ^* is a bilinear form. Furthermore,

$$\Phi^*(x + K, y + K) = \Phi(x, y) = \Phi(y, x) = \Phi^*(y + K, x + K),$$

so that Φ^* is a symmetric form.

We next show that $\{a_1 + K, \ldots, a_r + K\}$ is a basis of A/K. Let $\alpha_1, \ldots, \alpha_r$ be real numbers such that

$$\alpha_1(a_1 + K) + \cdots + \alpha_r(a_r + K) = 0 + K.$$

Since

$$\alpha_1(a_1 + K) + \cdots + \alpha_r(a_r + K) = \alpha_1 a_1 + \cdots + \alpha_r a_r + K,$$

we deduce that $\alpha_1 a_1 + \cdots + \alpha_r a_r + K = 0 + K$. Hence, there exist real numbers $\alpha_{r+1}, \ldots, \alpha_n$ such that

$$\alpha_1 a_1 + \cdots + \alpha_r a_r = \alpha_{r+1} a_{r+1} + \cdots + \alpha_n a_n$$

or

$$\alpha_1 a_1 + \cdots + \alpha_r a_r - \alpha_{r+1} a_{r+1} - \cdots - \alpha_n a_n = 0_A.$$

Since the elements $a_1, \ldots, a_n$ are linearly independent, Proposition 4.2.7 implies that

$$\alpha_1 = \cdots = a_r = \alpha_{r+1} = \cdots = \alpha_n = 0.$$

Applying Proposition 4.2.7 again, we see that the cosets $a_1 + K, \ldots, a_r + K$ are linearly independent. Now let $x = \sum_{1 \le j \le n} \xi_j a_j$ be an arbitrary element of A, where $\xi_j \in \mathbb{R}$, for $1 \le j \le n$. Then, since $a_j + K = K$ for $j \in \{r+1, \ldots, n\}$, we have

$$x + K = \sum_{1 \le j \le n} \xi_j a_j + K = \sum_{1 \le j \le n} \xi_j (a_j + K) = \sum_{1 \le j \le r} \xi_j (a_j + K),$$

which proves that the cosets $a_1 + K, \ldots, a_r + K$ form a basis of the quotient space A/K. Clearly this basis is orthogonal, by definition of Φ^*.

Employing the same arguments, we see that the cosets $b_1 + K, \ldots, b_s + K$ also form an orthogonal basis of the quotient space A/K.

Finally,

$$\begin{aligned}
\Phi^*(a_j + K, a_j + K) &= \Phi(a_j, a_j) > 0 \text{ for } j \in \{1, \ldots, k\}, \\
\Phi^*(a_j + K, a_j + K) &= \Phi(a_j, a_j) < 0 \text{ for } j \in \{k+1, \ldots, r\}, \\
\Phi^*(b_j + K, b_j + K) &= \Phi(b_j, b_j) > 0 \text{ for } j \in \{1, \ldots, t\}, \\
\Phi^*(b_j + K, b_j + K) &= \Phi(b_j, b_j) < 0 \text{ for } j \in \{t+1, \ldots, r\}.
\end{aligned}$$

This shows that Φ^* is nonsingular and Proposition 6.3.1 implies that $k = t$ and this now proves the result.

6.3.3. Definition. *Let A be a vector space over $\mathbb{R}$ and let Φ be a symmetric bilinear form on A. Let $\{a_1, \ldots, a_n\}$ be an orthogonal basis of A. Set*

$$\begin{aligned}
\rho &= \{j \mid 1 \le j \le n \text{ and } \Phi(a_j, a_j) > 0\}, \\
\nu &= \{j \mid 1 \le j \le n \text{ and } \Phi(a_j, a_j) < 0\}, \\
\xi &= \{j \mid 1 \le j \le n \text{ and } \Phi(a_j, a_j) = 0\}.
\end{aligned}$$

By Theorem 6.3.2, $|\rho|$, $|\nu|$, and $|\xi|$ are invariants of Φ. The number $|\rho| = \mathbf{pi}(\Phi)$ is called the positive index of inertia of Φ and the number $|\nu| = \mathbf{ni}(\Phi)$ is called the negative index of inertia of Φ.

A form Φ is called positive definite (respectively negative definite) if $\mathbf{pi}(\Phi) = \mathbf{dim}_F(A)$ (respectively, $\mathbf{ni}(\Phi) = \mathbf{dim}_F(A)$).

6.3.4. Corollary. *Let A be a vector space over $\mathbb{R}$ and let Φ be a symmetric bilinear form on A. Then, A has an orthogonal direct decomposition $A = A_{(+)} \oplus A_{(-)} \oplus K$ where K is the kernel of Φ, the restriction of Φ to $A_{(+)}$ is a positive definite form, and the restriction of Φ to $A_{(-)}$ is a negative definite form. In all such decompositions the dimensions of $A_{(+)}$ and $A_{(-)}$ are invariants.*

Proof. Let $\{a_1, \dots, a_n\}$ be an orthogonal basis of A. As above, we can suppose that $\Phi(a_j, a_j) > 0$ whenever $1 \le j \le k$, $\Phi(a_j, a_j) < 0$ whenever $k+1 \le j \le r$, and $\Phi(a_j, a_j) = 0$ whenever $r+1 \le j \le n$. Let $A_{(+)}$ denote the subspace generated by $a_1, \dots, a_k$, let $A_{(-)}$ denote the subspace generated by $a_{k+1}, \dots, a_r$ and let K be the subspace, generated by $a_{r+1}, \dots, a_n$. The result is now immediate.

Let $\{a_1, \dots, a_n\}$ be an orthogonal basis of A. Assume that $\Phi(a_j, a_j) > 0$ whenever $1 \le j \le k$, $\Phi(a_j, a_j) < 0$ whenever $k+1 \le j \le r$, and $\Phi(a_j, a_j) = 0$ whenever $r+1 \le j \le n$. Put $\Phi(a_j, a_j) = \alpha_j$, for $1 \le j \le n$. Then $\alpha_j > 0$ for $1 \le j \le k$, so $\sqrt{\alpha_j}$ is a real number. Further, $\alpha_j < 0$ for $k+1 \le j \le r$, so $\sqrt{-\alpha_j}$ is a real number. Now let

$$c_j = \frac{1}{\sqrt{\alpha_j}} a_j \text{ whenever } 1 \le j \le k,$$

$$c_j = \frac{1}{\sqrt{-\alpha_j}} a_j \text{ whenever } k+1 \le j \le r, \text{ and}$$

$$c_j = a_j \text{ whenever } r+1 \le j \le n.$$

Then, $\Phi(c_j, c_j) = \Phi((\frac{1}{\sqrt{\alpha_j}} a_j, \frac{1}{\sqrt{\alpha_j}} a_j) = \frac{1}{\sqrt{\alpha_j}} \cdot \frac{1}{\sqrt{\alpha_j}} \cdot \Phi(a_j, a_j) = \frac{1}{\alpha_j} \cdot \alpha_j = 1$ whenever $1 \le j \le k$ and, similarly,

$$\Phi(c_j, c_j) = -1 \text{ whenever } k+1 \le j \le r \text{ and}$$

$$\Phi(c_j, c_j) = 0 \text{ whenever } r+1 \le j \le n.$$

6.3.5. Definition. *Let A be a vector space over $\mathbb{R}$ and let Φ be a symmetric bilinear form on A. An orthogonal basis $\{a_1, \dots, a_n\}$ of A is called orthonormal, if $\Phi(a_j, a_j)$ is equal to one of the numbers $0, 1$ or -1.*

A consequence of the work above is the following fact.

6.3.6. Corollary. *Let A be a vector space over $\mathbb{R}$ and let Φ be a symmetric bilinear form on A. Then A has an orthonormal basis.*

Using an orthonormal basis allows us to write a symmetric bilinear form rather simply. To see this, let $\{a_1, \dots, a_n\}$ be an orthonormal basis of A. Assume that $\Phi(a_j, a_j) = 1$ whenever $1 \le j \le k$, $\Phi(a_j, a_j) = -1$ whenever $k+1 \le j \le r$, and $\Phi(a_j, a_j) = 0$ whenever $r+1 \le j \le n$. Let x, y be arbitrary elements of A, and let $x = \sum_{1\le j\le n} \xi_j a_j$ and $y = \sum_{1\le j\le n} \eta_j a_j$ be their decompositions relative to the basis $\{a_1, \dots, a_n\}$. Then it is easy to see that

$$\Phi(x, y) = \xi_1\eta_1 + \cdots + \xi_k\eta_k - \xi_{k+1}\eta_{k+1} \cdots - \xi_r\eta_r,$$

where $\xi_j, \eta_k \in \mathbb{R}$. This is sometimes called the canonical form of the symmetric bilinear form Φ.

The usual scalar product in $\mathbb{R}^3$ gives us an example of a positive definite symmetric form and the following theorem provides us with conditions for a symmetric form to be positive definite.

6.3.7. Proposition. *Let A be a vector space over $\mathbb{R}$ and let Φ be a symmetric bilinear form on A. Then, Φ is positive definite if and only if $\Phi(x, x) > 0$ for each nonzero element x of A.*

Proof. Suppose that Φ is positive definite. Let $\{a_1, \dots, a_n\}$ be an orthogonal basis of A. Then $\alpha_j = \Phi(a_j, a_j) > 0$ for all j, where $1 \le j \le n$. If $x = \sum_{1\le j\le n} \xi_j a_j$ is an arbitrary element of A, then

$$\Phi(x, x) = \xi_1^2\alpha_1 + \cdots + \xi_n^2\alpha_n > 0.$$

Conversely, assume that $\Phi(x, x) > 0$ for each nonzero element x. Let $\{a_1, \dots, a_n\}$ be an orthogonal basis of A. Then, our assumption implies that $\Phi(a_j, a_j) > 0$ for all j, where $1 \le j \le n$. This means that $\mathbf{pi}(\Phi) = n = \mathbf{dim}_F(A)$, so that Φ is positive definite.

6.3.8. Definition. *Let F be a field and let $S = [\sigma_{jt}] \in \mathbf{M}_n(F)$. The minors*

$$\sigma_{11}, \quad \mathbf{minor}\{1, 2; 1, 2\}, \quad \mathbf{minor}\{1, 2, 3; 1, 2, 3\}, \dots,$$
$$\mathbf{minor}\{1, 2, \dots, k; 1, 2, \dots, k\}, \dots, \quad \mathbf{minor}\{1, 2, \dots, n; 1, 2, \dots, n\}$$

are called the principal minors of the matrix S.

6.3.9. Theorem *(Sylvester). Let A be a vector space over $\mathbb{R}$ and let Φ be a symmetric bilinear form on A. If Φ is positive definite, then all principal minors of the matrix of Φ relative to an arbitrary basis are positive. Conversely, if there is a basis of A such that all principal minors of the matrix of Φ relative to this basis are positive, then the form Φ is positive definite.*

Proof. Suppose that Φ is positive definite and let $\{a_1, \dots, a_n\}$ be an arbitrary basis of A. We prove that the principal minors are positive by induction on n.

If $n = 1$, then the matrix of Φ relative to the basis $\{a_1\}$ has only one coefficient $\Phi(a_1, a_1)$ which is positive, since Φ is positive definite. Since this is the only principal minor, the result follows for $n = 1$.

Suppose now that $n > 1$ and that our assertion has been proved for spaces of dimension less than n. Let $S = [\sigma_{jt}] \in \mathbf{M}_n(\mathbb{R})$ denote the matrix of Φ relative to the basis $\{a_1, \ldots, a_n\}$. Let B denote the subspace generated by $a_1, \ldots, a_{n-1}$. Then, the matrix of the restriction of Φ to B relative to the basis $\{a_1, \ldots, a_{n-1}\}$ is

$$\begin{pmatrix} \sigma_{11} & \sigma_{12} & \cdots & \sigma_{1,n-1} \\ \sigma_{21} & \sigma_{22} & \cdots & \sigma_{2,n-1} \\ \vdots & \vdots & \ddots & \vdots \\ \sigma_{n-1,1} & \sigma_{n-1,2} & \cdots & \sigma_{n-1,n-1} \end{pmatrix}.$$

By Proposition 6.3.7, $\Phi(x, x) > 0$ for each nonzero element $x \in A$. In particular, this is valid for each element $x \in B$ and, by Proposition 6.3.7, the restriction of Φ to B is positive definite. The induction hypothesis shows that all principal minors of the matrix of this restriction relative to the basis $\{a_1, \ldots, a_{n-1}\}$ are positive. However, all these principal minors are principal minors of the matrix S. We still need to prove that the last principal minor of S is positive, which means that we need to prove that $\mathbf{det}(S) > 0$. Theorem 6.2.22 shows that A has an orthogonal basis $\{c_1, \ldots, c_n\}$ and we let $\Phi(c_j, c_j) = \gamma_j$, for $1 \leq j \leq n$. Since Φ is positive definite, $\gamma_j > 0$ for all $j \in \{1, \ldots, n\}$. The matrix L of Φ relative to $\{c_1, \ldots, c_n\}$ is diagonal, the diagonal entries being $\gamma_1, \gamma_2, \ldots, \gamma_n$ and, by Proposition 2.3.11, $\mathbf{det}(L) = \gamma_1\gamma_2 \ldots \gamma_n > 0$. Theorem 6.1.8 shows that $L = T^t ST$ where T is the transition matrix from the first basis to the second. By Theorem 2.5.1, $\mathbf{det}(L) = \mathbf{det}(T^t)\mathbf{det}(S)\mathbf{det}(T)$, whereas Proposition 2.3.3 shows that $\mathbf{det}(T^t) = \mathbf{det}(T)$. Therefore, $\mathbf{det}(L) = \mathbf{det}(T)^2\mathbf{det}(S)$ and since $\mathbf{det}(L) > 0$, we deduce that $\mathbf{det}(S) > 0$. This completes the first part of the proof.

We now assume that the space A has a basis $\{a_1, \ldots, a_n\}$ such that all principal minors of the matrix S of Φ relative to this basis are positive. Let $S = [\sigma_{jt}] \in \mathbf{M}_n(\mathbb{R})$. We again use induction on n to prove that Φ is positive definite. If $n = 1$, then the matrix of Φ relative to the basis $\{a_1\}$ has only one coefficient σ_{11}, which is its only principal minor. By our assumption, $\sigma_{11} = \Phi(a_1, a_1) > 0$, which means that Φ is positive definite so the result follows for $n = 1$.

Suppose now that $n > 1$ and that we have proved our assertion for spaces having dimensions less than n. Let B denote the subspace generated by the elements $a_1, \ldots, a_{n-1}$. Then, the matrix of the restriction of Φ on B relative to the basis $\{a_1, \ldots, a_{n-1}\}$ is

$$\begin{pmatrix} \sigma_{11} & \sigma_{12} & \cdots & \sigma_{1,n-1} \\ \sigma_{21} & \sigma_{22} & \cdots & \sigma_{2,n-1} \\ \vdots & \vdots & \ddots & \vdots \\ \sigma_{n-1,1} & \sigma_{n-1,2} & \cdots & \sigma_{n-1,n-1} \end{pmatrix}.$$

Clearly, the principal minors of this matrix are also principal minors of the matrix S and hence, all principal minors of this matrix are positive. By the induction hypothesis, it follows that the restriction of Φ to the subspace B is positive definite. Theorem 6.2.22 shows that B has an orthogonal basis $\{c_1, \ldots, c_{n-1}\}$ and we let $\Phi(c_j, c_j) = \gamma_j$, for $1 \leq j \leq n-1$. Since the restriction of Φ to B is positive definite, $\gamma_j > 0$ for all $j \in \{1, \ldots, n-1\}$.

The last principal minor $\mathbf{det}(S)$ is positive, so it is nonzero. Hence Φ is nonsingular. The subspace B is also nonsingular, so Theorem 6.2.14 implies that $A = B \oplus B^{\perp}$. Then, by Proposition 6.2.11, $\mathbf{dim}_F(B^{\perp}) = \mathbf{dim}_F(A) - \mathbf{dim}_F(B) = n - (n-1) = 1$. Let c_n be a nonzero element of $B^{\perp}$ so that $\{c_n\}$ is a basis of $B^{\perp}$. From the choice of c_n we have $\Phi(c_j, c_n) = 0$ for all $j \in \{1, \ldots, n-1\}$ and since $\{c_1, \ldots, c_{n-1}\}$ is an orthogonal basis of B, $\{c_1, \ldots, c_n\}$ is an orthogonal basis of A. We noted above that $\Phi(c_j, c_j) > 0$ for all $j \in \{1, \ldots, n-1\}$ and it remains to show that $\Phi(c_n, c_n) = \gamma_n > 0$, since this means that $\mathbf{pi}(\Phi) = n$ from which it follows that Φ is positive definite. To see that $\gamma_n > 0$, note that the matrix L of Φ relative to the basis $\{c_1, \ldots, c_n\}$ is diagonal, the diagonal entries being $\gamma_1, \gamma_2, \ldots, \gamma_n$. By Proposition 2.3.11, $\mathbf{det}(L) = \gamma_1\gamma_2 \ldots \gamma_n$. Also, Theorem 6.1.8 implies that $L = T^t S T$, where T is the transition matrix from the first basis to the second one. As above, we obtain $\mathbf{det}(L) = \mathbf{det}(T)^2\mathbf{det}(S)$. The last principal minor of S is $\mathbf{det}(S)$, so our conditions imply that $\mathbf{det}(S) > 0$. Hence, $\mathbf{det}(L) = \gamma_1\gamma_2 \ldots \gamma_n > 0$ and $\gamma_j > 0$ for all $j \in \{1, \ldots, n-1\}$ so that $\gamma_n > 0$, as required.

EXERCISE SET 6.3

6.3.1. Over the space $A = \mathbb{Q}^4$ a bilinear form with the matrix

$$\begin{pmatrix} 1 & 2 & -1 & 0 \\ 2 & -1 & 0 & 1 \\ -1 & 0 & 0 & 2 \\ 0 & 1 & 2 & -1 \end{pmatrix}$$

relative to the standard basis is given. Find an orthonormal basis of the space.

6.3.2. Over the space $A = \mathbb{Q}^4$ a bilinear form with the matrix

$$\begin{pmatrix} 3 & 1 & -1 & 2 \\ 1 & 0 & 2 & 4 \\ -1 & 2 & -1 & 3 \\ 2 & 4 & 3 & 0 \end{pmatrix}$$

relative to the standard basis is given. Find an orthonormal basis of the space.

6.3.3. Over the space $A = \mathbb{Q}^3$ a bilinear form with the matrix

$$\begin{pmatrix} 1 & 2 & -1 \\ 2 & -1 & 0 \\ -1 & 0 & 0 \end{pmatrix}$$

relative to the standard basis is given. Find an orthonormal basis of the space.

6.3.4. Over the space $A = \mathbb{Q}^3$ a bilinear form with the matrix

$$\begin{pmatrix} 3 & 1 & -1 \\ 1 & 0 & 2 \\ -1 & 2 & -1 \end{pmatrix}$$

relative to the standard basis is given. Find an orthonormal basis of the space.

6.3.5. Prove that a symmetric bilinear form over $\mathbb{R}$ is negative definite if and only if all principal minors of this form in some basis have alternating signs as their order grows and the first minor is negative.

6.3.6. Let Φ be a nonsingular bilinear symmetric form over $\mathbb{R}$ whose negative inertial index is 1, let a be an element with the property that $\Phi(a, a) < 0$ and let $B = \mathbb{R}a$. Prove that the reduction of this form over the space B is a nonsingular form.

6.3.7. Change the form $f(x) = x_1^2 - 2x_2^2 - 2x_3^2 - 4x_1x_2 + 4x_1x_3 + 8x_2x_3$ to its canonical form over the field of rational numbers.

6.3.8. Change the form $f(x) = x_1^2 + x_2^2 + 3x_3^2 + 4x_1x_2 + 2x_1x_3 + 2x_2x_3$ to its canonical form over the field of rational numbers.

6.3.9. Change the form $f(x) = x_1x_2 + x_1x_3 + x_1x_4 + x_2x_3 + x_2x_4 + x_3x_4$ to its canonical form over the field of rational numbers.

6.3.10. Change the form $f(x) = x_1^2 - 3x_3^2 - 2x_1x_2 + 2x_1x_3 - 6x_2x_3$ to its canonical form over the field of real numbers.

6.3.11. Change the form $f(x) = x_1^2 + 5x_2^2 - 4x_3^2 + 2x_1x_2 - 4x_1x_3$ to its canonical form over the field of real numbers.

6.3.12. Change the form $f(x) = 2x_1x_2 + 2x_3x_4$ to its canonical form over the field of rational numbers.

6.3.13. Change the form $f(x) = 2x_1x_2 + 2x_1x_3 - 2x_1x_4 - 2x_2x_3 + 2x_2x_4 + 2x_3x_4$ to its canonical form over the field of rational numbers.

6.3.14. Change the form $f(x) = \frac{1}{4}x_1^2 + x_2^2 + x_3^2 + x_4^2 + 2x_2x_4$ to its canonical form over the field of rational numbers.

6.3.15. Find all values of the parameter λ for which the quadratic form $f(x) = 5x_1^2 + x_2^2 + \lambda x_3^2 + 4x_1x_2 - 2x_1x_3 - 2x_2x_3$ is positive definite.

6.3.16. Find all values of the parameter λ for which the quadratic form $f(x) = x_1^2 + 4x_2^2 + x_3^2 + 2\lambda x_1x_2 + 10x_1x_3 + 6x_2x_3$ is positive definite.

6.4 EUCLIDEAN SPACES

In previous sections we considered bilinear forms as a generalization of the concept of a scalar product defined on $\mathbb{R}^3$. Although we considered different types of bilinear forms and their properties and characteristics, we did not cover some important metric characteristics of a space such as length, angle, area, and so on. The existence of a scalar product allows us to introduce the geometric characteristics mentioned above. In three dimensional geometric space, the scalar product is often introduced in calculus courses to define the length of a vector and to find angles between vectors. We can also proceed in the reverse direction and define the concepts of length and angles which then allow us to define the scalar product. In this section, we first extend the concept of a scalar product to arbitrary vector spaces over $\mathbb{R}$.

6.4.1. Definition. *Let A be a vector space over $\mathbb{R}$. We say that A is a Euclidean space if there is a mapping $\langle , \rangle : A \times A \longrightarrow \mathbb{R}$ satisfying the following properties:*

(E 1) $\langle x + y, z\rangle = \langle x, z\rangle + \langle y, z\rangle$;
(E 2) $\langle \alpha x, z\rangle = \alpha\langle x, z\rangle$;
(E 3) $\langle x, y\rangle = \langle y, x\rangle$;
(E 4) *if* $x \neq 0_A$, *then* $\langle x, x\rangle > 0$.

Also $\langle x, y\rangle$ is called the scalar or inner product of the elements $x, y \in A$.

If we write $\Phi(x, y) = \langle x, y\rangle$, then Φ is a positive definite, symmetric bilinear form defined on A. Thus, a Euclidean space is nothing more than a real vector space on which a positive definite, symmetric bilinear form is defined, using Proposition 6.3.7.

The main example of a Euclidean space is the space $\mathbb{R}^3$ with the usual scalar product $\langle x, y\rangle = |x||y| \cos\alpha$ where α is the angle between the vectors x and y. The following proposition shows that an arbitrary finite-dimensional vector space over $\mathbb{R}$ can always be made into a Euclidean space.

6.4.2. Proposition. *Let A be a finite-dimensional vector space over $\mathbb{R}$. Then, there is a scalar product defined on A such that A is Euclidean.*

Proof. Let $\{a_1, \dots, a_n\}$ be an arbitrary basis of A. In Corollary 6.1.7, we showed how to define a bilinear form on A using an arbitrary matrix S relative to this basis. Let S be the identity matrix I in that construction. By Corollary 6.1.7, the mapping $\langle\,,\rangle : A \times A \longrightarrow \mathbb{R}$ defined by $\langle x, y\rangle = \sum_{1\le j\le n} \xi_j\eta_j$ where $x = \sum_{1\le j\le n} \xi_j a_j$ and $y = \sum_{1\le t\le n} \eta_t a_t$, is a bilinear form on A. Since I is a symmetric matrix, Proposition 6.1.11 implies that this form is symmetric. Finally, Theorem 6.3.9 shows that this form is positive definite. Hence A is Euclidean, by our remarks above.

As another example, let $C_{[a,b]}$ be the set of all continuous real functions defined on a closed interval $[a, b]$. As we discussed in Section 4.1, $C_{[a,b]}$ is a subspace of $\mathbb{R}^{[a,b]}$. However, this vector space is not finite dimensional. For every pair of functions $x(t), y(t) \in C_{[a,b]}$ we put

$$\langle x(t), y(t)\rangle = \int_a^b x(t)y(t)\,\mathrm{d}t.$$

By properties of the definite integral, $\langle,\rangle$ is a symmetric bilinear form on $C_{[a,b]}$. If $x(t)$ is a nonzero continuous function, then

$$\langle x(t), x(t)\rangle = \int_a^b x(t)^2\,\mathrm{d}t.$$

We recall that the definite integral of a continuous nonnegative, nonzero function is positive. Hence, this bilinear form is positive definite and therefore, the vector space $C_{[a,b]}$ is an infinite-dimensional Euclidean space.

We note the following useful property of orthogonal subsets.

6.4.3. Proposition. *Let A be a Euclidean space and let $\{a_1, \dots, a_m\}$ be an orthogonal subset of nonzero elements. Then $\{a_1, \dots, a_m\}$ is linearly independent.*

Proof. Let $\alpha_1, \dots, \alpha_m$ be real numbers such that $\alpha_1 a_1 + \cdots + \alpha_m a_m = 0_A$. Then

$$\begin{aligned} 0 &= \langle 0_A, a_j\rangle = \langle \alpha_1 a_1 + \cdots + \alpha_m a_m, a_j\rangle \\ &= \alpha_1\langle a_1, a_j\rangle + \cdots + \alpha_{j-1}\langle a_{j-1}, a_j\rangle + \alpha_j\langle a_j, a_j\rangle + \alpha_{j+1}\langle a_{j+1}, a_j\rangle \\ &\quad + \cdots + \alpha_m\langle a_m, a_j\rangle = \alpha_j\langle a_j, a_j\rangle, \text{ for } 1 \le j \le m. \end{aligned}$$

Since $a_j \ne 0_A$, $\langle a_j, a_j\rangle > 0$ and so $\alpha_j = 0$, for $1 \le j \le m$. Proposition 4.2.7 shows that the elements $\{a_1, \dots, a_m\}$ are linearly independent.

6.4.4. Definition. *Let A be a Euclidean space over $\mathbb{R}$ and let x be an element of A. The number $+\sqrt{\langle x, x\rangle}$ is called the norm (or the length) of x and will be denoted by $\|x\|$.*

We note that $\|x\| \ge 0$ and $\|x\| = 0$ if and only if $x = 0_A$. If α is an arbitrary real number, we have

$$\|\alpha x\| = \sqrt{\langle \alpha x, \alpha x\rangle} = \sqrt{\alpha^2\langle x, x\rangle} = |\alpha|\sqrt{\langle x, x\rangle} = |\alpha|\,\|x\|\,.$$

6.4.5. Proposition *(The Cauchy–Bunyakovsky–Schwarz Inequality). Let A be a Euclidean space and let x, y be arbitrary elements of A. Then,* $|\langle x, y\rangle| \leq \|x\|\,\|y\|$ *and* $|\langle x, y\rangle| = \|x\|\,\|y\|$ *if and only if x, y are linearly dependent.*

Proof. We consider the scalar product $\langle x + \lambda y, x + \lambda y\rangle$, where $\lambda \in \mathbb{R}$. We have

$$\langle x + \lambda y, x + \lambda y\rangle = \langle x, x\rangle + 2\lambda\langle x, y\rangle + \lambda^2\langle y, y\rangle.$$

Since $\langle x + \lambda y, x + \lambda y\rangle \geq 0$, the latter quadratic polynomial in λ is always nonnegative and this implies that its discriminant is nonpositive. Thus, using the quadratic formula,

$$\langle x, y\rangle^2 - \langle x, x\rangle\langle y, y\rangle \leq 0.$$

It follows that

$$|\langle x, y\rangle| \leq \sqrt{\langle x, x\rangle}\sqrt{\langle y, y\rangle} \text{ or } |\langle x, y\rangle| \leq \|x\|\,\|y\|\,.$$

If $x = \lambda y$ for some real number λ, then

$$\begin{aligned} |\langle x, y\rangle| &= |\langle \lambda y, y\rangle| = |\lambda|\,|\langle y, y\rangle| = |\lambda|\,\|y\|^2 = |\lambda|\,\|y\|\,\|y\| \\ &= (|\lambda|\,\|y\|)\,\|y\| = \|\lambda y\|\,\|y\| = \|x\|\,\|y\|\,. \end{aligned}$$

If x, y are linearly independent, then $x + \lambda y$ is nonzero for all λ, so that $\langle x + \lambda y, x + \lambda y\rangle > 0$ and hence $|\langle x, y\rangle| < \|x\|\,\|y\|$.

6.4.6. Corollary *(The Triangle Inequality). Let A be a Euclidean space and let x, y be arbitrary elements of A. Then* $\|x + y\| \leq \|x\| + \|y\|$*, and* $\|x + y\| = \|x\| + \|y\|$ *if and only if* $y = \lambda x$ *where* $\lambda \geq 0$.

Proof. We have

$$\|x + y\|^2 = \langle x + y, x + y\rangle = \langle x, x\rangle + 2\langle x, y\rangle + \langle y, y\rangle.$$

By Proposition 6.4.5, $|\langle x, y\rangle| \leq \|x\|\,\|y\|$ and also $\langle x, x\rangle = \|x\|$. So we obtain

$$\|x + y\|^2 \leq \|x\|^2 + 2\,\|x\|\,\|y\| + \|y\|^2 = (\|x\| + \|y\|)^2$$

and hence $\|x + y\| \leq \|x\| + \|y\|$.

Again by Proposition 6.4.5, $\langle x, y\rangle = \|x\|\,\|y\|$ if and only if $y = \lambda x$ and $\lambda \geq 0$, so only in this case we have $\|x + y\| = \|x\| + \|y\|$.

We next describe the Gram–Schmidt process, which allows us to transform an arbitrary linearly independent subset $\{a_1, \dots, a_m\}$ of a Euclidean space into an orthogonal set $\{b_1, \dots, b_m\}$ of nonzero elements.

Put $b_1 = a_1$ and $b_2 = \xi_1 b_1 + a_2$, where ξ is some real number to be determined. Since the elements $b_1 (= a_1)$ and a_2 are linearly independent, b_2 is nonzero for all ξ_1. We choose the number ξ_1 such that b_2 and b_1 are orthogonal as follows:

$$0 = \langle b_1, b_2 \rangle = \langle b_1, \xi_1 b_1 + a_2 \rangle = \xi_1 \langle b_1, b_1 \rangle + \langle b_1, a_2 \rangle.$$

Since $\langle b_1, b_1 \rangle > 0$, we deduce that $\xi_1 = \frac{-\langle b_1, a_2 \rangle}{\langle b_1, b_1 \rangle}$.

Now suppose that we have inductively constructed the orthogonal subset $\{b_1, \dots, b_t\}$ of nonzero elements and that for every $j \in \{1, \dots, t\}$ the element b_j is a linear combination of $a_1, \dots, a_j$. This will be valid for b_{t+1} if we define $b_{t+1} = \xi_1 b_1 + \cdots + \xi_t b_t + a_{t+1}$. Then the element b_{t+1} is nonzero, because $\xi_1 b_1 + \cdots + \xi_t b_t \in \mathbf{Le}(\{b_1, \dots, b_t\}) = \mathbf{Le}(\{a_1, \dots, a_t\})$ and $a_{t+1} \notin \mathbf{Le}(\{a_1, \dots, a_t\})$. The coefficients $\xi_1, \dots, \xi_t$ are chosen to satisfy the condition that b_{t+1} must be orthogonal to all vectors $b_1, \dots, b_t$ so that

$$\begin{aligned} 0 &= \langle b_j, b_{t+1} \rangle = \langle b_j, \xi_1 b_1 + \cdots + \xi_t b_t + a_{t+1} \rangle \\ &= \xi_1 \langle b_j, b_1 \rangle + \cdots + \xi_t \langle b_j, b_t \rangle + \langle b_j, a_{t+1} \rangle, \text{ for } 1 \le j \le t. \end{aligned}$$

Since the subset $\{b_1, \dots, b_t\}$ is orthogonal,

$$\xi_j \langle b_j, b_j \rangle + \langle b_j, a_{t+1} \rangle = 0, \text{ for } 1 \le j \le t.$$

It follows that $\xi_j = -\langle b_j, a_{t+1} \rangle / \langle b_j, b_j \rangle$, for $1 \le j \le t$. Thus, in general, we have

$$b_{t+1} = a_{t+1} - \frac{\langle b_1, a_{t+1} \rangle}{\langle b_1, b_1 \rangle} b_1 - \cdots - \frac{\langle b_t, a_{t+1} \rangle}{\langle b_t, b_t \rangle} b_t.$$

Continuing this process, we construct the orthogonal subset $\{b_1, \dots, b_m\}$ of nonzero elements.

Employing this process to an arbitrary basis of a finite-dimensional Euclidean space of dimension n, we obtain an orthogonal subset consisting of n nonzero elements. Proposition 6.4.3 shows that this subset is linearly independent and it is therefore, a basis of the space. Thus, in this way we can obtain an orthogonal basis of a Euclidean space. Applying the remark related to the first step of the process of orthogonalization and taking into account that every nonzero element can be included in some basis of the space, we deduce that every nonzero element of a finite-dimensional Euclidean space belongs to some orthogonal basis.

As in Section 6.3, we can normalize an orthogonal basis, so that the vectors have norm 1. Let $\{a_1, \dots, a_n\}$ be an orthogonal basis. Put $c_j = \alpha_j a_j$ where $\alpha_j = \frac{1}{\sqrt{\langle a_j, a_j \rangle}}$, for $1 \le j \le n$. Then, we have $\langle c_j, c_k \rangle = 0$ whenever $j \ne k$, and $\langle c_j, c_j \rangle = 1$, for $1 \le j, k \le n$. This gives us an orthonormal basis. Hence we have proved.

6.4.7. Theorem. *Let A be a finite-dimensional Euclidean space. Then A has an orthonormal basis $\{c_1, \dots, c_n\}$. If $x = \sum_{1 \le j \le n} \xi_j a_j$ and $y = \sum_{1 \le j \le n} \eta_j a_j$ are arbitrary elements of A, where $\xi_j, \eta_j \in \mathbb{R}$, then $\langle x, y \rangle = \xi_1 \eta_1 + \cdots + \xi_n \eta_n$.*

The above formula will be familiar as the usual dot product defined on $\mathbb{R}^3$ when we use the standard basis $\{(1, 0, 0), (0, 1, 0), (0, 0, 1)\}$.

6.4.8. Corollary. *Let $\alpha_1, \dots, \alpha_n, \beta_1, \dots, \beta_n$ be real numbers. Then*

$$|\alpha_1\beta_1 + \cdots + \alpha_n\beta_n| \leq \sqrt{(\alpha_1^2 + \cdots + \alpha_n^2)}\sqrt{(\beta_1^2 + \cdots + \beta_n^2)}.$$

Proof. Let A be a finite-dimensional Euclidean space and choose an orthonormal basis $\{c_1, \dots, c_n\}$ of A. Let $x = \sum_{1\leq j\leq n} \alpha_j a_j$ and $y = \sum_{1\leq j\leq n} \beta_j a_j$, where $\alpha_j, \beta_j \in \mathbb{R}$, for $1 \leq j \leq n$. By Theorem 6.4.7

$$\langle x, y\rangle = \sum_{1\leq j\leq n} \alpha_j\beta_j, \ \|x\| = \sqrt{\langle x, x\rangle} = \sqrt{(\alpha_1^2 + \cdots + \alpha_n^2)},$$

$$\text{and } \ \|y\| = \sqrt{(\beta_1^2 + \cdots + \beta_n^2)}.$$

By Proposition 6.4.5, $\langle x, y\rangle \leq \|x\|\,\|y\|$; this implies the required inequality.

There are various standard mappings defined on Euclidean spaces; those preserving the scalar product are now the ones of interest since these are the ones which preserve length and angle.

6.4.9. Definition. *Let A, V be Euclidean spaces. A linear mapping $f : A \longrightarrow V$ is said to be metric if $\langle f(x), f(y)\rangle = \langle x, y\rangle$ for all elements $x, y \in A$. A metric bijective mapping is called an isometry.*

6.4.10. Theorem. *Let A, V be finite-dimensional Euclidean spaces. Then, there exists an isometry from A to V if and only if $\mathbf{dim}_{\mathbb{R}}(A) = \mathbf{dim}_{\mathbb{R}}(V)$.*

Proof. Assume that $f : A \longrightarrow V$ is an isometry. Then, f is an isomorphism from A to V and Corollary 5.1.9 shows that $\mathbf{dim}_{\mathbb{R}}(A) = \mathbf{dim}_{\mathbb{R}}(V)$.

Conversely, assume that $\mathbf{dim}_{\mathbb{R}}(A) = \mathbf{dim}_{\mathbb{R}}(V)$. By Theorem 6.4.7, the spaces A and V have orthonormal bases, say $\{c_1, \dots, c_n\}$ and $\{u_1, \dots, u_n\}$ respectively. Let $x = \sum_{1\leq j\leq n} \xi_j c_j$, be an arbitrary element of A, where $\xi_j \in \mathbb{R}$, for $1 \leq j \leq n$. Define a mapping $f : A \longrightarrow V$ by $f(x) = \sum_{1\leq j\leq n} \xi_j u_j$. By Proposition 5.1.12, f is a linear mapping. As in the proof of Theorem 5.1.13, we can show that f is a linear isomorphism. If $y = \sum_{1\leq j\leq n} \eta_j a_j$ is another element of A then, by Theorem 6.4.7, $\langle x, y\rangle = \xi_1\eta_1 + \cdots + \xi_n\eta_n$. We have $f(y) = \sum_{1\leq j\leq n} \eta_j u_j$ and again using Theorem 6.4.7 we obtain

$$\langle f(x), f(y)\rangle = \xi_1\eta_1 + \cdots + \xi_n\eta_n = \langle x, y\rangle.$$

Thus f is an isometry.

6.4.11. Definition. *Let A be a Euclidean space. A metric linear transformation f of A is called orthogonal.*

Here are some characterizations of orthogonal transformations.

6.4.12. Proposition. *Let A be a finite-dimensional Euclidean space and let f be a linear transformation of A.*

(i) *If f is an orthogonal transformation of A and $\{c_1, \dots, c_n\}$ is an arbitrary orthonormal basis of A, then $\{f(c_1), \dots, f(c_n)\}$ is an orthonormal basis of A. In particular, f is an automorphism of A.*
(ii) *Suppose that A has an orthonormal basis $\{c_1, \dots, c_n\}$ such that $\{f(c_1), \dots, f(c_n)\}$ is also an orthonormal basis of A. Then f and f^{-1} are orthogonal transformations.*
(iii) *If S is the matrix of an orthogonal transformation f relative to an arbitrary orthonormal basis $\{c_1, \dots, c_n\}$, then $S^t = S^{-1}$.*
(iv) *If f, g are orthogonal transformations of A, then $f \circ g$ is an orthogonal transformation.*

Proof.

(i) We have $\langle f(c_j), f(c_j)\rangle = \langle c_j, c_j\rangle = 1$ for all $j \in \{1, \dots, n\}$ and $\langle f(c_j), f(c_k)\rangle = \langle c_j, c_k\rangle = 0$ whenever $j \neq k$ and $j, k \in \{1, \dots, n\}$. It follows that $\{f(c_1), \dots, f(c_n)\}$ is an orthonormal basis of A. In particular, $\mathbf{Im} f = A$ and Proposition 5.2.14 proves that f is an automorphism.

(ii) Let $x = \sum_{1 \le j \le n} \xi_j c_j$ and $y = \sum_{1 \le j \le n} \eta_j c_j$ be arbitrary elements of A, where $\xi_j, \eta_j \in \mathbb{R}$, for $1 \le j \le n$. Since f is linear, $f(x) = \sum_{1 \le j \le n} \xi_j f(c_j)$ and $f(y) = \sum_{1 \le j \le n} \eta_j f(c_j)$. By Theorem 6.4.7, $\langle x, y\rangle = \xi_1\eta_1 + \cdots + \xi_n\eta_n$ and $\langle f(x), f(y)\rangle = \xi_1\eta_1 + \cdots + \xi_n\eta_n$, so that $\langle f(x), f(y)\rangle = \langle x, y\rangle$. This shows that f is an orthogonal transformation.

Since f^{-1} transforms the orthonormal basis $\{f(c_1), \dots, f(c_n)\}$ into the orthonormal basis $\{c_1, \dots, c_n\}$, f^{-1} is also an orthogonal transformation.

(iii) Let $S = [\sigma_{jt}] \in \mathbf{M}_n(\mathbb{R})$ denote the matrix of f relative to the basis $\{c_1, \dots, c_n\}$. By (i), the subset $\{f(c_1), \dots, f(c_n)\}$ is an orthonormal basis of A. Then, we can consider S as the transition matrix from the basis $\{c_1, \dots, c_n\}$ to the basis $\{f(c_1), \dots, f(c_n)\}$. Clearly, in every orthonormal basis the matrix of our bilinear form $\langle\, ,\rangle$ is E. Corollary 5.2.12 gives the equation $I = S^t I S = S^t S$, which implies that $S^t = S^{-1}$.

Since assertion (iv) is straightforward, the result follows.

6.4.13. Definition. *Let $S \in \mathbf{M}_n(\mathbb{R})$. The matrix S is called orthogonal if $S^t = S^{-1}$.*

The following proposition provides us with some basic properties of orthogonal matrices.

6.4.14. Proposition. *Let $S = [\sigma_{jt}] \in \mathbf{M}_n(\mathbb{R})$ be an orthogonal matrix. Then the following holds true:*

(i) $\sum_{1 \le t \le n} \sigma_{jt}\sigma_{kt} = \delta_{jk}, \sum_{1 \le t \le n} \sigma_{tj}\sigma_{tk} = \delta_{jk}$, *for* $1 \le j, k \le n$.
(ii) $\mathbf{det}(S) = \pm 1$.
(iii) S^{-1} *is also an orthogonal matrix.*
(iv) *If* R *is also orthogonal then* SR *is orthogonal.*
(v) *Let* A *be a Euclidean space of dimension* n. *Suppose that* $\{a_1, \dots, a_n\}$ *and* $\{c_1, \dots, c_n\}$ *are two orthonormal bases of* A. *If* T *is the transition matrix from* $\{a_1, \dots, a_n\}$ *to* $\{c_1, \dots, c_n\}$, *then* T *is orthogonal;*
(vi) *Let* A *be a Euclidean space of dimension* n *and let* f *be a linear transformation of* A. *Suppose that* $\{a_1, \dots, a_n\}$ *is an orthonormal basis of* A. *If* S *is the matrix of* f *relative to this basis, then* f *is an orthogonal linear transformation of* A.

Proof.

(i) follows from the definition of orthogonal matrix.
(ii) We have $SS^t = I$. Theorem 2.5.1 implies that $1 = \mathbf{det}(I) = \mathbf{det}(S)\mathbf{det}(S^t)$. By Proposition 2.3.3, $\mathbf{det}(S) = \mathbf{det}(S^t)$, so $\mathbf{det}(S)^2 = 1$. It follows that $\mathbf{det}(S) = \pm 1$.
(iii) We have $(S^{-1})^t = (S^t)^t = S = (S^{-1})^{-1}$.
(iv) By Theorem 2.1.10, we obtain $(SR)^t = R^t S^t = R^{-1}S^{-1} = (SR)^{-1}$. Hence SR is orthogonal.
(v) By Proposition 5.1.12, there exists a linear transformation f of A such that $f(a_j) = c_j$, for $1 \le j \le n$. By definition, the matrix of f relative to the basis $\{a_1, \dots, a_n\}$ is T. By Proposition 6.4.12(ii), f is an orthogonal transformation. Again using Proposition 6.4.12(iii), we see that T is an orthogonal matrix.
(vi) We have

$$\langle f(a_j), f(a_m)\rangle = \left\langle \sum_{1\le t\le n} \sigma_{tj} a_t, \sum_{1\le k\le n} \sigma_{km} a_k \right\rangle = \sum_{1\le t\le n}\sum_{1\le k\le n} \sigma_{tj}\sigma_{km}\langle a_t, a_k\rangle$$
$$= \sum_{1\le t\le n}\sum_{1\le k\le n} \sigma_{tj}\sigma_{km}\delta_{tk} = \sum_{1\le t\le n} \sigma_{tj}\sigma_{tm} = \delta_{jm},$$

using (i). Therefore, the set $\{f(a_1), \dots, f(a_n)\}$ is an orthonormal basis of A and Proposition 6.4.12(ii) shows that f is an orthogonal transformation.

Next we consider another important type of linear transformation of Euclidean spaces.

6.4.15. Definition. *Let* A *be a Euclidean space. A linear transformation* f *of* A *is called symmetric (or self-conjugate), if* $\langle f(x), y\rangle = \langle x, f(y)\rangle$ *for all elements* $x, y \in A$.

It is easy to see that if f and g are symmetric transformations, then $f+g$ is also a symmetric transformation. Also, if α is a real number then αf is also a symmetric transformation. There is, of course, a connection between these types of transformations and symmetric matrices.

6.4.16. Proposition. *Let A be a Euclidean space of dimension n.*

(i) *Let $\{a_1, \dots, a_n\}$ be an arbitrary orthonormal basis of A. If f is a symmetric transformation of A, then the matrix of f relative to the basis $\{a_1, \dots, a_n\}$ is symmetric.*

(ii) *Let f be a linear transformation of A. Suppose that $\{a_1, \dots, a_n\}$ is an orthonormal basis of A such that the matrix of f relative to this basis is symmetric. Then f is a symmetric linear transformation of A.*

Proof.

(i) Let $S = [\sigma_{jt}] \in \mathbf{M}_n(\mathbb{R})$ denote the matrix of f relative to the basis $\{a_1, \dots, a_n\}$. Since the basis is an orthonormal basis, we have

$$\langle f(a_j), a_m\rangle = \left\langle \sum_{1\le t\le n} \sigma_{tj} a_t, a_m \right\rangle = \sum_{1\le t\le n} \sigma_{tj}\langle a_t, a_m\rangle = \sigma_{mj} \text{ and,}$$

$$\langle a_j, f(a_m)\rangle = \left\langle a_j, \sum_{1\le k\le n} \sigma_{km} a_k \right\rangle \sum_{1\le k\le n} \sigma_{km}\langle a_j, a_k\rangle = \sigma_{jm},$$

for $1 \le j, m \le n$. Since $\langle f(a_j), a_m\rangle = \langle a_j, f(a_m)\rangle$, we have $\sigma_{mj} = \sigma_{jm}$, for $1 \le j, m \le n$, which shows that S is symmetric.

(ii) Let $x = \sum_{1\le j\le n} \xi_j a_j$ and $y = \sum_{1\le j\le n} \eta_j a_j$ be arbitrary elements of A, where $\xi_j, \eta_j \in \mathbb{R}$ for $1 \le j \le n$. Then

$$\begin{aligned} f(x) &= \sum_{1\le j\le n} \xi_j f(a_j) = \sum_{1\le j\le n} \xi_j \left(\sum_{1\le t\le n} \sigma_{tj} a_t \right) \\ &= \sum_{1\le t\le n} \left(\sum_{1\le j\le n} \xi_j \sigma_{tj} \right) a_t, \end{aligned}$$

and similarly,

$$f(y) = \sum_{1\le t\le n} \left(\sum_{1\le j\le n} \eta_j \sigma_{tj} \right) a_t.$$

It follows that

$$\begin{aligned}\langle f(x), y\rangle &= \left\langle \sum_{1\le t\le n}\left(\sum_{1\le j\le n}\xi_j\sigma_{tj}\right)a_t, \sum_{1\le k\le n}\eta_k a_k\right\rangle \\ &= \sum_{1\le t\le n}\sum_{1\le j\le n}\sum_{1\le k\le n}\xi_j\sigma_{tj}\eta_k\langle a_t, a_k\rangle = \sum_{1\le j\le n}\sum_{1\le k\le n}\xi_j\sigma_{kj}\eta_k,\end{aligned}$$

and

$$\begin{aligned}\langle x, f(y)\rangle &= \left\langle \sum_{1\le j\le n}\xi_j a_j, \sum_{1\le t\le n}\left(\sum_{1\le k\le n}\eta_k\sigma_{tk}\right)a_t\right\rangle \\ &= \sum_{1\le j\le n}\sum_{1\le t\le n}\sum_{1\le k\le n}\xi_j\eta_k\sigma_{tk}\langle a_j, a_t\rangle \\ &= \sum_{1\le j\le n}\sum_{1\le k\le n}\xi_j\eta_k\sigma_{jk} = \sum_{1\le j\le n}\sum_{1\le k\le n}\xi_j\sigma_{jk}\eta_k.\end{aligned}$$

Since S is symmetric, $\sigma_{jk} = \sigma_{kj}$, for $1 \le j, k \le n$, which implies that $\langle f(x), y\rangle = \langle x, f(y)\rangle$.

The following result gives an important property of symmetric matrices.

6.4.17. Theorem. *Let $S = [\sigma_{jt}] \in \mathbf{M}_n(\mathbb{R})$. If S is symmetric, then its characteristic polynomial $\chi_S(X)$ has only real roots. Thus the eigenvalues of a real symmetric matrix are real.*

Proof. Since the polynomial $\chi_S(X)$ has real coefficients, it follows from Theorem 7.5.14 (see the next chapter) that each of its roots is complex. Let λ_0 denote one such root. Thus $\mathbf{det}(S - \lambda_0 I) = 0$. Then the matrix of the system

$$\begin{aligned}(\sigma_{11} - \lambda_0)x_1 + \sigma_{12}x_2 + \cdots + \sigma_{1n}x_n &= 0_F \\ \sigma_{21}x_1 + (\sigma_{22} - \lambda_0)x_2 + \cdots + \sigma_{2n}x_n &= 0_F \\ &\vdots \\ \sigma_{n1}x_1 + \sigma_{n2}x_2 + \cdots + (\sigma_{nn} - \lambda_0)x_n &= 0_F\end{aligned} \tag{6.5}$$

is singular, and the results of Section 5.3 shows that this system has a nonzero solution $(\xi_1, \ldots, \xi_n)$. We remark that the numbers $\xi_1, \ldots, \xi_n$ are complex. So we have

$$\sum_{1\le j\le n}\sigma_{tj}\xi_j = \lambda_0\xi_t, \text{ for } 1 \le t \le n.$$

Multiplying both sides of the tth equation of Equation 6.5 by the conjugate $\overline{\xi}_t$ of ξ_t and adding these equations, we obtain

$$\sum_{1\le t\le n}\sum_{1\le j\le n}\sigma_{tj}\xi_j\overline{\xi}_t=\lambda_0\sum_{1\le t\le n}\xi_t\overline{\xi}_t.$$

The latter is a sum of nonnegative real numbers, at least one of which is positive and so it is nonzero. We will prove that λ_0 is a real number. For this, we will prove that $\sum_{1\le t\le n}\sum_{1\le j\le n}\sigma_{tj}\xi_j\overline{\xi}_t$ is real by showing that it coincides with its complex conjugate, $\overline{\sum_{1\le t\le n}\sum_{1\le j\le n}\sigma_{tj}\xi_j\overline{\xi}_t}$. To see this, we will use the symmetric property of the matrix S and the fact that S has real entries. We have

$$\begin{aligned}\overline{\sum_{1\le t\le n}\sum_{1\le j\le n}\sigma_{tj}\xi_j\overline{\xi}_t}&=\sum_{1\le t\le n}\sum_{1\le j\le n}\overline{\sigma_{tj}\xi_j\overline{\xi}_t}=\sum_{1\le t\le n}\sum_{1\le j\le n}\sigma_{tj}\overline{\xi}_j\xi_t\\&=\sum_{1\le t\le n}\sum_{1\le j\le n}\sigma_{jt}\overline{\xi}_j\xi_t=\sum_{1\le t\le n}\sum_{1\le j\le n}\sigma_{tj}\overline{\xi}_t\xi_j\\&=\sum_{1\le t\le n}\sum_{1\le j\le n}\sigma_{tj}\xi_j\overline{\xi}_t.\end{aligned}$$

The second-to-last equation is obtained by a simple juxtaposition of the summation indices t and j. The result follows.

6.4.18. Corollary. *Let f be a symmetric linear transformation of a finite-dimensional Euclidean space. Then, its characteristic polynomial $\chi_S(X)$ has only real roots.*

Now, we are in a position to formulate the following characterization of symmetric transformations of Euclidean space.

6.4.19. Theorem. *Let A be a Euclidean space of dimension n and let f be a transformation of A. Then, f is symmetric if and only if A has an orthonormal basis consisting of eigenvectors of this transformation.*

Proof. If A has an orthonormal basis $\{a_1,\dots,a_n\}$ such that $f(a_j)=\gamma_j a_j$, for $1\le j\le n$, then the matrix of f relative to the basis $\{a_1,\dots,a_n\}$ is diagonal. Since a diagonal matrix is clearly symmetric, Proposition 6.4.16 implies that f is symmetric.

Conversely, let f be a symmetric linear transformation of A. We will use induction on the dimension, n, of the space A. Certainly, if $n=1$, then every linear transformation of A will transform every vector a into some multiple of a. It follows that if $a\ne 0$, then a is an eigenvector for f and it is easily seen that each linear transformation of A is symmetric. By forming $a/\|a\|$, we obtain an orthonormal basis of A.

Suppose that the theorem is proved for all Euclidean spaces of dimension at most $n-1$. By Corollary 6.4.18, the characteristic polynomial $\chi_f(X)$ has a real root γ, which is an eigenvalue for f. Let a be an eigenvector corresponding to γ, so that $f(a)=\gamma a$. Let $\alpha=\langle a,a\rangle$ and let $a_1=\frac{1}{\sqrt{\alpha}}a$. Then clearly, $\langle a_1,a_1\rangle=1$ and $f(a_1)=\gamma a_1$.

Let A_1 be the subspace generated by the element a_1. Since $\langle a_1,a_1\rangle=1$, the restriction of the bilinear form on A_1 is nonsingular. By Theorem 6.2.14, there exists a decomposition $A=A_1\oplus A_1^{\perp}$. Let $x\in A_1^{\perp}$. Then

$$\langle a_1, f(x)\rangle = \langle f(a_1), x\rangle = \langle \gamma a_1, x\rangle = \gamma\langle a_1, x\rangle = 0.$$

Thus, $f(x)\in A_1^{\perp}$ and it follows that the restriction of f to $A_1^{\perp}$ is a linear transformation of $A_1^{\perp}$. Since f is symmetric, the restriction of f to $A_1^{\perp}$ is also symmetric and the induction hypothesis implies that $A_1^{\perp}$ has an orthonormal basis $\{a_2,\dots,a_n\}$, consisting of eigenvectors of f. All these elements are orthogonal to a_1, and therefore, $\{a_1,\dots,a_n\}$ is an orthonormal basis of the space A consisting of eigenvectors of the transformation f.

The results of this section can be extended to vector spaces over the complex numbers. Such a complex linear space, A, is called unitary if there is a scalar multiplication, axioms (**E 1**), (**E 2**), and (**E 4**) hold, with the proviso that $\langle x,y\rangle$ is a complex number and in the last axiom the scalar square of a nonzero vector is real and positive. Axiom (**E 3**) should be substituted by the axiom: $\overline{\langle x, y\rangle} = \langle \mathbf{y}, \mathbf{x}\rangle$. Almost all the results of this section can be extended to unitary spaces.

EXERCISE SET 6.4

6.4.1. Let A be a finite-dimensional Euclidean space and let K, L be subspaces of A. Suppose that $\dim_{\mathbb{R}}(L) < \dim_{\mathbb{R}}(K)$. Prove that K contains an element a, which is orthogonal to all elements of L.

6.4.2. Let $A=\mathbb{R}^4$ and let a_1,a_2,a_3,a_4 be an orthonormal basis of A. Let b, c be elements of A whose coordinates relative to this basis are $(1,-2,2,-3)$ and $(2,-3,2,4)$, respectively. Prove that b, c are orthogonal and find a complement to the set $\{b,c\}$ which, together with $\{b,c\}$, gives an orthogonal basis of A.

6.4.3. Let $A=\mathbb{R}^4$ and let a_1,a_2,a_3,a_4 be an orthonormal basis of A. Let b, c be elements of A, whose coordinates relative to this basis are $(1,1,1,2)$ and $(1,2,3,-3)$, respectively. Prove that b, c are orthogonal and find a complement to the set $\{b,c\}$ which, together with $\{b,c\}$, gives an orthogonal basis of A.

6.4.4. Let $A = \mathbb{R}^3$ and let a_1, a_2, a_3 be an orthonormal basis of A. Let b, c be elements of A, whose coordinates relative to this basis are $(2/3, 1/3, 2/3)$ and $(1/3, 2/3, -2/3)$, respectively. Prove that b, c are orthogonal and find a complement to the set $\{b, c\}$, which together with $\{b, c\}$, gives an orthogonal basis.

6.4.5. Let $A = \mathbb{R}^4$ and let a_1, a_2, a_3, a_4 be an orthonormal basis of A. Let b, c be elements of A, whose coordinates relative to this basis are $(1/2, 1/2, 1/2, 1/2)$ and $(1/2, 1/2, -1/2, -1/2)$. Prove that b, c are orthogonal and find a complement to the set $\{b, c\}$, which together with $\{b, c\}$, gives an orthogonal basis.

6.4.6. Let $A = \mathbb{R}^4$ and let a_1, a_2, a_3, a_4 be an orthonormal basis of A. Let b, c, d be elements of A, whose coordinates relative to this basis are $(1, 2, 2, -1)$, $(3, 2, 8, -7)$, and $(1, 1, -5, 3)$, respectively. Construct an orthonormal basis of the linear envelope of the set $\{b, c, d\}$.

6.4.7. Let $A = \mathbb{R}^4$ and let a_1, a_2, a_3, a_4 be an orthonormal basis of A. Let b, c, d be elements of A, whose coordinates relative to this basis are $(1, 1, -1, -2)$, $(5, 8, -2, -3)$, and $(3, 9, 3, 8)$, respectively. Construct an orthonormal basis of the linear envelope of the set $\{b, c, d\}$.

6.4.8. Let $A = \mathbb{R}^4$ and let a_1, a_2, a_3, a_4 be an orthonormal basis of A. Let b, c, d, u be elements of A, whose coordinates relative to this basis are $(2, 1, 3, -1)$, $(7, 4, 3, -3)$, $(1, 1 - 6, 0)$, and $(5, 7, 7, 8)$, respectively. Construct an orthonormal basis of the linear envelope of the set $\{b, c, d, u\}$.

6.4.9. Let A be a finite-dimensional Euclidean space and L be a subspace of A. Prove that every element x of A can be uniquely represented as $x = y + z$ where $y \in L$ and z is orthogonal to L. The element y is called the orthogonal projection of x on the subspace L and the element z is called the orthogonal component of x relative to L.

6.4.10. Let $A = \mathbb{R}^4$ and let a_1, a_2, a_3, a_4 be an orthonormal basis of A. Let b, c, d be elements of A, whose coordinates relative to this basis are $(1, 1, 1, 1)$, $(1, 2, 2, -1)$, and $(1, 0, 0, 3)$. Let x be an element whose coordinates relative to the given basis are $(4, -1, -3, 4)$. Find the orthogonal projection y and the orthogonal component z of the element x on the linear envelope L of the subset $\{b, c, d\}$.

6.4.11. Let $A = \mathbb{R}^4$ and let a_1, a_2, a_3, a_4 be an orthonormal basis of A. Let b, c, d be elements of A, whose coordinates relative to this basis are $(2, 1, 1, -1)$, $(1, 1, 3, 0)$, and $(1, 2, 8, 1)$. Let x be an element whose coordinates relative to the given basis are $(5, 2, -2, 2)$. Find the orthogonal projection y and the orthogonal component z of the element x on the linear envelope L of the subset $\{b, c, d\}$.

6.4.12. Let $A = \mathbb{R}^4$ and let a_1, a_2, a_3, a_4 be an orthonormal basis of A. Let x be an element whose coordinates relative to the given basis are $(7, -4, -1, 2)$. Let L be a subspace, whose coordinates relative to $\{a_1, a_2, a_3, a_4\}$ satisfy the following system of equations:

$$\begin{aligned} 2x_1 + x_2 + x_3 + 3x_4 &= 0, \\ 3x_1 + 2x_2 + 2x_3 + x_4 &= 0, \\ x_1 + 2x_2 + 2x_3 - 9x_4 &= 0. \end{aligned}$$

Find the orthogonal projection y and the orthogonal component z of the element x on L.

6.4.13. Let $A = \mathbb{R}^3$ and let a_1, a_2, a_3, be an orthonormal basis of A. Let f be a linear transformation of A, whose matrix relative to the given basis is

$$\begin{pmatrix} 11 & 2 & -8 \\ 2 & 2 & 10 \\ -8 & 10 & 5 \end{pmatrix}.$$

Find the orthonormal basis consisting of eigenvectors of the linear transformation f and the matrix of f relative to this basis.

6.4.14. Let A be a finite-dimensional Euclidean space and let f be a symmetric linear transformation of A. Then f is called positively (respectively nonnegatively) defined, if its eigenvalues are positive (respectively nonnegative). Show that f is positively (respectively nonnegatively) defined, if $\langle f(a), a \rangle > 0$ (respectively $\langle f(a), a \rangle \geq 0$) for every nonzero element a of the space A.

CHAPTER 7

RINGS

7.1 RINGS, SUBRINGS, AND EXAMPLES

In mathematics, we often deal with sets that involve several algebraic operations that are connected to each other in some way, typically by some kind of distributive law. Natural examples of this, with which the reader is familiar, include the sets of numbers $\mathbb{Z}$, $\mathbb{Q}$, and $\mathbb{R}$, the set of real functions, and the sets of matrices $\mathbf{M}_n(\mathbb{Z})$, $\mathbf{M}_n(\mathbb{Q})$, $\mathbf{M}_n(\mathbb{R})$. These sets have some important common properties that enable us to classify them as a type of algebraic structure called a ring. Rings are one of the most important algebraic structures. In Chapter 3 we discussed fields, which are very special types of rings. However, a very basic and motivating example of a ring is the ring of integers. The study of natural extensions of the ring of integers, such as rings of "algebraic numbers", began the subject of ring theory. Another fundamental type of ring that has received a lot of attention is the ring of polynomials.

A German mathematician, Richard Dedekind (1831–1916), introduced the concept of a ring. The term *ring* (Zahlring) is due to another great German mathematician, David Hilbert (1862–1943). The axiomatic approach to the study of rings is due to Adolph Fraenkel (1891–1965) and the German born mathematician, Emmy Noether (1882–1935), known for her groundbreaking contributions to abstract algebra and theoretical physics. She was described by David Hilbert, Albert Einstein, and others as the most important woman in the history of mathematics. She revolutionized the theories of rings, algebras, and fields.

7.1.1. Definition. *A set R together with two algebraic binary operations, which we shall call addition and multiplication, is called a ring if the following properties hold:*

(R 1) *The operation of addition, denoted by $+$, has the properties*

(i) *addition is commutative, so $a + b = b + a$, for all $a, b \in R$;*

(ii) *addition is associative, so $a + (b + c) = (a + b) + c$, for all $a, b, c, \in R$;*

(iii) *there exists a zero element $0_R \in R$ such that $a + 0_R = a$, for all $a \in R$;*

(iv) *each element $a \in R$ has an additive inverse, $-a \in R$, called an opposite (or negative), such that $a + (-a) = 0_R$.*

(R 2) *Addition and multiplication are connected by the distributive laws: $(a + b)c = ac + bc$ and $a(b + c) = ab + ac$ for all $a, b, c \in R$.*

We call R, together with just the operation of addition, the additive group of the ring R.

The existence of negative elements allows us to introduce the operation of subtraction in R by making the definition that $a - b = a + (-b)$.

There are a number of elementary consequences, which follow from the definition of a ring and which we now list.

7.1.2. Proposition. *Let R be a ring and let a, b, c be elements of R. Then the following properties hold:*

(i) $a \cdot 0_R = 0_R \cdot a = 0_R$,

(ii) $a(-b) = (-a)b = -ab$,

(iii) $a(b - c) = ab - ac$ *and* $(a - b)c = ac - bc$.

In particular $(-a)(-b) = ab$.

Proof. For each $b \in R$ we have $b + 0_R = b$. By the distributive law,

$$ab = a \cdot (b + 0_R) = ab + a \cdot 0_R.$$

Since the element $ab \in R$ has a negative, $-ab$, we can add it to both sides and the equality becomes

$$-ab + ab = -ab + ab + a \cdot 0_R,$$

so $0_R = a \cdot 0_R$, since $-ab + ab = 0_R$ and, likewise, $0_R \cdot a = 0_R$, by a similar argument. Therefore (i) follows.

To prove (ii), from the definition of the negative element and the distributive laws we obtain

$$0_R = a \cdot 0_R = a(b + (-b)) = ab + a(-b) \text{ and}$$
$$0_R = 0_R \cdot b = (a + (-a)) \cdot b = ab + (-a) \cdot b.$$

These equations show that $a(-b)$ is the negative of ab and $(-a)b$ is also the negative of ab and hence (ii) follows. Also by replacing a by $-a$ we see that $(-a)(-b) = (-(-a))b = ab$ since the negative of $-a$ is a itself.

These equations show that subtraction and multiplication are also connected by the distributive laws since

$$a(b - c) = a(b + (-c)) = ab + a(-c) = ab - ac$$

and likewise $(a - b)c = ac - bc$.

7.1.3. Definition. *Let R be a ring.*

- **(R 3)** *R is called associative if the multiplication in R is associative so $a(bc) = (ab)c$ for all $a, b, c \in R$.*
- **(R 4)** *R is called commutative, if the multiplication in R is commutative so $ab = ba$ for all $a, b \in R$.*
- **(R 5)** *R is a ring with identity if R has an identity element e relative to the operation of multiplication so $ae = ea = a$ for all $a \in R$.*

While we consider only associative rings in this book, it is worth noting that the theory of some nonassociative rings that are, in some sense, close to associative rings has been studied in some depth. We here mention some of the most important examples of these types of rings. As usual we denote the product $a \cdot a$ by a^2.

A ring R is called a Lie ring, if it satisfies the conditions:

- **(LR 1)** $a^2 = 0_R$ for each $a \in R$.
- **(LR 2)** $(ab)c + (bc)a + (ca)b = 0_R$ for all $a, b, c \in R$.

The condition **(LR 2)** is called the Jacobi identity.

A ring R is called a Jordan ring, if it is commutative and satisfies the condition

$$(\mathbf{JR})((a \cdot a)b)a = (a \cdot a)(b \cdot a) \text{ for all elements } a, b \in R.$$

Finally, a ring R is called an alternative ring, if it satisfies the conditions:

- **(AR 1)** $(aa)b = a(ab)$ and
- **(AR 2)** $(ba)a = b(aa)$, for arbitrary elements $a, b \in R$.

Further important information concerning nonassociative rings can be found in the classical books of Bahturin (1986), Bourbaki (2004), and Serre (1992).

[Bahturin Yu. Identical relations in Lie algebras. Utrecht, Netherlands: Brill Academic Publishers; 1986.]

[Bourbaki N. Lie groups and Lie algebras. Berlin: Springer; 2004. Chapters 1–9.]

[Serre JP. Lie algebras and Lie groups. New York: Springer; 1992.]

Since our goal is to understand something about the theory of associative rings, we, from this point on, use the term *ring* to mean an associative ring. Thus, we assume that our rings satisfy properties **(R 1)**, **(R 2)**, and **(R 3)** and use the term *ring* in this respect without necessarily mentioning that the ring is an associative ring.

Let R be a ring with identity and suppose that $0_R = e$. Then we have

$$a = ae = a0_R = 0_R,$$

for each $a \in R$. Hence, a ring in which the zero and identity elements coincide consists only of the zero element. We term the ring $\{0_R\}$ the trivial ring and we consider only nontrivial rings below.

From the definition, it follows that a ring R with identity is a semigroup with identity under multiplication. Therefore, R contains the subset $\mathbf{U}(R)$ consisting of all invertible elements of this semigroup. By Corollary 3.1.15, this subset is stable and we shall see later that $\mathbf{U}(R)$ is a group under multiplication, when this multiplication is restricted to $\mathbf{U}(R)$. We note that $0_R \notin \mathbf{U}(R)$.

7.1.4. Definition. *A nonzero element a of a ring R is called a left (respectively right) zero divisor, if there is a nonzero element b such that $ab = 0_R$ (respectively $ba = 0_R$).*

For commutative rings, if $ab = 0_R$ then $ba = 0_R$, so every left zero-divisor is a right zero-divisor and conversely, and we just talk about zero-divisors in this case. If $ab = 0_R$ and $a \in \mathbf{U}(R)$ then a^{-1} exists and

$$0_R = a^{-1} \cdot 0_R = a^{-1}(ab) = (a^{-1}a)b = e \cdot b = b.$$

Thus, an invertible element cannot be a zero-divisor.

7.1.5. Proposition. *Let R be a ring and let $a, x, y \in R$.*

(i) *If a is not a left zero-divisor and if $ax = ay$ then $x = y$.*

(ii) *If a is not a right zero-divisor and if $xa = ya$ then $x = y$.*

Proof. Since the proofs of (i) and (ii) are similar, we merely prove (i). If $ax = ay$, it follows that $ax - ay = 0_R$. By the distributive law, we obtain $a(x - y) = 0_R$. Since a is not a left zero-divisor, this implies that $x - y = 0_R$ and hence $x = y$.

Proposition 7.1.5 is called the left (or the right) cancellation law. We obtain the following immediate consequence.

7.1.6. Corollary. *Let R be a ring with no zero-divisors. If $a, x, y \in R$ and $ax = ay$ (respectively $xa = ya$) then $x = y$.*

We here remind the reader that a ring means an associative ring. The ring of integers has some very standard properties that are common to certain types of rings, which we now introduce.

7.1.7. Definition. *A ring R is called an integral domain if R is commutative, has a multiplicative identity, and has no zero-divisors.*

We are naturally interested in subsets of rings that are themselves rings.

7.1.8. Definition. *Let R be a ring. A subset H of R is called a subring, if H is stable under the operations of addition and multiplication and H is also a ring under the restrictions of these operations to the set H. When H is a subring of R, we write $H \leq R$.*

We next give a criterion for a nonempty subset of a ring to be a subring. It is very similar to the criterion for a subset of a vector space to be a subspace.

7.1.9. Theorem. *Let R be a ring. If H is a subring of R, then H satisfies the following conditions:*

(SR 1) *if $x, y \in H$, then $x - y \in H$;*
(SR 2) *if $x, y \in H$, then $xy \in H$.*

Conversely, suppose that H is a nonempty subset of R. If H satisfies the conditions (**SR 1**) *and* (**SR 2**)*, then H is a subring of R.*

Proof. Let H be a subring of R. Certainly H is a stable subset under addition and multiplication and it also follows that H has a zero element 0_H. Thus, $x + 0_H = x$ for each element $x \in H$. By Definition 7.1.1, there is an element $-x \in R$ and we have $-x + x + 0_H = -x + x$ so that $0_R + 0_H = 0_R$. It follows that $0_H = 0_R$. By Definition 7.1.1, for each element $x \in H$ there is an element $y \in H$ such that $x + y = 0_H$ and since $0_H = 0_R$ it follows that y is a negative of x. As seen in Section 3.2, each negative element is uniquely determined and hence $y = -x$. In particular, $-x \in H$. Now if x, y are arbitrary elements of H then $-y \in H$ and, since H is a stable subset under addition, we have

$$x - y = x + (-y) \in H.$$

Hence H satisfies the condition **(SR 1)**.

Since H is a stable subset under multiplication, $xy \in H$, so that H satisfies **(SR 2)**.

Conversely, suppose that H is not empty and satisfies **(SR 1)** and **(SR 2)**. If $u \in H$ then, by **(SR 1)**, $0_R = u - u \in H$. If x is an arbitrary element of H then, by **(SR 1)**, $-x = 0_R - x \in H$. If also y is an arbitrary element of H, then $-y \in H$ and, using **(SR 1)**, we obtain $x + y = x - (-y) \in H$. Hence, H is a stable subset under addition. The condition **(SR 2)** shows that H is a stable subset under multiplication. Thus, the restrictions of addition and multiplication to H are binary operations on H. All the other conditions of **(R 1)–(R 3)** are valid for H because they are valid for all elements of R.

We note that even when a ring contains a multiplicative identity, as is true in the ring $\mathbb{Z}$, the subrings need not contain a multiplicative identity, as seen when considering the subring $2\mathbb{Z} = \{2r | r \in \mathbb{Z}\}$ of $\mathbb{Z}$. Indeed, it is possible for a ring and a subring to contain different multiplicative identities, as we shall see later.

A subring H of a ring R with multiplicative identity is called unitary if H contains the identity of the ring R.

Every ring R always contains at least two subrings, the subset $\{0_R\}$ and the entire ring R. It is very easy to generate further generic examples of subrings.

7.1.10. Corollary. *Let R be a ring and let $\mathfrak{S}$ be a family of subrings of R. The intersection $\cap\mathfrak{S}$ of all subrings of this family is also a subring in R.*

Proof. Let $S = \cap\mathfrak{S}$. Since $0_R \in U$ for all $U \in \mathfrak{S}$ it follows that $S \neq \emptyset$. Let $x, y \in S$. Then $x - y, xy \in U$, for all subrings $U \in \mathfrak{S}$ and therefore $x - y, xy$ belongs to the intersection of all such subrings. Thus, $x - y, xy \in S$ and Theorem 7.1.9 implies that S is a subring of R.

We note that a union of subrings is not necessarily a subring. For example, if n is a fixed positive integer then the subset

$$n\mathbb{Z} = \{nk \mid k \in \mathbb{Z}\}$$

satisfies both conditions **(SR 1)** and **(SR 2)**, and therefore is a subring. In particular, the subsets $2\mathbb{Z}$ and $3\mathbb{Z}$ are subrings, but $2\mathbb{Z} \cup 3\mathbb{Z}$ does not contain the integer $5 = 2 + 3$, and therefore is not a subring. However, there are some instances when unions of subrings are again subrings, as we see next. We recall that $\mathfrak{L}$ is a local family if whenever $H, K \in \mathfrak{L}$ there is a subring $L \in \mathfrak{L}$ such that $H, K \leq L$.

7.1.11. Corollary. *Let R be a ring and let $\mathfrak{L}$ be a local family of subrings of R. Then the union, $\cup\mathfrak{L}$, is a subring of R.*

Proof. Let $V = \cup\mathfrak{L}$ and let $x, y \in V$. There are subrings $H, K \in \mathfrak{L}$ such that $x \in H$, $y \in K$. We choose a subring $L \in \mathfrak{L}$ with the property that $H, K \leq L$. Then $x, y \in L$. Since L is a subring, $x - y, xy \in L$, by Theorem 7.1.9, and therefore $x - y, xy \in V$. Now we apply Theorem 7.1.9 to deduce that V is a subring.

There are a number of further special cases that we mention here.

7.1.12. Corollary. *Let R be a ring and let $\mathfrak{L}$ be a linearly ordered family of subrings of R. Then the union $\cup\mathfrak{L}$ of all subrings from this family is a subring of R.*

7.1.13. Corollary. *Let R be a ring and let*

$$H_1 \leq H_2 \leq \cdots \leq H_n \leq \cdots$$

be an ascending system of subrings of R. Then the union $\bigcup_{n\in N} H_n$ is a subring of R.

Let M be a subset of the ring R and let $\mathfrak{S}$ be the family of all subrings which contains the subset M. By Corollary 7.1.10, the intersection $r(M) = \cap\mathfrak{S}$ is a subring of R.

7.1.14. Definition. *The subring $r(M)$ is called the subring generated by the subset M, and the subset M is called a set of generators for $r(M)$. In particular, if $r(M) = R$, then we say that M generates the ring R.*

A ring R is called finitely generated if there exists a finite subset M such that $r(M) = R$. The following lemma is very easy to deduce.

7.1.15. Lemma. *Let R be a ring and let H be a subring of R. If $M \subseteq H$ then $r(M) \leq H$.*

If H is a subring containing the set M then Lemma 7.1.15 implies that H also contains $r(M)$. In this sense $r(M)$ is the minimal subring containing the subset M. Clearly, if M is a subring of a ring R then $r(M) = M$.

We now consider a very useful construction, the Cartesian (or direct) product of finitely many rings. Let $R_1, \dots, R_n$ be rings and let $D = R_1 \times \cdots \times R_n$ be the Cartesian product of the underlying sets $R_1, \dots, R_n$. We define an addition and multiplication on D by

$$(y_1, \dots, y_n) + (x_1, \dots, x_n) = (y_1 + x_1, \dots, y_n + x_n) \text{ and}$$
$$(y_1, \dots, y_n)(x_1, \dots, x_n) = (y_1x_1, \dots, y_nx_n),$$

where $y_j, x_j \in R_j$, for $1 \leq j \leq n$. Thus, we add and multiply componentwise.

Since y_j, x_j are elements of the ring R_j, their sum and product are defined in the ring R_j, where $1 \leq j \leq n$. Thus, the addition and multiplication of D are binary operations on D. The addition is associative since

$$\begin{aligned}
&\big((w_1, \dots, w_n) + (x_1, \dots, x_n)\big) + (y_1, \dots, y_n) \\
&= \big((w_1 + x_1) + y_1, \dots, (w_n + x_n) + y_n\big) \\
&= \big(w_1 + (x_1 + y_1), \dots, w_n + (x_n + y_n)\big) \\
&= (w_1, \dots, w_n) + \big((x_1, \dots, x_n) + (y_1, \dots, y_n)\big).
\end{aligned}$$

Using the same type of argument, we can prove that multiplication is also associative in D.

The addition is commutative since we have, using the commutativity of the addition in each of the component rings,

$$(w_1, \ldots, w_n) + (x_1, \ldots, x_n) = (w_1 + x_1, \ldots, w_n + x_n)$$
$$= (x_1 + w_1, \ldots, x_n + w_n) = (x_1, \ldots, x_n) + (w_1, \ldots, w_n).$$

There is a zero element, namely $(0_1, \ldots, 0_n)$, where 0_j is the zero element of the ring R_j, for $1 \le j \le n$. In addition, the negative of $(y_1, \ldots, y_n)$ is

$$-(y_1, \ldots, y_n) = (-y_1, \ldots, -y_n).$$

If R_j is a ring with identity and e_j is the identity element of R_j, for $1 \le j \le n$, then clearly $(e_1, \ldots, e_n)$ is the identity element of D.

Finally, we show that addition and multiplication are connected by the distributive laws. Indeed,

$$\begin{aligned}
&\big((y_1, \ldots, y_n) + (x_1, \ldots, x_n)\big)(z_1, \ldots, z_n)\\
&\quad = (y_1 + x_1, \ldots, y_n + x_n)(z_1, \ldots, z_n)\\
&\quad = \big((y_1 + x_1)z_1, \ldots, (y_n + x_n)z_n\big) = (y_1z_1 + x_1z_1, \ldots, y_nz_n + x_nz_n)\\
&\quad = (y_1z_1, \ldots, y_nz_n) + (x_1z_1, \ldots, x_nz_n)\\
&\quad = (y_1, \ldots, y_n)(z_1, \ldots, z_n) + (x_1, \ldots, x_n)(z_1, \ldots, z_n).
\end{aligned}$$

The other distributive law can be proved similarly. Thus, all the ring axioms are valid for D and D, together with the operations just defined, is called the Cartesian (or the direct) product of the rings $R_1, \ldots, R_n$. It is also possible to define the Cartesian product of infinitely many rings, by using similar (componentwise) laws of addition and multiplication. This Cartesian product is sometimes called the unrestricted direct product, since there is also a slightly different "restricted direct product." In the case of finitely many rings these different concepts (which we have only loosely discussed here) coincide. If the rings $R_1, \ldots, R_n$ are commutative, then it is very easy to see, using the type of methods used above, that their Cartesian product is also commutative. We note that $\mathbb{Z} \times \mathbb{Z}$ is a ring with multiplicative identity $(1, 1)$ that contains the subring $\mathbb{Z} \times \{0\}$ that has identity element $(1, 0)$, thus justifying our earlier claim that a ring and a subring can have different multiplicative identities.

We next consider the structure of $\mathbf{U}(D)$, the set of invertible elements of D, in the case when the rings R_i contain multiplicative identities. Let $(y_1, \ldots, y_n) \in \mathbf{U}(D)$. Then there exists $(z_1, \ldots, z_n)$ such that

$$\begin{aligned}
(y_1, \ldots, y_n)(z_1, \ldots, z_n) &= (y_1z_1, \ldots, y_nz_n)\\
&= (e_1, \ldots, e_n) = (z_1, \ldots, z_n)(y_1, \ldots, y_n).
\end{aligned}$$

Thus, $y_j z_j = e_j = z_j y_j$, so that $y_j \in \mathbf{U}(R_j)$ for each j, where $1 \leq j \leq n$. Conversely if $y_j \in \mathbf{U}(R_j)$ for every j, where $1 \leq j \leq n$ we can repeat the previous arguments in the reverse order, to see that $(y_1, \ldots, y_n) \in \mathbf{U}(D)$ and hence we obtain the equality

$$\mathbf{U}(R_1 \times \cdots \times R_n) = \mathbf{U}(R_1) \times \cdots \times U(R_n).$$

We next consider some important examples of rings.

Number rings

From the definitions it is clear that a field is a very special type of commutative ring. Thus, the set $\mathbb{R}$ of all real numbers is a commutative ring, since $\mathbb{R}$ is a field, as seen in Chapter 3. The set $\mathbb{Q}$ of all rational numbers is a subfield of $\mathbb{R}$, and the set $\mathbb{Z}$ of all integers is a unitary subring. The subring $\mathbb{Z}$ is finitely generated, generated by the number 1. We have already noted that, for every $n \geq 0$, the subset $n\mathbb{Z} = \{nk \mid k \in \mathbb{Z}\}$ is a subring of $\mathbb{Z}$. Conversely, let H be a subring of $\mathbb{Z}$. If $H = \{0\}$, then $H = 0_{\mathbb{Z}}$. Suppose now that H contains nonzero elements. Note that if $x \in H$ then $-x \in H$ and hence H must contain positive integers. Let n be the least positive element of H. Then $2n = n + n \in H$, $3n = 2n + n \in H$, and similarly, $kn \in H$ for each positive integer k. If $k < 0$ then $-k > 0$ and hence $(-k)n \in H$. Since H is a subring, $kn = -(-k)n \in H$ and from this it follows that $n\mathbb{Z} \leq H$. Next, let m be an arbitrary element of H. By Theorem 1.4.1, there are integers q, r such that $m = qn + r$ where $0 \leq r < n$. Since $n\mathbb{Z} \leq H$, we have $r = m - qn \in H$. Since n is the least positive element of H and, since $0 \leq r < n$, we deduce that $r = 0$. Thus, $m = qn \in n\mathbb{Z}$, which proves that $H = n\mathbb{Z}$. Thus, the subrings of $\mathbb{Z}$ are precisely of the form $n\mathbb{Z}$, where n is some fixed, but arbitrary, element of $\mathbb{Z}$.

Next, let p be a prime and let $\mathbb{Q}_p = \left\{\frac{m}{p^k} \mid m, k \in \mathbb{Z}\right\}$. A fraction of the type $\frac{m}{p^k}$ is called a p-adic fraction. The equation

$$\frac{m}{p^k} - \frac{r}{p^s} = \frac{mp^s - rp^k}{p^{k+s}}$$

shows that $\mathbb{Q}_p$ satisfies the condition **(SR 1)**. Since $\mathbb{Q}_p$ clearly satisfies **(SR 2)**, it is a unitary subring of $\mathbb{Q}$ by Theorem 7.1.9, the ring of p-adic fractions. These examples constitute just a small fraction of rings of numbers.

Rings of Matrices

In what follows, we use similar notation to that used in Chapter 2. Let R be an integral domain. We let $\mathbf{M}_n(R)$ denote the set of all square matrices of dimension n whose entries belong to the ring R. If $A = [a_{ij}]$ and $B = [b_{ij}]$ are two matrices from $\mathbf{M}_n(R)$, then the sum $A + B$ is a matrix $C = [c_{ij}]$, whose elements are defined by $c_{ij} = a_{ij} + b_{ij}$ for each pair of indices i, j, where $1 \leq i, j \leq n$.

The product AB of these matrices is the matrix $D = [d_{ij}]$, whose entries are defined by

$$d_{ij} = a_{i1}b_{1j} + a_{i2}b_{2j} + \cdots + a_{in}b_{nj} = \sum_{1 \le k \le n} a_{ik}b_{kj}$$

for each pair of indices i, j, where $1 \le i, j \le n$.

Thus, we keep the rules of addition and multiplication introduced earlier for matrices with numerical entries.

The same arguments that we used in Section 2.1 show that $\mathbf{M}_n(R)$ is a ring, using the operations that we introduced above. Note that as with numerical matrices, this ring is also noncommutative.

We define the determinant of a matrix in the same way as we did for numerical matrices. Thus, if $A = [a_{ij}] \in \mathbf{M}_n(R)$ then

$$\mathbf{det}(A) = \sum_{\pi \in S_n} \mathbf{sign}\,\pi \; a_{1,\pi(1)}a_{2,\pi(2)} \ldots a_{n,\pi(n)}.$$

It is easy to see that all properties of determinants that we proved in Sections 2.3, 2.4 and Theorem 2.5.1 are valid for an integral domain R. Let $A \in \mathbf{U}(\mathbf{M}_n(R))$ and let $a = \mathbf{det}(A)$. Since

$$e = \mathbf{det}(I) = \mathbf{det}(AA^{-1}) = \mathbf{det}(A)\mathbf{det}(A^{-1}) = a\mathbf{det}(A^{-1}),$$

it follows that $a \in \mathbf{U}(R)$. Conversely, let A be a matrix such that $\mathbf{det}(A) \in \mathbf{U}(R)$ and let $B = [b_{ij}] \in \mathbf{M}_n(R)$, be the matrix whose entries are defined by

$$b_{ij} = A_{ji}/\mathbf{det}(A) \text{ for } 1 \le i, j \le n,$$

where A_{ji} is the cofactor corresponding to a_{ij} It is easy to see that $AB = BA = I$, so B is the inverse of the matrix A. Therefore,

$$\mathbf{U}(\mathbf{M}_n(R)) = \{A \in \mathbf{M}_n(R) \mid \mathbf{det}(A) \in \mathbf{U}(R)\}.$$

As with numerical matrices, we call the group $\mathbf{U}(\mathbf{M}_n(R))$ the general linear group of degree n over the ring R. In particular, if R is a field, then $\mathbf{U}(\mathbf{M}_n(R))$ is again the set of all nonsingular matrices.

We let E_{km} denote the matrix whose (k, m) entry is the multiplicative identity of R and in which all other entries are zero. As with numerical matrices we can easily prove that

$$E_{km}E_{rs} = \begin{cases} E_{ks}, & \text{if } m = r, \\ O, & \text{if } m \ne r. \end{cases}$$

For example, $E_{12}E_{11} = O$, the zero matrix, and $E_{11}E_{12} = E_{12}$, so it follows that the ring $\mathbf{M}_n(R)$ is noncommutative if $n \ge 2$. Furthermore, this example shows that $\mathbf{M}_n(R)$ has zero-divisors.

We let $T_n^{\circ}(R)$ denote the subset of $\mathbf{M}_n(R)$, that consists of all upper triangular matrices. It is clear that the difference of two upper triangular matrices is also upper triangular. Furthermore, let $A, B \in T_n^{\circ}(R)$, where $A = [a_{ij}]$, and $B = [b_{ij}]$. Set $C = AB = [c_{ij}]$ and suppose that $i > j$. Then we have

$$c_{ij} = a_{i1}b_{1j} + a_{i2}b_{2j} + \cdots + a_{i,i-1}b_{i-1,j} + a_{ii}b_{ij} + a_{i,i+1}b_{i+1,j} + \cdots + a_{in}b_{nj}.$$

Since

$$a_{i1} = a_{i2} = \cdots = a_{i,i-1} = 0 \text{ and } b_{ij} = b_{i+1,j} = \cdots = b_{nj} = 0,$$

it follows that $c_{ij} = 0$ for $i > j$ and hence $AB \in T_n^{\circ}(R)$. Observe that

$$\begin{aligned} c_{ii} &= a_{i1}b_{1i} + a_{i2}b_{2i} + \cdots + a_{i,i-1}b_{i-1,i} + a_{ii}b_{ii} + a_{i,i+1}b_{i+1,i} + \cdots + a_{in}b_{ni} \\ &= a_{ii}b_{ii}. \end{aligned}$$

Using Theorem 7.1.9 we see that $T_n^{\circ}(R)$ is a subring of $\mathbf{M}_n(R)$.

An upper triangular matrix is called zero triangular if all its entries on the main diagonal are zeros.

The set of all zero triangular matrices of dimension n with entries belonging to the ring R is denoted by $\mathbf{NT}_n(R)$. It is easy to see that the subset $\mathbf{NT}_n(R)$ satisfies the conditions **(SR 1)** and **(SR 2)**, so that Theorem 7.1.9 implies that $\mathbf{NT}_n(R)$ is a subring of $\mathbf{M}_n(R)$.

We let $D_n^{\circ}(R)$ denote the subset of all diagonal matrices of dimension n with entries belonging to the ring R and set

$$RI = \{\lambda I \mid \lambda \in R\}.$$

Thus, RI is the subset of $\mathbf{M}_n(R)$ consisting of the set of all scalar matrices. Using Theorem 7.1.9 it is possible to show that $D_n^{\circ}(R)$ and RI are unitary subrings of $\mathbf{M}_n(R)$. Of course $T_n^{\circ}(R)$ is also unitary. Finally, let

$$RE_{ii} = \{\lambda E_{ii} \mid \lambda \in R\}.$$

We have

$$\lambda E_{ii} - \mu E_{ii} = (\lambda - \mu)E_{ii} \text{ and } \lambda E_{ii} \cdot \mu E_{ii} = \lambda\mu E_{ii}.$$

Theorem 7.1.9 implies that RE_{ii} is a subring of $\mathbf{M}_n(R)$. Furthermore,

$$\lambda E_{ii} \cdot eE_{ii} = eE_{ii} \cdot \lambda E_{ii} = \lambda E_{ii}.$$

This equation shows that $eE_{ii} = E_{ii}$ is the multiplicative identity element of the subring RE_{ii}, so RE_{ii} is nonunitary, but has its own multiplicative identity element. Furthermore, if $\lambda \in U(R)$ then λE_{ii} has an inverse in RE_{ii}, namely $\lambda^{-1}E_{ii}$, even though λE_{ii} is not invertible in $\mathbf{M}_n(R)$.

Rings of Functions

In what follows we use the fact that two functions are equal precisely when they both take the same value at every point in the domain of the function. Let R be an arbitrary ring and let M be an arbitrary set. As usual we let R^M denote the set of all functions $f : M \longrightarrow R$ and define a sum and product on R^M by the rules

$$(f+g)(a) = f(a) + g(a) \text{ and } (fg)(a) = f(a)g(a),$$

for every element $a \in M$. We note that the product here is *not* defined using composition of mappings.

If $f, g \in R^M$, then

$$(f+g)(a) = f(a) + g(a) = g(a) + f(a) = (g+f)(a),$$

for each element $a \in M$. It follows that $f + g = g + f$, so that addition is commutative. Similarly, for $f, g, h \in R^M$ we have, using the associative law of addition in R,

$$\begin{aligned}\big(f + (g+h)\big)(a) &= f(a) + (g+h)(a) = f(a) + \big(g(a) + h(a)\big)\\ &= \big(f(a) + g(a)\big) + h(a) = (f+g)(a) + h(a)\\ &= \big((f+g) + h\big)(a),\end{aligned}$$

for an arbitrary element $a \in M$. Thus, $f + (g+h) = (f+g) + h$ and the associative law of addition holds in R^M

Let ϑ be the mapping defined by $\vartheta(a) = 0_R$ for each $a \in M$. Then

$$(f+\vartheta)(a) = f(a) + \vartheta(a) = f(a) + 0_R = f(a).$$

Thus, $f + \vartheta = f$ for every $f \in R^M$ and hence ϑ is the zero element of R^M. In addition, we define the mapping $-f$ by $(-f)(a) = -f(a)$, for all $a \in M$. Then

$$\big(f + (-f)\big)(a) = f(a) + (-f)(a) = f(a) - f(a) = 0_R,$$

for all $a \in M$ so that $f + (-f) = \vartheta$.

Next, let $f, g, h \in R^M$ and let $a \in M$. Then, using the distributive law of R,

$$\begin{aligned}\big(f(g+h)\big)(a) &= f(a)\big((g+h)\big)(a) = f(a)\big(g(a)+h(a)\big) = f(a)g(a)+f(a)h(a)\\ &= (fg)(a)+(fh)(a) = (fg+fh)(a),\end{aligned}$$

so $f(g+h) = fg + fh$. Similarly, $(f+g)h = fh + gh$, so the distributive laws hold in R^M.

Multiplication of functions is associative since

$$\begin{aligned}\big(f(gh)\big)(a) &= f(a)(gh)(a) = f(a)\big(g(a)h(a)\big)\\ &= \big(f(a)g(a)\big)h(a) = (fg)(a)h(a) = \big((fg)h\big)(a),\end{aligned}$$

using the associative law of R.

The function I defined by the rule $I(a) = e$, for all $a \in M$ serves as a multiplicative identity since

$$(fI)(a) = f(a)I(a) = f(a)e = f(a), \text{ for all } a \in M.$$

Thus, $fI = f$ and, similarly, $If = f$, which proves our claim.

It follows that R^M, together with the operations of addition and multiplication as defined above, is a ring. Furthermore, if R is commutative then the ring R^M is also commutative since

$$(fg)(a) = f(a)g(a) = g(a)f(a) = (gf)(a).$$

Next we consider $\mathbf{U}(R^M)$. This set consists of those functions f such that $f(a) \in \mathbf{U}(R)$ for all $a \in M$. In this case the inverse g of f is the function defined by $g(a) = \big(f(a)\big)^{-1}$; to avoid confusion with certain other notation, we so not denote the inverse of f here by f^{-1}.

The ring R^M has zero-divisors. To see this, let L be a proper nonempty subset of M and define functions f, g by the rules:

$$f(a) = \begin{cases} e, & \text{if } a \in L \\ 0_R, & \text{if } a \notin L, \end{cases} \quad \text{and } g(a) = \begin{cases} 0_R, & \text{if } a \in L \\ e, & \text{if } a \notin L. \end{cases}$$

The functions f, g are nonzero, but $f(a)g(a) = 0_R$ for every $a \in M$ and hence $fg = \vartheta$.

In calculus courses we usually deal with the situation when the set M is either $\mathbb{R}$ or an interval $[a, b] \subseteq \mathbb{R}$ and $R = \mathbb{R}$ is the field of real numbers. Thus, $R^M = \mathbb{R}^{\mathbb{R}}$ (respectively $R^M = \mathbb{R}^{[a,b]}$) is the ring of all functions defined on $\mathbb{R}$ (respectively on $[a, b]$) whose range is a subset of the field of real numbers. The ring $\mathbb{R}^{\mathbb{R}}$ has numerous well-known unitary subrings, including the subring of all continuous function, the subring of all differentiable functions, the subring of all twice differentiable functions and so on.

Boolean Rings

Let M be a set and, as usual, let $A = \mathfrak{B}(M)$ denote its Boolean. Thus, A is the set of all subsets of M. If $a, b \in A$, then put

$$a + b = (a \cup b)\backslash(a \cap b) \text{ and } ab = a \cap b.$$

There are some easy, but at times tedious, computations required to verify that A is a ring. The zero element of A is $\emptyset$ and the multiplicative identity of A is M. Thus, M is the only invertible element of this ring. Moreover, for each element a we have $a^2 = a$, $2a = \emptyset = 0_A$. Rings with these properties are called Boolean rings.

The Center of a Ring

If R is a ring, then we let $\zeta(R)$ denote the set of all elements of R that commute with every element of R. Thus,

$$\zeta(R) = \{x \in R | xy = yx, \text{ for all } y \in R\}.$$

Clearly $0_R \in \zeta(R)$. If $a, b \in \zeta(R)$ and $x \in R$ then

$$(a - b)x = ax - bx = xa - xb = x(a - b) \text{ and}$$
$$(ab)x = a(bx) = a(xb) = (ax)b = (xa)b = x(ab).$$

These equations show that $a - b, ab \in \zeta(R)$ and, by Theorem 7.1.9, $\zeta(R)$ is therefore a subring of R. The subring $\zeta(R)$ is called *the center of the ring* R.

If R is a ring with identity e, let $\mathbb{Z}e = \{ne \mid n \in \mathbb{Z}\}$, where $ne = \underbrace{e + e + \cdots + e}_{n}$, if $n > 0$, $0e = 0_R$ and $ne = -(-ne)$ if $n < 0$. Then, for all $k, n \in \mathbb{Z}$ we have

$$ne - ke = (n - k)e \text{ and } (ne)(ke) = (nk)e.$$

By Theorem 7.1.9, $\mathbb{Z}e$ is therefore a subring of R.

Since

$$(ne)a = n(ea) = na = a(ne),$$

it follows that $ne \in \zeta(R)$ and hence $\mathbb{Z}e \leq \zeta(R)$.

Nilpotent Elements in Rings

Let R be a ring. An element a is called *nilpotent* if $a^n = 0_R$, for some $n \in \mathbb{N}$. If R is a commutative ring, then the subset of all nilpotent elements is a subring. To see this, we note that the binomial theorem of algebra still holds, namely, for all $a, b \in R$, it is the case that, for all natural numbers m, $(a + b)^m = \sum_{j=0}^{m} \binom{m}{j} a^j b^{m-j}$. This can be proved by induction on m. Then, if $a^n = 0_R, b^t = 0_R$ for some $n, t \in \mathbb{N}$ we consider $(a - b)^{n+t}$. This element is a linear combination of the elements $a^k b^m$ where $k + m = n + t$. Hence if $k \geq n$, then $a^k b^m = 0_R$. If $k < n$, then $m > t$ and again $a^k b^m = 0_R$. Hence,

$$(a - b)^{n+t} = 0_R.$$

Further,

$$(ab)^{n+t} = a^{n+t}b^{n+t} = 0_R.$$

The Ring of Endomorphisms of an Abelian Group

Let A be an abelian group under addition. A homomorphism f from A to itself is called an endomorphism of A. Thus, by Definition 3.1.20, this means that $f(x + y) = f(x) + f(y)$ for each element $x \in A$. We let $\mathbf{End}(A)$ denote the set of all endomorphisms of A and if $f, g \in \mathbf{End}(A)$ we define addition of endomorphisms by $(f + g)(a) = f(a) + g(a)$ for every element $a \in A$.

We have, using the facts that f, g are endomorphisms and that A is abelian:

$$\begin{aligned}(f + g)(a + b) &= f(a + b) + g(a + b) = f(a) + f(b) + g(a) + g(b)\\ &= f(a) + g(a) + f(b) + g(b) = (f + g)(a) + (f + g)(b).\end{aligned}$$

It follows that $f + g$ is an endomorphism of A. Hence, the mapping $(f, g) \longmapsto f + g$, where $f, g \in \mathbf{End}(A)$ is a binary operation on the set $\mathbf{End}(A)$. The set $\mathbf{End}(A)$ is an abelian group under the operation defined above since

$$(f + g)(a) = f(a) + g(a) = g(a) + f(a) = (g + f)(a).$$

Also,

$$\begin{aligned}\big(f + (g + h)\big)(a) &= f(a) + (g + h)(a) = f(a) + \big(g(a) + h(a)\big),\\ \big((f + g) + h\big)(a) &= \big(f(a) + g(a)\big) + h(a) = \big(f(a) + g(a)\big) + h(a),\end{aligned}$$

for each element $a \in A$. It follows that

$$f + g = g + f \text{ and } f + (g + h) = (f + g) + h,$$

for arbitrary $f, g, h \in \mathbf{End}(A)$.

As above, we define a mapping $\vartheta : A \longrightarrow A$ by $\vartheta(a) = 0_A$ for each element $a \in A$. Then

$$(f + \vartheta)(a) = f(a) + \vartheta(a) = f(a), \text{ so } f + \vartheta = f,$$

for arbitrary $f \in \mathbf{End}(A)$. This shows that ϑ is the zero element of $\mathbf{End}(A)$.

Finally, put $(-f)(a) = -f(a)$ for each $a \in A$. Clearly $f + (-f) = \vartheta$ and all the axioms for an abelian group hold.

Next for $f, g \in \mathbf{End}(A)$ and $a, b \in A$ we have

$$\begin{aligned}(f \circ g)(a + b) &= f\big(g(a + b)\big) = f\big(g(a) + g(b)\big) = f\big(g(a)\big) + f\big(g(b)\big)\\ &= (f \circ g)(a) + (f \circ g)(b).\end{aligned}$$

This equation shows that $f \circ g$ is an endomorphism of A and hence the mapping

$$(f, g) \longmapsto f \circ g, \text{ where } f, g \in \mathbf{End}(A)$$

is a binary operation on A.

Let $f, g, h \in \mathbf{End}(A)$ and $a \in A$. We have

$$\begin{aligned}(f \circ (g+h))(a) &= f((g+h)(a)) = f(g(a)+h(a)) = f(g(a)) + f(h(a)) \\ &= (f \circ g)(a) + (f \circ h)(a) = (f \circ g + f \circ h)(a).\end{aligned}$$

It follows that

$$f \circ (g+h) = f \circ g + f \circ h.$$

Similarly we can prove that $(f+g) \circ h = f \circ h + g \circ h$. As we know from Theorem 1.3.2, multiplication of mappings is associative. The mapping ε_A, defined by the rule $\varepsilon_A(a) = a$ for all $a \in A$, is the multiplicative identity under multiplication and hence $\mathbf{End}(A)$ is a ring called the ring of endomorphisms of the abelian group A.

EXERCISE SET 7.1

Justify your work by writing a proof or by giving a counterexample.

7.1.1. On the set $R = \mathbb{Z} \times \mathbb{Z}$ we define operations of addition and multiplication by $(a, b) + (a_1, b_1) = (a + a_1, b + b_1)$, $(a, b)(a_1, b_1) = (0, 0)$. Is R a ring under these operations? If yes, does R have an identity element or zero divisors?

7.1.2. On the set $R = \mathbb{Z} \times \mathbb{Z}$ we define operations of addition and multiplication by $(a, b) + (a_1, b_1) = (a + a_1, b + b_1)$, $(a, b)(a_1, b_1) = (a + a_1 + b + b_1, 0)$. Is R a ring under these operations? If yes, does R have an identity element or zero-divisors?

7.1.3. On the set $R = \mathbb{Z} \times \mathbb{Z}$ we define operations of addition and multiplication by $(a, b) + (a_1, b_1) = (aa_1, bb_1)$, $(a, b)(a_1, b_1) = (a + a_1, b + b_1)$. Is R a ring under these operations? If yes, does R have an identity element or zero-divisors?

7.1.4. On the set $R = \mathbb{Z} \times \mathbb{Z}$ we define operations of addition and multiplication by $(a, b) + (a_1, b_1) = (a + a_1, b + b_1)$, $(a, b)(a_1, b_1) = (aa_1 + 3bb_1, ab_1 + ba_1)$. Is R a ring under these operations? If yes, does R have an identity element or zero-divisors?

7.1.5. On the set $R = \mathbb{Z} \times \mathbb{Z}$ we define operations of addition and multiplication by $(a, b) + (a_1, b_1) = (a + a_1, b + b_1)$, $(a, b)(a_1, b_1) = (aa_1 - 3bb_1, ab_1 + ba_1)$. Is R a ring under these operations? If yes, does R have an identity element or zero-divisors?

7.1.6. On the set $R = \mathbb{Z} \times \mathbb{Z}$ we define operations of addition and multiplication by $(a, b) + (a_1, b_1) = (a + a_1, b + b_1)$, $(a, b)(a_1, b_1) = (aa_1, bb_1)$. Prove that R is a ring with identity and find its zero-divisors.

7.1.7. Let R be a ring and let $a, b \in R$. Prove that the equation $a + x = b$ has a unique solution.

7.1.8. Let R be a ring, let $a, b \in \mathbf{U}(R)$ and let $a + b \neq 0_R$. Does the element $a + b$ have an inverse?.

7.1.9. Let A be an abelian group with an additive operation. If we define multiplication by one of the rules: (i) $ab = a - b$; (ii) $ab = b$; and (iii) $ab = 2a - b$, then is A a ring?

7.1.10. Let K be the ring of all real functions and let $f \in K$. Suppose that f is not a zero-divisor. Prove that $f \in \mathbf{U}(K)$.

7.1.11. Let $A \in \mathbf{M}_2(\mathbb{R})$ and suppose that A is not a zero-divisor. Prove that the matrix A is nonsingular.

7.1.12. Let R be a finite ring and let $a \in R$. Prove that either $a \in \mathbf{U}(R)$ or a is a zero-divisor.

7.1.13. Let R be a ring and let H_1, H_2 be subrings of R. Find necessary and sufficient condition for $H_1 \cup H_2$ to be a subring.

7.1.14. Find the subring of $\mathbb{C}$, generated by the subset $\mathbb{Z} \cup \{\sqrt{-5}\}$. Find all invertible elements of this subring.

7.1.15. Find the subring of $\mathbb{R}$, generated by the subset $\mathbb{Z} \cup \{\sqrt{7}\}$. Find all invertible elements of this subring.

7.1.16. Find the subring of $\mathbb{Q}$, generated by the subset $\mathbb{Z} \cup \{\frac{1}{5}\}$. Find all invertible elements of this subring.

7.1.17. Let L be the subset of $\mathbf{M}_2(\mathbb{Q})$ consisting of all matrices of the form

$$\begin{pmatrix} a & b \\ -b & a-b \end{pmatrix}.$$

Prove that L is a subring of $\mathbf{M}_2(\mathbb{Q})$. Is L commutative?

7.1.18. Give examples of zero-divisors in the ring $\mathbf{M}_2(\mathbb{Z})$.

7.1.19. Let R be a commutative ring and suppose that R has no zero-divisors. Prove that if R is finite, then R is a field.

7.2 EQUIVALENCE RELATIONS

The idea of a correspondence and the particular case that is the concept of a relation are some of the most commonly used ideas in mathematics. One of the most useful concepts here, that of an equivalence relation, is discussed in this section.

We recall from Chapter 1 that a binary relation on a set A is a subset of the Cartesian product $A \times A$. If Φ is a relation on A and if $\alpha = (x, y) \in \Phi$, then we say that elements x and y (in this given order) correspond to each other via Φ. Instead of the notation $(x, y) \in \Phi$ it is often usual to write $x \, \Phi \, y$ and we shall adhere to this convention. The notation $x \, \Phi \, y$ is meant to convey the idea that x is related to y via Φ. This form of notation is called infix.

At its most basic level the study of binary relations is actually the study of subsets of a Cartesian product. In particular, this means that it is possible to include one binary relation in another, or we may take the union or intersection of binary relations and so on.

Here are the most important properties of binary relations.

7.2.1. Definition. *Let A be a set with a binary relation Φ.*

(i) Φ *is called reflexive if* $(a, a) \in \Phi$ *(or* $a \, \Phi \, a$*), for each* $a \in A$*;*

(ii) Φ *is called transitive if, whenever* $a, b, c \in A$ *and* $(a, b), (b, c) \in \Phi$*, then* $(a, c) \in \Phi$ *(or, alternatively,* $a \, \Phi \, b$ *and* $b \, \Phi \, c$ *imply that* $a \, \Phi \, c$*);*

(iii) Φ *is called symmetric if, whenever* $a, b \in A$ *and* $(a, b) \in \Phi$*, then* $(b, a) \in \Phi$ *(or, alternatively,* $a \, \Phi \, b$ *implies* $b \, \Phi \, a$*);*

(iv) Φ *is called antisymmetric if, whenever* $a, b \in A$ *and* $(a, b), (b, a) \in \Phi$ *then* $a = b$ *(or, alternatively,* $a \, \Phi \, b$ *and* $b \, \Phi \, a$ *imply* $a = b$*).*

Since a relation is a certain type of set, we can use the same notation for relations that we use for sets. We mention here some notation that is often used for relations on finite sets. If A is a finite set we make a pair of perpendicular axes and label the axes with points representing the elements of A. If $a, b \in A$ and $(a, b) \in \Phi$, then we can plot the point (a, b), as we do in the usual regular coordinate system, by finding the point on the horizontal axis labeled a and the point on the vertical axis labelled b and putting a mark (cross or circle) at the place where the lines drawn from these points would intersect.

Another well-known method of representing a relation on a finite set is based on the use of oriented graphs. Here we represent the elements of A by the vertices of a directed graph, and if $a, b \in A$ then we represent the fact that $a \, \Phi \, b$ by drawing an arrow from the vertex a to the vertex b.

It is also useful to represent a relation on a finite set by means of square matrices. In this case the matrix will be an $n \times n$ matrix, where $A = \{a_1, \dots, a_n\}$ has n elements. Let Φ denote a binary relation on A and let the matrix $M(\Phi) = [\alpha_{jk}]$ be the matrix corresponding to this binary relation, defined as follows. Let

$$\alpha_{jk} = \begin{cases} 1, & \text{if } (a_j, a_k) \in \Phi \\ 0, & \text{if } (a_j, a_k) \notin \Phi. \end{cases}$$

For example, let $A = \{1, 2, 3, 4, 5\}$ and let

$$\Phi = \{(1, 1), (1, 3), (2, 2), (2, 4), (4, 1), (5, 3)\}$$

Then we can describe the relation Φ with the help of the matrix

$$\begin{pmatrix} 1 & 0 & 1 & 0 & 0 \\ 0 & 1 & 0 & 1 & 0 \\ 0 & 0 & 0 & 0 & 0 \\ 1 & 0 & 0 & 0 & 0 \\ 0 & 0 & 1 & 0 & 0 \end{pmatrix}.$$

Such a matrix is often called a Boolean matrix to underline the fact that all its elements are 0 and 1.

If a relation is represented by a graph and if the relation is reflexive, then there will be a loop at each vertex of the graph, from the vertex to itself. The corresponding Boolean matrix of a reflexive relation will have a main diagonal whose entries are all 1.

There are many examples of reflexive relations and here we give only a few: If A is the set of all straight lines in the plane then the relation of "being parallel" is certainly reflexive; the relation "looks alike" on a certain set of people is clearly reflexive since everyone looks alike themselves; the relation of "having the same gender" on a set of animals is certainly reflexive, and so on.

In the graph of a symmetric relation, for every arc from the vertex a to the vertex b, there is a corresponding arc from b to a. Thus, for a symmetric relation, we can use nondirected graphs (that is, graphs without arrows) to illustrate the relation. The matrix of a symmetric relation clearly will have a line of symmetry along the main diagonal and the matrix itself will be an example of a symmetric matrix. Thus, the matrix $A = [a_{ij}]$ represents a symmetric relation on a finite set precisely when $a_{ij} = a_{ji}$, for all i, j such that $1 \leq i, j \leq n$. As an example we note that the relation "x is the brother of y" is a symmetric relation on the set of all male humans, but is not symmetric on the set of all humans. It is important to be precise as to the set involved, as these examples make clear.

Next we note some examples of transitive relations: the relation "to be divisible by" on the set of whole numbers; the relation "to be greater than" on the set of real numbers; the relation "to be older than" on a set of people; the relation "to have the same color as" on the set of toys, and so on.

Equivalence relations are closely connected to the idea of a partition of a set, which we now describe.

7.2.2. Definition. *A family $\mathfrak{S}$ of subsets of a set A is called a covering if $A = \cup\mathfrak{S}$ (thus, for each $x \in A$ there exists $S \in \mathfrak{S}$ such that $x \in S$). A covering $\mathfrak{S}$ is called a partition of the set A, if $X \cap Y = \emptyset$, whenever $X, Y \in \mathfrak{S}$ and $X \neq Y$; thus, all pairs of distinct subsets of the partition have empty intersection.*

Let $\mathfrak{S}$ be a partition of the set A and define a binary relation $\Gamma(\mathfrak{S})$ on A by the rule that $(x, y) \in \Gamma(\mathfrak{S})$ if and only if the elements x and y belong to the same set S from the family $\mathfrak{S}$. The relation $\Gamma(\mathfrak{S})$ has various properties which we now exhibit.

Since $A = \cup\mathfrak{S}$ then, for each element $x \in A$, there exists a subset $S \in \mathfrak{S}$ such that $x \in S$. Thus, $(x, x) \in \Gamma(\mathfrak{S})$ and hence the relation $\Gamma(\mathfrak{S})$ is reflexive. It is clear that the relation $\Gamma(\mathfrak{S})$ is symmetric. Finally, let $(x, y), (y, z) \in \Gamma(\mathfrak{S})$. It follows that there exist subsets $S, R \in \mathfrak{S}$ such that $x, y \in S$ and $y, z \in R$. In particular, $y \in S \cap R$ and using the definition of a partition, we see that $S = R$. Hence the elements x, z belongs to S which is an element of the partition $\mathfrak{S}$. Thus, $(x, z) \in \Gamma(\mathfrak{S})$ so the relation $\Gamma(\mathfrak{S})$ is transitive.

7.2.3. Definition. *A binary relation Φ on a set A is called an equivalence relation or an equivalence if it is reflexive, symmetric and transitive.*

We give some examples next. First we say that two polygons are equivalent if they have the same number of vertices. Thus, for example, under this relation all triangles are equivalent, and it is easy to see that this relation is an equivalence relation. The family of all triangles can itself be partitioned into the subsets of acute, right angled, and obtuse triangles and this partition helps to define an equivalence relation on the set of all triangles; the relation "the sides of the triangle are all equal" is also an equivalence relation on the set of all triangles, as is the relation "the figure A is similar to the figure B" on the set of all geometric figures.

Every equivalence relation on a set is very closely connected to a partition of that set into classes which we call equivalence classes. One main reason for studying equivalence relations is that such relations allow us to construct new mathematical objects quite rigorously. For example, the relation of colinearity of rays is a partition of the plane or space into classes of colinear rays. Each of these classes is called a direction or a path. In this way, we can transform the intuitive idea of direction into a rigorously defined concept. In a similar way given a collection of figures we can define a relation on this set of figures by saying that figure A is related to figure B if and only if A has the same shape as B. Although the concept of an equivalence relation is quite deep there are many elementary examples. For example, one common type of exercise that young children undertake is to classify a set of toys by their color. This classification involves setting up a partition of the toys into different colors, which in turn can be obtained by the equivalence relation that toy A is related to toy B if and only if A has the same color as B.

We now develop a rigorous relationship between partitions and equivalence relations.

7.2.4. Definition. *Let Φ be an equivalence relation on the set A and let $x \in A$. The subset $[x]_\Phi = \{y \in A \mid (x, y) \in \Phi\}$ is called the equivalence class of x.*

It is important to note that each equivalence class is uniquely defined by each of its elements. Indeed, let $y \in [x]_\Phi$ so that $(x, y) \in \Phi$. If $z \in [y]_\Phi$, then $(y, z) \in \Phi$

also. Since the equivalence relation is transitive it follows that $(x, z) \in \Phi$ also and hence $z \in [x]_\Phi$. Thus, $[y]_\Phi \subseteq [x]_\Phi$. Because equivalence relations are symmetric we also have $[x]_\Phi \subseteq [y]_\Phi$ and hence $[x]_\Phi = [y]_\Phi$.

Since $(x, x) \in \Phi$, it follows that $x \in [x]_\Phi$ and hence the family of all equivalence classes forms a covering set of A. Next we consider the intersection, $[x]_\Phi \cap [y]_\Phi$, of two equivalence classes and suppose that this intersection is not empty. Let $z \in [x]_\Phi \cap [y]_\Phi$. Then, as we noted above, $[z]_\Phi = [y]_\Phi$ and $[z]_\Phi = [x]_\Phi$ from which it follows that $[x]_\Phi = [y]_\Phi$. Therefore every pair of distinct equivalence classes has empty intersection and we deduce that the family of all equivalence classes is a partition, $\mathbf{P}(\Phi)$, of the set A.

7.2.5. Theorem. *Let A be a set. Then the mapping $\rho : \Phi \longmapsto \mathbf{P}(\Phi)$ is a bijection from the set of all equivalence relations defined on A to the set of all partitions of A.*

Proof. We have just observed that every equivalence relation defined on A does indeed give rise to a partition of A, via the equivalence classes of the relation. To show that ρ is injective let Φ, Θ be two distinct equivalence relations on A. Then we may assume that there exists a pair (x, y), with $x, y \in A$, such that $(x, y) \in \Phi \setminus \Theta$. Since $(x, y) \in \Phi$, $y \in [x]_\Phi$ and, since $(x, y) \notin \Theta$, $y \notin [x]_\Theta$. Assume, for a contradiction, that $\mathbf{P}(\Phi) = \mathbf{P}(\Theta)$. Then there exists an element $z \in A$ such that $[z]_\Theta = [x]_\Phi$. In particular, $x \in [z]_\Theta$. Since an equivalence class is represented by each of its elements, we have $[z]_\Theta = [x]_\Theta$. Hence $[x]_\Theta = [x]_\Phi$. However, $y \notin [x]_\Theta$ so $y \notin [x]_\Phi$ and we obtain the desired contradiction. This contradiction shows that $\mathbf{P}(\Phi) \neq \mathbf{P}(\Theta)$ and hence the mapping ρ is injective.

We next prove that ρ is surjective and it will then follow that ρ is bijective. To this end, let $\mathfrak{S}$ be an arbitrary partition of A. As seen above, each partition $\mathfrak{S}$ gives rise to an equivalence relation $\Phi(\mathfrak{S})$. If $x \in A$ then $x \in S$ for some subset S of the partition $\mathfrak{S}$ and the definition of $\Phi(\mathfrak{S})$ then implies that S is a subset of the equivalence class of x. Conversely, let $y \in A$ be an element of the equivalence class of x. Thus, $(x, y) \in \Phi(\mathfrak{S})$ which means that there is a subset Q of $\mathfrak{S}$ such that $x, y \in Q$. In this case $x \in S \cap Q$ and, since distinct subsets of a partition are disjoint, we see that $Q = S$. Hence $y \in S$. Thus, the equivalence class of x in $\Phi(\mathfrak{S})$ coincides with an element of the partition of $\mathfrak{S}$. This shows that $\mathbf{P}(\Phi) = \mathfrak{S}$ and the result follows.

We now consider some further examples of equivalence relations.

If A is an arbitrary set then there are two extreme cases, namely, the case when $\Phi = A \times A$, which is the largest equivalence relation on the set A and the case when $\Phi = \{(x, x) \mid x \in A\}$ (the diagonal of the Cartesian product $A \times A$), which is the smallest equivalence relation on the set A. All other equivalence relations on A are situated between these two extreme cases.

Other examples of equivalence relations, which can easily be checked, are as follows:

(i) the relation of "being parallel" on the set of all straight lines of a plane;

(ii) the relation of similarity on the set of all geometrical figures;
(iii) the relation "to be equivalent equations" on the set of equations;
(iv) the relation "to belong to the same species" on the set of animals;
(v) the relation "to be relatives" on the set of people;
(vi) the relation "to be the same height" on the set of people;
(vii) the relation "to live in the same city" on the set of people;
(viii) the relation "has the same birthday as" on the set of all people;
(ix) the relation "is similar to" or " is congruent to" on the set of all triangles;
(x) the relation "has the same image" on the elements of the domain of a fixed function.

We often use the symbols $\approxeq, \equiv, \approx, \sim$ and others to help denote equivalence in an equivalence relation. Thus, $x \sim y$ denotes that x is equivalent to y.

Here are some more mathematical examples.

(i) Let $a > 0$ and in the first quadrant $P = \{(x, y) \mid x > 0, y > 0\}$ consider the set of all hyperbolas $G_a = \{(x, y) \mid xy = a\}$. As a is allowed to vary, the system $\{G_a \mid a > 0\}$ gives a partition of the set P and hence defines an equivalence relation on P, using Theorem 7.2.5.

(ii) Let M be the set of all sequences $\mathbf{s} = (x_n)_{n \in \mathbb{N}}$ of rational numbers. Consider the relation Φ on M defined by the rule: $(\mathbf{s}, \mathbf{r}) \in \Phi$ if and only if

$$\lim_{n \longrightarrow \infty} (x_n - y_n) = 0.$$

Here $\mathbf{r} = (y_n)_{n \in \mathbb{N}}$. It easy to see that Φ is an equivalence relation.

(iii) Let $M = [0, 1]$. Define a relation P on M by $(x, y) \in P$ if and only if $x - y$ is a rational number. It easy to see that P is an equivalence relation.

There is one further very important example, which we consider now.

Let m be a fixed, but arbitrary, integer. Two integers a, b are said to be congruent modulo m if $a - b$ is divisible by m. Congruence modulo m is denoted by $a \equiv b \pmod{m}$. Thus, $a \equiv b \pmod{m}$ if and only if $m|(a - b)$. This relation is easily seen to be an equivalence relation, using the properties of divisibility and it will be considered in detail later.

We consider next the very important concept of factorization of mappings.

7.2.6. Definition. *Let Φ be an equivalence relation on the set A. The set A/Φ of all equivalence classes of A by the relation Φ is called the factor-set of A by Φ.*

We may define a mapping

$$\sigma_\Phi : A \longrightarrow A/\Phi.$$

by

$$\sigma_\Phi(a) = [a]_\Phi\,, \text{ for all } a \in A.$$

The mapping σ_Φ is called an infinite surjection of A on the factor-set A/Φ.

Let A, B be sets and let $f : A \longrightarrow B$ be a mapping. We connect this mapping with a binary relation $\Delta(f)$ defined by

$$(x, y) \in \Delta(f) \text{ if and only if } f(x) = f(y), \text{ where } x, y \in A.$$

Clearly, the relation $\Delta(f)$ is reflexive since $f(x) = f(x)$. The relation is symmetric since the equations $f(x) = f(y)$ and $f(y) = f(x)$ are equivalent. In addition, the relation is transitive since the equations $f(x) = f(y)$ and $f(y) = f(z)$ imply $f(x) = f(z)$, for all $x, y, z \in A$. Hence, $\Delta(f)$ is an equivalence relation on the set A.

Now consider the factor-set $A/\Delta(f)$ and define the mapping

$$\psi_f : A/\Delta(f) \longrightarrow \mathbf{Im} f$$

by $\psi_f([a]_{\Delta(f)}) = f(a)$ for each equivalence class $[a]_{\Delta(f)} \in A/\Delta(f)$.

First we note that ψ_f is well defined. This means that ψ_f does not depends on the choice of representative of the equivalence class. Indeed, let c be an element of the set A such that $[c]_{\Delta(f)} = [a]_{\Delta(f)}$. Then by the definition of the relation $\Delta(f)$ we have $f(a) = f(c)$ and it follows that ψ_f is well-defined.

The mapping ψ_f is bijective. If $b \in \mathbf{Im} f$ there exists an element $a \in A$ such that $b = f(a)$, so we have $\psi_f([a]_{\Delta(f)}) = f(a) = b$, and ψ_f is surjective. If ψ_f $([a]_{\Delta(f)}) = \psi_f([c]_{\Delta(f)})$, then by the definition of ψ_f we have $f(a) = f(c)$, so $(a, c) \in \Delta(f)$ and therefore $[c]_{\Delta(f)} = [a]_{\Delta(f)}$. Thus, ψ_f is injective and therefore bijective.

Finally, we consider the following product of the mappings $\mathbf{j}_D \circ \psi_f \circ \sigma_{\Delta(f)}$, where $D = \mathbf{Im} f$ and $\mathbf{j}_D$ is the canonical injection of D into B. For an arbitrary element $a \in A$ we obtain

$$\mathbf{j}_D \circ \psi_f \circ \sigma_{\Delta(f)}(a) = \mathbf{j}_D(\psi_f(\sigma_{\Delta(f)}(a))) = \mathbf{j}_D(\psi_f([a]_{\Delta(f)})) = \mathbf{j}_D(f(a)) = f(a).$$

Since the domain of the mapping $\mathbf{j}_D \circ \psi_f \circ \sigma_{\Delta(f)}$ is A, and the range of $\mathbf{j}_D \circ \psi_f \circ \sigma_{\Delta(f)}$ is B, we see that $f = \mathbf{j}_D \circ \psi_f \circ \sigma_{\Delta(f)}$ and hence obtain the following theorem.

7.2.7. Theorem. *Let A, B be sets and let $f : A \longrightarrow B$ be a mapping. Then the following assertions hold:*

(i) *The relation $\Delta(f)$ defined by*

$$(x, y) \in \Delta(f) \textit{ if and only if } f(x) = f(y), \textit{ where } x, y \in A,$$

is an equivalence relation on the set A.

(ii) *The mapping $\psi_f : A/\Delta(f) \longrightarrow \mathbf{Im} f$, defined by the rule*

$$\psi_f([a]_{\Delta(f)}) = f(a) \textit{ for each equivalence class } [a]_{\Delta(f)} \in A/\Delta(f)$$

is bijective.

(iii) $f = \mathbf{j}_D \circ \psi_f \circ \sigma_{\Delta(f)}$.

In particular, an arbitrary mapping is a product of a surjection, a bijection, and an injection. The decomposition of the mapping in Theorem 7.2.7 is called canonical.

EXERCISE SET 7.2

Justify your work with a proof or a counterexample.

7.2.1. For $a, b \in \mathbb{R}$ define $a \simeq b$ to mean that $ab = 0$. Prove or disprove each of the following:

(a) The relation $\simeq$ is reflexive.

(b) The relation $\simeq$ is symmetric.

(c) The relation $\simeq$ is transitive.

7.2.2. For $a, b \in \mathbb{R}$ define $a \simeq b$ to mean that $ab \neq 0$. Prove or disprove each of the following:

(a) The relation $\simeq$ is reflexive.

(b) The relation $\simeq$ is symmetric.

(c) The relation $\simeq$ is transitive.

7.2.3. For $a, b \in \mathbb{R}$ define $a \simeq b$ to mean that $|a - b| < 7$. Prove or disprove each of the following:

(a) The relation $\simeq$ is reflexive.

(b) The relation $\simeq$ is symmetric.

(c) The relation $\simeq$ is transitive.

7.2.4. Define a mapping $f : \mathbb{R} \longrightarrow \mathbb{R}$ by the rule $f(x) = x^2 + 1$, where $x \in \mathbb{R}$. For $a, b \in \mathbb{R}$ define $a \simeq b$ to mean that $f(a) = f(b)$.

(a) Prove that $\simeq$ is an equivalence relation on $\mathbb{R}$.

(b) List all elements in the set $\{x \in \mathbb{R} \mid x \simeq 5\}$.

7.2.5. For points $(a, b), (c, d) \in \mathbb{R}^2$ define $(a, b) \simeq (c, d)$ to mean that $a^2 + b^2 = c^2 + d^2$.

(a) Prove that $\simeq$ is an equivalence relation on $\mathbb{R}^2$.

(b) List all elements in the set $\{(x, y) \in \mathbb{R}^2 \mid (x, y) \simeq (0, 0)\}$.

(c) List five distinct elements in the set $\{(x, y) \in \mathbb{R}^2 \mid (x, y) \simeq (1, 0)\}$.

7.2.6. Determine whether the relations represented by the following sets of ordered pairs are reflexive, symmetric, or transitive. Which are equivalence relations?

(a) $\{(1, 1), (2, 1), (2, 2), (3, 1), (3, 2), (3, 3)\}$

(b) $\{(1, 2), (1, 3), (2, 3), (2, 1), (3, 2), (3, 1)\}$

(c) $\{(1, 1), (1, 3), (2, 2), (3, 2), (1, 2)\}$.

7.2.7. Fractions are numbers of the form $\frac{a}{b}$ where a and b are integers and $b \neq 0$. Fraction equality is defined by $\frac{a}{b} = \frac{c}{d}$ if and only if $ad = bc$. Determine whether fraction equality is an equivalence relation.

7.2.8. Determine whether the relations represented by the following sets and descriptions are reflexive, symmetric, or transitive. Which are equivalence relations?

(a) "Has the same shape as" on the set of all triangles.

(b) "Is a factor of" on the set $\{1, 2, 3, 4, ..\}$.

(c) "Has the primary residence in the same state as" on the set of all people in the United States.

7.2.9. The equivalence class of an element $x \in S$ is the set of all elements that are equivalent to $x : [x] = \{y \in S \mid y \approx x\}$.

(a) Suppose that $\approx$ is an equivalence relation on a set S.
Show that for every x and y in S, $[x] = [y]$ if $x \approx y$ and $[x] \cap [y] = \emptyset$ otherwise.

(b) Show that $\bigcup_{x \in S}[x] = S$.

7.2.10. Suppose that S is a nonempty set. Show that equality $(=)$ is an equivalence relation on S and that $[x] = \{x\}$ for each $x \in S$.
The trivial relation $\approx$ on S is defined by $x \approx y$ for all x and y in S. Show that $\approx$ is an equivalence relation on S with only one equivalence class, namely S itself.

7.2.11. Define a relation on $\mathbb{Z} \times \mathbb{N}$ by $(j, k) \approx (m, n)$ if and only if $jn = km$.

(a) Show that $\approx$ is an equivalence relation.

(b) Define $\frac{m}{n}$ to be the equivalence class generated by (m, n). Show that this definition agrees with the usual notion of equality of rational numbers.

(c) Show that the usual definitions for addition and multiplication of rational numbers are consistent. That is, these definitions are independent of the particular representatives used for the equivalence classes.

7.2.12. Let $\mathbf{M}_{m \times n}(\mathbb{R})$ denote the set of $m \times n$ matrices with real entries.
The following are called row operations on a matrix:

1. Multiply a row by a nonzero real number.
2. Interchange two rows.
3. Add a multiple of a row to another row.

If A and B are $m \times n$ matrices, then A and B are said to be row equivalent if A can be transformed into B by a finite sequence of row operations. Show that row equivalence is an equivalence relation on $\mathbf{M}_{m \times n}(\mathbb{R})$. **Hint**: Note that each row operation can be reversed by another row operation.

7.2.13. Two $n \times n$ matrices A and B are said to be similar if there exists an invertible $n \times n$ matrix P such that $P^{-1}AP = B$. Show that similarity is an equivalence relation on $\mathbf{M}_{n\times n}(\mathbb{R})$.

7.3 IDEALS AND QUOTIENT RINGS

The concept of an ideal arose in the study of rings in which uniqueness of factorization into products of prime elements did not hold. The search to discover some form of "weak uniqueness" property led mathematicians to the concept of an ideal. Ideals were first introduced by Dedekind in 1876 in the third edition of his book Vorlesungen über Zahlentheorie (Lectures on Number Theory). The theory of ideals, a generalization of the concept of ideal numbers developed by Ernest Kummer, was later expanded by David Hilbert and especially Emmy Noether.

7.3.1. Definition. *A subring H of a ring R is called an ideal of R if, for each element $x \in R$ and every element $h \in H$, both products xh and hx lie in H.*

The concept of an ideal is often developed using more general notions as follows. A subring H of a ring R is called a left ideal (respectively a right ideal), if for each element $x \in R$ and every element $h \in H$ the product xh (respectively hx) lies in H. Every subset of the ring, which is both a left and a right ideal simultaneously, is an ideal. Thus, one can talk about the "two-sided" ideals of a ring. If R is a commutative ring, then all these concepts coincide.

Using Theorem 7.1.9 we obtain the following criterion for a subset to be an ideal. This is often used as the working definition of ideal.

7.3.2. Proposition. *Let R be a ring. A non empty subset H of a ring R is an ideal if and only if the following conditions hold:*

(I 1) *if $x, y \in H$, then $x - y \in H$;*
(I 2) *if $x \in R$ and $h \in H$ then the products xh and hx both belong to H.*

In every ring R the subsets $\{0_R\}$ and R are always ideals, as is easily verified.

7.3.3. Definition. *A ring R is called simple if its only ideals are $\{0_R\}$ and R.*

The following corollaries are analogs of the corresponding assertions for subrings and we therefore omit their proofs.

7.3.4. Corollary. *Let R be a ring and let $\mathfrak{S}$ be a family of ideals of R. The intersection $\cap\mathfrak{S}$ of all ideals from this family is also an ideal of R.*

7.3.5. Corollary. *Let R be a ring and let $\mathfrak{L}$ be a local family of ideals of R. Then their union $\cup\mathfrak{L}$ is also an ideal of R.*

7.3.6. Corollary. *Let R be a ring and let $\mathfrak{L}$ be a linearly ordered family of ideals of R. Then the union $\cup\mathfrak{L}$ of all ideals from this family is also an ideal of R.*

7.3.7. Corollary. *Let R be a ring and let*

$$H_1 \leq H_2 \leq \cdots \leq H_n \leq \cdots$$

be an ascending system of ideals of R. Then the union $\bigcup_{n\in\mathbb{N}} H_n$ is also an ideal of R.

If M is a subset of a ring R and $\mathfrak{S}$ is the family of all ideals of R containing M then the intersection, $\mathbf{id}(M) = \cap\mathfrak{S}$, is an ideal of R, by Corollary 7.3.4.

7.3.8. Definition. *The ideal $\mathbf{id}(M)$ is called the ideal that is generated by the subset M, and M is called a set of generators for this ideal.*

If $M = \{a\}$, then we write $\mathbf{id}(a)$ instead of $\mathbf{id}(\{a\})$ for the ideal that is generated by a. Such ideals are called the principal ideals of the ring R.

Certainly if $a, x \in R$ then $ax \in \mathbf{id}(a)$. Let

$$aR = \{ax \mid x \in R\} \text{ and } Ra = \{xa \mid x \in R\}.$$

It is easy to see that aR is a right ideal of R and that Ra is a left ideal of R. We next prove that when R is commutative then $aR = \mathbf{id}(a)$. We already know that $aR \leq \mathbf{id}(a)$, so it remains to prove that $\mathbf{id}(a) \leq aR$. To show this we prove that aR is an ideal of R and since it contains a it must then contain $\mathbf{id}(a)$ also. Hence, we must prove that aR satisfies the conditions **(I 1)** and **(I 2)**. To this end, let $u, v \in aR$. Then $u = ay$ and $v = az$ for certain elements $y, z \in R$. We now have

$$u - v = ay - az = a(y - z) \in aR \text{ and } gu = ug = (ay)g = a(yg) \in aR$$

for all elements g of the ring R. Thus, aR is an ideal of R. Since $a = ae$, we have $a \in aR$ and therefore $\mathbf{id}(a) \leq aR$. This proves that $aR = \mathbf{id}(a)$.

We have already shown that an arbitrary subring of the ring $\mathbb{Z}$ is of the form $n\mathbb{Z}$ where $n \geq 0$ is fixed. Hence every subring of $\mathbb{Z}$ is a principal ideal.

If R is a ring and H, K are subrings of R let

$$H + K = \{x + y \mid x \in H, y \in K\}.$$

The subset $H + K = \{h + k \mid h \in H, k \in K\}$ is called the sum of the subrings H and K. From the definition it follows that $H, K \subseteq H + K$. We note that a sum of two subrings is not always a subring. However, if one of the component subrings, H say, is an ideal then $H + K$ is a subring. Indeed, let $a, b \in H + K$.

Then $a = x + y$ and $b = x_1 + y_1$ for certain elements $x, x_1 \in H, y, y_1 \in K$. We have

$$a - b = (x + y) - (x_1 + y_1) = (x - x_1) + (y - y_1) \in H + K \text{ and}$$
$$ab = (x + y)(x_1 + y_1) = xx_1 + xy_1 + yx_1 + yy_1.$$

Since H is an ideal, then $xx_1 + xy_1 + yx_1 \in H$, so that

$$(xx_1 + xy_1 + yx_1) + yy_1 \in H + K.$$

Hence $H + K$ satisfies the conditions **(SR 1)** and **(SR 2)** of Theorem 7.1.9 so that $H + K$ is a subring of R. Moreover, if H, K are both ideals of R, then $H + K$ is also an ideal of R. For, if $x \in H, y \in K, z \in R$, then

$$z(x + y) = zx + zy \in H + K \text{ and } (x + y)z = xz + yz \in H + K,$$

so that $H + K$ satisfies the conditions **(I 1), (I 2)** and, by Proposition 7.3.2, it is an ideal of R.

7.3.9. Proposition. *Let R be a ring and let H an ideal of R. If H contains an invertible element of R, then $H = R$.*

Proof. Suppose that $x \in H \cap \mathbf{U}(R)$. Since H is an ideal, $e = x^{-1}x \in H$ and if a is an arbitrary element of R, then $a = ae \in H$. This means that $H = R$.

7.3.10. Corollary. *Every division ring, and hence every field, is a simple ring.*

However, the following theorem shows that the converse assertion is not true. Not every simple ring is a division ring.

7.3.11. Theorem. *The ring $\mathbf{M}_n(F)$ over the field F is simple.*

Proof. We have to show that the only nonzero ideal of $\mathbf{M}_n(F)$ is the ring itself. To this end, let L be a nonzero ideal of the ring $\mathbf{M}_n(F)$ and suppose that $B = [b_{ij}]$ is a nonzero matrix that is an element of L. Then there are indices k, m such that $b_{km} \neq 0_F$. We write $B = \sum_{1 \leq i,j \leq n} b_{ij}E_{ij}$ and note that

$$E_{st}B = E_{st}\left(\sum_{1 \leq ij \leq n} b_{ij}E_{ij}\right) = \sum_{1 \leq i,j \leq n} b_{ij}E_{st}E_{ij} = \sum_{1 \leq j \leq n} b_{tj}E_{sj}.$$

Hence

$$E_{st}BE_{qr} = \left(\sum_{1 \leq j \leq n} b_{tj}E_{sj}\right)E_{qr} = \sum_{1 \leq j \leq n} b_{tj}E_{sj}E_{qr} = b_{tq}E_{sr}.$$

Since L is an ideal of $\mathbf{M}_n(F)$, it follows that $E_{st}BE_{qr} \in L$ for any s, t, q, r. Now let $A = [a_{ij}]$ be an arbitrary matrix. Since $E_{ik}BE_{mj} = b_{km}E_{ij} \in L$ we see that

$$a_{ij}E_{ij} = (a_{ij}b_{km}^{-1}I)(E_{ik}BE_{mj}) \in L$$

for every pair of indices i, j and this proves that $A = \sum_{1 \le i,j \le n} a_{ij}E_{ij} \in L$. We deduce that $L = \mathbf{M}_n(F)$.

As we have already seen in Section 7.1, $\mathbf{M}_n(F)$ has zero-divisors if $n \ge 2$ and it cannot therefore be a division ring. However, as the following theorem shows, the converse statement to Corollary 7.3.10 is true for commutative rings.

7.3.12. Theorem. *Every simple commutative ring is a field.*

Proof. Let a be an arbitrary nonzero element of R. The nonzero ideal aR contains a and hence $aR = R$, since R is simple. Therefore, there is an element $x \in R$ such that $ax = e$, so a is invertible. Consequently every nonzero element is invertible, so that R is a field.

Let R be a ring, let H be a subring of R and let $x, y \in R$. We define a relation $\sum_H$ on R by $(x, y) \in \sum_H$ if and only if $x - y \in H$.

The relation $\sum_H$ is reflexive since $x - x = 0_R \in H$, which means that $(x, x) \in \sum_H$. The relation $\sum_H$ is symmetric. For, if $(x, y) \in \sum_H$ then $x - y \in H$ and, since H is a subring, it follows that H contains $-(x - y) = y - x$, so that $(y, x) \in \sum_H$. The relation $\sum_H$ is transitive. For, if $(x, y), (y, z) \in \sum_H$ then $x - y, y - z \in H$ and since H is a subring it contains $(x - y) + (y - z) = x - z$. Thus, $(x, z) \in \sum_H$. Hence $\sum_H$ is an equivalence relation on R.

We now determine the equivalence class of the element $x \in R$ under the relation $\sum_H$. If $(x, y) \in \sum_H$, then $x - y = h \in H$ and it follows that $y = x + (-h)$. Let $x + H = \{x + u \mid u \in H\}$. The subset $x + H$ is called the coset of the element x relative to the subring H and x is called a representative of this coset. Thus, every element equivalent to x (under $\sum_H$) belongs to the coset $x + H$. Conversely, if $z \in x + H$, then $z = x + u$ for some element $u \in H$. Then we have

$$x - z = -u \in H,$$

and this means that $(x, z) \in \sum_H$. Consequently, the equivalence classes under the relation $\sum_H$ are exactly the cosets relative to the subring H. It follows that the coset $x + H$ is defined by each of its elements. Thus, if $y \in x + H$ then $y + H = x + H$ and, by the argument following Definition 7.2.4, two cosets either coincide or have empty intersection. Furthermore, the ring R is the union of all the cosets. Thus, the family of all cosets under H is a partition of R. If $H = \{0_R\}$, then $x + H = \{x\}$ for each element $x \in R$ and we obtain the largest partition of R consisting of all one-element subsets; if $H = R$, then we obtain the smallest partition consisting only of the set R.

Now let H be an ideal of the ring R. We define addition and multiplication on the set of all cosets of H by the rules:

$$(x + H) + (y + H) = x + y + H \text{ and}$$
$$(x + H)(y + H) = xy + H.$$

These operations are well defined, as we now show. If also x_1, y_1 are elements of the ring R such that $x + H = x_1 + H$ and $y + H = y_1 + H$, then $x_1 = x + u$, $y_1 = y + v$ for some elements $u, v \in H$. Hence

$$x_1 + y_1 = (x + u) + (y + v) = (x + y) + (u + v),$$
$$x_1 y_1 = (x + u)(y + v) = xy + uy + xv + uv.$$

Since H is an ideal, it follows that $(u + v), uy, xv, uv \in H$. Therefore,

$$x + y + H = x_1 + y_1 + H \text{ and } xy + H = x_1 y_1 + H,$$

which shows that the operations are well defined.

Next,

$$\begin{aligned}(x + H) + (y + H) &= x + y + H = y + x + H\\ &= (y + H) + (x + H) \text{ and}\\ (x + H) + \big((y + H) + (z + H)\big) &= (x + H) + (y + z + H)\\ &= x + (y + z) + H\\ &= \big((x + y) + z\big) + H\\ &= (x + y + H) + (z + H)\\ &= \big((x + H) + (y + H)\big) + (z + H),\end{aligned}$$

which shows that the operation of addition of cosets is commutative and associative. Also

$$(x + H) + (0_R + H) = x + 0_R + H = x + H$$

and hence $0_R + H = H$ is the zero element under addition of cosets. Clearly,

$$(x + H) + (-x + H) = (x + (-x)) + H = 0_R + H = H,$$

and hence

$$-(x + H) = (-x) + H.$$

Furthermore,

$$\begin{aligned}(x+H)(y+H+z+H) &= (x+H)(y+z+H) = x(y+z)+H \\ &= xy+xz+H \\ &= xy+H+xz+H \\ &= (x+H)(y+H)+(x+H)(z+H)\end{aligned}$$

and, similarly,

$$(x+H+y+H)(z+H) = (x+H)(z+H)+(y+H)(z+H).$$

Also

$$\begin{aligned}&(x+H)\big((y+H)(z+H)\big) = (x+H)(yz+H) = x(yz)+H \\ &= (xy)z+H = (xy+H)(z+H) = \big((x+H)(y+H)\big)(z+H) \text{ and} \\ &(e+H)(x+H) = ex+H = x+H = xe+H = (x+H)(e+H),\end{aligned}$$

so under the operations of addition and multiplication the set of cosets is a ring.

7.3.13. Definition. *Let R be a ring and let H be an ideal of R. The set of all cosets of the ideal H is called the quotient (or factor) ring of R over H and is denoted by R/H.*

We observe that if the ring R is commutative, then every quotient ring of R is also commutative.

As an important example we consider the quotient rings of the ring $\mathbb{Z}$ of all integers. Let H be an ideal of $\mathbb{Z}$. We have already observed that an arbitrary ideal of $\mathbb{Z}$ is of the form $n\mathbb{Z}$, where $n > 0$ is fixed. When $H = \{0\}$ the quotient ring is again $\mathbb{Z}$, so we suppose that $n > 0$. Consider the set of cosets

$$\mathfrak{N} = \{n\mathbb{Z}, 1+n\mathbb{Z}, \ldots, (n-1)+n\mathbb{Z}\},$$

and suppose that among these there are two cosets that are the same. Thus, $k+n\mathbb{Z} = t+n\mathbb{Z}$, for some k, t where $0 \leq k, t \leq n-1$. Let us assume, without loss of generality, that $k \geq t$. Then $k \in t+n\mathbb{Z}$ so $k = t+nm$, for some integer m. It follows that $k-t = nm$. Since $k-t \geq 0$, we have $m \geq 0$. However, $k-t \leq n-1$, which implies that $m = 0$ and $k = t$. Thus, the cosets of the family $\mathfrak{N}$ are distinct.

Next, let s be an arbitrary integer. By Theorem 1.4.1, $s = qn+r$ where $0 \leq r < n$ and it follows that $s+n\mathbb{Z} = r+n\mathbb{Z}$. This means that the set of all cosets of $n\mathbb{Z}$ coincides with $\mathfrak{N}$. Consequently, when $n > 0$ the quotient ring $\mathbb{Z}/n\mathbb{Z}$ is finite and indeed $|\mathbb{Z}/n\mathbb{Z}| = n$.

Finally, if R is a ring and H is an ideal, consider the mapping $\sigma_H : R \longrightarrow R/H$, defined by the rule:

$$\sigma_H(x) = x + H \text{ where } x \in R.$$

Then

$$\sigma_H(x + y) = x + y + H = x + H + y + H = \sigma_H(x) + \sigma_H(y) \text{ and}$$
$$\sigma_H(xy) = xy + H = (x + H)(y + H) = \sigma_H(x)\sigma_H(y).$$

This observation leads us to consider mappings of rings, which preserve the operations of addition and multiplication.

EXERCISE SET 7.3

Justify your work with a proof or a counterexample where required.

7.3.1. On the set $R = \mathbb{F}_3 \times \mathbb{Z}$ we define operations of addition and multiplication by $(a, b) + (a_1, b_1) = (a + a_1, b + b_1)$, $(a, b)(a_1, b_1) = (aa_1, bb_1)$. Prove that R is a ring with identity. Find all ideals of R.

7.3.2. Let R be a commutative ring and let $n \in \mathbb{N}$. Is the set $nR = \{nx \mid x \in R\}$ an ideal of R?

7.3.3. Write the multiplication table for the ring $\mathbb{Z}_{16} = \mathbb{Z}/16\mathbb{Z}$.

7.3.4. Write the multiplication table for the ring $\mathbb{Z}_{14} = \mathbb{Z}/14\mathbb{Z}$.

7.3.5. Let $M = \{2k + 2ti \mid k, t \in \mathbb{Z}\}$. Prove that M is an ideal of the ring $\mathbb{Z}[i]$. Find all elements of the quotient ring $\mathbb{Z}[i]/M$.

7.3.6. Let $M = \{2k + 2ti \mid k, t \in \mathbb{Z}\}$. Find all zero-divisors of the quotient ring $\mathbb{Z}[i]/M$.

7.3.7. Let $M = 3\mathbb{Z}[i]$. Prove that the quotient ring $\mathbb{Z}[i]/M$ is a field of order 9.

7.3.8. Let $M = n\mathbb{Z}[i]$. Prove that the quotient ring $\mathbb{Z}[i]/M$ is a field if and only if n is a prime and n is not equal to the sum of the squares of two integers.

7.3.9. Let F be a field and let M be the subset of all polynomials from the ring $F[X_1, X_2]$ having zero constant term. Prove that M is an ideal of the ring $F[X_1, X_2]$. Prove that M is a not a principal ideal of $F[X_1, X_2]$.

7.4 HOMOMORPHISMS OF RINGS

In this section, we consider mappings of rings that respect (or preserve) the operations. In Section 3.1 we introduced the general concept of homomorphism and applying this to rings we obtain the following definition.

7.4.1. Definition. *Let R, S be rings. The mapping $f : R \longrightarrow S$ is called a ring homomorphism if it satisfies the conditions*

$$f(x+y) = f(x) + f(y) \text{ and } f(xy) = f(x)f(y)$$

for all elements $x, y \in R$.

In this section we only consider rings, so the term *homomorphism* will always be understood to mean "ring homomorphism."

As usual an injective homomorphism is called a monomorphism, a surjective homomorphism is called an epimorphism and a bijective homomorphism is called an isomorphism.

If $f : R \longrightarrow S$ is an isomorphism then, as in Section 3.1, the natural mapping $f^{-1} : S \longrightarrow R$ is also an isomorphism.

7.4.2. Definition. *Let R, S be rings. Then R and S are called isomorphic if there exists an isomorphism from R to S or, equivalently, from S to R. This will be denoted by $R \cong S$.*

The easiest example of an isomorphism is the identity permutation $\varepsilon_R : R \longrightarrow R$. It is also easy to show that if $f : R \longrightarrow S$ and $g : S \longrightarrow U$ are homomorphisms, then their product $g \circ f$ is likewise a homomorphism.

7.4.3. Proposition. *Let R, S be rings and let $f : R \longrightarrow S$ be a homomorphism. Then the following properties hold:*

(i) *$f(0_R) = 0_S$;*
(ii) *$f(-x) = -f(x)$ for every element $x \in R$;*
(iii) *$f(x-y) = f(x) - f(y)$ for all $x, y \in R$;*
(iv) *if H is a subring of R, then its image $f(H)$ is a subring of S. In particular, $f(R) = \mathbf{Im}\, f$ is a subring of S;*
(v) *if V is a subring of S, then its preimage $f^{-1}(V)$ is a subring of R;*
(vi) *if V is an ideal of S, then its preimage $f^{-1}(V)$ is an ideal of R. In particular,*

$$\mathbf{Ker}\, f = \{x \in R \mid f(x) = 0_S\} = f^{-1}(0_S)$$

is an ideal of R;
(vii) *if R has a multiplicative identity, e, then $f(e)$ is the identity element of the subring $\mathbf{Im}\, f$;*
(viii) *if R is commutative, then $\mathbf{Im}\, f$ is commutative.*

Proof.
(i) We have $x + 0_R = x$ for each $x \in R$. Then

$$f(x) + f(0_R) = f(x + 0_R) = f(x).$$

Since the element $f(x)$ has a negative in S, we may add this negative to both sides of these equations to obtain

$$0_S = -f(x) + f(x) = -f(x) + f(x) + f(0_R) = 0_S + f(0_R) = f(0_R).$$

(ii) From the definition of the negative element we have $x + (-x) = 0_R$, so that

$$0_S = f(0_R) = f(x + (-x)) = f(x) + f(-x).$$

This equation show that $f(-x)$ is the negative of $f(x)$.

(iii) We have

$$f(x - y) = f(x + (-y)) = f(x) + f(-y) = f(x) + (-f(y)) = f(x) - f(y).$$

(iv) Let $x, y \in H$ and let $a = f(x), b = f(y)$. Then

$$a + b = f(x) + f(y) = f(x + y) \in f(H) \text{ and}$$
$$ab = f(x)f(y) = f(xy) \in f(H).$$

It follows from Theorem 7.1.9 that $f(H)$ is a subring of S.

(v) Let $x, y \in f^{-1}(V)$. Then $f(x), f(y) \in V$. Since V is a subring of S, $f(x) - f(y) = f(x - y) \in V$, and $f(x)f(y) = f(xy) \in V$, which imply that $x - y, xy \in f^{-1}(V)$. From Theorem 7.1.9 it follows that $f^{-1}(V)$ is a subring of R.

(vi) As in the proof of (v), if $x, y \in f^{-1}(V)$ then $x - y \in f^{-1}(V)$. If also $r \in R$ then $f(xr) = f(x)f(r) \in V$, since V is an ideal of S and hence $xr \in f^{-1}(V)$. Likewise $rx \in f^{-1}(V)$ and it follows that $f^{-1}(V)$ is an ideal of R. Of course $\{0_S\}$ is an ideal of S, so the statement concerning $\mathbf{Ker} f$ is clear.

(vii) If $a \in \mathbf{Im} f$, then $a = f(x)$ for some element $x \in R$. Therefore

$$a = f(x) = f(xe) = f(x)f(e) = af(e) \text{ and } a = f(x) = f(ex)$$
$$= f(e)f(x) = f(e)a,$$

which means that $f(e)$ is the identity element of the subring $\mathbf{Im} f$.

(viii) Finally, let R be a commutative ring and let $a, b \in \mathbf{Im} f$. Then $a = f(x)$ and $b = f(y)$ for some elements $x, y \in R$. We now have

$$ab = f(x)f(y) = f(xy) = f(yx) = f(y)f(x) = ba.$$

The ideal $\mathbf{Ker} f$ is called the kernel of the homomorphism f.

7.4.4. Theorem *(The Theorem on Monomorphisms). Let R, S be rings. A homomorphism $f : R \longrightarrow S$ is a monomorphism if and only if* $\mathbf{Ker} f = \{0_R\}$. *If $f : R \longrightarrow S$ is a monomorphism, then $R \cong$* $\mathbf{Im} f$.

Proof. If f is a monomorphism, and if $x \neq 0_F$ then $f(x) \neq f(0_R) = 0_S$. This means that no nonzero element x belongs to $\mathbf{Ker}\, f$ and hence $\mathbf{Ker}\, f = \{0_R\}$. Conversely, let $\mathbf{Ker}\, f = \{0_R\}$ and let x, y be elements of R such that $f(x) = f(y)$. Then

$$f(x - y) = f(x) - f(y) = 0_S,$$

and hence $x - y \in \mathbf{Ker}\, f$. It follows that $x - y = 0_R$, so that $x = y$. Thus, f is an injective homomorphism and hence it is a monomorphism.

7.4.5. Theorem *(The First Isomorphism Theorem, Version 1). Let R, S be rings and let $f : R \longrightarrow S$ be an epimorphism. Then S is isomorphic to $R/\mathbf{Ker}\, f$.*

Proof. For the sake of convenience put $H = \mathbf{Ker}\, f$. As in Section 7.2 we define an equivalence relation $\Delta(f)$ on R by $(x, y) \in \Delta(f)$ if and only if $f(x) = f(y)$. As in the previous theorem, this is equivalent to $f(x - y) = 0$. Thus, $(x, y) \in \Delta(f)$ if and only if $x - y \in H$. Consequently $\Delta(f) = \sum_H$ and, by the results of Section 7.2, we see that $[x]_{\Delta(f)} = x + H$ for each $x \in R$, and therefore $R/\Delta(f) = R/H$. We now consider the mapping $\Psi_f : R/H \longrightarrow S$, defined by $\Psi_f(x + H) = f(x)$. By Theorem 7.2.7, Ψ_f is a bijection so the proof will be complete once we prove that Ψ_f is a homomorphism. We have

$$\begin{aligned}\Psi_f(x + H + y + H) &= \Psi_f(x + y + H) = f(x + y) = f(x) + f(y)\\ &= \Psi_f(x + H) + \Psi_f(y + H),\end{aligned}$$

and

$$\begin{aligned}\Psi_f((x + H)(y + H)) &= \Psi_f(xy + H) = f(xy) = f(x)f(y)\\ &= \Psi_f(x + H)\Psi_f(y + H).\end{aligned}$$

The first isomorphism theorem now follows.

7.4.6. Theorem *(The First Isomorphism Theorem, Version 2). Let R, S be rings and let $f : R \longrightarrow S$ be a homomorphism. Then $R/\mathbf{Ker}\, f \cong \mathbf{Im}\, f \leq S$.*

Proof. The restriction of f to the mapping $R \longrightarrow \mathbf{Im}\, f$ is an epimorphism and hence, from Theorem 7.4.5, we see that $\mathbf{Im}\, f \cong R/\mathbf{Ker}\, f$. Finally, by Proposition 7.4.3, $\mathbf{Im}\, f$ is a subring of S.

We now consider some applications of these results.

The Characteristic of a Ring

Let R be a ring and let $f : \mathbb{Z} \longrightarrow R$ be the mapping defined by $f(n) = ne$, where $n \in \mathbb{Z}$. Clearly,

$$f(n + k) = (n + k)e = ne + ke = f(n) + f(k) \text{ and}$$
$$f(nk) = (nk)e = n(ke) = (ne)(ke) = f(n)f(k),$$

for all $n, k \in \mathbb{Z}$. By Proposition 7.4.3,

$$\mathbf{Im}f = \{ne \mid n \in \mathbb{Z}\} = \mathbb{Z}e$$

is a subring of R. By Theorem 7.4.6, $\mathbb{Z}e = \mathbf{Im}f \cong \mathbb{Z}/\mathbf{Ker}f$. By Proposition 7.4.3, $\mathbf{Ker}f$ is an ideal of $\mathbb{Z}$. From the description of ideals of the ring $\mathbb{Z}$ we have $\mathbf{Ker}f = n\mathbb{Z}$, for some fixed, but arbitrary, $n \geq 0$.

If $\mathbf{Ker}f = \{0\}$, then $\mathbb{Z}e \cong \mathbb{Z}$. In this case we say that the ring R has characteristic 0 and write $\mathbf{char}R = 0$. If $\mathbf{Ker}f = n\mathbb{Z}$, where $n > 0$, then $\mathbb{Z}e \cong \mathbb{Z}/n\mathbb{Z}$ and in this case we say that the ring R has characteristic $n > 0$ and write $\mathbf{char}R = n$. In this case if a is an arbitrary element of R, then

$$na = (ne)a = 0_R a = 0_R.$$

If n is not prime, then $n = kt$, where $1 < k, t < n$. Certainly $k, t \notin n\mathbb{Z}$. Therefore $ke \neq 0_R$ and $te \neq 0_R$. However,

$$(ke)(te) = (kt)e = ne = 0_R.$$

So if $\mathbf{char}R$ is not prime then the ring R has zero-divisors.

7.4.7. Proposition. *If a ring R has no zero-divisors then either* $\mathbf{char}R = 0$ *or* $\mathbf{char}R = p$ *for some prime p. In particular, if R is a division ring or an integral domain, then either* $\mathbf{char}R = 0$ *or* $\mathbf{char}R = p$ *for some prime p.*

In Section 3.2 we considered examples of prime fields. These were the field $\mathbb{Q}$ of all rational numbers and the field $\mathbb{F}_p$, for some prime p. Now we are in a position to give a different definition of the field $\mathbb{F}_p$. It turns out that this is none other than the quotient ring $\mathbb{Z}/p\mathbb{Z}$ where p is a prime.

We show first that the quotient ring $\mathbb{Z}/p\mathbb{Z}$ is a field. Let $p\mathbb{Z} \neq x + p\mathbb{Z} \in \mathbb{Z}/p\mathbb{Z}$. Since $x \notin p\mathbb{Z}$ the integers x and p are relatively prime and hence there exist integers k, r such that $xk + pr = 1$. We have

$$1 + p\mathbb{Z} = xk + pr + p\mathbb{Z} = xk + p\mathbb{Z} = (x + p\mathbb{Z})(k + p\mathbb{Z}).$$

This shows that every nonzero coset $x + p\mathbb{Z}$ is invertible and therefore $\mathbb{F}_p = \mathbb{Z}/p\mathbb{Z}$ is a field. As we proved in Section 7.3, $|\mathbb{F}_p| = p$. The fact that $\mathbb{F}_p$ is a prime field has been proved in Section 3.2.

7.4.8. Theorem. *Let F be a field and let F_0 be the prime subfield of F.*

(i) *If* $\mathbf{char}F = p$ *is a prime then* $F_0 \cong \mathbb{F}_p$.
(ii) *If* $\mathbf{char}F = 0$, *then* $F_0 \cong \mathbb{Q}$.

Proof. Consider the mapping $f : \mathbb{Z} \longrightarrow F$ defined by $f(n) = ne$, where $n \in \mathbb{Z}$. As we proved above, this mapping is a homomorphism. If $\mathbf{char}F = p$ is a prime, then $\mathbf{Ker} f = p\mathbb{Z}$ and, as in the proof of the first isomorphism theorem, the mapping $\psi : \mathbb{Z}/p\mathbb{Z} \longrightarrow F$ defined by the rule $\psi(n + p\mathbb{Z}) = f(n)$, where $n \in \mathbb{Z}$, is also a homomorphism and this homomorphism is nonzero, because $\psi(1 + p\mathbb{Z}) = f(1) = e \neq 0_F$. Clearly, $\mathbf{Im}\,\psi = \mathbf{Im}\, f = \mathbb{Z}e$ and we deduce the result in this case by applying Theorem 3.2.16.

Suppose now that $\mathbf{char} F = 0$. In this case, $\mathbf{Ker} f = \{0\}$ and Theorem 7.4.4 implies that f is a monomorphism. Since $e \in F_0$ it follows that $\mathbb{Z}e$ is a subring of F_0. We extend the mapping f to a mapping $f_1 : \mathbb{Q} \longrightarrow F$ in the following way. Let $\frac{m}{n}$ be an arbitrary element of $\mathbb{Q}$. Then $n \neq 0$ and hence $f(n) = ne \neq 0_F$. Since F is a field, its nonzero element ne is invertible in F. We now set

$$f_1\left(\frac{m}{n}\right) = (me)(ne)^{-1}.$$

The function f_1 is well defined. For, if $\frac{k}{t} = \frac{m}{n}$ then we have

$$kn = tm \text{ so } (ke)(ne) = (te)(me) \text{ and hence}$$
$$(me)(ne)^{-1} = (ke)(te)^{-1} = (te)^{-1}(ke).$$
$$\text{Thus } f_1(m/n) = (me)(ne)^{-1} = (ke)(te)^{-1} = f_1(k/t),$$

and that f_1 is well defined follows. The mapping f_1 is an extension of f since if $n \in \mathbb{Z}$, then $n = \frac{n}{1}$ so that

$$f_1(n) = f_1\left(\frac{n}{1}\right) = (ne)(e)^{-1} = (ne)(e)(e)^{-1} = ne = f(n).$$

The mapping f_1 is a homomorphism since

$$\begin{aligned} f_1\left(\frac{m}{n} + \frac{k}{t}\right) &= f_1\left(\frac{mt + kn}{nt}\right) = \big((mt + kn)e\big)\big((nt)e\big)^{-1} \\ &= \big((mt)e + (kn)e\big)\big((ne)(te)\big)^{-1} \\ &= \big((me)(te) + (ke)(ne)\big)(ne)^{-1}(te)^{-1} \\ &= (me)(te)(ne)^{-1}(te)^{-1} + (ke)(ne)(ne)^{-1}(te)^{-1} \\ &= (me)(ne)^{-1} + (ke)(te)^{-1} = f_1\left(\frac{m}{n}\right) + f_1\left(\frac{k}{t}\right). \end{aligned}$$

Also

$$\begin{aligned} f_1\left(\left(\frac{m}{n}\right)\left(\frac{k}{t}\right)\right) &= f_1\left(\frac{mk}{nt}\right) = \big((mk)e\big)\big((nt)e\big)^{-1} = (me)(ke)\big((ne)(te)\big)^{-1} \\ &= (me)(ke)(ne)^{-1}(te)^{-1} = (me)(ne)^{-1}(ke)(te)^{-1} \\ &= f_1\left(\frac{m}{n}\right) f_1\left(\frac{k}{t}\right). \end{aligned}$$

By Theorem 3.2.16, $\mathbb{Q} \cong \mathbf{Im} f_1$. Since $\mathbf{Im} f_1$ is a subfield of F, it follows that $\mathbf{Im} f_1 \geq F_0$. On the other hand, $ne \in F_0$ for every $n \in \mathbb{Z}$ and if $n \neq 0$, then $(ne)^{-1} \in F_0$. Therefore $(me)(ne)^{-1} \in F_0$ for every pair $m, n \in \mathbb{Z}$, where $n \neq 0$. Thus, $\mathbf{Im} f_1 \leq F_0$, and then $\mathbf{Im} f_1 = F_0$.

The following theorem allows us to construct new ring extensions.

7.4.9. Theorem. *Let R, K be rings and let $f : R \longrightarrow K$ be a monomorphism. Then there exists a ring S such that S is isomorphic to K and R is a subring of S. If e_R, e_K are the multiplicative identities of R and K respectively, and if $f(e_R) = e_K$, then e_R is the multiplicative identity of S. Finally, if R and K are fields, then R is a subfield of S.*

Proof. We may assume that $K \cap R = \emptyset$. Indeed, if this is not true we can replace K by an isomorphic image having empty intersection with R. For example, we can put $K_1 = K \times \{1\}$ and define operations by the rules

$$(x, 1) + (y, 1) = (x + y, 1) \text{ and } (x, 1)(y, 1) = (xy, 1).$$

Let $U = \mathbf{Im} f$. By Theorem 7.4.4, U is a subring of K and $U \cong R$. Let $A = K \setminus U$ and $S = R \cup A$. Let $g : S \longrightarrow K$ be the mapping defined by

$$g(x) = \begin{cases} f(x) \text{ if } x \in R, \\ x \text{ if } x \in A. \end{cases}$$

It is easy to check that g is a bijection from S onto K. We now define operations of addition, $\oplus$, and multiplication, $\otimes$, on S by

$$x \oplus y = g^{-1}(g(x) + g(y)) \text{ and } x \otimes y = g^{-1}(g(x)g(y)),$$

for arbitrary $x, y \in S$. Then,

$$g(x \oplus y) = g(g^{-1}(g(x) + g(y))) = g(x) + g(y) \text{ and}$$
$$g(x \otimes y) = g(g^{-1}(g(x)g(y))) = g(x)g(y).$$

We have to show that S is a ring and to this end we have

$$x \oplus y = g^{-1}(g(x) + g(y)) = g^{-1}(g(y) + g(x)) = y \oplus x,$$

so $\oplus$ is commutative. Also

$$\begin{aligned}(x \oplus y) \oplus z &= g^{-1}\big(g(x \oplus y) + g(z)\big) = g^{-1}\big((g(x) + g(y)) + g(z)\big) \\ &= g^{-1}\big(g(x) + (g(y) + g(z))\big) = g^{-1}\big(g(x) + g(y \oplus z)\big) \\ &= x \oplus (y \oplus z);\end{aligned}$$

so $\oplus$ is associative. Further, if $x \in R$ then

$$x \oplus 0_R = g^{-1}\big(g(x) + g(0_R)\big) = g^{-1}(g(x) + 0_K) = g^{-1}(g(x)) = x.$$

Also, if $x \in R$ then

$$x \oplus (-x) = g^{-1}\big(g(x) + g(-x)\big) = g^{-1}\big(f(x + (-x))\big) = g^{-1}\big(f(0_R)\big) = 0_R.$$

If $x \in A$, then $-x \notin U$, because U is a subring. Thus,

$$x \oplus (-x) = g^{-1}\big(g(x) + g(-x)\big) = g^{-1}\big(x + (-x)\big) = g^{-1}(0_K) = 0_R,$$

and it follows that S is an abelian group under the operation $\oplus$.

Moreover,

$$\begin{aligned}(x \oplus y) \otimes z &= g^{-1}\big(g(x \oplus y)g(z)\big) = g^{-1}\big((g(x) + g(y))g(z)\big)\\ &= g^{-1}\big(g(x)g(z) + g(y)g(z)\big) = g^{-1}\big(g(x \otimes z) + g(y \otimes z)\big)\\ &= (x \otimes z) \oplus (y \otimes z).\end{aligned}$$

We can prove the equation

$$x \otimes (y \oplus z) = (x \otimes y) \oplus (x \otimes z)$$

in a similar manner. Also, $\otimes$ is associative since

$$\begin{aligned}(x \otimes y) \otimes z &= g^{-1}\big(g(x \otimes y)g(z)\big) = g^{-1}\big((g(x)g(y))g(z)\big)\\ &= g^{-1}\big(g(x)(g(y)g(z))\big) = g^{-1}\big(g(x)g(y \otimes z)\big) = x \otimes (y \otimes z).\end{aligned}$$

Furthermore,

$$\begin{aligned}&x \otimes g^{-1}(e_K) = g^{-1}\big(g(x)g(g^{-1}(e_K))\big) = g^{-1}(g(x)e_K) = g^{-1}\big(g(x)\big) = x \text{ and}\\ &g^{-1}(e_K) \otimes x = g^{-1}\big(g(g^{-1}(e_K))g(x)\big) = g^{-1}\big(e_K g(x)\big) = g^{-1}\big(g(x)\big) = x.\end{aligned}$$

so that $g^{-1}(e_K)$ is the identity element of S. In particular, if $f(e_R) = g(e_R) = e_K$, then e_R is the identity element of S. It follows that S is a ring. We have already proved that

$$g(x \oplus y) = g(x) + g(y) \text{ and } g(x \otimes y) = g(x)g(y),$$

which shows that g is an isomorphism.

If $x, y \in R$, then

$$x \oplus y = g^{-1}\big(g(x) + g(y)\big) = g^{-1}\big(f(x) + f(y)\big)$$
$$= g^{-1}\big(f(x+y)\big) = g^{-1}\big(g(x+y)\big) = x + y \text{ and}$$
$$x \otimes y = g^{-1}\big(g(x)g(y)\big) = g^{-1}\big(f(x)f(y)\big) = g^{-1}\big(f(xy)\big) = g^{-1}\big(g(xy)\big) = xy$$

Hence, the restriction of these operations to R give rise to the original operations on R. Thus, by Theorem 7.1.9, R is a subring of S.

Finally suppose that R, K are fields. By Theorem 3.2.6 it is sufficient to check the condition **(SF 2)** only. If x is a nonzero element of R, then it has a multiplicative inverse in R. Since the multiplicative identity of R is also the identity element in S, the inverse of x in R is the inverse element to x in the field S.

This theorem shows that to construct an extension E of the ring R it is sufficient to construct a monomorphism of R into some ring E isomorphic to the ring K.

We next consider some applications of the results we obtained above. Recall that in Section 7.1 we decided to consider associative rings only. The following remarkable theorem, sometimes called the Dorroh extension Theorem, shows that every ring is a subring of some ring with identity.

7.4.10. Theorem. *For every ring R there exists a ring K with multiplicative identity such that R is a subring of K.*

Proof. If R has an identity element then we take $K = R$. Therefore suppose that R does not have a multiplicative identity. By Theorem 7.4.9, it is enough to construct a monomorphism $f : R \longrightarrow K$ where K is a ring with identity. Let $K = R \times \mathbb{Z}$. Define operations on K by

$$(x, n) + (y, k) = (x + y, n + k) \text{ and } (x, n)(y, k) = (xy + kx + ny, nk),$$

for all $x, y \in R$ and $n, k \in \mathbb{Z}$.

Since the addition on K is defined componentwise, using the addition in R and $\mathbb{Z}$, K inherits all the properties of addition from R and $\mathbb{Z}$. Thus, K is an abelian group under addition. Also it is easy to see that

$$0_K = (0_R, 0) \text{ and } -(x, n) = (-x, -n).$$

Furthermore, multiplication is distributive over addition since,

$$\big((x, n) + (y, k)\big)(z, m) = (x + y, n + k)(z, m)$$
$$= \big((x + y)z + (n + k)z + m(x + y), (n + k)m\big)$$
$$= (xz + yz + nz + kz + mx + my, nm + km)$$

and

$$\begin{aligned}(x, n)(z, m) + (y, k)(z, m) &= (xz + nz + mx, nm) + (yz + kz + my, km)\\ &= (xz + nz + mx + yz + kz + my, nm + km),\end{aligned}$$

so that

$$\big((x, n) + (y, k)\big)(z, m) = (x, n)(z, m) + (y, k)(z, m).$$

Similarly,

$$(x, n)\big((y, k) + (z, m)\big) = (x, n)(y, k) + (x, n)(z, m).$$

The multiplication in K is associative since,

$$\begin{aligned}\big((x, n)(y, k)\big)(z, m) &= (xy + ny + kx, nk)(z, m)\\ &= \big((xy + ny + kx)z + m(xy + ny + kx) + (nk)z, (nk)m\big)\\ &= \big((xy)z + n(yz) + k(xz) + m(xy) + (mn)y\\ &\quad + (mk)x + (nk)z, (nk)m\big)\end{aligned}$$

whereas

$$\begin{aligned}(x, n)\big((y, k)(z, m)\big) &= (x, n)(yz + kz + my, km)\\ &= \big(x(yz + kz + my) + n(yz + kz + my) + (km)x, n(km)\big)\\ &= \big(x(yz) + k(xz) + m(xy) + n(yz) + (nk)z\\ &\quad + (nm)y + (km)x, n(km)\big).\end{aligned}$$

Since the multiplications in R and $\mathbb{Z}$ are associative, it follows that

$$\big((x, n)(y, k)\big)(z, m) = (x, n)\big((y, k)(z, m)\big).$$

Consequently K is a ring. Finally,

$$(x, n)(0_R, 1) = (x, n) = (0_R, 1)(x, n),$$

so that $(0_R, 1)$ is the multiplicative identity element of K. Thus, K is a unitary ring.

We next define the mapping $f : R \longrightarrow K$ by $f(x) = (x, 0)$, where $x \in R$. Clearly,

$$\begin{aligned}f(x + y) &= (x + y, 0) = (x, 0) + (y, 0) = f(x) + f(y) \text{ and}\\ f(xy) &= (xy, 0) = (x, 0)(y, 0) = f(x)f(y),\end{aligned}$$

for all $x, y \in R$. Evidently, $\mathbf{Ker} f = \{0_R\}$ and, by Theorem 7.4.4, f is a monomorphism. The result follows.

The following theorem indicates that there are some types of "universal" objects for rings.

7.4.11. Theorem. *Let R be a ring with identity. Then there exists a monomorphism from R into the endomorphism ring of some abelian group.*

Proof. By Theorem 7.4.10, we may assume that R has a multiplicative identity. Let $a \in R$ and consider the mapping $\tau_a : (R, +) \longrightarrow (R, +)$ defined by $\tau_a(x) = ax$, where $x \in R$. We have

$$\tau_a(x + y) = a(x + y) = ax + ay = \tau_a(x) + \tau_a(y)$$

and hence τ_a is an endomorphism of the additive group of the ring R. Of course $\mathbf{End}(R, +)$ is a ring.

We next consider the mapping $f : R \longrightarrow \mathbf{End}(R, +)$, defined by the rule $f(a) = \tau_a$ for each $a \in R$. If $x \in R$, then

$$\tau_{a+b}(x) = (a + b)x = ax + bx = \tau_a(x) + \tau_b(x) = (\tau_a + \tau_b)(x)$$

and

$$\tau_{ab}(x) = (ab)x = a(bx) = \tau_a(bx) = \tau_a(\tau_b(x)) = \tau_a \circ \tau_b(x).$$

From this it follows that $\tau_{a+b} = \tau_a + \tau_b$ and $\tau_{ab} = \tau_a \circ \tau_b$, for all $a, b \in R$. We therefore have

$$f(a + b) = \tau_{a+b} = \tau_a + \tau_b = f(a) + f(b) \text{ and}$$
$$f(ab) = \tau_{ab} = \tau_a \circ \tau_b = f(a) \circ f(b)$$

for all $a, b \in R$. These equations show that f is a ring homomorphism. Finally, if $a \neq b$, then

$$\tau_a(e) = ae = a \neq b = be = \tau_b(e),$$

so that $f(a) = \tau_a \neq \tau_b = f(b)$ and hence f is a monomorphism.

EXERCISE SET 7.4

7.4.1. On the set $\mathbb{R} \times \mathbb{R} \times \mathbb{R}$ we define operations of addition and multiplication by $(a, b, c) + (a_1, b_1, c_1) = (a + a_1, b + b_1, c + c_1)$, $(a, b, c)(a_1, b_1, c_1) = (aa_1 - bb_1, ab_1 + ba_1, ac_1 + ca_1)$. Is $\mathbb{R} \times \mathbb{R} \times \mathbb{R}$ a ring? If yes, is there a monomorphism from the field of complex numbers to this ring?

7.4.2. On the set $\mathbb{R} \times \mathbb{R}$ we define operations of addition and multiplication by $(a, b) + (c, d) = (a + c, b + d)$; $(a, b)(c, d) = (ac - 3bd, ad + 2bd +$

bc). Is $\mathbb{R} \times \mathbb{R}$ a field? If yes, is this field isomorphic to the field of complex numbers?

7.4.3. Let $K = \mathbb{Z}[\sqrt{2}] = \{a + b\sqrt{2} \mid a, b \in \mathbb{Z}\}$ and let L be the subset of $\mathbf{M}_2(\mathbb{Z})$ consisting of all matrices of the form

$$\begin{pmatrix} a & b \\ 2b & a \end{pmatrix}.$$

Prove directly that K is a subring of $\mathbb{R}$, that L is a subring of $\mathbf{M}_2(\mathbb{Z})$, and that K and L are isomorphic.

7.4.4. Let $P_1 = \{x + y\sqrt{2} \mid x, y \in \mathbb{Q}\}$, $P_2 = \{x + y\sqrt{5} \mid x, y \in \mathbb{Q}\}$. Prove directly that P_1 and P_2 are subrings of $\mathbb{R}$. Is the map $f : P_1 \longrightarrow P_2$ defined by the rule $f(x + y\sqrt{2}) = x + y\sqrt{5}$ an isomorphism of P_1 to P_2?

7.4.5. Let $K = \mathbb{Z}[i\sqrt{3}] = \{a + ib\sqrt{3} \mid a, b \in \mathbb{Z}\}$. Prove directly that K is a subring of $\mathbb{C}$. Define the mapping $f : K \longrightarrow \mathbf{M}_2(\mathbb{Z})$ by the rule

$$f(a + ib\sqrt{3}) = \begin{pmatrix} a & -3b \\ b & a \end{pmatrix}.$$

Is f a monomorphism?

7.4.6. Prove that the rings $(\mathbb{Z}/6\mathbb{Z})/(3\mathbb{Z}/6\mathbb{Z})$ and $\mathbb{Z}/3\mathbb{Z}$ are isomorphic.

7.4.7. Prove that the rings $(\mathbb{Z}/8\mathbb{Z})/(4\mathbb{Z}/8\mathbb{Z})$ and $\mathbb{Z}/4\mathbb{Z}$ are isomorphic.

7.4.8. Let R be an integral domain. Suppose that the additive subgroup $\langle e \rangle$ is finite. Prove that $|\langle e \rangle|$ is a prime.

7.4.9. Let R be an integral domain and let $a, b \in R$. Prove that the additive cyclic subgroups $\langle a \rangle$, $\langle b \rangle$ have the same orders.

7.4.10. Let $R = \{a + bi \mid a, b \in \mathbb{Z}\} = \mathbb{Z}[i]$ and let $H = 4iR$. Find the order of the quotient ring R/H. Find all zero-divisors and invertible elements of R/H.

7.4.11. Let $R = \{a + bi \mid a, b \in \mathbb{Z}\} = \mathbb{Z}[i]$ and let $H = 3R$. Find the order of the quotient ring R/H. Find all zero-divisors and invertible elements of R/H.

7.4.12. On the set $F = \mathbb{R} \times \mathbb{R}$ we define operations of addition and multiplication by $(a, b) + (c, d) = (a + c, b + d)$, $(a, b) \cdot (c, d) = (ac - 3bd, ad + 2bd + bc)$. Is F a field? If yes, is this field isomorphic to the field of complex numbers? Find a root of the polynomial $X^2 + 1$ in F.

7.4.13. Let $\mathbb{Q}[\sqrt{p}\,] = \{x + y\sqrt{p} \mid x, y \in \mathbb{Q}\}$ where p is a prime. Is $\mathbb{Q}[\sqrt{p}\,]$ a subring of $\mathbb{R}$? Is $\mathbb{Q}[\sqrt{p}\,]$ a subfield of $\mathbb{R}$? Are $\mathbb{Q}[\sqrt{7}]$ and $\mathbb{Q}[\sqrt{5}]$ isomorphic?

7.4.14. Prove that every field F of characteristic 0 contains one and only one field isomorphic to $\mathbb{Q}$.

7.4.15. Let R be a commutative ring. Prove that R is a field if and only if every epimorphism $f : R \longrightarrow L$ where L is a nonzero ring is an isomorphism.

7.4.16. Let F be a field and let M be the subset of all polynomials from the ring $F[X_1, X_2]$ having zero constant term. Prove that the quotient ring $F[X_1, X_2]/M$ is isomorphic to the field F.

7.4.17. Let F be a field and let M be the ideal of the polynomial ring $F[X_1, X_2]$, generated by $X_1 - X_2$. Prove that the quotient ring $F[X_1, X_2]/M$ is isomorphic to the ring $F[X_1]$.

7.5 RINGS OF POLYNOMIALS AND FORMAL POWER SERIES

Of all rings, the ring of polynomials is one of the most important and plays a fundamental role in commutative ring theory. Many important mathematical problems have been solved using this theory. Here we consider one example that requires us to consider the ring of polynomials.

Let K be a commutative ring with identity and let R be a unitary subring of K, containing no zero-divisors. Let M be a subset of K and consider the family

$$\mathfrak{M} = \{H \mid H \text{ is a subring of } K \text{ containing both } R \text{ and } M\}.$$

Put $R[M] = \cap\mathfrak{M}$, which we call the subring of K generated by R and M, or the subring generated over R by M. By Corollary 7.1.10 $R[M]$ is a subring. By its definition $R[M]$ is the least subring, which contains both the subring R and the subset M.

The simplest case to consider here is the case when M consists of one element y. In this case, we write $R[y]$ instead of $R[\{y\}]$ and we next determine the elements of $R[y]$. By Theorem 7.1.9 we note that $y^n \in R[y]$ for all $n \in \mathbb{N}$ and hence, for $a_0, a_1, \dots a_n \in R$, all possible sums of the type

$$a_0 + a_1 y + a_2 y^2 + \cdots + a_n y^n$$

belong to $R[y]$. Such an element is nothing more than a polynomial in y with coefficients in R. Moreover, the subring $R[y]$ consists of all these sums. Indeed, for two arbitrary such sums $a_0 + a_1 y + a_2 y^2 + \cdots + a_n y^n$ and $b_0 + b_1 y + b_2 y^2 + \cdots + b_k y^k$ where $n \geq k$ we have

$$\begin{aligned}(a_0 + a_1 y + \ldots + a_n y^n) + (b_0 + b_1 y + \cdots + b_k y^k) = (a_0 + b_0) + (a_1 + b_1)y \\ + \cdots (a_k + b_k)y^k \\ + a_{k+1} y^{k+1} + \cdots + a_n y^n;\end{aligned}$$

and

$$\begin{aligned}(a_0 + a_1 y + \cdots + a_n y^n)(b_0 + b_1 y + \cdots + b_k y^k) = a_0 b_0 &+ (a_0 b_1 + a_1 b_0) y \\ &+ (a_0 b_2 + a_1 b_1 + a_2 b_0) y^2 \\ &+ \cdots + a_n b_k y^{n+k}.\end{aligned}$$

By Theorem 7.1.9, the set of all sums of the type $a_0 + a_1 y + \cdots + a_n y^n$ forms a subring. Since this subring contains R and y, it coincides with $R[y]$. Note that y is an element of the ring K and, for this reason, some polynomials in y that look different can actually give the same polynomial, which is clearly inconvenient. In order to overcome this indeterminacy, we need to remove multiple ways of writing an element somehow. The term *variable* arises in this way and needs to be denoted by some symbol. There is a better way to proceed here, where the main idea lies in using the operations defined above to construct polynomials. For example, every polynomial can be defined by its coefficients and if we formally define operations between coefficients then it is possible to formally retrieve the ring of polynomials. Here we will be a little bit more general and define polynomials as a certain subring of some larger ring, namely, the ring of formal power series. Let R be an integral domain and consider the set $R[[X]]$ of all sequences

$$(a_n)_{n \in \mathbb{N}_0} = (a_0, a_1, \ldots, a_n, \ldots),$$

where $a_i \in R$. We shall often write $(a_n)_{n \in \mathbb{N}_0}$ as simply (a_n) for brevity.

Two sequences $(a_n)_{n \in \mathbb{N}_0}$ and $(b_n)_{n \in \mathbb{N}0}$ are called equal, if $a_n = b_n$ for each $n \in \mathbb{N}_0$. We define the following operations of addition and multiplication in $R[[X]]$:

$$(a_n) + (b_n) = (a_n + b_n)$$

and

$$(a_n)(b_n) = (d_n)$$

where

$$d_n = a_0 b_n + a_1 b_{n-1} + \cdots + a_n b_0 = \sum_{0 \le j \le n} a_j b_{n-j} = \sum_{j+k=n} a_j b_k,$$

for each $n \in \mathbb{N}_0$.

In this way addition in $R[[X]]$ is reduced to the addition of corresponding elements of the ring R, so it is commutative and associative. Clearly, the zero element is the sequence

$$0 = (0_R, 0_R, \ldots, 0_R, \ldots).$$

Every sequence (a_n) has a negative, $(-a_n)$, which is the additive inverse of (a_n). So, this set of sequences is an abelian group under addition.

It is easy to prove that multiplication of sequences is commutative, so we need to prove the distributivity rule:

$$((a_n) + (b_n))(c_n) = (a_n)(c_n) + (b_n)(c_n).$$

Let

$$(u_n) = ((a_n) + (b_n))(c_n) \text{ and } (v_n) = (a_n)(c_n) + (b_n)(c_n).$$

Then, for arbitrary $n \in \mathbb{N}_0$, we have

$$u_n = \sum_{j+k=n} (a_j + b_j)c_k = \sum_{j+k=n} a_j c_k + \sum_{j+k=n} b_j c_k \text{ and}$$

$$v_n = \sum_{j+k=n} a_j c_k + \sum_{j+k=n} b_j c_k.$$

Hence, $u_n = v_n$ for each $n \in \mathbb{N}_0$ so the distributive law follows.

The multiplication is associative. To see this let

$$((a_n)(b_n))(c_n) = (w_n) \text{ and } (a_n)((b_n)(c_n)) = (z_n).$$

For arbitrary $n \in \mathbb{N}_0$ we have

$$w_n = \sum_{m+t=n} \left(\sum_{j+k=m} a_j b_k \right) c_t = \sum_{j+k+t=n} (a_j b_k) c_t \text{ and}$$

$$z_n = \sum_{j+m=n} a_j \left(\sum_{k+t=m} b_k \right) c_t = \sum_{j+k+t=n} a_j (b_k c_t).$$

Hence $w_n = z_n$ for each $n \in \mathbb{N}_0$, so the multiplication is associative.

The element $e = (e, 0_R, 0_R, \ldots, 0_R, \ldots)$ is the multiplicative identity, as is easily seen.

Consequently, $R[[X]]$ is a commutative ring with identity and, indeed, $R[[X]]$ is an integral domain. To see this latter fact, let $(a_n) \neq 0$ and $(b_n) \neq 0$ be two sequences in $R[[X]]$. Then there are indices j, k such that $a_j \neq 0_R$, $b_k \neq 0_R$ but $a_n = 0_R$ for $0 \leq n \leq j$, and $b_n = 0_R$ for $0 \leq n \leq k$. If

$$(a_n)(b_n) = (d_n)$$

then

$$d_{j+k} = a_0 b_{j+k} + \cdots + a_{j-1} b_{k+1} + a_j b_k + a_{j+1} b_{k-1} + \cdots + a_{j+k} b_0 = a_j b_k \neq 0_R,$$

which proves that $R[[X]]$ is an integral domain.

7.5.1. Definition. *A sequence $(a_n)_{n\in\mathbb{N}_0}$ is called a polynomial of degree $m \geq 1$, if there exists a positive integer m such that $a_n = 0_R$ whenever $n > m$, but $a_m \neq 0_R$. In this case the coefficient a_m is called the leading coefficient of the polynomial.*

If $a_n = 0_R$ for all $n \in \mathbb{N}_0$, then the polynomial is called the zero polynomial. Sometimes the zero polynomial is considered to be a polynomial of infinite degree. We denote the subset of $R[[X]]$ consisting of all polynomials by $R[X]$. Next, let $(a_n), (b_n)$ be two polynomials of degree m, t, respectively. Then $a_n = b_n = 0_R$ for $n > \mathbf{max}\{m, t\}$ and

$$a_0b_n + a_1b_{n-1} + \cdots + a_jb_{n-j} + \cdots + a_nb_0 = 0_R,$$

for $n > m + t$. Theorem 7.1.9 therefore shows that $R[X]$ is a subring of the ring $R[[X]]$ and clearly $R[X]$ is a unitary subring.

Now let $X = (0_R, e, 0_R, \ldots, 0_R, \ldots)$. By the definition of multiplication, we obtain

$$X^n = (\underbrace{0_R, 0_R, \ldots, 0_R}_{n}, e, 0_R, \ldots, 0_R, \ldots).$$

For an arbitrary element $a \in R$ we let

$$\overline{a} = (a, 0_R, 0_R, \ldots, 0_R, \ldots).$$

Then

$$\overline{a}X^n = (\underbrace{0_R, 0_R, \ldots, 0_R}_{n}, a, 0_R, \ldots, 0_R, \ldots).$$

We next let $(a_n)_{n\in\mathbb{N}_0}$ be a polynomial of degree m. We can write this polynomial in the form

$$\begin{aligned}(a_n) &= (a_0, 0_R, 0_R, \ldots, 0_R, 0_R, 0_R, \ldots) + (0_R, a_1, 0_R, \ldots, 0_R, 0_R, 0_R, \ldots)\\ &\quad + \cdots (0_R, 0_R, 0_R, \ldots, 0_R, a_m, 0_R, \ldots)\\ &= \overline{a}_0 + \overline{a}_1X + \cdots + \overline{a}_mX^m.\end{aligned}$$

Furthermore,

$$\begin{aligned}\overline{a} + \overline{b} &= (a, 0_R, \ldots, 0_R, \ldots) + (b, 0_R, \ldots, 0_R, \ldots)\\ &= (a + b, 0_R, \ldots, 0_R, \ldots) = \overline{a + b} \text{ and likewise}\\ \overline{a}\overline{b} &= (ab, 0_R, \ldots, 0_R, \ldots) = \overline{ab}.\end{aligned}$$

These equations show that the mapping $a \longmapsto \overline{a}$, where $a \in R$, is a homomorphism of R into $R[[X]]$, which is clearly injective. In particular, R is isomorphic to its image in $R[[X]]$. In algebra, we often consider isomorphic objects as identical, so we identify R with its image and instead of $\overline{a}$ we just write a. Thus, $a \in R$ is identified with the corresponding element $\overline{a}$ of $R[[X]]$. In this way, we can write a polynomial in the form

$$f(X) = a_0 + a_1 X + \cdots + a_m X^m,$$

which, of course, is the normal way of writing a polynomial and we have now formally justified this process. We denote the degree of the polynomial $f(X)$ by $\mathbf{deg} f(X)$. From the properties mentioned above, we observe the following relations:

$$\mathbf{deg}(f(X) \pm g(X)) \leq \mathbf{max}\{\mathbf{deg} f(X), \mathbf{deg}\, g(X)\} \text{ and}$$
$$\{\mathbf{deg}(f(X)g(X)) = \mathbf{deg} f(X) + \mathbf{deg} g(X)\}.$$

From the second equation, we deduce that $\mathbf{U}(R[X]) = \mathbf{U}(R)$ and that $R[X]$ has no zero-divisors if R has no such elements.

We now return to formal power series. Just as with polynomials we can write them as a sum of powers of X; however, the way we write such power series involves infinite sums. Since we cannot usually talk about limits of sequences in arbitrary rings, we must agree on some rules for adding an infinite set of sequences.

A collection of sequences, $\{A_j \mid j \in \mathbb{N}_0\}$, where $A_j = (a_{jn})_{n\in\mathbb{N}_0}$ and $a_{jn} \in R$, is called summable if the set $\{a_{jn} \mid j \in \mathbb{N}_0\}$ contains only a finite set of nonzero elements, for each $n \in \mathbb{N}_0$. In this case, the sum $\sum_{j\in\mathbb{N}_0} A_j$ of the collection of sequences $\{A_j \mid j \in \mathbb{N}_0\}$ is the sequence $(b_n)_{n\in\mathbb{N}_0}$, where b_n is the sum of all the nonzero elements of the sequence $\{a_{jn} \mid j \in \mathbb{N}_0\}$.

We now show that the collection $\{a_n X^n \mid n \in \mathbb{N}_0\}$ is summable and that its sum is equal to $(a_n)_{n\in\mathbb{N}_0}$. To see this we write these sequences one under another and check if every column of the resulting infinite matrix contains only finitely many nonzero elements. By adding the sequences, we will add finitely many elements in each column. We have

$$(a_n)_{n\in\mathbb{N}_0} = \begin{array}{llllllllll} (a_0, & 0_R, & 0_R, & 0_R, & \ldots, & 0_R, & 0_R, & 0_R, & \ldots) & + \\ (0_R, & a_1, & 0_R, & 0_R, & \ldots, & 0_R, & 0_R, & 0_R, & \ldots) & + \\ \ldots & \ldots & \ldots & \ldots & \ldots & \ldots & \ldots & \ldots & \ldots & \ldots . \\ (0_R, & 0_R, & 0_R, & 0_R, & \ldots & 0_R, & a_m & 0_R, & \ldots) & + \\ \ldots & \ldots & \ldots & \ldots & \ldots & \ldots & \ldots & \ldots & \ldots & \ldots \end{array}$$

In this way, the formal power series $(a_n)_{n\in\mathbb{N}_0}$ can be written in the more regular form as $\sum_{n\in\mathbb{N}_0} a_n X^n$.

Finally, we will find which series lie in the group $\mathbf{U}(R[[X]])$. Let $f(X) = \sum_{n\in\mathbb{N}_0} a_n X^n$, $g(X) = \sum_{n\in\mathbb{N}_0} b_n X^n$ and suppose that $f(X)g(X) = e$. It follows that $a_0 b_0 = e = b_0 a_0$, since R is commutative and hence $a_0 \in \mathbf{U}(R)$.

Conversely, suppose that $a_0 \in \mathbf{U}(R)$. Suppose that $f(X)g(X) = e = g(X)f(X)$ and let us use this equation to determine the coefficients of $g(X)$ in terms of those of $f(X)$. In this case it will follow that $f(X)$ has a multiplicative inverse and then $f(X) \in \mathbf{U}(R[\![X]\!])$. We have $a_0b_0 = e$, so that $b_0 = a_0^{-1}$. Further $a_0b_1 + a_1b_0 = 0_R$. It follows that $b_1 = a_1a_0^{-2}$. If the coefficients $b_0, b_1, \ldots, b_n$ have been defined then the coefficient b_{n+1} can be found from the equation

$$a_0b_{n+1} + a_1b_n + \cdots + a_nb_1 + a_{n+1} = 0_R,$$

and we see that

$$b_{n+1} = (-a_1b_n - \cdots - a_nb_1 - a_{n+1})a_0^{-1}.$$

Thus, the group $\mathbf{U}(R[\![X]\!])$ consists of all series of the form $\sum_{n\in\mathbb{N}_0} a_nX^n$ where $a_0 \in \mathbf{U}(R)$. In particular, if R is a field, then $\mathbf{U}(R[\![X]\!])$ consists of series of the type $\sum_{n\in\mathbb{N}_0} a_nX^n$, in which $a_0 \neq 0_R$.

We next determine the ideals in the ring of formal power series. The following important theorems describe their structure. We first prove a version of the division algorithm of $\mathbb{Z}$ in $F[X]$, in the case when F is a field.

7.5.2. Theorem. *Let F be a field, let $f(X), g(X) \in F[X]$ and suppose that $g(X) \neq 0_F$. Then there exist polynomials $q(X), r(X) \in F[X]$ such that $f(X) = q(X)g(X) + r(X)$, where either $r(X) = 0_F$ or* $\mathbf{deg}\, r(X) < \mathbf{deg}\, g(X)$. *This representation is unique.*

Proof. Let

$$f(X) = a_0 + a_1X + \cdots + a_nX^n \text{ and } g(X) = b_0 + b_1X + \cdots + b_kX^k,$$

where $b_k \neq 0_F$. We will use induction on n to prove the theorem. If $\mathbf{deg}\, f(X) < \mathbf{deg}\, g(X)$ then put $r(X) = f(X)$ and $q(X) = 0_F$. Thus, we may assume that $\mathbf{deg} f(X) \geq \mathbf{deg} g(X)$. If $\mathbf{deg} f(X) = 0$ then we set $r(X) = 0_F$, $q(X) = a_0b_0^{-1}$. Suppose now that $n > 0$ and suppose that our theorem has been proved for all polynomials of degree less than n. The polynomial $a_nb_k^{-1}X^{n-k}g(X)$ has degree n and its leading coefficient is a_n. Then the degree of the polynomial $f(X) - a_nb_k^{-1}X^{n-k}g(X) = f_1(X)$ is less than n and, by induction,

$$f_1(X) = q_1(X)g(X) + r(X),$$

where either $r(X) = 0_F$ or $\mathbf{deg}\, r(X) < \mathbf{deg}\, g(X)$. We now have

$$f(X) = a_nb_k^{-1}X^{n-k}g(X) + f_1(X) = q(X)g(X) + r(X),$$

where

$$q(X) = q_1(X) + a_nb_k^{-1}X^{n-k}$$

and the first part follows.

Suppose also that

$$f(X) = q_2(X)g(X) + r_2(X),$$

where either $r_2(X) = 0_F$ or $\mathbf{deg}\, r_2(X) < \mathbf{deg}\, g(X)$. Then

$$q(X)g(X) + r(X) = q_2(X)g(X) + r_2(X),$$

and it follows that

$$g(X)(q(X) - q_2(X)) = r_2(X) - r(X).$$

The polynomial $r_2(X) - r(X)$ is either zero or its degree is less than $\mathbf{deg}\, g(X)$. On the other hand, if $q(X) - q_2(X) \neq 0_F$, then

$$\mathbf{deg}(g(X)(q(X) - q_2(X)) = \mathbf{deg}\, g(X) + \mathbf{deg}(q(X) - q_2(X)) \geq \mathbf{deg}\, g(X),$$

which gives a contradiction if $r_2(X) - r(X) \neq 0_F$. Thus, $r_2(X) - r(X) = 0_F$, which implies that $q(X) - q_2(X) = 0_F$ and hence $r_2(X) = r(X)$. This establishes the uniqueness portion of the result.

7.5.3. Corollary. *Let F be a field. Then every ideal of the ring $F[X]$ is principal.*

Proof. Let H be an ideal of the ring $F[X]$. If $H = \{0_F\}$, then $H = 0_F F[X]$, which is a principal ideal. Assume next that H contains nonzero polynomials. Among these we choose a polynomial $g(X)$ of least degree. Let $f(X)$ be an arbitrary element of H. By Theorem 7.5.2, we know that

$$f(X) = q(X)g(X) + r(X),$$

where either $r(X) = 0_F$ or $\mathbf{deg}\, r(X) < \mathbf{deg}\, g(X)$.

If we suppose that $r(X) \neq 0_F$, then $r(X) = f(X) - q(X)g(X) \in H$, which contradicts the choice of the polynomial $g(X)$. This contradiction proves that $r(X) = 0_F$, so that $f(X) = q(X)g(X)$. It follows that $H = g(X)F[X]$ and again H is a principal ideal. The result follows.

Here is a further example of a ring all of whose ideals are principal.

7.5.4. Theorem. *Let F be a field. Then every ideal of the ring $F[\![X]\!]$ is principal.*

Proof. Let H be an ideal of $F[\![X]\!]$. If $H = \{0_F\}$, then $H = 0_F F[\![X]\!]$ is a principal ideal.

Assume next that H contains a nonzero formal power series. If $\sum_{n \in \mathbb{N}_0} a_n X^n \in H$ and $a_0 \neq 0_F$, then, as mentioned earlier,

$$\sum_{n \in \mathbb{N}_0} a_n X^n \in \mathbf{U}(F[\![X]\!]).$$

By Proposition 7.3.9, $H = F[\![X]\!]$, so that $H = eF[\![X]\!]$, a principal ideal. Next consider the case when the constant term a_0 of every power series $\sum_{n\in\mathbb{N}_0} a_n X^n$ in H is zero. This means, that every element $f(X)$ of H can be written as $Xg(X)$ for some $g(X) \in F[\![X]\!]$. We choose a natural number m such that for some $f(X) \in H$ we have $f(X) = X^m g(X)$, where $g(X)$ is a series with nonzero constant term. Thus, $g(X) \in U(F[\![X]\!])$. Let $h(X)$ be a power series such that $g(X)h(X) = e$. Since H is an ideal of the ring $F[\![X]\!]$, then

$$f(X)h(X) = X^m g(X)h(X) = X^m \in H. \tag{7.1}$$

Hence $X^m F[\![X]\!] \leq H$. We choose m least such that $X^m \in H$. If $f(X) \in H$ then $f(X) = X^n k(X)$ for some $n \geq m$ and $k(X) \in F[\![X]\!]$. Thus, $f(X) \in X^m F[\![X]\!]$ and hence $H = X^m F[\![X]\!]$.

We now return to the ring $R[y]$ which we discussed at the start of this section. Let K be a commutative ring and let R be a unitary subring of K. Suppose that R has no zero-divisors and that $y \in K$. If $f(X) \in R[X]$ then

$$f(X) = a_0 + a_1 X + \cdots + a_n X^n, \quad \text{where } a_0, a_1, \ldots, a_n \in R.$$

Let

$$f(y) = a_0 + a_1 y + \cdots + a_n y^n.$$

Clearly, $f(y) \in R[y]$.

7.5.5. Definition. *An element y is said to be a root, or a zero, of a polynomial $f(X)$, if $f(y) = 0_R$.*

A root might or might not be an element of the ring R. For example, the polynomial $X^2 - 2 \in \mathbb{Q}[X]$ has no rational roots; its roots are real numbers and belong to the field $\mathbb{R}$; the polynomial $X^2 + 1 \in \mathbb{Q}[X]$ has no real roots. It is easy to check the following.

7.5.6. Proposition. *Let K be a commutative ring and let R be a unitary subring of K. Suppose that R is an integral domain. If y is a fixed element of K, then the mapping $\eta : R[X] \longrightarrow K$, defined by $\eta(f(X)) = f(y)$ for each $f(X) \in R[X]$, is a homomorphism, sometimes called the evaluation homomorphism.*

Consider now $\mathbf{Im}\,\eta$. By Proposition 7.4.3, $\mathbf{Im}\,\eta$ is a subring of K. If $f(X) = a$ is a polynomial of zero degree, then $\eta(f(X)) = a$. It follows from this that $R \leq \mathbf{Im}\,\eta$. On the other hand,

$$\mathbf{Im}\,\eta = \{\eta(f(X)) \mid f(X) \in R[X]\} = \{f(y) \mid f(X) \in R[X]\},$$

and, as we noted above, $f(y) \in R[y]$ for any $f(X) \in R[X]$. This means that $\mathbf{Im}\,\eta = R[y] = \{f(y) \mid f(X) \in R[X]\}$.

Next we consider $\mathbf{Ker}\,\eta$.

7.5.7. Definition. *An element y is said to be transcendental over the ring R, if there is no polynomial with coefficients in R for which y is a root of that polynomial.*

This means that if an element y is transcendental over R, then $\mathbf{Ker}\,\eta = \{0_F\}$. Theorem 7.4.4 shows that η is then a monomorphism and in this case $R[X] \cong R[y]$.

We next consider the case when $\mathbf{Ker}\,\eta \neq \{0_F\}$. By Proposition 7.4.3, $\mathbf{Ker}\,\eta$ is an ideal of the ring $R[X]$ and $\mathbf{Ker}\,\eta$ consists only of those polynomials $f(X)$ for which $f(y) = 0_F$. Thus, $\mathbf{Ker}\,\eta$ consists of all polynomials that have y as a root.

7.5.8. Definition. *An element y is said to be algebraic over the ring R if there exists a polynomial $f(X) \in R[X]$ such that y is a root of $f(X)$.*

If R is a field then, from Corollary 7.5.3 we deduce that $\mathbf{Ker}\,\eta = h(X)R[X]$ for some polynomial $h(X) \in R[X]$. If $r(X)$ is another polynomial with the property that $\mathbf{Ker}\,\eta = r(X)R[X]$, then

$$h(X) = r(X)u(X) \text{ and } r(X) = h(X)v(X),$$

for some $u(X), v(X) \in R[X]$. It follows that

$$h(X) = r(X)u(X) = h(X)v(X)u(X),$$

so $v(X)u(X) = e$. Since $\mathbf{U}(R[X]) = \mathbf{U}(R)$, this means that $v(X) = c \in \mathbf{U}(R)$ and $u(X) = c^{-1}$. Hence $r(X) = ch(X)$ where $c \in \mathbf{U}(R)$.

Among all the polynomials generating the ideal $\mathbf{Ker}\,\eta$, we choose the one whose leading coefficient is the identity element and we denote this polynomial by $\mathbf{m}_y(X)$. The polynomial $\mathbf{m}_y(X)$ is called the minimal polynomial of y over the field R. By Theorem 7.4.6, we have

$$R[y] = \mathbf{Im}\,\eta \cong R[X]/\mathbf{Ker}\,\eta \text{ and } \mathbf{Ker}\,\eta = \mathbf{m}_y(X)R[X].$$

Let F be a field and let $y \in F$. From Theorem 7.5.2 we obtain the decomposition $f(X) = q(X)(X - y) + r(X)$ where either $r(X) = 0_F$ or $\mathbf{deg}\,r(X) < \mathbf{deg}\,(X - y) = 1$. Hence, in any case, $r(X) \in F$, so $r(X) = b \in F$. By Proposition 7.5.6,

$$f(y) = q(y)(y - y) + b = b.$$

Thus, $f(X) = (X - y)q(X) + f(y)$.

We say that a polynomial $f(X)$ is divisible by a polynomial $g(X)$ if there exists a polynomial $q(X)$ such that $f(X) = g(X)q(X)$. We have now proved.

7.5.9. Proposition. *Let F be a field and let $f(X) \in F[X]$. The element $y \in F$ is a root of a polynomial $f(X)$ if and only if $(X - y)$ divides $f(X)$.*

7.5.10. Corollary. *Let F be a field and let $f(X) \in F[X]$. Then*

$$f(X) = (X - c_1)^{k_1} \dots (X - c_m)^{k_m} q(X),$$

where the polynomial $q(X)$ has no roots in the field F, $c_j \neq c_t$ whenever $j \neq t$ and $k_1, \dots, k_m \in \mathbb{N}$.

7.5.11. Corollary. *Let F be a field, let $f(X) \in F[X]$ and let $M = \{a \in F \mid a$ is a root of a polynomial $f(X)\}$. Then $|M| \leq \mathbf{deg}\, f(X)$.*

This draws attention to certain fields in which every polynomial has a root. We make the following definition.

7.5.12. Definition. *A field F is called algebraically closed if every polynomial $f(X) \in F[X]$ has a root in F.*

7.5.13. Corollary. *Let F be a field. Then F is algebraically closed if and only if, for any polynomial $f(X) \in F[X]$,*

$$f(X) = a(X - c_1)^{k_1} \dots (X - c_m)^{k_m},$$

where a is the leading coefficient of the polynomial $f(X)$, $c_1, \dots, c_m \in F$, $c_j \neq c_t$ whenever $j \neq t$ and $k_1, \dots, k_m \in \mathbb{N}$.

From this corollary, we obtain the following set of equations, collectively known as the Viète formulas. They reflect important relations between the roots and the coefficients of a polynomial.

Let

$$f(X) = a_0 + a_1 X + \dots + a_n X^n,$$

where $a_n \neq 0_F$ and also suppose that

$$f(X) = a(X - c_1)(X - c_2) \dots (X - c_n).$$

Then we have

$$\begin{aligned}
a_0 &= (-1)^n a c_1 c_2 \dots c_n;\\
a_1 &= (-1)^n a (c_1 c_2 \dots c_{n-1} + c_1 c_3 \dots c_{n-1} c_n + \dots + c_2 c_3 \dots c_{n-1} c_n);\\
&\vdots\\
a_{n-2} &= a(c_1 c_2 + c_1 c_3 + \dots + c_1 c_n + c_2 c_3 + \dots + c_{n-1} c_n);\\
a_{n-1} &= -a(c_1 + c_2 + \dots + c_n);\\
a_n &= a.
\end{aligned}$$

The following important theorem gives us our first example of an algebraically closed field, but its proof is omitted.

7.5.14. Theorem *(Gauss). The field $\mathbb{C}$ of complex numbers is algebraically closed.*

A complex number α is called algebraic if α is an algebraic element over the field $\mathbb{Q}$. A complex number α is called transcendental if α is a transcendental element over the field $\mathbb{Q}$.

It is easy to find examples of algebraic numbers. However, it is not easy to prove that a given number is transcendental.

In the nineteenth century, Lindeman proved that e is a transcendental number. In the same century, Liouville proved that π is transcendental. The Russian mathematician, Gelfond (1906–1968), developed an advanced theory, allowing us to determine the transcendency of a wide class of numbers appearing in analysis. Many of the proofs of these facts use powerful analytic methods, so we do not consider them here. However, we illustrate how to construct certain examples of transcendental numbers using a technique due to Liouville. The following result is the basis of this method.

7.5.15. Proposition. *Let α be an algebraic number and let $f(X) \in \mathbb{Q}[X]$ be the minimal polynomial of α. Suppose that $n = \mathbf{deg}\, f(X)$. Then there exists a number $\mu = \mu(n)$, independent of p, q, such that for each rational number $\frac{p}{q} \neq \alpha$ the following inequality holds*

$$\left|\alpha - \frac{p}{q}\right| > \frac{\mu}{q^n}.$$

Proof. Let

$$f(X) = a_0 + a_1X + \cdots + a_nX^n \in \mathbb{Q}[X].$$

Without loss of generality, we can suppose that $a_0, a_1, \ldots, a_n \in \mathbb{Z}$ and we note also that $f(\frac{p}{q}) \neq 0$, since $f(X)$ is irreducible over $\mathbb{Q}$. Also we note that

$$\begin{aligned} f\left(\frac{p}{q}\right) &= a_0 + a_1\left(\frac{p}{q}\right) + \cdots + a_n\left(\frac{p}{q}\right)^n \\ &= \frac{a_0q^n + a_1pq^{n-1} + \cdots + a_np^n}{q^n}, \end{aligned}$$

so that $|f(\frac{p}{q})| \geq \frac{1}{q^n}$, since $a_0q^n \ldots a_np^n \in \mathbb{Z}$. If $|\frac{p}{q} - \alpha| > 1$ then we take $\mu = 1$. Since $f(X)$ has α as a root we have $f(X) = (X - \alpha)f_1(X)$ so that $f(\frac{p}{q}) = (\frac{p}{q} - \alpha)f_1(\frac{p}{q})$. If $|\frac{p}{q} - \alpha| \leq 1$ then let

$$\lambda > \mathbf{Max}\{f_1(x) \mid \text{ where } |x - \alpha| \leq 1\}.$$

In this case,

$$\frac{1}{q^n} \leq \left| f(\frac{p}{q}) \right| \leq \left| \frac{p}{q} - \alpha \right| \lambda.$$

It follows that

$$\left| \frac{p}{q} - \alpha \right| > \frac{1}{\lambda} \cdot \frac{1}{q^n}.$$

If $\lambda < 1$ then $\frac{1}{q^n} < \left| \frac{p}{q} - \alpha \right|$ and again we take $\mu = 1$. If $\lambda \geq 1$ then

$$\left| \frac{p}{q} - \alpha \right| \geq \frac{1}{\lambda} \cdot \frac{1}{q^n} > \frac{1}{2\lambda} \cdot \frac{1}{q^n},$$

so we may take $\mu = \frac{1}{2\lambda}$ in this case.

One consequence of Proposition 7.5.15 is that it is not possible, for all rationals p/q, for $|\alpha - \frac{p}{q}|$ to be less than K/q^{n+1} for some fixed K independent of p, q. Otherwise, we would have $K/q^{n+1} > \mu/q^n$, which implies that $K > \mu q$ for all integers q, which is false.

Now it is easy to observe, for example, that the number $\gamma = \sum_{m \geq 1} \frac{1}{10^{m!}}$ is transcendental. Suppose that γ is algebraic. Let $\gamma_r = \sum_{m=1}^{r} \frac{1}{10^{m!}}$. Then $\gamma_r = \frac{p}{10^{r!}}$, say, a rational number. We have

$$\begin{aligned} 0 < \gamma - \gamma_r &= \frac{1}{10^{(r+1)!}} + \frac{1}{10^{(r+2)!}} + \cdots \\ &< \frac{1}{10^{(r+1)!}} + \frac{1}{10 \cdot 10^{(r+1)!}} \frac{1}{10^2 \cdot 10^{(r+1)!}} \cdots \\ &< \frac{2}{10^{(r+1)!}} \end{aligned} \tag{7.2}$$

However, if n is arbitrary and $r > n$ then $10^{r!n} < 10^{r!r} < 10^{(r+1)!}$ so that $\frac{\mu}{10^{r!n}} > \frac{\mu}{10^{(r+1)!}}$. Hence (7.2) implies that

$$\gamma - \gamma_r = \gamma - \frac{p}{q} < \frac{2}{q^n},$$

where $q = 10^{r!}$. By our remark above, it now follows that if the minimal polynomial of γ is of degree $n - 1$, then we obtain a contradiction. Since n is arbitrary this also gives a contradiction. Consequently, γ is transcendental.

EXERCISE SET 7.5

7.5.1. Prove that the polynomials $f(X), g(X) \in \mathbb{F}_5[X]$ are equal, where $f(X) = (X+3)^5$ and $g(X) = X^5 + 3$.

7.5.2. For which values of a, b, c are the following polynomials $f(X), g(X) \in \mathbb{Z}[X]$, $f(X) = aX^2(X+1) + b(X^2+1)(X-6) + cX(X^2+1)$, $g(X) = X^2 + 5X + 6$ equal?

7.5.3. For which values of a, b, c are the following polynomials $f(X), g(X) \in \mathbb{Z}[X]$, $f(X) = aX^2(X+3) + b(X-1)(X-6) + c(X+1)$, $g(X) = 2X^3 + 5X^2 + 8X + 7$ equal?

7.5.4. For which values of a is the polynomial $f(X) \in \mathbb{Z}[X]$, $f(X) = X^4 + 6X^3 + 11X^2 + aX + 1$ the square of some polynomial $g(X) \in \mathbb{Z}[X]$. Find all such $g(X)$.

7.5.5. For which values of a, b is the polynomial $f(X) \in \mathbb{Z}[X]$, $f(X) = X^4 + 6X^3 + 11X^2 + aX + 1$ the cube of some polynomial $g(X) \in \mathbb{Z}[X]$.

7.5.6. Can the polynomial $f(X) = 8X^6 - 36aX^5 + 66a^2X^4 - 63a^3X^3 + 33a^4X^2 - 9a^5X + a^6 \in \mathbb{R}[X]$ be the cube of a polynomial $g(X) \in \mathbb{Z}[X]$, where $a \in \mathbb{R}$?

7.5.7. Let F be a finite field and $\rho : F \longrightarrow F$ be a mapping. Prove that there exists a polynomial $f(X) \in F[X]$ such that $\rho(a) = f(a)$ for each $a \in F$.

7.5.8. Prove that the ring $\mathbb{Z}[X]$ does not contain a polynomial $f(X)$ such that $f(7) = 11$ and $f(11) = 13$.

7.5.9. For which values of a does the polynomial $g(X) = X^2 - a \in \mathbb{Z}[X]$ divide the polynomial $f(X) = 3X^4 - 2X^2 - 5 \in \mathbb{Z}[X]$?

7.5.10. For which values of a does the polynomial $g(X) = X^2 - a \in \mathbb{Q}[X]$ divide the polynomial $f(X) = 3X^4 - 2X^2 - 5 \in \mathbb{Q}[X]$?

7.5.11. For which values of a, b, c does the polynomial $g(X) = X^2 + a - 1 \in \mathbb{R}[X]$ divide the polynomial $f(X) = X^3 - bX - c \in \mathbb{R}[X]$?

7.5.12. Divide the polynomial $f(X) = 4X^5 - 6X^3 + 2X^2 - 4$ by $g(X) = 2X^2 - 5X + 1$ in the ring $\mathbb{Q}[X]$.

7.5.13. Divide the polynomial $f(X) = (2i+3)X^3 - 4iX + i - 24$ by $g(X) = X^2 + i$ in the ring $\mathbb{C}[X]$.

7.5.14. Divide the polynomial $f(X) = 4X^3 + 2X^2 - X + 1$ by $g(X) = 2X + 3$ in the ring $\mathbb{F}_5[X]$.

7.5.15. For which values of a, b do the polynomials $g_1(X) = X - 1, g_2(X) = X + 1 \in \mathbb{R}[X]$ simultaneously divide the polynomial $f(X) = X^5 - a^2 X^2 + bX + 1 \in \mathbb{R}[X]$?

7.5.16. Let R be an integral domain, let $a \in R$ and let $f(X) = X^n - a^n, n \in \mathbb{N}$. Prove that $f(X)$ is divisible by $g(X) = X - a$.

7.5.17. The subset M of the ring $\mathbb{Z}[X]$ contains the number 0 and all polynomials of the form $f(X) = a_0X^2 + a_1X^3 + \cdots + a_{n-2}X^n$. Prove that M is an ideal of $\mathbb{Z}[X]$. Is M a principal ideal?

7.5.18. Let F be a subfield of a field P and let $\alpha \in P$ be algebraic over F. Let $H = \{f(X) \in F[X] \mid f(\alpha) = 0_F\}$. Prove that H is an ideal of $F[X]$.

7.5.19. Let α be an irrational root of the polynomial $pX^2 + qX + r$, where $p, q, r \in \mathbb{R}$ and $p \neq 0$. Is the subset $\mathbb{Q}[\alpha] = \{x + y\alpha \mid x, y \in \mathbb{Q}\}$ a subfield of $\mathbb{R}$?

7.5.20. If α is a root of the polynomial $X^n - 1 \in \mathbb{C}[X]$, then prove that $\alpha^{n-1} + \alpha^{n-2} + \cdots + \alpha + 1 = 0$.

7.6 RINGS OF MULTIVARIABLE POLYNOMIALS

In Section 7.5 we constructed the ring of polynomials in one variable over a commutative ring R. In this section, we extend this construction, quite naturally, to the case of multivariable polynomials. As in Section 7.5, we suppose that the commutative ring R does not have zero-divisors.

Let $K = R[X]$ be the ring of polynomials in one variable X, as constructed in Section 7.5 and recall that K also has no zero-divisors. In turn, we can consider the ring $Q = K[Y]$ of polynomials in the variable Y over the ring K. Here the variable Y plays the same role relative to K as the variable X played relative to R. Elements of Q can be written in the form:

$$b_0 + b_1Y + \cdots + b_mY^m, \text{ where } b_j \in K, \text{ for } 0 \leq j \leq m,$$

and this form is unique. Every element b_j can be written in the form

$$b_j = a_{j0} + a_{j1}X + \cdots + a_{j,n(j)}X^{n(j)}.$$

It follows from the construction that the variable permutes with each element of the ring over which the polynomials have been constructed. In particular, the variables X and Y permute. Therefore, every element of Q has a presentation in the form

$$\sum_{1\leq j\leq m} \sum_{1\leq k\leq n(j)} a_{jk}X^kY^j, \text{ where } a_{jk} \in R, \text{ for } 1 \leq k \leq n(j) \text{ and } 1 \leq j \leq m.$$

This ring, $Q = R[X, Y]$, is called the ring of polynomials in the two (independent) commuting variables X and Y over the ring R.

We may repeat this argument as many times as we need to obtain the ring $R[X_1, \ldots, X_n]$ of polynomials in the (independent) commuting variables

$X_1, \dots, X_n$ over R. An element $f \in R[X_1, \dots, X_n]$ is a sum of elements of the kind $a_{(k_1,\dots,k_n)} X_1^{k_1} \dots X_n^{k_n}$ where $a_{(k_1,\dots,k_n)}$ are elements of the ring R. The expressions $a_{(k_1,\dots,k_n)} X_1^{k_1} \dots X_n^{k_n}$ are called monomials. The number k_j is called the degree of the monomial relative to the variable X_j and the sum $k_1 + \dots + k_n$ is called the complete degree of this monomial.

Two polynomials in $R[X]$ are equal precisely when the corresponding coefficients are equal. Then two polynomials $f, g \in R[X_1, \dots, X_n]$ are equal precisely when the sets of monomials making the decompositions of f and g are the same. Thus, if $a_{(k_1,\dots,k_n)} X_1^{k_1} \dots X_n^{k_n}$ is a monomial in the decomposition of f then $a_{(k_1,\dots,k_n)} X_1^{k_1} \dots X_n^{k_n}$ is the corresponding monomial in g and conversely. In particular, both polynomials contain a monomial of the form $X_1^{k_1} \dots X_n^{k_n}$ and the coefficient of this monomial is $a_{(k_1,\dots,k_n)}$ in both cases.

7.6.1. Definition

(i) *The degree of a polynomial $f \in R[X_1, \dots, X_n]$ relative to the variable X_j is the maximal degree relative to X_j of all nonzero monomials from the decomposition of f. This number is denoted by* $\mathbf{deg}_j f(X_1, \dots, X_n)$ *or* $\mathbf{deg}_j f$ *for short.*

(ii) *The complete degree of a polynomial $f \in R[X_1, \dots, X_n]$ is the maximal complete degree of the nonzero monomials from the decomposition of f. We denote this number by* $\mathbf{deg} f(X_1, \dots, X_n)$ *or* $\mathbf{deg}\, f$ *for short.*

As in the case of polynomials in one variable, we do not prescribe any degree to the zero element. Sometimes for the sake of convenience, the zero element is considered as a polynomial of infinite degree. It no longer makes sense to talk about the leading coefficient of a multivariable polynomial since the decomposition may include several monomials all of the same maximal degree.

7.6.2. Definition. *A polynomial $f \in R[X_1, \dots, X_n]$ is called homogeneous (or uniform) or a form of degree t, if each monomial in its decomposition has the same complete degree t.*

Linear polynomials are of degree one, whereas quadratic and cubic polynomials are of degrees two and three, respectively. By combining monomials of the same complete degree, we can write the polynomial $f(X_1, \dots, X_n)$ as a sum of forms of different degrees, say

$$f(X_1, \dots, X_n) = f_0(X_1, \dots, X_n) + f_1(X_1, \dots, X_n) + \dots + f_k(X_1, \dots, X_n),$$

where $k = \mathbf{deg} f(X_1, \dots, X_n)$ and f_i is homogeneous of degree i. We note that this representation is unique.

In Section 7.5 we proved that the ring $R[X]$ of polynomials over an integral domain has no zero divisors and hence that $R[X]$ is also an integral domain. It follows, by induction, that the ring $R[X_1, \dots, X_n]$ of multivariable polynomials over an integral domain R also has no zero-divisors and hence is itself an integral domain. However, the following more general result holds.

7.6.3. Theorem. *Let R be an integral domain and let $f, g \in R[X_1, \dots, X_n]$. Then*

$$\mathbf{deg}_j(fg) = \mathbf{deg}_j f + \mathbf{deg}_j g$$

for every j, where $1 \le j \le n$ and

$$\mathbf{deg}(fg) = \mathbf{deg}\, f + \mathbf{deg}\, g.$$

Proof. The first assertion was proved in Section 7.5.

Let $f = f_0 + f_1 + \cdots + f_k$ be a decomposition of f as a sum of forms where $k = \mathbf{deg}\, f$ and let $g = g_0 + g_1 + \cdots + g_m$ be a decomposition of g as a sum of forms where $m = \mathbf{deg}\, g$. Thus,

$$fg = f_0g_0 + (f_0g_1 + f_1g_0) + (f_0g_2 + f_1g_1 + f_2g_0) + \cdots + f_kg_m.$$

Forms f_k, g_m are nonzero and, since the ring $R[X_1, \dots, X_n]$ has no zero-divisors, the polynomial f_kg_m is nonzero. It is clearly the case that every monomial of the decomposition has complete degree less than $k + m$ and it follows that

$$\mathbf{deg}(fg) = k + m = \mathbf{deg}\, f + \mathbf{deg}\, g.$$

7.6.4. Definition. *Let K be a commutative ring and let R be a unitary subring of K. If $f \in R[X_1, \dots, X_n]$, where $f = \sum a_{(k_1,\dots,k_n)} X_1^{k_1} \dots X_n^{k_n}$, and if $y_1, \dots, y_n \in K$ then let*

$$f(y_1, \dots, y_n) = \sum a_{(k_1,\dots,k_n)} y_1^{k_1} \dots y_n^{k_n}.$$

The element $f(y_1, \dots, y_n)$ of the ring K is called the value of the polynomial $f(X_1, \dots, X_n)$ at the point $(y_1, \dots, y_n)$, or its value at $X_1 = y_1, \dots, X_n = y_n$. We say that $(y_1, \dots, y_n)$ is a root of $f(X_1, \dots, X_n)$, if $f(y_1, \dots, y_n) = 0_R$.

As in the case of one variable, we have the following result.

7.6.5. Proposition. *Let K be a commutative ring, let R be a unitary subring of K and let $y_1, \dots, y_n \in K$. Then the mapping*

$$\eta[y_1, \dots, y_n] : R[X_1, \dots, X_n] \longrightarrow K,$$

defined by

$$\eta[y_1, \dots, y_n](f) = f(y_1, \dots, y_n), \text{ whenever } f(X_1, \dots, X_n) \in R[X_1, \dots, X_n],$$

is a homomorphism.

This result follows from Proposition 7.5.6 and with the help of a simple induction argument. As in the case of a single variable, $\mathbf{Im}\,\eta[y_1, \ldots, y_n] = R[y_1, \ldots, y_n]$ is a subring, generated by the elements $y_1, \ldots, y_n$ over R.

7.6.6. Definition. *The elements* $y_1, \ldots, y_n \in K$ *are called algebraically dependent over* R, *if* $\mathbf{Ker}\,\eta[y_1, \ldots, y_n] \neq \{0_R\}$, *and, algebraically independent, if* $\mathbf{Ker}\,\eta[y_1, \ldots, y_n] = \{0_R\}$.

In particular, if the elements $y_1, \ldots, y_n$ are algebraically independent over R, then the subring $R[y_1, \ldots, y_n]$ is isomorphic to the polynomial ring $R[X_1, \ldots, X_n]$. Proposition 7.6.5 shows the universality of the polynomial ring.

We consider now the standard form of writing a multivariable polynomial which is often called the lexicographic form, since it is reminiscent of the way a dictionary works.

7.6.7. Definition. *A monomial* $a_{(k_1,\ldots,k_n)}X_1^{k_1}\ldots X_n^{k_n}$ *is higher than the monomial* $a_{(m_1,\ldots,m_n)}X_1^{m_1}\ldots X_n^{m_n}$ *if there is a number* d *such that* $k_1 = m_1, \ldots, k_{d-1} = m_{d-1}$, *but* $k_d > m_d$. *We call* d *the height of the monomial.*

We point out that if one monomial is higher than another this does not mean that its corresponding degree must be bigger. Thus, the lexicographic order is not connected with the polynomial degree. For each pair of monomials in a given polynomial, one is always higher than the other and the relationship of "being higher than" is a transitive relation. In this way all members of the decomposition of a polynomial f can be situated in lexicographic order. The monomial that is highest in this lexicographic order is called the highest member of the polynomial f.

7.6.8. Theorem. *Let* R *be an integral domain and let* $f, g \in R[X_1, \ldots, X_n]$. *Then the highest member of the polynomial* fg *is the product of the highest members of* f *and* g.

Proof. Let $a_{(k_1,\ldots,k_n)}X_1^{k_1}\ldots X_n^{k_n}$ and $b_{(t_1,\ldots,t_n)}X_1^{t_1}\ldots X_n^{t_n}$ be the highest monomials in the polynomials f and g, respectively. Let

$$a_{(m_1,\ldots,m_n)}X_1^{m_1}\ldots X_n^{m_n} \text{ and } b_{(r_1,\ldots,r_n)}X_1^{r_1}\ldots X_n^{r_n}$$

be arbitrary monomials from the corresponding decompositions of f and g. Then there exist positive integers d (respectively q) such that $k_1 = m_1, \ldots, k_{d-1} = m_{d-1}$, but $k_d \geq m_d$ and $t_1 = r_1, \ldots, t_{q-1} = r_{d-1}$, but $t_q \geq r_q$. We compare the products

$$(a_{(k_1,\ldots,k_n)}X_1^{k_1}\ldots X_n^{k_n})(b_{(t_1,\ldots,t_n)}X_1^{t_1}\ldots X_n^{t_n}) = a_{(k_1,\ldots,k_n)}b_{(t_1,\ldots,t_n)}X_1^{k_1+t_1}\ldots X_n^{k_n+t_n}$$

and

$$(a_{(m_1,\ldots,m_n)}X_1^{m_1}\ldots X_n^{m_n})(b_{(r_1,\ldots,r_n)}X_1^{r_1}\ldots X_n^{r_n})$$
$$= a_{(m_1,\ldots,m_n)}b_{(r_1,\ldots,r_n)}X_1^{m_1+r_1}\ldots X_n^{m_n+r_n}.$$

We may assume, without loss of generality, that $d \geq q$. Then we have

$$k_1 + t_1 = m_1 + r_1, \ldots, k_{d-1} + t_{d-1} = m_{d-1} + r_{d-1},$$

but

$$k_d + t_d \geq m_d + r_d.$$

Thus,

$$a_{(k_1,\ldots,k_n)}b_{(t_1,\ldots,t_n)}X_1^{k_1+t_1}\ldots X_n^{k_n+t_n}$$

is higher than

$$a_{(m_1,\ldots,m_n)}b_{(r_1,\ldots,r_n)}X_1^{m_1+r_1}\ldots X_n^{m_n+r_n},$$

which means that the product

$$a_{(k_1,\ldots,k_n)}b_{(t_1,\ldots,t_n)}X_1^{k_1+t_1}\ldots X_n^{k_n+t_n}$$

is the highest monomial of fg.

Next we consider some important specific types of polynomial.

7.6.9. Definition. *Let R be a commutative ring, let π be a permutation of degree n and let*

$$f(X_1,\ldots,X_n) \in R[X_1,\ldots,X_n], \ \textit{where}$$
$$f(X_1,\ldots,X_n) = \sum a_{(k_1,\ldots,k_n)}X_1^{k_1}\ldots X_n^{k_n}.$$

Define the mapping

$$S_\pi : R[X_1,\ldots,X_n] \longrightarrow R[X_1,\ldots,X_n]$$

by

$$S_\pi(f(X_1,\ldots,X_n)) = \sum a_{(k_1,\ldots,k_n)}X_{\pi(1)}^{k_1}\ldots X_{\pi(n)}^{k_n}.$$

A polynomial $f(X_1,\ldots,X_n)$ is called symmetric, if

$$S_\pi(f(X_1,\ldots,X_n)) = f(X_1,\ldots,X_n)$$

for every permutation $\pi \in S_n$.

Clearly the polynomials that do not change upon permuting any of the variables are symmetric. Also every element of R is a symmetric polynomial, since it does not depend on any variable. Here are some examples of symmetric polynomials

$$\begin{aligned}
\sigma_1(X_1, \dots, X_n) &= X_1 + \cdots + X_n, \\
\sigma_2(X_1, \dots, X_n) &= X_1X_2 + X_1X_3 + \cdots + X_{n-1}X_n, \\
\sigma_3(X_1, \dots, X_n) &= X_1X_2X_3 + X_1X_2X_4 + \cdots + X_{n-2}X_{n-1}X_n, \\
&\vdots \\
\sigma_{n-1}(X_1, \dots, X_n) &= X_1X_2 \dots X_{n-1} + X_1X_2 \dots X_{n-2}X_n + \cdots + X_2X_3 \dots X_n, \\
\sigma_n(X_1, \dots, X_n) &= X_1X_2 \dots X_{n-1}X_n.
\end{aligned}$$

These polynomials are called the elementary symmetric polynomials. By Theorem 2.2.7, every permutation is a product of certain transpositions. Therefore, in order to check that a polynomial is symmetric, it is sufficient to check that the polynomial does not change under the action of any transposition.

7.6.10. Proposition. *Let R be an integral domain. Then the subset of all symmetric polynomials of $R[X_1, \dots, X_n]$ is a subring of $R[X_1, \dots, X_n]$.*

This assertion is left to the reader. It is quite straightforward to prove.

Note that all the variables $X_1, \dots, X_n$ are included in the decomposition of each symmetric polynomial and all variables must have the same degree. By the Viete formulas, we deduce that the coefficients of a polynomial in $R[X]$ with leading coefficient one are elementary symmetric polynomials of the polynomial roots.

Proposition 7.6.10 implies the following.

7.6.11. Corollary. *Let R be an integral domain and let $f(X_1, \dots, X_n)$ denote an arbitrary polynomial over R. Then the polynomial*

$$f(\sigma_1(X_1, \dots, X_n), \dots, \sigma_n(X_1, \dots, X_n))$$

is symmetric.

Conversely, the following theorem holds.

7.6.12. Theorem. *Let R be an integral domain and let $f(X_1, \dots, X_n)$ be an arbitrary symmetric polynomial over R. Then there exists a polynomial $g(X_1, \dots, X_n) \in R[X_1, \dots, X_n]$ such that*

$$f(X_1, \dots, X_n) = g(\sigma_1(X_1, \dots, X_n), \dots, \sigma_n(X_1, \dots, X_n)).$$

Proof. Let $a_0X_1^{k_1}\dots X_n^{k_n}$ be the highest monomial in the lexicographic ordering of the polynomial $f(X_1,\dots,X_n)$. First we show that $k_1 \geq k_2 \geq \dots \geq k_n$. Suppose the contrary and let $k_j < k_{j+1}$ for some j. Since $f(X_1,\dots,X_n)$ is a symmetric polynomial it has a monomial $a_0X_1^{k_1}\dots X_j^{k_{j+1}}X_{j+1}^{k_j}\dots X_n^{k_n}$. However, this latter monomial is higher than $a_0X_1^{k_1}\dots X_j^{k_{j+1}}X_{j+1}^{k_j}\dots X_n^{k_n}$, a contradiction which shows that $k_1 \geq k_2 \geq \dots \geq k_n$.

We next consider the polynomial

$$h_1 = a_0\sigma_1^{k_1-k_2}\sigma_2^{k_2-k_3}\dots\sigma_{n-1}^{k_{n-1}-k_n}\sigma_n^{k_n}.$$

By Corollary 7.6.11, h_1 is a symmetric polynomial in the variables $X_1,\dots,X_n$. The highest monomials of the polynomials $\sigma_1,\sigma_2,\dots,\sigma_n$ are $X_1, X_1X_2, X_1X_2X_3,\dots,X_1\dots X_n$, respectively, and therefore by, Theorem 7.6.8, the highest monomial of the polynomial h_1 is

$$\begin{aligned}&a_0X_1^{k_1-k_2}(X_1X_2)^{k_2-k_3}\dots(X_1X_2\dots X_{n-1})^{k_{n-1}-k_n}(X_1X_2\dots X_n)^{k_n}\\&= a_0X_1^{k_1}\dots X_n^{k_n}.\end{aligned}$$

It follows from this that the highest member of the symmetric polynomial $f - h_1 = f_1$ is lower than the monomial $a_0X_1^{k_1}\dots X_n^{k_n}$, the highest member of the polynomial f. Repeating the same argument for the polynomial f_1, we see that $f_1 = h_2 + f_2$, where h_2 is a product of elementary symmetric polynomials with coefficient in R and f_2 is a symmetric polynomial whose highest member is lower than the highest member of f_1. We now have $f = h_1 + h_2 + f_2$.

As we continue this process, at some stage the process terminates, which is to say that for some positive integer t we shall have $f_t = 0_R$ and then

$$f(X_1,\dots,X_n) = h_1 + h_2 + \dots + h_t,$$

where h_j is a product of elementary symmetric polynomials with coefficient in R, for $1 \leq j \leq t$. Indeed, if this process does not terminate then we would obtain an infinite series of symmetric polynomials $\{f_n \mid n \in \mathbb{N}\}$ in which every highest member of each of these polynomials will be lower than the highest members of the preceding polynomials and therefore will be lower than $a_0X_1^{k_1}\dots X_n^{k_n}$. However, if $bX_1^{m_1}\dots X_n^{m_n}$ is the highest member of f_j, then as we showed above $m_1 \geq m_2 \geq \dots \geq m_n$. On the other hand, since $a_0X_1^{k_1}\dots X_n^{k_n}$ is higher than $bX_1^{m_1}\dots X_n^{m_n}$, we have $k_1 \geq m_1$. However, it is easy to observe that there are only finitely many systems of whole numbers $m_1, m_2,\dots,m_n$ with the properties $m_1 \geq m_2 \geq \dots \geq m_n$ and $k_1 \geq m_1$. It follows from this that our sequence of polynomials is finite.

The following result is a natural complement to Theorem 7.6.12.

7.6.13. Theorem. *Let R be an integral domain and let $f(X_1, \ldots, X_n)$ be an arbitrary symmetric polynomial over R. Then f can be uniquely expressed as a polynomial in the elementary symmetric polynomials.*

Proof. If the polynomial $f(X_1, \ldots, X_n)$ has two distinct representations

$$f(X_1, \ldots, X_n) = g(\sigma_1, \sigma_2, \ldots, \sigma_n) = h(\sigma_1, \sigma_2, \ldots, \sigma_n),$$

then the difference

$$r(\sigma_1, \sigma_2, \ldots, \sigma_n) = g(\sigma_1, \sigma_2, \ldots, \sigma_n) - h(\sigma_1, \sigma_2, \ldots, \sigma_n)$$

would be a nonzero polynomial in $\sigma_1, \sigma_2, \ldots, \sigma_n$. At the same time the substitution in this polynomial of the variables $\sigma_1, \sigma_2, \ldots, \sigma_n$ in terms of $X_1, \ldots, X_n$ would lead us to the zero polynomial in the ring $R[X_1, \ldots, X_n]$. We have therefore only to prove that if the polynomial $r(\sigma_1, \sigma_2, \ldots, \sigma_n)$ is nonzero, then the polynomial $q(X_1, \ldots, X_n)$ obtained from $r(\sigma_1, \sigma_2, \ldots, \sigma_n)$ by substituting for $\sigma_1, \sigma_2, \ldots, \sigma_n$ in terms of $X_1, \ldots, X_n$ is also nonzero.

If $a\sigma_1^{k_1} \ldots \sigma_n^{k_n}$ is one of the members of $r(\sigma_1, \sigma_2, \ldots, \sigma_n)$, and $a \neq 0_R$ then, after substitution of all $\sigma_1, \sigma_2, \ldots, \sigma_n$ by their expressions we will get a polynomial in $X_1, \ldots, X_n$, the highest member of which is, by Theorem 7.6.8,

$$aX_1^{k_1}(X_1X_2)^{k_2} \ldots (X_1 \ldots X_j)^{k_j} \ldots (X_1 \ldots X_n)^{k_n} = aX_1^{m_1} \ldots X_j^{m_{j+1}} X_{j+1}^{m_j} \ldots X_n^{m_n},$$

where

$$\begin{aligned} m_1 &= k_1 + k_2 + \cdots + k_n, \\ m_2 &= k_2 + \cdots + k_n, \\ &\vdots \\ m_{n-1} &= k_{n-1} + k_n, \\ m_n &= k_n. \end{aligned}$$

It follows that

$$k_n = m_n \text{ and } k_j = m_j - m_{j+1}, \text{ for } 1 \leq j \leq n-1.$$

Thus, if we know the exponents $m_1, \ldots, m_n$ we can deduce the exponents $k_1, \ldots, k_n$ of the original polynomial $r(\sigma_1, \sigma_2, \ldots, \sigma_n)$.

Now consider all members of the polynomial $r(\sigma_1, \sigma_2, \ldots, \sigma_n)$ and for each of them find the highest member in its representation as a polynomial in $X_1, \ldots, X_n$. Select the highest of these monomials. This member does not have any like terms among other members and that is why having nonzero coefficient it appears only one time. This implies that not all coefficients of $q(X_1, \ldots, X_n)$ are zeros, so this polynomial is not the zero element of $R[X_1, \ldots, X_n]$. This completes the proof.

EXERCISE SET 7.6

Justify your work, giving a proof or counterexample where necessary.

7.6.1. Let F be a field and let M be the subset of all homogeneous polynomials of the ring $F[X_1, \dots, X_n]$. Prove that M is a subring of $F[X_1, \dots, X_n]$.

7.6.2. How many homogeneous polynomials of degree 2 are there in the ring $\mathbb{F}_2[X_1, X_2, X_3]$?

7.6.3. How many homogeneous polynomials of degree 2 are there in the ring $\mathbb{F}_3[X_1, X_2, X_3]$?

7.6.4. Order the following polynomial lexicographically: $f(X_1, X_2, X_3) = 2X_1^3 (X_2 + X_3) - 3(X_1^2 + X_3^2)X_1^2X_2^2 + 5X_1^4X_2X_3^2 \in \mathbb{Z}[X_1, X_2, X_3]$.

7.6.5. Does the polynomial $g(X_1, X_2, X_3, X_4) = (X_1 - X_4)(X_2 + X_3)$ divides the polynomial $f(X_1, X_2, X_3, X_4) = (X_1X_2 - X_3X_4)^5 + (X_1X_3 - X_2 X_4)^5$ in the ring $\mathbb{Z}[X_1, X_2, X_3, X_4]$?

7.6.6. Is the polynomial $f(X_1, X_2, X_3) = X_1^2X_2 + X_1^2X_3 + X_2^2X_1 + X_2^2X_3 + X_1 + X_2 + X_3$ symmetric?

7.6.7. Is the polynomial $f(X_1, X_2, X_3) = 2X_1^3 + 2X_1^3 + 2X_3^3 + X_1X_2X_3 - 1$ symmetric?

7.6.8. Is the polynomial $f(X_1, X_2, X_3, X_4) = (X_1X_2 + X_3X_4)(X_1X_4 + X_2X_3) (X_1X_3 + X_2X_4)$ symmetric?

7.6.9. Using the minimal number of monomials complete the polynomial $f(X_1, X_2) = X_2^2 + 2X_2$ to a symmetric one.

7.6.10. Using the minimal number of monomials complete the polynomial $f(X_1, X_2, X_3) = X_1^3 + 2X_1X_2 + 2X_2X_3 + 5$ to a symmetric one.

7.6.11. Using the minimal number of monomials complete the polynomial $f(X_1, X_2, X_3) = (X_1 + X_2)^2 + 2X_1X_3 + X_1X_2X_3$ to a symmetric one.

7.6.12. Express the polynomial $f(X_1, X_2, X_3) = X_1^3X_2 + X_1X_2^3 + 2X_1^2 + 2X_2^2$ in terms of elementary symmetric polynomials.

7.6.13. Express the polynomial $f(X_1, X_2, X_3) = 2X_1^4X_2 - 5X_1^2X_2 + 2X_1X_1^4 - 5X_1X_2^2$ in terms of elementary symmetric polynomials.

7.6.14. Express the polynomial $f(X_1, X_2, X_3) = 2X_1^3 + X_2^3 + X_3^3 - X_1 - X_2 - X_3$ in terms of elementary symmetric polynomials.

7.6.15. Express the polynomial $f(X_1, X_2, X_3) = X_1^5X_2X_3 + X_1X_2^5X_3 + X_1X_2X_3^5 + 2X_1X_2X_3$ in terms of elementary symmetric polynomials.

7.6.16. Express the polynomial $f(X_1, X_2, X_3) = (X_1 - X_2)^2 + (X_1 - X_3)^2 + (X_2 - X_3)^2$ in terms of elementary symmetric polynomials.

7.6.17. Let $S_n(X_1, X_2) = X_1^n + X_2^n$. Prove that for each $k > 2$ we have $S_k(X_1, X_2) = \sigma_1(X_1, X_2)S_{k-1}(X_1, X_2) - \sigma_2(X_1, X_2)S_{k-2}$.

7.6.18. Find real solutions of the following system:

$$x_1^3 + x_2^3 = 35,$$
$$x_1 + x_2 = 5.$$

7.6.19. Find real solutions of the following system:

$$x_1^3 + x_2^3 + x_3^3 = 1,$$
$$x_1^2 + x_2^2 + x_3^2 = 9,$$
$$x_1 + x_2 + x_3 = 1.$$

CHAPTER 8

GROUPS

8.1 GROUPS AND SUBGROUPS

In Section 8.1, we briefly introduce the concept of a group and give some examples of groups and subgroups. The term group belongs to the great French mathematician, Evariste Galois. The theory of permutations had its beginnings in the investigation of roots of algebraic equations, which had been developed by the likes of Lagrange, Vandermonde, Gauss, Ruffini, Cauchy and it is this idea that led to the concept of an abstract group. Some initial results in group theory were obtained by these mathematicians.

However, Evariste Galois is considered to be the founder of group theory, since he reduced the study of algebraic equations to the study of permutation groups. He introduced the concept of a normal subgroup and understood its importance. He also considered groups having special given properties and introduced the idea of a "linear presentation" of a group that is very close to the concept of homomorphism. His brilliant work was not understood for a long period of time; his ideas were disseminated by Serre and Jordan, later, after his tragic death.

Results by Galois anticipated the study of finite groups of permutations. In this initial stage, group theory was represented using groups of permutations only. The definition of an abstract group was introduced by Cayley (1821–1895) in the middle of the nineteenth century. However, for a long period of time, the investigation of abstract groups was considered as a part of permutation group theory.

Only at the end of the nineteenth century, did the real development of group theory begin. The development of geometry, topology, differential equations, and other mathematical disciplines required the study of groups of transformations. It took quite a long time to understand the relationship between groups and the ideas of invariance and symmetry. Everywhere in which a key role is played by a symmetrical property of an object (algebraic or differential equations, crystal lattices, geometric figures, and so on), a group of transformations (movements, substitutions of variables, permutations of indices, and so on) appears. Groups are some kind of measure of the symmetry of an object, which is why they are so important for the classification of such objects. These are the main reasons why groups are of vital importance in different branches of mathematics, physics, chemistry, and so on.

We will repeat the definition of a group for the sake of completeness.

8.1.1. Definition. *A group is a set G, together with a given binary operation*

$$(x, y) \longmapsto xy, \text{ where } x, y \in G,$$

satisfying the properties (the group axioms)

(G 1) *the operation is associative, so for all elements $x, y, z \in G$, the equation*

$$x(yz) = (xy)z \text{ holds;}$$

(G 2) *G has an identity element, an element e having the property*

$$xe = ex = x,$$

for all $x \in G$;

(G 3) *every element $x \in G$ has an inverse, $x^{-1} \in G$, an element such that*

$$xx^{-1} = x^{-1}x = e.$$

8.1.2. Definition. *Let G be a group. If the group operation is commutative, then the group is called abelian. Often, an additive notation is used for abelian groups so, in this case, the group axioms take the following form:*

(AG 1) *the operation is commutative, so*

$$x + y = y + x$$

for all $x, y \in G$;

(AG 2) *the operation is associative,*

$$x + (y + z) = (x + y) + z$$

for all $x, y, z \in G$;

(AG 3) *G has a zero element, an element 0_G having the property that*

$$x + 0_G = 0_G + x = x$$

for all $x \in G$;

(AG 4) *every element $x \in G$ has an opposite element, $-x \in G$, an element such that*

$$x + (-x) = (-x) + x = 0_G.$$

We can weaken the definition of a group somewhat, as follows.

8.1.3. Theorem. *Let G be a semigroup. Then, G is a group if and only if G satisfies*

(i) *G has a right identity element, an element $e_r \in G$ such that $xe_r = x$ for each element $x \in G$;*

(ii) *for each element $x \in G$ there exists a right inverse, an element $x_r \in G$ such that $xx_r = e_r$.*

Proof. If G is a group, then its identity element is a right identity element, and the inverse of x is a right inverse element. Conversely, let the semigroup G satisfy conditions (i) and (ii). We have

$$e_r(xx_r) = e_re_r = e_r = xx_r.$$

Multiplying both sides of the equation $e_rxx_r = xx_r$ on the right by the right inverse of x_r, we obtain $e_rxe_r = xe_r$, so that $e_rx = x$, because $xe_r = x$. Thus, e_r is a left identity element also and hence e_r is the unique right and left identity of G.

From the equation $xx_r = e_r$, we obtain $x_rxx_r = x_re_r = x_r$. We multiply both sides of the last equation on the right by the right inverse of x_r and obtain $x_rxe_r = e_r$. Thus, $x_rx = e_r$ and hence x_r is a left inverse of x, from which it follows that $x_r = x^{-1}$.

8.1.4. Theorem. *Let G be a semigroup. Then, G is a group if and only if for arbitrary elements $a, b \in G$ the equations $ax = b$ and $xa = b$ have solutions.*

Proof. If G is a group, then the equation $ax = b$ has the solution $x = a^{-1}b$ and the equation $xa = b$ has the solution $x = ba^{-1}$. Conversely, let G be a semigroup in which the given equations have solutions. In particular, the equation $ax = a$ has a solution e. Let b be an arbitrary element of G and let c be a solution of the equation $xa = b$. Then we have

$$be = (ca)e = c(ae) = ca = b,$$

so e is a right identity for G. For an arbitrary element $a \in G$, the solution of the equation $xa = e$ is a right inverse element and the result follows by Theorem 8.1.3.

We note that the solution, x, of $ax = b$ is uniquely determined.

A group G is called finite if it contains only finitely many elements and a group that is not finite is called infinite. We let $|G|$ denote the number of elements in the finite group G and call this number the order of G.

8.1.5. Theorem. *Let G be a finite semigroup. Then, G is a group if and only if for arbitrary elements $a, b, c \in G$, the equations $ab = ac$ and $ba = ca$ together imply $b = c$.*

Proof. If G is a group, then the element a has an inverse. From the equation $ab = ac$, we obtain

$$b = eb = (a^{-1}a)b = a^{-1}(ab) = a^{-1}(ac) = (a^{-1}a)c = ec = c,$$

and similarly we can prove that if $ba = ca$ then again $b = c$, so the given conclusions hold.

Conversely, let $G = \{g_1, \dots, g_n\}$. From the hypotheses, it follows that

$$ag_1, \dots, ag_n$$

are n distinct elements of G and hence $G = \{ag_1, \dots, ag_n\}$. Thus, $b \in \{ag_1, \dots, ag_n\}$ and hence the equation $ax = b$ has a solution in the group G. Similarly, the equation $xa = b$ also has a solution in G and Theorem 8.1.4 implies the result.

Now, we will consider the very important concept of a subgroup, an idea that was introduced in Section 3.2.

8.1.6. Definition. *Let G be a group. A stable subset H of a group G is called a subgroup of G if H is a group relative to the operation given in G. The fact that H is a subgroup of G will be denoted by $H \le G$.*

As usual there is a short method for determining whether a nonempty subset of a group is a subgroup, which we present in the next theorem.

8.1.7. Theorem *(subgroup criterion). Let G be a group. If H is a subgroup of G, then H satisfies the conditions*

(SG 1) *if $x, y \in H$, then $xy \in H$;*

(SG 2) *if $x \in H$, then $x^{-1} \in H$.*

Conversely, if H is a nonempty subset of G satisfying conditions **(SG 1)** *and* **(SG 2)**, *then H is a subgroup of G.*

Proof. If H is a subgroup, then condition **(SG 1)** follows from the fact that the operation is closed. Let e_H be the identity element of H. Then, for an arbitrary element $x \in H$, we have $xe_H = x$. Since the element x has an inverse in G, we have, multiplying on the left by this inverse,

$$x^{-1}xe_H = x^{-1}x = e.$$

Hence $e \in H$ and $e = e_H$. Also, it follows that the inverse elements to x in H and in G coincide. In particular, condition **(SG 2)** holds.

Conversely, let the nonempty subset H satisfy conditions **(SG 1)** and **(SG 2)**. From condition **(SG 1)**, it follows that H is a stable subset of G. In particular, the operation of G induced on H is a binary operation on H. This operation is associative, since the original operation on G is associative. If $x \in H$ then, by **(SG 2)**, $x^{-1} \in H$. By **(SG 1)**, $e = xx^{-1} \in H$ and hence e is an identity element for H. Finally, every element of H has a multiplicative inverse, by **(SG 2)**.

We can reduce the number of conditions to check still further as follows. It is important to realize that we must also check that $H \neq \emptyset$.

8.1.8. Corollary. *Let G be a group. If H is a subgroup of G, then H satisfies the following condition:*

(SG3) *If $x, y \in H$, then $xy^{-1} \in H$.*

Conversely, if H is a nonempty subset of G satisfying condition **(SG3)**, *then H is a subgroup of G.*

Proof. We will show that condition **(SG 3)** is equivalent to conditions **(SG 1)** and **(SG 2)**. Clearly, **(SG 3)** is a consequence of **(SG 1)** and **(SG 2)**.

If **(SG 3)** holds and $x \in H$, then $e = xx^{-1} \in H$. Furthermore, $x^{-1} = ex^{-1} \in H$, so that **(SG 2)** holds. Finally, let $y \in H$. In this case, we have already proved that $y^{-1} \in H$. Therefore, $xy = x(y^{-1})^{-1} \in H$. Hence **(SG 1)** holds.

Let G be an abelian group with additive notation. We define an operation of subtraction on G by $x - y = x + (-y)$ and, in this case, Corollary 8.1.8 is as follows:

A nonempty subset H of an additive group G is a subgroup if and only if the condition

(SG 3) if $x, y \in H$, then $x - y \in H$ holds.

8.1.9. Corollary. *Let G be a group and let H be a subgroup of G. A subset K of H is a subgroup of G if and only if K is a subgroup of H.*

We use Corollary 8.1.9 to deduce that an intersection of subgroups is again a subgroup.

8.1.10. Corollary. *Let G be a group and let $\mathfrak{S}$ be a family of subgroups of G. Then the intersection $\bigcap \mathfrak{S}$ of the subgroups of this family is a subgroup of G.*

Proof. Let $T = \bigcap \mathfrak{S}$, let $x, y \in T$, and let U be an arbitrary subgroup of the family $\mathfrak{S}$. Then $xy^{-1} \in U$, by **(SG 3)**, and therefore $xy^{-1} \in T$. Corollary 8.1.8 gives the result.

One important means of constructing subgroups is provided by the next definition.

8.1.11. Definition. *Let G be a group, let H be a subgroup of G, and let M a subset of G. Set*

$$C_H(M) = \{x \in H \mid xg = gx \text{ for every } g \in M\}.$$

The subset $C_H(M)$ is called the centralizer of the subset M in the subgroup H. The subset $\zeta(G) = C_G(G)$ is called the center of G. If $M = \{a\}$, then instead of $C_H(\{a\})$, we will write $C_H(a)$.

8.1.12. Corollary. *Let G be a group, let H be a subgroup of G, and let M be a subset of G. The centralizer, $C_H(M)$, is a subgroup of G. In particular, the center $\zeta(G)$ is also a subgroup of G.*

Proof. Certainly, $e \in C_H(M) \neq \emptyset$. Next, let $x, y \in C_H(M)$ and let $g \in M$. By the associative law,

$$(xy)g = x(yg) = x(gy) = (xg)y = (gx)y = g(xy)$$

and hence $xy \in C_H(M)$. Next $xg = gx$ so, multiplying on the right and left sides by x^{-1}, we have $x^{-1}xgx^{-1} = x^{-1}gxx^{-1}$, which implies that $gx^{-1} = x^{-1}g$ since $x^{-1}x = xx^{-1} = e$. Thus $x^{-1} \in C_H(M)$ and Theorem 8.1.7 completes the proof.

Unions of subgroups need not be subgroups in general, but in certain cases, unions may sometimes be subgroups. We give some examples in the next few corollaries.

8.1.13. Corollary. *Let G be a group and let $\mathfrak{L}$ be a local family of subgroups of the group G. Then the union, $\bigcup \mathfrak{L}$, of the subgroups of this family is a subgroup of G.*

Proof. Let $V = \bigcup \mathfrak{L}$ and let $x, y \in V$. There exist subgroups H, K, belonging to the family $\mathfrak{L}$, such that $x \in H$, $y \in K$. We choose a subgroup $F \in \mathfrak{L}$, containing both subgroups H and K. Then $x, y \in F$. Since F is a subgroup, $xy^{-1} \in F$ by Corollary 8.1.8. Hence $xy^{-1} \in V$ and Corollary 8.1.8 finishes the proof.

8.1.14. Corollary. *Let G be a group and let $\mathfrak{L}$ be a family of subgroups of G, linearly ordered by inclusion. Then the union, $\bigcup \mathfrak{L}$, of the subgroups of this family is a subgroup of G.*

8.1.15. Corollary. *Let G be a group and let*

$$H_1 \leq H_2 \leq \cdots \leq H_n \leq \cdots$$

be an ascending series of subgroups of G. Then the union, $\bigcup_{n\in\mathbb{N}} H_n$, of the subgroups of this series is a subgroup of G.

Another standard way to construct subgroups is via generating sets. To explain this idea, let M be a subset of a group G and let $\mathfrak{S}$ be the family of all subgroups containing M. Then the intersection $\langle M \rangle = \bigcap \mathfrak{S}$ is a subgroup, by Corollary 8.1.8.

8.1.16. Definition. *The intersection $\langle M \rangle = \bigcap \mathfrak{S}$ is a subgroup that is called the subgroup generated by the subset M, and the subset M is called a system of generators of the subgroup $\langle M \rangle$. In particular, if $\langle M \rangle = G$, then we say that M generates the group G. A group G is called finitely generated, if there is a finite subset M such that $G = \langle M \rangle$.*

The following proposition is quite easy to prove and is left for the reader.

8.1.17. Proposition. *Let G be a group and let H be a subgroup of G. If $M \subseteq H$, then $\langle M \rangle \leq H$.*

If H is a subgroup of G, containing a subset M, then H also contains $\langle M \rangle$. This means that $\langle M \rangle$ is the smallest subgroup of all the subgroups containing M. It is clear that if M is a subgroup of a group G, then $\langle M \rangle = M$.

It is worth understanding what exactly characterizes the elements of $\langle M \rangle$. The simplest situation occurs when M consists of a single element x, so let x be an element of a group G and let H be a subgroup containing x. It follows from Theorem 8.1.7 and induction that $x^n, e = x^0, x^{-n} \in H$ for arbitrary $n \in \mathbb{N}$, which means that

$$\{x^n \mid n \in \mathbb{Z}\} \subseteq H.$$

On the other hand, $x^n x^{-m} = x^{n-m}$, by Proposition 3.1.16. By Corollary 8.1.8 this implies that $\{x^n \mid n \in \mathbb{Z}\}$ is a subgroup of the group G. This means that $\{x^n \mid n \in \mathbb{Z}\} = \langle x \rangle$ is the subgroup generated by the subset $\{x\}$, or the subgroup generated by the element x. Thus, subgroups generated by a single element are quite transparent.

8.1.18. Definition. *The subgroup $\{x^n \mid n \in \mathbb{Z}\} = \langle x \rangle$ is called a cyclic subgroup. An element y with the property that $\langle y \rangle = \langle x \rangle$ is called a generator of $\langle x \rangle$. A group G is called cyclic if it coincides with at least one of its cyclic subgroups.*

Later, we shall see that several different elements may generate a cyclic group. We now consider the subgroup $\langle x \rangle$ further. There are two cases:

(i) $x^n \neq x^m$, whenever $n \neq m$.
(ii) There exist integers n, m, such that $n \neq m$ but $x^n = x^m$.

In case (i), $\langle x \rangle$ is an infinite group. We say that the element x has infinite order. In case (ii), suppose that $n > m$. From the equation $x^n = x^m$, we obtain $x^{n-m} = e$, which means that some positive power of x is the identity element. Let t be the least positive integer for which $x^t = e$ and let n be an arbitrary integer. By Theorem 1.4.1, $n = tq + r$ where $0 \leq r < t$ and we have

$$x^n = x^{tq+r} = x^{tq}x^r = (x^t)^q x^r = x^r \text{ since } x^t = e.$$

Similarly, we can prove that

$$e = x^0, x = x^1, x^2, \ldots, x^{t-1}$$

are distinct and it follows, in this case, that

$$\langle x \rangle = \{x^n \mid n \in \mathbb{Z}\} = \{x^0 = e, x = x^1, x^2, x^{t-1}\}.$$

In this case, we say that the element x has finite order. The order of the element x is the least positive integer t such that $x^t = e$. We will denote this by $|x|$ and note that here $|x| = t$. Also, by definition, $|x| = |\langle x \rangle|$, so the two meanings of order coincide in this case. Observe, that $|e| = 1$.

8.1.19. Definition.

(i) *A group G is called periodic if each of its elements have finite order.*
(ii) *A group G is called torsion free, if each of its nontrivial elements has infinite order.*
(iii) *A group G is called mixed, if it has nontrivial elements of both finite and infinite orders.*

8.1.20. Theorem. *Let $G = \langle g \rangle$ be a cyclic group and let $y \in G$.*

(i) *If G is an infinite group, then $\langle y \rangle = G$ if and only if $y = g$ or $y = g^{-1}$.*
(ii) *If G is a finite group of order t, then the element $y = g^m$ has order $t/\mathbf{GCD}(m, t)$.*
(iii) *If G is finite and $|G| = t$, then $\langle y \rangle = G$ if and only if $y = g^m$ where $\mathbf{GCD}(m, t) = 1$.*

Proof.

(i) First let $G = \langle g \rangle$ be an infinite cyclic group and let y also generate G. Since $y \in \langle g \rangle$, then $y = g^m$ for some $m \in \mathbb{Z}$. On the other hand, $\langle y \rangle = G$, so that $g = y^d$, for some $d \in \mathbb{Z}$. Thus we have

$$g = y^d = (g^m)^d = g^{md}.$$

Multiplying both sides of this equation by g^{-1}, we obtain $g^{md-1} = e$. Since $\langle g \rangle$ is an infinite cyclic group, this means that $md - 1 = 0$, so that $m, d \in \{1, -1\}$. Therefore, we have only two possibilities, that $y = g$ or $y = g^{-1}$.

(ii) Now, let G be finite of order t. Let $y = g^m$ and let $\mathbf{GCD}(m, t) = k$. Then

$$y^{t/k} = (g^m)^{t/k} = (g^t)^{m/k} = e.$$

Thus y has order at most t/k. Next suppose that $y^r = e$. Then $g^{mr} = e$. By Theorem 1.4.1, there exist integers a, b such that $mr = at + b$, where $0 \leq b < t$. Then

$$e = g^{mr} = g^{at+b} = (g^t)^a g^b = g^b.$$

Since g has order t, it follows that $b = 0$. Hence $mr = at$ and $\frac{m}{k}r = a\frac{t}{k}$. Since $\mathbf{GCD}(\frac{m}{k}, \frac{t}{k}) = 1$, it follows that $\frac{m}{k}$ divides a so that $a = c(\frac{m}{k})$ for some $c \geq 1$ and $(\frac{m}{k})r = c(\frac{m}{k})(tk)$. Thus $r = c(\frac{t}{k}) \geq \frac{t}{k}$. It now follows that $y = g^m$ has the stated order.

(iii) This part of the theorem follows easily from part (ii).

8.1.21. Theorem. *Let $G = \langle g \rangle$ be a cyclic group. If H is a subgroup of G, then H is also cyclic.*

Proof. Since H is a subgroup, $e \in H$. If $H = \{e\}$, then $H = \langle e \rangle$ is cyclic. Suppose now that H contains nontrivial elements. All elements of G are powers of g. Therefore, there exists an integer m such that $e \neq g^m \in H$. If $m < 0$, then by condition **(SG 2)**, we have $(g^m)^{-1} = g^{-m} \in H$. This means that H contains positive powers of g that are nontrivial. Let

$$\Omega = \{k > 0 \mid e \neq g^k \in H\}.$$

We let d to be the least natural number in Ω. In particular, $g^d \in H$ and, by Proposition 8.1.17, $\langle g^d \rangle \leq H$. We show that in fact $H = \langle g^d \rangle$. Let x be an arbitrary element of H. Then $x = g^n$ for some $n \in \mathbb{Z}$. By Theorem 1.4.1, $n = dq + r$, where $0 \leq r < d$ and it follows that

$$x = g^n = g^{dq+r} = (g^d)^q(g^r) \text{ and } g^r = x(g^{dq})^{-1}.$$

By **(SG 3)**, $g^r \in H$ and therefore $r \in \Omega$ if $r \neq 0$, which contradicts the choice of d. Thus, $r = 0$ and hence $x = g^n = (g^d)^q$. This implies that $H \leq \langle g^d \rangle$ and, since $H \geq \langle g^d \rangle$, $H = \langle g^d \rangle$, a cyclic group.

We now consider generating sets a bit more generally.

8.1.22. Theorem. *Let G be a group and let M be a subset of G. Then the subgroup $\langle M \rangle$ consists of elements of the form $x_1x_2 \ldots x_n$, where $x_j \in M$ or $x_j^{-1} \in M$, for $1 \leq j \leq n$ and n is arbitrary.*

Proof. Let Y be the subset of all elements of the form $x_1x_2 \ldots x_n$ where $x_j \in M$ or $x_j^{-1} \in M$, for $1 \leq j \leq n$. By Theorem 8.1.7, $Y \subseteq \langle M \rangle$. On the other hand, we have $M \subseteq Y$ and next we show that Y satisfies **(SG 3)**. Then Corollary 8.1.8 will imply that Y is a subgroup.

To this end, suppose that $g_1, g_2 \in Y$. Then $g_1 = x_1x_2 \ldots x_n$, $g_2 = x_{n+1}x_{n+2} \ldots x_m$, where $x_j \in M$ or $x_j^{-1} \in M$, for $1 \leq j \leq m$. We have

$$g_1g_2^{-1} = (x_1x_2 \ldots x_n)(x_{n+1}x_{n+2} \ldots x_m)^{-1} = x_1x_2 \ldots x_nx_m^{-1} \ldots x_{n+1}^{-1} \in Y.$$

We note that, of course, $x_j^{-1} \in M$ or $x_j = (x_j^{-1})^{-1} \in M$ and hence Y is a subgroup containing M. It follows that $\langle M \rangle \leq Y$. Hence $Y = \langle M \rangle$.

A group G is called finitely generated, if there exists a finite subset M of G such that $\langle M \rangle = G$. Finite and finitely generated groups were the very first subjects of investigation in group theory. Very important problems were formulated by Burnside at the start of the twentieth century concerning finitely generated groups. These problems, although easy to be stated, proved to be extremely difficult and for many group theorists, these problems of Burnside played a role similar to the role that Fermat's Last Theorem played for number theorists. These problems played a significant role, not only in group theory but also in the development of the theory of associative algebras and the theory of Lie algebras.

The first "general" Burnside problem can be stated very simply: is every periodic finitely generated group finite? The negative answer to this problem was obtained by Golod in 1964. After that, some other examples of infinite periodic finitely generated groups were constructed, most notably the elegant and clear examples constructed by Grigorchuk.

The "second" Burnside problem is more limiting still. Although the examples of Golod and Grigorchuk are periodic, there is no bound on the orders of the elements occurring. Thus the second Burnside problem asks if G is a finitely generated group and if there is a positive integer m such that $g^m = e$ for each element $g \in G$, then is G finite? In 1968, Adyan and Novikov obtained a negative solution to this problem. Thus, among the groups with condition $g^m = e$, there are infinite groups. So the following "restricted" Burnside problem is very natural. Let G be a finite group, generated by the finite subset M, suppose that $|M| \leq k$ and that there is a positive integer m such that $g^m = e$ for each element $g \in G$. Is there a function $b(k, m)$ such that $|G| \leq b(k, m)$? This restricted Burnside problem was finally solved affirmatively by Zelmanov and, for this achievement, in 1996, he was awarded the highest mathematical honor, the Fields Medal.

Finally, we consider the useful construction of the Cartesian (direct) product of a finite set of groups. Let $G_1, \ldots, G_n$ be groups and let $D = G_1 \times \cdots \times G_n$ be its

Cartesian product as sets. On the set D, we define an operation of multiplication by the rule

$$(g_1, \dots, g_n)(x_1, x_2, \dots, x_n) = (g_1x_1, \dots, g_nx_n),$$

where $g_j, x_j \in G_j$, for $1 \le j \le n$. Since g_j, x_j are elements of G_j, their product is also an element of G_j, for $1 \le j \le n$. Therefore, this is a binary operation defined on D. This operation is associative since

$$((g_1, \dots, g_n)(x_1, x_2, \dots, x_n))(y_1, \dots, y_n) = ((g_1x_1)y_1, \dots, (g_nx_n)y_n)$$
$$= (g_1(x_1y_1), \dots, g_n(x_ny_n)) = (g_1, \dots, g_n)((x_1, \dots, x_n)(y_1, \dots, y_n)).$$

If e_j is the identity element of G_j, for $1 \le j \le n$, then the tuple $(e_1, \dots, e_n)$ is easily seen to be the identity element for D and it is also easy to see that

$$(g_1, \dots, g_n)^{-1} = (g_1^{-1}, \dots, g_n^{-1}).$$

Therefore, all the group axioms hold for D and D is called the Cartesian product of the groups $G_1, \dots, G_n$. The group D is also called the direct product of the groups $G_1, \dots, G_n$. In general, when infinitely many groups G_i are concerned, the constructions of Cartesian and direct products are different and lead to different groups, but for finitely many groups $G_1, \dots, G_n$, the two concepts coincide.

If all groups $G_1, \dots, G_n$ are abelian, then their Cartesian product is also an abelian group since

$$(g_1, \dots, g_n)(x_1, x_2, \dots, x_n) = (g_1x_1, \dots, g_nx_n)$$
$$= (x_1g_1, \dots, x_ng_n) = (x_1, \dots, x_n)(g_1, \dots, g_n).$$

If the operation is additive in each of $G_1, \dots, G_n$, then we use additive notation for the operation in the direct product, which is sometimes called a direct sum in this case, and we write

$$(g_1, \dots, g_n) + (x_1, x_2, \dots x_n) = (g_1 + x_1, \dots, g_n + x_n).$$

EXERCISE SET 8.1

8.1.1. Let $G = \{a + bi\sqrt{5} \mid a, b \in \mathbb{Q}, a^2 + b^2 \neq 0\}$. Is G a subgroup of $\mathbf{U}(\mathbb{C})$?

8.1.2. Define an operation on $G = \mathbb{Z} \times \mathbb{Z}$ by $(n_1, m_1) * (n_2, m_2) = (n_1 + n_2, (-1)^{n_2}m_1 + m_2)$. Is this operation commutative? Is G a group?

8.1.3. Let $G = \langle g \rangle$ be a group and let $|g| = 32$. Find all elements $x \in G$ with the property $G = \langle x \rangle$.

8.1.4. Is the set of complex numbers $\{\alpha \mid |\alpha| = r\}$ a subgroup of $\mathbf{U}(\mathbb{C})$? Is the set of complex numbers $\{\alpha \mid 0 \neq |\alpha| \leq r\}$ a subgroup of $\mathbf{U}(\mathbb{C})$? Is the set of complex numbers $\{\alpha \mid 0 \neq |\alpha| \geq r\}$ a subgroup of $\mathbf{U}(\mathbb{C})$?

8.1.5. Let z be a 600th root of unity in $\mathbb{C}$ of order 6 in the multiplicative group $\mathbf{U}(\mathbb{C})$. Find all the possibilities for z.

8.1.6. Suppose that $g^2 = e$ for all elements g of a group G. Prove that G is abelian.

8.1.7. Let G be an abelian group. Prove that the subset of all elements of G having finite order is a subgroup.

8.1.8. The mapping $\varphi : x + iy \longrightarrow x - iy$ is an isomorphism of $\mathbb{C}$ onto $\mathbb{C}$. Find the order of φ as an element of $S(\mathbb{C})$.

8.1.9. On the set $G = \mathbb{Z} \times \{-1, 1\}$, we define the operation $\star$ by $(m, a) \star (n, b) = (m + an, ab)$. Is G a group? Is this operation commutative?

8.1.10. On the set $G = \mathbb{Z} \times \{-1, 1\}$, we define the operation $\diamond$ by $(m, a) \diamond (n, b) = (bm + n, ab)$. Is G a group? Is this operation commutative?

8.1.11. On the set $G = \mathbb{Z} \times \{-1, 1\}$, we define the operation $\circ$ by $(m, a) \circ (n, b) = (m + n, ab)$. Is G a group? Is this operation commutative?

8.1.12. On the set $G = \mathbb{Z} \times \{-1, 1\}$, we define the operation $\boxtimes$ by $(m, a) \boxtimes (n, b) = (bm + an, ab)$. Is G a group? Is this operation commutative?

8.1.13. On the set $G = \mathbb{Z} \times \mathbb{Z}$, we define the operation $\dagger$ by $(a, b) \dagger (c, d) = (a + c, (-1)^c b + d)$. Is G a group? Is this operation commutative? Are the sets $H = \{(a, 0) \mid a \in \mathbb{Z}\}$ and $K = \{(0, a) \mid a \in \mathbb{Z}\}$ subgroups of G?

8.1.14. On the set $G = \mathbb{Z} \times \mathbb{Q}$, we define the operation $\ddagger$ by $(a, b) \ddagger (c, d) = (a + c, 2^c b + d)$. Is G a group? Is this operation commutative? Are the sets $H = \{(a, 0) \mid a \in \mathbb{Z}\}$ and $K = \{(0, a) \mid a \in \mathbb{Q}\}$ subgroups of G?

8.1.15. On a set with four elements, define commutative and associative operations each having an identity element.

8.1.16. Let $M = \{x, y, z\}$. Define an algebraic binary operation on M such that M becomes a semigroup with identity but not a group.

8.2 EXAMPLES OF GROUPS AND SUBGROUPS

Groups Consisting of Numbers

We begin this section by giving some examples, by no means all, of the groups occurring in the complex number system. The set $\mathbb{R}$ of all real numbers is a group under the operation of addition. The additive identity element is 0, in this case. This group is torsion free, since if $x \in \mathbb{R}$ and $n \in \mathbb{Z}$ then $nx = 0$ implies immediately that $x = 0$. The subsets $\mathbb{Q}$, of all rational numbers and $\mathbb{Z}$, of all integers, are clearly subgroups of $\mathbb{R}$, under the naturally induced operation of addition since, for example, the difference of two rational numbers is again a rational number. The group $\mathbb{Z}$ is an example of an infinite cyclic group, generated by the number 1. If $n \geq 0$ is a fixed integer, the subset $n\mathbb{Z} = \{nk \mid k \in \mathbb{Z}\}$ is also a subgroup of $\mathbb{Z}$. Indeed, if $k, t \in \mathbb{Z}$, then $nk - nt = n(k - t)$, which shows that $n\mathbb{Z}$ satisfies condition **(SG 3)** and Corollary 8.1.8 implies that $n\mathbb{Z}$ is a subgroup of $\mathbb{Z}$. Furthermore, Theorem 8.1.21 shows that each subgroup of $\mathbb{Z}$ coincides with one of the subsets $n\mathbb{Z}$, for some $n \geq 0$.

Now, let p be a prime and let

$$\mathbb{Q}_p = \left\{ \frac{m}{p^k} \mid m, k \in \mathbb{Z} \right\}.$$

A fraction of the type $\frac{m}{p^k}$ is called p-adic. The equation

$$\frac{m}{p^k} - \frac{r}{p^s} = \frac{mp^s - rp^k}{p^{k+s}}$$

shows that $\mathbb{Q}_p$, under addition, satisfies condition **(SG 3)**. Corollary 8.1.8 implies that $\mathbb{Q}_p$ is a subgroup of $\mathbb{Q}$ called the additive group of p-adic fractions.

The set $\mathbb{R}\backslash\{0\} = \mathbb{R}^\times$ of all nonzero real numbers is a group under the operation of multiplication. The multiplicative identity is the number 1, in this case. The subset, $\mathbb{Q}\backslash\{0\} = \mathbb{Q}^\times$, of all nonzero rational numbers and the subset $\{1, -1\}$ are subgroups of $\mathbb{R}^\times$. Moreoever, $\{1, -1\}$ is an example of a finite cyclic group. We note that the only two elements of $\mathbb{R}^\times$ that have finite order are 1 and -1, since these are the only real solutions of the equation $x^n = 1$.

Boolean Sets

Let A be an arbitrary set and let $\mathfrak{B}(A)$ be the Boolean of A. From Theorem 1.1.10, we see that $\mathfrak{B}(A)$ is a group under the operation Δ defined by $C\Delta B = (C\backslash B) \cup (B\backslash C))$. This group is abelian with identity element $\emptyset$ and the equation $B\Delta B = \emptyset$ shows that every element of this group has order 2.

Groups of Invertible Elements of a Semigroup

Many important examples of groups have their origins in semigroups. Let S be a semigroup with identity. As in Section 3.1, we let $\mathbf{U}(S)$ denote the subset of all invertible elements of S. By Corollary 3.1.15, $\mathbf{U}(S)$ is stable in S. Thus the restriction of the binary operation given on S to the set $\mathbf{U}(S)$ is a binary operation on $\mathbf{U}(S)$. This operation is associative because it is associative in S. By definition, the identity element belongs to $\mathbf{U}(S)$ and every element has an inverse in $\mathbf{U}(S)$. Therefore, axioms **(G1)–(G3)** hold in $\mathbf{U}(S)$.

The group $\mathbf{U}(S)$ is called the group of invertible elements of S.

In Section 3.1, we noted that the set $\mathbf{P}(A)$ of all transformations of a set A is a semigroup with identity. In particular, since $\mathbf{U}(\mathbf{P}(A)) = \mathbf{S}(A)$, the set $\mathbf{S}(A)$, the set of all permutations of A, is a group under the operation of multiplication of functions. This group is called the group of permutations of the set A. For the group $\mathbf{S}_n$, we reserve a special term, namely, the symmetric group of degree n.

In Section 2.5, we observed that $\mathbf{U}(M_n(\mathbb{R})) = GL_n(\mathbb{R})$. In particular, the subset $GL_n(\mathbb{R})$ of all nonsingular square matrices of dimension n with real entries is a group under the operation of matrix multiplication, called the general linear group of degree n over $\mathbb{R}$. The subset $GL_n(\mathbb{Q})$ is easily seen to be a subgroup of $GL_n(\mathbb{R})$. Indeed, by definition, the product of two matrices with rational entries is a matrix with rational entries. If $A \in GL_n(\mathbb{Q})$, then Theorem 2.5.3 shows that $A^{-1} \in GL_n(\mathbb{Q})$. Hence $GL_n(\mathbb{Q})$ satisfies conditions **(SG 1)–(SG 2)** and Theorem 8.1.7 implies that $GL_n(\mathbb{Q})$ is a subgroup of $GL_n(\mathbb{R})$.

Permutation Groups

Let A be an arbitrary set. We observed above that the set $\mathbf{S}(A)$ of all permutations of A is a group. We now consider some important examples of subgroups of this group.

The Stabilizer of Subset

Let B be a subset of A and let

$$\mathbf{St}(B) = \{\pi \in \mathbf{S}(A) \mid \pi(b) = b \text{ for each element } b \in B\}.$$

If $\pi, \sigma \in \mathbf{St}(B)$ then, for every element $b \in$ B, we have

$$\pi \circ \sigma(b) = \pi(\sigma(b)) = \pi(b) = b,$$

so that $\pi \circ \sigma \in \mathbf{St}(B)$ and also

$$b = \varepsilon_A(b) = (\pi^{-1} \circ \pi)(b) = (\pi^{-1}(\pi(b)) = \pi^{-1}(b).$$

Thus $\pi^{-1} \in \mathbf{St}(B)$ also. Consequently, the subset $\mathbf{St}(B)$ satisfies both conditions **(SG 1)** and **(SG 2)** and, by Theorem 8.1.7, it is a subgroup of $\mathbf{S}(A)$. In particular, the subset $\mathbf{St}(a)$ is a subgroup for each element $a \in A$. We also note that $\mathbf{St}(B) = \bigcap_{b\in B} \mathbf{St}(b)$.

Now, let

$$\mathbf{Inv}(B) = \{\pi \in \mathbf{S}(A) \mid \pi(B) = B\}.$$

This set is the set of permutations leaving B invariant. If $\pi, \sigma \in \mathbf{Inv}(B)$ and $b \in B$, then $\sigma(b) = b_1 \in B$. Therefore,

$$\pi \circ \sigma(b) = \pi(\sigma(b)) = \pi(b_1) = b_2 \in B,$$

so that $\pi \circ \sigma \in \mathbf{Inv}(B)$. Since $\pi(B) = B$, there exists an element $b_3 \in B$ such that $b = \pi(b_3)$. Then

$$b_3 = \varepsilon_A(b_3) = (\pi^{-1} \circ \pi)(b_3) = (\pi^{-1}(\pi(b_3)) = \pi^{-1}(b),$$

so $\pi^{-1} \in \mathbf{Inv}(B)$. Consequently, the subset $\mathbf{Inv}(B)$ satisfies conditions **(SG 1)** and **(SG 2)**. By Theorem 8.1.7, $\mathbf{Inv}(B)$ is a subgroup of $\mathbf{S}(A)$. We note that $\mathbf{St}(B) \leq \mathbf{Inv}(B)$ while $\mathbf{Inv}(a) = \mathbf{St}(a)$ for each $a \in A$.

Finitary Permutations

Let $\pi \in \mathbf{S}(A)$. The subset

$$\mathbf{Supp}\,\pi = \{a \in A \mid \pi(a) \neq a\}$$

is called the support of the permutation π and its complement $A \backslash \mathbf{Supp}\,\pi$ is called the set of fixed points of π. A permutation π is called finitary if its support, $\mathbf{Supp}\,\pi$, is finite. We let $\mathbf{FS}(A)$ denote the subset of all finitary permutations of the set A. If $\pi, \sigma \in \mathbf{FS}(A)$ and $b \in A \backslash (\mathbf{Supp}\,\pi \cup \mathbf{Supp}\,\sigma)$, then

$$\pi \circ \sigma(b) = \pi(\sigma(b)) = \pi(b) = b,$$

so that $\mathbf{Supp}(\pi \circ \sigma) \subseteq \mathbf{Supp}\,\pi \cup \mathbf{Supp}\,\sigma$. Thus, $\mathbf{Supp}(\pi \circ \sigma)$ is finite and $\pi \circ \sigma \in \mathbf{FS}(A)$. Furthermore, if $b \in A \backslash \mathbf{Supp}\,\pi$, then

$$b = \varepsilon_A(b) = (\pi^{-1} \circ \pi)(b) = (\pi^{-1}(\pi(b)) = \pi^{-1}(b),$$

which implies that $\mathbf{Supp}\,\pi^{-1} \subseteq \mathbf{Supp}\,\pi$. Since $\pi = (\pi^{-1})^{-1}$, $\mathbf{Supp}\,\pi \subseteq \mathbf{Supp}\,\pi^{-1}$ also and hence $\mathbf{Supp}\,\pi^{-1} = \mathbf{Supp}\,\pi$. In particular, $\mathbf{Supp}\,\pi^{-1}$ is finite and hence $\pi^{-1} \in \mathbf{FS}(A)$. Consequently, the subset $\mathbf{FS}(A)$ satisfies both conditions **(SG 1)** and **(SG 2)** and Theorem 8.1.7 implies that it is a subgroup of $\mathbf{S}(A)$. The group $\mathbf{FS}(A)$ is called the group of finitary permutations of A.

Groups of Symmetries

Let $E = \mathbb{R}^n$, and let $d(x, y)$ denote the distance between the points x, y of the space E. As we saw in Chapter 6, a bijective mapping $f \in \mathbf{S}(E)$ is called an isometry of the space E, if f fixes the distance between the points of E, which means that

$$d(f(x), f(y)) = d(x, y),$$

for every pair of points x, y of E. The set of all isometries of E is denoted by $\mathbf{Isom}(E)$. If $f, g \in \mathbf{Isom}(E)$ and $x, y \in E$, then

$$d(f \circ g(x), f \circ g(y)) = d(f(g(x)), f(g(y))) = d(g(x), g(y)) = d(x, y),$$

so that $f \circ g \in \mathbf{Isom}(E)$. Furthermore,

$$d(x, y) = d(f(f^{-1}(x)), f(f^{-1}(y))) = d(f^{-1}(x), f^{-1}(y))$$

and hence $f^{-1} \in \mathbf{Isom}(E)$. Consequently, the subset $\mathbf{Isom}(E)$ satisfies both conditions **(SG 1)** and **(SG 2)** and, by Theorem 8.1.7, it is a subgroup of $\mathbf{S}(E)$.

Now, let M be a subset of E and set

$$\mathbf{Sym}(M) = \mathbf{Isom}(E) \cap \mathbf{Inv}(M).$$

Thus, $\mathbf{Sym}(M)$ consists of those isometries of E that transform the set M into itself. By Corollary 8.1.10, $\mathbf{Sym}(M)$ is a subgroup of $\mathbf{Isom}(E)$. This subgroup is called the symmetry group of M.

Figuratively speaking, the group $\mathbf{Sym}(M)$ measures the level of symmetry of M. For example, the fact that an isosceles triangle looks more symmetrical than a scalene triangle can be interpreted as follows. The group of symmetries of an isosceles triangle consists of the identity permutation and a reflection in its axis of symmetry while the group of symmetries of a scalene triangle consists of only the identity permutation. The group of symmetries of an equilateral triangle consists of six isometries, the identity rotations through 120° and 240°, about the center of the triangle and three reflections in its three axes of symmetry. Thus the size of the symmetry group tends to reflect the amount of symmetry (or lack thereof) that the particular figure has.

The symmetry of a molecule is an isometry of the space that transforms each atom of the molecule to an atom of the same type and keeps all valency connections between atoms. For example, the molecule of phosphor consists of four atoms situated in the vertices of a regular tetrahedron. An important role of symmetry lies in crystallography. Here by symmetry of a crystal, we understand an isometry of the space that keeps the disposition of atoms of the crystal and all their connections and transforms every atom into an atom of the same element.

Symmetry is also important in certain laws of physics. In this case, symmetries are transformations of the coordinates that keep the law invariant. Thus the laws of mechanics should be kept under translation from one inertial system to another. The transformation of coordinates in Galileo–Newtonian mechanics for movement in a line look like

$$x' = x - vt, \quad t' = t,$$

whereas in the mechanics of special relativity, they are

$$x' = \frac{x - vt}{\sqrt{1 - \left(\frac{v}{c}\right)^2}}, \quad t' = \left(t - \frac{v}{c^2}x\right)\sqrt{1 - \left(\frac{v}{c}\right)^2},$$

where c is the speed of light. The group of symmetries is called the group of Galileo–Newtonian mechanics in the first case and the Lorentz group in the second.

Symmetry groups of finite objects can be connected with subgroups of a permutation group of finite degree. For example, let $\mathbf{P}_3$ be an equilateral triangle whose vertices are labeled as 1, 2, and 3. Each symmetry transforms the vertices in some way and we can assign a permutation of degree 3 to this symmetry. The identity transformation corresponds to the identity permutation

$$\varepsilon = \begin{pmatrix} 1 & 2 & 3 \\ 1 & 2 & 3 \end{pmatrix}.$$

The (counterclockwise) rotation of the triangle by 120° about the center corresponds to the permutation

$$\pi_1 = \begin{pmatrix} 1 & 2 & 3 \\ 2 & 3 & 1 \end{pmatrix}, \quad \text{whereas}$$

the rotation of the triangle by 240° about the center corresponds to the permutation

$$\pi_2 = \begin{pmatrix} 1 & 2 & 3 \\ 3 & 1 & 2 \end{pmatrix}.$$

Each of the reflections in a line through a vertex and the midpoint of the opposite side corresponds to one of the following permutations:

$$\pi_3 = \begin{pmatrix} 1 & 2 & 3 \\ 1 & 3 & 2 \end{pmatrix}, \pi_4 = \begin{pmatrix} 1 & 2 & 3 \\ 3 & 2 & 1 \end{pmatrix}, \pi_5 = \begin{pmatrix} 1 & 2 & 3 \\ 2 & 1 & 3 \end{pmatrix}.$$

From this it follows that the group $\mathbf{S}_3$ is the group of symmetries of an equilateral triangle.

We also observe that the group of symmetries of a square consists of the identity permutation ε as follows:

(i) counterclockwise rotations about the center of the square through angles $90^\circ, 180^\circ, 270^\circ$—with corresponding permutations

$$\pi_1 = \begin{pmatrix} 1 & 2 & 3 & 4 \\ 2 & 3 & 4 & 1 \end{pmatrix}, \pi_2 = \begin{pmatrix} 1 & 2 & 3 & 4 \\ 3 & 4 & 1 & 2 \end{pmatrix}, \pi_1 = \begin{pmatrix} 1 & 2 & 3 & 4 \\ 4 & 1 & 2 & 3 \end{pmatrix};$$

(ii) two reflections in the diagonals of the square—with corresponding permutations

$$\pi_4 = \begin{pmatrix} 1 & 2 & 3 & 4 \\ 1 & 4 & 3 & 2 \end{pmatrix}, \pi_5 = \begin{pmatrix} 1 & 2 & 3 & 4 \\ 3 & 2 & 1 & 4 \end{pmatrix};$$

(iii) two reflections through lines connecting midpoints of opposite sides—with corresponding permutations

$$\pi_6 = \begin{pmatrix} 1 & 2 & 3 & 4 \\ 2 & 1 & 4 & 3 \end{pmatrix}, \pi_7 = \begin{pmatrix} 1 & 2 & 3 & 4 \\ 4 & 3 & 2 & 1 \end{pmatrix}.$$

We can write the multiplication table for $\mathbf{Sym}(\mathbf{P}_4)$ as a matrix. In the first row and in the first column, we write the group elements and at the intersection of the row beginning with an element x and the column beginning with an element y, we will write the element xy. This table is the Cayley table of the group and for the symmetry group of the square, it is as follows:

	ε	π_1	π_2	π_3	π_4	π_5	π_6	π_7
ε	ε	π_1	π_2	π_3	π_4	π_5	π_6	π_7
π_1	π_1	π_2	π_3	ε	π_7	π_6	π_4	π_5
π_2	π_2	π_3	ε	π_1	π_5	π_4	π_7	π_6
π_3	π_3	ε	π_1	π_2	π_6	π_7	π_5	π_4
π_4	π_4	π_6	π_5	π_7	ε	π_2	π_1	π_3
π_5	π_5	π_7	π_4	π_6	π_2	ε	π_3	π_1
π_6	π_6	π_5	π_7	π_4	π_3	π_1	ε	π_2
π_7	π_7	π_4	π_6	π_5	π_1	π_3	π_2	ε

Observe that the element $\sigma = \pi_1$ has order 4, that $\pi_2 = \sigma^2$ and that $\pi_3 = \sigma^3$. Let $\iota = \pi_4$. Then $\iota^2 = \varepsilon$ and from the Cayley table, it follows that $\mathbf{Sym}(\mathbf{P}_4) = \langle \sigma, \iota \rangle$. The elements σ, ι do not commute since $\iota\sigma = \sigma^3\iota$. This group is called the dihedral group of order 8. Here, we have considered only the most basic examples of symmetry groups.

Linear Groups

Linear groups (more precisely finite-dimensional linear groups) are subgroups (including the group itself) of the group $\mathbf{GL}_n(\mathbb{R})$. We next consider some important subgroups of $\mathbf{GL}_n(\mathbb{R})$. As we have already noted, the set $\mathbf{GL}_n(\mathbb{Q})$ of all invertible matrices of degree n with rational coefficients is a subgroup of $\mathbf{GL}_n(\mathbb{R})$.

Next let

$$\mathbf{GL}_n(\mathbb{Z}) = \{A \in \mathbf{M}_n(\mathbb{Z}) \mid \mathbf{det}(A) \in \{1, -1\}\}.$$

By definition, the product of two matrices with coefficients in $\mathbb{Z}$ again has integer coefficients. If $A \in \mathbf{GL}_n(\mathbb{Z})$, then since $\mathbf{det}(A) = \pm 1$, Theorem 2.5.3 implies that $A^{-1} \in \mathbf{GL}_n(\mathbb{Z})$. Consequently, $\mathbf{GL}_n(\mathbb{Z})$ satisfies conditions **(SG1)** and **(SG2)** and, by Theorem 8.1.7, it is a subgroup of $\mathbf{GL}_n(\mathbb{R})$.

For the next example, let

$$\mathbf{SL}_n(\mathbb{R}) = \{A \in \mathbf{M}_n(\mathbb{R}) \mid \mathbf{det}(A) = 1\}$$

and let $A, B \in \mathbf{SL}_n(\mathbb{R})$. Theorem 2.5.1 shows that

$$\mathbf{det}(AB) = \mathbf{det}(A)\mathbf{det}(B) = 1.$$

It follows that $AB \in \mathbf{SL}_n(\mathbb{R})$. By the same reasoning,

$$1 = \mathbf{det}(I) = \mathbf{det}(AA^{-1}) = \mathbf{det}(A)\mathbf{det}(A^{-1}),$$

and hence $\mathbf{det}(A^{-1}) = 1$ so that $A^{-1} \in \mathbf{SL}_n(\mathbb{R})$. Consequently, $\mathbf{SL}_n(\mathbb{R})$ satisfies conditions **(SG 1)** and **(SG 2)** and, by Theorem 8.1.7, it is a subgroup of $\mathbf{GL}_n(\mathbb{R})$, called the special linear group of degree n over $\mathbb{R}$. In a similar fashion, the subgroups $\mathbf{SL}_n(\mathbb{Q})$ and $\mathbf{SL}_n(\mathbb{Z})$ can also be defined.

We denote by $\mathbf{T}_n(\mathbb{R})$, the set of all nonsingular triangular (more precisely, upper triangular) matrices of degree n with real entries. Let $A, B \in \mathbf{T}_n(\mathbb{R})$, where $A = [a_{ij}]$, $B = [b_{ij}]$ and $C = AB = [c_{ij}]$. Suppose that A, B are upper triangular so that if $i > j$, then $a_{ij} = b_{ij} = 0$. We have

$$c_{ij} = a_{i1}b_{1j} + a_{i2}b_{2j} + \cdots + a_{i,i-1}b_{i-1,j} + a_{ii}b_{ij} + a_{i,i+1}b_{i+1,j} + \cdots + a_{in}b_{nj}.$$

Since

$$a_{i1} = a_{i2} = \cdots = a_{i,i-1} = 0 \text{ and } b_{ij} = b_{i+1,j} = \cdots = b_{nj} = 0,$$

it follows that $c_{ij} = 0$. Hence $AB \in \mathbf{T}_n(\mathbb{R})$. We note that

$$\begin{aligned} c_{ii} &= a_{i1}b_{1i} + a_{i2}b_{2i} + \cdots + a_{i,i-1}b_{i-1,i} + a_{ii}b_{ii} + a_{i,i+1}b_{i+1,i} + \cdots + a_{in}b_{ni} \\ &= a_{ii}b_{ii}. \end{aligned}$$

Next, we consider the matrix $A^{-1} = [x_{ij}]$. Theorem 2.5.3 shows that $x_{ij} = \frac{A_{ji}}{\det(A)}$, for $1 \leq i, j \leq n$, where $A_{ji} = (-1)^{i+j}\mathbf{det}(S_{ji})$ and $S_{ji} = [y_{km}]$ is the submatrix of A constructed by eliminating the jth row and the ith column. Let $i > j$ and consider an arbitrary column of the matrix S_{ji}, say the kth column. If $k < j$, then the kth column of S_{ji} consists of the coefficients

$$y_{1k} = a_{1k}, \ldots, y_{k-1,k} = a_{k-1,k}, y_{kk} = a_{kk}, y_{k+1,k} = 0, \ldots, y_{n-1,k} = 0.$$

If $j = k$, then the kth column consists of the elements

$$y_{1k} = a_{1k}, \ldots, y_{k-1,k} = a_{k-1,k}, y_{kk} = 0, y_{k+1,k} = 0, \ldots, y_{nk} = 0.$$

If $i > k > j$, then the kth column consists of the coefficients

$$\begin{aligned} &y_{1k} = a_{1k}, \ldots, y_{j-1,k} = a_{j-1,k}, y_{jk} = a_{j+1,k}, \ldots, \\ &y_{k-1,k} = a_{kk}, y_{kk} = 0, \ldots, y_{n-1,k} = 0. \end{aligned}$$

If $k \geq i$, then the kth column consists of the coefficients

$$y_{1k} = a_{1,k+1}, \ldots, y_{j-1,k} = a_{j-1,k+1}, y_{jk} = a_{j+1,k+1}, \ldots,$$
$$y_{k-1,k} = a_{k,k+1}, y_{kk} = a_{k+1,k+1}, y_{k+1,k} = 0, \ldots, y_{n-1,k} = 0.$$

This means that the matrix S_{ji} is upper triangular. If $j < n - 1$ then $y_{jj} = 0$ and, by Proposition 2.3.11, we see that $A_{ji} = (-1)^{j+i}\mathbf{det}(S_{ji}) = 0$. If $j = n - 1$, then $i = n$; and in this case, all entries of the last row are zero. By Corollary 2.3.6, $A_{ji} = (-1)^{j+i}\mathbf{det}(S_{ji}) = 0$. So, if $i > j$, then $x_{ij} = \frac{A_{ji}}{\det(A)} = 0$, which means that $A^{-1} \in \mathbf{T}_n(\mathbb{R})$. Consequently, $\mathbf{T}_n(\mathbb{R})$ satisfies both conditions **(SG 1)** and **(SG 2)**. Thus, $\mathbf{T}_n(\mathbb{R})$ is a subgroup of $\mathbf{GL}_n(\mathbb{R})$, by Theorem 8.1.7. It then follows that $\mathbf{T}_n(\mathbb{Q}) = \mathbf{GL}_n(\mathbb{Q}) \cap \mathbf{T}_n(\mathbb{R})$ and $\mathbf{T}_n(\mathbb{Z}) = \mathbf{GL}_n(\mathbb{Z}) \cap \mathbf{T}_n(\mathbb{R})$ are also subgroups.

Next, let $\mathbf{UT}_n(\mathbb{R})$ denote the set of all unitriangular matrices of degree n with real entries. If $A, B \in \mathbf{UT}_n(\mathbb{R})$, where $A = [a_{ij}]$, $B = [b_{ij}]$, and $C = AB = [c_{ij}]$ then the arguments above show that C is a triangular matrix. Since $c_{ii} = a_{ii}b_{ii}$, it follows that $c_{ii} = 1$ for each i, where $1 \leq i \leq n$. From this, it also follows that the inverse of a unitriangular matrix is unitriangular. Consequently, $\mathbf{UT}_n(\mathbb{R})$ satisfies both conditions **(SG 1)** and **(SG 2)** and, by Theorem 8.1.7, it is a subgroup of $\mathbf{GL}_n(\mathbb{R})$. Then $\mathbf{UT}_n(\mathbb{Q}) = \mathbf{GL}_n(\mathbb{Q}) \cap \mathbf{UT}_n(\mathbb{R})$ and $\mathbf{UT}_n(\mathbb{Z}) = \mathbf{GL}_n(\mathbb{Z}) \cap \mathbf{UT}_n(\mathbb{R})$ are also subgroups.

We denote the set of all nonsingular diagonal matrices of degree n with real entries by $\mathbf{D}_n(\mathbb{R})$. If $A, B \in \mathbf{D}_n(\mathbb{R})$, where $A = [a_{ij}]$, $B = [b_{ij}]$, and if $C = AB = [c_{ij}]$, then the previous arguments show that C is a diagonal matrix. Also, $c_{ii} = a_{ii}b_{ii}$ for every i, where $1 \leq i \leq n$ and hence $AB = BA$. From this equation, we see that the inverse of a diagonal matrix is also diagonal. Consequently, $\mathbf{D}_n(\mathbb{R})$ satisfies both conditions **(SG 1)** and **(SG 2)** and, by Theorem 8.1.7, it is an abelian subgroup of $\mathbf{GL}_n(\mathbb{R})$. Then $\mathbf{D}_n(\mathbb{Q}) = \mathbf{GL}_n(\mathbb{Q}) \cap \mathbf{D}_n(\mathbb{R})$ and $\mathbf{D}_n(\mathbb{Z}) = \mathbf{GL}_n(\mathbb{Z}) \cap \mathbf{D}_n(\mathbb{R})$ are both subgroups. If $A \in \mathbf{D}_n(\mathbb{Z})$, then $a_{ii} \in \{1, -1\}$ for each i, where $1 \leq i \leq n$ and it follows that $\mathbf{D}_n(\mathbb{Z})$ is a finite abelian group of order 2^n.

We recall from Chapter 6 that a matrix $A \in \mathbf{M}_n(\mathbb{R})$ is called orthogonal if $AA^t = I$, so that $A^{-1} = A^t$. If $AA^t = I$, Proposition 2.3.3 and Theorem 2.5.1 imply that

$$1 = \mathbf{det}(I) = \mathbf{det}(AA^t) = \mathbf{det}(A)\mathbf{det}(A^t) = (\mathbf{det}(A))^2.$$

Thus, an orthogonal matrix is always nonsingular. Let $\mathbf{O}_n(\mathbb{R})$ denote the subset of $\mathbf{M}_n(\mathbb{R})$ consisting of all orthogonal matrices. If $A, B \in \mathbf{O}_n(\mathbb{R})$ then, by Theorem 2.1.10,

$$(AB)^t = B^tA^t = B^{-1}A^{-1} = (AB)^{-1} \text{ and } (A^{-1})^t = (A^t)^t = A = (A^{-1})^{-1}.$$

From this, we see that $AB \in \mathbf{O}_n(\mathbb{R})$ and $A^{-1} \in \mathbf{O}_n(\mathbb{R})$, so Theorem 8.1.7 implies that $\mathbf{O}_n(\mathbb{R})$ is a subgroup of $\mathbf{GL}_n(\mathbb{R})$.

Quasicyclic (or Prüfer) p groups

Let p be a prime and let

$$C_{p^\infty} = \{x \in \mathbb{C} \mid x^{p^k} = 1 \text{ for some } k \in \mathbb{N}_0\}.$$

We show that C_{p^∞} is a subgroup of the multiplicative group $\mathbf{U}(\mathbb{C})$. Let $x, y \in C_{p^\infty}$ and let k, t be positive integers such that $x^{p^k} = 1$, $y^{p^t} = 1$. If $m = \mathbf{max}\{k, t\}$ then $(\frac{x}{y})^{p^m} = 1$ and we deduce from Corollary 8.1.8 that C_{p^∞} is a subgroup of $\mathbf{U}(\mathbb{C})$.

Let $x \in C_{p^\infty}$ and let k be the least natural number for which $x^{p^k} = 1$, so that x, of order p^k, is a primitive p^kth root of unity. Thus $C_{p^k} = \langle x \rangle$ is a cyclic group of order p^k. Clearly

$$C_p \leq C_{p^2} \leq \cdots \leq C_{p^k} \leq \cdots \leq \bigcup_{k \in \mathbb{N}} C_{p^k} = C_{p^\infty}.$$

Now let H be a subgroup of C_{p^∞}. Then either the orders of the elements of H are bounded or they are not. In the former case, there is a number k, such that $y^{p^k} = 1$ for every $y \in H$ and H contains an element x, such that $|\langle x \rangle| = p^k$. This means that $H \leq C_{p^k} = \langle x \rangle$, so $H = C_{p^k}$.

In the latter case, for each $k \in \mathbb{N}$, there exists an element $z \in H$ such that $|\langle z \rangle| = p^m$, where $m \geq k$. From $|\langle z \rangle| = p^m$, it follows that $\langle z \rangle = C_{p^m}$. Since $m \geq k$, $C_{p^k} \leq C_{p^m} = \langle z \rangle \leq H$. So for each $k \in \mathbb{N}_0$, the subgroup H contains C_{p^k} and this means that $H = C_{p^\infty}$. Thus, every proper subgroup of C_{p^∞} is finite and cyclic.

This group C_{p^∞} is called a quasicyclic or a Prüfer p group. As shown above, C_{p^∞} is an infinite group, but all its proper subgroups are finite. In 1938, Schmidt raised the problem of describing the infinite groups all of whose proper subgroups are finite. Chernikov proved that for many important types of infinite groups, only the Prüfer groups have all proper subgroups finite. However, in 1978, the first example of such a group distinct from the Prüfer group was constructed by Olshanskii.

EXERCISE SET 8.2

Justify your work with a proof or a counterexample when necessary.

8.2.1. Which of the following sets are subgroups in $\mathbf{S}(\mathbb{R} \times \mathbb{R})$:

$$A = \{\iota_\alpha \mid \iota_\alpha : (x, y) \longrightarrow (x + \alpha, y + \alpha), \alpha \in \mathbb{R}\},\ B = \{\iota_\alpha \mid \alpha \in \mathbb{Q}\},$$
$$C = \{\iota_\alpha \mid \alpha \in \mathbb{Z}\},\ D = \{j_\alpha \mid j_\alpha : (x, y) \longrightarrow (\alpha x, \alpha y), \alpha \in \mathbb{R}\},$$
$$E = \{j_\alpha \mid \alpha \in \mathbb{Q}, \alpha \neq 0\}?$$

8.2.2. Prove that the set of matrices

$$\left\{\begin{pmatrix}1 & 0\\0 & 1\end{pmatrix}, \begin{pmatrix}1 & 0\\0 & -1\end{pmatrix}, \begin{pmatrix}-1 & 0\\0 & 1\end{pmatrix}, \begin{pmatrix}-1 & 0\\0 & -1\end{pmatrix}\right\}$$

is a subgroup of $\mathbf{GL}_2(\mathbb{Z})$.

8.2.3. Consider the following permutations of the set $\mathbb{R}\backslash\{0, 1\}$: $u_1(x) = x$, $u_2(x) = \frac{1}{x}$, $u_3(x) = 1 - x$, $u_4(x) = \frac{x}{x-1}$, $u_5(x) = \frac{(x-1)}{x}$, $u_6(x) = \frac{1}{1-x}$. Is the set $\{u_1, u_2, u_3, u_4, u_5, u_6\}$ a subgroup of $\mathbf{S}(\mathbb{R}\backslash\{0, 1\})$?

8.2.4. Let $H = \{\pi \mid \pi \in A_5, \pi(1) = 1\}$. Prove that H is a subgroup of A_5. Find $|H|$.

8.2.5. Find the order of the element

$$A = \begin{pmatrix}1 & 0\\0 & -1\end{pmatrix}$$

in the group $\mathbf{GL}_2(\mathbb{Q})$.

8.2.6. Find the center of the group A_4.

8.2.7. Prove that the group A_4 has no subgroups of order 6.

8.2.8. Let $x, y \in \mathbb{Q}$. Prove directly that $\langle x, y\rangle$ is a group under addition.

8.2.9. Let A be a vector space over a field F. A linear transformation f of a space A is called finitary, if the subspace $A(f - 1) = \{a(f - 1) \mid a \in A\}$ has finite dimension. Is the subset $\mathbf{FGL}(F, A)$ of all finitary isomorphisms of A a subgroup in the group $\mathbf{GL}(F, A)$ of all isomorphisms from A onto A?

8.2.10. Are the following mappings isometries of $\mathbb{R}$: $\alpha : x \longrightarrow x + 2$, $\beta : x \longrightarrow nx$, $n \notin \{1, -1\}$, $\gamma : x \longrightarrow x^2$, $\eta : x \longrightarrow -x$?

8.3 COSETS

In this section, we consider some important equivalence relations on groups analogous to those used in ring theory. To this end, let G be a group and let H be a subgroup of G. We define a relation Σ_H by the rule:

Let $x, y \in G$. Then $(x, y) \in \Sigma_H$ if and only if $xy^{-1} \in H$.

The relation Σ_H is reflexive since $xx^{-1} = e \in H$, so that $(x, x) \in \Sigma_H$. The relation Σ_H is symmetric since if $(x, y) \in \Sigma_H$ then $xy^{-1} \in H$ so $(xy^{-1})^{-1} = (y^{-1})^{-1}x^{-1} = yx^{-1} \in H$, because H is a subgroup of G. Thus $(y, x) \in \Sigma_H$. Finally, the relation Σ_H is transitive. If $xy^{-1}, yz^{-1} \in H$ then, since H is a subgroup, it contains the product, $xy^{-1}yz^{-1} = xz^{-1}$, of these two elements. Hence, if $(x, y), (y, z) \in \Sigma_H$ then, $(x, z) \in \Sigma_H$. Consequently, Σ_H is an equivalence relation.

If $(x, y) \in \Sigma_H$, then $xy^{-1} = h \in H$. Multiplying both sides of this equation on the left by h^{-1} and on the right by y gives $y = h^{-1}x$. Let

$$Hx = \{ux \mid u \in H\}.$$

8.3.1. Definition. *The subset Hx is called a right coset of H in G, or a right H-coset, and the element x is called its coset representative.*

Thus, each element equivalent to x (relative to Σ_H) belongs to Hx. Conversely, if $z \in Hx$ then, $z = ux$, for some element $u \in H$ and we have

$$xz^{-1} = x(ux)^{-1} = xx^{-1}u^{-1} = u^{-1} \in H.$$

Therefore $(x, z) \in \Sigma_H$, so the equivalence classes of Σ_H are precisely the right cosets of H. This implies that the right coset Hx is defined by each of its representatives x; thus, if $y \in Hx$, then $Hy = Hx$. Because they are equivalence classes, two right cosets either coincide or have an empty intersection and the group G is the union of all its right cosets. Thus the family of all right cosets of G by a subgroup H is a partition of G. If $H = \langle e \rangle$, then $Hx = \{x\}$ for each element $x \in G$, so we obtain the largest partition of G, consisting of one element sets. If $H = G$, then we obtain the smallest partition consisting of only one set, G.

Similarly, now let

$$xH = \{xu \mid u \in H\}.$$

8.3.2. Definition. *The subset xH is called a left coset of H in G, and the element x is called its left coset representative.*

As stated above, we can define an equivalence relation Λ_H on G by defining $(x, y) \in \Lambda_H$ if and only if $y^{-1}x \in H$. In this case, the left cosets form the corresponding equivalence classes. Thus the left coset xH is defined by each of its elements x in the sense that if $y \in xH$, then $yH = xH$. Therefore, two left cosets either coincide or have an empty intersection and G is the union of all the left cosets. Hence the family of all left cosets of H in G is a partition of G.

As an example of such a partition we consider the group $\mathbf{S}_n$. Let

$$P_i = \{\pi \in \mathbf{S}_n \mid \pi(n) = i\}.$$

It is easy to prove that $P_i \cap P_j = \emptyset$ whenever $i \neq j$ and that

$$\mathbf{S}_n = \bigcup_{1 \leq i \leq n} P_i = P_1 \cup P_2 \cup \cdots \cup P_n.$$

Since $P_n = \mathbf{St}(n)$ is the stabilizer of n, it is a subgroup of $\mathbf{S}_n$. Also, it is not hard to prove that $P_i = \tau_{in} \circ P_n$, where τ_{in} is the transposition interchanging i and n.

As one might expect, there is a connection between left and right cosets.

8.3.3. Proposition. *Let G be a group and let H be a subgroup of G. The mapping*

$$\nu : Hx \longmapsto x^{-1}H$$

is a bijection from the set of all right cosets of H in G onto the set of all left cosets of H in G.

Proof. First, we must show that ν is a mapping, in the sense that it does not depend on the choice of the representative x. To this end, let y be another representative of the coset Hx. Then $Hx = Hy$, so $y = ux$ for some element $u \in H$. By Proposition 3.1.16, $y^{-1} = x^{-1}u^{-1} \in x^{-1}H$, and so $y^{-1}H = x^{-1}H$.

Furthermore, the mapping ν is injective. For, if Hx, Hy are right cosets and if $\nu(Hx) = \nu(Hy)$ then

$$x^{-1}H = \nu(Hx) = \nu(Hy) = y^{-1}H.$$

Then $y^{-1} = x^{-1}v$ for some element $v \in H$, and $y = v^{-1}x \in Hx$. It follows that $Hx = Hy$ and that ν is injective follows.

Finally, ν is surjective since if zH is a left coset then

$$\nu(Hz^{-1}) = (z^{-1})^{-1}H = zH.$$

Hence ν is a bijective mapping, as required.

Let G be a group and let H be a subgroup of G. In every left (respectively right) coset of G by H, we choose a representative and let $\mathbf{lt}(G, H)$ (respectively, $\mathbf{rt}(G, H)$) denote the sets of all these selected representatives.

8.3.4. Definition. *The subset $\mathbf{lt}(G, H)$ (respectively $\mathbf{rt}(G, H)$) is called a left transversal (respectively right transversal) to H in G.*

It follows that $G = \bigcup_{x \in \mathbf{lt}(G,H)} xH$ (and also that $G = \bigcup_{x \in \mathbf{rt}(G,H)} Hx$). Furthermore, the equation $xH = yH$ (respectively $Hx = Hy$) for $x, y \in \mathbf{lt}(G, H)$ (respectively, $x, y \in \mathbf{rt}(G, H)$) implies that $x = y$.

8.3.5. Definition. *Let G be a group and let H be a subgroup of G. The number of distinct right cosets of H in G is called the index of H in G and it is denoted by $|G : H|$. If the set of all right cosets of H in G is finite then the subgroup H is said to have finite index in G. We say that H has infinite index in G if the set of all right cosets of H in G is infinite.*

Proposition 8.3.3 implies that the index of H in G is also equal to the number of distinct left cosets of H in G. This simply means that

$$|G : H| = |\mathbf{rt}(G, H)| = |\mathbf{lt}(G, H)|.$$

Clearly, in the case when $H = G$, $|G : H| = 1$. If G is a finite group G and $H = \langle e \rangle$, then $|G : H| = |G|$.

8.3.6. Theorem. *Let G be a group, let H, K be subgroups of G and suppose that $K \leq H$. Put $T = \mathbf{lt}(G, H)$ and $U = \mathbf{lt}(H, K)$. Then the subset*

$$R = \{tu \mid x \in T, u \in U\}$$

is a left transversal to K in G.

Proof. We have

$$G = \bigcup_{t \in T} tH, H = \bigcup_{u \in U} uK.$$

If x is an arbitrary element of G, then $x \in tH$ for some element $t \in T$, so that $x = th$ where $h \in H$. Also, there exists an element $u \in U$ such that $h \in uK$, so $h = uv$ for some $v \in K$. Thus, $x = tuv$ which implies that

$$G = \bigcup_{t \in T, u \in U} tuK = \bigcup_{tu \in R} tuK.$$

Suppose now that $tuK = zwK$, where $t, z \in T$ and $u, w \in U$. Then $tu = zwv$ for some element $v \in K$. Since $u, w, v \in H$, the cosets tH and zH have nonempty intersection and therefore coincide, since they both contain the element tu. By definition of T, it follows that $t = z$. Multiplying both sides of the equation $tu = zwv$ on the left by t^{-1}, we obtain $u = wv$. This means that $u \in wK$ and, since $u \in uK$, the cosets uK and wK have nonempty intersection. Therefore, $uK = wK$ and from the definition of U, it follows that $u = w$. Thus the equation $tuK = zwK$ is true if and only if $t = z$ and $u = w$ and together with the equation $G = \bigcup_{y \in R} yK$ it follows that the subset R is a left transversal to the subgroup K in G.

We place the following important rephrasing of Theorem 8.3.6 on record.

8.3.7. Corollary. *Let G be a group, let H, K be subgroups of G and let $K \leq H$. Then, $|G : K|$ is finite if and only if the indices $|G : H|$ and $|H : K|$ are both finite. In this case, $|G : K| = |G : H|\,|H : K|$.*

Next we have one of the most important theorems in finite group theory.

8.3.8. Corollary *(Lagrange's Theorem). Let G be a finite group and let H be a subgroup of G. Then $|G| = |G : H|\,|H|$. In particular, the order of a subgroup of a finite group is a divisor of the order of the group.*

To see this, let $K = \langle e \rangle$ in Theorem 8.3.6 so that $|H : K| = |H|$.

8.3.9. Corollary. *Let G be a finite group and let x be an element of G. Then the order of x is a divisor of the order of G.*

To see this, we note that the order of an element is the order of the cyclic group that this element generates. Lagrange's theorem can then be used to prove the result.

8.3.10. Corollary. *Let G be a finite group. If $|G|$ is a prime, then G is a cyclic group.*

Proof. To see this, let $e \neq g \in G$. Then $\langle g \rangle$ has at least one nontrivial element, so that $|\langle g \rangle| > 1$. By Corollary 8.3.8, we see that the equation $|\langle g \rangle| = |G|$ follows since $|G|$ is prime. This implies that $G = \langle g \rangle$.

8.3.11. Corollary. *Let G be a group and let H, K be subgroups of G. If the indices $|G : H|$ and $|G : K|$ are finite then the index $|G : H \cap K|$ is also finite. Moreover, $|G : H \cap K| \leq |G : H||G : K|$.*

Proof. Let $L = H \cap K$, let $T = \mathbf{lt}(H, L)$, and let $x, y \in T$, where $x \neq y$. If $xK = yK$, then $x^{-1}y \in K$. Since $x, y \in H$, we have $x^{-1}y \in H \cap K = L$, so that $xL = yL$. From the choice of x and y, it follows that $x = y$. Thus if $xL \neq yL$ then $xK \neq yK$. Since $|G : K|$ is finite, this implies that $|H : L|$ is finite and that $|H : L| \leq |G : K|$. Corollary 8.3.7 implies that

$$|G : H \cap K| = |G : H||H : H \cap K| \leq |G : H||G : K|.$$

This has the following interesting consequence.

8.3.12. Corollary *(Poincare's Theorem). The intersection of a finite set of subgroups each having finite index in a group is a subgroup of finite index.*

As we proved in Theorem 8.1.21, every subgroup of a cyclic group is cyclic. Now we are ready for some further details.

8.3.13. Theorem. *Let $G = \langle g \rangle$ be a cyclic group and let H be a subgroup of G.*

(i) *If G is infinite and $H = \langle g^n \rangle$, where $n \neq 0$, then H is infinite and $|G : H| = n$; if $n = 0$, then the index $|G : H|$ is infinite.*
(ii) *If G has finite order r, then $|H|$ is a divisor of r. Conversely, for each divisor d of the number r, there is one and only one subgroup of order d and it coincides with $\langle g^{\frac{r}{d}} \rangle$.*

Proof.

(i) Let x be an arbitrary element of G. Then, $x = g^k$ for some $k \in \mathbb{Z}$ and Theorem 1.4.1 implies that $k = nq + s$ where $0 \leq s < n$. We have

$$g^k = g^{nq+s} = (g^n)^q g^s \in g^s H.$$

It follows that

$$G = \bigcup_{0 \leq s < n} g^s H.$$

Next if $g^s H = g^t H$ where $0 \leq t \leq s < n$ then we have $g^t = g^s u$, for some element $u \in H$. Since $u = g^{nm}$, for some $m \in \mathbb{Z}$, we have

$$g^t = g^s g^{nm} = g^{s+nm}.$$

Since G is an infinite cyclic group, g has infinite order so $t = s + nm$. From the choice of s, t we conclude that $m = 0$ and hence $s = t$. Therefore, the cosets $g^s H$ are distinct from each other for $0 \leq s < n$. It follows that $|G : H| = n$.

(ii) Now let $|G| = r < \infty$. By Lagrange's Theorem, Corollary 8.3.8, $|H|$ is a divisor of r. Conversely, let d be a divisor of r, say $r = db$, where $b \in \mathbb{N}$. If we suppose that $|g^b| = u < d$, then $g^{bu} = e$ and $bu < bd = r$, which contradicts the fact that g has order r. Thus $|g^b| = d$. Let $\langle g^v \rangle$ be another subgroup of order d. Then, $g^{vd} = e$ and therefore vd is divisible by $r = db$, so v is divisible by b. Thus $\langle g^v \rangle \leq \langle g^b \rangle$. However, $d = |\langle g^v \rangle|$ and $d = |\langle g^b \rangle|$, so that $\langle g^v \rangle = \langle g^b \rangle$.

8.3.14. Corollary. *A group G has only two subgroups ($\langle e \rangle$ and G) if and only if $|G|$ is a prime.*

Proof. Let g be a nontrivial element of G. If G has only two subgroups then $\langle g \rangle = G$ and G is a cyclic group. If we suppose that $|g|$ is infinite, then

$$\langle e \rangle \neq \langle g^2 \rangle \neq \langle g \rangle,$$

a contradiction, which shows that $|g|$ is finite. By Theorem 8.3.13, $|G|$ is a prime number.

EXERCISE SET 8.3

8.3.1. Prove that a group G is a cyclic group of order p^2 where p is a prime if and only if G has exactly three subgroups.

8.3.2. Prove that a finite group of even order has an odd number of elements of order exactly 2.

8.3.3. Prove that if A and B are finite subgroups of a group G and if $\mathbf{GCD}(|A|, |B|) = 1$ then $A \cap B = \{e\}$.

8.3.4. Let A, B be subgroups of a group G. Give an example to show that $AB = \{ab | a \in A, b \in B\}$ need not be a subgroup of G.

8.3.5. Let A, B be subgroups of a group G. Define $f : A \times B \longrightarrow AB$ by $f(a, b) = ab$. Prove that if $ab \in AB$ then $\phi^{-1}(ab) = \{(ac, c^{-1}b) | c \in A \cap B\}$. Deduce that $|AB| \cdot |A \cap B| = |A| \cdot |B|$. Note that AB need not be a group.

8.3.6. Compute the set of left cosets and the set of right cosets for the subgroup generated by the cycle $(1\,2\,3\,4)$ in the group $\mathbf{S}_4$.

8.3.7. Compute the set of left cosets and the set of right cosets for the subgroup $\{e, (1\,2)(3\,4), (1\,3)(2\,4), (1\,4)(2\,3)\}$ in the group $\mathbf{S}_4$.

8.3.8. Let G be a group and let H, K be subgroups such that $\mathbf{GCD}(|G : H|, |G : K|) = 1$. Prove that $G = HK$.

8.3.9. Prove that if H, K are subgroups of a group G, then $H \cup K$ is a subgroup of G if and only if either $H \leq K$ or $K \leq H$. Hence show that no group is a union of two proper subgroups.

8.3.10. Give an example of a group that is a union of three proper subgroups.

8.3.11. Find all groups of order at most 5.

8.4 NORMAL SUBGROUPS AND FACTOR GROUPS

In Section 8.3, we considered partitions of a group into left and right cosets. The natural questions arise as to whether the partitions are significantly different and whether they can coincide. Thus, is it ever the case that Σ_H and Λ_H are the same and when does this happen? For what types of subgroups H does this happen? Of course, for abelian groups G, it is always the case that $xh = hx$ for all elements $x \in G$ and $h \in H$. Therefore, in this case, $xH = Hx$ for all $x \in G$ and hence $\Sigma_H = \Lambda_H$ for all subgroups H of an abelian group G.

The following proposition gives a further special case when $\Sigma_H = \Lambda_H$.

8.4.1. Proposition. *Let G be a group and let H be a subgroup of G. If $|G : H| = 2$, then $xH = Hx = G \backslash H$, for each $x \notin H$ and $xH = Hx = H$ for each $x \in H$.*

Proof. Indeed, we have $G = H \cup gH$ for some element $g \in G$. It follows that $gH = G \backslash H$. If $x \notin H$ then $xH \neq H$ and $xH = gH = G \backslash H$. A similar argument is valid for right cosets. Thus $xH = Hx = G \setminus H$ when $x \notin H$. On the other hand, if $x \in H$ then $xH = H = Hx$ and the result follows.

We now consider the example of the group $\mathbf{S}_3$ and a couple of its subgroups. We let

$$H = \mathbf{A}_3 = \left\{\varepsilon, \pi = \begin{pmatrix} 1 & 2 & 3 \\ 2 & 3 & 1 \end{pmatrix}, \pi^2 = \begin{pmatrix} 1 & 2 & 3 \\ 3 & 1 & 2 \end{pmatrix}\right\}, \text{ and } K = \langle \iota_{12} \rangle.$$

First we note that, since $|H| = 3$, Corollary 8.3.8 implies that $|\mathbf{S}_3 : H| = 2$. Consequently, the family of all left cosets of $\mathbf{A}_3$ in $\mathbf{S}_3$ coincides with the family of right cosets of this subgroup and hence $\Sigma_{\mathbf{A}_3} = \Lambda_{\mathbf{A}_3}$. However, for the subgroup K, there are the following three right cosets, obtained by direct calculation,

$$K, K \circ \iota_{13} = \left\{\iota_{13}, \pi^2\right\}, K \circ \iota_{23} = \{\iota_{23}, \pi\},$$

and the following three left cosets,

$$K, \iota_{13} \circ K = \{\iota_{13}, \pi\}, \iota_{23} \circ K = \left\{\iota_{23}, \pi^2\right\}.$$

These two sets are different of course, so in this case $\Sigma_K \neq \Lambda_K$.

Let G be a group and let H be a subgroup of G such that the family of all left cosets of H in G coincides with the family of all right cosets of H in G. If $x \in G$, then there exists an element $y \in G$ such that $xH = Hy$. Since $x \in xH$, it follows that $x \in Hy \cap Hx$ and since right cosets are equal or disjoint, we have that $Hy = Hx$ so $xH = Hx$.

8.4.2. Definition. *Let G be a group. The subgroup H is called normal in G, if $xH = Hx$ for each element $x \in G$. We denote the fact that H is normal in G by $H \lhd G$.*

8.4.3. Definition. *Let G be a group and let A, B be two subsets of G. The product AB is defined to be the subset $\{ab \mid a \in A, b \in B\}$.*

In the case, when A consists of the single element a, we write aB instead of $\{a\}B$ and a similar convention holds for Ab, whenever $B = \{b\}$. In a similar fashion, we can define the product of any finite set of subsets of a group. Since the operation in a group is associative, this subset multiplication is also an associative operation.

8.4.4. Definition. *Let G be a group and let A, B be two subgroups of G. The subgroups A, B are said to be permutable if $AB = BA$.*

8.4.5. Proposition. *Let G be a group and let A, B be two subgroups of G. The product AB is a subgroup if and only if the subgroups A, B are permutable and in this case,*

$$AB = \langle A \cup B \rangle = \langle A, B \rangle.$$

Proof. Assume that AB is a subgroup of G. Since $A = Ae \leq AB$ and $B = eB \leq AB$, we have $BA \subseteq \langle A, B \rangle \leq AB$. Let x be an arbitrary element of AB. Since AB is a subgroup of G, then by Theorem 8.1.7, $x^{-1} \in AB$ and hence $x^{-1} = uv$ for elements $u \in A, v \in B$. We now have

$$x = (x^{-1})^{-1} = (uv)^{-1} = v^{-1}u^{-1}.$$

Again by Theorem 8.1.7, $u^{-1} \in A$ and $v^{-1} \in B$. Hence $v^{-1}u^{-1} \in BA$, which proves that $AB \subseteq BA$ and therefore $AB = BA = \langle A, B \rangle$.

Conversely, suppose that $AB = BA$. If $x, y \in AB$, then $x = a_1b_1$ and $y = a_2b_2$, for some elements $a_1, a_2 \in A, b_1, b_2 \in B$. We have

$$xy^{-1} = (a_1b_1)(a_2b_2)^{-1} = (a_1b_1)(b_2^{-1}a_2^{-1}) = a_1(b_1b_2^{-1})a_2^{-1}.$$

Since B is a subgroup, Corollary 8.1.8 implies that $b_1b_2^{-1} \in B$ and hence $(b_1b_2^{-1})a_2^{-1} \in BA = AB$. Therefore $(b_1b_2^{-1})a_2^{-1} = a_3b_3$, where $a_3 \in A$ and $b_3 \in B$. Hence

$$xy^{-1} = a_1(a_3b_3) = (a_1a_3)b_3 \in AB$$

and AB is a subgroup, by Corollary 8.1.8.

The following proposition is very useful.

8.4.6. Proposition *(Dedekind Modular Law). Let G be a group and let A, B, C be subgroups of G. Suppose that $A \leq C$. Then $(AB) \cap C = A(B \cap C)$.*

Proof. In fact, $A(B \cap C) \subseteq AB$, and $A(B \cap C) \subseteq AC = C$. Hence

$$A(B \cap C) \subseteq (AB) \cap C.$$

Conversely, let $x \in (AB) \cap C$. Then $x = ab$, where $a \in A$ and $b \in B$. We have $b = a^{-1}x \in AC = C$, so that $b \in B \cap C$ and hence $x \in A(B \cap C)$. This implies that $(AB) \cap C \subseteq A(B \cap C)$ and we have

$$A(B \cap C) = (AB) \cap C,$$

which completes the proof.

Note that from the inclusion $A \leq C$, we have $A \cap C = A$, so we can rewrite the equation in the statement of Proposition 8.4.6 in the form

$$(AB) \cap C = (A \cap C)(B \cap C).$$

However, $A(B \cap C) \subseteq (AB) \cap C$ is not valid in general.

8.4.7. Definition. *Let G be a group. A subgroup P is called permutable (or quasi-normal) in G, if $PH = HP$ for each subgroup H of G.*

The most important examples of permutable subgroups are normal subgroups. Here are some criteria for normality.

8.4.8. Proposition. *Let G be a group and let H be a subgroup of G. The following are equivalent:*

(i) *H is a normal subgroup of G.*
(ii) *$(xH)(yH) = xyH$ for all $x, y \in G$.*
(iii) *$x^{-1}Hx \subseteq H$ for every element $x \in G$.*
(iv) *$x^{-1}Hx = H$ for every element $x \in G$.*

Proof.
(i) $\Longrightarrow$ (ii). We have

$$(xH)(yH) = x(Hy)H = x(yH)H = (xy)(HH) = xyH.$$

This latter equality follows from the fact that if $u, v \in H$, then Theorem 8.1.7 implies that $uv \in H$, so $HH \subseteq H$. On the other hand, since $e \in H$, $H = eH \subseteq HH$, and hence $HH = H$.

(ii) $\Longrightarrow$ (iii). For each $x \in G$, we have

$$x^{-1}Hx = (x^{-1}Hx)e \subseteq (x^{-1}Hx)H = (x^{-1}H)(xH) = x^{-1}xH = eH = H,$$

using the given hypothesis.

(iii) $\Longrightarrow$ (iv). Let $x \in G$. Then $x^{-1}Hx \subseteq H$, by hypothesis. Also $(x^{-1})^{-1}Hx^{-1} = xHx^{-1} \subseteq H$. Therefore

$$H = (x^{-1}x)H(x^{-1}x) = x^{-1}(xHx^{-1})x \subseteq x^{-1}Hx,$$

and so $x^{-1}Hx = H$.

(iv) $\Longrightarrow$ (i). We have

$$xH = x(x^{-1}Hx) = (xx^{-1})Hx = eHx = Hx.$$

We next consider some examples of normal subgroups. Certainly, $\langle e \rangle \lhd G$ and $G \lhd G$.

8.4.9. Definition. *A group G is called simple, if it has only two normal subgroups, namely, $\langle e \rangle$ and G.*

By contrast, a group of all whose subgroups are normal is essentially the direct opposite of a simple group. Such groups are called Dedekind groups and their structure was described (in the case of finite groups) in the following paper of Dedekind (1897).

[Dedekind Über R. Gruppen deren sammtliche Teiler Normalteiler sind. Math Annalen 1897;48:548–561.]

Clearly, every abelian group is a Dedekind group. Every simple abelian group is a finite group of prime order, by Corollary 8.3.14.

We also note that every subgroup of the center $\zeta(G)$ of a group G is normal in G. To see this, note that if $h \in \zeta(G)$ and $g \in G$, then $g^{-1}hg = hg^{-1}g = h \in \zeta(G)$ so if $H \leq \zeta(G)$ then $g^{-1}Hg \subseteq H$.

8.4.10. Proposition. *Let G be a group and let $\mathfrak{S}$ be a family of normal subgroups of G. Then the intersection, $\cap\mathfrak{S}$, of all subgroups of this family is also normal in G.*

Proof. Let $S = \cap\mathfrak{S}$. By Corollary 8.1.10, S is a subgroup of G. Let $x \in S$ and $g \in G$. If U is an arbitrary subgroup of the family $\mathfrak{S}$ then $x \in U$ and, since U is normal in G, it contains the element $g^{-1}xg$. It follows that $g^{-1}xg$ belongs to the intersection of all subgroups of the family $\mathfrak{S}$ and hence to S. Proposition 8.4.8 completes the proof.

8.4.11. Corollary. *Let G be a group and let $\mathfrak{L}$ be a local family of normal subgroups of G. Then the union, $\cup\mathfrak{L}$, of subgroups of this family is normal in G.*

Proof. Let $V = \cup\mathfrak{L}$. By Corollary 8.1.13, V is a subgroup. Let $x \in V, g \in G$ and choose a subgroup $L \in \mathfrak{L}$, which contains x. Since L is a normal subgroup, it also contains $g^{-1}xg$. It follows that $g^{-1}xg \in V$ and, by Proposition 8.4.8, V is normal in G.

8.4.12. Corollary. *Let G be a group and let $\mathfrak{L}$ be a linearly ordered family of normal subgroups of G. Then the union, $\cup\mathfrak{L}$, of subgroups of this family is normal in G.*

8.4.13. Corollary. *Let G be a group and let*

$$H_1 \leq H_2 \leq \cdots \leq H_n \leq \cdots$$

be an ascending series of normal subgroups of G. Then the union, $\bigcup_{n\in\mathbb{N}} H_n$, of the subgroups from this series is a normal subgroup of G.

Let M be a subset of a group G and let $\mathfrak{S}$ be the family of all normal subgroups that contain M. Then the intersection, $\langle M\rangle^G = \cap\mathfrak{S}$, is normal in G, by Proposition 8.4.10.

8.4.14. Definition. *Let G be a group and let M be a subset of G. The subgroup $\langle M\rangle^G$ is called the normal subgroup generated by the subset M or the normal closure of M in G.*

If H is a normal subgroup containing the subset M then, by definition, H contains the normal subgroup $\langle M\rangle^G$ and, in this sense, $\langle M\rangle^G$ is the least normal subgroup containing M. It is clear that if M is a normal subgroup of G, then $\langle M\rangle^G = M$.

If H is a subgroup of G, then the subset $x^{-1}Hx$ is a subgroup for each element $x \in G$. Indeed, let $a, b \in x^{-1}Hx$. Then

$$a = x^{-1}ux \text{ and } b = x^{-1}vx$$

for some elements $u, v \in H$. We now have

$$ab^{-1} = x^{-1}ux(x^{-1}vx)^{-1} = x^{-1}ux(x^{-1}v^{-1}x) = x^{-1}uv^{-1}x.$$

Since H is a subgroup, $uv^{-1} \in H$. Therefore $x^{-1}Hx$ satisfies **(SG 3)** and, by Corollary 8.1.8, $x^{-1}Hx$ is a subgroup. We say that the subgroups H and $x^{-1}Hx$ are conjugate. By Proposition 8.4.8, a subgroup H is normal if and only if H coincides with each of its conjugates.

8.4.15. Proposition. *Let G be a group and let H be a subgroup of G. Then*

$$\bigcap_{x \in G} x^{-1}Hx = \mathbf{Core}_G(H)$$

is a normal subgroup of G. If H contains a subgroup K that is normal in G then $K \le \mathbf{Core}_G(H)$. Thus, $\mathbf{Core}_G(H)$ is the largest normal subgroup of G contained in H.

Proof. We proved above that $x^{-1}Hx$ is a subgroup of G for every $x \in G$. By Corollary 8.1.10, $\mathbf{Core}_G(H)$ is also a subgroup. Now let $u \in \mathbf{Core}_G(H)$ and let $x, g \in G$. Then $u \in (xg^{-1})^{-1}H(xg^{-1}) = gx^{-1}Hxg^{-1}$, so $u = gx^{-1}vxg^{-1}$ for some element $v \in H$. This implies that $g^{-1}ug = x^{-1}vx \in x^{-1}Hx$. Since x is an arbitrary element of G, $g^{-1}ug \in \mathbf{Core}_G(H)$. By Proposition 8.4.8, $\mathbf{Core}_G(H)$ is a normal subgroup of G. Furthermore, since the subgroup K is normal, $K = x^{-1}Kx \le x^{-1}Hx$. Thus, $K \le \bigcap_{x \in G} x^{-1}Hx = \mathbf{Core}_G(H)$.

8.4.16. Definition. *Let G be a group. We say that the elements g, y are conjugate in G if there exists an element $u \in G$ such that $g = u^{-1}yu$. More precisely, we shall say that the elements y and $g = u^{-1}yu$ are conjugate in the group G with the help of the element u.*

Note that this relation of conjugacy is an equivalence relation. Indeed, $g = e^{-1}ge$ and this relation is therefore reflexive. If $g = u^{-1}yu$, then

$$(u^{-1})^{-1}gu^{-1} = ugu^{-1} = u(u^{-1}yu)u^{-1} = (uu^{-1})y(uu^{-1}) = y.$$

Thus, y and g are conjugate with the help of u^{-1} and hence, conjugacy is a symmetric relation. Finally, if $g = u^{-1}yu$ and $y = v^{-1}zv$, then

$$g = u^{-1}yu = u^{-1}(v^{-1}zv)u = u^{-1}v^{-1}zvu = (vu)^{-1}z(vu),$$

and therefore conjugacy is a transitive relation.

8.4.17. Definition. *Let G be a group. If x is an element of G then the equivalence class of x under the relation of conjugacy is called the conjugacy class of x and is denoted by x^G. Thus $x^G = \{g^{-1}xg \mid g \in G\}$.*

Since conjugacy is an equivalence relation, the conjugacy classes form a partition of G.

8.4.18. Proposition. *Let G be a group and let $x, u, v \in G$. The elements $u^{-1}xu$ and $v^{-1}xv$ coincide if and only if $uv^{-1} \in C_G(x)$.*

Proof. Indeed, if $u^{-1}xu = v^{-1}xv$, then we obtain $vu^{-1}xuv^{-1} = x$ or $(uv^{-1})^{-1}x(uv^{-1}) = x$. This means that $uv^{-1} \in C_G(x)$. The converse can be established by arguing in reverse.

8.4.19. Corollary. *Let G be a group and let $x \in G$. There is a bijection from x^G onto the set $\{gC_G(x) \mid g \in G\}$.*

Proof. Indeed, consider the mapping $\phi : x^G \longrightarrow \{gC_G(x) \mid g \in G\}$, defined by the rule

$$\phi : u^{-1}xu \longmapsto uC_G(x), \text{ where } u \in G.$$

The mapping ϕ is injective since if $uC_G(x) = vC_G(x)$, then $uv^{-1} \in C_G(x)$, and Proposition 8.4.18 implies that $u^{-1}xu = v^{-1}xv$. The fact that the mapping ϕ is surjective is clear.

8.4.20. Corollary. *Let G be a group and let $x \in G$. If the subset x^G is finite, then the index $|G : C_G(x)|$ is also finite. Moreover, $\left|x^G\right| = |G : C_G(x)|$.*

Let G be a group. Put

$$\mathbf{FC}(G) = \{x \in G \mid x^G \text{ is finite}\}.$$

8.4.21. Corollary. *Let G be a group. The subset $\mathbf{FC}(G)$ is a subgroup of G.*

Proof. Indeed, from $(g^{-1}xg)^{-1} = g^{-1}x^{-1}g$, it follows that $(x^{-1})^G = (x^G)^{-1}$. Thus, if x^G is finite then $(x^{-1})^G$ is also finite. Furthermore, $g^{-1}(xy)g = g^{-1}xgg^{-1}yg$, which implies that $(xy)^G \subseteq x^G y^G$. Therefore, if x^G and y^G are finite, then $(xy)^G$ is finite. The fact that $\mathbf{FC}(G)$ is a subgroup follows by Theorem 8.1.7.

We next consider the family $\mathfrak{L}(G)$, of all subgroups of the group G. We can prove that the relation "the subgroups H and K are conjugate in G" is an equivalence relation on $\mathfrak{L}(G)$. Here, to say that H is conjugate to K means that there exists $g \in G$ such that $g^{-1}Hg = K$. As with our discussion concerning the conjugacy of elements, we consider equivalence classes of conjugate subgroups. The set of subgroups conjugate to a subgroup H in a group G, will be denoted by $\mathbf{cl}_G(H)$. Thus $\mathbf{cl}_G(H) = \{g^{-1}Hg | g \in G\}$. By Proposition 8.4.8, the subgroup H is normal if and only if $\mathbf{cl}_G(H) = \{H\}$.

8.4.22. Definition. *Let G be a group and let H be a subgroup of G. The subset*

$$N_G(H) = \{x \in G \mid x^{-1}Hx = H\}$$

is called the normalizer of H in G.

8.4.23. Proposition. *Let G be a group and let H be a subgroup of G. The normalizer $N_G(H)$ is a subgroup of G.*

Proof. Let $x \in N_G(H)$. Then, $x^{-1}Hx = H$ and hence, $H = xHx^{-1}$ follows easily by premultiplying the equation by x and postmultiplying it by x^{-1}. Thus $x^{-1} \in N_G(H)$. Next, if $x, y \in N_G(H)$, we have

$$(xy)^{-1}H(xy) = y^{-1}x^{-1}Hxy = y^{-1}(x^{-1}Hx)y = y^{-1}Hy = H.$$

Hence $x^{-1}, xy \in N_G(H)$ for all $x, y \in H$ and $N_G(H)$ is therefore a subgroup of G, by Theorem 8.1.7

We note that $N_G(H)$ is the largest subgroup of G in which H is normal.

8.4.24. Proposition. *Let G be a group and let H be a subgroup of G. The subgroups $u^{-1}Hu$ and $v^{-1}Hv$ coincide if and only if $uv^{-1} \in N_G(H)$.*

The proof of this is similar to the proof of Proposition 8.4.18, so we leave the proof to the reader.

8.4.25. Corollary. *Let G be a group and let H be a subgroup of G. There is a bijection from the set $\mathbf{cl}_G(H)$ to the set $\{N_G(H)g \mid g \in G\}$.*

Proof. We let $\phi : \mathbf{cl}_G(H) \longrightarrow \{N_G(H)g \mid g \in G\}$ be the mapping defined by

$$\phi : u^{-1}Hu \longmapsto N_G(H)u, \text{ where } u \in G.$$

The mapping ϕ is injective, since if $N_G(H)u = N_G(H)v$, then $uv^{-1} \in N_G(H)$ and Proposition 8.4.24 implies that $u^{-1}Hu = v^{-1}Hv$. It is clear that ϕ is surjective and hence bijective.

We can also deduce the following result analogous to Corollary 8.4.20

8.4.26. Corollary. *Let G be a group and let H be a subgroup of G. If the set $\mathbf{cl}_G(H)$ is finite, then the index $|G : N_G(H)|$ is also finite and, in this case, $|\mathbf{cl}_G(H)| = |G : N_G(H)|$.*

In Proposition 8.4.8, we saw that the left cosets of a normal subgroup H of a group G satisfy the equation $xHyH = xyH$. If we take different coset representatives x_1H, y_1H of xH and yH, respectively, then $x = x_1h$ and $y = y_1k$ for elements $h, k \in H$. Then

$$xHyH = x_1hHy_1kH = x_1hy_1kH = x_1hy_1H = x_1y_1(y_1^{-1}hy_1)H = x_1y_1H,$$

because H is normal, so $y_1^{-1}hy_1 \in H$. Proposition 8.4.8(ii) therefore implies that the set of all left cosets of a normal subgroup H of G is stable relative to the operation of multiplication of subsets, since the argument given above shows that this operation of multiplication is well defined. Moreover, when H is normal in

G, the set of left H-cosets forms a group under this operation of multiplication defined on the left H-cosets. Indeed, as we mentioned above, the operation of multiplication of subsets is associative. The identity element is H itself, since

$$(xH)H = xHH = xH \text{ and } H(xH) = (Hx)H = (xH)H = xH.$$

The reciprocal element to xH is $x^{-1}H$, since

$$(xH)(x^{-1}H) = (xx^{-1})H = eH = (x^{-1}x)H = (x^{-1}H)(xH).$$

8.4.27. Definition. *Let G be a group and let H be a normal subgroup of G. The group of all left H-cosets is called the factor group of G by H and it is denoted by G/H.*

It is interesting to note that some properties of a group are inherited by its factor groups. For example, if G is an abelian group, then each of its factor groups is also abelian. To see this, note that if H is normal in G and if $x, y \in G$ then $xHyH = xyH = yxH = yHxH$. However, some properties of a group are not inherited by factor groups as happens, for example, with the property of being infinite. For example, consider the additive group $\mathbb{Z}$ of all integers and its subgroup $n\mathbb{Z}$, where n is a fixed integer. In Section 7.3 we showed that $|\mathbb{Z}/n\mathbb{Z}| = n$, which implies that the infinite group $\mathbb{Z}$ has finite factor groups. On the other hand, it is obvious that every factor group of a finite group is finite. We use the notion of factor groups a lot in our study of groups. This concept is very important in group theory.

There is another question related to factor groups. If $H = \langle e \rangle$, then for each element $x \in G$, we have $xH = x\langle e \rangle = \{x\}$, and $xHyH = \{x\}\{y\} = \{xy\}$. This means that the factor group $G/\langle e \rangle$ is no different from G. In particular, the algebraic properties of G and $G/\langle e \rangle$ are identical.

8.4.28. Definition. *The factor group G/H is called proper, if H is a nontrivial normal subgroup.*

More than 50 years ago, the study of the influence of properties of proper factor groups on a group was started. A summary of the main results obtained in this area can be found in the book by Kurdachenko *et al*. (2002) [Kurdachenko LA, Otal J, Subbotin IY. Groups with prescribed quotient groups and associated module theory. New Jersey: World Scientific Publishing Company; 2002.]

8.4.29. Definition. *Let G be a group and let x, y be elements of G. The element $[x, y] = x^{-1}y^{-1}xy$ is called the commutator of the elements x, y.*

If $xy = yx$, then $x^{-1}y^{-1}xy = e$. Performing these arguments in reverse, we obtain $xy = yx$ from $[x, y] = e$. Thus, x and y commute if and only if $[x, y] = e$, which is to say that two elements commute if and only if their commutator is the identity. We note that $[x, y]^{-1} = [y, x]$. Thus the inverse of a commutator is also a commutator.

8.4.30. Definition. *Let G be a group. The subgroup of G generated by the subset $\{[x, y] \mid x, y \in G\}$ is called the derived subgroup or the commutator subgroup of the group G and is denoted by $[G, G]$ or G'. An element of $[G, G]$ is therefore a product of commutators.*

Observe that

$$\begin{aligned}[g^{-1}xg, g^{-1}yg] &= (g^{-1}xg)^{-1}(g^{-1}yg)^{-1}(g^{-1}xg)(g^{-1}yg)\\ &= g^{-1}x^{-1}gg^{-1}y^{-1}gg^{-1}xgg^{-1}yg\\ &= g^{-1}x^{-1}y^{-1}xyg = g^{-1}[x, y]g\end{aligned}$$

and from this it follows that $[G, G]$ is a normal subgroup of G.

8.4.31. Proposition. *Let G be a group. Then*

(i) *The factor group $G/[G, G]$ is abelian.*
(ii) *If H is a normal subgroup of G such that G/H is abelian, then $[G, G] \le H$.*

Proof.
(i) To see this, let $D = [G, G]$ and consider the cosets xD, yD. We have

$$\begin{aligned}[xD, yD] &= (xD)^{-1}(yD)^{-1}(xD)(yD) = (x^{-1}D)(y^{-1}D)(xD)(yD)\\ &= x^{-1}y^{-1}xyD = [x, y]D = D,\end{aligned}$$

which shows that G/D is abelian, since the commutator of two arbitrary elements of G/D is the identity element of G/D.

(ii) Let H be a normal subgroup of G such that G/H is abelian. This means that for all elements $x, y \in G$, we have $xHyH = yHxH$ and hence $[x, y]H = H$. However, this means that $[x, y] \in H$. Thus, every commutator is in H and so $[G, G] \le H$ follows.

EXERCISE SET 8.4

8.4.1. Prove that every factor group of a cyclic group is also cyclic.

8.4.2. Let G be a group and let N be a normal subgroup of G. Prove that if $N \le H \le G$ then H/N is a subgroup of G/N and that every subgroup of G/N arises in this way. Thus, also prove that if X is a subgroup of G/N then $X = H/N$, for some subgroup H of G containing N.

8.4.3. Prove that if G has a normal subgroup N of index k then $g^k \in N$ for all $g \in G$.

8.4.4. Prove that if N is a normal subgroup of G and if H is a subgroup of G then HN is a subgroup of G.

8.4.5. In problem 8.4.4, further prove that if H is also normal in G then HN is normal in G.

8.4.6. Give an example of a nonabelian group G which has an abelian factor group.

8.4.7. Let G be a group and let $H \leq G$ be such that $|G : H| = 2$. Prove that H is a normal subgroup of G.

8.4.8. Let G be a finite group of odd order and let x be the product of all the elements of G, in some order. Prove that $x \in G'$.

8.4.9. Prove that $\mathbf{A}_n$ is a normal subgroup of $\mathbf{S}_n$. Describe the group $\mathbf{S}_n/\mathbf{A}_n$.

8.4.10. Prove that if G is a group, N is a normal subgroup of G, and H is a subgroup of G then $H \cap N$ is a normal subgroup of H.

8.4.11. Let N be a cyclic normal subgroup of G. Prove that every subgroup of N is normal in G.

8.4.12. Give an example to show that if $A \leq B \leq C$ and if A is normal in B and B is normal in C then A need not be normal in C. (Thus normality is not a transitive relation.)

8.4.13. Let N be a normal subgroup of the finite group G and let $\mathbf{GCD}(|N|, |G : N|) = 1$. Prove that if $|N| = k$ and $x \in G$, the equation $x^k = e$ implies that $x \in N$.

8.4.14. Prove that $\mathbf{SL}_n(\mathbb{R})$ is a normal subgroup of $\mathbf{GL}_n(\mathbb{R})$ and describe the group $\mathbf{GL}_n(\mathbb{R})/\mathbf{SL}_n(\mathbb{R})$.

8.4.15. Prove that every subgroup of the center of a group is normal in the group.

8.4.16. Show that the converse of Lagrange's theorem is false by showing that $\mathbf{A}_4$ has no subgroup of order 6.

8.5 HOMOMORPHISMS OF GROUPS

We next consider homomorphisms of groups. We begin with the elementary properties of homomorphisms. For the convenience of the reader, we recall the definition of a group homomorphism.

8.5.1. Definition. *Let G, H be groups and let $f : G \longrightarrow H$ be a mapping. Then f is called a homomorphism if*

$$f(xy) = f(x)f(y) \textit{ for all } x, y \in G.$$

An injective homomorphism is called a monomorphism, a surjective homomorphism is called an epimorphism and a bijective homomorphism is called an isomorphism. If $f : G \longrightarrow H$ is an isomorphism then, as we saw in Section 3.1, the mapping $f^{-1} : H \longrightarrow G$ is also an isomorphism.

8.5.2. Definition. *Let G, H be groups. Then, G and H are called isomorphic if there exists an isomorphism from G onto H, and we write this as $G \cong H$.*

It is easy to see that the identity mapping $\varepsilon_G : G \longrightarrow G$ is an isomorphism and also that if $f : G \longrightarrow H$ and $g : H \longrightarrow K$ are homomorphisms then their product $g \circ f$ is also a homomorphism.

8.5.3. Proposition. *Let G, U be groups and let $f : G \longrightarrow U$ be a homomorphism. Then*

(i) *$f(e) = e_U$ is the identity element of U;*
(ii) *if $f(x) = u$, then $f(x^{-1}) = u^{-1}$;*
(iii) *if H is a subgroup of G then its image, $f(H)$, is a subgroup of U; in particular, $f(G) = \mathbf{Im} f$ is a subgroup of U;*
(iv) *if V is a subgroup of U then its preimage, $f^{-1}(V)$, is a subgroup of G;*
(v) *if V is a normal subgroup of U then its preimage, $f^{-1}(V)$, is a normal subgroup of G; in particular, $f^{-1}(\langle e \rangle)$ is a normal subgroup of G.*

Proof.

(i) By definition of the identity element, $ex = x$ for every $x \in G$ and hence, $ee = e$. It follows that

$$f(e) = f(ee) = f(e)f(e).$$

Since $f(e)$ has an inverse, we obtain, multiplying on the right (or left) by $f(e)^{-1}$,

$$e_U = e_U f(e) = f(e).$$

(ii) By definition of inverses, we have $xx^{-1} = e = x^{-1}x$. It follows that

$$f(x)f(x^{-1}) = f(xx^{-1}) = f(e) = e_U = f(e) = f(x^{-1}x) = f(x^{-1})f(x).$$

Thus $uf(x^{-1}) = e = f(x^{-1})u$ so $u^{-1} = f(x^{-1})$.

(iii) Let $u, v \in f(H)$. Then there exist elements $a, b \in H$ such that $u = f(a)$ and $v = f(b)$. We have

$$uv^{-1} = f(a)f(b^{-1}) = f(ab^{-1}) \in f(H),$$

because H is a subgroup of G. By Corollary 8.1.8, $f(H)$ is a subgroup of U.

(iv) Let $x, y \in f^{-1}(V)$. Then $f(x), f(y) \in V$ and so $f(x)f(y)^{-1} \in V$. Thus

$$f(xy^{-1}) = f(x)f(y^{-1}) = f(x)(f(y))^{-1} \in V,$$

and hence $xy^{-1} \in f^{-1}(V)$. By Corollary 8.1.8, $f^{-1}(V)$ is a subgroup of G.

(v) Let g be an arbitrary element of G and let $x \in f^{-1}(V)$. Then $f(x) \in V$ and

$$f(g^{-1}xg) = f(g^{-1})f(x)f(g) = (f(g))^{-1}f(x)f(g) \in V,$$

because V is normal in U. It follows that $g^{-1}xg \in f^{-1}(V)$ and by Proposition 8.4.8, $f^{-1}(V)$ is a normal subgroup of G.

We note that, by contrast with (v), the image of a normal subgroup need not be normal in U unless the map f is an epimorphism. We saw in (v) that $f^{-1}(\langle e \rangle)$ is a normal subgroup of G whenever $f : G \longrightarrow U$ is a homomorphism. This important subgroup is known as the kernel, which we now formally define in a similar fashion to that used for rings.

8.5.4. Definition. *Let G, U be groups and let $f : G \longrightarrow U$ be a homomorphism. The normal subgroup* $\mathbf{Ker}\, f = \{x \in G \mid f(x) = e\}$ *is called the kernel of the homomorphism f. The subgroup* $\mathbf{Im}\, f = \{f(x) \mid x \in G\}$ *is called the image of f.*

We next give a number of classical theorems on homomorphisms similar to those appearing in Chapter 7. We have supplied the proofs, even though they are very similar to the corresponding proofs in the previous chapter.

8.5.5. Theorem *(The Theorem on Monomorphisms). Let G, U be groups. Then, a homomorphism $f : G \longrightarrow U$ is a monomorphism if and only if* $\mathbf{Ker}\, f = \langle e \rangle$. *In this case,* $G \cong \mathbf{Im}\, f$.

Proof. Indeed, if f is a monomorphism, then $x \neq e$ implies that $f(x) \neq f(e) = e_U$. This means that no nontrivial element x belongs to $\mathbf{Ker}\, f$ and hence $\mathbf{Ker}\, f = \langle e \rangle$.

Conversely, let $\mathbf{Ker}\, f = \langle e \rangle$, and assume that x, y are elements of G such that $f(x) = f(y)$. Then $f(x)f(y)^{-1} = f(y)f(y)^{-1} = e_U$ and

$$e_U = f(x)f(y)^{-1} = f(x)f(y^{-1}) = f(xy^{-1}),$$

so $xy^{-1} \in \mathbf{Ker}\, f = \langle e \rangle$. This means that $xy^{-1} = e$ and hence $x = y$. Therefore f is an injective mapping.

The next theorem is analogous to Theorem 7.4.5 and could be proved in a similar fashion, but here, we give an alternative rendition of this proof.

8.5.6. Theorem *(First Isomorphism Theorem, Version 1). Let G, U be groups and let $f : G \longrightarrow U$ be an epimorphism. Then U is isomorphic to the factor group $G/\mathbf{Ker} f$.*

Proof. We let $N = \mathbf{ker} f$ and define a mapping $\psi_f : G/N \longrightarrow U$ by $\psi_f(xN) = f(x)$. First, we must show that this mapping is well defined. If $xN = yN$ then $y = xn$, for some $n \in N$ and we have

$$\psi_f(yN) = f(y) = f(xn) = f(x)f(n) = f(x)e = f(x) = \psi_f(xN).$$

Now ψ_f is a homomorphism since

$$\psi_f(xHyH) = \psi_f(xyH) = f(xy) = f(x)f(y) = \psi_f(xH)\psi_f(yH).$$

Also ψ_f is an epimorphism since f is. Furthermore, if $\psi_f(xN) = e_U$ then the definition of $\overline{f}$ shows that $f(x) = e_U$ and hence $x \in \mathbf{ker} f = N$. By Theorem 8.5.5, ψ_f is a monomorphism and hence ψ_f is an isomorphism.

8.5.7. Theorem *(First Isomorphism Theorem, Version 2). Let G, U be groups and let $f : G \longrightarrow U$ be a homomorphism. Then $G/\mathbf{Ker} f \cong \mathbf{Im} f \leq U$.*

Proof. The restriction of f to the mapping $G \longrightarrow \mathbf{Im} f$ is an epimorphism. Then by Theorem 8.5.6, we deduce that $\mathbf{Im} f \cong G/\mathbf{Ker} f$. By Proposition 8.5.3, $\mathbf{Im} f$ is a subgroup of U.

As the first application of the above theorems, we describe all cyclic groups.

8.5.8. Theorem. *Let $G = \langle g \rangle$ be a cyclic group.*

(i) *If G is infinite then G is isomorphic to the additive group $\mathbb{Z}$ of all integers.*
(ii) *If G is finite and $|G| = m$, then $G \cong \mathbb{Z}/m\mathbb{Z}$.*

Proof. Let $f : \mathbb{Z} \longrightarrow G$ be the mapping defined by $f(n) = g^n$, where $n \in \mathbb{Z}$. We have

$$f(n + k) = g^{n+k} = g^n g^k = f(n)f(k),$$

so that f is a homomorphism. Since every element of G is an integer power of g, f is an epimorphism. Suppose that G is infinite. Then $n \neq k$ implies that $f(n) = g^n \neq g^k = f(k)$, so the mapping f is an injection and therefore it is an isomorphism.

Suppose now that G is a finite group. In this case, $|g| = m$ and hence $g^m = e$. Thus, $m \in \mathbf{Ker} f$ and Theorem 8.1.7 implies that $\langle m \rangle = m\mathbb{Z} \leq \mathbf{Ker} f$. By Theorem 8.1.21, $\mathbf{Ker} f = \langle t \rangle = t\mathbb{Z}$, for some $t \in \mathbb{Z}$ and since $m \in t\mathbb{Z}$, we see that $m = ts$, for some $s \in \mathbb{Z}$. By Theorem 8.5.6, $G \cong \mathbb{Z}/\mathbf{Ker} f$ and hence

$|G/\mathbf{Ker} f| = m$. On the other hand, Theorem 8.3.13 implies that $|\mathbb{Z}/m\mathbb{Z}| = m$ and $|\mathbb{Z}/t\mathbb{Z}| = t$, which show that $t = m$. Consequently, $G \cong \mathbb{Z}/m\mathbb{Z}$.

Here are some further interesting examples.

8.5.9. Example. Let a be a real number, where $a > 1$. Define the mapping $f : \mathbb{R} \longrightarrow \mathbb{R}^\times$ by $f(x) = a^x$, for each $x \in \mathbb{R}$. We have

$$f(x + y) = a^{x+y} = a^x a^y = f(x)f(y),$$

so that f is a homomorphism. Clearly, f is injective and $\mathbf{Im} f$ consists of all positive real numbers. Thus, f is an isomorphism from the group of additive real numbers onto the multiplicative group of all positive real numbers. Of course, the inverse map in this case is $\log_a : \mathbb{R}^\times \longrightarrow \mathbb{R}$.

8.5.10. Example. Define the mapping $f : \mathbf{GL}_n(\mathbb{R}) \longrightarrow \mathbb{R}^\times$ by $f(A) = \mathbf{det}(A)$, whenever $A \in \mathbf{GL}_n(\mathbb{R})$. By Theorem 2.5.1,

$$f(AB) = \mathbf{det}(AB) = \mathbf{det}(A)\mathbf{det}(B) = f(A)f(B),$$

so f is a homomorphism. Also,

$$\mathbf{Ker}\, f = \{A \in GL_n(\mathbb{R}) \mid \mathbf{det}(A) = 1\} = \mathbf{SL}_n(\mathbb{R}),$$

and hence $\mathbf{SL}_n(\mathbb{R})$ is normal in $\mathbf{GL}_n(\mathbb{R})$. It is clear that the mapping f is surjective so, by Theorem 8.5.6,

$$\mathbf{GL}_n(\mathbb{R})/\mathbf{SL}_n(\mathbb{R}) \cong \mathbb{R}^\times.$$

Similarly,

$$\mathbf{GL}_n(\mathbb{Q})/\mathbf{SL}_n(\mathbb{Q}) \cong \mathbb{Q}^\times \text{ and } \mathbf{GL}_n(\mathbb{Z})/\mathbf{SL}_n(\mathbb{Z}) \cong \{1, -1\}.$$

8.5.11. Example. Define the mapping $f : \mathbf{T}_n(\mathbb{R}) \longrightarrow \mathbf{D}_n(\mathbb{R})$ by

$$\begin{pmatrix} a_{11} & a_{12} & a_{13} & \dots & a_{1n-1} & a_{1n} \\ 0 & a_{22} & a_{23} & \dots & a_{2n-1} & a_{2n} \\ 0 & 0 & a_{33} & \dots & a_{3n-1} & a_{3n} \\ \vdots & \vdots & \vdots & \vdots & \vdots & \vdots \\ 0 & 0 & 0 & \dots & 0 & a_{nn} \end{pmatrix} \overset{f}{\longmapsto} \begin{pmatrix} a_{11} & 0 & 0 & \dots & 0 & 0 \\ 0 & a_{22} & 0 & \dots & 0 & 0 \\ 0 & 0 & a_{33} & \dots & 0 & 0 \\ \vdots & \vdots & \vdots & \vdots & \vdots & \vdots \\ 0 & 0 & 0 & \dots & 0 & a_{nn} \end{pmatrix}.$$

From Section 8.2 it follows that f is a homomorphism (sometimes called an erasing homomorphism). The mapping f is clearly surjective and $\mathbf{Ker} f = \mathbf{UT}_n(\mathbb{R})$. Thus, $\mathbf{UT}_n(\mathbb{R})$ is a normal subgroup of $\mathbf{T}_n(\mathbb{R})$ and, by Theorem 8.5.6, $\mathbf{T}_n(\mathbb{R})/\mathbf{UT}_n(\mathbb{R}) \cong \mathbf{D}_n(\mathbb{R})$. However, we note that $\mathbf{UT}_n(\mathbb{R})$ is not normal in $\mathbf{GL}_n(\mathbb{R})$.

8.5.12. Example. Let G be a group. A homomorphism $\varphi : G \longrightarrow G$ is called an endomorphism of G. The set of all endomorphisms of G is denoted by $\mathbf{End}(G)$. A bijective endomorphism φ of G (thus φ is also a permutation of G) is called an automorphism of the group G and we denote the set of all automorphisms of G by $\mathbf{Aut}(G)$. We note that $\mathbf{Aut}(G)$ is a subgroup of $\mathbf{S}(G)$, the group of permutations of the set G. To see this, let $\varphi, \psi \in \mathbf{Aut}(G)$. Then

$$\begin{aligned}(\varphi \circ \psi)(xy) &= \varphi(\psi(xy)) = \varphi(\psi(x)\psi(y)) = \varphi(\psi(x))\varphi(\psi(y)) \\ &= (\varphi \circ \psi)(x)(\varphi \circ \psi)(y).\end{aligned}$$

This means that $\varphi \circ \psi \in \mathbf{Aut}(G)$. In Section 3.1 we showed that the inverse of an automorphism is again an automorphism. Thus, $\mathbf{Aut}(G)$ satisfies Theorem 8.1.7 and hence it is a subgroup of $\mathbf{S}(G)$.

In a group G we choose an arbitrary element g and define the mapping $\mathbf{in}_g : G \longrightarrow G$ by $\mathbf{in}_g(x) = g^{-1}xg$ for each $x \in G$. We have

$$\mathbf{in}_g(xy) = g^{-1}xyg = g^{-1}xgg^{-1}yg = \mathbf{in}_g(x)\mathbf{in}_g(y),$$

so that $\mathbf{in}_g$ is an endomorphism. If x is an arbitrary element of G then $\mathbf{in}_g(gxg^{-1}) = g^{-1}(gxg^{-1})g = x$ which implies that $\mathbf{in}_g$ is surjective. If

$$\mathbf{in}_g(x) = \mathbf{in}_g(y), \text{ then } g^{-1}xg = g^{-1}yg.$$

Multiplying these equations on the right by g^{-1} and on left by g, we obtain $x = y$. Thus, $\mathbf{in}_g$ is injective and hence is an automorphism of G. The mapping $\mathbf{in}_g$ is called the inner automorphism of G induced by g.

Next, we define a further mapping $\Phi : G \longrightarrow \mathbf{Aut}(G)$ by $\Phi(g) = \mathbf{in}_{g^{-1}}$ for each element $g \in G$. From the equations

$$\begin{aligned}\mathbf{in}_{gh}(x) &= (gh)^{-1}x(gh) = h^{-1}g^{-1}xgh = h^{-1}(g^{-1}xg)h = \mathbf{in}_h(g^{-1}xg) \\ &= \mathbf{in}_h(\mathbf{in}_g(x)) = (\mathbf{in}_h \circ \mathbf{in}_g)(x),\end{aligned}$$

it follows that $\mathbf{in}_{gh} = \mathbf{in}_h \circ \mathbf{in}_g$. Also

$$\Phi(gh) = \mathbf{in}_{(gh)^{-1}} = \mathbf{in}_{h^{-1}g^{-1}} = \mathbf{in}_{g^{-1}} \circ \mathbf{in}_{h^{-1}} = \Phi(g) \circ \Phi(h),$$

and this shows that Φ is a homomorphism. If $u \in \mathbf{Ker}\,\Phi$, then $\Phi(u) = \mathbf{in}_{u^{-1}} = \varepsilon_G$. Hence

$$\mathbf{in}_{u^{-1}}(x) = uxu^{-1} = \varepsilon_G(x) = x,$$

for each element $x \in G$. Note that $uxu^{-1} = x$ is equivalent to $ux = xu$, for each $x \in G$, and thus $u \in \zeta(G)$, so $\mathbf{Ker}\,\Phi \leq \zeta(G)$. Since it is clear that $\zeta(G) \leq \mathbf{Ker}\,\Phi$, we have $\mathbf{Ker}\Phi = \zeta(G)$. We will denote $\mathbf{Im}\,\Phi$ by $\mathbf{Inn}(G)$. By Proposition 8.5.3, $\mathbf{Inn}(G)$ is a subgroup of $\mathbf{Aut}(G)$ which we call the group of inner automorphisms of G. By Theorem 8.5.7, $\mathbf{Inn}(G) \cong G/\zeta(G)$.

8.5.13. Example. Let G be a group, let H be a subgroup of G and let $\mathfrak{R}_H = \{xH \mid x \in G\}$ be a partition of G into the left H-cosets. If $g \in G$ then there is a mapping $\iota_g : \mathfrak{R}_H \longrightarrow \mathfrak{R}_H$ defined by $\iota_g(xH) = (gx)H$. This mapping is easily seen to be well defined. The mapping ι_g is surjective since $xH = g(g^{-1}xH) = \iota_g(g^{-1}xH)$ and injective since if $\iota_g(xH) = \iota_g(yH)$, then $gxH = gyH$, from which it follows easily that $xH = yH$. Consequently, ι_g is a permutation of the set $\mathfrak{R}_H$. We next consider the mapping $\Psi : G \longrightarrow \mathbf{S}(\mathfrak{R}_H)$, defined by $\Psi(g) = \iota_g$ for each $g \in G$. We have

$$\iota_{gv}(xH) = (gv)(xH) = g(vxH) = \iota_g(vxH) = \iota_g(\iota_v(xH)) = \iota_g \circ \iota_v(xH),$$

so $\iota_{gv} = \iota_g \circ \iota_v$. Thus

$$\Psi(gv) = \iota_{gv} = \iota_g \circ \iota_v = \Psi(g) \circ \Psi(v),$$

so that Ψ is a homomorphism. Let $g \in K = \mathbf{Ker}\,\Psi$. Then $\Psi(g) = \iota_g = \varepsilon$, where ε is the identity permutation of the set $\mathfrak{R}_H$. Thus $\iota_g(xH) = gxH = xH$ for every $x \in G$. Hence $gx = xh$ for some $h \in H$ dependent upon x. Multiplying both sides on the right by x^{-1} we see that $g = xhx^{-1} \in xHx^{-1}$, for all $x \in G$. Hence $g \in \bigcap_{x \in G} xHx^{-1} = \mathbf{Core}_G(H)$. It follows that $K \leq \mathbf{Core}_G(H)$. Since the above chain of reasoning can be reversed, we also have $\mathbf{Core}_G(H) \leq K$ and hence $\mathbf{Ker}\,\Psi = \mathbf{Core}_G(H)$.

If $H = \langle e \rangle$, then $K = \langle e \rangle$ and, by Theorem 8.5.5, Ψ is a monomorphism. We have proved the following results.

8.5.14. Theorem *(Cayley's Theorem). Every group G is isomorphic to some subgroup of a permutation group on some set.*

Cayley's Theorem is particularly pleasing in the finite case since it suggests that studying and understanding the finite symmetric groups enables us to get further information about all finite groups.

8.5.15. Corollary. *If G is a finite group of order n then G is isomorphic to some subgroup of the symmetric group, $\mathbf{S}_n$.*

Finally, we state the following result. It shows that if G has a subgroup H of finite index n, then the set of cosets of H is permuted by the elements of G in the manner described in Example 8.5.13. Thus, G acts on this set of cosets as an appropriate element of a small permutation group. There is a homomorphism from G into $\mathbf{S}_n$ in this case, where $n = |G : H|$.

8.5.16. Corollary. *Let G be a group and let H be a subgroup of G. Suppose that H has finite index in G and that $|G : H| = n$. Then, H contains a subgroup K such that K is normal in G and $|G/K| \leq n!$.*

EXERCISE SET 8.5

Show your work by exhibiting a proof or a counterexample where necessary.

8.5.1. Let $G = \{\alpha_{ab} \mid \alpha_{ab} : \mathbb{R} \longrightarrow \mathbb{R}\}$, where the mapping α_{ab} is defined by $\alpha_{ab}(x) = ax + b, a \neq 0, a, b \in \mathbb{R}$.

(a) Prove that G is a subgroup of $\mathbf{S}(\mathbb{R})$.

(b) Prove that the mapping $\theta : \alpha_{ab}\ \alpha_{a0}$ is an endomorphism of G. Find $\mathbf{Im}\,\theta$ and $\mathbf{Ker}\,\theta$.

8.5.2. Define the mapping $f_{ab} : \mathbb{R} \longrightarrow \mathbb{R}$ by the rule $f_{ab}(x) = ax + b$ where $a, b \in \mathbb{R}, a \neq 0$. Prove that the subset $G = \{f_{ab} \mid a, b \in \mathbb{R}, a \neq 0\}$ is a subgroup of $\mathbf{S}(\mathbb{R})$. Put $H = \{f_{1b} \mid b \in \mathbb{R}\}$. Prove that H is a normal subgroup of G and that $G/H \cong \mathbf{U}(\mathbb{R}) \cong \{f_{a0} \mid a \neq 0, a \in \mathbb{R}\}$.

8.5.3. Define the mapping $\Theta : Z \longrightarrow \mathbf{U}(\mathbb{Q})$ by the rule

$$\Theta(x) = \begin{Bmatrix} 1, & \text{if } x \text{ is even} \\ -1, & \text{if } x \text{ is odd.} \end{Bmatrix}.$$

Prove that Θ is a homomorphism. Find $\mathbf{Im}\,\Theta$ and $\mathbf{Ker}\,\Theta$.

8.5.4. Prove Corollary 8.5.16 in detail.

8.5.5. Decide which of the following mappings constitute group homomorphisms and find the kernels for those which are. Which of them are isomorphisms?

(a) $f : G \longrightarrow G$ defined by $f(x) = x^3$, where G is the group of nonzero real numbers under multiplication.

(b) $f : \mathbb{R} \longrightarrow \mathbb{R}$ defined by $f(x) = x^2$.

(c) $f : \mathbf{M}_n(\mathbb{R}) \longrightarrow \mathbb{R}$ defined by $f(A) = \mathbf{det}(A)$.

(d) $f : G \times G \longrightarrow G$ defined by $f(a, b) = a + b$, where G is an abelian group, written additively.

(e) $f : G \longrightarrow G/N$ defined by $f(x) = xN$, where G is a group and N is a normal subgroup of G.

8.5.6. Let $G = \mathbb{C}^*$ denote the group of complex numbers under multiplication and let $\mathbb{R}^+$ denote the group of positive real numbers under multiplication. Let $N = \{x + iy | x^2 + y^2 = 1\}$. Prove that $G/N \cong \mathbb{R}^+$.

8.5.7. Use the first isomorphism theorem to prove the second isomorphism theorem, that if G is a group, $H \leq G$, and N is a normal subgroup of G then $HN/N \cong H/(H \cap N)$.

8.5.8. Use the first isomorphism theorem to prove the third isomorphism theorem, that if G is a group and N, K are normal subgroups of G such that $N \leq K$ then $(G/N)/(K/N) \cong G/K$.

8.5.9. Let G be the additive group of $\mathbb{Z}[X]$ and let H be the multiplicative group $\mathbb{Q}^+$ of all positive rationals. Prove, using the fundamental theorem of arithmetic that $G \cong H$.

8.5.10. Determine all the homomorphisms from $\mathbb{Z}_{12}$ to itself and decide which ones are actually isomorphisms.

8.5.11. Verify that the mapping $\mathbf{sign} : \mathbf{S}_n \longrightarrow \{-1, 1\}$ is a homomorphism and find its kernel.

8.5.12. Let $f : \mathbb{Q} \longrightarrow \mathbb{C}^*$ be defined by $f\left(\frac{m}{n}\right) = e^{2\pi i m/n}$. Prove that f is a homomorphism and find the kernel of f. Determine the group with which $\mathbb{Q}/\mathbf{ker}\, f$ is isomorphic.

8.5.13. Let p be a prime and let $\mathbf{Q}_p = \{m/p^k | m \in \mathbb{Z}, k \in \mathbb{N}_0\}$. Prove that the mapping $f : \mathbb{Q}_p \longrightarrow \mathbb{C}^*$ defined by $f(m/p^k) = e^{2\pi i m/p^k}$ is a homomorphism and find its kernel and image.

8.5.14. Let G, H be groups. Prove that $\pi : G \times H \longrightarrow G$ defined by $\pi(g, h) = g$ is a homomorphism, find its kernel and use the first isomorphism theorem to deduce what $G \times H/\mathbf{ker}\, f$ is isomorphic to.

8.5.15. Let G be a group containing two normal subgroups H, K such that $H \cap K = \{e\}$. Let $HK = \{hk | h \in H, k \in K\}$. Prove that $HK \cong H \times K$.

8.5.16. Prove that if $G = GL_n(\mathbb{R})$ then $G' \geq SL_n(\mathbb{R})$.

8.5.17. Let G be a finite group and let p be the least prime dividing the order of G. Prove that if H is a subgroup of G and if $|G : H| = p$ then H is a normal subgroup of G.

8.5.18. Let G be a group and let H, K be normal subgroups of G. Prove that $G/(H \cap K)$ is isomorphic to a subgroup of $G/H \times G/K$.

CHAPTER 9

ARITHMETIC PROPERTIES OF RINGS

9.1 EXTENDING ARITHMETIC TO COMMUTATIVE RINGS

This section is concerned with extending the usual arithmetic in the ring $\mathbb{Z}$ of integers to other types of commutative rings. The rings that we consider here are very special and form a small subset of rings in general. However, among them there are very useful and important types of rings such as certain rings of polynomials. Additionally, mathematicians working in other branches of mathematics often need to work with polynomial rings and other types of rings considered here. Such rings are, therefore, an essential part of any algebra text.

We begin with the notion of divisibility and observe how this concept is studied in rings other than $\mathbb{Z}$.

9.1.1. Definition. *Let R be a commutative ring and suppose that R has no zero divisors. Let $a, b \in R$. We say that a divides b or b is divisible by a or that b is a multiple of a, if there exists $c \in R$ such that $b = ac$. We denote this by $a \mid b$.*

Let R be a commutative ring with no zero divisors. Suppose that $a \mid b$ and $b \mid a$, so $b = ac$ and $a = bd$ for some elements $c, d \in R$. Then $a = (ac)d = a(cd)$, so that $0_R = a - acd = a(e - cd)$ and hence $cd = e$. Thus, $c, d \in \mathbf{U}(R)$. We shall sometimes say that a has been cancelled or that a has been divided out in this sort of situation.

9.1.2. Definition. *Elements a, b of the ring R are called associates (in R), if $b = au$ for some element $u \in \mathbf{U}(R)$.*

9.1.3. Theorem. *Let R be an integral domain. The relation "to be associate elements" is an equivalence relation on R.*

Proof. Indeed, the equation $a = ae$ shows that the relation is reflexive. If $b = au$ where $u \in \mathbf{U}(R)$, then $a = bu^{-1}$ and also $u^{-1} \in \mathbf{U}(R)$, which shows that the relation is symmetric. Finally, also let $c = bv$, where $v \in \mathbf{U}(R)$. Then $c = (au)v = a(uv)$ and $uv \in \mathbf{U}(R)$, so that the relation is transitive.

The equivalence classes of the relation "to be associate elements" are called the associate classes of R. If $u \in \mathbf{U}(R)$, then u is a divisor of each element a since $a = u(u^{-1}a)$. Also note that $u^{-1}a$ is an associate of a.

9.1.4. Definition. *For the integral domain R and the element $a \in R$, the associates of a and the elements of $\mathbf{U}(R)$ are called improper divisors of a. Also, if $a = bc$, where $b, c \in R \setminus \mathbf{U}(R)$, then the elements b, c are called proper divisors of the element a.*

We observe the following properties of divisibility.

9.1.5. Proposition. *Let R be an integral domain and let $a, b, c \in R$.*

(i) *If a divides b and b divides c, then a divides c.*
(ii) *If a divides both b and c, then a divides $b + c$.*
(iii) *If a divides b, then a divides bc.*
(iv) *If a divides each of the elements $b_1, \dots, b_n$ and $c_1, \dots, c_n$ are arbitrary elements of R, then a divides $b_1c_1 + \cdots + b_nc_n$.*

The proofs of these results follow using the definitions. For example, (i) works because we can write $b = au$ and $c = bv$ for some $u, v \in R$ so that $c = auv$. Since $uv \in R$ it follows that a divides c.

The notion of divisibility can be translated into the language of ideals, an idea that will be useful later.

9.1.6. Proposition. *Let R be an integral domain and let $a, b \in R$. Then a divides b if and only if $bR \leq aR$.*

Proof. If $a \mid b$, then $b = ac$ for some element $c \in R$ and hence $b \in aR$. Since aR is an ideal, it follows that $bR \leq aR$. Conversely, if $bR \leq aR$, then $b \in aR$ and, therefore, $b = ac$ for some element $c \in R$.

9.1.7. Corollary. *Let R be an integral domain and let $a, b \in R$.*

(i) *a, b are associates if and only if $bR = aR$.*
(ii) *If a is a proper divisor of b, then $bR \lneqq aR$.*

We now extend the notion of the greatest common divisor to arbitrary integral domains.

9.1.8. Definition. *Let R be an integral domain and let $a, b \in R$. The element d of R is called the greatest common divisor (or the highest common factor) of a, b if d satisfies the conditions:*

(**GCD 1**) *d divides a and d divides b;*

(**GCD 2**) *if c is a common divisor of both a and b then c divides d.*

We denote the fact that d is a greatest common divisor of a and b by $d =$ **GCD**(a, b).

Clearly every associate of d also satisfies (**GCD 1**) and (**GCD 2**). Conversely, if an element $d_1 \in R$ satisfies the conditions (**GCD 1**) and (**GCD 2**) then d, d_1 divide each other and therefore they are associates. Thus, the greatest common divisors of a, b are precisely the associates of one another. It is quite common language to refer to *the* greatest common divisor.

The concept of the least common multiple is dual to the concept of the greatest common divisor.

9.1.9. Definition. *Let R be an integral domain and let $a, b \in R$. The element m of R is called a least common multiple of a and b, if m satisfies the conditions:*

(**LCM 1**) *both a and b divide m;*

(**LCM 2**) *if c is a common multiple of both a and b then m divides c.*

We denote the fact that m is a least common multiple of a and b by $m =$ **LCM** (a, b).

Clearly, every associate of m also satisfies (**LCM 1**) and (**LCM 2**). Conversely, if an element $m_1 \in R$ satisfies (**LCM 1**) and (**LCM 2**), then m, m_1 divide each other, and therefore they are associates. Thus, the least common multiples of a, b are precisely the associates of one another. Again it is common language to refer to *the* least common multiple. In an arbitrary commutative ring, not all pairs of elements need have a greatest common divisor or a least common multiple. In this section, we consider rings where the existence of the greatest common divisors and the least common multiples is always guaranteed, although there may not be an easy way to find such in general.

9.1.10. Definition. *An integral domain R is called a principal ideal domain (PID) if every ideal of R is principal; thus, if $I \triangleleft R$ then $I = aR$ for some $a \in I$.*

9.1.11. Theorem. *Let R be a principal ideal domain. For every pair, a, b, of elements of R there is a greatest common divisor and a least common multiple. Moreover,*

(i) $d = \mathbf{GCD}(a, b)$ *if and only if* $dR = aR + bR$;
(ii) $m = \mathbf{LCM}(a, b)$ *if and only if* $mR = aR \cap bR$.

Proof. In Section 7.3 we observed that the sum of two ideals is an ideal, so that $aR + bR$ is an ideal of R. Since R is a PID, there is an element $d \in R$ such that $dR = aR + bR$. Of course, $aR \leq dR$ and $bR \leq dR$ so Proposition 9.1.6 shows that $d \mid a$ and $d \mid b$. Next if c is a common divisor of a and b, then, again by Proposition 9.1.6, we see that $aR \leq cR$ and $bR \leq cR$. Thus, $dR = aR + bR \leq cR$, so that c is a divisor of d. Thus, the element d satisfies the conditions (**GCD 1**), and (**GCD 2**). Conversely, if d_1 is a greatest common divisor of a and b, then, as mentioned above, the elements d and d_1 are associates. In particular, Corollary 9.1.7 shows that $d_1R = dR = aR + bR$.

For the least common multiple, since every ideal of R is principal, there exists an element $m \in R$ such that $mR = aR \cap bR$. The inclusions $mR \leq aR$ and $mR \leq bR$ together with Proposition 9.1.6 show that $a \mid m$ and $b \mid m$. If c is another common multiple of the elements a, b then Proposition 9.1.6 this time implies that $cR \leq aR$ and $cR \leq bR$, so $cR \leq aR \cap bR = mR$. Therefore, m is a divisor of c and m satisfies the conditions (**LCM 1**) and (**LCM 2**). Conversely, if m_1 is another least common multiple of the elements a, b, then, as mentioned above, the elements m and m_1 are associates. In particular, Corollary 9.1.7 proves that $m_1R = mR = aR \cap bR$.

The following rather long result is really quite elementary and summarizes many of the properties of greatest common divisors and least common multiples.

9.1.12. Corollary. *Let R be a principal ideal domain and let a, b be elements of R. Then the following assertions hold:*

(i) $a = \mathbf{GCD}(a, b)$ *if and only if* $a \mid b$;
(ii) $\mathbf{GCD}(a, O_R) = a$;
(iii) $\mathbf{GCD}(ax, bx) = x\mathbf{GCD}(a, b)$;
(iv) $\mathbf{GCD}(\mathbf{GCD}(a, b), c) = \mathbf{GCD}(a, \mathbf{GCD}(b, c))$;
(v) $a = \mathbf{LCM}(a, b)$ *if and only if* $b \mid a$;
(vi) $\mathbf{LCM}(a, 0_R) = 0_R$;
(vii) $\mathbf{LCM}(ax, bx) = x\mathbf{LCM}(a, b)$;
(viii) $\mathbf{LCM}(\mathbf{LCM}(a, b), c) = \mathbf{LCM}(a, \mathbf{LCM}(b, c))$;
(ix) *if* $d = \mathbf{GCD}(a, b)$, *then there exist* $x, y \in R$ *such that* $d = ax + by$;
(x) *if a, b are both nonzero then* $\mathbf{GCD}(a, b)$ *and* $\mathbf{LCM}(a, b)$ *are also nonzero.*

Proof. These assertions are mostly clear. We illustrate by giving a proof of the last one. Let $d = \mathbf{GCD}(a, b)$ and let $m = \mathbf{LCM}(a, b)$. By Theorem 9.1.11, $dR = aR + bR$ and $mR = aR \cap bR$. It is therefore immediate that dR is nonzero, since, for example, $a \in aR \leq dR$. If $m = 0_R$ then $aR \cap bR = \{0_R\}$. However,

$ab \in aR \cap bR$ so that $ab = 0_R$. This is a contradiction since R is an integral domain and $a, b \neq 0_R$. This contradiction shows that the element m must be nonzero.

We say that elements a, b of a PID R are relatively prime if the multiplicative identity is one of the greatest common divisors.

9.1.13. Corollary. *Let R be a principal ideal domain. The elements a, b of R are relatively prime if and only if there exist elements $x, y \in R$ such that $e = ax + by$ (where e is the multiplicative identity of the ring R).*

Proof. Corollary 9.1.12 implies the necessity of this assertion. Suppose now that the ring R has elements x, y with the property that $e = ax + by$. Then $e \in aR + bR$. By Proposition 7.3.9 $eR = R = aR + bR$ and Theorem 9.1.11 implies that $e = \mathbf{GCD}(a, b)$.

9.1.14. Corollary. *Let R be a principal ideal domain, a, b be nonzero elements of R, $d = \mathbf{GCD}(a, b)$, and $a = da_1, b = db_1$. Then, the elements a_1, b_1 are relatively prime.*

Proof. By Corollary 9.1.12, there exist elements $x, y \in R$ such that $d = ax + by = da_1x + da_2y$. Since R has no zero divisors and since d is nonzero, by Corollary 9.1.12, we can divide both sides of this equation by d and obtain $e = a_1x + b_1y$. Corollary 9.1.13 implies that a_1 and b_1 are relatively prime.

The following properties hold for relatively prime elements.

9.1.15. Proposition. *Let R be a principal ideal domain and let $a, b, c \in R$.*

(i) *If a, b are relatively prime and a, c are also relatively prime, then a and bc are relatively prime.*
(ii) *If a divides bc and a, b are relatively prime, then a divides c.*

Proof.
(i) By Corollary 9.1.13, there exist elements x_1, x_2, y_1, y_2 in R such that $e = ax_1 + bx_2$ and $e = ay_1 + cy_2$. Now we have

$$e = (ax_1 + bx_2)(ay_1 + cy_2) = a(ax_1y_1 + bx_2y_1 + cx_1y_2) + (bc)(x_2y_2).$$

By Corollary 9.1.13 the elements a and bc are relatively prime.
(ii) By Corollary 9.1.13, there exist $x_1, x_2 \in R$ such that $e = ax_1 + bx_2$. We have $c = ce = cax_1 + cbx_2$ and since a divides bc and ca, it follows that a divides c.

Next we indicate the relationship between the greatest common divisor and the least common multiple.

9.1.16. Theorem. *Let R be a principal ideal domain and let a, b be nonzero elements of R.*

(i) *If* $m = \mathbf{LCM}(a, b)$ *and* $ab = md$ *then* $d = \mathbf{GCD}(a, b)$.

(ii) *If* $d = \mathbf{GCD}(a, b)$, *where* $a = da_1$ *and* $b = db_1$, *then* $m = da_1b_1$ *is a least common multiple of* a *and* b.

Proof.

(i) Since m is a common multiple of a and b, we have $m = a_2a$, $m = b_2b$ for elements a_2, b_2 of R. Then $ab = md = aa_2d$ and, dividing by the nonzero element a, we obtain $b = a_2d$, which shows that b is divisible by d. Similarly, we can prove that d divides a. Next suppose that c is an arbitrary common divisor of a and b, so $a = a_3c$, $b = b_3c$ for some $a_3, b_3 \in R$. The element a_3b_3c is therefore a common multiple of both a and b and, hence, by the definition of least common multiple, there exists an element $z \in R$ such that $a_3b_3c = mz$. Hence, by hypothesis, $md = ab = (a_3c)(b_3c) = mzc$. By Corollary 9.1.12, m is nonzero, so we can divide both sides by m to obtain $d = zc$. Hence, $c \mid d$ and, consequently, d satisfies the conditions (**GCD 1**), (**GCD 2**).

(ii) It is clear that m divides a and b. Suppose that c is an arbitrary common multiple of a and b, say $c = aa_4$, $c = bb_4$ for some $a_4, b_4 \in R$. Then,

$$c = aa_4 = da_1a_4 \text{ and } c = bb_4 = db_1b_4,$$

so $da_1a_4 = db_1b_4$. By Corollary 9.1.12, d is nonzero and dividing both sides by d we obtain $a_1a_4 = b_1b_4$. By Corollary 9.1.14, a_1 and b_1 are relatively prime, so Proposition 9.1.15 implies that a_1 divides b_4, say $b_4 = a_1x$, where $x \in R$. Hence, $c = db_1b_4 = db_1a_1x = mx$, since $m = db_1a_1$. Consequently, m divides c and hence m satisfies the conditions (**LCM 1**) and (**LCM 2**).

9.1.17. Definition. *Let* R *be an integral domain. A nonzero element* p *of* R *is called a prime element if* $p \notin \mathbf{U}(R)$ *and* p *cannot be written as a product of two proper divisors. Thus, an element* p *is a prime if and only if in every decomposition* $p = ab$ *at least one of the factors* a, b *is invertible.*

Prime elements of integral domains are closely connected to maximal ideals. An ideal M is called a maximal ideal of the ring R if for each ideal H that is situated between M and R (so $M \leq H \leq R$) either $M = H$ or $H = \text{R}$.

9.1.18. Lemma. *Let* R *be a ring. An ideal* M *is maximal in* R *if and only if* R/M *is a simple ring. In particular, an ideal* M *of a commutative ring* R *is maximal if and only if* R/M *is a field.*

Proof. Let M be a maximal ideal of R and let Q be an arbitrary ideal of R/M. Let H be the preimage of Q under the natural homomorphism $\sigma_M : R \longrightarrow R/M$. Then H is an ideal of R containing $M = \mathbf{Ker}\sigma_M$, by Proposition 7.4.3. It follows that either $M = H$ or $H = R$. In the first case, $Q = \sigma_M(H) = H/M = M/M$ is the zero ideal of R/M, whereas in the latter case, $Q = \sigma_M(R) = R/M$. This means that R/M is a simple ring.

Conversely, let R/M be a simple ring and let H be an ideal, situated between M and R. Since σ_M is an epimorphism, Proposition 7.4.3 implies that $\sigma_M(H)$ is an ideal of R/M. It follows that either $\sigma_M(H)$ is the zero ideal of R/M or $\sigma_M(H) = R/M$. In the first case, $H \leq \mathbf{Ker}\sigma_M = M$ and hence $H = M$. For the latter case, note that $\sigma_M(H) = H/M = R/M$ so $H = R$. Hence M is a maximal ideal of R. When the ring R is commutative, we use Theorem 7.3.12 to deduce that R/M is a field.

We next consider some important properties of prime elements.

9.1.19. Lemma. *Let R be a principal ideal domain.*

(i) *If p is a prime element of R and a is an arbitrary element of R, then either p divides a or a and p are relatively prime.*
(ii) *The element p of R is prime if and only if the ideal pR is maximal in R.*

Proof.
(i) Let $d = \mathbf{GCD}(a, p)$. Since d is a divisor of p, either the elements p and d are associates or $d \in \mathbf{U}(R)$. In the first case, $p = \mathbf{GCD}(a, p)$ so p divides a. In the second case, p and a are relatively prime.

(ii) Let p be a prime of R and let H be an ideal of R such that $pR \leq H \leq R$. Since R is a PID, choose $y \in R$ such that $H = yR$. Then $pR \leq yR$ so $p = yx$ for some $x \in R$. Since p is a prime element either y and p are associates (when $pR = yR = H$), or $y \in \mathbf{U}(R)$ (when $H = yR = R$). Thus, pR is a maximal ideal. Conversely, let pR be a maximal ideal of R. If $p = zw$ for certain elements $z, w \in R$ then, by Proposition 9.1.6, we have $pR \leq zR$. It follows that either $pR = zR$ or $zR = R$. In the first case, p and z are associates. In the latter case $z \in \mathbf{U}(R)$. Hence p is a prime element of R.

We can now establish the main results of this section. The following result is sometimes known as the Fundamental Theorem of Arithmetic.

9.1.20. Theorem. *Let R be a principal ideal domain and let $0_R \neq a \in R$.*

(i) *If $a \notin \mathbf{U}(R)$ then a can be written as a product of primes of R.*
(ii) *If $a \notin \mathbf{U}(R)$ and $a = p_1 \dots p_n = q_1 \dots q_m$ are two decompositions of a into products of prime elements of R, then $n = m$ and we can renumber the elements of the second decomposition so that $q_j = u_j p_j$ for some $u_j \in \mathbf{U}(R)$, for all $1 \leq j \leq n$.*

Proof.
(i) Suppose, for a contradiction, that this is not true and let D denote the subset of nonzero elements of $R \backslash \mathbf{U}(R)$ that cannot be decomposed as a product of primes. By our assumption, $D \neq \emptyset$.

Suppose first that there is an ascending chain of ideals

$$d_1 R \leq d_2 R \leq \cdots \leq d_n R \leq \cdots$$

such that $d_n \in D$ for each $n \in \mathbb{N}$. By Corollary 7.3.7, $H = \bigcup_{n \in \mathbb{N}} d_n R$ is an ideal of R and, since R is a PID, there is an element $d \in R$ such that $H = dR$. By the choice of H, there exists $k \in \mathbb{N}$ such that $d \in d_k R$, which implies that $dR \leq d_k R$. On the other hand, $d_k R \leq H = dR$, so that $d_k R = dR = d_j R$, for all $j \geq k$. Thus, every such chain is a finite chain and terminates in finitely many steps.

Now let $a = a_1$ be an arbitrary element of D. Then a_1 is not a product of primes and so, in particular, a_1 is not prime. Hence, it is possible to write $a_1 = bc$ where $b, c \in R \setminus \mathbf{U}(R)$. Note that this also means, by Proposition 9.1.6 and Corollary 9.17, that $a_1 R \lneqq bR$ and $a_1 R \lneqq cR$. Now if b, c are both not in D then both b and c are products of primes and hence a_1 is also a product of primes, so that $a_1 \notin D$, contrary to the choice of a_1. Hence either $b \in D$ or $c \in D$ and in any case there is an element $a_2 \in D$ such that $a_1 R \lneqq a_2 R$. However, this argument can now be repeated for the element a_2 and consequently there is an element $a_3 \in R$ such that $a_2 R \lneqq a_3 R$. In this way, we can construct an infinite ascending chain of ideals

$$a_1 R \lneqq a_2 R \lneqq \cdots \lneqq a_n R \lneqq \cdots$$

such that $a_n \in D$ for each $n \in \mathbb{N}$, which contradicts our initial argument. Hence, D is empty and assertion (i) follows.

(ii) For this assertion we use induction on n. If $n = 1$, then $a = p_1$ is a prime element. By the definition of a prime, one of the factors of the second decomposition, say q_1, is an associate of p_1. By cancelling the nonzero element p_1, it follows that there is an element $u \in \mathbf{U}(R)$ such that $e = uq_2 \ldots q_m$. Hence each remaining q_i is an element of $\mathbf{U}(R)$, contrary to the definition of a prime. Hence the result holds when $n = 1$.

Suppose now that $n > 1$ and assume that our assertion is true for all elements that decompose into a product of fewer than n primes. Then $m \geq n$ and we have

$$p_1 \ldots p_n = q_1 \ldots q_m. \tag{9.1}$$

By Lemma 9.1.19, either the elements p_1 and q_1 are associates or these elements are relatively prime. In the latter case, Proposition 9.1.15 implies that p_1 divides the product $q_2 \ldots q_m$. By repeating this argument we see that either p_1 and q_2 are associates or p_1 divides $q_3 \ldots q_m$. Since $m \geq n$ we see upon repeating the argument sufficiently often that p_1 is an associate of some q_i or that p_1 divides q_m, in which case p_1 is an associate of q_m. In any case, p_1 is an associate of some q_i and, by renumbering if necessary, we may assume that p_1, q_1 are associates, say $q_1 = u_1 p_1$, where $u_1 \in \mathbf{U}(R)$. By cancelling p_1 in Equation 9.1 we see that

$$p_2 \ldots p_n = u_1 q_2 q_3 \ldots q_m.$$

Now we can apply the induction hypothesis to the element $p_2 \ldots p_n$. Hence $n - 1 = m - 1$, so $n = m$ and after some renumbering we obtain $q_j = u_j p_j$ for some $u_j \in \mathbf{U}(R)$, where $2 \leq j \leq n$. The result follows.

We often say that the decomposition of Theorem 9.1.20 is unique in the sense that the primes occurring in the decomposition are unique up to the order that the factors are written in and to within multiplication by invertible elements of R. Moreover, the prime factors occurring need not be distinct, so we can write the product more succinctly in the form

$$a = up_1^{k_1} \dots p_m^{k_m},$$

where $p_1, \dots, p_m$ are pairwise different primes and $k_1, \dots, k_m$ are nonzero integers.

For the ring R, we let $\sigma(R)$ denote a set of representatives of prime elements, one for each associate class.

9.1.21. Corollary. *Let R be a principal ideal domain, let a, b be nonzero elements of R, let $a = up_1^{k_1} \dots p_m^{k_m}$, $b = vp_1^{t_1} \dots p_m^{t_m}$ where $p_1, \dots, p_m$ are distinct elements of the set $\sigma(R)$ and $u, v \in \mathbf{U}(R)$.*

(i) *The element a divides the element b if and only if $k_j \leq t_j$ for each j, where $1 \leq j \leq m$.*

(ii) $\mathbf{GCD}(a, b) = p_1^{d_1} \dots p_m^{d_m}$ *where $d_j = \mathbf{min}\{k_j, t_j\}$ for each j, such that $1 \leq j \leq m$.*

(iii) $\mathbf{LCM}(a, b) = p_1^{q_1} \dots p_m^{q_m}$ *where $q_j = \mathbf{max}\{k_j, t_j\}$ for each j, such that $1 \leq j \leq m$.*

In particular, the elements a and b are relatively prime if none of the prime factors of a are prime factors of b.

Although the previous corollary gives us a method in principle for finding the greatest common divisors, it is dependent upon being able to factor an element of a principle ideal domain into products of primes and this can be an exceedingly difficult task to perform in general, even in the ring of integers.

9.1.22. Definition. *Let R be an integral domain. Then R is said to be a unique factorization domain or a factorial domain if it satisfies the conditions:*

(i) *If a is a nonzero element of R and $a \notin \mathbf{U}(R)$, then a is a product of prime elements of R.*

(ii) *If $a = p_1 \dots p_n = q_1 \dots q_m$ are two decompositions of a into a product of prime elements of R, then $n = m$ and we can renumber the elements of the second decomposition in such way that $q_j = u_j p_j$ for some $u_j \in \mathbf{U}(R)$, where $1 \leq j \leq n$.*

By Theorem 9.1.20, every PID is a unique factorization domain. If R is a unique factorization domain, then for every pair of its elements (and hence for each finite set of elements) there exists a greatest common divisor and a least common multiple, which can be obtained as in Corollary 9.1.21 (ii) and (iii).

Now we show that not every unique factorization domain is a PID and in passing we consider some important examples of such domains. In order to do this, we consider polynomial rings in some more detail. We begin with the following generalization of Theorem 7.5.2.

9.1.23. Theorem. *Let R be an integral domain, let $f(X), g(X) \in R[X]$ and let $g(X) \neq 0_R$. Let*

$$k = \mathbf{max}\{\mathbf{deg}\, f(X) - \mathbf{deg}\, g(X) + 1, 0\},$$

and let d be the leading coefficient of the polynomial $g(X)$. Then there exist polynomials $q(X), r(X) \in R[X]$ such that $d^k f(X) = q(X)g(X) + r(X)$ where either $r(X) = 0_R$ or $\mathbf{deg}\, r(X) < \mathbf{deg}\, g(X)$. This presentation of the polynomial $d^k f(X)$ is unique.

Proof. Let

$$f(X) = a_0 + a_1 X + \cdots + a_m X^m \text{ and } g(X) = b_0 + b_1 X + \cdots + b_n X^n,$$

where $a_m = c \neq 0_F, b_n = d \neq 0_F$. We apply induction on m. If $\mathbf{deg}\, g(X) > \mathbf{deg}\, f(X)$, then put $r(X) = f(X)$ and $q(X) = 0_F$. Thus, we may assume that $\mathbf{deg}\, f(X) \geq \mathbf{deg}\, g(X)$. If $\mathbf{deg}\, f(X) = 0$ then set $r(X) = 0_R$ and $q(X) = c$. Suppose next that $m > 0$ and suppose, inductively, that the theorem is true for all polynomials of degree less than m. The degree of the polynomial $cX^{m-n}g(X)$ is m and the coefficient of X^m is cd. Therefore, the degree of the polynomial $df(X) - cX^{m-n}g(X)$ is less than m and the induction hypothesis implies that there are polynomials $q_1(X), r(X) \in R[X]$ such that

$$d^{(m-1)-n+1}(df(X) - cX^{m-n}g(X)) = q_1(X)g(X) + r(X),$$

where either $r(X) = 0_R$ or $\mathbf{deg}\, r(X) < \mathbf{deg}\, g(X)$. Now we have

$$d^k f(X) = q(X)g(X) + r(X),$$

where

$$q(X) = cd^{m-n}X^{m-n} + q_1(X).$$

The uniqueness of this presentation can be proved in the same way as in Theorem 7.5.2.

9.1.24. Definition. *Let R be a unique factorization domain. A polynomial $f(X) \in R[X]$ is called primitive, if the greatest common divisor of all its coefficients is the multiplicative identity element.*

Every polynomial $f(X) \in R[X]$ can be represented in the form $f(X) = cg(X)$, where $c \in R$, $g(X) \in R[X]$ and $g(X)$ is a primitive polynomial. Here we take c to be a greatest common divisor of all the coefficients of the polynomial $f(X)$. Conversely, if $f(X) = c_1 g_1(X)$ where $g_1(X)$ is a primitive polynomial, then clearly c_1 is also a greatest common divisor of all coefficients of the polynomial $f(X)$. Thus, the elements c and c_1 are associates. The element c is called the content of $f(X)$ and is denoted by $\mathbf{c}(f(X))$ or $\mathbf{c}(f)$. Thus, the content is just the greatest common divisor of the coefficients. As a greatest common divisor, the content of a polynomial is not unique, since it can only be determined up to multiplication by an element of $\mathbf{U}(R)$. A polynomial $f(X)$ is primitive if and only if $\mathbf{c}(f) \in \mathbf{U}(R)$.

9.1.25. Proposition *(Gauss's Lemma). Let R be a unique factorization domain and let $f(X), g(X) \in R[X]$. Then $\mathbf{c}(fg) = \mathbf{c}(f)\mathbf{c}(g)$. Hence if $f(X)$ and $g(X)$ are both primitive then $f(X)g(X)$ is also a primitive polynomial.*

Proof. Let

$$f(X) = a_0 + a_1 X + \cdots + a_m X^m,\ g(X) = b_0 + b_1 X + \cdots + b_n X^n, \text{ and}$$
$$f(X)g(X) = c_0 + c_1 X + \cdots + c_{m+n} X^{m+n}.$$

First suppose that $f(X), g(X)$ are primitive polynomials and assume that the product is not primitive. Choose a prime element q that is a divisor of $\mathbf{c}(fg)$. Then q divides every coefficient of $f(X)g(X)$, so $q \mid c_j$ for each j, where $0 \le j \le m+n$. Since $f(X)$ is primitive, there exists a positive integer k such that q divides $a_0, a_1, \ldots, a_{k-1}$, but q does not divide a_k and, similarly, there exists a positive integer t such that q divides $b_0, b_1, \ldots, b_{t-1}$ but q does not divide b_t. We have

$$c_{k+t} = a_0 b_{k+t} + a_1 b_{k+t-1} + \cdots + a_{k-1} b_{t+1} + a_k b_t + a_{k+1} b_{t-1} + \cdots + a_{k+t} b_0.$$

Then

$$a_k b_t = c_{k+t} - a_0 b_{k+t} - a_1 b_{k+t-1} - \cdots - a_{k-1} b_{t+1} - a_{k+1} b_{t-1} - \cdots - a_{k+t} b_0$$

and, since q divides each term on the right-hand side, q divides $a_k b_t$. Since q is prime and q does not divide a_k, Lemma 9.1.19 implies that q and a_k are relatively prime. Hence q must divide b_t, by Proposition 9.1.15, which is a contradiction to the choice of t . This proves that $f(X)g(X)$ is a primitive polynomial.

More generally,

$$f(X) = \mathbf{c}(f) f_1(X) \text{ and } g(X) = \mathbf{c}(g) g_1(X),$$

where $f_1(X), g_1(X)$ are primitive polynomials. Furthermore, $f(X)g(X) = \mathbf{c}(f)\mathbf{c}(g)f_1(X)g_1(X)$. Since $f_1(X)g_1(X)$ is a primitive polynomial, it follows that $\mathbf{c}(f)\mathbf{c}(g)$ is the content of $f(X)g(X)$.

9.1.26. Lemma. *Let R be a unique factorization domain and suppose that $f(X), g(X) \in R[X]$. Suppose that $g(X)$ is a primitive polynomial. If $g(X)$ divides $df(X)$ for some nonzero element $d \in R$, then $g(X)$ divides $f(X)$.*

Proof. We have $df(X) = g(X)h(X)$ for some polynomial $h(X) \in R[X]$. By Proposition 9.1.25, $d\mathbf{c}(f) = \mathbf{c}(g)\mathbf{c}(h)$. Since $\mathbf{c}(g) \in \mathbf{U}(R)$ it follows that $\mathbf{c}(h) = du\mathbf{c}(f)$ where $u = (\mathbf{c}(g))^{-1}$. Furthermore, $h(X) = \mathbf{c}(h)h_1(X)$ where $h_1(X)$ is a primitive polynomial. We now have

$$df(X) = g(X)h(X) = g(X)\mathbf{c}(h)h_1(X) = g(X)du\mathbf{c}(f)h_1(X).$$

Since R is an integral domain and $d \neq 0$ we have, upon cancelling d,

$$f(X) = \mathbf{c}(f)ug(X)h_1(X)$$

and the result follows.

We next need a technical lemma.

9.1.27. Lemma. *Let R be a unique factorization domain and let $a, b \in R$. Suppose that $p \in R$ is prime. If p divides ab then either p divides a or p divides b.*

Proof. We know that there is an element $c \in R$ such that $ab = cp$. If $a = p_1 \dots p_k$ and $b = q_1 \dots q_l$ as products of primes, then it follows that $p_1 \dots p_k q_1 \dots q_l = cp$. However, R is a unique factorization domain so $p = up_i$, for some i or $p = uq_j$ for some j and some invertible element u. If $p = up_i$ then $a = u^{-1}p_1 \dots p_{i-1}pp_{i+1}$ and hence p divides a. Otherwise, by a similar argument, p divides b.

We are now led to the following very pleasing result, which enables us to construct other unique factorization domains.

9.1.28. Theorem. *Let R be a unique factorization domain. Then $R[X]$ is a unique factorization domain.*

Proof. To begin we prove that every nonzero and noninvertible element $f(X)$ of the ring $R[X]$ can be written as a product of prime elements. If $\mathbf{deg}\, f(X) = m = 0$, then $f(X)$ is an element of R and the hypothesis implies that $f(X)$ can be so written. Let $m > 0$ and assume that the theorem is valid for all polynomials of degree less than m. Rewrite the polynomial $f(X)$ as $f(X) = \mathbf{c}(f)f_1(X)$ where $f_1(X)$ is a primitive polynomial. If $f_1(X)$ is a product of primes, then the result holds. Therefore, suppose that $f_1(X) = g(X)h(X)$. If

deg $g(X) =$ **deg** $f_1(X)$ then **deg** $h(X) = 0$ and hence $h(X)$ is an element of R. Thus, since $f_1(X)$ is primitive, $h(X)$ is invertible. So we may assume that **deg** $g(X) <$ **deg** $f_1(X)$ and **deg** $h(X) <$ **deg** $f_1(X)$. However, we may now apply the induction hypothesis to $g(X)$ and $h(X)$, which implies that they are products of prime elements of $R[X]$. Hence, $f_1(X)$ and $f(X)$ are also products of primes in $R[X]$.

Uniqueness can be proved with the help of Lemma 9.1.27 using the same arguments that were developed in the proof of Theorem 9.1.20.

A simple induction allows us to extend this result easily as follows.

9.1.29. Corollary. *Let R be a unique factorization domain. Then $R[X_1, \dots, X_n]$ is a unique factorization domain.*

In particular, every field is, by default, a unique factorization domain, so we have the following corollary.

9.1.30. Corollary. *If F is a field, then $F[X_1, \dots, X_n]$ is a unique factorization domain.*

$\mathbb{Z}$ is the standard example of a unique factorization domain, so we also can state the following corollary.

9.1.31. Corollary. *$\mathbb{Z}[X_1, \dots, X_n]$ is a unique factorization domain.*

Using the results that we have recently obtained, it is easy to show that there are unique factorization domains that are not PIDs. To see this let F be a field and note that by Corollary 9.1.30 $F[X, Y]$, the polynomial ring in two variables X and Y, is a unique factorization domain. Consider the ideal $H = XF[X, Y] + YF[X, Y]$ and suppose that $H = f(X, Y)F[X, Y]$, for some polynomial $f(x, y)$ of degree at least 1. Then we have $X = h(X, Y)f(X, Y)$ and $Y = g(X, Y)f(X, Y)$, where $h(X, Y), g(X, Y) \in F[X, Y]$. By Theorem 7.6.3, **deg** $h(X, Y) = 0 =$ **deg** $g(X, Y)$, so that $h(X, Y) = u$ and $g(X, Y) = v$ are nonzero elements of the field F. Thus, $f(X, Y) = u^{-1}X = v^{-1}Y$, which contradicts the fact that X and Y are independent variables. Hence, $F[X, Y]$ is a unique factorization domain that is not a PID.

Finally, we construct an example of a ring in which the decomposition into prime factors exists but which is not unique. To do this, we consider the following construction of quadratic fields and rings. Let r be an integer with the property that $\sqrt{r} \notin \mathbb{Q}$ and let

$$\mathbb{Q}[\sqrt{r}] = \{a + b\sqrt{r} \mid a, b \in \mathbb{Q}\}, \mathbb{Z}[\sqrt{r}] = \{a + b\sqrt{r} \mid a, b \in \mathbb{Z}\}.$$

Let α, β be arbitrary elements of $\mathbb{Q}[\sqrt{r}]$, say $\alpha = a + b\sqrt{r}, \beta = a_1 + b_1\sqrt{r}$. Then

$$\alpha - \beta = (a - a_1) + \sqrt{r}(b - b_1) \text{ and } \alpha\beta = (aa_1 + bb_1r) + \sqrt{r}(ab_1 + ba_1).$$

Thus, $\alpha - \beta, \alpha\beta \in \mathbb{Q}[\sqrt{r}]$ and, if $\alpha, \beta \in \mathbb{Z}[\sqrt{r}]$, then $\alpha - \beta, \alpha\beta \in \mathbb{Z}[\sqrt{r}]$. By Theorem 7.1.9, $\mathbb{Q}[\sqrt{r}], \mathbb{Z}[\sqrt{r}]$ are subrings of $\mathbb{C}$. Clearly, $1 \in \mathbb{Z}[\sqrt{r}]$. If $\alpha \neq 0$ then it is easy to see that

$$\alpha^{-1} = \frac{a}{a^2 - rb^2} + \sqrt{r}\left(\frac{-b}{a^2 - rb^2}\right) \in \mathbb{Q}[\sqrt{r}],$$

and Theorem 3.2.6 shows that $\mathbb{Q}[\sqrt{r}]$ is a subfield of $\mathbb{C}$. If $r > 0$, then $\mathbb{Q}[\sqrt{r}]$ is called a real quadratic field; if $r < 0$, then $\mathbb{Q}[\sqrt{r}]$ is called an imaginary quadratic field.

If $\alpha = a + b\sqrt{r} \in \mathbb{Q}[\sqrt{r}]$, then the rational number $\mathbf{N}(\alpha) = a^2 - rb^2$ is called the norm of α. If $\mathbf{N}(\alpha) = 0$, then $\frac{a^2}{b^2} = r$. Since $\sqrt{r} \notin \mathbb{Q}$, we conclude that $a = 0$, and hence $b = 0$. Then $\mathbf{N}(\alpha) = 0$ if and only if $\alpha = 0$.

Simple calculations show that $\mathbf{N}(\alpha\beta) = \mathbf{N}(\alpha)\mathbf{N}(\beta)$.

In particular,

$$1 = \mathbf{N}(1) = \mathbf{N}(\alpha\alpha^{-1}) = \mathbf{N}(\alpha)\mathbf{N}(\alpha^{-1}),$$

so that

$$\mathbf{N}(\alpha^{-1}) = \frac{1}{\mathbf{N}(\alpha)}.$$

We note that for elements of the ring $\mathbb{Z}[\sqrt{r}]$ the norm is an integer; moreover in the case when $r < 0$, we have $\mathbf{N}(\alpha) \geq 0$ for each element $\alpha \in \mathbb{Z}[\sqrt{r}]$.

9.1.32. Proposition. *Let $r \in \mathbb{Z}$ and $r < 0$.*

(i) $\mathbf{U}(\mathbb{Z}[\sqrt{r}]) = \{1, -1\}$ *if $r \neq -1$ and* $\mathbf{U}(\mathbb{Z}[i]) = \{1, -1, i, -i\}$.

(ii) *Let α be a nonzero element of $\mathbb{Z}[\sqrt{r}]$. If $\alpha \notin \mathbf{U}(\mathbb{Z}[\sqrt{r}])$, then α is a product of prime elements.*

Proof.

(i) Let $d = -r$, where $d > 0$. If $\alpha = a + b\sqrt{r} \in \mathbf{U}(\mathbb{Z}(\sqrt{r}))$ then, since $\mathbf{N}(\alpha)$ is an integer, the equation $1 = \mathbf{N}(\alpha)\mathbf{N}(\alpha^{-1})$ implies that $\mathbf{N}(1) = 1$. However, the equation $a^2 + db^2 = 1$ has the following integer solutions:

$a = 0, b = 1$; $a = 0, b = -1$; $a = 1, b = 0$; $a = -1, b = 0$ in the case when $d = 1$; and $a = 1, b = 0$; $a = -1, b = 0$, in the case when $d \neq 1$.

(ii) We use induction on $\mathbf{N}(\alpha)$. Since $\alpha \notin \mathbf{U}(\mathbb{Z}(\sqrt{r}))$, $\mathbf{N}(\alpha) > 1$. The equation $\mathbf{N}(\alpha) = 2$ is valid only if $d = 1$ or 2. If $d = 1$, then

$$\alpha \in \{1 + i, 1 - i, -1 + i, -1 - i\}.$$

If $d = 2$, then

$$\alpha \in \{i\sqrt{2}, -i\sqrt{2}\}.$$

The equation $\mathbf{N}(\alpha) = 3$ is valid only if $d = 2$ or 3. If $d = 2$, then

$$\alpha \in \{1 + i\sqrt{2}, 1 - i\sqrt{2}, -1 + i\sqrt{2}, -1 - i\sqrt{2}\}.$$

If $d = 3$, then

$$\alpha \in \{i\sqrt{3}, -i\sqrt{3}\}.$$

Consider the equation $\mathbf{N}(\alpha) = 4$. If $d = 1$, then

$$\alpha \in \{2, -2, 2i, -2i\};$$

if $d = 2$, then

$$\alpha \in \{2, -2\};$$

if $d = 3$, then

$$\alpha \in \{2, -2\};$$

if $d = 4$, then

$$\alpha \in \{2, -2, i, -i\};$$

if $d > 4$, then again

$$\alpha \in \{2, -2\}.$$

It is easy to check that when $\alpha \in \mathbb{Z}[\sqrt{r}]$ and $\mathbf{N}(\alpha) \leq 4$ then α can be written as a product of primes, so now let $\alpha \in \mathbb{Z}[\sqrt{r}]$ be such that $\mathbf{N}(\alpha) \geq 4$. Assume inductively that the result holds for all $\beta \in \mathbb{Z}[\sqrt{r}]$ for which $\mathbf{N}(\beta)$ is less than some natural number k and let α be such that $\mathbf{N}(\alpha) = k$. If α is a prime then certainly α is a product of primes. Therefore, we may assume that $\alpha = \beta\gamma$, where β, γ are proper divisors of α. Then $\beta, \gamma \notin \mathbf{U}(\mathbb{Z}[\sqrt{r}])$ so $\mathbf{N}(\beta) > 1$ and $\mathbf{N}(\gamma) > 1$. It follows from the equation $\mathbf{N}(\beta\gamma) = \mathbf{N}(\beta)\mathbf{N}(\gamma)$, that $\mathbf{N}(\beta) < \mathbf{N}(\alpha)$ and $\mathbf{N}(\gamma) < \mathbf{N}(\alpha)$. By the induction hypothesis, β, γ, and hence α can be decomposed into a product of prime elements. This completes the proof.

Now we find a concrete value of r for which, in the ring $\mathbb{Z}[\sqrt{r}]$, there exist elements having two distinct decompositions into primes. One such value is $r = -5$. Indeed,

$$9 = 3 \times 3 = (2 + i\sqrt{5})(2 - i\sqrt{5}).$$

By Proposition 9.1.32, the elements 3 and $(2 + i\sqrt{5})$, 3 and $(2 - i\sqrt{5})$ are not associates. Also, $3, 2 + i\sqrt{5}, 2 - i\sqrt{5}$ are primes in the ring $\mathbb{Z}[\sqrt{-5}]$. To see this let

$$\alpha \in \{3, 2 + i\sqrt{5}, 2 - i\sqrt{5}\},$$

and suppose that $\alpha = \beta\gamma$. We have $9 = \mathbf{N}(\alpha) = \mathbf{N}(\beta\gamma) = \mathbf{N}(\beta)\mathbf{N}(\gamma)$. This implies that $\mathbf{N}(\beta) = 3 = \mathbf{N}(\gamma)$. However, if $\beta = x + y\sqrt{-5}$ then $\mathbf{N}(\alpha) = x^2 + 5y^2$ and the equation $x^2 + 5y^2 = 3$ has no integer solutions. Consequently 9 has two different prime factorizations in the ring $\mathbb{Z}[\sqrt{-5}]$.

EXERCISE SET 9.1

In each of the problems justify your reasoning, using a proof or counterexample.

9.1.1. Prove that $30 \mid n^5 - n$ for all positive integers n.

9.1.2. Prove that $42 \mid n^7 - n$ for all positive integers n.

9.1.3. Prove that $8 \mid n^2 - 1$ for all odd positive integers n.

9.1.4. Suppose that $3 \nmid n$. Prove that $6 \mid n^2 - 1$.

9.1.5. Find an element that generates the ideal $4\mathbb{Z} + 6\mathbb{Z} + 8\mathbb{Z} + 10\mathbb{Z} + 15\mathbb{Z} + 20\mathbb{Z}$.

9.1.6. Find integers n, k such that $5\mathbb{Z} + 7\mathbb{Z} = n\mathbb{Z}$, $5\mathbb{Z} \cap 7\mathbb{Z} = k\mathbb{Z}$.

9.1.7. Find integers n, k such that $-3\mathbb{Z} + 12\mathbb{Z} = n\mathbb{Z}$, $-3\mathbb{Z} \cap 12\mathbb{Z} = k\mathbb{Z}$.

9.1.8. Find integers n, k such that $5\mathbb{Z} + 11\mathbb{Z} = n\mathbb{Z}$, $5\mathbb{Z} \cap 11\mathbb{Z} = k\mathbb{Z}$.

9.1.9. In the ring $\mathbb{Z}[i]$ find the prime decomposition of the element $5 + 9i$.

9.1.10. In the ring $\mathbb{Z}[i]$ find the prime decomposition of the element $12 + 5i$.

9.1.11. In the ring $\mathbb{Z}[i]$ find the prime decomposition of the element $3 - 2i$.

9.1.12. In the ring $\mathbb{Z}[i]$ find the prime decomposition of the element $2 + 5i$.

9.1.13. In the ring $\mathbb{Z}[i]$ find the prime decomposition of the element $-2 + 11i$.

9.1.14. In the ring $\mathbb{Z}[i]$ find the prime decomposition of the element $2 + 9i$.

9.1.15. Let $\alpha \in \mathbb{Z}[i]$ and suppose that $\mathbf{N}(\alpha) = p$ is a prime. Is α a prime element of the ring $\mathbb{Z}[i]$?

9.1.16. Let $x, y \in \mathbb{Z}$. Suppose that $\mathbf{GCD}(x, y) = 1$. Prove that $\mathbf{GCD}(x + y, x - y) = 1$ or 2.

9.1.17. Let $x, y, z \in \mathbb{Z}$. Suppose that $\mathbf{GCD}(x, y) = 1$ and $\mathbf{GCD}(x, z) = 1$. Prove that $\mathbf{GCD}(x, yz) = 1$.

9.1.18. Let $x, y, z \in \mathbb{Z}$. Prove that $\mathbf{GCD}(zx, zy) = z\mathbf{GCD}(x, y)$.

9.1.19. Let $x, y, z \in \mathbb{Z}$. Prove that $\mathbf{GCD}(\mathbf{GCD}(x, y), z) = \mathbf{GCD}(x, \mathbf{GCD}(y, z))$.

9.1.20. In the ring $\mathbb{Z}[i\sqrt{11}] = \{x + yi\sqrt{11} \mid x, y \in \mathbb{Z}\}$ find an element that does not have a unique prime factorization.

9.2 EUCLIDEAN RINGS

In the previous section, we extended some common arithmetic concepts and results to PIDs. In particular, we proved the existence of greatest common divisors and least common multiples of elements of such rings. We also showed that in such rings elements can always be written as a product of primes in a unique way. However, there is a difference between the existence of some object and obtaining an algorithm that allows us to construct such an object. In this section, we consider a smaller class of rings for which we can construct such algorithms.

9.2.1. Definition. *Let R be an integral domain. Then R is said to be a Euclidean ring if there exists a function $\delta : R\backslash\{0_R\} \longrightarrow \mathbb{N}$ satisfying the following conditions:*

*(**E 1**) $\delta(xy) \geq \delta(x)$ for all $x, y \in R\backslash\{0_R\}$;*

*(**E 2**) for all $x, y \in R$ where $y \neq 0_R$ there exist $w, z \in R$ such that $x = wy + z$ and either $z = 0_R$ or $\delta(z) < \delta(y)$.*

We often call w the quotient and z the remainder. Note that if $\delta(xy) = \delta(x)\delta(y)$ then $\delta(xy) \geq \delta(x)$ since $\delta(y) \geq 1$. As we have seen in Section 1.4, the first natural example of a Euclidean ring is the ring $\mathbb{Z}$ of all integers, where the function δ is taken to be the absolute value. The reason for the term *Euclidean* here lies in the fact that in his famous book, "The Elements," Euclid proved that the set of integers form what we now call a Euclidean ring by proving property (**E 2**) in that case. By Theorem 7.5.2, the ring of polynomials in one variable over a field is Euclidean. We now consider some other examples.

9.2.2. Proposition. *The ring $\mathbb{Z}[i]$ is Euclidean. The function δ is defined by $\delta(\alpha) = \mathbf{N}(\alpha)$ for every element $\alpha \in \mathbb{Z}[i]$.*

Proof. In Section 9.1 we proved, in a more general setting, that $\mathbb{Z}[i]$ is a subring of the field $\mathbb{C}$ and hence $\mathbb{Z}[i]$ is an integral domain. If $\alpha = a + bi$, where $a, b \in \mathbb{Z}$, then $\delta(\alpha) = \mathbf{N}(\alpha) = a^2 + b^2$. In Section 9.1, we saw that $\mathbf{N}(\alpha\beta) = \mathbf{N}(\alpha)\ \mathbf{N}(\beta)$ holds in $\mathbb{Z}[i]$ and this implies that (**E 1**) holds.

Now let $\alpha = a + bi$ and $\beta = c + di \neq 0$ where $a, b, c, d \in \mathbb{Z}$. Then $\frac{\alpha}{\beta} = x + yi$, where $x, y \in \mathbb{Q}$ and hence there exist integers u, v such that $|x - u| \leq \frac{1}{2}$ and

$|y - v| \leq \frac{1}{2}$. Let $\gamma = u + vi$ and $\rho = \alpha - \beta\gamma$. By definition, $\gamma, \rho \in \mathbb{Z}[i]$ and either $\rho = 0$ or

$$\begin{aligned}\mathbf{N}(\rho) &= \mathbf{N}(\alpha - \beta\gamma) = \mathbf{N}\left(\beta\left(\frac{\alpha}{\beta} - \gamma\right)\right) = \mathbf{N}(\beta)\mathbf{N}\left(\frac{\alpha}{\beta} - \gamma\right) \\ &= \mathbf{N}(\beta)\mathbf{N}((x-u) + (y-v)i) = \mathbf{N}(\beta)((x-u)^2 + (y-v)^2) \\ &\leq \mathbf{N}(\beta)\left(\frac{1}{4} + \frac{1}{4}\right) = \frac{1}{2}\mathbf{N}(\beta) < \mathbf{N}(\beta).\end{aligned}$$

This proves (**E 2**) from which it follows that $\mathbb{Z}[i]$ is a Euclidean ring, called the *ring of Gaussian integers*, named after Gauss who introduced this ring and studied its properties

We now discuss another example. Let $\varpi = \frac{-1+\sqrt{-3}}{2}$, which is a root of the polynomial $X^2 + X + 1 \in \mathbb{Z}[X]$. Hence $\varpi^2 + \varpi + 1 = 0$ and from the equation $(X-1)(X^2 + X + 1) = X^3 - 1$ we see that ϖ is a primitive third root of unity. Notice also that

$$\varpi^2 = -\varpi - 1 = \frac{-1 - \sqrt{-3}}{2}.$$

By the results of Section 7.5, the set $\mathbb{Z}[\varpi]$ consists of all numbers of the type $x + y\varpi$, where $x, y \in \mathbb{Z}$. Furthermore, $\mathbb{Z}[\varpi]$ is a subring of $\mathbb{C}$, so it is an integral domain and therefore has no zero divisors. Also note that $\mathbb{Z}[\varpi]$ is closed under complex conjugation. Indeed if $\mathbf{\Delta}$ denotes complex conjugation, then

$$\mathbf{\Delta}(\sqrt{-3}) = \mathbf{\Delta}(i\sqrt{3}) = -i\sqrt{3} = -\sqrt{-3},$$

and therefore $\mathbf{\Delta}(\varpi) = \varpi^2$. Thus, if $\alpha = x + y\varpi$, then

$$\mathbf{\Delta}(\alpha) = x + y\mathbf{\Delta}(\varpi) = x + y\varpi^2 = (x - y) - y\varpi \in \mathbb{Z}[\varpi],$$

since $\varpi^2 = -\varpi - 1$.

For $\alpha = a + b\varpi \in \mathbb{Z}[\varpi]$, we let $\mathbf{N}(\alpha) = \alpha\mathbf{\Delta}(\alpha) = (a + b\varpi)(a + b\varpi^2) = a^2 - ab + b^2$, since $1 + \varpi + \varpi^2 = 0$. We define $\delta : \mathbb{Z}[\varpi] \setminus \{0\} \longrightarrow \mathbb{N}$ by $\delta(\alpha) = \mathbf{N}(\alpha)$.

9.2.3. Proposition. *$\mathbb{Z}[\varpi]$ is a Euclidean ring with this definition of δ.*

Proof. We have already noted that $\mathbb{Z}[\varpi]$ is an integral domain. If $\alpha = a + b\varpi$, where $a, b \in \mathbb{Z}$, then we know that $\mathbf{N}(\alpha) = a^2 - ab + b^2$ and also that $\mathbf{N}(\alpha\beta) = \mathbf{N}(\alpha)\mathbf{N}(\beta)$ so (**E 1**) follows.

Now let $\alpha = a + b\varpi$ and $\beta = c + d\varpi \neq 0$ where $a, b, c, d \in \mathbb{Z}$. Since $\beta\mathbf{\Delta}(\beta) = \mathbf{N}(\beta) \in \mathbb{N}$ and $\alpha\mathbf{\Delta}(\beta) \in \mathbb{Z}[\varpi]$ it follows that

$$\frac{\alpha\mathbf{\Delta}(\beta)}{\beta\mathbf{\Delta}(\beta)} = \frac{\alpha}{\beta} = x + y\varpi,$$

where $x, y \in \mathbb{Q}$. There are integers u, v such that $|x - u| \leq \frac{1}{2}$ and $|y - v| \leq \frac{1}{2}$. Put $\gamma = u + v\varpi$ and $\rho = \alpha - \beta\gamma$. By definition, $\gamma, \rho \in \mathbb{Z}[\varpi]$ and either $\rho = 0$ or

$$\begin{aligned}
\mathbf{N}(\rho) &= \mathbf{N}(\alpha - \beta\gamma) = \mathbf{N}\beta(\alpha/\beta - \gamma) = \mathbf{N}(\beta)\mathbf{N}\left(\frac{\alpha}{\beta} - \gamma\right) \\
&= \mathbf{N}(\beta)\mathbf{N}((x - u) + (y - v)\varpi) \\
&= \mathbf{N}(\beta)((x - u)^2 - (x - u)(y - v) + (y - v)^2) \\
&\leq \mathbf{N}(\beta)\left(\frac{1}{4} + \frac{1}{4} + \frac{1}{4}\right) = \frac{3}{4}\mathbf{N}(\beta) < \mathbf{N}(\beta).
\end{aligned}$$

Thus, (**E2**) is also valid.

In the book (LeVeque, 1956), the reader can find the proof of the following interesting theorem: The ring $\mathbb{Z}[\sqrt{r}]$ is Euclidean if and only if r is one of the numbers from the set

$$\{-11, -7, -3, -2, -1, 2, 3, 5, 6, 7, 11, 13, 17, 19, 21, 29, 33, 37, 41, 57, 73, 97\}.$$

[LeVeque WJ. Topics in Number Theory. Vol. 2. Addison-Wesley: Reading, MA, 1956.]

The following property of Euclidean rings is fundamental.

9.2.4. Theorem. *Every Euclidean ring is a principal ideal domain.*

Proof. Let R be a Euclidean ring and let H be an ideal of R. If H is the zero ideal, then H is generated by the zero element, so that H is principal. Therefore, we need to consider only the case when H is nonzero. Let

$$\Delta(H) = \{\delta(x) \mid 0_R \neq x \in H\}.$$

Since $\Delta(H)$ is a subset of $\mathbb{N}$, $\Delta(H)$ has a least element m. In H we choose an element y with the property that $\delta(y) = m$. Let x be an arbitrary element of H. By (**E 2**), there exist elements w, z such that $x = wy + z$, where either $z = 0_R$ or $\delta(z) < \delta(y)$. Since H is an ideal, $wy \in H$ and hence $z = x - wy \in H$. If we suppose that $z \neq 0_R$ then $\delta(z) < \delta(y)$, which contradicts the choice of y. Thus, $z = 0_R$ and therefore $x = wy$. Hence $H \leq yR$ and since $yR \leq H$ in any case, we have $yR = H$. The result follows.

By Theorem 9.1.20 we see that Euclidean rings are also unique factorization domains.

9.2.5. Corollary. *Let R be a Euclidean ring and suppose that $0_R \neq a \in R$.*

(i) *If $a \notin \mathbf{U}(R)$, then a can be written as a product of prime elements of the ring R.*
(ii) *If $a \notin \mathbf{U}(R)$ and $a = p_1 \dots p_n = q_1 \dots q_m$ are two decompositions of a into products of prime elements of the ring R, then $n = m$ and we can renumber the elements of the second decomposition such that $q_j = u_j p_j$ for some $u_j \in \mathbf{U}(R)$, where $1 \leq j \leq n$.*

The next assertion follows from Theorems 9.1.11 and 9.2.4

9.2.6. Corollary. *Let R be a Euclidean ring. Every pair of elements of R has a greatest common divisor and a least common multiple in R.*

For any two elements a, b of a PID, we proved the existence of $\mathbf{GCD}(a, b)$ and $\mathbf{LCM}(a, b)$. A practical algorithm for finding $\mathbf{GCD}(a, b)$ is not readily available in PIDs in general, because it is generally not easy to factor. However, for Euclidean rings, the Euclidean algorithm that follows represents one technique for finding such greatest common divisors.

Let R be a Euclidean ring and let a, b be arbitrary elements of R. If $a = 0_R$, then $\mathbf{GCD}(a, b) = b$. Therefore, we can assume that a, b are both nonzero. We divide a by b to obtain $a = bq_1 + r_1$ where either $r_1 = 0_R$ or $\delta(r_1) < \delta(b)$ and $q_1, r_1 \in R$. Next, if $r_1 \neq 0_R$ we divide b by r_1 to obtain $b = r_1 q_2 + r_2$ where either $r_2 = 0_R$ or $\delta(r_2) < \delta(r_1)$. If $r_2 \neq 0_R$, then we divide r_1 by r_2 to obtain $r_1 = r_2 q_3 + r_3$ where either $r_3 = 0_R$ or $\delta(r_3) < \delta(r_2)$. Continuing this process, in the general case, if $r_j \neq 0_R$, we divide r_{j-1} by r_j to obtain a quotient q_{j+1} and remainder r_{j+1}. Since we have $\delta(r_{j+1}) < \delta(r_j)$ and, since $\delta(r_j)$ is a natural number, this process must terminate in finitely many steps. This means that at some stage the corresponding remainder r_{k+1} is the zero element. Thus, we have the following chain of equations.

$$
\begin{aligned}
a &= bq_1 + r_1, \\
b &= r_1 q_2 + r_2, \\
r_1 &= r_2 q_3 + r_3, \\
&\vdots \\
r_{k-3} &= r_{k-2} q_{k-1} + r_{k-1}, \\
r_{k-2} &= r_{k-1} q_k + r_k, \\
r_{k-1} &= r_k q_{k+1}.
\end{aligned}
\tag{9.2}
$$

We claim that $r_k = \mathbf{GCD}(a, b)$. To see this note that

$$r_{k-2} = r_{k-1}q_k + r_k = r_k q_{k+1} q_k + r_k = r_k(q_{k+1}q_k + e),$$

so that r_k divides r_{k-2}. Furthermore,

$$\begin{aligned} r_{k-3} &= r_{k-2}q_{k-1} + r_{k-1} = r_k(q_{k+1}q_k + e)q_{k-1} + r_k q_{k+1} \\ &= r_k(q_{k+1}q_k q_{k-1} + q_{k-1} + q_{k+1}), \end{aligned}$$

so that r_k divides r_{k-3}. Continuing in this manner, working up Equation 9.2 and employing a suitable induction argument if necessary, we see that r_k divides both a and b. Thus, r_k is a common divisor of a and b.

Next, let u be an arbitrary common divisor of a and b. From the equation $r_1 = a - bq_1$, we see that u divides r_1. From the equation $r_2 = b - r_1 q_2$, u divides r_2 and continuing in this way we obtain, finally, that r_k is divisible by u. It follows that r_k is the greatest common divisor of a and b. Having obtained $\mathbf{GCD}(a, b)$, we can find $\mathbf{LCM}(a, b)$ using Theorem 9.1.16.

From Corollary 9.1.12 it follows that, for the element $d = \mathbf{GCD}(a, b)$, there are elements x, y of the Euclidean ring R such that $d = ax + by$. The Euclidean algorithm also helps us to find these elements x, y. Indeed from the chain (Eq. 9.2), we have

$$d = r_k = r_{k-2} - r_{k-1}q_k \text{ and}$$
$$r_{k-1} = r_{k-3} - r_{k-2}q_{k-1},$$

so that

$$\begin{aligned} d &= r_{k-2} - (r_{k-3} - r_{k-2}q_{k-1})q_k = r_{k-2} - r_{k-3}q_k + r_{k-2}q_{k-1}q_k \\ &= r_{k-2}(e + q_{k-1}q_k) - r_{k-3}q_k = r_{k-2}y_1 - r_{k-3}x_1. \end{aligned}$$

Continuing in this way, going up the chain (Eq. 9.2) we finally obtain $d = ax + by$.

EXERCISE SET 9.2

Justify your work.

9.2.1. Find $\mathbf{GCD}(-1 + i, 2 - i)$, $\mathbf{LCM}(-1 + i, 2 - i)$.

9.2.2. Find $\mathbf{GCD}(-3 - i, 2 + 7i)$, $\mathbf{LCM}(-3 - i, 2 + 7i)$.

9.2.3. Find $\mathbf{GCD}(5 - \varpi, 7 + 2\varpi)$, $\mathbf{LCM}(5 - \varpi, 7 + 2\varpi)$.

9.2.4. Find $\mathbf{GCD}(7 + i, 3 + 5i)$, $\mathbf{LCM}(7 + i, 3 + 5i)$.

9.2.5. Find **GCD**$(5 + 9\varpi, 15 + 8\varpi)$, **LCM**$(5+9\varpi, 15+8\varpi)$, if $\varpi = \frac{-1+\sqrt{-3}}{2}$.

9.2.6. Find **GCD**$(2 + i, 3 - i)$, **LCM**$(2 + i, 3 - i)$.

9.2.7. Find **GCD**$(3 + 2\varpi, 1 - 3\varpi)$, **LCM**$(3 + 2\varpi, 1 - 3\varpi)$, if $\varpi = \frac{-1+\sqrt{-3}}{2}$.

9.2.8. Let $f(X)=X^4 + 5X^2 + 6, g(X)=4X^3 + 3 \in \mathbb{F}_7[X]$ where $\mathbb{F}_7 = \mathbb{Z}/7\mathbb{Z}$ and n denotes $n + 7\mathbb{Z}$. Find a polynomial that generates the ideal $f(X)\mathbb{F}_7[X] \cap g(X)\mathbb{F}_7[X]$.

9.2.9. Let $f(X) = X^3 + 4X^2 + 3, g(X) = 3X^3 + 2X + 4 \in \mathbb{F}_5[X]$ where $\mathbb{F}_5 = \mathbb{Z}/5\mathbb{Z},\ n = n + 5\mathbb{Z}$. Find a polynomial that generates the ideal $f(X)\mathbb{F}_5[X] + g(X)\mathbb{F}_5[X]$.

9.2.10. If $X - d$ divides $f(X) = a_0 + a_1X + \cdots + a_nX^n$, then prove that d divides a_0.

9.2.11. Find the reminder when $f(X)$ is divided by $(X - a)(X - b)$.

9.2.12. Without using the Euclidian algorithm find **GCD**$(f(X), g(X))$ where $f(X) = X^3 + 3X^2 - 2, g(X) = X^3 + 3X^2 - X - 3$.

9.2.13. Prove that the polynomials $f(X) = a_0 + a_1X + a_2X^2 + \cdots + a_nX^n$ and $g(X) = a_1X + a_2X^2 + \cdots + a_nX^n$ are relatively prime if $a_0 \neq 0$.

9.2.14. Use the Euclidean algorithm to find **GCD**$(f(X), g(X))$, if $f(X), g(X) \in \mathbb{Q}[X]$ where $f(X) = X^3 + X^2 - 4X - 6, g(X) = X^3 + X^2 - 10X - 6$.

9.2.15. Use the Euclidean algorithm to find **GCD**$(f(X), g(X))$, if $f(X), g(X) \in \mathbb{Q}[X]$ where $f(X) = 3X^4 - 3X^3 + 4X^2 - X + 1, g(X) = 2X^3 - X^2 + X + 1$.

9.2.16. Use the Euclidean algorithm to find **GCD**$(f(X), g(X))$, if $f(X), g(X) \in \mathbb{C}[X]$ where $f(X) = X^4 + 2iX^3 - 2X^2 - 2iX + 1, g(X) = X^3 + (i + 1)X^2 + iX$.

9.2.17. Find **LCM**$(f(X), g(X))$, if $f(X), g(X) \in \mathbb{Q}[X]$ where $f(X) = X^4 - 4X^3 + 4X^2 - 5X - 2, g(X) = X^2 - X - 2$.

9.2.18. Find **LCM**$(f(X), g(X))$, if $f(X), g(X) \in \mathbb{Q}[X]$ where $f(X) = 2X^3 + 7X^2 + 4X - 3,\ g(X) = X^3 + X^2 - 3X + 1$.

9.2.19. Find **LCM**$(f(X), g(X))$, if $f(X), g(X) \in \mathbb{C}[X]$ where $f(X) = X^4 + 2iX^3 - 2X^2 - 2iX + 1, g(X) = X^3 + (i + 1)X^2 + iX$.

9.2.20. Let $f(X), g(X) \in \mathbb{Q}[X]$ where $f(X) = X^3 + 5X^2 + 6X + 2,\ g(X) = X^2 + 6X + 5$. Find the polynomials $u(X), v(X) \in \mathbb{Q}[X]$ such that **GCD**$(f(X), g(X)) = u(X)f(X) + v(X)g(X)$.

9.3 IRREDUCIBLE POLYNOMIALS

As we mentioned in the previous section, the polynomial ring $F[X]$, over a field F, is Euclidean. From Corollary 9.2.6 we obtain the following.

9.3.1. Proposition. *Let F be a field. Then all pairs of polynomials $f(X), g(X) \in F[X]$ have a greatest common divisor and a least common multiple belonging to $F[X]$.*

9.3.2. Definition. *Let R be an integral domain. A polynomial $f(X)$ of degree at least 1 with coefficients belonging to R is called irreducible or indecomposable over R, if it is not possible to represent this polynomial as a product $f(X) = u(X)v(X)$ of two polynomials $u(X), v(X) \in R[X]$, satisfying the conditions $0 < \mathbf{deg}\, u(X) < \mathbf{deg}\, f(X)$ and $0 < \mathbf{deg}\, v(X) < \mathbf{deg}\, f(X)$. A polynomial that is not irreducible is called reducible. Thus, a reducible polynomial can be factored as a product of two other polynomials of smaller degree, but of degree at least 1.*

It follows from this definition that every polynomial of first degree is irreducible. We need to point out that irreducibility is determined relative to the ring under consideration. Thus, for example, the polynomial $X^2 - 2$ is irreducible over the field $\mathbb{Q}$, while over the field $\mathbb{R}$ it can be represented as a product of two polynomials of first degree, namely, $X - \sqrt{2}$ and $X + \sqrt{2}$. Likewise, the polynomial $X^2 + 1$ is irreducible over $\mathbb{R}$, whereas over $\mathbb{C}$ it is reducible into the form $(X + i)(X - i)$. The polynomial $X^4 + 4$ is reducible over $\mathbb{Q} : X^4 + 4 = (X^2 + 2X + 2)(X^2 - 2X + 2)$, while both its factors are irreducible not only over $\mathbb{Q}$ but also over $\mathbb{R}$.

In Section 7.5 we proved that $\mathbf{U}(F[X]) = \mathbf{U}(F)$. Thus, a polynomial of degree at least 1 in $F[X]$ is never invertible in $F[X]$. Hence, the irreducible polynomials of $F[X]$ are precisely the prime elements of $F[X]$. From Theorem 9.1.28 we know that the polynomial ring $F[X]$, over a field F, is a unique factorization domain and hence every polynomial of degree at least 1 with coefficients in F can be written as a product of irreducible polynomials with coefficients in F. Since this is such an important result, we have decided to write the polynomial version of Theorem 9.1.20 next. For $F[X]$ we usually take the set $\sigma(F[X])$, which consists of a set of representatives of the various associate classes, to be a set of monic polynomials, where a polynomial is monic if its leading coefficient is equal to e, the multiplicative identity of F.

9.3.3. Proposition. *Let F be a field and let $f(X)$ be a polynomial of degree at least 1. Then*

$$f(X) = a(p_1(X))^{k_1} \ldots (p_m(X))^{k_m},$$

where a is the leading coefficient of the polynomial $f(X)$, and $p_1(X), \ldots, p_m(X)$ are distinct irreducible polynomials whose leading coefficients belong to $\mathbf{U}(F)$.

This representation is unique, up to the order that the polynomials are written in and to within multiplication by elements of $\mathbf{U}(F)$.

The following property is also important and its proof is reminiscent of the proof of the theorem that $\mathbb{Z}$ has infinitely many primes.

9.3.4. Theorem. *Let F be a field. The set of associate classes of irreducible polynomials in $F[X]$ is infinite.*

Proof. If the field F is infinite, the polynomials of first degree of the form $X - a$, are irreducible, and are not associates, as a is allowed to vary over F. The theorem therefore follows in this case. Thus, suppose that F is finite. Assume that we have already found m distinct irreducible polynomials $p_1(X), \dots, p_m(X)$. We may assume that each of these is monic. Let $q(X) = p_1(X) \dots p_m(X) + e$. By Proposition 9.3.3, $q(X)$ has an irreducible monic divisor $u(X)$. Assume that $u(X)$ coincides with one of the polynomials above, so $u(X) = p_j(X)$, say. The equation $e = q(X) - p_1(X) \dots p_m(X)$ and Corollary 9.1.13 imply that $q(X)$ is relatively prime to $p_1(X), \dots, p_m(X)$. Thus, $u(X)$ is relatively prime to each polynomial from the set $\{p_1(X), \dots, p_m(X)\}$. Hence, given a finite set of irreducible polynomials, there is always an irreducible polynomial that is relatively prime with each of the selected polynomials. This means that the set of irreducible monic polynomials of the ring $F[X]$ is infinite.

One interesting fact that arises from the proof of the previous result is the following corollary.

9.3.5. Corollary. *Let F be a finite field. The degrees of the irreducible polynomials of the ring $F[X]$ are unbounded.*

We show that Corollary 9.3.5 is also true in the ring $\mathbb{Q}[X]$. For this we will consider irreducible polynomials with rational coefficients.

First observe that if $f(X) \in \mathbb{Q}[X]$ has some noninteger coefficients, then multiplying $f(X)$ by the least common multiple $\mathbf{s}$ of the denominators of all the coefficients of $f(X)$, we obtain a polynomial $\mathbf{s}f(X)$, with all integer coefficients. It is clear that $f(X)$ and $\mathbf{s}f(X)$ have the same roots and we note that, in addition, if one of them is irreducible over $\mathbb{Q}$ then the second is also irreducible over this field. On the other hand, if $f(X)$ is a polynomial with integer coefficients that is irreducible over the ring $\mathbb{Z}$, then it is irreducible over the field $\mathbb{Q}$. However, the following is also true.

9.3.6. Theorem. *A polynomial $f(X) \in \mathbb{Z}[X]$ is irreducible over the ring of integers $\mathbb{Z}$ if and only if it is irreducible over the field $\mathbb{Q}$ of all rational numbers.*

Proof. Obviously, if $f(X)$ is irreducible over $\mathbb{Q}$, then it is irreducible over $\mathbb{Z}$. Conversely, suppose that $f(X)$ is irreducible over $\mathbb{Z}$, but not over $\mathbb{Q}$, so that $f(X) = f_1(X)f_2(X)$, where $f_1(X), f_2(X) \in \mathbb{Q}[X]$ and $0 < \mathbf{deg}\, f_1(X), \mathbf{deg}\, f_2(X) < \mathbf{deg}\, f(X)$. Multiplying both sides by the least common multiple of the denominators of the coefficients of $f_1(X)$ and $f_2(X)$, we have $af(X) = f_3(X)f_4(X)$ where $f_3(X), f_4(X) \in \mathbb{Z}[X]$ and $\mathbf{deg}\, f_1(X) = \mathbf{deg}\, f_3(X)$, $\mathbf{deg}\, f_2(X) = \mathbf{deg}\, f_4(X)$. Furthermore, $af(X) = bf_5(X)f_6(X)$, where b is the content of $f_3(X)\, f_4(X)$ and $f_5(X), f_6(X) \in \mathbb{Z}[X]$, are primitive with $\mathbf{deg}\, f_5(X) = \mathbf{deg}\, f_3(X)$ and $\mathbf{deg}\, f_6(X) = \mathbf{deg}\, f_4(X)$. By Lemma 9.1.26, $f_5(X)$ divides $f(X)$. In addition, $\mathbf{deg}\, f_1(X) = \mathbf{deg}\, f_5(X)$, and $0 < \mathbf{deg}\, f_5(X) < \mathbf{deg}\, f(X)$. Thus, $f(X)$ has the factor $f_5(X)$, which contradicts the fact that $f(X)$ is irreducible over $\mathbb{Z}[X]$. Consequently, $f(X)$ is irreducible over $\mathbb{Q}$.

Thus, all questions regarding the irreducibility of polynomials over $\mathbb{Q}$ become questions as to whether they are irreducible over $\mathbb{Z}$. The question regarding the irreducibility of a given polynomial $f(X) \in \mathbb{Z}[X]$ over $\mathbb{Q}$ is quite complicated. There is a method, due to Kronecker, which allows us to determine if $f(X)$ is irreducible over $\mathbb{Q}$, but this method is rather cumbersome and quite limited. However, there are many results giving sufficient conditions for irreducibility over $\mathbb{Q}$, one standard one being the following, usually called Eisenstein's criterion.

9.3.7. Theorem. *Let $f(X) = a_0 + a_1X + \cdots + a_nX^n \in \mathbb{Z}[X]$ and suppose that there exists a prime p such that the following conditions hold:*

(i) *the leading coefficient a_n is not divisible by p;*
(ii) *all other coefficients of $f(X)$ are divisible by p;*
(iii) *a_0 is not divisible by p^2.*

Then the polynomial $f(X)$ is irreducible over $\mathbb{Q}$.

Proof. Assume, for a contradiction, that $f(X)$ is reducible over $\mathbb{Q}$. Theorem 9.3.6 shows that then $f(X)$ is reducible over $\mathbb{Z}$, so $f(X) = f_1(X)f_2(X)$ where $f_1(X), f_2(X) \in \mathbb{Z}[X]$ and $0 < \mathbf{deg}\, f_1(X), \mathbf{deg}\, f_2(X) < \mathbf{deg}\, f(X)$. Let $f_1(X) = b_0 + b_1X + \cdots + b_kX^k$ and $f_2(X) = c_0 + c_1X + \cdots + c_tX^t$, where $b_i, c_i \in \mathbb{Z}$.

We have $a_0 = b_0c_0$. By conditions (ii), (iii), either p divides b_0 and p does not divide c_0, or p divides c_0 and p does not divide b_0. Without loss of generality, assume that p divides i b_0. Now, upon multiplying and equating coefficients we see that, for each l,

$$a_l = b_lc_0 + b_{l-1}c_1 + \cdots + b_0c_l,$$

where we set $b_r = 0$ if $r > k$ and $c_s = 0$ if $s > t$. Since p does not divide a_n and $a_n = b_kc_t$, it follows that p does not divide b_k either. Hence, there is a least integer r such that p divides b_r but p does not divide b_{r+1}. However,

$$a_{r+1} = b_{r+1}c_0 + b_rc_1 + \cdots + b_0c_{r+1}.$$

Since p divides $a_{r+1} - (b_r c_1 + \cdots + b_0 c_{r+1})$ we have that p divides $b_{r+1}c_0$ and, since p does not divide b_{r+1} it follows that p divides c_0. We conclude that p^2 divides $b_0 c_0 = a_0$, contrary to hypothesis (iii). Thus, $f(X)$ is irreducible.

Since, for all natural numbers n and all primes p, $X^n + p$ is irreducible by Eisenstein's criterion we have the following.

9.3.8. Corollary. *The degrees of the monic irreducible polynomials over $\mathbb{Q}$ are unbounded.*

By contrast, a classical theorem of Gauss shows that the field $\mathbb{C}$ of complex numbers is algebraically closed, which means that every irreducible polynomial in $\mathbb{C}[X]$ has degree 1 and one consequence is that every irreducible polynomial in $\mathbb{R}[X]$ has degree at most 2.

9.3.9. Corollary. *Let p be a prime number. The polynomial*

$$f_p(X) = 1 + X + \cdots + X^{p-1} \in \mathbb{Z}[X]$$

is irreducible over $\mathbb{Q}$.

Proof. We note that $X^p - 1 = (X^{p-1} + X^{p-2} + \cdots + X + 1)(X - 1)$ so the roots of $X^{p-1} + \cdots + X + 1$ are complex pth roots of unity. Together with 1, these roots lie on the unit circle in the complex plane, which is the reason why the polynomial $f_p(X)$ is sometimes called a cyclotomic polynomial.

Consider $g(X) = f_p(X + 1)$. If we suppose that

$$f_p(X) = g_1(X)g_2(X), \text{ where } g_1(X), g_2(X) \in \mathbb{Z}[X]$$

and

$$0 < \mathbf{deg}\, g_1(X), \mathbf{deg}\, g_2(X) < \mathbf{deg}\, f_p(X)$$

then

$$g(X) = f_p(X + 1) = g_1(X + 1)g_2(X + 1) = g_3(X)g_4(X)$$

with

$$g_3(X), g_4(X) \in \mathbb{Z}[X] \text{ and } 0 < \mathbf{deg}\, g_3(X), \mathbf{deg}\, g_4(X) < \mathbf{deg}\, g(X),$$

Hence, $f_p(X)$ is irreducible if and only if $g(X)$ is irreducible. We have

$$\begin{aligned} g(X) = f_p(X + 1) &= \frac{(X + 1)^p - 1}{X + 1 - 1} \\ &= X^{p-1} + C_1^p X^{p-2} + \cdots + C_{p-2}^p X + C_{p-1}^p, \end{aligned}$$

where $C_k^p = \frac{p!}{k!(p-k)!}$ are binomial coefficients. Now

$$p! = k!(p-k)!C_k^p.$$

However, p divides $p!$ but not $k!(p-k)!$ (since the factors, which are all less than p, are relatively prime to p); so, C_k^p is divisible by p. The last coefficient is $C_{p-1}^p = p$ and hence $g(X)$ satisfies the conditions of Theorem 9.3.7. Hence $g(X)$ and, therefore, $f_p(X)$ is irreducible over $\mathbb{Q}$.

By Proposition 9.3.3, every polynomial decomposes into a product of powers of irreducible polynomials. As we noted already, there are no truly satisfactory ways of determining the irreducibility of polynomials with rational coefficients. The question of obtaining a decomposition of a polynomial into a product of powers of irreducible polynomials is not easy. One technique of interest involves the concept of the derivative of a polynomial.

9.3.10. Definition. *Let R be a commutative ring and let $f(X) \in R[X]$. If*

$$f(X) = a_0 + a_1X + \cdots + a_nX^n,$$

then the polynomial

$$f'(X) = a_1 + 2a_2X + \cdots + na_nX^{n-1}$$

is called the derivative of the polynomial $f(X)$.

Note that this is a purely formal definition. We are not taking limits of any kind. As in calculus, there is a product rule and a chain rule, among other derivative rules.

9.3.11. Lemma. *Let R be a commutative ring and let $f(X), g(X)$ be arbitrary polynomials over R. Then*

(i) $(f(X) + g(X))' = f'(X) + g'(X)$;
(ii) $(f(X)g(X))' = f'(X)g(X) + f(X)g'(X)$.

Proof. Let

$$f(X) = a_0 + a_1X + \cdots + a_nX^n \text{ and } g(X) = b_0 + b_1X + \cdots + b_kX^k.$$

To prove (i) we may assume that $n = k$, by making some coefficients 0_R, if necessary. We have

$$\begin{aligned}(f(X) + g(X))' &= \big((a_0 + b_0) + (a_1 + b_1)X + \cdots + (a_n + b_n)X^n\big)' \\ &= (a_1 + b_1) + 2(a_2 + b_2)X + \cdots + n(a_n + b_n)X^{n-1} \\ &= a_1 + 2a_2X + \cdots + na_nX^{n-1} + b_1 + 2b_2X + \cdots + nb_nX^{n-1} \\ &= f'(X) + g'(X).\end{aligned}$$

To prove (ii) we use induction on k, the degree of $g(X)$. If $g(X) = 0_R$ then (ii) is clear. If $k = 0$ then

$$f(X)g(X) = b_0a_0 + b_0a_1X + \cdots + b_0a_nX^n,$$

and

$$\begin{aligned}(f(X)g(X))' &= b_0a_1 + 2b_0a_2X + \cdots + nb_0a_nX^{n-1} = f'(X)b_0 \\ &= f'(X)g(X) + f(X)g'(X).\end{aligned}$$

Suppose now that $k > 0$ and assume that the lemma is true for all polynomials with degree less than k. Let

$$h(X) = b_0 + b_1X + \cdots + b_{k-1}X^{k-1}.$$

Then $g(X) = h(X) + b_kX^k$ and, therefore,

$$f(X)g(X) = f(X)(h(X) + b_kX^k) = f(X)h(X) + f(X)b_kX^k.$$

This equation and (i) imply that

$$(f(X)g(X))' = (f(X)h(X) + f(X)b_kX^k)' = (f(X)h(X))' + (f(X)b_kX^k)'.$$

By the induction hypothesis, $(f(X)h(X))' = f'(X)h(X) + f(X)h'(X)$. Furthermore, we have

$$f(X)b_kX^k = b_ka_0X^k + b_ka_1X^{k+1} + \cdots + b_ka_nX^{k+n},$$

so

$$(f(X)b_kX^k)' = (b_ka_0X^k)' + (b_ka_1X^{k+1})' + \cdots + (b_ka_nX^{k+n})'.$$

The derivative of an arbitrary member of the last sum is

$$\begin{aligned}(b_ka_jX^{k+j})' &= (k + j)b_ka_jX^{k+j-1} = kb_ka_jX^{k+j-1} + jb_ka_jX^{k+j-1} \\ &= (kb_kX^{k-1})a_jX^j + b_kX^k(ja_jX^{j-1}).\end{aligned}$$

Hence

$$\begin{aligned}(f(X)b_kX^k)' &= (kb_kX^{k-1})a_0 + (k+1)b_ka_1X^k + \cdots + (k+n)b_ka_nX^{k+n-1}\\ &= (a_1 + 2a_2X + \cdots + na_nX^{n-1})b_kX^k + (a_0 + a_1X + \cdots\\ &\quad + a_nX^n)(kb_kX^{k-1})\\ &= f'(X)b_kX^k + f(X)(b_kX^k)'.\end{aligned}$$

Finally,

$$\begin{aligned}(f(X)g(X))' &= (f(X)h(X))' + (f(X)b_kX^k)'\\ &= f'(X)h(X) + f(X)h'(X) + f'(X)b_kX^k + f(X)(b_kX^k)'\\ &= f'(X)(h(X) + b_kX^k) + f(X)(h'(X) + (b_kX^k)')\\ &= f'(X)(h(X) + b_kX^k) + f(X)(h(X) + b_kX^k)'\\ &= f'(X)g(X) + f(X)g'(X).\end{aligned}$$

A further induction allows us to deduce the "chain rule."

9.3.12. Corollary. *Let R be a commutative ring and let $f_1(X), \ldots, f_n(X) \in R[X]$. Then*

$$(f_1(X)\ldots f_n(X))' = \sum_{1\le j\le n} f_1(X)\ldots f_{j-1}(X)f_j'(X)f_{j+1}(X)\ldots f_n(X)$$

and, in particular,

$$(f(X)^n)' = n(f(X)^{n-1})f'(X).$$

9.3.13. Definition. *Let R be an integral domain and let $f(X) \in R[X]$. Suppose that $f(X)$ is divisible by the irreducible polynomial $p(X)$. Then $p(X)$ has multiplicity m in $f(X)$ if $p(X)^m$ divides $f(X)$ but $p(X)^{m+1}$ does not divide $f(X)$. In particular if, for some $a \in F$, $p(X) = X - a$ has multiplicity m then a is said to be a root of $f(X)$ of multiplicity m.*

9.3.14. Proposition. *Let F be a field of characteristic zero and let $f(X) \in F[X]$. If $p(X)$ is an irreducible divisor of $f(X)$ with multiplicity m then $p(X)$ is a divisor of $f'(X)$ with multiplicity $m - 1$.*

Proof. We have $f(X) = (p(X))^m u(X)$ where $p(X)$ does not divide $u(X)$. By Lemma 9.1.19, $u(X)$ and $p(X)$ are relatively prime. Corollary 9.3.12 shows that $f'(X) = (p(X))^{m-1}(mp'(X)u(X) + p(X)u'(X))$. If we suppose that $p(X)$ divides $(mp'(X)u(X) + p(X)u'(X))$, then $p(X)$ must divide $mp'(X)u(X)$. Since **char** $F = 0$, then **deg** $p'(X) =$ **deg** $p(X) - 1$, so $p'(X)$ is a nonzero polynomial and this also shows that $p(X)$ does not divide $p'(X)$. Then, by Lemma 9.1.19,

$p'(X)$ and $p(X)$ are relatively prime. Using Proposition 9.1.15 we see that $p(X)$ and $mp'(X)u(X)$ are relatively prime, which proves that $p(X)$ divides $f'(X)$ with multiplicity $m - 1$.

We now give some applications of these results. Let F be a field of characteristic zero and suppose that $f(X) \in F[X]$. By Proposition 9.3.3,

$$f(X) = a(p_1(X))^{k_1} \dots (p_m(X))^{k_m},$$

where a is the leading coefficient of the polynomial $f(X)$, and $p_1(X), \dots, p_m(X)$ are distinct monic irreducible polynomials. By Proposition 9.3.14 and Corollary 9.1.21, we have

$$g(X) = \mathbf{GCD}(f(X), f'(X)) = (p_1(X))^{k_1 - 1} \dots (p_m(X))^{k_m - 1}.$$

We can apply the Euclidean algorithm to find the polynomial $g(X)$. Furthermore, $f(X) = g(X)w(X)$ where, clearly, $w(X) = ap_1(X) \dots p_m(X)$. Thus, dividing the polynomial $f(X)$ by $g(X)$, we obtain the polynomial $w(X)$, whose decomposition includes all the irreducible factors of the polynomial $f(X)$ but with multiplicity 1. This can sometimes be used as a tool for finding the decomposition of $f(X)$, since $w(X)$ will typically have somewhat smaller degree than $f(X)$. Thus, essentially, we need to find only the decomposition of $w(X)$.

Note that for fields of characteristic p, for the prime p, Proposition 9.3.14 does not hold. For example, suppose that the field F has prime characteristic $p > 0$ and that $f(X) = X^p$. Then

$$f'(X) = pX^{p-1} = (pe)X^{p-1} = 0_F.$$

In other words, a polynomial of degree larger than 1 could have zero derivative. However, in the general case, the following theorem holds.

9.3.15. Theorem. *Let F be a field and let $f(X) \in F[X]$. An element $c \in F$ is a multiple root of $f(X)$ if and only if $f(c) = 0_F$ and $f'(c) = 0_F$.*

Proof. Let b be an arbitrary element of F. By Theorem 7.5.2, $f(X) = (X - b)^2 q(X) + r(X)$ where either $r(X) = 0_F$ or $\mathbf{deg}\, r(X) < 2$. In particular, the polynomial $r(X)$ is of the form $r(X) = d(X - b) + w$ for some elements $d, w \in F$. We note that $r(b) = f(b) = w$. Hence

$$f(X) = (X - b)^2 q(X) + d(X - b) + w.$$

Using Lemma 9.3.11 and Corollary 9.3.12, we obtain

$$f'(X) = (X - b)(2q(X) + (X - b)q'(X)) + d,$$

and we note that $d = f'(b)$. So, finally,

$$f(X) = (X - b)^2 q(X) + f'(b)(X - b) + f(b)$$

and

$$f'(X) = (X - b)(2q(X) + (X - b)q'(X)) + f'(b).$$

If c is a multiple root of $f(X)$, then $(X - c)^2$ divides $f(X)$. It follows that $f'(c) = f(c) = 0_F$. Conversely, if $f'(c) = f(c) = 0_F$, then the equations above show that $(X - c)^2$ divides $f(X)$.

To conclude this section, we turn again to polynomials with rational coefficients and clarify the question of finding the rational roots, using a process known as the rational root test.

As we mentioned above we can always multiply a polynomial $f(X) \in \mathbb{Q}[X]$ by the least common multiple **s** of the denominators of the coefficients. In this way, we obtain the polynomial $\mathbf{s}f(X)$ with all integer coefficients and clearly the polynomials $f(X)$ and $\mathbf{s}f(X)$ have the same roots. So we need to explore the question of finding rational roots of a polynomial with integer coefficients. Let

$$f(X) = a_0 + a_1 X + \cdots + a_n X^n \in \mathbb{Z}[X],$$

and let c be an integer root of the polynomial $f(X)$. By Proposition 7.5.9,

$$f(X) = (X - c)q(X) \text{where } q(X) = b_0 + b_1 X + \cdots + b_{n-1} X^{n-1}.$$

Thus,

$$a_n = b_{n-1}, a_{n-1} = b_{n-2} - cb_{n-1}, \ldots, a_1 = b_0 - cb_1, a_0 = (-c)b_0.$$

These equations show that all the coefficients $b_0, b_1, \ldots, b_{n-1}$ are integers and that c is a divisor of a_0. Hence, if the polynomial $f(X)$ has an integer root then the possibilities for such roots must be the integer divisors of a_0, where both negative and positive divisors are used. We then need to check which of these is a root by evaluating $f(X)$ at the proposed root. If the evaluation gives 0_F then we have located a root. In order to simplify this procedure, we can proceed as follows.

Of course, 1 and -1 are divisors of a_0. Thus, we always need to find $f(1)$ and $f(-1)$. So

$$f(1) = (1 - c)q(1) \text{ and } f(-1) = (-1 - c)q(-1).$$

Since the coefficients of $q(X)$ are integers, $q(1)$ and $q(-1)$ are also integers. Thus, we need to check only such divisors d of a_0 for which $\frac{f(1)}{1-d}$ and $\frac{f(-1)}{1+d}$ are integers.

Now we can apply the following

9.3.16. Theorem. *Let $f(X) = a_0 + a_1X + \cdots + a_{n-1}X^{n-1} + X^n \in \mathbb{Z}[X]$. If c is a rational root of $f(X)$, then c is an integer.*

Proof. Suppose, for a contradiction, that c is not an integer. Then $c = \frac{m}{k}$ and we can assume that the integers m and k are relatively prime. We have

$$0 = a_0 + a_1\left(\frac{m}{k}\right) + \cdots + a_{n-1}\left(\frac{m}{k}\right)^{n-1} + \left(\frac{m}{k}\right)^n,$$

and therefore,

$$\frac{m^n}{k} = -a_0k^{n-1} - a_1mk^{n-2} - \cdots - a_{n-1}m^{n-1}.$$

Since m and k are relatively prime, m^n and k are also relatively prime. This means that on the left-hand side of the last equation we have a rational noninteger, while on the right-hand side we have an integer, which is the contradiction sought.

Now let $f(X) = a_0 + a_1X + \cdots + a_nX^n \in \mathbb{Z}[X]$. Consider the polynomial $g(X) = a_n^{n-1}f(X)$, so

$$g(X) = a_n^{n-1}a_0 + a_n^{n-1}a_1X + \cdots + a_n^nX^n.$$

Put $Y = a_nX$, then

$$g(Y) = a_n^{n-1}a_0 + a_n^{n-2}a_1Y + \cdots + a_{n-1}Y^{n-1} + Y^n.$$

By Theorem 9.3.16, each rational root of $g(Y)$ must be an integer and we saw earlier how to find all integer roots of $g(Y)$. Finally, if c is an integer root of $g(Y)$ then $\frac{c}{a_n}$ is a root of the polynomial $f(X)$.

EXERCISE SET 9.3

9.3.1. In the ring $R = \mathbb{F}_5[X], \mathbb{F}_5 = \mathbb{Z}/5\mathbb{Z}$ find the prime factorization of the polynomial $3X^3 + 2X^2 + X + 1$ where $n = n + 5\mathbb{Z}$.

9.3.2. In the ring $R = \mathbb{F}_3[X]$, where $\mathbb{F}_3 = \mathbb{Z}/3\mathbb{Z}$, find the prime factorization of the polynomial $X^4 + 2X^3 + 1$ where $n = n + 3\mathbb{Z}$.

9.3.3. Prove that the polynomial $f(X) = X^3 - 2$ is irreducible over the field $\mathbb{Q}$.

9.3.4. Prove that the polynomial $f(X) = X^2 + X + 1$ is irreducible over the field $\mathbb{Q}$.

9.3.5. Prove that the polynomial $f(X) = X^4 + 1$ is irreducible over the field $\mathbb{Q}$.

9.3.6. Prove that the polynomial $f(X) = X^2 + X + 1$ is irreducible over the field $\mathbb{F}_5$.

9.3.7. Using Eisenstein's criterion, prove that the polynomial $f(X) = X^4 - 8X^3 + 12X^2 - 6X + 2$ is irreducible over the field $\mathbb{Q}$.

9.3.8. Using Eisenstein's criterion, prove that the polynomial $f(X) = X^4 - X^3 + 2X + 1$ is irreducible over the field $\mathbb{Q}$.

9.3.9. In the ring $\mathbb{Q}[X]$ factor $2X^5 - X^4 - 6X^3 + 3X^2 + 4X - 2$ into irreducible factors.

9.3.10. Prove that the polynomial $f(X) = X^5 - X^2 + 1 \in \mathbb{Z}[X]$ is irreducible over $\mathbb{Z}$.

9.3.11. Prove that the polynomial $f(X) = X^3 - X^2 + X + 1 \in \mathbb{Z}[X]$ is irreducible over $\mathbb{Z}$.

9.3.12. Prove that the polynomial $f(X) = X^{2n} + X^n + 1 \in \mathbb{Z}[X]$ is irreducible over $\mathbb{Z}$ for each $n \in \mathbb{N}$.

9.3.13. Find the rational roots of the polynomial $X^4 + 2X^3 - 13X^2 - 38X - 24$.

9.3.14. Find the rational roots of the polynomial $2X^3 + 3X^2 + 6X - 4$.

9.3.15. Find the rational roots of the polynomial $X^4 - 2X^3 - 8X^2 + 13X - 24$.

9.3.16. Let $f(X) = X^3 + X^2 + aX + 3 \in \mathbb{R}[X]$. For which real number a does this polynomial have multiple roots?

9.3.17. Let $f(X) = X^3 + 3X^2 + 3aX - 4 \in \mathbb{R}[X]$. For which real number a does this polynomial have multiple roots?

9.3.18. Let $f(X) = X^3 + 3X^2 + 4 \in \mathbb{R}[X]$. Find all the irreducible factors of $f(X)$, together with their multiplicity.

9.3.19. Let $f(X) = X^5 + 4X^4 + 7X^3 + 8X^2 + 2 \in \mathbb{R}[X]$. Find all the irreducible factors of the polynomial $f(X)$, together with their multiplicity.

9.3.20. Let $f(X) = X^5 - iX^4 + 5X^3 - iX^2 + 4i \in \mathbb{C}[X]$. Find the irreducible factors of the polynomial $f(X)$.

9.4 ARITHMETIC FUNCTIONS

In this section, we consider some important *number-theoretic functions* that are of the form $f : \mathbb{N} \longrightarrow \mathbb{C}$, whose domain is the set of natural numbers.

9.4.1. Definition. *A number-theoretic function whose domain is $\mathbb{N}$ and whose range is a subset of the complex numbers is called an arithmetical function or an arithmetic function.*

We first consider some important examples of number-theoretic functions.

9.4.2. Definition. *Let $\nu : \mathbb{N} \longrightarrow \mathbb{N}$ be the function defined in the following way: $\nu(1) = 1$, and $\nu(n)$ is the number of all positive divisors of n if $n > 1$.*

9.4.3. Definition. *Let $\sigma : \mathbb{N} \longrightarrow \mathbb{N}$ be the function defined by $\sigma(1) = 1$ and $\sigma(n)$ is the sum of all the positive divisors of n if $n > 1$.*

The next proposition provides us with certain formulae, allowing us to find the values of these functions.

9.4.4. Proposition. *Let n be a positive integer and suppose that $n = p_1^{k_1} p_2^{k_2} \dots p_t^{k_t}$ is its prime decomposition where p_j is a prime for $1 \leq j \leq t$ and $p_k \neq p_j$ whenever $k \neq j$. Then*

(i) $\nu(n) = (k_1 + 1) \dots (k_t + 1)$;

(ii) $\sigma(n) = \dfrac{(p_1^{k_1+1} - 1)}{(p_1 - 1)} \dots \dfrac{(p_t^{k_t+1} - 1)}{(p_t - 1)}$.

Proof. Let m be an arbitrary divisor of n. Then $m = p_1^{s_1} p_2^{s_2} \dots p_t^{s_t}$, where $0 \leq s_j \leq k_j$, for $1 \leq j \leq t$. Since $\mathbb{Z}$ is a unique factorization domain, the decompositions of m and n are unique. It follows that the mapping

$$m \longmapsto (s_1, \dots, s_t), \text{ where } 0 \leq s_j \leq k_j, \text{ for } 1 \leq j \leq t$$

is bijective. Consequently, the number of all divisors of n is equal to the number of all t-tuples $(s_1, \dots, s_t)$ of positive integers s_j, where $0 \leq s_j \leq k_j$, and $1 \leq j \leq t$. It is evident that for each j there are $k_j + 1$ choices for s_j and hence a total number of $(k_1 + 1) \dots (k_{t-1} + 1)(k_t + 1)$ choices for the tuple $(s_1, \dots, s_t)$. Therefore,

$$\nu(n) = (k_1 + 1) \dots (k_t + 1).$$

To justify the formula for $\sigma(n)$ we will use induction on t. If $t = 1$, so $n = p_1^{k_1}$, then

$$\sigma(n) = 1 + p_1 + p_1^2 + \dots + p_1^{k_1} = \frac{(p_1^{k_1+1} - 1)}{(p_1 - 1)}.$$

Suppose that $t > 1$, let $r = p_1^{k_1} p_2^{k_2} \dots p_{t-1}^{k_{t-1}}$, and suppose we have already proved that

$$\sigma(r) = \frac{(p_1^{k_1+1} - 1)}{(p_1 - 1)} \cdots \frac{(p_{t-1}^{k_{t-1}+1} - 1)}{(p_{t-1} - 1)}.$$

Let S be the set of all t-tuples $(s_1, \dots, s_t)$ of positive integers s_j, where $0 \le s_j \le k_j$, for $1 \le j \le t$, and let M be the set of all $(t-1)$-tuples $(s_1, \dots, s_{t-1})$ of positive integers s_j, where $0 \le s_j \le k_j$, for $1 \le j \le t-1$. Then we have

$$\begin{aligned}
\sigma(n) &= \sum_{(s_1,\dots,s_t)\in S} p_1^{s_1} p_2^{s_2} \dots p_{t-1}^{s_{t-1}} p_t^{s_t} \\
&= \sum_{(s_1,\dots,s_{t-1})\in M} p_1^{s_1} p_2^{s_2} \dots p_{t-1}^{s_{t-1}} + \sum_{(s_1,\dots,s_{t-1})\in M} p_1^{s_1} p_2^{s_2} \dots p_{t-1}^{s_{t-1}} p_t \\
&\quad + \sum_{(s_1,\dots,s_{t-1})\in M} p_1^{s_1} p_2^{s_2} \dots p_{t-1}^{s_{t-1}} p_t^2 + \dots + \sum_{(s_1,\dots,s_{t-1})\in M} p_1^{s_1} p_2^{s_2} \dots p_{t-1}^{s_{t-1}} p_t^{k_t} \\
&= \left(\sum_{(s_1,\dots,s_{t-1})\in M} p_1^{s_1} p_2^{s_2} \dots p_{t-1}^{s_{t-1}} \right) (1 + p_t + p_t^2 + \dots + p_t^{k_t}) \\
&= \frac{(p_1^{k_1+1} - 1)}{(p_1 - 1)} \cdots \frac{(p_{t-1}^{k_t+1} - 1)}{(p_{t-1} - 1)} \frac{(p_t^{k_t+1} - 1)}{(p_t - 1)},
\end{aligned}$$

using the induction hypothesis.

There is a very interesting number-theoretical problem connected with the function $\sigma(n)$. A positive integer n is called *perfect*, if $\sigma(n) = 2n$. For example, the positive integers 6 and 28 are perfect. Proposition 9.4.4 implies that if $2^{k+1} - 1$ is a prime, then $n = 2^k(2^{k+1} - 1)$ is perfect. Euler proved that every even perfect number has such a form. Thus, the problem of finding all even perfect numbers is reduced to finding primes of the form $2^{k+1} - 1$.

9.4.5. Definition. *A prime p is called a Mersenne prime if $p = 2^k - 1$ for some positive integer k.*

The following two important problems about perfect numbers remain unsolved at the time of writing:

1. *Are there infinitely many perfect numbers?*
2. *Is there an odd perfect number?*

We want a method for constructing further number-theoretic functions from given ones. There are many ways to do this but one particularly interesting method is as follows.

9.4.6. Definition. *The Dirichlet product of two number-theoretic functions f and g is the function $f \boxtimes g$ defined by*

$$(f \boxtimes g)(n) = \sum_{kt=n} f(k)g(t).$$

We note that $f \boxtimes g$ is again a number-theoretic function. The following proposition lists some of the important properties of the Dirichlet product.

9.4.7. Proposition.

(i) *Dirichlet multiplication of number-theoretic functions is commutative.*
(ii) *Dirichlet multiplication of number-theoretic functions is associative.*
(iii) *Dirichlet multiplication of number-theoretic functions has an identity element. This is the function $\boldsymbol{E}$ defined by the rule $\boldsymbol{E}(1) = 1$ and $\boldsymbol{E}(n) = 0$ for $n > 1$.*

Proof. Since multiplication of complex numbers is commutative, assertion (i) is easy. To prove (ii) we consider the products $(f \boxtimes g) \boxtimes h$ and $f \boxtimes (g \boxtimes h)$. We have

$$((f \boxtimes g) \boxtimes h)(n) = \sum_{kt=n} (f \boxtimes g)(k)h(t)$$

$$= \sum_{kt=n} \left(\sum_{uv=k} f(u)g(v) \right) h(t) = \sum_{(uv)t=n} (f(u)g(v))h(t) \text{ and}$$

$$(f \boxtimes (g \boxtimes h))(n) = \sum_{um=n} f(u)(g \boxtimes h)(m)$$

$$= \sum_{um=n} f(u) \left(\sum_{vt=m} g(v)h(t) \right) = \sum_{u(vt)=n} f(u)(g(v)h(t)).$$

Since multiplication of complex numbers is associative,

$$((f \boxtimes g) \boxtimes h)(n) = (f \boxtimes (g \boxtimes h))(n), \text{ for each } n \in \mathbb{N},$$

which means that $(f \boxtimes g) \boxtimes h = f \boxtimes (g \boxtimes h)$.

Finally,

$$(f \boxtimes \boldsymbol{E})(n) = \sum_{kt=n} f(k)\boldsymbol{E}(t) = f(n)\boldsymbol{E}(1) = f(n) \text{ for each } n \in \mathbb{N},$$

and hence $f \boxtimes \boldsymbol{E} = f$.

Another important number-theoretic function $\mathcal{S}$ is defined by the rule $\mathcal{S}(n) = 1$ *for each* $n \in \mathbb{N}$. We have

$$(f \boxtimes \mathcal{S})(n) = \sum_{kt=n} f(k)\mathcal{S}(t) = \sum_{k|n} f(k), \quad \text{for each } n \in \mathbb{N},$$

where the sum is taken over all divisors k of n.

9.4.8. Definition. *The function $f \boxtimes \mathcal{S}$ is said to be* the summator function for the function f.

The next important number-theoretic function that we consider is the *Möbius function* μ.

9.4.9. Definition. *Let $n = p_1^{k_1} p_2^{k_2} \dots p_t^{k_t}$, where $p_1, \dots, p_t$ are distinct primes. The Möbius function μ is defined as follows.*

$$\mu(n) = \begin{cases} 1, \textit{ if } n = 1; \\ 0, \textit{ if there exists } j \textit{ such that } k_j \geq 2; \\ (-1)^t, \textit{ if } k_j \leq 1 \textit{ for all } j. \end{cases}$$

The summator function for μ turns out to be $\boldsymbol{E}$ as our next proposition shows.

9.4.10. Proposition. $\mu \boxtimes \mathcal{S} = \boldsymbol{E}$.

Proof. We have $\mu \boxtimes \mathcal{S}(1) = \mu(1) = 1$. Next, let $n = p_1^{k_1} p_2^{k_2} \dots p_t^{k_t} > 1$ be the prime decomposition of n where $p_i \neq p_j$ whenever $i \neq j$. Choose an arbitrary divisor m of n. Then

$$m = p_1^{s_1} p_2^{s_2} \dots p_{t-1}^{s_{t-1}} p_t^{s_t}, \quad \text{where } 0 \leq s_j \leq k_j, \text{ for } 1 \leq j \leq t.$$

If there exists an index j such that $s_j \geq 2$, then $\mu(m) = 0$. Let T denote the set of all tuples $(s_1, \dots, s_t)$ of length t such that $0 \leq s_j \leq 1$, for $1 \leq j \leq t$. Then

$$\mu \boxtimes \mathcal{S}(n) = \sum_{(s_1,\dots,s_t)\in T} \mu(p_1^{s_1} p_2^{s_2} \dots p_{t-1}^{s_{t-1}} p_t^{s_t}).$$

Let $\mathbf{Supp}(s_1, \dots, s_t) = \{j \mid 1 \leq j \leq t, s_j = 1\}$. By the definition of the Möbius function

$$\mu(p_1^{s_1} p_2^{s_2} \dots p_{t-1}^{s_{t-1}} p_t^{s_t}) = \begin{cases} 1 \text{ if } |\mathbf{Supp}(s_1, \dots, s_t)| \text{ is even;} \\ -1 \text{ if } |\mathbf{Supp}(s_1, \dots, s_t)| \text{ is odd.} \end{cases}$$

We prove that the number of tuples $(s_1, \ldots, s_t)$ of length t [where $(0 \leq s_j \leq 1)$] such that $|\mathbf{Supp}(s_1, \ldots, s_t)|$ is even, coincides with the number of tuples for which $|\mathbf{Supp}(s_1, \ldots, s_t)|$ is odd. It will follow that $\mu(n) = 0$.

To prove the required assertion, we apply induction on t. First let $t = 2$. In this case, $|\mathbf{Supp}(0, 0)|$ and $|\mathbf{Supp}(1, 1)|$ are even and $|\mathbf{Supp}(0, 1)|$ and $|\mathbf{Supp}(1, 0)|$ are odd. Hence for $t = 2$, the result holds. Suppose next that $t > 2$ and that our assertion is true for all tuples of length $t - 1$. Let U denote the set of all tuples of numbers $(s_1, \ldots, s_t)$ of length t where $0 \leq s_j \leq 1$ such that $s_t = 0$ and let V denote the corresponding set when $s_t = 1$. Clearly, $|U| = |V|$ and is equal to the number of $(t-1)$-tuples $(s_1, \ldots, s_{t-1})$, where $0 \leq s_j \leq 1$ and $1 \leq j \leq t - 1$. Hence, the number of t-tuples $(s_1, \ldots, s_t)$ of the subset U such that $|\mathbf{Supp}(s_1, \ldots, s_t)|$ is even coincides with the number of tuples $(s_1, \ldots, s_{t-1})$ of length $t - 1$ such that $|\mathbf{Supp}(s_1, \ldots, s_{t-1})|$ is also even. Similarly, the number of t-tuples $(s_1, \ldots, s_t)$ of the subset U such that $|\mathbf{Supp}(s_1, \ldots, s_t)|$ is odd coincides with the number of $(t-1)$-tuples $(s_1, \ldots, s_{t-1})$ such that $|\mathbf{Supp}(s_1, \ldots, s_{t-1})|$ is also odd. Also, the number of t-tuples $(s_1, \ldots, s_t)$ of the subset V such that $|\mathbf{Supp}(s_1, \ldots, s_t)|$ is even coincides with the number of $(t-1)$-tuples $(s_1, \ldots, s_{t-1})$ such that $|\mathbf{Supp}(s_1, \ldots, s_{t-1})|$ is also odd. Finally, the number of t-tuples $(s_1, \ldots, s_t)$ of the subset V such that $|\mathbf{Supp}(s_1, \ldots, s_t)|$ is odd coincides with the number of $(t-1)$-tuples $(s_1, \ldots, s_{t-1})$ such that $|\mathbf{Supp}(s_1, \ldots, s_{t-1})|$ is even. Using the induction hypothesis, we deduce that the number of t-tuples $(s_1, \ldots, s_t)$ of the subset U (respectively V) such that $|\mathbf{Supp}(s_1, \ldots, s_t)|$ is even coincides with the number of t-tuples $(s_1, \ldots, s_t)$ such that $|\mathbf{Supp}(s_1, \ldots, s_t)|$ is odd. This proves the assertion.

The following important result can now be obtained.

9.4.11. Theorem *(The Möbius Inversion Formula). Let f be a number-theoretic function and let F be the summator function for f. Then*

$$f(n) = \sum_{k|n} \mu(k) F(\frac{n}{k})$$

for each $n \in \mathbb{N}$.

Proof. We have $F = f \boxtimes \mathcal{S}$. Proposition 9.4.10 implies that

$$F \boxtimes \mu = (f \boxtimes S) \boxtimes \mu = f \boxtimes (S \boxtimes \mu) = f \boxtimes (\mu \boxtimes S) = f \boxtimes \boldsymbol{E} = f.$$

The next number-theoretic function plays a very important role in many areas of mathematics.

9.4.12. Definition. *The Euler function φ is defined by $\varphi(1) = 1$ and if $n > 1$ then*

$$\varphi(n) = \{k \mid k \in \mathbb{N}, 1 \leq k < n, \mathbf{GCD}(n, k) = 1\}.$$

The next theorem provides us with an alternative approach to the Euler Function, which shows one of its important use in group theory.

9.4.13. Theorem.

(i) *Let G be a finite cyclic group of order n. Then the number of generators of G coincides with $\varphi(n)$.*

(ii) *If k is a positive integer, then $k + n\mathbb{Z} \in \mathbf{U}(\mathbb{Z}/n\mathbb{Z})$ if and only if $\mathbf{GCD}(k, n) = 1$. In particular, $\varphi(n) = |\mathbf{U}(\mathbb{Z}/n\mathbb{Z})|$.*

Proof.

(i) Since G is cyclic, $G = \langle g \rangle$ for some element $g \in G$. By Theorem 8.1.20, an element $y = g^k \in G$ is a generator for G if and only if $\mathbf{GCD}(k, n) = 1$. From Section 8.1, it follows that

$$\langle g \rangle = \left\{g^0, g^1, g^2, \ldots, g^{n-1}\right\}.$$

Consequently, by its definition, $\varphi(n)$ is equal to the number of generators of G.

(ii) Let $k + n\mathbb{Z} \in \mathbf{U}(\mathbb{Z}/n\mathbb{Z})$. Then there exists a coset $s + n\mathbb{Z}$ such that

$$(ks + n\mathbb{Z}) = (k + n\mathbb{Z})(s + n\mathbb{Z}) = 1 + n\mathbb{Z}.$$

Thus, $ks + nr = 1$ for some $r \in \mathbb{Z}$ and Corollary 1.4.7 implies that $\mathbf{GCD}(k, n) = 1$. Conversely, if $\mathbf{GCD}(k, n) = 1$ we can reverse these arguments to see that $k + n\mathbb{Z} \in \mathbf{U}(\mathbb{Z}/n\mathbb{Z})$. It follows that $|\mathbf{U}(\mathbb{Z}/n\mathbb{Z})| = \varphi(n)$.

9.4.14. Corollary *(Euler's Theorem). Let n be a positive integer and suppose that k is an integer such that $\mathbf{GCD}(k, n) = 1$. Then $k^{\varphi(n)} \equiv 1 \pmod{n}$.*

Proof. Since $\mathbf{GCD}(k, n) = 1$, it follows that $k + n\mathbb{Z} \in \mathbf{U}(\mathbb{Z}/n\mathbb{Z})$. Corollary 8.3.9 and Theorem 9.4.13 together imply that $(k + n\mathbb{Z})^{\varphi(n)} = 1 + n\mathbb{Z}$. However,

$$(k + n\mathbb{Z})^{\varphi(n)} = k^{\varphi(n)} + n\mathbb{Z},$$

so the result follows.

9.4.15. Corollary *(Fermat's Little Theorem). Let p be a prime and let k be an integer. If p does not divide k then $k^{p-1} \equiv 1 \pmod{p}$.*

Proof. Since p is a prime, $\varphi(p) = p - 1$ and we can apply Corollary 9.4.14.

Corollary 9.4.15 can also be written in the following form:

Let p be a prime and k be an integer. Then $k^p \equiv k (mod\ p)$.

We next consider the summator function for the Euler function. Let G be a group. Define the binary relation $\circ$ on G by the rule: $x \circ y$ if and only if $|x| = |y|$, for $x, y \in G$. It is easy to check that $\circ$ is an equivalence relation and, if n is a positive integer, let

$$G_n = \{x \mid x \in G \text{ and } |x| = n\}.$$

The subset G_n is an equivalence class under the relation $\circ$. Let G_∞ denote the subset of all elements of G whose orders are infinite. By Theorem 7.2.5 the family of subsets $\{G_n \mid n \in \mathbb{N} \cup \{\infty\}\}$ is a partition of G. We note that G_n can be empty for some positive integer n or $n = \infty$. In particular, if G is a finite group, then $G_\infty = \emptyset$. Moreover, if y is an arbitrary element of a finite group G, then Corollary 8.3.9 implies that $|y|$ is a divisor of $|G|$. Thus, the family of subsets $\{G_k | k \text{ is a divisor of } |G|\}$ is a partition of the finite group G. It follows that if $|G| = n$ then $G = \bigcup_{k|n} G_k$. It will often be the case that G_k will also be empty. However, the one exception here is that of a cyclic group. In fact G_k is nonempty for all $k \mid |G|$ if and only if G is a finite cyclic group.

These observations allows us to obtain the following interesting identity.

9.4.16. Theorem. $\sum_{k|n} \varphi(k) = n.$

Proof. Let $G = \langle g \rangle$ be a cyclic group of order n. We noted above that $G = \bigcup_{k|n} G_k$. Let k be a divisor of n and put $d = \frac{n}{k}$. Also let $x = g^d$. We have $x^k = (g^d)^k = g^{dk} = g^n = e$. If we suppose that $x^t = e$ for some positive integer $t < k$, then $e = x^t = (g^d)^t = g^{dt}$. Since $dt < n$, we obtain a contradiction to the fact that $|g| = |G| = n$. Thus, $|x| = k$. Hence, the subset G_k is not empty for every divisor k of n.

Let z be an element of G having order k. Then $z = g^m$ for some positive integer m, and $e = z^k = (g^m)^k = g^{mk}$. It follows that $n = dk$ divides mk and hence d divides m, so $m = ds$ for some positive integer s. We have

$$z = g^m = g^{ds} = (g^d)^s = x^s,$$

which proves that $z \in \langle x \rangle$. Furthermore, $|\langle z \rangle| = |z| = |x| = k$, so that $\langle z \rangle = \langle x \rangle$. Thus, every element of order k is a generator for the subgroup $\langle x \rangle$. By Theorem 9.4.13, the number of all such elements is equal to $\varphi(k)$. Hence for each divisor k of n, we have $|G_k| = \varphi(k)$. Therefore,

$$n = |G| = \sum_{k|n} |G_k| = \sum_{k|n} \varphi(k).$$

9.4.17. Corollary. *Let G be a finite group of order n. If k is a divisor of n, then let $G[k] = \{x \mid x \in G \text{ and } x^k = e\}$. Suppose that $|G[k]| \leq k$ for each divisor k of n. Then the group G is cyclic.*

Proof. We have already noted above that the family of subsets $\{G_k \mid k|n\}$ is a partition of a finite group G. This implies that $G = \bigcup_{k|n} G_k$ and $|G| = \sum_{k|n} G_k$.

Suppose that the subset G_k is not empty for some divisor k of n and let $y \in G_k$. By Corollary 9.3.10, if $z \in \langle y \rangle$ then $|z|$ divides k, and therefore $z^k = e$. Now, the equations $|y| = |\langle y \rangle| = k$ together with the conditions of this corollary imply that $G_k \subseteq \langle y \rangle$. Moreover, each element of order k is a generator for a subgroup $\langle y \rangle$. Applying Theorem 9.4.13, we deduce that the number of all such elements is equal to $\varphi(k)$. Consequently, if the subset G_k is not empty for some divisor k of n, then $|G_k| = \varphi(k)$. Comparing the following two equations $n = |G| = \sum_{k|n} |G_k|$ and $n = \sum_{k|n} \varphi(k)$, we see that G_k is nonempty for each divisor k of n. In particular, $G_n \neq \emptyset$. Let $g \in G_n$. The equation $n = |g| = |\langle g \rangle|$ implies that $\langle g \rangle = G$.

9.4.18. Corollary. *Let F be a field and let G be a finite subgroup of $U(F)$. Then G is cyclic.*

Proof. Let $|G| = n$ and let k be a divisor of n. If an element y of G satisfies the condition $y^k = e$, then y is a root of $X^k - e \in F[X]$. By Corollary 7.5.11, this polynomial has at most k roots. By Corollary 9.4.17, G is a cyclic group.

9.4.19. Corollary. *Let F be a finite field. Then its multiplicative group $U(F)$ is cyclic.*

We know already that $\mathbb{Z}/p\mathbb{Z}$ is a field whenever p is a prime, so we deduce the following fact.

9.4.20. Corollary. *Let p be a prime. Then $U(\mathbb{Z}/p\mathbb{Z})$ is cyclic.*

We showed in Theorem 9.4.16 that the identity permutation of the set $\mathbb{N}$ is the summator function for the Euler function. This allows us to obtain a formula for the values of the Euler function. Employing Theorems 9.4.11 and 9.4.16 we have

$$\varphi(n) = \sum_{k|n} \mu(k)\frac{n}{k} = n \sum_{k|n} \frac{\mu(k)}{k},$$

for each $n \in \mathbb{N}$. Now let $n = p_1^{k_1} p_2^{k_2} \dots p_t^{k_t}$ be the prime decomposition of n where $p_i \neq p_j$ whenever $i \neq j$. If m is a divisor of n, then $m = p_1^{s_1} p_2^{s_2} \dots p_t^{s_t}$ where $0 \leq s_j \leq k_j$, $1 \leq j \leq t$. If there exists j such that $s_j \geq 2$, then $\mu(m) = 0$. Hence

$$\sum_{k|n} \frac{\mu(k)}{k} = 1 - \sum_{1 \leq j \leq t} \frac{1}{p_j} + \sum_{1 \leq j < m \leq t} \frac{1}{p_j p_m} - \dots + \frac{(-1)^t}{p_1 p_2 \dots p_t}$$

$$= \left(1 - \frac{1}{p_1}\right)\left(1 - \frac{1}{p_2}\right)\dots\left(1 - \frac{1}{p_t}\right).$$

Consequently,

$$\begin{aligned}\varphi(n) &= n\left(1-\frac{1}{p_1}\right)\left(1-\frac{1}{p_2}\right)\dots\left(1-\frac{1}{p_t}\right)\\ &= p_1^{k_1-1}(p_1-1)p_2^{k_2-1}(p_2-1)\dots p_t^{k_t-1}(p_t-1).\end{aligned}$$

In particular, if p is a prime and $k \in \mathbb{N}$, then $\varphi(p^k) = p^{k-1}(p-1) = p^k - p^{k-1}$.
Moreover, from the above formula we deduce that

$$\varphi(nk) = \varphi(n)\varphi(k), \text{ whenever } \mathbf{GCD}(k, n) = 1.$$

9.4.21. Definition. *The number-theoretic function f is called multiplicative if*

(i) *there is a positive integer n such that $f(n) \neq 0$;*
(ii) *if* $\mathbf{GCD}(k, n) = 1$, *then* $f(nk) = f(n)f(k)$.

Thus, φ is a multiplicative function. The next theorem gives us an important property of multiplicative functions.

9.4.22. Theorem. *If a number-theoretic function f is multiplicative, then the summator function for f is also multiplicative.*

Proof. Let k and t be positive integers such that $\mathbf{GCD}(k, t) = 1$. If d is a divisor of kt, then clearly $d = uv$ where $u|k$, $v|t$. Let $F = (f \boxtimes \mathcal{S})$ be the summator function for f. Then

$$\begin{aligned}F(kt) &= \sum_{u|k, v|t} f(uv) = \sum_{u|k, v|t} f(u)f(v)\\ &= \left(\sum_{u|k} f(u)\right)\left(\sum_{v|t} f(v)\right) = F(k)F(t).\end{aligned}$$

At the end of this section, we discuss some applications of the above results. Recently the Euler function has found applications in cryptography. A giant leap forward occurred in cryptography in the second half of the twentieth century, with the invention of public key cryptography. The main idea is the concept of a trapdoor function—a function that has an inverse, but whose inverse is very difficult to calculate. In 1976, Rivest, Shamir, and Adleman succeeded in finding such a class of functions. It turns out that if you take two very large numbers and multiply them together, a machine can quickly compute the answer. However, if you give the machine the answer and ask it for the two factors, the factorization cannot be computed in a useful amount of time. The public key system, built upon these ideas, is now known as the RSA (Rivest, Shamir, and Adelman) key after the three men who created it. The main idea in RSA cryptography is as follows. One chooses two arbitrary primes p and q and calculates $n = pq$ and $\varphi(n) = (p-1)(q-1)$. Next, one picks an arbitrary number $k < \varphi(n)$ that is relatively prime with $\varphi(n)$. As we can see by the proof of Theorem 9.4.13, $k + \varphi(n)\mathbb{Z} \in \mathbf{U}(\mathbb{Z}/\varphi(n)\mathbb{Z})$. Consequently, there exists a positive integer t such that $kt \equiv 1 \ (\mathbf{mod}\ \varphi(n))$. We can find t using the Euclidian algorithm, which has been described in Section 9.2. the numbers n and k determine the coding method. They are not secret and they form the open (or public) key. Only the primes p, q and the number t are kept secret. First, the message should be written in numerical form with the help of ordinary digits 0, 1, 2, 3, 4, 5, 6, 7, 8, 9. After that, one divides this message into blocks $M_1, \dots M_s$ of a certain length w. The number m should satisfy the restriction $10w < m$. Usually one chooses the numbers p and q up to 100 or more digits each. Each block M_j can be considered as a representative of the coset $M_j + n\mathbb{Z}$. Encryption of the block M_j is done by substituting it by the block $E(M_j)$, where $E(M_j) \equiv M_j^k \ (\mathbf{mod}\ n)$. Since the number n is large, only a computer can perform this computation. Decryption is done using the following procedure. The choice of the number t requires that $kt \equiv 1 + r\varphi(n)$. Now one can apply Euler's theorem (Corollary 9.4.14). In this case, M_j and n are required to be relatively prime. Nevertheless, we can show that this application is valid in any case. Let m be an arbitrary positive integer. If $\mathbf{GCD}(m, n) = 1$, then Corollary 9.4.14 shows that $m^{\varphi(n)} \equiv 1(\mathbf{mod}\ n)$, and

$$m^{kt} = m^{1+r\varphi(n)} = m(m^{\varphi(n)})^r \equiv m(\mathbf{mod}\ n).$$

Suppose now that $\mathbf{GCD}(m, n) \neq 1$. Since $n = pq$, then either p divides m and q does not divide m, or conversely, q divides m and p does not divide m. Consider the first case; the consideration of the second case is similar.

We have

$$m = pu \text{ and } m^{kt} - m = (pu)^{kt} - pu.$$

On the other hand,

$$m^{kt} = m^{1+r\varphi(n)} = m^{1+r(p-1)(q-1)} = m(m^{q-1})^{r(p-1)}.$$

Since $\mathbf{GCD}(m, q) = 1$, Corollary 9.4.15 leads us to $m^{(q-1)} \equiv 1(\mathbf{mod}\ q)$. Therefore,

$$m^{kt} \equiv m(m^{q-1})^{r(p-1)}(\mathbf{mod}\ q) \equiv m(1^{q-1})^{r(p-1)}(\mathbf{mod}\ q) \equiv m(\mathbf{mod}\ q).$$

Hence q divides $m^{kt} - m$. As we have seen above, p divides $m^{kt} - m$, so that $n = pq$ divides $m^{kt} - m$. Consequently, in any case we have $m^{kt} \equiv m(\mathbf{mod}\ n)$. For block M_j we have

$$E(M_j)^t = (M_j^k)^t = M_j^{kt} = M_j^{1+r\varphi(n)} = M_j(Mj^{\varphi(n)})^r.$$

So we can write $E(M_j)^t \equiv M_j(\mathbf{mod}\ n)$. Recall that the number M_j satisfies the condition $1 \leq M_j < n$, and therefore it is uniquely determined by the congruence $E(M_j)^t \equiv M_j(\mathbf{mod}\ n)$.

The problem of reliability of the RSA code is reduced to the question: Can the block $E(M_j)$ be decoded? For this one needs to solve the congruence $x^k \equiv E(M_j)(\mathbf{mod}\ n)$ without knowing the number t. In the next section, we will study congruences of the type $x^k \equiv a(\mathbf{mod}\ n)$. For now, we can only state that there is no general method for the solution of such congruences. In reality, it could be done by the examination of all cases. From the choice of w, it follows that this sorting requires the consideration of 100^{100} cases, which is not quite realistic.

Therefore, to crack the RSA code one needs to find t if n and k given. If the decomposition $n = pq$ is known, then this is not difficult. In turn, knowing the numbers n, k, and t, one can find the decomposition $n = pq$. So the finding of t requires the same efforts as is needed for finding the decomposition $n = pq$. However, in the case when each of the factors has 100 digits, this is not realistic yet.

We attach some worked examples of exercises supplied with solutions.

Some Worked Exercises

9.4.23. *Find a positive integer n such that n has exactly 14 positive divisors including 12.*

Solution. Let $n = p_1^{k_1} p_2^{k_2} \dots p_t^{k_t}$ be the primary decomposition of n where p_j is a prime, for $1 \leq j \leq n$ and $p_k \neq p_j$ whenever $k \neq j$. By Proposition 9.4.4, $\nu(n) = (k_1 + 1) \dots (k_t + 1) = 14 = 2 \cdot 7$. It follows that $(k_1 + 1) = 2, (k_2 + 1) = 7, t = 2$. On the other hand, 12 divides n, so that $p_1 = 3$, $p_2 = 2$, $k_1 = 1$, and $k_2 = 6$. Hence $n = 2^6 \times 3 = 192$.

9.4.24. *Let n be a positive integer. Find the product of all positive divisors of n.*

Solution. Let $D(n)$ be the set of all positive divisors of n. Let $P(n)$ be the set of all nonordered pairs $\{d, \frac{n}{d}\}$ where $d \in D(n)$. If $n \neq k^2$ for some positive integer k, then clearly $|P(n)| = \frac{|D(n)|}{2} = \frac{\nu(n)}{2}$. Since $d \cdot \frac{n}{d} = n$, the product of all elements of $D(n)$ is equal $n^{\frac{\nu(n)}{2}}$. If $n = k^2$ for some positive integer k, then clearly $|P(n)| = \frac{|D(n)|}{2} = \frac{\nu(n)-1}{2}) + 1$. So in this case, the product of all elements of $D(n)$ is equal to $k^{\frac{\nu(n)-1}{2}} = n^{\frac{1}{2}} \cdot n^{\frac{\nu(n)-1}{2}} = n^{\frac{\nu(n)}{2}}$.

9.4.25. *Find a positive integer n such that the product of all positive divisors of n is* 810 000.

Solution. We have $810\,000 = 2^4 \times 3^4 \times 5^4$. Therefore, n has the form $n = 2^a \times 3^b \times 5^c$. By example 9.4.24, $810\,000 = n^{\frac{\nu(n)}{2}}$. By Proposition 9.4.4,

$$\nu(n) = \nu(2^a \times 3^b \times 5^c) = (a + 1)(b + 1)(c + 1).$$

Since $\frac{\nu(n)}{2} \leq 4$, $(a+1)(b+1)(c+1) \leq 8$. Since $a \geq 1$, $b \geq 1$, $c \geq 1$, we obtain that $a = 1$, $b = 1$, $c = 1$, that is, $n = 30$.

9.4.26. *Let n be a positive integer and let $\{d_1, \ldots d_k\}$ be the set of all positive divisors of n. Prove that*

$$n = \frac{d_1 + \cdots + d_k}{\frac{1}{d_1} + \cdots + \frac{1}{d_k}}.$$

Solution. Without loss of generality we may suppose that $d_1 < \cdots < d_k$. Then

$$d_1 = \frac{n}{d_k}, d_2 = \frac{n}{d_{k-1}}, \ldots, d_k = \frac{n}{d_1}.$$

It follows that

$$\frac{d_1 + \cdots + d_k}{\frac{1}{d_1} + \cdots + \frac{1}{d_k}} = \frac{n(d_1 + \cdots + d_k)}{n(\frac{1}{d_1} + \cdots + \frac{1}{d_k})} = \frac{n(d_1 + \cdots + d_k)}{\frac{n}{d_1} + \cdots + \frac{n}{d_k}} = \frac{n(d_1 + \cdots + d_k)}{d_1 + \cdots + d_k} = n.$$

9.4.27. *Let n be a positive integer. Prove that $\varphi(n)$ divides $n!$.*

Solution. Let $n = p_1^{k_1} p_2^{k_2} \ldots p_t^{k_t}$ be the primary decomposition of n, where p_j is a prime, for $1 \leq j \leq n$ and $p_k \neq p_j$ whenever $k \neq j$. It was already shown above that

$$\varphi(n) = (p_1 - 1)(p_2 - 1) \ldots (p_t - 1) p_1^{k_1 - 1} p_2^{k_2 - 1} \ldots p_t^{k_t - 1}.$$

Since $p_k \neq p_j$, $p_k - 1 \neq p_j - 1$ whenever $k \neq j$. Since $p_j - 1 < n$ for each j, $1 \leq j \leq n$, $(p_1 - 1)(p_2 - 1) \ldots (p_t - 1)$ divides $(n-1)!$. Furthermore, $p_1^{k_1 - 1} p_2^{k_2 - 1} \ldots p_t^{k_t - 1}$ divides n. Therefore, $\varphi(n)$ divides $n!$.

9.4.28. *Let n be a positive integer and let*

$$\Phi_n = \{k \mid k \in \mathbb{N}, 1 \leq k < n, \mathbf{GCD}(n, k) = 1\}.$$

Find the sum of all the elements of Φ_n.

Solution. If $k < n$, then $k + (n - k) = n$. Suppose that d is a divisor of k and n. The above equation shows that d is a divisor of $n - k$ and n. Conversely, if d is a divisor of $n - k$ and n, then d is a divisor of k and n. It follows that $\mathbf{GCD}(n, k) = 1$ if and only if $\mathbf{GCD}(n, n - k) = 1$. Therefore, we can divide the set Φ_n into pairs $\{k, n - k\}$. The amount of all these pairs is equal to $\frac{\varphi(n)}{2}$. Since $k + (n - k) = n$, the sum of all elements of Φ_n is equal to $n \cdot \frac{\varphi(n)}{2}$.

9.4.29. *Let a be a positive integers. Find a positive integer n satisfying the equation $\varphi(n) = \frac{1}{a} \cdot n$.*

Solution. If $a = 1$, then clearly $n = 1$. Suppose that $n > 1$ and let $n = p_1^{k_1} p_2^{k_2} \dots p_t^{k_t}$ be the primary decomposition of n where p_j is a prime, for $1 \leq j \leq n$ and $p_k \neq p_j$ whenever $k \neq j$. The equation $\varphi(n) = \frac{1}{a} \cdot n$ is equivalent to $\frac{n}{\varphi(n)} = a$. We have

$$\frac{n}{\varphi(n)} = \frac{p_1^{k_1} p_2^{k_2} \dots p_t^{k_t}}{(p_1 - 1)(p_2 - 1) \dots (p_t - 1) p_1^{k_1 - 1} p_2^{k_2 - 1} \dots p_t^{k_t - 1}}$$
$$= \frac{p_1 p_2 \dots p_t}{(p_1 - 1)(p_2 - 1) \dots (p_t - 1)}.$$

Since $\frac{n}{\varphi(n)} = a$ is a positive integer, $(p_1 - 1)(p_2 - 1) \dots (p_t - 1)$ divides $p_1 p_2 \dots p_t$. If a prime p_j is odd, then $p_j - 1$ is even. Only one of the numbers $p_1, p_2, \dots, p_t$ can be equal to 2. If follows that either $n = 2^{k_1}$ or $n = 2^{k_1} p_2^{k_2}$. In the first case, $\frac{n}{\varphi(n)} = \frac{2}{2-1} = 2$. Hence for $a = 2$ the solutions of the equation $\varphi(n) = \frac{1}{a} \cdot n$ have the form 2^k.

In the second case,

$$\frac{n}{\varphi(n)} = \frac{2p_2}{(2 - 1)(p_2 - 1)} = \frac{2p_2}{p_2 - 1}.$$

The number $\frac{2p_2}{p_2 - 1}$ is a positive integer. Since p_2 is a prime and $p_2 - 1 < p_2$, $\mathbf{GCD}(p_2 - 1, p_2) = 1$. Therefore, if $p_2 - 1 = 2$, $\frac{2p_2}{p_2 - 1}$ is an integer, that is, $p_2 = 3$. Hence in the second case, $n = 2^k 3^t$, and in this case $a = 3$.

EXERCISE SET 9.4

9.4.1. Find $\mu(1086)$.

9.4.2. Find $\mu(3001)$.

9.4.3. Find $\varphi(1226)$.

9.4.4. Find $\varphi(1137)$.

9.4.5. Find $\varphi(2137)$.

9.4.6. Find $\varphi(1989)$.

9.4.7. Find $\varphi(1789)$.

9.4.8. Find $\varphi(1945)$.

9.4.9. Use Euler's theorem to find the remainder when we divide 197^{157} by 35.

9.4.10. Use Euler's theorem to find the remainder when we divide 500^{810000} by each of 5, 7, 11, 13.

9.4.11. Find the remainder when we divide 204^{41} by 111.

9.4.12. Find the remainder when $10^{100} + 40^{100}$ is divided by 7.

9.4.13. Find the remainder when $5^{70} + 7^{50}$ is divided by 12.

9.4.14. Find the last two digits of the number 3^{100}.

9.4.15. Find the last two digits of the number 903^{1294}.

9.4.16. Prove that 42 divides $a^7 - a$ for each $a \in$ N.

9.4.17. Prove that 100 divides $a^{42} - a^2$ for each $a \in \mathbb{N}$.

9.4.18. Prove that 65 divides $a^{12} - b^{12}$ whenever GCD$(a, 65) =$ **GCD**$(b, 65) =$ 11.

9.4.19. Prove that $p^{q-1} + q^{p-1} \equiv 1$ (**mod** pq) where p, q are primes and $p \neq q$.

9.4.20. Prove that if $a_1 + a_2 + a_3 \equiv 0$ (**mod** 30), then $a_1^5 + a_2^5 + a_3^5 \equiv 0$ (**mod** 30).

9.5 CONGRUENCES

In this section, we consider the process of solving congruences. This is a special case of the problem of finding roots of polynomials over commutative rings. In Section 7.4, we introduced polynomials over a commutative ring R and considered the specific case when R is an integral domain. Polynomials over an arbitrary commutative ring can also be discussed but in this more general situation some of the standard properties of polynomials are lost.

A most natural first case to consider is the ring $R = \mathbb{Z}/n\mathbb{Z}$, where n is fixed. We use the notation that if k is an integer then $\hat{k}$ will denote the coset $k + n\mathbb{Z}$. Let

$$f(X) = \hat{a}_0 + \hat{a}_1 X + \cdots + \hat{a}_n X^n \in (\mathbb{Z}/n\mathbb{Z})[X].$$

Then the equation

$$\hat{a}_0 + \hat{a}_1 X + \cdots + \hat{a}_n X^n = \hat{0}$$

leads to the congruence

$$a_0 + a_1 X + \cdots + a_n X^n \equiv 0 \pmod{n}.$$

9.5.1. Definition. *An integer t is called a solution of the congruence*

$$a_0 + a_1 X + \cdots + a_n X^n \equiv 0 \pmod{n}$$

if

$$a_0 + a_1t + \cdots + a_nt^n \equiv 0 \pmod{n}.$$

The following more precise definition is needed.

9.5.2. Definition. *The solutions t and s of*

$$a_0 + a_1X + \cdots + a_nX^n \equiv 0 \pmod{n}$$

are called equivalent, if $t + n\mathbb{Z} = s + n\mathbb{Z}$ or $t \equiv s \pmod{n}$.

So, in order to find a complete set of solutions of the equation

$$\hat{a}_0 + \hat{a}_1X + \cdots + \hat{a}_nX^n = \hat{0},$$

it is necessary to find the complete set of inequivalent solutions of the congruence

$$a_0 + a_1X + \cdots + a_nX^n \equiv 0 \pmod{n}.$$

A natural first step here is the case when $f(X)$ has degree 1. Thus, we have the equation $\hat{c} + \hat{a}x = \hat{0}$ or $\hat{a}x = \hat{b}$ where $\hat{b} = -\hat{c}$. This equation leads to the congruence $ax \equiv b \pmod{n}$.

9.5.3. Theorem. *Let n be a positive integer. The congruence $ax \equiv b \pmod{n}$ has a solution if and only if $d = \mathbf{GCD}(a, n)$ divides b. If c is a solution of $ax \equiv b$* mod *n and if $m = \frac{n}{d}$, then*

$$\{c, c + m, \ldots, c + (d - 1)m\}$$

is a complete set of solutions of this congruence.

Proof. Let u be a solution of $ax \equiv b \pmod{n}$ so

$$(a + n\mathbb{Z})(u + n\mathbb{Z}) = au + n\mathbb{Z} = b + n\mathbb{Z}.$$

We have $b = au + nz$ for some $z \in \mathbb{Z}$. Since d divides $au + nz$, d divides b. Conversely, let d divide b, say $b = b_1d$, where $b_1 \in \mathbb{Z}$. By Corollary 1.4.6, there exist integers v, w such that $av + nw = d$. Multiplying both sides of this equation by b_1, we obtain $avb_1 + nwb_1 = db_1$ and hence $a(vb_1) + n(wb_1) = b$. The last equation shows that the integer $u = vb_1$ is a solution of $ax \equiv b \pmod{n}$. Note also that our reasoning above shows how to find the solution of the congruence $ax \equiv b \pmod{n}$, as well as showing the sufficiency of the condition that d divides b. Indeed, we can find v, w with the aid of the Euclidean algorithm, as in Section 9.2.

Finally, let c, y be two solutions of $ax \equiv b \pmod{n}$. We have

$$ac + n\mathbb{Z} = b + n\mathbb{Z} = ay + n\mathbb{Z},$$

so $ay - ac + n\mathbb{Z} = n\mathbb{Z}$. Thus, n divides $ac - ay$. Let $a = a_1 d$, where $a_1 \in \mathbb{Z}$. Then $md \mid (da_1 c - da_1 y)$ so $m \mid a_1(c - y)$. By Corollary 1.4.8, the integers a_1 and m are relatively prime, so Corollary 1.4.9 shows that $m \mid (y - c)$. Hence $y = c + mk$ for some $k \in \mathbb{Z}$.

Conversely, it is not difficult to see that the integer $y = c + mk$ is a solution of $ax \equiv b \pmod{n}$. It is clear that the solutions $c, c + m, \dots, c + (d - 1)m$ are not equivalent. For the arbitrary solution $y = c + mk$ of $ax \equiv b \pmod{n}$, divide k by d, say $k = ds + r$, where $0 \le r < d$. Then

$$y = c + mk = c + m(ds + r) = c + mr + mds = c + mr + ns,$$

so that the solution y is equivalent to $c + mr$. This proves that the family of integers $\{c, c + m, \dots, c + (d - 1)m\}$ is a complete set of solutions of the congruence $ax \equiv b \pmod{n}$.

There are a couple of special cases that are worth pointing out.

9.5.4. Corollary. *Let n be a positive integer. If a and n are relatively prime then $ax \equiv b \pmod{n}$ always has solutions. In this case, all solutions are equivalent, so the solution of this congruence is unique. In this case $x \equiv a^{-1}b \pmod{n}$.*

9.5.5. Corollary. *Let p be a prime and let a be a positive integer such that p does not divide a. Then $ax \equiv b \pmod{p}$ always has solutions. The solutions are all equivalent so there is a unique solution.*

We require the following elementary result next.

9.5.6. Lemma. *Let $b_1, \dots, b_t$ be integers and let n be a positive integer. If $\mathbf{GCD}(b_j, n) = 1$ for all $1 \le j \le t$, then $\mathbf{GCD}(b_1 \dots b_t, n) = 1$.*

Proof. By Theorem 9.4.13, $b_j + n\mathbb{Z} \in \mathbf{U}(\mathbb{Z}/n\mathbb{Z})$ for each j, where $1 \le j \le t$. Corollary 3.1.15 shows that $\mathbf{U}(\mathbb{Z}/n\mathbb{Z})$ is a stable subset of $\mathbb{Z}/n\mathbb{Z}$, and therefore

$$(b_1 + n\mathbb{Z}) \dots (b_t + n\mathbb{Z}) = b_1 \dots b_t + n\mathbb{Z} \in \mathbf{U}(\mathbb{Z}/n\mathbb{Z}).$$

Again using Theorem 9.4.13, we see that $\mathbf{GCD}(b_1 \dots b_t, n) = 1$.

If a positive integer n has a form $n = n_1 \dots n_t$, where $\mathbf{GCD}(n_j, n_k) = 1$ whenever $j \ne k$, then very often we can reduce the congruence modulo n to a system of congruences modulo $n_1, \dots, n_t$. The classical Chinese Remainder Theorem that follows plays a key role here.

9.5.7. Theorem. *Let n be a positive integer and suppose that $n = n_1 \dots n_t$, where* $\mathbf{GCD}(n_j, n_k) = 1$ *whenever $j \neq k$. If $b_1, \dots, b_t$ are arbitrary integers, then the system of congruences*

$$x \equiv b_1 \quad (\mathbf{mod}\ n_1),$$
$$\vdots$$
$$x \equiv b_t \quad (\mathbf{mod}\ n_t),$$

has solutions. Moreover, if c, d are two solutions of this system of congruences, then $c \equiv d$ $(\mathbf{mod}\ n)$.

Proof. Let $m_j = \frac{n}{n_j}$, for $1 \leq j \leq t$. By Lemma 9.5.6, $\mathbf{GCD}(n_j, m_j) = 1$ for each j, where $1 \leq j \leq t$. By Corollary 1.4.7, there exist integers r_j, s_j such that $r_j n_j + s_j m_j = 1$, for $1 \leq j \leq t$. Let $e_j = s_j m_j$. Then $e_j + n_j\mathbb{Z} = 1 + n_j\mathbb{Z}$ and $e_j \in n_k\mathbb{Z}$ whenever $j \neq k$. Let

$$c = b_1 e_1 + \cdots + b_t e_t.$$

Then

$$\begin{aligned} c + n_j\mathbb{Z} &= b_j e_j + \sum_{k \neq j} b_k e_k + n_j\mathbb{Z} = (b_j e_j + n_j\mathbb{Z}) + \left(\sum_{k \neq j} b_k e_k + n_j\mathbb{Z}\right) \\ &= (b_j + n_j\mathbb{Z})(e_j + n_j\mathbb{Z}) + \left(\sum_{k \neq j} (b_k + n_j\mathbb{Z})(e_k + n_j\mathbb{Z})\right) \\ &= (b_j + n_j\mathbb{Z})(1 + n_j\mathbb{Z}) = (b_j + n_j\mathbb{Z}). \end{aligned}$$

In other words $c \equiv b_j$ $(\mathbf{mod}\ n_j)$, for $1 \leq j \leq t$.

If d is another solution of this system of congruences, then

$$c + n_j\mathbb{Z} = b_j + n_j\mathbb{Z} = d + n_j\mathbb{Z},$$

and then $c - d \in n_j\mathbb{Z}$, so $n_j \mid (c - d)$ for each j, where $1 \leq j \leq t$. Using induction and Corollary 1.4.9, we deduce that $n = n_1 \dots n_t$ divides $(c - d)$. Thus, $c \equiv d$ $(\mathbf{mod}\ n)$.

This result has several interesting consequences, which we now deduce.

9.5.8. Corollary. *Let n be a positive integer and suppose that $n = n_1 \dots n_t$, where* $\mathbf{GCD}(n_j, n_k) = 1$ *whenever $j \neq k$. Then*

$$\mathbb{Z}/n\mathbb{Z} \cong \mathbb{Z}/n_1\mathbb{Z} \times \cdots \times \mathbb{Z}/n_t\mathbb{Z},$$

and

$$\mathbf{U}(\mathbb{Z}/n\mathbb{Z}) \cong \mathbf{U}(\mathbb{Z}/n_1\mathbb{Z}) \times \cdots \times \mathbf{U}(\mathbb{Z}/n_t\mathbb{Z}).$$

Proof. Let R be the ring $\mathbb{Z}/n_1\mathbb{Z} \times \cdots \times \mathbb{Z}/n_t\mathbb{Z}$ and consider the mapping

$$f : \mathbb{Z} \longrightarrow R,$$

defined by

$$f(k) = (k + n_1\mathbb{Z}, \ldots, k + n_t\mathbb{Z}) \text{ for each } k \in \mathbb{Z}.$$

The mapping f is a ring homomorphism. Indeed,

$$\begin{aligned} f(k+s) &= (k + s + n_1\mathbb{Z}, \ldots, k + s + n_t\mathbb{Z}) \\ &= (k + n_1\mathbb{Z} + s + n_1\mathbb{Z}, \ldots, k + n_t\mathbb{Z} + s + n_t\mathbb{Z}) \\ &= (k + n_1\mathbb{Z}, \ldots, k + n_t\mathbb{Z}) + (s + n_1\mathbb{Z}, \ldots, s + n_t\mathbb{Z}) \\ &= f(k) + f(s) \end{aligned}$$

and

$$\begin{aligned} f(ks) &= (ks + n_1\mathbb{Z}, \ldots, ks + n_t\mathbb{Z}) \\ &= \big((k + n_1\mathbb{Z})(s + n_1\mathbb{Z}), \ldots, (k + n_t\mathbb{Z})(s + n_t\mathbb{Z})\big) \\ &= (k + n_1\mathbb{Z}, \ldots, k + n_t\mathbb{Z})(s + n_1\mathbb{Z}, \ldots, s + n_t\mathbb{Z}) = f(k)f(s). \end{aligned}$$

Now let $(k_1 + n_1\mathbb{Z}, \ldots, k_t + n_t\mathbb{Z})$ be an arbitrary element of R. By Theorem 9.5.7, there exists $k \in \mathbb{Z}$ such that $k + n_j\mathbb{Z} = k_j + n_j\mathbb{Z}$ for each j, where $1 \le j \le t$. Thus,

$$(k_1 + n_1\mathbb{Z}, \ldots, k_t + n_t\mathbb{Z}) = (k + n_1\mathbb{Z}, \ldots, k + n_t\mathbb{Z}) = f(k).$$

Hence f is an epimorphism and, by Theorem 7.4.5, $R \cong \mathbb{Z}/\mathbf{Ker}\, f$. To determine $\mathbf{Ker}\, f$, let $k \in \mathbf{Ker}\, f$. Then

$$f(k) = (k + n_1\mathbb{Z}, \ldots, k + n_t\mathbb{Z}) = (n_1\mathbb{Z}, \ldots, n_t\mathbb{Z}),$$

so that $k \in n_j\mathbb{Z}$ for each j, where $1 \le j \le t$. Hence $n_j|k$ for every j, where $1 \le j \le t$. By induction and Corollary 1.4.9, we deduce that $n = n_1 \ldots n_t$ divides k, so that $k \in n\mathbb{Z}$.

Conversely, if $k \in n\mathbb{Z}$ then $k \in n_j\mathbb{Z}$, for $1 \le j \le t$ so $k \in \mathbf{Ker}\, f$. Consequently,

$$\mathbf{Ker}\, f = n\mathbb{Z} \text{ and } \mathbb{Z}/n\mathbb{Z} \cong R = \mathbb{Z}/n_1\mathbb{Z} \times \cdots \times \mathbb{Z}/n_t\mathbb{Z}.$$

Using the results of Section 7.1 which describes the group of invertible elements of a Cartesian product of rings, we have

$$\mathbf{U}(\mathbb{Z}/n\mathbb{Z}) \cong \mathbf{U}(\mathbb{Z}/n_1\mathbb{Z}) \times \cdots \times \mathbf{U}(\mathbb{Z}/n_t\mathbb{Z}).$$

The following important special case arises.

9.5.9. Corollary. *Let n be a positive integer and suppose that $n = p_1^{k_1} p_2^{k_2} \dots p_t^{k_t}$ is its decomposition into primes, where $p_i \neq p_j$ whenever $i \neq j$. Then*

$$\mathbb{Z}/n\mathbb{Z} \cong \mathbb{Z}/p_1^{k_1}\mathbb{Z} \times \cdots \times \mathbb{Z}/p_t^{k_t}\mathbb{Z},$$

and

$$\mathbf{U}(\mathbb{Z}/n\mathbb{Z}) \cong \mathbf{U}(\mathbb{Z}/p_1^{k_1}\mathbb{Z}) \times \cdots \times \mathbf{U}(\mathbb{Z}/p_t^{k_t}\mathbb{Z}).$$

Corollary 9.5.9 reduces the description of $\mathbb{Z}/n\mathbb{Z}$ and $\mathbf{U}(\mathbb{Z}/n\mathbb{Z})$ to the case when $n = p^k$ is a power of a prime. The description of $\mathbf{U}(\mathbb{Z}/p^k\mathbb{Z})$ requires separate consideration of the cases when p is odd and when $p = 2$.

9.5.10. Definition. *Let n be a positive integer. We say that n has a primitive root if the group $\mathbf{U}(\mathbb{Z}/n\mathbb{Z})$ is cyclic. We say that an integer k is a primitive root modulo n, if $\langle k + n\mathbb{Z}\rangle = \mathbf{U}(\mathbb{Z}/n\mathbb{Z})$.*

From Theorem 9.4.13 it follows that if k is a primitive root modulo n then $\mathbf{GCD}(k, n) = 1$. Also since $|k + n\mathbb{Z}| = |\langle k + n\mathbb{Z}\rangle|$, by Theorem 9.4.13 we deduce that $|k + n\mathbb{Z}| = \varphi(n)$.

9.5.11. Definition. *Let n be a positive integer and let k be an integer. Suppose that $\mathbf{GCD}(k, n) = 1$. The positive integer t is called the order of k modulo n, if t is the order of $k + n\mathbb{Z}$ in the group $\mathbf{U}(\mathbb{Z}/n\mathbb{Z})$.*

9.5.12. Proposition. *Let n be a positive integer. The integer k is a primitive root modulo n if and only if the order of k modulo n is $\varphi(n)$.*

Proof. If k is a primitive root modulo n then, as seen above, $|k + n\mathbb{Z}| = \varphi(n)$. Conversely, let k have order $\phi(n)$ modulo n. Then $\varphi(n) = |k + n\mathbb{Z}| = |\langle k + n\mathbb{Z}\rangle|$. Theorem 9.4.13 shows that $\varphi(n) = |\mathbf{U}(\mathbb{Z}/n\mathbb{Z})|$, so that

$$|\langle k + n\mathbb{Z}\rangle| = |\mathbf{U}(\mathbb{Z}/n\mathbb{Z})|,$$

and hence

$$\langle k + n\mathbb{Z}\rangle = \mathbf{U}(\mathbb{Z}/n\mathbb{Z}).$$

Thus, $\mathbf{U}(\mathbb{Z}/n\mathbb{Z})$ is a cyclic group.

Observe, that not every natural number has a primitive root. For example, it is easy to see that

$$\mathbf{U}(\mathbb{Z}/8\mathbb{Z}) = \{1 + 8\mathbb{Z}, 3 + 8\mathbb{Z}, 5 + 8\mathbb{Z}, 7 + 8\mathbb{Z}\},$$

and

$$(3 + 8\mathbb{Z})^2 = 9 + 8\mathbb{Z} = 1 + 8\mathbb{Z},$$

$$(5 + 8\mathbb{Z})^2 = 25 + 8\mathbb{Z} = 1 + 8\mathbb{Z},$$

$$(7 + 8\mathbb{Z})^2 = 49 + 8\mathbb{Z} = 1 + 8\mathbb{Z}.$$

However, $\varphi(8) = |\mathbf{U}(\mathbb{Z}/8\mathbb{Z})| = 4$. Consequently, the group $\mathbf{U}(\mathbb{Z}/8\mathbb{Z})$ is not cyclic, so that 8 has no primitive roots.

The case $p \neq 2$ is different, however.

9.5.13. Lemma. *Let t be a positive integer, let a, b be integers and let p be a prime. If $a \equiv b \pmod{p^t}$, then $a^p \equiv b^p \pmod{p^{t+1}}$.*

Proof. We have $a = b + cp^t$ for some integer c. It follows, from the Binomial theorem, that

$$a^p = (b + cp^t)^p = b^p + C_1^p b^{p-1} cp^t + d,$$

where

$$d = C_2^p b^{p-2}(cp^t)^2 + \cdots + C_{p-1}^p b(cp^t)^{p-1} + (cp^t)^p,$$

and $C_k^p = \frac{p!}{k!(p-k)!}$ are binomial coefficients. As seen in Section 9.3, p divides C_k^p, for $1 \leq k \leq p$. This implies that if $2 \leq k < p$ then $C_k^p b^{p-k}(cp^t)^k$ is divisible by $pp^{tk} = p^{tk+1} \geq p^{t+1}$. Also, $(cp^t)^p$ is divisible by p^{t+1} since

$$pt - (t + 1) = pt - t - 1 = t(p - 1) - 1 \geq t - 1 \geq 0.$$

Hence p^{t+1} divides d, and we have

$$a^p = b^p + pb^{p-1}cp^t + d = b^p + p^{t+1}b^{p-1}c + p^{t+1}d_1,$$

so that $a^p \equiv b^p \pmod{p^{t+1}}$.

9.5.14. Lemma. *Let t be a positive integer, let a, b be integers, and let p be an odd prime. If $t \geq 2$ and $s = p^{t-2}$, then $(1 + ap)^s \equiv 1 + ap^{t-1} \pmod{p^t}$.*

Proof. We will use induction on t. If $t = 2$, then the assertion is clear, so suppose that $t \geq 2$ and, inductively, assume the result true for t. By the induction hypothesis, $(1 + ap)^s \equiv 1 + ap^{t-1} \pmod{p^t}$ and from Lemma 9.5.13, we have

$$(1 + ap)^{sp} \equiv (1 + ap^{t-1})^p \pmod{p^{t+1}}.$$

However, $sp = p^{t-2}p = p^{t-1}$ whereas, by the Binomial theorem,

$$(1 + ap^{t-1})^p = 1 + C_1^p ap^{t-1} + d,$$

where

$$d = C_2^p (ap^{t-1})^2 + \cdots + C_{p-1}^p (ap^{t-1})^{p-1} + (ap^{t-1})^p,$$

and C_k^p are binomial coefficients. As in Section 9.3, p divides C_k^p, for $1 \leq k < p$. For $2 \leq k < p$ we have

$$1 + tk - k - t - 1 = tk - k - t = t(k-1) - k \geq 2k - 2 - k = k - 2 \geq 0,$$

so that $pp^{(t-1)k} \geq p^{t+1}$. The last term $(ap^{t-1})^p$ of the decomposition of d is divisible by p^{t+1} because

$$(t-1)p - t - 1 \geq 3(t-1) - t - 1 = 2t - 4 \geq 0,$$

so that $(p^{t-1})^p \geq p^{t+1}$. Hence p^{t+1} divides d, and we have

$$(1 + ap^{t-1})^p = 1 + pap^{t-1} + d = 1 + ap^t + p^{t+1}d_1,$$

for some $d_1 \in \mathbb{Z}$ so that $(1 + ap)^m \equiv 1 + ap^t \pmod{p^{t+1}}$, where $m = p^{t-1}$. This completes the induction and the proof.

9.5.15. Corollary. *Let t be a positive integer, let a be an integer, and let p be an odd prime. If p does not divide a then the order of $1 + ap$ modulo p^t is equal to p^{t-1}.*

Proof. Let $m = p^{t-1}$. From Lemma 9.5.14 we deduce that

$$(1 + ap)^m \equiv 1 + ap^t \qquad \pmod{p^{t+1}},$$

so that $(1 + ap)^m \equiv 1 \pmod{p^t}$. Thus, the order of the element $1 + ap + p^t\mathbb{Z}$ in $\mathbf{U}(\mathbb{Z}/p^t\mathbb{Z})$ divides p^{t-1}. On the other hand, if $s = p^{t-2}$ then, again by Lemma 9.5.14,

$$(1 + ap)^s \equiv 1 + ap^{t-1} \qquad \pmod{p^t}.$$

Since p does not divide a, $(1 + ap)^s + p^t\mathbb{Z} \neq 1 + p^t\mathbb{Z}$ and hence

$$\left|1 + ap + p^t\mathbb{Z}\right| = p^{t-1}.$$

9.5.16. Theorem. *Let t be a positive integer and let p be an odd prime. Then $\mathbf{U}(\mathbb{Z}/p^t\mathbb{Z})$ is a cyclic group. Thus, p^t has primitive roots.*

Proof. From Corollary 9.4.20, we see that $\mathbf{U}(\mathbb{Z}/p\mathbb{Z})$ is a cyclic group so there exists an integer m with the property that $\langle m + p\mathbb{Z}\rangle = \mathbf{U}(\mathbb{Z}/p\mathbb{Z})$. Then m is not divisible by p. Suppose that $m^{p-1} \equiv 1 \pmod{p^2}$ and consider $m + p$. We have, by the binomial theorem,

$$(m + p)^{p-1} = m^{p-1} + (p-1)m^{p-2}p + p^2 d,$$

for some $d \in \mathbb{Z}$, as before. Since p does not divide $(p-1)m^{p-2}$, p^2 does not divide $(m+p)^{p-1} - 1$. However, $m + p + p\mathbb{Z} = m + p\mathbb{Z}$. Consequently, if $m^{p-1} \equiv 1 \pmod{p^2}$, we can replace m by $m + p$ and therefore assume that $m^{p-1} - 1$ is not divisible by p^2. Suppose next that the order of the element $m + p^t\mathbb{Z}$ in $\mathbf{U}(\mathbb{Z}/p^t\mathbb{Z})$ is n. By the above, we may assume that $m^{p-1} = 1 + ap$, where p does not divide a. Furthermore,

$$\begin{aligned}(1 + ap + p^t\mathbb{Z})^n &= (m^{p-1} + p^t\mathbb{Z})^n = ((m + p^t\mathbb{Z})^{p-1})^n \\ &= ((m + p^t\mathbb{Z})^n)^{p-1} = (1 + p^t\mathbb{Z})^{p-1} = 1 + p^t\mathbb{Z}.\end{aligned}$$

Corollary 9.5.15 shows that $1 + ap + p^t\mathbb{Z}$, as an element of the group $\mathbf{U}(\mathbb{Z}/p^t\mathbb{Z})$, has order p^{t-1}. It follows that $s = p^{t-1}$ divides n, say $n = p^{t-1}k = sk$ so $(m + p\mathbb{Z})^n = (m + p\mathbb{Z})^{sk}$. By Corollary 9.4.15, $(m + p\mathbb{Z})^{p-1} = 1 + p\mathbb{Z}$ and therefore

$$(m + p\mathbb{Z})^p = (m + p\mathbb{Z})^{p-1}(m + p\mathbb{Z}) = (1 + p\mathbb{Z})(m + p\mathbb{Z}) = m + p\mathbb{Z}.$$

Using an induction argument, we conclude that

$$(m + p\mathbb{Z})^s = (m + p\mathbb{Z}).$$

Hence,

$$(m + p\mathbb{Z})^n = (m + p\mathbb{Z})^{sk} = (m + p\mathbb{Z})^k.$$

However, $m + p^t\mathbb{Z}$ has order n so $(m + p^t\mathbb{Z})^n = 1 + p^t\mathbb{Z}$ and we see that $(m + p\mathbb{Z})^k = 1 + p\mathbb{Z}$. Since

$$\langle m + p\mathbb{Z}\rangle = \mathbf{U}(\mathbb{Z}/p\mathbb{Z})$$

we deduce that

$$|m + p\mathbb{Z}| = |\langle m + p\mathbb{Z}\rangle| = |\mathbf{U}(\mathbb{Z}/p\mathbb{Z})|.$$

By Theorem 9.4.13, $|\mathbf{U}(\mathbb{Z}/p\mathbb{Z})| = \varphi(p) = p - 1$. Thus, we have $|m + p\mathbb{Z}| = p - 1$. The equation $(m + p\mathbb{Z})^k = 1 + p\mathbb{Z}$ shows that $p - 1$ divides k, so $p^{t-1}(p-1) = \varphi(p^t)$ divides $n = p^{t-1}k$. On the other hand, by Corollary 8.3.9, n divides $\left|\mathbf{U}(\mathbb{Z}/p^t\mathbb{Z})\right| = \varphi(p^t)$. Hence $n = \varphi(p^t)$. Proposition 9.5.12 completes the proof.

The proof of this theorem is constructive since it shows how to find a primitive root. Moreover, it shows how to do it if at least one of the primitive roots modulo a prime number is obtained. In the general case the last task can be quite complicated. The following theorem helps reduce the possible selection significantly.

9.5.17. Theorem. *Let q be an odd prime and let $q-1 = p_1^{k_1} p_2^{k_2} \dots p_t^{k_t}$ be the primary decomposition of $q-1$ where $p_i \neq p_j$ whenever $i \neq j$. An integer g with the property* $\mathbf{GCD}(q, g) = 1$ *is a primitive root modulo q if and only if it is not a solution of any of the congruences $x^{\frac{q-1}{p_j}} \equiv 1 \pmod{q}$, for $1 \leq j \leq t$.*

Proof. In fact, if g is a primitive root modulo q then, by Proposition 9.5.12, $|g + q\mathbb{Z}| = q - 1$. Thus, if p is an arbitrary prime divisor of $q-1$, then $(g + q\mathbb{Z})^{\frac{q-1}{p}} \neq 1 + q\mathbb{Z}$, so g satisfies none of the given congruences.

Conversely, let g be an integer such that $\mathbf{GCD}(g, q) = 1$ and suppose that g satisfies none of the congruences $x^{\frac{q-1}{p_j}} \equiv 1 \pmod{q}$, for $1 \leq j \leq t$. Suppose that g is not a primitive root modulo q. Thus, $\langle g + q\mathbb{Z}\rangle \neq \mathbf{U}(\mathbb{Z}/q\mathbb{Z})$. From Corollary 8.3.9 we see that $|g + q\mathbb{Z}| = m$ where m is a proper divisor of $q-1$. However, then $m = p_1^{r_1} p_2^{r_2} \dots p_t^{r_t}$ where $r_j \leq k_j$, for $1 \leq j \leq t$, and there exists an index s such that $r_s \neq k_s$. In particular, m is a divisor of $\frac{q-1}{p_s}$, and therefore

$$1 + q\mathbb{Z} = (g + q\mathbb{Z})^m = (g + q\mathbb{Z})^{\frac{q-1}{p^s}}.$$

Then g satisfies the congruence $x^{\frac{q-1}{p_s}} \equiv 1 \pmod{q}$, a contradiction from which we conclude that g is a primitive root modulo q.

We now consider the case $p = 2$.

9.5.18. Theorem. *Let t be a positive integer. Then $\mathbf{U}(\mathbb{Z}/2\mathbb{Z})$ and $\mathbf{U}(\mathbb{Z}/4\mathbb{Z})$ are cyclic groups. If $t \geq 3$, then*

$$\mathbf{U}(\mathbb{Z}/2^t\mathbb{Z}) = \{(-1)^a 5^b + 2^t\mathbb{Z} \mid a \in \{0, 1\}, 0 \leq b \leq 2^{t-2}\}.$$

In particular, $\mathbf{U}(\mathbb{Z}/2^t\mathbb{Z})$ is generated by $-1 + 2^t\mathbb{Z}$ and $5 + 2^t\mathbb{Z}$.

Proof. It is clear that $\mathbf{U}(\mathbb{Z}/2\mathbb{Z}) = \{1 + 2\mathbb{Z}\} = \langle 1 + 2\mathbb{Z}\rangle$, and $\mathbf{U}(\mathbb{Z}/4\mathbb{Z}) = \{1 + 4\mathbb{Z}, 3 + 4\mathbb{Z}\} = \langle 3 + 4\mathbb{Z}\rangle$ so suppose that $t \geq 3$. We show that $5^s \equiv 1 + 2^{t-1} \pmod{2^t}$ where $s = 2^{t-3}$. For $t = 3$ this is clear so we assume inductively that the result is true for some integer $t \geq 3$ and prove that the result is valid for $t + 1$. To this end, since $5^s \equiv 1 + 2^{t-1} \pmod{2^t}$, a simple computation shows that $(5^s)^2 \equiv (1 + 2^{t-1})^2 \equiv 1 + 2 \times 2^{t-1} + 2^{2t-2} \pmod{2^{t+1}}$. However,

$t \geq 3$ so $2t - 2 \geq t + 1$ and $2s = 2^{t-2}$ so $5^m \equiv 1 + 2^t \pmod{2^{t+1}}$, where $m = 2^{t-2}$. Now we have

$$(5 + 2^t\mathbb{Z})^m = 5^m + 2^t\mathbb{Z} = 1 + 2^t\mathbb{Z}.$$

Thus, the order of the element $5 + 2^t\mathbb{Z}$ in $\mathbf{U}(\mathbb{Z}/2^t\mathbb{Z})$ divides 2^{t-2}. On the other hand, for $s = 2^{t-3}$, we have $5^s \equiv 1 + 2^{t-1} \pmod{2^t}$ and $5^s + 2^t\mathbb{Z} \neq 1 + 2^t\mathbb{Z}$. This means that $|5 + 2^t\mathbb{Z}| = 2^{t-2}$.

We next consider the set

$$\{(-1)^a 5^b + 2^t\mathbb{Z} \mid a \in \{0, 1\}, 0 \leq b \leq 2^{t-2}\}.$$

Suppose that $(-1)^a 5^b + 2^t\mathbb{Z} = (-1)^c 5^d + 2^t\mathbb{Z}$ for some $a, c \in \{0, 1\}, 0 \leq b, d \leq 2^{t-2}$. Then, since $5 \equiv 1 \pmod 4$ we have $(-1)^a + 4\mathbb{Z} = (-1)^c + 4\mathbb{Z}$, which is possible only if $a = c$. This implies that $5^b + 2^t\mathbb{Z} = 5^d + 2^t\mathbb{Z}$, and then $(5 + 2^t\mathbb{Z})^{b-d} = 1 + 2^t\mathbb{Z}$. Since $|5 + 2^t\mathbb{Z}| = 2^{t-2}$, it follows that 2^{t-2} divides $b - d$, and by the choice of b, d we conclude that $b - d = 0$ and $b = d$. Consequently, all elements of the set

$$\{(-1)^a 5^b + 2^t\mathbb{Z} \mid a \in \{0, 1\}, 0 \leq b \leq 2^{t-2}\}$$

are distinct and therefore

$$|\{(-1)^a 5^b + 2^t\mathbb{Z} \mid a \in \{0, 1\}, 0 \leq b \leq 2^{t-2}\}| = 22^{t-2} = 2^{t-1} = \varphi(2^t).$$

By Theorem 9.4.13, $|\mathbf{U}(\mathbb{Z}/2^t\mathbb{Z})| = \varphi(2^t)$, and hence

$$\mathbf{U}(\mathbb{Z}/2^t\mathbb{Z}) = \{(-1)^a 5^b + 2^t\mathbb{Z} \mid a \in \{0, 1\}, 0 \leq b \leq 2^{t-2}\}.$$

We next consider congruences of higher powers, beginning with congruences of the type $x^k \equiv a \pmod n$.

9.5.19. Definition. *Let n, k be positive integers and let a be an integer with the property* $\mathbf{GCD}(a, n) = 1$. *We say that a is a k-power residue modulo n, if the congruence $x^k \equiv a \pmod n$ has a solution y such that* $\mathbf{GCD}(y, n) = 1$. *In the special case when $k = 2$ we call "a" a quadratic residue modulo n.*

9.5.20. Theorem. *Let n, k be positive integers and let a be an integer such that* $\mathbf{GCD}(a, n) = 1$. *Suppose that n has primitive roots. The integer a is a k-power residue modulo n if and only if $a^{\frac{\varphi(n)}{d}} \equiv 1 \pmod n$ where $d =$* $\mathbf{GCD}(k, \varphi(n))$. *Moreover, if the congruence $x^k \equiv a \pmod n$ has solutions, then it has exactly d solutions.*

Proof. Let g be a primitive root modulo n, so $\langle g + n\mathbb{Z}\rangle = \mathbf{U}(\mathbb{Z}/n\mathbb{Z})$ and let y be a solution of the congruence $x^k \equiv a \pmod{n}$ for which $\mathbf{GCD}(y, n) = 1$. By Theorem 9.4.13, $y + n\mathbb{Z}, a + n\mathbb{Z} \in \mathbf{U}(\mathbb{Z}/n\mathbb{Z})$. Then

$$a + n\mathbb{Z} = (g + n\mathbb{Z})^b = g^b + n\mathbb{Z}$$

and

$$y + n\mathbb{Z} = (g + n\mathbb{Z})^z = g^z + n\mathbb{Z}$$

for some $b, z \in \mathbb{Z}$. We have

$$g^b + n\mathbb{Z} = a + n\mathbb{Z} = (y + n\mathbb{Z})^k = (g^z + n\mathbb{Z})^k = g^{zk} + n\mathbb{Z}.$$

Thus,

$$(g + n\mathbb{Z})^b = (g + n\mathbb{Z})^{kz} \text{and } 1 + n\mathbb{Z} = (g + n\mathbb{Z})^{kz-b} = g^{kz-b} + n\mathbb{Z}.$$

Since $\langle g + n\mathbb{Z}\rangle = \mathbf{U}(\mathbb{Z}/n\mathbb{Z})$, Theorem 9.4.13 implies that $|g + n\mathbb{Z}| = \varphi(n)$, from which we deduce that $\varphi(n)$ divides $kz - b$. Thus, $kz \equiv b \pmod{\varphi(n)}$. By Theorem 9.5.3, d must divide b, say $b = cd$ for some positive integer c. We have

$$(a + n\mathbb{Z})^{\frac{\varphi(n)}{d}} = (g + n\mathbb{Z})^{b\frac{\varphi(n)}{d}} = (g + n\mathbb{Z})^{\varphi(n)c} = ((g + n\mathbb{Z})^{\varphi(n)})^c = 1 + n\mathbb{Z}.$$

From this we have $a^{\frac{\varphi(n)}{d}} \equiv 1 \pmod{n}$.

Conversely, let $a^{\frac{\varphi(n)}{d}} \equiv 1 \pmod{n}$ be valid. Then $(g + n\mathbb{Z})^{b\frac{\varphi(n)}{d}} = 1 + n\mathbb{Z}$. Since $|g + n\mathbb{Z}| = \varphi(n)$, $\varphi(n)$ divides $b\frac{\varphi(n)}{d}$, and it follows that d divides b. By Theorem 9.5.3, $kz \equiv b \pmod{\varphi(n)}$ has exactly d solutions. Using the above arguments, we can show that there is a one-to-one correspondence between distinct solutions of the congruence $kz \equiv b \pmod{\varphi(n)}$ and distinct solutions of the congruence $x^k \equiv a \pmod{n}$. So the latter congruence has exactly d solutions. The proof is complete.

Let n be a positive integer and let $n = 2^{k_1} p_2^{k_2} \dots p_t^{k_t}$ be its prime decomposition, where $p_i \neq p_j$ whenever $i \neq j$. Let m be a positive integer and let a be an integer with the property $\mathbf{GCD}(a, n) = 1$. Consider the system of congruences

$$x^m \equiv a \pmod{2^{k_1}}, x^m \equiv a \pmod{p_2^{k_2}}, \dots, x^m \equiv a \pmod{p_t^{k_t}}.$$

Suppose that these congruences have solutions $y_1, \dots, y_t$ respectively. By Theorem 9.5.7, there exists an integer y such that

$$y + 2^{k_1}\mathbb{Z} = y_1 + 2^{k_1}\mathbb{Z}, y + p_2^{k_2}\mathbb{Z} = y_2 + p_2^{k_2}\mathbb{Z}, \dots, y + p_t^{k_t}\mathbb{Z} \equiv y_t + p_t^{k_t}\mathbb{Z}.$$

Then

$$y^m + 2^{k_1}\mathbb{Z} = (y + 2^{k_1}\mathbb{Z})^m = (y_1 + 2^{k_1}\mathbb{Z})^m = y_1^m + 2^{k_1}\mathbb{Z} = a + 2^{k_1}\mathbb{Z},$$

and, similarly,

$$y^m + p_j^{k_j}\mathbb{Z} = a + p_j^{k_j}\mathbb{Z} \text{ for } 2 \le j \le t.$$

In other words,

$$y^m - a \in 2^{k_1}\mathbb{Z} \text{ and } y^m - a \in p_j^{k_j}\mathbb{Z} \text{ for } 2 \le j \le t,$$

which means that 2^{k_1} divides $y^m - a$ and $p_j^{k_j}$ divides $y^m - a$, for $2 \le j \le t$. Hence, the product $n = 2^{k_1} p_2^{k_2} \dots p_t^{k_t}$ divides $y^m - a$. Thus, we can see that y is a solution of the congruence $x^m \equiv a \pmod{n}$. However, it is clear that each solution of the congruence

$$x^m \equiv a \pmod{n}$$

is also a solution of the system of congruences

$$x^m \equiv a \pmod{2^{k_1}}, x^m \equiv a \pmod{p_2^{k_2}}, \dots, x^m \equiv a \pmod{p_t^{k_t}}$$

Consequently, we can reduce the problem of finding solutions of the congruence $x^m \equiv a \pmod{n}$ to the special case when $n = p^k$ for some prime p. If $p \neq 2$, then Theorem 9.5.16 implies that p^k has a primitive root, so that we can apply Theorem 9.5.20. The case $p = 2$ again requires separate consideration. If $k \le 2$, then the number 2^k has primitive roots, so that we can apply Theorem 9.5.20. So the important case is the case when $k \ge 3$.

9.5.21. Theorem. *Let n, t be positive integers, let $t \ge 3$ and let a be an odd integer. If n is odd, then the congruence $x^n \equiv a$ $(\mathbf{mod}\ 2^t)$ has a solution and this solution is unique. If n is even, then a is an n–power residue modulo 2^t if and only if $a \equiv 1 \pmod{4}$ and $a^{\frac{s}{d}} \equiv 1 \pmod{2^t}$ where $s = 2^{t-2}, d = \mathbf{GCD}(n, 2^{t-2})$. Moreover, if the given congruence has a solution it has exactly $2d$ solutions.*

Proof. Suppose first that n is odd. The results of Section 9.4 imply that $m = \varphi(2^t) = 2^{t-1}(2-1) = 2^{t-1}$. By Corollary 1.4.7, there exist integers u, v such that $un + mv = 1$. For an arbitrary integer x we let $\hat{x} = x + 2^t\mathbb{Z}$. Then

$$\hat{a} = \hat{a}^1 = \hat{a}^{nu+mv} = (\hat{a}^{un})(\hat{a}^{mv}) = (\hat{a}^u)^n(\hat{a}^m)^v.$$

By Corollary 9.4.14, $(\hat{a}^m) = \hat{1}$. So $y = a^u$ is a solution of the congruence $x^n \equiv a$ $(\mathbf{mod}\ 2^t)$. We can find the number u using the Euclidian algorithm of Section 9.2.

Now suppose that n is even and that y is an odd integer, which is a solution of $x^n \equiv a \pmod{2^t}$. By Theorem 9.4.13,

$$y + 2^t\mathbb{Z}, a + 2^t\mathbb{Z} \in \mathbf{U}(\mathbb{Z}/2^t\mathbb{Z}).$$

Taking account of Theorem 9.5.18, we deduce that

$$a + 2^t\mathbb{Z} = (-1)^b 5^c + 2^t\mathbb{Z} \text{ and } y + 2^t\mathbb{Z} = (-1)^r 5^z + 2^t\mathbb{Z}$$

for certain $b, c, z, r \in \mathbb{Z}$. Now we have

$$\begin{aligned}(-1)^b 5^c + 2^t\mathbb{Z} = a + 2^t\mathbb{Z} = (y + 2^t\mathbb{Z})^n &= ((-1)^r 5^z + 2^t\mathbb{Z})^n \\ &= (-1)^{rn} 5^{zn} + 2^t\mathbb{Z} = 5^{zn} + 2^t\mathbb{Z},\end{aligned}$$

since n is even. In particular, $b = 0$, so

$$a \equiv 1 \pmod 4 \text{ and } a + 2^t\mathbb{Z} = 5^c + 2^t\mathbb{Z}.$$

Let $n = dw$. Then

$$\begin{aligned}(a + 2^t\mathbb{Z})^{\frac{s}{d}} &= (5^c + 2^t\mathbb{Z})^{\frac{s}{d}} = (5^{zn} + 2^t\mathbb{Z})^{\frac{s}{d}} = (5 + 2^t\mathbb{Z})^{\frac{zns}{d}} \\ &= (5 + 2^t\mathbb{Z})^{\frac{zsdw}{d}} = (5 + 2^t\mathbb{Z})^{szw}.\end{aligned}$$

From the proof of Theorem 9.5.18, it follows that 5 has order s, modulo 2^t, so that

$$(a + 2^t\mathbb{Z})^{\frac{s}{d}} = 1 + 2^t\mathbb{Z}, \text{ and hence } a^{\frac{s}{d}} \equiv 1 \pmod{2^t}.$$

The equations

$$(5 + 2^t\mathbb{Z})^c = 5^c + 2^t\mathbb{Z} = 5^{zn} + 2^t\mathbb{Z} = (5 + 2^t\mathbb{Z})^{zn}$$

imply $nz \equiv c \pmod{2^{t-2}}$. By Theorem 9.5.3, this congruence has exactly d solutions. Note that y and $-y$ are both solutions of $x^n \equiv a \pmod{2^t}$, when n is even.

Conversely, let $a \equiv 1 \pmod 4$ and $a^{\frac{s}{d}} \equiv 1 \pmod{2^t}$. Then, $b = 0$ and $a + 2^t\mathbb{Z} = 5^c + 2^t\mathbb{Z}$. Thus,

$$1 + 2^t\mathbb{Z} = (a + 2^t\mathbb{Z})^{\frac{s}{d}} = (5^c + 2^t\mathbb{Z})^{\frac{s}{d}} = (5 + 2^t\mathbb{Z})^{\frac{cs}{d}}.$$

From Theorem 9.5.18 we know that $\left|5 + 2^t\mathbb{Z}\right| = 2^{t-2} = s$. Therefore $\frac{cs}{d}$ is divisible by s and hence $d \mid c$. In this case, by Theorem 9.5.3, the congruence $nz \equiv c \pmod{2^{t-2}}$ has a solution. As above, this implies that an integer y (together with $-y$) for which $y + 2^t\mathbb{Z} = 5^z + 2^t\mathbb{Z}$ is a solution of the congruence $x^n \equiv a$ $(\textbf{mod } 2^t)$.

We next consider an important special case of the congruence $x^m \equiv a$ $(\textbf{mod } n)$, namely the case when $n = p$ is an odd prime and $m = 2$, so we have $x^2 \equiv a \ (\textbf{mod } p)$. Since $p \neq 2$, then

$$-1 + p\mathbb{Z} = (p - 1) + p\mathbb{Z} \neq 1 + p\mathbb{Z},$$

and hence $-\hat{1} = -1 + p\mathbb{Z}$ and $\hat{1} = 1 + p\mathbb{Z}$ are the only solutions of the equation $X^2 = \hat{1}$. By Corollary 9.4.15, for an arbitrary integer a that is relatively prime to p, we have

$$((a + p\mathbb{Z})^{\frac{p-1}{2}})^2 = (a + p\mathbb{Z})^{p-1} = 1 + p\mathbb{Z},$$

and therefore either

$$(a + p\mathbb{Z})^{\frac{p-1}{2}} = -1 + p\mathbb{Z},$$

or

$$(a + p\mathbb{Z})^{\frac{p-1}{2}} = 1 + p\mathbb{Z}.$$

Let p be an odd prime and let a be an integer such that p does not divide a. We define the Legendre symbol $\left(\frac{a}{p}\right)$ by the rule

$$\left(\frac{a}{p}\right) = \begin{cases} 1 \text{ if } a \text{ is a quadratic residue modulo } p, \\ -1 \text{ if } a \text{ is a nonquadratic residue modulo } p. \end{cases}$$

The symbol $\left(\frac{a}{p}\right)$ is also called the quadratic character.

9.5.22. Theorem. *Let p be an odd prime and let a, b be integers such that* $\textbf{GCD}(a, p) = \textbf{GCD}(b, p) = 1$.

(i) *a is a quadratic residue modulo p if and only if $a^{\frac{p-1}{2}} \equiv 1 \ (\textbf{mod } p)$;*

(ii) $a^{\frac{p-1}{2}} \equiv \left(\frac{a}{p}\right) \ (\textbf{mod } p)$;

(iii) $\left(\frac{ab}{p}\right) \equiv \left(\frac{a}{p}\right)\left(\frac{b}{p} \ (\textbf{mod } p)\right)$.

Proof. By Corollary 9.4.20, p has primitive roots, so that for the proof of (i) we can apply Theorem 9.5.20. Assertion (i) together with the arguments proceeding this theorem prove (ii). Since the quotient ring $\mathbb{Z}/p\mathbb{Z}$ is commutative, then

$$((a + p\mathbb{Z})(b + p\mathbb{Z}))^k = (a + p\mathbb{Z})^k(b + p\mathbb{Z})^k$$

for every integer k. In particular,

$$\left(\frac{ab}{p}\right) = (ab + p\mathbb{Z})^{\frac{p-1}{2}} = ((a + p\mathbb{Z})(b + pZ))^{\frac{p-1}{2}}$$
$$= (a + p\mathbb{Z})^{\frac{p-1}{2}}(b + p\mathbb{Z})^{\frac{p-1}{2}} = \left(\frac{a}{p}\right)\left(\frac{b}{p}\right),$$

which proves (iii).

Now we consider the solution of general quadratic congruences, namely congruences of the type $ax^2 + bx + c \equiv 0 \pmod{p}$, where p is an odd prime. If p divides a, then the given congruence becomes a linear congruence of the type considered earlier. Hence, we may suppose that p does not divide a. Since a and p are then relatively prime, a^{-1} exists and by multiplying $ax^2 + bx + c \equiv 0 \pmod{p}$ by a^{-1} we obtain a congruence of the form $x^2 + b_1x + c_1 \equiv 0 \pmod{p}$.

Since p is odd, the congruence $2x \equiv b_1 \pmod{p}$ has one solution b_2, by Theorem 9.5.3. Thus, we obtain the congruence $x^2 + 2b_2x + c_1 \equiv 0 \pmod{p}$. We now complete the square, in a familiar fashion. This gives

$$x^2 + 2b_2x + c_1 = x^2 + 2b_2x + b_2^2 + (c_1 - b_2^2) = (x + b_2)^2 + c_2,$$

where $c_2 = (c_1 - b_2^2)$ and we let $y = x + b_2$, $c_3 = -c_2$ to obtain the congruence $y^2 \equiv c_3 \pmod{p}$.

Using Theorem 9.5.22 we can determine if c_3 is a quadratic residue modulo p. If c_3 is a quadratic residue then we can find a solution of this congruence using the methods of Theorem 9.5.20.

We end this section with a further well-known result. This remarkable theorem was first discovered by Bhaskara I, and much later disseminated in Europe by Ibn al-Haytham (circa 1000 AD), but it is named after John Wilson who announced this result in 1770, although he could not prove it. Lagrange gave the first proof in 1773. The resulting theorem became known as Wilson's theorem, despite its history and is a primality test of sorts.

9.5.23. Theorem *(Wilson's Theorem).*

(i) *If p is a prime, then $(p-1)! \equiv -1 \pmod{p}$.*
(ii) *If n is not prime, then either $n = 4$ or $(n-1)! \equiv 0 \pmod{n}$.*

Proof.

(i) If $p = 2$, then since $1 \equiv -1 \pmod{2}$, the result follows. Therefore we can suppose that p is an odd prime. As above, for an arbitrary integer x, we let $\hat{x} = x + p\mathbb{Z}$. From Corollary 9.4.15 we know that $a^{p-1} \equiv 1 \pmod{p}$ for each integer a, relatively prime to p. Thus, $\hat{a}$ is a root of the polynomial $X^{p-1} - \hat{1}$.

Since $|\mathbf{U}(\mathbb{Z}/p\mathbb{Z})| = p - 1$, all elements of $\mathbf{U}(\mathbb{Z}/p\mathbb{Z})$ are roots of the polynomial $X^{p-1} - \hat{1}$. From Proposition 7.5.9 we deduce that

$$X^{p-1} - \hat{1} = (X - \hat{1})(X - \hat{2})\dots(X - \widehat{(p-1)}).$$

Setting $X = \hat{0}$ we have $-\hat{1} = (-1)^{p-1}\widehat{(p-1)}! = \widehat{(p-1)}!$. The last expression can be rewritten in congruences as $(p-1)! \equiv -1 \pmod{p}$.

(ii) Now suppose that n is not prime. If $n = 4$ then $(4-1)! \equiv 2 \pmod{4}$, so we may suppose that $n \geq 6$. If n is not the square of a prime then $n = km$, where $k \neq m$ and $1 < k, m < n$. Then k, m are both factors of $(n-1)!$ so n divides $(n-1)!$. If $n = q^2$ for some prime q then $(n-1)!$ has q and $2q$ as factors, when $q \neq 2$, so once again n divides $(n-1)!$. This completes the proof.

EXERCISE SET 9.5

9.5.1. Let p be a prime and let $a^2 \equiv b^2 \pmod{p}$. Prove that $a \equiv b \pmod{p}$ or $a \equiv -b \pmod{p}$.

9.5.2. Let $x, y \in \mathbb{Z}$ and let $z = \mathbf{GCD}(x, y) \neq 1$. Prove that $ax \equiv bx \pmod{y}$ implies $a \equiv b \pmod{\frac{y}{z}}$.

9.5.3. Let $x, y \in \mathbb{Z}$, $\mathbf{GCD}(x, y) = 1$. Prove that $ax \equiv bx \pmod{y}$ implies $a \equiv b \pmod{y}$.

9.5.4. Find the order of the number 8 modulo 31.

9.5.5. Solve $8x \equiv 11 \pmod{83}$.

9.5.6. Solve $8x \equiv 17 \pmod{19}$.

9.5.7. Find the solutions of the system

$$3x \equiv 5 \pmod{7},$$
$$2x \equiv 1 \pmod{5}.$$

9.5.8. Find the solutions of the system

$$x \equiv 2 \pmod{7},$$
$$x \equiv 5 \pmod{9},$$
$$x \equiv 11 \pmod{15}.$$

9.5.9. For which values of a does the system

$$x \equiv 3 \pmod{11},$$
$$x \equiv 11 \pmod{20},$$

$$x \equiv 1 \pmod{15},$$
$$x \equiv a \pmod{8}.$$

have a solution.

9.5.10. Find the positive integers a, for which the congruence $x^3 \equiv a \pmod{13}$ is soluble.

9.5.11. Find the positive integers a, for which the congruence $x^3 \equiv a \pmod{15}$ is soluble.

9.5.12. Find the number of solutions of the congruence $x^2 \equiv 3 \pmod{101}$.

9.5.13. Find $\left(\frac{15}{23}\right)$.

9.5.14. Find $\left(\frac{14}{37}\right)$.

9.5.15. Find $\left(\frac{7}{41}\right)$.

9.5.16. Solve $x^5 \equiv 10 \pmod{11}$.

9.5.17. Prove that the congruence $x^2 \equiv 1 \pmod{2^k}$ has one solution for $k = 1$, two solutions for $k = 2$, and four solutions for $k \geq 3$.

9.5.18. Solve $x^4 \equiv 4 \pmod{17}$.

9.5.19. Solve $3x^2 + 5x + 1 \equiv 0 \pmod{17}$.

9.5.20. Solve $5x^2 + 2x + 3 \equiv 15 \pmod{51}$.

CHAPTER 10

THE REAL NUMBER SYSTEM

This chapter is dedicated to the development of the most important systems of numbers, namely, the systems of the natural numbers, the integers, the rational numbers, and the real numbers. We have so far been using these systems quite informally, assuming their properties without question. In this chapter, we take a more formal look at the real number system. One could argue that since the formal development of these systems does not require any special knowledge and we have already used their properties in the previous chapters, this chapter logically belongs at the beginning of the book. However, the rigorous construction of these number systems requires some experience and perhaps mathematical maturity, which we hope that the reader has now attained, but perhaps did not have before. We believe that the reader has gained this experience working with this book. Since the theme of numbers is so very important, this experience will play a key role here.

10.1 THE NATURAL NUMBERS

The notion of a natural number is one of the most fundamental and most important in mathematics. The system of natural numbers was the first abstract scientific concept created by man. Having dealt in everyday life, with certain quantities of real things, people noted certain general properties of numbers and developed the notion of counting numbers. This apparently simple concept is in some ways so profound that it has prompted some people to believe that this concept comes directly from God. A great German number theorist, Leopold Kronecker (1823–1891) said: "Die ganzen Zahlen hat der liebe Gott gemacht, alles andere ist Menschenwerk (God made the natural numbers, all else is the work of man)." [Heinrich Weber. Leopold Kronecker. Jahresberichte DMV 1893; 2:5–31.] Creating the notion of a natural number is a first step not only in mathematics, but in the development of all sciences.

We will not touch upon the great and interesting history of the development of this concept, since such a task would bring us far beyond the scope of this book. We proceed directly to the modern axiomatic theory of natural numbers. This theory was developed at the end of the nineteenth century and named in honor of a famous Italian mathematician, Giuseppe Peano (1858–1932), whose input in the axiomatization of natural numbers was of exceptional mathematical and philosophical value.

We already noted a lack of agreement in the interpretation of the number 0 as a nonnatural number, so we will now give the axioms for the set $\mathbb{N}_0$.

10.1.1. Definition. *The set $\mathbb{N}_0$ is a nonempty set and for all $a \in \mathbb{N}_0$, there is a uniquely defined element a', called the immediate successor of a and for which the following axioms hold:*

(P 1) *$a = b$ implies that $a' = b'$.*

(P 2) *There is an element 0 (the natural number 0) such that 0 is not the immediate successor of any element of $\mathbb{N}_0$. Thus $0 \neq a'$ for all elements $a \in \mathbb{N}_0$.*

(P 3) *If $a, b \in \mathbb{N}_0$ and $a' = b'$, then $a = b$.*

(P 4) *(the induction axiom). Let M be a subset of $\mathbb{N}_0$ satisfying the conditions:*

(i) $0 \in M$;

(ii) if $a \in M$, then $a' \in M$.

Then $M = \mathbb{N}_0$.

Axiom **(P 4)** is a law, which states that if a set is a subset of the set $\mathbb{N}_0$ and contains 0, and if for each number in the given set the succeeding natural number is in the set, then the given set is identical to the set $\mathbb{N}_0$. This is the basis for a very important instrument for establishing proofs; namely, the principle of mathematical induction, which we discussed in Section 1.4 and which we have used repeatedly throughout this book.

At once the following two questions arise, namely, is there any set satisfying axioms **(P 1)**–**(P 4)** and, if so, is this set unique? The answer to the first question is obtained once we have built a set theoretical model of the set of natural numbers. Here, as already occured in Section 1.1, the question of the level of rigor arises. Absolute rigor in the development of the theory of the natural numbers can be achieved with the aid of some additional important set of theoretical axioms such as, for example, the axiom of the universum. Such a level of exposition is far beyond the scope of this book and requires significant mathematical maturity. It will be enough for us to just show how the elements of the set $\mathbb{N}_0$ are defined. For this, we will proceed by defining the elements of $\mathbb{N}_0$ in the following way:

$$0 = \emptyset, 1 = \{\emptyset\}, 2 = 1 \cup \{1\} = \{0, 1\}, 3 = 2 \cup \{2\} = \{0, 1, 2\}, \ldots.$$

In general, if the natural number n has been defined, then we define its immediate successor n' by

$$n' = n + 1 = n \cup \{n\} = \{0, 1, \ldots, n\}.$$

The question of uniqueness requires a broader approach. There are distinct sets satisfying Definition 10.1.1, but all of them have absolutely identical properties with respect to the statement "a' succeeds a." In other words, all of them are isomorphic in some sense. The complete answer to this question does not require special knowledge, but has a technical nature. Unfortunately, again, we shall not say more on this topic.

As consequences of axioms **(P 1)**–**(P 4)**, we can obtain all the well-known properties of the natural numbers. Let $a, b \in \mathbb{N}_0$. If $b = a'$, then we say that a *precedes* b. By axiom **(P 2)**, 0 has no immediate predecessor and the following statement shows that 0 is the unique number with this property.

10.1.2. Proposition. *Let $a, b \in \mathbb{N}_0$ and suppose that $a \neq 0$. Then a has only one predecessor.*

Proof. Let

$$M = \{x \mid x \in \mathbb{N}_0 \text{ and } x = y' \text{ for some } y \in \mathbb{N}_0\} \cup \{0\}.$$

It follows that $0 \in M$. Also, if $a \in M$, then $a' \in M$ and, by axiom **(P 4)**, $M = \mathbb{N}_0$. If we suppose that $a = b'$ and $a = c'$, then $b' = c'$ and, by axiom **(P 3)**, $b = c$. Thus, every element of $\mathbb{N}_0$ other than 0 has a unique predecessor.

Addition of natural numbers is defined inductively as follows.

10.1.3. Definition. *Let n be a fixed natural number. Then*

(i) *$n + 0 = n$ and $n + 1 = n'$;*
(ii) *if $n + k$ has been defined then set $n + k' = (n + k)'$.*

Using this definition, we obtain the following familiar properties.

10.1.4. Theorem. *Let $a, b, c \in \mathbb{N}_0$. The following assertions hold:*

(i) *$a + (b + c) = (a + b) + c$ (the associative property);*

(ii) $a + b = b + a$ *(the commutative property);*
(iii) *if* $a \neq 0$, *then* $a + b \neq 0$.

Proof.
(i) Let $a, b \in \mathbb{N}_0$ be fixed and let

$$M = \{c \in \mathbb{N}_0 \mid a + (b + c) = (a + b) + c\}.$$

We have $a + (b + 0) = a + b = (a + b) + 0$, using Definition 10.1.3(i), and $(a + b) + 1 = (a + b)' = a + b' = a + (b + 1)$, by Definition 10.1.3(ii), so that $0, 1 \in M$. Suppose now that $c \in M$, that is, $(a + b) + c = a + (b + c)$. We now have, again by Definition 10.1.3(ii),

$$(a + b) + c' = ((a + b) + c)' = (a + (b + c))' = a + (b + c)' = a + (b + c').$$

By axiom **(P 4)**, $M = \mathbb{N}_0$ and the result now follows by the principle of mathematical induction.

(ii) Let a be an arbitrary element of $\mathbb{N}_0$. We use the principle of mathematical induction and first prove that $a + 0 = 0 + a$. Let

$$M_1 = \{a \in \mathbb{N}_0 \mid a + 0 = 0 + a\}.$$

If $a = 0$, then $0 + 0 = 0 = 0 + 0$, so $0 \in M_1$. Suppose next that $a \neq 0$ and $a \in M_1$. Then, by Definition 10.1.3(i),

$$a' + 0 = a' \text{ and } 0 + a' = (0 + a)' = a',$$

so that $a' \in M_1$. By axiom **(P 4)**, $M_1 = \mathbb{N}_0$ and hence $a + 0 = 0 + a$ for all $a \in \mathbb{N}_0$.

Next, using induction, we shall prove that $a + 1 = 1 + a$. Let

$$M_2 = \{a \in \mathbb{N}_0 \mid a + 1 = 1 + a\}.$$

We already know that $0 \in M_2$ by the previous argument. Suppose that $a \neq 0$ and $a \in M_2$. Then

$$a' + 1 = (a + 1) + 1 = (1 + a) + 1 = (1 + a)' = 1 + a',$$

so that $a' \in M_2$. By axiom **(P 4)**, we have $M_2 = \mathbb{N}_0$.

We prove now that $a + b = b + a$ for all $b \in \mathbb{N}_0$, by using induction. We let a be fixed and let

$$M_3 = \{b \in \mathbb{N}_0 \mid a + b = b + a\}.$$

We have already proved that $0, 1 \in M_3$. Suppose that $b \in M_3$. Then $a + b = b + a$, and

$$\begin{aligned} a + b' &= (a + b)' = (b + a)', \text{ since } b \in M_3 \\ &= b + (a + 1) = b + (1 + a) = (b + 1) + a = b' + a, \end{aligned}$$

since $a \in M_3$, so that $b' \in M_3$. By axiom **(P 4)**, $M_3 = \mathbb{N}_0$ and (ii) follows.

(iii) Since $a \neq 0$, we have $a = u'$ for some $u \in \mathbb{N}_0$, using Proposition 10.1.2. Then

$$a + b = b + a = b + u' = (b + u)'$$

and axiom **(P 2)** shows that $a + b \neq 0$. The result follows.

Now we will inductively define the operation of multiplication of natural numbers.

10.1.5. Definition. *Let n be a fixed natural number. Then*

(i) *$n \cdot 0 = 0 \cdot n = 0$ and $n \cdot 1 = n$;*
(ii) *if the product $n \cdot k$ has already been defined then put $n \cdot k' = n \cdot k + n$.*

Now we are ready to prove some well-known properties of multiplication. As usual we shall often omit the multiplication sign "$\cdot$".

10.1.6. Theorem. *Let $a, b, c \in \mathbb{N}_0$. The following assertions hold:*

(i) *$(a + b)c = ac + bc$ and $a(b + c) = ab + ac$ (the distributive property);*
(ii) *$ab = ba$ (the commutative property);*
(iii) *$(ab)c = a(bc)$ (the associative property).*

Proof.

(i) Our proof will proceed in a now familiar manner. Let a, b be fixed natural numbers and let

$$M = \{c \in \mathbb{N}_0 \mid (a + b)c = ac + bc\}.$$

We have

$$(a + b)0 = 0 = 0 + 0 = a0 + b0,$$

using Definition 10.1.5 and, likewise,

$$(a + b)1 = a + b = a1 + b1.$$

So $0, 1 \in M$. Suppose now that $c \in M$, that is, $(a + b)c = ac + bc$. We have

$$(a + b)c' = (a + b)c + (a + b), \text{ by definition}$$

$$= ac + bc + (a + b), \text{ by induction}$$
$$= (ac + a) + (bc + b) = ac' + bc',$$

by several applications of Theorem 10.1.4. Thus $c' \in M$ and hence, by axiom **(P 4)**, $M = \mathbb{N}_0$. This result follows by the principle of mathematical induction. We prove the second part of (i) using part (ii).

(ii) Let a be an arbitrary element of $\mathbb{N}_0$. If $a = 0$, then $ab = 0b = 0 = b0 = ba$, by Definition 10.1.5. Therefore, we may suppose further that $a \neq 0$. Let

$$M_1 = \{b \in \mathbb{N}_0 \mid ab = ba\}.$$

We have $a0 = 0 = 0a$, so $0 \in M_1$, and also $a1 = a$. Using induction on a, we prove that $1a = a$ and note that $1 \times 0 = 0$, by the above. Since $a \neq 0, a = d'$ for some element $d \in \mathbb{N}_0$, by Proposition 10.1.2, so we can assume inductively that $1d = d$. Then

$$1a = 1d' = 1d + 1 = d + 1 = d' = a.$$

It follows that $0, 1 \in M_1$. Suppose next that $b \neq 0$ and $b \in M_1$. Then

$$1b' = 1b + b = b1 + b = b + 1 = b' = b'1,$$

so that $b' \in M_1$. By axiom **(P 4)**, $M_1 = \mathbb{N}_0$. Thus, (ii) is proved and the second part of (i) follows since, using $(a + b)c = ac + bc$, it follows that

$$a(b + c) = (b + c)a = ba + ca = ab + ac.$$

(iii) If one of the elements a, b, c is zero, then the result holds. Therefore, we may assume that a, b, c are nonzero natural numbers. Let a, b be fixed and put

$$M_2 = \{c \in \mathbb{N}_0 \mid (ab)c = a(bc)\}.$$

We have, by Definition 10.1.5,

$$(ab)1 = ab = a(b1),$$

so that $0, 1 \in M_2$. Suppose now that $c \in M$, so $(ab)c = a(bc)$, for all $a, b \in \mathbb{N}_0$. We have now

$$(ab)c' = (ab)c + ab = a(bc) + ab = a(bc + b) = a(bc'),$$

so that $c \in M_2$. By axiom **(P 4)**, $M_2 = \mathbb{N}_0$ and the result follows.

Our definition of the natural numbers was based on the main relation "b succeeds a." This choice of the word "succeeds" shows that there is some natural order on $\mathbb{N}_0$.

10.1.7. Definition. *Let $k, n \in \mathbb{N}_0$. If there exists a natural number m such that $n = k + m$, then we say that "n is greater than or equal to k," which is denoted by $n \geq k$. Alternatively, we may say that "k is less than or equal to n," which is denoted by $k \leq n$. If, in this case, $n \neq k$, then we say that "n is (strictly) greater than k" which is denoted by $n > k$ or that "k is (strictly) less than n" and denote this by $k < n$.*

10.1.8. Theorem. *Let $a, b \in \mathbb{N}_0$.*

(i) *If a is nonzero then $a + b \neq b$.*
(ii) *One and only one from the following assertions is valid: $a = b$, $a < b$, or $a > b$.*

Proof.

(i) Let a be nonzero. First we prove that $a + b \neq b$ for each $b \in \mathbb{N}_0$. This is valid for $b = 0$, because $a + 0 = a \neq 0$. If $b = 1$, then $a + b = a + 1 = a' \neq 1$, by Proposition 10.1.2 and the fact that $0' = 1$. Suppose that we have already proved that $a + b \neq b$ and consider $a + b'$. We have $a + b' = (a + b)'$, by definition. However, if $(a + b)' = b'$ then $a + b = b$, by Proposition 10.1.2, contrary to our induction hypothesis. Consequently, $a + b' = (a + b)' \neq b'$ and, by the principle of mathematical induction, $a + b \neq b$ for each $b \in \mathbb{N}_0$.

(ii) First, we show that if $a, b \in \mathbb{N}_0$ are arbitrary then at most one of the assertions $a = b$, $a < b$, and $b < a$ is true. If $a > b$ or $a < b$ then, by definition, $a \neq b$. Suppose that $a > b$ and $b > a$. Then $a = b + k$ and $b = a + m$ for some k, $m \in \mathbb{N}_0$. Moreover, k and m are nonzero and, by Theorem 10.1.4(iii), $m + k \neq 0$. In this case,

$$a = b + k = (a + m) + k = a + (m + k).$$

Since $m + k \neq 0$, it follows from (i) that $a + (m + k) \neq a$, and we obtain a contradiction, which shows that $a > b$ and $a < b$ cannot both happen simultaneously.

We next have to show that if $a, b \in \mathbb{N}_0$, then one of $a = b$, $a < b$, or $a > b$ is true. To this end if, for the elements a, b, only one of the relations $a = b$, $a > b$, or $b > a$ is true then we will say that these elements *are comparable*. Let a be fixed and put

$$M = \{b \in \mathbb{N}_0 \mid a \text{ and } b \text{ are comparable}\}.$$

If $a = 0$ then $b = 0 + b$, which implies that $0 \leq b$ for each $b \in \mathbb{N}_0$ so that 0 is comparable to b for each $b \in \mathbb{N}_0$. Therefore, we can suppose that $a \neq 0$. In this case, we have again $0 \leq a$, so that $0 \in M$. Since $a \neq 0$, it follows from Proposition 10.1.2 that $a = u'$ for some $u \in \mathbb{N}_0$. We have that $a = u + 1$ and hence by definition $1 \leq a$, so that $1 \in M$.

Suppose that $b \in M$ and consider the element b'. If $a = b$, then $b' = b + 1 = a + 1$ and $a \leq b'$. If $a < b$ then $b = a + c$ for some $0 \neq c \in \mathbb{N}_0$ and

$$b' = b + 1 = (a + c) + 1 = a + (c + 1),$$

so that $a \leq b'$. Clearly $c + 1 \neq 0$, which shows that $a < b'$.

Finally, suppose that $a > b$. Then $a = b + v$ for some $0 \neq v \in \mathbb{N}_0$. If $v = 1$, then $a = b'$ and the result follows. If $v \neq 0, 1$ then, by Proposition 10.1.2, $v = w'$ so $v = w + 1$ for some $0 \neq w \in \mathbb{N}_0$. In this case,

$$a = b + v = b + (w + 1) = b + (1 + w) = (b + 1) + w = b' + w.$$

Since $w \neq 0$, $a > b'$ and it follows that $b' \in M$. By axiom **(P 4)**, $M = \mathbb{N}_0$ and the result follows by the principle of mathematical induction.

10.1.9. Proposition. *Let $a, b, c \in \mathbb{N}_0$. Then the following assertions hold:*

(i) $a \leq a$;
(ii) $a < b$ *and* $b < c$ *imply* $a < c$;
(iii) $a \leq b$ *and* $b < c$ *imply* $a < c$;
(iv) $a < b$ *and* $b \leq c$ *imply* $a < c$;
(v) $a \leq b$ *and* $b \leq c$ *imply* $a \leq c$;
(vi) $a \leq b$ *and* $b \leq a$ *imply* $a = b$.

Proof. We have $a = a + 0$, so (i) follows.

(ii) We have $b = a + u$ and $c = b + v$, for some $u, v \in \mathbb{N}_0$. Since $a \neq b$ and $b \neq c$, u, v are nonzero. Hence

$$c = b + v = (a + u) + v = a + (u + v),$$

so that $a \leq c$. By Theorem 10.1.4(iii), $u + v \neq 0$, so that $a + (u + v) \neq a$ by Theorem 10.1.8(i). Therefore $a < c$.

(iii) If $a \neq b$, then we may apply (ii). Suppose that $a = b$. Since $c = b + v = a + v$ for some $v \in \mathbb{N}_0$, with $v \neq 0$ it follows that $a \leq c$. By Theorem 10.1.8(i), $a + v \neq a$, so that $a \neq c$, and $a < c$.

(iv) The proof is similar.

(v) Follows from (ii)–(iv).

(vi) We have $b = a + u$ and $a = b + v$, for some $u, v \in \mathbb{N}_0$. Suppose that $u \neq 0$; then

$$a = b + v = (a + u) + v = a + (u + v).$$

By Theorem 10.1.4(iii) $u + v \neq 0$, and using Theorem 10.1.8(i), we obtain

$$a + (u + v) = (u + v) + a \neq a.$$

This contradiction proves that $u = 0$ and therefore $a = b$.

10.1.10. Theorem *(Laws of Monotonicity of Addition). Let $a, b, c \in \mathbb{N}_0$. Then the following properties hold:*

(i) *if $a + c = b + c$ then $a = b$;*
(ii) *if $a < b$ then $a + c < b + c$;*
(iii) *if $a + c < b + c$ then $a < b$;*
(iv) *if $a \leq b$ then $a + c \leq b + c$;*
(v) *if $a + c \leq b + c$ then $a \leq b$.*

Proof.
(i) We proceed by induction on c. If $c = 0$, then $a = a + 0 = b + 0 = b$. If $c = 1$ then, by hypothesis, $a' = a + 1 = b + 1 = b'$ and axiom **(P 3)** implies that $a = b$. Next, suppose inductively that we have already proved that $a + c = b + c$ implies $a = b$ and suppose that $a + c' = b + c'$. We have $(a + c)' = a + c' = b + c' = (b + c)'$. We again apply axiom **(P 3)** and deduce that $a + c = b + c$. By the induction hypothesis, it follows that $a = b$.

(ii) Since $a < b$, we know that $b = a + u$ for some $0 \neq u \in \mathbb{N}_0$. Then

$$b + c = (a + u) + c = a + (u + c) = a + (c + u) = (a + c) + u,$$

and so $a + c \leq b + c$. However, $u \neq 0$ so

$$(a + c) + u \neq a + c,$$

by Theorem 10.1.8. This implies that $a + c \neq b + c$.

(iii) Suppose that $a + c < b + c$. By Theorem 10.1.8(ii), we know that either $a = b, a < b$, or $a > b$. If $a = b$ then clearly $a + c = b + c$, contrary to the hypothesis. If $b < a$ then, by (ii), $b + c < a + c$, which is also impossible. So $a < b$.

(iv) follows from (ii) and (i).

(v) follows from (iii) and (i).

10.1.11. Theorem *(Laws of Monotonicity of Multiplication). Let $a, b, c \in \mathbb{N}_0$. Then the following properties hold:*

(i) *if a, b are nonzero then $ab \neq 0$;*
(ii) *if $a < b$ and $c \neq 0$ then $ac < bc$;*
(iii) *if $ac = bc$ and $c \neq 0$ then $a = b$;*
(iv) *if $ac < bc$ and $c \neq 0$ then $a < b$;*
(v) *if $a \leq b$ and $c \neq 0$ then $ac \leq bc$;*
(vi) *if $ac \leq bc$ and $c \neq 0$, then $a \leq b$;*
(vii) *if $a = bc$ and $c \neq 0$ then $a \geq b$;*
(viii) *if $bc = 1$ then $b = c = 1$.*

Proof.

(i) Since a, b are nonzero, $a = u'$, $b = v'$ for some $u, v \in \mathbb{N}_0$. We have

$$ab = av' = a(v + 1) = av + a = av + u' = (av + u)'.$$

Using axiom **(P 2)**, we deduce that ab is nonzero.

(ii) Since $a < b$, we have $b = a + u$ for some $0 \neq u \in \mathbb{N}_0$. Then

$$bc = (a + u)c = ac + uc,$$

so that $ac \leq bc$. By (i), $uc \neq 0$, so Theorem 10.1.8(i) implies that $ac + uc \neq ac$. Therefore $ac < bc$.

(iii) By Theorem 10.1.8(ii), precisely one of the following holds: namely, $a = b$, $a < b$, or $a > b$. If $a < b$ then, by (ii), $ac < bc$. Since $ac = bc$ we have a contradiction and a similar argument shows that $a > b$ is also impossible. Thus $a = b$.

(iv) The proof is similar.

(v) Follows from (ii).

(vi) Follows from (iii) and (iv).

(vii) If $b = 0$ then the result is clear, so assume $b > 0$. Since $b \neq 0$, we have $b \geq 1$. Then, by (v), $a = bc \geq b$.

(viii) By (vii), $b \leq 1$ and $c \leq 1$. Clearly, $b, c \neq 0$, so that $b = c = 1$.

10.1.12. Corollary. *Let $a, b \in \mathbb{N}_0$ and suppose that $b \neq 0$. Then there is a natural number c such that $bc > a$.*

Proof. If $a = 0$ then $b1 = b > 0 = a$, so suppose that $a \neq 0$. Since $b \neq 0$, $b = u'$ for some $u \in \mathbb{N}_0$, so $b = u + 1$. This means that $b \geq 1$. Put $c = a + 1$ so $c > a$. Applying Theorem 10.1.11(ii), we deduce that $bc > ab$ and $ab \geq a1 = a$, and Proposition 10.1.9(iii) implies that $bc > a$.

10.1.13. Corollary. *Let $a \in \mathbb{N}_0$. If $b \in \mathbb{N}_0$ is a number such that $a \leq b \leq a + 1$, then either $b = a$, or $b = a + 1$. In particular, if $c > a$, then $c \geq a + 1$, and if $d < a + 1$, then $d \leq a$.*

Proof. We have $b = a + u$ for some $u \in \mathbb{N}_0$. If $b \neq a$ then $u \neq 0$ and then there is an element $v \in \mathbb{N}_0$ such that $u = v'$. Thus $u = v + 1$, so that $u \geq 1$. Using Theorem 10.1.10(iv), we deduce that $b = a + u \geq a + 1$. Together with Proposition 10.1.9, this gives $b = a + 1$.

10.1.14. Corollary. *Let S be a nonempty subset of $\mathbb{N}_0$. Then S has a least element.*

Proof. If $0 \in S$, then 0 is the least element of S. If $0 \notin S$ but $1 \in S$, then 1 is the least element of S. Indeed, if $u \in S$ then $u \neq 0$, since $0 \notin S$. In this case, there is an element $v \in \mathbb{N}_0$ such that $u = v'$, so $u = v + 1$, which means that $u \geq 1$. Suppose now that $1 \notin S$. Let

$$M = \{n \in \mathbb{N}_0 \mid n \leq k \text{ for each element } k \in S\}.$$

By the above assumption, $0, 1 \in M$. If we suppose that for every $a \in M$, the number $a + 1$ also belongs to M then, by axiom **(P 4)**, $M = \mathbb{N}_0$. In this case, $S = \emptyset$ and we obtain a contradiction. This contradiction shows that there is an element $b \in M$ such that $b + 1 \notin M$. Since $b \in M$, then $b \leq k$ for each element $k \in S$. Suppose that $b \notin S$. Then $b < k$ for each $k \in S$. Then, Corollary 10.1.13 shows that $b + 1 \leq k$ for each element $k \in S$. It follows that $b + 1 \in M$, and we obtain a contradiction to the definition of b. Hence $b \in S$, and therefore b is the least element of S.

10.2 THE INTEGERS

The concept of numbers has a long history of development especially as a means of counting. From the dawn of civilization, natural numbers found applications as a useful tool in counting. Some documents (as, for example, the Nine Chapters on the Mathematical Art by Jiu Zhang Suan-Shu) show that certain types of negative numbers appeared in the period of the Han Dynasty (202 BC–220 AD).

The use of negative numbers in situations like mathematical problems of debt was known in early India (the 7th century AD). The Indian mathematician Brahmagupta, in Brahma-Sphuta-Siddhanta (written in 628 AD), discussed the use of negative numbers to produce the general form of the quadratic formula that remains in use today. Arabian mathematicians brought these concepts to Europe. Most of the European mathematicians did not use negative numbers until the seventeenth century, and even into the eighteenth century, it was common practice to exclude any negative results derived from equations, on the assumption that they were meaningless.

Natural numbers serve as a basis for the constructive development of all other number systems. Consequently, we shall construct the integers, the rationals, and the real numbers as extensions that satisfy some given key additional proprieties compared with the original set. Here there is much more to be done than simply adjoining numbers to a given set. We would like to obtain an extension S of a set of numbers M so that certain operations or relations between the elements of M are enhanced on S and so that certain operations or relations that generally have no validity in M will be valid in the extended set S.

For example, subtraction of natural numbers is not always feasible within the system of natural numbers, since if a, b are natural numbers $a - b$ does not always make sense. For integers, however, the process of subtraction is always feasible. However, division of one integer by some other nonzero integer also does not make sense in the system of integers, but such a division works perfectly well in the system of rational numbers. For rational numbers, limit operations are not always valid, but for real numbers, it is always possible to attempt to take limits. In real numbers, we are not able to take a root of an even power of a negative number, but we can always do it working in the set of complex numbers. Thus, at each stage the number system is extended to include operations that are always allowed.

There is one other important restriction we need to be aware of in extending number systems. The extension S should be *minimal* with respect to the given properties that we would like to keep. The integers form such an extension of the natural numbers in that the new set keeps all the properties of the operations of addition and multiplication and the original ordering, but now we are able to subtract any two integers. Subtraction requires the existence of the set of opposites, or additive inverses, for all numbers in the set.

We note that many pairs of numbers (u, v) are related in the sense that the difference $u - v$ is invariant (for example, $2 = 3 - 1 = 5 - 3 = 121 - 119 = -5 - (-7)$, and so on, all have difference 2). These pairs of numbers are therefore related and this leads us to the following construction.

10.2.1. Definition. *We say that the pairs* $(a, b), (k, n) \in \mathbb{N}_0 \times \mathbb{N}_0$ *are equivalent and we write* $(a, b)R(k, n)$ *if* $a + n = b + k$. *Put*

$$\mathbf{c}(a, b) = \{(k, n) \in \mathbb{N}_0 \times \mathbb{N}_0 \mid a + n = b + k\}.$$

We note that the relation R is an equivalence relation on $\mathbb{N}_0 \times \mathbb{N}_0$ with equivalence classes the elements $\mathbf{c}(a, b)$. To see this, we note that $(a, b)R(a, b)$ since $a + b = b + a$ and if $(a, b)R(k, n)$ then

$$a + n = b + k = k + b = n + a,$$

so $(k, n)R(a, b)$. Thus R is both reflexive and symmetric. Finally, R is transitive: if $(a, b)R(k, n)$ and $(k, n)R(u, v)$ then $a + n = b + k$ and $k + v = n + u$; so, omitting parentheses, using the associative and commutative laws of $\mathbb{N}_0$, we have

$$a + v + n = a + n + v = b + k + v = b + n + u = b + u + n.$$

Thus $a + v = b + u$, using Theorem 10.1.10(i) and that R is an equivalence relation follows. By definition, the equivalence class of (a, b) is the set of all pairs (n, k) that are equivalent to (a, b) and this is precisely the set $\mathbf{c}(a, b)$. Of course, the equivalence classes of an equivalence relation form a partition; in this case, the subsets $\mathbf{c}(a, b)$ form a partition of $\mathbb{N}_0 \times \mathbb{N}_0$.

We can now use this construction to formally define the set of integers.

10.2.2. Definition. *Let*

$$\mathbb{Z} = \{\mathbf{c}(a, b) \mid (a, b) \in \mathbb{N}_0 \times \mathbb{N}_0\}.$$

There are natural operations of addition and multiplication on $\mathbb{Z}$.

10.2.3. Definition. *The operations of addition and multiplication are defined on* $\mathbb{Z}$ *by the rules*

$$\mathbf{c}(a, b) + \mathbf{c}(c, d) = \mathbf{c}(a + c, b + d);$$
$$\mathbf{c}(a, b)\mathbf{c}(c, d) = \mathbf{c}(ac + bd, ad + bc).$$

First we must be sure that these operations are well defined, which is to say that they are independent of the equivalence class representative chosen. To see this, let $(k, n) \in \mathbf{c}(a, b)$ and $(t, m) \in \mathbf{c}(c, d)$. This means that $a + n = b + k$ and $c + m = d + t$. We need to show that $\mathbf{c}(a + c, b + d) = \mathbf{c}(k + t, n + m)$. Using the properties of natural number addition, we have

$$(a + c) + (n + m) = a + n + c + m = b + k + d + t = (b + d) + (k + t),$$

and it follows that the pairs $(a + c, b + d)$ and $(k + t, n + m)$ are equivalent, so that the addition is well defined.

We also want $\mathbf{c}(a, b)\mathbf{c}(c, d) = \mathbf{c}(k, n)\mathbf{c}(t, m)$, which is to say that we require $\mathbf{c}(ac + bd, ad + bc) = \mathbf{c}(kt + nm, km + nt)$. We do this in two steps. From $a + n = b + k$, we obtain $ac + nc = bc + kc$ and $bd + kd = ad + nd$, which imply

$$\begin{aligned} ac + bd + kd + nc &= ac + nc + bd + kd = bc + kc + ad + nd \\ &= (ad + bc) + (kc + nd). \end{aligned}$$

It follows that the pairs

$$(ac + bd, ad + bc) \text{ and } (kc + nd, kd + nc)$$

are equivalent. Using the same arguments, we deduce that the pairs

$$(kc + nd, kd + nc) \text{ and } (kt + nm, km + nt)$$

are equivalent. By transitivity, we see that the pairs

$$(ac + bd, ad + bc) \text{ and } (kt + nm, km + nt)$$

are equivalent, so multiplication is also well defined.

Now we obtain the basic properties of these operations. The following theorem shows that $\mathbb{Z}$ is a commutative ring.

10.2.4. Theorem. *Let* $a, b, c, d, u, v \in \mathbb{N}_0$. *The following properties hold:*

(i) $\mathbf{c}(a, b) + \mathbf{c}(c, d) = \mathbf{c}(c, d) + \mathbf{c}(a, b)$, *so addition is commutative;*

(ii) $\mathbf{c}(a, b) + (\mathbf{c}(c, d) + \mathbf{c}(u, v)) = (\mathbf{c}(a, b) + \mathbf{c}(c, d)) + \mathbf{c}(u, v)$, *so addition is associative;*

(iii) $\mathbf{c}(a, b) + \mathbf{c}(0, 0) = \mathbf{c}(a, b)$, *so addition has a zero element;*
(iv) $\mathbf{c}(a, b) + \mathbf{c}(b, a) = \mathbf{c}(0, 0)$, *so every element of* $\mathbb{Z}$ *has an additive inverse;*
(v) $\mathbf{c}(a, b)\mathbf{c}(c, d) = \mathbf{c}(c, d)\mathbf{c}(a, b)$, *so multiplication is commutative;*
(vi) $\mathbf{c}(a, b)(\mathbf{c}(c, d)\mathbf{c}(u, v)) = (\mathbf{c}(a, b)\mathbf{c}(c, d))\mathbf{c}(u, v)$, *so multiplication is associative;*
(vii) $\mathbf{c}(a, b)\mathbf{c}(1, 0) = \mathbf{c}(a, b)$, *so multiplication has an identity element;*
(viii) $\mathbf{c}(a, b)(\mathbf{c}(c, d) + \mathbf{c}(u, v)) = \mathbf{c}(a, b)\mathbf{c}(c, d) + \mathbf{c}(a, b)\mathbf{c}(u, v)$, *so multiplication is distributive over addition.*

Proof. The properties we are asserting for $\mathbb{Z}$ generally follow from the corresponding property in $\mathbb{N}_0$, to which we shall not specifically refer.

(i) We have

$$\mathbf{c}(a, b) + \mathbf{c}(c, d) = \mathbf{c}(a + c, b + d) = \mathbf{c}(c + a, d + b) = \mathbf{c}(c, d) + \mathbf{c}(a, b).$$

(ii) Next

$$\begin{aligned}\mathbf{c}(a, b) + (\mathbf{c}(c, d) + \mathbf{c}(u, v)) &= \mathbf{c}(a, b) + \mathbf{c}(c + u, d + v)\\ &= \mathbf{c}(a + (c + u), b + (d + v))\\ &= \mathbf{c}((a + c) + u, (b + d) + v)\\ &= \mathbf{c}(a + c, b + d) + \mathbf{c}(u, v)\\ &= (\mathbf{c}(a, b) + \mathbf{c}(c, d)) + \mathbf{c}(u, v).\end{aligned}$$

(iii) Clearly $\mathbf{c}(0, 0)$ is the additive identity since

$$\mathbf{c}(a, b) + \mathbf{c}(0, 0) = \mathbf{c}(a + 0, b + 0) = \mathbf{c}(a, b).$$

We also note that $\mathbf{c}(0, 0) = \mathbf{c}(n, n)$ for each $n \in \mathbb{N}_0$, since $0 + n = n + 0$.

(iv) Then

$$\mathbf{c}(a, b) + \mathbf{c}(b, a) = \mathbf{c}(a + b, b + a) = \mathbf{c}(a + b, a + b) = \mathbf{c}(0, 0),$$

so $\mathbf{c}(b, a)$ is the additive inverse of $\mathbf{c}(a, b)$.

(v) The multiplicative properties are only slightly more cumbersome to prove.

$$\mathbf{c}(a, b)\mathbf{c}(c, d) = \mathbf{c}(ac + bd, ad + bc) = \mathbf{c}(ca + db, cb + da) = \mathbf{c}(c, d)\mathbf{c}(a, b).$$

(vi) For associativity, we have

$$\begin{aligned}&\mathbf{c}(a, b)(\mathbf{c}(c, d)\mathbf{c}(u, v))\\ &\quad = \mathbf{c}(a, b)\mathbf{c}(cu + dv, cv + du)\\ &\quad = \mathbf{c}(a(cu + dv) + b(cv + du), a(cv + du) + b(cu + dv))\\ &\quad = \mathbf{c}(acu + adv + bcv + bdu, acv + adu + bcu + bdv)\end{aligned}$$

and

$$
\begin{aligned}
&(\mathbf{c}(a, b)\mathbf{c}(c, d))\mathbf{c}(u, v)) \\
&\quad = \mathbf{c}(ac + bd, ad + bc)\mathbf{c}(u, v) \\
&\quad = \mathbf{c}((ac + bd)u + (ad + bc)v, (ac + bd)v + (ad + bc)u) \\
&\quad = \mathbf{c}(acu + bdu + adv + bcv, acv + bdv + adu + bcu).
\end{aligned}
$$

Comparing these expressions, we see that

$$\mathbf{c}(a, b)(\mathbf{c}(c, d)\mathbf{c}(u, v)) = (\mathbf{c}(a, b)\mathbf{c}(c, d))\mathbf{c}(u, v).$$

(vii) Next, $\mathbf{c}(1, 0)$ is the multiplicative identity since

$$\mathbf{c}(a, b)\mathbf{c}(1, 0) = \mathbf{c}(a1 + b0, a0 + b1) = \mathbf{c}(a, b).$$

(viii) For the distributive property, we have

$$
\begin{aligned}
\mathbf{c}(a, b)(\mathbf{c}(c, d) + \mathbf{c}(u, v)) &= \mathbf{c}(a, b)\mathbf{c}(c + u, d + v) \\
&= \mathbf{c}(a(c + u) + b(d + v), a(d + v) + b(c + u)) \\
&= \mathbf{c}(ac + au + bd + bv, ad + av + bc + bu)
\end{aligned}
$$

and

$$
\begin{aligned}
\mathbf{c}(a, b)\mathbf{c}(c, d) + \mathbf{c}(a, b)\mathbf{c}(u, v) &= \mathbf{c}(ac + bd, ad + bc) + \mathbf{c}(au + bv, av + bu) \\
&= \mathbf{c}(ac + bd + au + bv, ad + bc + av + bu).
\end{aligned}
$$

Close inspection shows that $\mathbf{c}(a, b)(\mathbf{c}(c, d) + \mathbf{c}(u, v)) = \mathbf{c}(a, b)\mathbf{c}(c, d) + \mathbf{c}(a, b)$ $\mathbf{c}(u, v)$.

Now we consider the mapping $\iota : \mathbb{N}_0 \longrightarrow \mathbb{Z}$, defined by $\iota(n) = \mathbf{c}(n, 0)$, where $n \in \mathbb{N}_0$. If $\iota(n) = \iota(m)$ then $\mathbf{c}(n, 0) = \mathbf{c}(m, 0)$, so $n + 0 = 0 + m$ which gives $n = m$. Thus the mapping ι is injective. Furthermore,

$$
\begin{aligned}
\iota(n) + \iota(k) &= \mathbf{c}(n, 0) + \mathbf{c}(k, 0) = \mathbf{c}(n + k, 0) = \iota(n + k) \text{ and} \\
\iota(n)\iota(k) &= \mathbf{c}(n, 0)\mathbf{c}(k, 0) = \mathbf{c}(nk + 0, n0 + 0k) = \mathbf{c}(nk, 0) = \iota(nk)
\end{aligned}
$$

for every $n, k \in \mathbb{N}_0$. It follows that ι induces a bijection between $\mathbb{N}_0$ and $\mathbf{Im}\,\iota$ that respects the operations of addition and multiplication and so is a type of *isomorphism*. (Of course, $\mathbb{N}_0$ is not a group or ring.) Thus, we may identify $n \in \mathbb{N}_0$ with its image $\mathbf{c}(n, 0)$ and we shall write $\mathbb{N}_0 \equiv \mathbf{Im}\,\iota$. We set $\mathbf{c}(n, 0) = \mathbf{n}$ for each $n \in \mathbb{N}_0$. Then we have $\mathbf{c}(n, k) = \mathbf{c}(n, 0) + \mathbf{c}(0, k)$. By Theorem 10.2.4, the element $\mathbf{c}(0, k)$ is the additive inverse of the element $\mathbf{c}(k, 0) = \mathbf{k}$. As usual, for the additive inverse of $\mathbf{k}$, we write $-\mathbf{k}$. Thus if $\mathbf{c}(k, 0) = \mathbf{k}$, then $\mathbf{c}(0, k) = -\mathbf{k}$.

So, for the element $\mathbf{c}(n, k)$, we obtain

$$\mathbf{c}(n, k) = \mathbf{c}(n, 0) + \mathbf{c}(0, k) = \mathbf{n} + (-\mathbf{k}).$$

The existence of additive inverses allows us to define *subtraction*. We define the difference of two integers $\mathbf{n}, \mathbf{k}$ by

$$\mathbf{n} - \mathbf{k} = \mathbf{n} + (-\mathbf{k}), \text{ where } \mathbf{n}, \mathbf{k} \in \mathbb{Z}.$$

Thus we have $\mathbf{c}(n, k) = \mathbf{n} + (-\mathbf{k}) = \mathbf{n} - \mathbf{k}$. Immediately we observe the following properties of subtraction:

$$-(-\mathbf{k}) = \mathbf{k},$$
$$\mathbf{n}(-\mathbf{k}) = \mathbf{c}(n, 0)\mathbf{c}(0, k) = \mathbf{c}(n0 + 0k, nk + 00) = \mathbf{c}(0, nk) = -\mathbf{nk} = (-\mathbf{n})\mathbf{k}$$

and

$$\mathbf{n}(\mathbf{k} - \mathbf{m}) = \mathbf{n}(\mathbf{k} + (-\mathbf{m})) = nk + \mathbf{n}(-\mathbf{m}) = \mathbf{nk} + (-\mathbf{nm}) = \mathbf{nk} - \mathbf{nm}.$$

Next we see that $\mathbb{Z} = \{\pm\, \mathbf{n} \mid n \in \mathbb{N}_0\}$. Indeed, for every pair of numbers n and $k \in \mathbb{N}_0$, Theorem 10.1.8 implies that $n = k$, or $n > k$, or $n < k$. In the first case, the pair (n, n) is equivalent to $(0, 0)$, so $\mathbf{c}(n, k) = \mathbf{c}(0, 0) = \mathbf{0}$. In the second case, for some $m \in \mathbb{N}_0$, we have $n = k + m$. Therefore $(n, k) = (k + m, k)$ which is equivalent to the pair $(m, 0)$. However, $\mathbf{c}(m, 0) = \mathbf{m}$, so $\mathbf{c}(n, k) = \mathbf{m}$. In the last case $k = n + t$, for some $t \in \mathbb{N}_0$, so $(n, k) = (n, n + t)$ is equivalent to the pair $(0, t)$ and so $\mathbf{c}(n, k) = -\mathbf{t}$.

Let $\mathbf{n}, \mathbf{k} \in \mathbf{Im}\ \iota$ and suppose that $\mathbf{n} - \mathbf{k} \in \mathbf{Im}\ \iota$. We have $\mathbf{n} = \mathbf{c}(n, 0)$ and $\mathbf{k} = \mathbf{c}(k, 0)$, where $n, k \in \mathbb{N}_0$. Then $\mathbf{n} - \mathbf{k} = \mathbf{c}(n, 0) + \mathbf{c}(0, k) = \mathbf{c}(n, k)$. By Theorem 10.1.8, $n = k$, or $n > k$, or $n < k$. In the first case, $n = n + 0$. In the second case, we have $n = k + m$ for some $m \in \mathbb{N}_0$, so $(n, k) = (k + m, k)$ is equivalent to the pair $(m, 0) = \mathbf{m}$. In the last case, $k = n + t$ for some $t \in \mathbb{N}_0$. Therefore $(n, k) = (n, n + t)$ is equivalent to the pair $(0, t)$, which does not belong to $\mathbb{N}_0$. This allows us to define the difference of two numbers n and k of $\mathbb{N}_0$ by the following rule: the difference $n - k$ is defined if and only if $n \geq k$, which is to say that $n = k + m$, and in this case, we put $n - k = m$.

We can extend the existing order on $\mathbb{N}_0$ to the set $\mathbb{Z}$ in the following way.

10.2.5. Definition. *Let* $\mathbf{k}, \mathbf{n} \in \mathbb{Z}$. *If* $\mathbf{n} - \mathbf{k} \in \mathbb{N}_0$, *then we say that* $\mathbf{n}$ is greater than or equal to $\mathbf{k}$ *and denote this by* $\mathbf{n} \geq \mathbf{k}$, *or we say that* $\mathbf{k}$ *is less than or equal to* $\mathbf{n}$ *and denote this by* $\mathbf{k} \leq \mathbf{n}$. *If in addition* $\mathbf{n} \neq \mathbf{k}$, *then we say that* $\mathbf{n}$ *is greater than* $\mathbf{k}$ *and denote this by* $\mathbf{n} > \mathbf{k}$, *or we say that* $\mathbf{k}$ *less that* $\mathbf{n}$ *and denote this by* $\mathbf{k} < \mathbf{n}$.

Let $n, k \in \mathbb{N}_0$ and suppose that $\mathbf{n} = \mathbf{c}(n, 0) \geq \mathbf{k} = \mathbf{c}(k, 0)$. Then

$$\mathbf{n} - \mathbf{k} = \mathbf{c}(n, 0) + \mathbf{c}(0, k) = \mathbf{c}(n, k) = \mathbf{m}, \text{ say.}$$

Thus, $\mathbf{m} = \mathbf{n} - \mathbf{k}$, so $\mathbf{n} = \mathbf{m} + \mathbf{k}$. Since $\mathbf{m} \in \mathbf{Im}\ \iota$, we have $\mathbf{m} = \mathbf{c}(m, 0)$, for some $m \in \mathbb{N}_0$. Then $\mathbf{c}(m, 0) = \mathbf{c}(n, k)$, which implies that $m + k = n + 0 = n$. Hence

$n \geq k$. This shows that the order induced on $\mathbf{Im}\ \iota \equiv \mathbb{N}_0$ by the order we established on $\mathbb{Z}$ coincides with the order that we established on $\mathbb{N}_0$ in Section 10.1.

10.2.6. Theorem. *Let* $\mathbf{a}, \mathbf{b} \in \mathbb{Z}$. *Then one and only one of the following relations holds:* $\mathbf{a} = \mathbf{b}$, $\mathbf{a} < \mathbf{b}$ *or* $\mathbf{a} > \mathbf{b}$.

Proof. Let $\mathbf{a} = \mathbf{c}(n, k)$, $\mathbf{b} = \mathbf{c}(t, m)$ and suppose that $\mathbf{a} \neq \mathbf{b}$. If $\mathbf{a} - \mathbf{b} \in \mathbf{Im}\,\iota$ then, by definition, $\mathbf{a} \geq \mathbf{b}$ and, since $\mathbf{a} \neq \mathbf{b}$, we conclude that $\mathbf{a} > \mathbf{b}$.

Suppose now that $\mathbf{a} - \mathbf{b} \notin \mathbf{Im}\ \iota$. Then

$$\mathbf{a} - \mathbf{b} = \mathbf{c}(n, k) - \mathbf{c}(t, m) = \mathbf{c}(n, k) + \mathbf{c}(m, t) = \mathbf{c}(n + m, k + t).$$

We must have $k + t > n + m$ otherwise $\mathbf{c}(n + m, k + t) = \mathbf{c}(n + m - k - t, 0) \in \mathbf{Im}\,\iota$, contrary to our assumption. In this case,

$$\mathbf{b} - \mathbf{a} = \mathbf{c}(k + t, n + m) = \mathbf{c}((k + t) - (n + m), 0) \in \mathbf{Im}\ \iota$$

It follows that $\mathbf{b} \geq \mathbf{a}$ and since $\mathbf{a} \neq \mathbf{b}$, we deduce that $\mathbf{b} > \mathbf{a}$.

10.2.7. Proposition. *Let* $\mathbf{a}, \mathbf{b}, \mathbf{c} \in \mathbb{Z}$. *Then*

(i) $\mathbf{a} \leq \mathbf{a}$;
(ii) $\mathbf{a} \leq \mathbf{b}$ *and* $\mathbf{b} \leq \mathbf{c}$ *imply* $\mathbf{a} \leq \mathbf{c}$;
(iii) $\mathbf{a} < \mathbf{b}$ *and* $\mathbf{b} \leq \mathbf{c}$ *imply* $\mathbf{a} < \mathbf{c}$;
(iv) $\mathbf{a} \leq \mathbf{b}$ *and* $\mathbf{b} < \mathbf{c}$ *imply* $\mathbf{a} < \mathbf{c}$;
(v) $\mathbf{a} < \mathbf{b}$ *and* $\mathbf{b} < \mathbf{c}$ *imply* $\mathbf{a} < \mathbf{c}$;
(vi) $\mathbf{a} \leq \mathbf{b}$ *and* $\mathbf{b} \geq \mathbf{a}$ *imply* $\mathbf{a} = \mathbf{b}$.

Proof. We have $\mathbf{a} = \mathbf{a} + \mathbf{0}$, which implies (i).

(ii) We have $\mathbf{b} - \mathbf{a}, \mathbf{c} - \mathbf{b} \in \mathbf{Im}\,\iota$ and hence

$$\mathbf{c} - \mathbf{a} = (\mathbf{c} - \mathbf{b}) + (\mathbf{b} - \mathbf{a}) \in \mathbf{Im}\,\iota$$

It follows that $\mathbf{a} \leq \mathbf{c}$. If $\mathbf{c} \neq \mathbf{b}$ or $\mathbf{b} \neq \mathbf{a}$, then $\mathbf{c} - \mathbf{b} \neq \mathbf{0}$ or $\mathbf{b} - \mathbf{a} \neq \mathbf{0}$. Then $\mathbf{c} - \mathbf{a} = (\mathbf{c} - \mathbf{b}) + (\mathbf{b} - \mathbf{a}) \neq \mathbf{0}$, so $\mathbf{c} > \mathbf{a}$, which implies (iii)–(v).

(vi) We have $\mathbf{b} - \mathbf{a}$ and $\mathbf{a} - \mathbf{b} \in \mathbf{Im}\,\iota$. Let $\mathbf{a} = \mathbf{c}(n, k)$, $\mathbf{b} = \mathbf{c}(t, m)$. Then

$$\mathbf{a} - \mathbf{b} = \mathbf{c}(n, k) + \mathbf{c}(m, t) = \mathbf{c}(n + m, k + t),$$

and

$$\mathbf{b} - \mathbf{a} = \mathbf{c}(t, m) + \mathbf{c}(k, n) = \mathbf{c}(t + k, n + m).$$

Since $\mathbf{a} - \mathbf{b} \in \mathbf{Im}\,\iota \equiv \mathbb{N}_0$, we have $n + m \geq k + t$ and since $\mathbf{b} - \mathbf{a} \in \mathbf{Im}\,\iota$, we have $k + t \geq n + m$. By Proposition 10.1.9, $k + t = n + m$ and it follows that $\mathbf{b} - \mathbf{a} = \mathbf{0}$ so $\mathbf{a} = \mathbf{b}$.

10.2.8. Theorem. *Let* $\mathbf{a}, \mathbf{b}, \mathbf{c} \in \mathbb{Z}$.

(i) *if* $\mathbf{a} \leq \mathbf{b}$, *then* $\mathbf{a} + \mathbf{c} \leq \mathbf{b} + \mathbf{c}$;

(ii) *if* $\mathbf{a} < \mathbf{b}$, *then* $\mathbf{a} + \mathbf{c} < \mathbf{b} + \mathbf{c}$;
(iii) *if* $\mathbf{a} + \mathbf{c} \leq \mathbf{b} + \mathbf{c}$, *then* $\mathbf{a} \leq \mathbf{b}$;
(iv) *if* $\mathbf{a} + \mathbf{c} < \mathbf{b} + \mathbf{c}$, *then* $\mathbf{a} < \mathbf{b}$.

Proof.
(i) We have $\mathbf{b} - \mathbf{a} \in \mathbf{Im}\ \iota \equiv \mathbb{N}_0$. Note that

$$\mathbf{b} - \mathbf{a} = \mathbf{b} + \mathbf{0} - \mathbf{a} = \mathbf{b} + \mathbf{c} - \mathbf{c} - \mathbf{a} = (\mathbf{b} + \mathbf{c}) - (\mathbf{a} + \mathbf{c}),$$

so that $(\mathbf{b} + \mathbf{c}) - (\mathbf{a} + \mathbf{c}) \in \mathbf{Im}\,\iota$. It follows that $\mathbf{a} + \mathbf{c} \leq \mathbf{b} + \mathbf{c}$.
(ii) If $\mathbf{b} \neq \mathbf{a}$, then $\mathbf{b} - \mathbf{a} \neq \mathbf{0}$. In this case, $(\mathbf{b} + \mathbf{c}) - (\mathbf{a} + \mathbf{c}) \neq \mathbf{0}$, so that $\mathbf{a} + \mathbf{c} < \mathbf{b} + \mathbf{c}$.
(iii) If $\mathbf{a} + \mathbf{c} \leq \mathbf{b} + \mathbf{c}$, then by (i), we have

$$\mathbf{a} = \mathbf{a} + \mathbf{c} + (-\mathbf{c}) \leq \mathbf{b} + \mathbf{c} + (-\mathbf{c}) = \mathbf{b}.$$

(iv) The proof is similar.

10.2.9. Theorem. *Let* $\mathbf{a}, \mathbf{b}, \mathbf{c} \in \mathbb{Z}$ *and* $\mathbf{c} \neq \mathbf{0}$.

(i) *if* $\mathbf{a}, \mathbf{b} \neq \mathbf{0}$ *then* $\mathbf{ab} \neq \mathbf{0}$;
(ii) *if* $\mathbf{ac} = \mathbf{bc}$ *then* $\mathbf{a} = \mathbf{b}$;
(iii) *if* $\mathbf{a} \leq \mathbf{b}$ *and* $\mathbf{c} > \mathbf{0}$, *then* $\mathbf{ac} \leq \mathbf{bc}$;
(iv) *if* $\mathbf{a} < \mathbf{b}$ *and* $\mathbf{c} > \mathbf{0}$, *then* $\mathbf{ac} < \mathbf{bc}$;
(v) *if* $\mathbf{a} \leq \mathbf{b}$ *and* $\mathbf{c} < \mathbf{0}$, *then* $\mathbf{ac} \geq \mathbf{bc}$;
(vi) *if* $\mathbf{a} < \mathbf{b}$ *and* $\mathbf{c} < \mathbf{0}$, *then* $\mathbf{ac} > \mathbf{bc}$;
(vii) *if* $\mathbf{ac} < \mathbf{bc}$ *and* $\mathbf{c} > \mathbf{0}$, *then* $\mathbf{a} < \mathbf{b}$;
(viii) *if* $\mathbf{ac} \leq \mathbf{bc}$ *and* $\mathbf{c} > \mathbf{0}$, *then* $\mathbf{a} \leq \mathbf{b}$;
(ix) *if* $\mathbf{ac} < \mathbf{bc}$ *and* $\mathbf{c} < \mathbf{0}$, *then* $\mathbf{a} > \mathbf{b}$;
(x) *if* $\mathbf{ac} \leq \mathbf{bc}$ *and* $\mathbf{c} > \mathbf{0}$, *then* $\mathbf{a} \leq \mathbf{b}$.

Proof.
(i) Suppose, for a contradiction, that $\mathbf{ab} = \mathbf{0}$. If $\mathbf{a} = \mathbf{c}(n, k)$ and $\mathbf{b} = \mathbf{c}(t, m)$ then

$$\mathbf{ab} = \mathbf{c}(n, k)\mathbf{c}(t, m) = \mathbf{c}(nt + km, nm + kt) = \mathbf{c}(0, 0).$$

It follows that $nt + km = mn + kt$. Since $\mathbf{a} \neq 0$, $n \neq k$ so, by Theorem 10.1.8, either $n < k$ or $n > k$. Suppose first that $n < k$. Then $k = n + s$ for some $s \in \mathbb{N}_0$, where $s \neq 0$. We have

$$nt + km = nt + (n + s)m = nt + nm + sm, \text{ and}$$
$$nm + kt = nm + (n + s)t = nm + nt + st.$$

It follows that

$$nt + nm + sm = nm + nt + st,$$

and hence, by Theorem 10.1.10, $sm = st$. Since $s \neq 0$, Theorem 10.1.11 implies that $m = t$. However, in this case,

$$\mathbf{b} = \mathbf{c}(t, m) = \mathbf{c}(t, t) = \mathbf{0},$$

and we obtain the desired contradiction. In the case, when $n > k$ the proof is similar.

(ii) Since $\mathbf{ac} = \mathbf{bc}$, then $\mathbf{0} = \mathbf{bc} - \mathbf{ac} = (\mathbf{b} - \mathbf{a})\mathbf{c}$. Since $\mathbf{c} \neq \mathbf{0}$, (i) implies that $\mathbf{b} - \mathbf{a} = \mathbf{0}$ and hence $\mathbf{a} = \mathbf{b}$.

(iii) Since $\mathbf{a} \leq \mathbf{b}$, we have $\mathbf{b} - \mathbf{a} \in \mathbf{Im}\ \iota$. Also $\mathbf{c} \in \mathbf{Im}\,\iota$ and hence $(\mathbf{b} - \mathbf{a})\mathbf{c} = \mathbf{bc} - \mathbf{ac} \in \mathbf{Im}\,\iota$. It follows that $\mathbf{ac} \leq \mathbf{bc}$.

(iv) By (ii), $\mathbf{ac} \leq \mathbf{bc}$. Since $\mathbf{a} < \mathbf{b}$, we have $\mathbf{b} - \mathbf{a} \neq \mathbf{0}$ and, by (i), $\mathbf{bc} - \mathbf{ac} = (\mathbf{b} - \mathbf{a})\mathbf{c} \neq \mathbf{0}$, so that, $\mathbf{ac} < \mathbf{bc}$.

(v) We have $\mathbf{0} > \mathbf{c}$, so that $\mathbf{0} - \mathbf{c} = -\mathbf{c} \in \mathbf{Im}\,\iota$. It follows that $-\mathbf{c} > \mathbf{0}$. Then, by (iii), $\mathbf{a}(-\mathbf{c}) \leq \mathbf{b}(-\mathbf{c})$ and we deduce that

$$-\mathbf{bc} - (-\mathbf{ac}) = \mathbf{ac} - \mathbf{bc} \in \mathbf{Im}\,\iota,$$

which proves that $\mathbf{ac} \geq \mathbf{bc}$.

(vi) As in (iv), $(\mathbf{b} - \mathbf{a})\mathbf{c} \neq \mathbf{0}$ and, together with (v), this gives $\mathbf{ac} > \mathbf{bc}$.

(vii) By Theorem 10.1.12, there is one and only one possibility, namely, that $\mathbf{a} = \mathbf{b}$, $\mathbf{a} < \mathbf{b}$, or $\mathbf{a} > \mathbf{b}$. If $\mathbf{a} > \mathbf{b}$ then, by (iv), $\mathbf{ac} > \mathbf{bc}$, which is contrary to the hypothesis. Likewise, if $\mathbf{a} = \mathbf{b}$ then $\mathbf{ac} = \mathbf{bc}$, again contrary to $\mathbf{ac} < \mathbf{bc}$. It follows that $\mathbf{a} < \mathbf{b}$.

For assertions (viii)–(x), the proofs are similar.

10.2.10. Corollary. *Let* $\mathbf{a}, \mathbf{b} \in \mathbb{Z}$ *and suppose that* $\mathbf{b} > \mathbf{0}$. *Then there is a number* $\mathbf{c} \in \mathbb{Z}$ *such that* $\mathbf{bc} > \mathbf{a}$.

Proof. If $\mathbf{a} \leq \mathbf{0}$, then $\mathbf{b1} = \mathbf{b} > \mathbf{0} > \mathbf{a}$. Suppose that $\mathbf{a} > \mathbf{0}$. Then $\mathbf{a} = \mathbf{c}(a, 0)$ and $\mathbf{b} = \mathbf{c}(b, 0)$, where $a, b \in \mathbb{N}_0$. By Corollary 10.1.12, there exists $c \in \mathbb{N}_0$ such that $bc > a$. Then $\mathbf{a} = \mathbf{c}(a, 0) < \mathbf{c}(bc, 0) = c(b, 0)\mathbf{c}(c, 0) = \mathbf{bc}$.

10.2.11. Corollary. *Let* $\mathbf{a} \in \mathbb{Z}$. *If* $\mathbf{b}$ *is an integer such that* $\mathbf{a} \leq \mathbf{b} \leq \mathbf{a} + \mathbf{1}$, *then either* $\mathbf{b} = \mathbf{a}$ *or* $\mathbf{b} = \mathbf{a}+\mathbf{1}$.

Proof. Since $\mathbf{0} < \mathbf{1}$, it follows from Theorem 10.2.8 that $\mathbf{a} < \mathbf{a} + \mathbf{1}$. Suppose that $\mathbf{b} \neq \mathbf{a}$. Then $\mathbf{a} < \mathbf{b} \leq \mathbf{a} + \mathbf{1}$. Again using Theorem 10.2.8, we deduce that $\mathbf{0} < \mathbf{b} - \mathbf{a} \leq \mathbf{1}$. However, this means that $\mathbf{b} - \mathbf{a} \in \mathbf{Im}\,\iota \equiv \mathbb{N}_0$. We noted above that the order introduced on $\mathbb{Z}$ induces the same order in $\mathbb{N}_0$ that was introduced originally in Section 10.1. Therefore, we can apply Corollary 10.1.13 and deduce that $\mathbf{b} = \mathbf{a} + \mathbf{1}$.

In this way, we obtain all the common properties of the integers. Furthermore, we identify the set $\mathbb{N}_0$ with its image $\mathbf{Im}\,\iota \equiv \mathbb{N}_0$ in $\mathbb{Z}$ and hence we assume that $\mathbb{N}_0$ is a subset of $\mathbb{Z}$. Also, for all integers we will use a common font, so, for example, we will write n instead of $\mathbf{n}$.

We next recall the notion of the absolute value of an integer.

Let n be an integer and set

$$|n| = \begin{cases} n, & \text{if } n \in \mathbb{N}_0, \\ -n, & \text{if } n \notin \mathbb{N}_0. \end{cases}$$

It is not hard to prove that $|nm| = |n||m|$ and other properties of absolute value can also be obtained.

Finally, we consider the question concerning the uniqueness of $\mathbb{Z}$. Let G be an additive group containing $\mathbb{N}$. Suppose that 0 is the zero element of G and consider the subgroup X, generated by $\mathbb{N}$. If $x, y \in \mathbb{N}$ then, by Corollary 8.1.8, $x - y \in X$ and we let $Y = \{x - y \mid x, y \in \mathbb{N}\}$. If $u, v \in Y$, then $u = x_1 - y_1$, $v = x_2 - y_2$ and we have

$$\begin{aligned} u - v &= (x_1 - y_1) - (x_2 - y_2) \\ &= x_1 - y_1 - x_2 + y_2 = (x_1 + y_2) - (y_1 + x_2) \in Y. \end{aligned}$$

Again by Corollary 8.1.8, we see that Y is a subgroup of G. For each element $x \in \mathbb{N}$, we have $x = x - 0 \in Y$, so that $\mathbb{N}_0 \subseteq Y$ and hence $X \leq Y$. On the other hand, we remarked above that $Y \leq X$, so $X = Y = \{x - y \mid x, y \in \mathbb{N}\}$. Now consider the mapping $\zeta : \mathbb{N} \times \mathbb{N} \longrightarrow X$, defined by $\zeta(n, k) = n - k$, where $n, k \in \mathbb{N}$.

We consider the equivalence relation $\Delta(\zeta)$ as defined in Section 7.2. Then $(a, b), (k, n) \in \Delta(\zeta)$ if and only if $\zeta(a, b) = \zeta(k, n)$, which means that $a - b = k - n$. Thus, $\Delta(\zeta)$ is the equivalence relation of Definition 10.2.1. By Theorem 7.2.7, there is a bijection ψ_ζ of $\mathbb{Z}$ onto $\mathbf{Im}\,\zeta = X$, defined by

$$\psi_\zeta(\mathbf{c}(n, k)) = \zeta(n, k) = n - k.$$

We have

$$\begin{aligned} \psi_\zeta(\mathbf{c}(a, b) + \mathbf{c}(c, d)) &= \psi_\zeta(\mathbf{c}(a + c, b + d)) = \zeta(a + c, b + d) \\ &= (a + c) - (b + d) = (a - b) + (c - d) \text{ and} \\ \psi_\zeta(\mathbf{c}(a, b)) + \psi_\zeta(\mathbf{c}(c, d)) &= \zeta(a, b) + \zeta(c, d) = (a - b) + (c - d). \end{aligned}$$

Likewise,

$$\begin{aligned} \psi_\zeta(\mathbf{c}(a, b)\mathbf{c}(c, d)) &= \psi_\zeta(\mathbf{c}(ac + bd, ad + bc)) = \zeta(ac + bd, ad + bc) \\ &= (ac + bd) - (ad + bc), \text{ whereas} \\ \psi_\zeta(\mathbf{c}(a, b))\psi_\zeta(\mathbf{c}(c, d)) &= \zeta(a, b)\zeta(c, d) = (a - b)(c - d) \\ &= ac - bc + bd - ad = (ac + bd) - (ad + bc). \end{aligned}$$

Hence

$$\psi_\zeta(\mathbf{c}(a, b) + \mathbf{c}(c, d)) = \psi_\zeta(\mathbf{c}(a, b)) + \psi_\zeta(\mathbf{c}(c, d))$$

and

$$\psi_\zeta(\mathbf{c}(a, b)\mathbf{c}(c, d)) = \psi_\zeta(\mathbf{c}(a, b))\psi_\zeta(\mathbf{c}(c, d)).$$

In other words, ψ_ζ is a bijection of $\mathbb{Z}$ onto X, which respects addition and multiplication, so ψ_ζ is an isomorphism. Consequently, if an arbitrary additive group G contains $\mathbb{N}_0$, then the subgroup generated by $\mathbb{N}$ is isomorphic to $\mathbb{Z}$ which proves the uniqueness of $\mathbb{Z}$.

10.3 THE RATIONALS

In this section, we are going to construct the set of rational numbers. We shall use the same ideas that helped us in Section 10.2, when we extended the set of natural numbers to the set of integers. In Section 10.2, the leading idea was concerned with enhancing the properties of addition, while in the current section, we will mainly focus on the multiplicative properties. Thus, in the set of integers, the inverse operation to multiplication does not work all the time in the sense that for all integers (other than 1 and -1) the multiplicative inverse (or the reciprocal) is not an integer itself. We would like to extend the set $\mathbb{Z}$ of integers to a set $\mathbb{Q}$ in which addition and multiplication possess the same properties and, additionally, division by nonzero elements is also defined. Of course, we are looking for such an extension that is minimal with respect to the properties that we would like to keep.

As in Section 10.2, we construct an equivalence relation this time defined on the set $\mathbb{Z} \times (\mathbb{Z}\backslash\{0\})$ and we shall then obtain a partition of this set. The equivalence classes of this partition will be called *the rational numbers*.

This partition is natural once we recall that a fraction $\frac{a}{b}$ is not defined in a unique way. Remember that the fractions $\frac{a}{b}$ and $\frac{c}{d}$ are equal if and only if $ad = bc$. Thus, with each fraction we associate some infinite set of pairs of integers.

10.3.1. Definition. *We say that the pairs $(a, b), (c, d) \in \mathbb{Z}\times(\mathbb{Z} \setminus \{0\})$ are equivalent, if $ad = bc$. For $(a, b) \in \mathbb{Z} \times (\mathbb{Z} \setminus \{0\})$, we define*

$$\frac{a}{b} = \{(c, d) \in \mathbb{Z} \times (\mathbb{Z}\backslash\{0\}) \mid ad = bc\}.$$

The set $\frac{a}{b}$ is called a fraction; a is called the numerator of the fraction and b is called the denominator of the fraction.

It is easy to see that the relation R defined by $(a, b)R(c, d)$ if and only if $ad = bc$ is an equivalence relation, as we now show. We use various properties of the integers. Clearly, $(a, b)R(a, b)$ since $ab = ba$ in $\mathbb{Z}$ so R is reflexive. Also R is symmetric, since if $(a, b)R(c, d)$ then $ad = bc$, so $cb = da$ which is to say that $(c, d)R(a, b)$ and the relation is symmetric. Finally, if $(a, b)R(c, d)$ and $(c, d)R(u, v)$ then $ad = bc$ and $cv = du$. Multiplying these equations by the nonzero integers v and b, respectively, we see that

$$adv = bcv = cvb = bdu,$$

and from this we have $(av - bu)d = 0$, using commutativity. Since $b \neq 0$, Theorem 10.2.9(ii) implies that $av = bu$, which means that $(a, b)R(u, v)$ and R is transitive. Thus R is an equivalence relation. It follows from Definition 10.3.1 that the equivalence class of (a, b) is precisely the fraction $\frac{a}{b}$. Since this equivalence relation partitions the set $\mathbb{Z} \times (\mathbb{Z} \setminus \{0\})$ into equivalence classes, it follows that if $\frac{a}{b} \cap \frac{c}{d} \neq \emptyset$ then $\frac{a}{b} = \frac{c}{d}$.

10.3.2. Definition. *The set* $\mathbb{Q}$ *is defined to be*

$$\mathbb{Q} = \left\{ \frac{a}{b} \mid (a, b) \in \mathbb{Z} \times (\mathbb{Z} \setminus \{0\}) \right\}.$$

10.3.3. Definition. *The operations of addition and multiplication are defined on* $\mathbb{Q}$ *by*

$$\frac{a}{b} + \frac{c}{d} = \frac{ad + bc}{bd} \ \textit{and} \ \frac{a}{b} \cdot \frac{c}{d} = \frac{ac}{bd}.$$

As in Section 10.2, we must be sure that these operations are well defined so that they are independent of the choice of elements from the equivalence class. Let $(k, n) \in \frac{a}{b}$ and $(t, m) \in \frac{c}{d}$. This means that $an = bk$ and $cm = dt$. Then

$$\begin{aligned}(ad + bc)(nm) &= ad \cdot nm + bc \cdot nm = (an)(dm) + (cm)(bn) \\ &= bk \cdot dm + dt \cdot bn = (km + nt)bd.\end{aligned}$$

It follows that the pairs $(ad + bc, bd)$ and $(km + nt, nm)$ are equivalent, so that the addition is well defined.

Furthermore,

$$(ac)(nm) = (an)(cm) = bk \cdot dt = (bd)(kt),$$

which implies that the pairs (ac, bd) and (kt, nm) are equivalent. Therefore multiplication is also well defined.

Now we obtain the basic properties of these operations.

10.3.4. Theorem. *Let* $a, b, c, d, u, v \in \mathbb{Z}$, *where* b, d, v *are nonzero. The following properties hold:*

(i) $\frac{a}{b} + \frac{c}{d} = \frac{c}{d} + \frac{a}{b}$, *so addition is* commutative*;*

(ii) $\left(\frac{a}{b} + \frac{c}{d}\right) + \frac{u}{v} = \frac{a}{b} + \left(\frac{c}{d} + \frac{u}{v}\right)$, *so addition is* associative*;*

(iii) $\frac{a}{b} + \frac{0}{d} = \frac{a}{b}$, *so the fraction* $\frac{0}{d}$ *is the* zero element for addition*;*

(iv) *the fraction* $\frac{-a}{b}$ *is an additive* inverse *to* $\frac{a}{b}$, *so every element of* $\mathbb{Q}$ *has an additive* inverse*;*

(v) $\frac{a}{b} \cdot \frac{c}{d} = \frac{c}{d} \cdot \frac{a}{b}$, *so multiplication is* commutative*;*

(vi) $\frac{a}{b} \cdot \left(\frac{c}{d} \cdot \frac{u}{v}\right) = \left(\frac{a}{b} \cdot \frac{c}{d}\right) \cdot \frac{u}{v}$, *so multiplication is associative;*

(vii) *the fraction* $\frac{d}{d}$ *is the multiplicative identity for each nonzero* $d \in \mathbb{Z}$*;*

(viii) *if* $a \neq 0$, *then* $\frac{a}{b} \cdot \frac{b}{a}$ *is the multiplicative identity, so every nonzero fraction has a* reciprocal, *or multiplicative inverse;*

(ix) $\frac{a}{b}\left(\frac{c}{d} + \frac{u}{v}\right) = \frac{a}{b} \cdot \frac{c}{d} + \frac{a}{b} \cdot \frac{u}{v}$, *so multiplication is distributive over addition.*

Proof.

(i) We have

$$\frac{a}{b} + \frac{c}{d} = \frac{ad + bc}{bd} = \frac{cb + da}{db} = \frac{c}{d} + \frac{a}{b}.$$

(ii) Next,

$$\frac{a}{b} + \left(\frac{c}{d} + \frac{u}{v}\right) = \frac{a}{b} + \frac{cv + du}{dv} = \frac{a(dv) + (cv + du)b}{b(dv)}.$$

On the other hand,

$$\left(\frac{a}{b} + \frac{c}{d}\right) + \frac{u}{v} = \frac{ad + bc}{bd} + \frac{u}{v} = \frac{(ad + bc)v + (bd)u}{(bd)v}.$$

Since

$$\begin{aligned} a(dv) + (cv + du)b &= a(dv) + (cv)b + (du)b = a(dv) + b(cv) + b(du) \\ &= (ad)v + (bc)v + (bd)u = (ad + bc)v + (bd)u, \end{aligned}$$

using commutativity, associativity, and distributivity in $\mathbb{Z}$ and since $(bd)v = b(dv)$ (ii) follows.

(iii) Next,

$$\frac{a}{b} + \frac{0}{d} = \frac{ad + b0}{bd} = \frac{ad}{bd}.$$

Since $a(bd) = (ab)d = (ba)d = b(ad)$, the fractions $\frac{a}{b}$ and $\frac{ad}{bd}$ coincide so $\frac{a}{b} + \frac{0}{d} = \frac{a}{b} = \frac{0}{d} + \frac{a}{b}$. We note that

$$\frac{0}{d} = \frac{0}{v}$$

for all nonzero $d, v \in \mathbb{Z}$, so $\frac{0}{d}$ is the zero element.

(iv) Also,

$$\frac{a}{b} + \frac{-a}{b} = \frac{ab + b(-a)}{b^2} = \frac{ab + (-ba)}{b^2} = \frac{ab - ba}{b^2} = \frac{0}{b^2} = \frac{-a}{b} + \frac{a}{b},$$

so $\frac{-a}{b}$ is the negative of $\frac{a}{b}$.

(v) For the commutative property, we see

$$\frac{a}{b} \cdot \frac{c}{d} = \frac{ac}{bd} = \frac{ca}{db} = \frac{c}{d} \cdot \frac{a}{b}.$$

(vi) can be proved in a similar manner to (v), using the associative property of integers.

(vii) To show that $\frac{d}{d}$ is the multiplicative identity, we note that

$$\frac{a}{b} \cdot \frac{d}{d} = \frac{ad}{bd} = \frac{d}{d} \cdot \frac{a}{b}$$

and since $(ad)b = (bd)a$, we have

$$\frac{ad}{bd} = \frac{a}{b}.$$

(viii) To see that $\frac{a}{b}$ has a multiplicative inverse, we note that when $a \neq 0$, $\frac{b}{a} \in \mathbb{Q}$ and

$$\frac{a}{b} \cdot \frac{b}{a} = \frac{ab}{ba} = \frac{ab}{ab} = \frac{b}{a} \cdot \frac{a}{b}.$$

By (vii), $\frac{ab}{ab}$ is the identity element.

(ix) For the distributive property, notice that

$$\frac{a}{b}\left(\frac{c}{d} + \frac{u}{v}\right) = \frac{a}{b} \cdot \left(\frac{cv + du}{dv}\right) = \frac{a(cv + du)}{bdv} = \frac{acv + adu}{bdv}.$$

On the other hand,

$$\frac{a}{b} \cdot \frac{c}{d} + \frac{a}{b} \cdot \frac{u}{v} = \frac{ac}{bd} + \frac{au}{bv} = \frac{ac(bv) + bd(au)}{bdbv} = \frac{b(acv + dau)}{b(dbv)}.$$

By (vii), we see

$$\frac{acv + dau}{dbv} = \frac{b(acv + adu)}{b(dbv)},$$

so the distributive property follows. The proof is complete.

The existence of additive inverses allows us to define the operation *of subtraction* of fractions. We define the difference of two fractions $\frac{n}{k}$ and $\frac{c}{d}$ by

$$\frac{n}{k} - \frac{c}{d} = \frac{n}{k} + \left(-\frac{c}{d}\right) = \frac{nd + k(-c)}{kd} = \frac{nd - kc}{kd}.$$

It is easy to check that this operation respects the properties of subtraction already obtained in $\mathbb{Z}$. Next we show how to embed $\mathbb{Z}$ into $\mathbb{Q}$. To this end, let $n, u, v \in \mathbb{Z}$ and let $u \neq 0$. Since $(nu)v = u(nv)$, the pairs (nu, u) and (nv, v) are equivalent, so that $\frac{nu}{u} = \frac{nv}{v}$. We define the mapping $\iota : \mathbb{Z} \longrightarrow \mathbb{Q}$, defined by $\iota(n) = \frac{nu}{u}$, where $n, u \in \mathbb{Z}$ and $u \neq 0$. The above argument shows that ι is independent of the choice of u, so ι is well defined. Suppose that $n \neq k$, but $\frac{nu}{u} = \frac{kv}{v}$, where $0 \neq u, v \in \mathbb{Z}$. Then

$$n(uv) = (nu)v = u(kv) = (kv)u = k(vu) = k(uv).$$

By Theorem 10.2.9(i), $uv \neq 0$, and applying Theorem 10.2.9(ii) we deduce that $n = k$. This contradiction shows that $\iota(n) \neq \iota(k)$ and hence ι is injective. Furthermore, for every $n, k \in \mathbb{Z}$, we have

$$\begin{aligned}\iota(n) + \iota(k) &= nu/u + kv/v = ((nu)v + u(kv))/uv = (n(uv) + k(uv))/uv \\ &= (n + k)uv/uv = \iota(n + k) \text{ and} \\ \iota(n)\iota(k) &= \frac{nu}{u} \cdot \frac{kv}{v} = \frac{(nu)(kv)}{uv} = \frac{(nk)(uv)}{uv} = \iota(nk).\end{aligned}$$

It follows that ι induces a bijection between $\mathbb{Z}$ and **Im** ι, which respects the operations of addition and multiplication. Thus, ι is a homomorphism from the ring $\mathbb{Z}$ to the ring $\mathbb{Q}$ and we can identify the integer n with its image $\iota(n) = \frac{nu}{u}$, in a manner familiar from Section 10.2 and we write $\frac{nu}{u} = n$ for each $u \in \mathbb{Z}$, where $u \neq 0$.

Then,

$$\frac{n}{k} = \frac{n \cdot 1 \cdot 1}{1 \cdot 1 \cdot k} = \frac{n1}{1} \cdot \frac{1}{1k} = n\left(\frac{1}{k}\right).$$

We write the reciprocal of $k \in \mathbb{Z}$ as k^{-1}. By Theorem 10.3.4, the fraction $\frac{1}{k}$ is the reciprocal to the element $\frac{k1}{1} = k$. So for the fraction $\frac{n}{k}$, we have

$$\frac{n}{k} = \frac{n1}{1} \cdot \frac{1}{k} = nk^{-1}.$$

The existence of reciprocals allows us to define the operation of division on nonzero fractions. We write $\left(\frac{u}{v}\right)^{-1}$ for the reciprocal of $\frac{u}{v}$ and note that by Theorem 10.3.4, $\left(\frac{u}{v}\right)^{-1} = \frac{v}{u}$. We define the quotient of two fractions $\frac{n}{k}$ and $\frac{u}{v}$, where $k, u, v \neq 0$ by

$$\frac{n}{k} \div \frac{u}{v} = \frac{\frac{n}{k}}{\frac{u}{v}} = \frac{n}{k} \cdot \left(\frac{u}{v}\right)^{-1} = \frac{n}{k} \cdot \frac{v}{u} = \frac{nuv}{ku}.$$

Theorem 10.3.4 implies that $\mathbb{Q}$ is a field and, as we saw in Section 7.4, it is a prime field.

The construction we used here is a very general one. Instead of using $\mathbb{Z}$, we could have used any integral domain R and exactly the same arguments could be used to extend R to a field F, known as the field of fractions of R. In this case, we use an equivalence relation defined on $R \times (R \setminus \{0\})$, using the same definition as given above and, if $r, s \in R$ and $s \neq 0$ then we define rs^{-1} to be the equivalence class of ordered pairs $\{(u, v) \in R \times (R \setminus \{0\}) | rv = su\}$. The field F then consists of the elements rs^{-1}, where $r \in R$ and $s \in R \setminus \{0\}$.

The fact that $\mathbb{Q}$ can be uniquely obtained in this way follows from Theorem 7.4.8.

Our next goal is to formally obtain some of the important properties of rational numbers. First, we define an order on the set $\mathbb{Q}$ in the following way.

10.3.5. Definition. *Let $\frac{n}{k} \in \mathbb{Q}$. If the integer nk is nonnegative, so $nk \geq 0$, or positive ($nk > 0$), then we say that $\frac{n}{k}$ is nonnegative (respectively positive). In this case, we will write $\frac{n}{k} \geq 0$ (respectively $\frac{n}{k} > 0$). Let $\frac{n}{k}, \frac{m}{t} \in \mathbb{Q}$. If $\frac{n}{k} - \frac{m}{t}$ is nonnegative (respectively positive), then we say that "$\frac{n}{k}$ is greater than or equal to $\frac{m}{t}$" or that "$\frac{m}{t}$ is less than or equal to $\frac{n}{k}$" and write these in the former case as $\frac{n}{k} \geq \frac{m}{t}$ and in the latter case as $\frac{m}{t} \leq \frac{n}{k}$. If, in this case, $\frac{n}{k} \neq \frac{m}{t}$, we say that "$\frac{n}{k}$ is greater then $\frac{m}{t}$" or that "$\frac{m}{t}$ is less then $\frac{n}{k}$" and write these in the former case as $\frac{n}{k} > \frac{m}{t}$ and in the latter case as $\frac{m}{t} < \frac{n}{k}$.*

Let $n, k, \in \mathbb{Z}$, $u, v \in \mathbb{Z} \setminus \{0\}$ and consider $\frac{nu}{u} - \frac{kv}{v}$. We have

$$\frac{nu}{u} - \frac{kv}{v} = \frac{n(uv) - u(kv)}{uv} = \frac{n(uv) - k(uv)}{uv} = \frac{(n-k)uv}{uv}.$$

By Theorem 10.2.9(xi), $(uv)^2 \geq 0$ and, by Theorem 10.2.9(i), $(uv)^2 \neq 0$, so that $(uv)^2 > 0$. Therefore, if $(n - k)(uv)^2 \geq 0$ (respectively, $(n - k)(uv)^2 > 0$) then, by Theorem 10.2.9(viii), $n - k \geq 0$ and $n \geq k$ (or respectively, by Theorem 10.2.9(vii), $n - k > 0$ and $n > k$). This shows that the order induced on $\mathbf{Im}\,\iota \cong \mathbb{Z}$ from $\mathbb{Q}$ coincides with the one introduced on $\mathbb{Z}$ in Section 10.2.

10.3.6. Theorem. *Let $x, y \in \mathbb{Q}$. Then one and only one of the relations $x = y$, $x < y$, or $x > y$ is valid.*

Proof. Let $x = \frac{n}{k}$, $y = \frac{m}{t}$, and suppose that $x \neq y$. We have

$$x - y = \frac{n}{k} - \frac{m}{t} = \frac{nt - km}{kt}.$$

Since $(nt - km)kt$ is an integer, Theorem 10.1.12 implies that $(nt - km)kt = 0$, or $(nt - km)kt > 0$, or $(nt - km)kt < 0$. Since $k, t \neq 0$, Theorem 10.2.9(i) shows that $kt \neq 0$. If $(nt - km)kt = 0$ then, using Theorem 10.2.9(i) again, we deduce that $nt = km$. This implies that $\frac{n}{k} = \frac{m}{t}$. If $(nt - km)kt > 0$, then, similarly, we can obtain that $\frac{n}{k} > \frac{m}{t}$, whereas if $(nt - km)kt < 0$ then $\frac{n}{k} < \frac{m}{t}$.

10.3.7. Proposition. *Let $x, y, z \in \mathbb{Q}$. Then the following properties hold:*

(i) $x \leq x$.

(ii) *If x and y are nonnegative, then $x + y$ and xy are nonnegative. Furthermore, if one of x and y is positive, then $x + y$ is positive. If both x and y are positive, then xy is positive.*

(iii) *$x \leq y$ and $y \leq z$ imply $x \leq z$.*

(iv) *$x < y$ and $y \leq z$ imply $x < z$.*

(v) *$x \leq y$ and $y < z$ imply $x < z$.*

(vi) *$x < y$ and $y < z$ imply $x < z$.*

(vii) *$x \leq y$ and $y \leq x$ imply $x = y$.*

Proof. Put $x = \frac{n}{k}$, $y = \frac{m}{t}$, and $z = \frac{r}{s}$.

(i) We have to show that $x - x \geq 0$ and to this end, we consider $\frac{n}{k} - \frac{n}{k} = \frac{nk-kn}{k^2}$. However, $(nk - kn)k^2 \geq 0$ which implies (i).

(ii) We have $nk, mt \geq 0$. Then $x + y = \frac{n}{k} + \frac{m}{t} = \frac{nt+km}{kt}$. To prove that $x + y \geq 0$, we must prove that $(nt + km)kt \geq 0$. However, $(nt + km)kt = nkt^2 + mtk^2$. By Theorem 10.2.9(xi), $t^2 \geq 0$, so $t^2 > 0$ because $t \neq 0$. The same arguments shows that $k^2 > 0$. Hence Theorem 10.2.9(iii) implies that $nkt^2, mtk^2 \geq 0$. By Theorem 10.2.8(i), $nkt^2 + mtk^2 \geq mtk^2$ and, since $mtk^2 \geq 0$, Proposition 10.2.7(ii) implies that $nkt^2 + mtk^2 \geq 0$. Thus $x + y \geq 0$. For the product xy, we have $xy = \frac{nm}{kt}$. Then, by Theorem 10.2.9(iii), $(nm)(kt) = (nk)(mt) \geq 0$, so $xy \geq 0$.

Furthermore, suppose that $x > 0$. Then $nk > 0$ and Theorem 10.2.9(iv) implies that $nkt^2 > 0$. Since $mtk^2 \geq 0$, Theorem 10.2.8(ii) shows that $nkt^2 + mtk^2 > mtk^2$ and using Proposition 10.2.7(iii), we deduce that $nkt^2 + mtk^2 > 0$, so that $x + y > 0$.

If $x, y > 0$ then, by Theorem 10.2.9(iv), $(nm)(kt) = (nk)(mt) > 0$ and we deduce that $xy > 0$.

(iii) Since $y \geq x$ and $z \geq y$ then, by definition, $y - x \geq 0$ and $z - y \geq 0$. By (ii) $z - x = (z - y) + (y - x) \geq 0$ and it follows that $z \geq x$.

(iv)–(vi) The proofs are similar.

(vii) We have

$$(mk - tn)tk = mtk^2 - nkt^2 \geq 0,$$

$$\text{and } (nt - km)tk = nkt^2 - mtk^2 \geq 0.$$

By Proposition 10.2.7(vi), $(nt-km)tk=nkt^2 - mtk^2 = 0$. Since $tk>0$, Theorem 10.2.9(i) implies that $nt - km = 0$ and this means that $x = \frac{n}{k} = \frac{m}{t} = y$.

10.3.8. Theorem. *Let $x, y, z \in \mathbb{Q}$. Then the following properties hold:*

(i) *if $x \leq y$, then $x + z \leq y + z$;*
(ii) *if $x < y$, then $x + z < y + z$;*
(iii) *if $x + z \leq y + z$, then $x \leq y$;*
(iv) *if $x + z < y + z$, then $x < y$.*

Proof. Here, we need to repeat the arguments of the proof of Theorem 10.2.8.

10.3.9. Theorem. *Let $x, y, z \in \mathbb{Q}$ and $z \neq 0$. Then the following properties hold:*

(i) *if x, y are nonzero then $xy \neq 0$;*
(ii) *if $xz = yz$, then $x = y$;*
(iii) *if $x \leq y$ and $z > 0$, then $xz \leq yz$;*
(iv) *if $x < y$ and $z > 0$, then $xz < yz$;*
(v) *if $x \leq y$ and $z < 0$, then $xz \geq yz$;*
(vi) *if $x < y$ and $z < 0$, then $xz > yz$;*
(vii) *if $xz < yz$ and $z > 0$, then $x < y$;*
(viii) *if $xz \leq yz$ and $z > 0$, then $x \leq y$;*
(ix) *if $xz < yz$ and $z < 0$, then $x > y$;*
(x) *if $xz \leq yz$ and $z < 0$, then $x \geq y$;*
(xi) *if $0 < x \leq y$ (respectively, $x \leq y < 0$), then $x^{-1} \geq y^{-1}$;*
(xii) *if $0 < x < y$ (respectively, $x < y < 0$), then $x^{-1} > y^{-1}$.*

Proof. Put $x = \frac{n}{k}$, $y = \frac{m}{t}$, and $z = \frac{r}{s}$.

(i) We have $xy = \frac{nm}{kt}$. If $xy = 0$, then $nm = 0$. However, $x \neq 0$ and $y \neq 0$ so $n, m \neq 0$, which contradicts Theorem 10.2.9(i). Thus $xy \neq 0$.

(ii) Since $xz = yz, 0 = yz - xz = (y - x)z$. By (i), we deduce that $y - x = 0$, because $z \neq 0$. Hence $y = x$.

(iii) Since $x \leq y$, $y - x$ is nonnegative. By Proposition 10.3.7(ii), the product $(y - x)z = yz - xz$ is also nonnegative and it follows that $xz \leq yz$.

(iv) Since $x < y$, $y - x$ is positive. By Proposition 10.3.7(ii), the product $(y - x)z = yz - xz$ is also positive and it follows that $xz < yz$.

(v) We have $0 > z$, so that $0 - z = -z > 0$. Then by (iii),

$$-xz = x(-z) \leq y(-z) = -yz.$$

It follows that

$$-yz - (-xz) = xz - yz \geq 0,$$

which proves the inequality $xz \geq yz$.

(vi) As in (iv), $(y - x)z \neq 0$, which together with (v) gives $xz > yz$.

(vii) By Theorem 10.3.6, $x = y$, $x < y$, or $x > y$. If $x < y$ then, by (iv), $xz < yz$. If $y < x$, then by (iv), $yz < xz$. Again by Theorem 10.3.6, $xz = yz$, $xz < yz$, or $xz > yz$. By hypothesis, the cases $xz = yz$ and $xz > yz$ do not occur and it follows that $x < y$.

(viii)–(x) The proofs are similar.

(xi) Clearly, the pair (n, k) is equivalent to the pair $(-n, -k)$. Thus, if $x = \frac{n}{k}$ and $y = \frac{m}{t}$, we may assume that $k, t > 0$. Then, by (iii),

$$\frac{nk}{k} = \frac{n}{k}k \leq \frac{m}{t}k = \frac{mk}{t} \text{ and } \frac{nkt}{k} \leq \frac{mkt}{t}.$$

It follows that

$$\frac{mkt}{t} - \frac{nkt}{k} \geq 0, \text{ or, equivalently }, \frac{mtk^2 - nkt^2}{kt} \geq 0.$$

This means that $(mk - nt)kt \geq 0$ and since $kt > 0$, we also have $mk - nt \geq 0$, by (viii). By the choice of x and y, either both integers n and m are nonnegative, or they are both nonpositive. In the first case, Proposition 10.3.7(ii) shows that nm is nonnegative. If $n, m \leq 0$ then, by Proposition 10.2.7(vii), $-n, -m \geq 0$, and $(-n)(-m) \geq 0$. On the other hand, $(-n)(-m) = -(-n)m = nm$. Thus, in either case, $nm \geq 0$ and by (iii) we deduce that $(mk - nt)nm \geq 0$. In turn, it follows that $\frac{t}{m} \leq \frac{k}{n}$.

(xii) The proof is similar.

10.3.10. Corollary. *Let $x, y \in \mathbb{Q}$ and suppose that $y > 0$. Then there is a natural number m such that $ym > x$.*

Proof. If $x \leq 0$, then $y1 = y > 0 > x$. Thus suppose that $x > 0$ and let $\frac{x}{y} = \frac{n}{k} > 0$. As in Theorem 10.3.9(xi) we may assume that $n, k > 0$, and by Corollary 10.3.13 that $k \geq 1$. Using Theorem 10.3.9(iii), we deduce that $n = k\frac{n}{k} \geq \frac{x}{y}$. Since $n + 1 > n$, Proposition 10.3.7(iv) implies that $n + 1 > \frac{x}{y}$. Using Theorem 10.3.9(iv), we deduce that $(n + 1)y > x$ so we can set $m = n + 1$.

As for integers, we can define the absolute value of a rational number. Let x be a rational number and let

$$|x| = \begin{cases} x, & \text{if } x \geq 0; \\ -x, & \text{if } x < 0. \end{cases}$$

It is not hard to prove that the following assertions hold for arbitrary $x, y \in \mathbb{Q}$.

$|x| \geq 0$, and $|x| = 0$ if and only if $x = 0$;

$|xy| = |x|\,|y|$;

$|x + y| \leq |x| + |y|$.

10.4 THE REAL NUMBERS

In this section, we construct the set of real numbers. In advanced calculus courses, the real numbers are usually introduced using the idea of a Dedekind Cut. Here we will use another approach which was introduced by G. Cantor. This approach is based on the notion of a Cauchy sequence, named after Augustin Louis Cauchy (1789–1857), a great French mathematician, whose work was influential in infinitesimal calculus and analysis in general. In our opinion, this approach has a very strong algebraic spirit and is very close to the one we used in the construction of the sets $\mathbb{Z}$ and $\mathbb{Q}$. However, we shall not describe the well-known properties of real numbers in detail since this is usually done in advanced calculus courses.

A long time ago, people observed that in many cases the set of rational numbers was not sufficient for measurement. Thus they arrived at the concept of incommensurable line segments. If KT and AB are two line segments and if AB and KT are commensurable, then there is a line segment CD such that KT contains it exactly r times and AB contains it s times. Then $\frac{|KT|}{|AB|} = \frac{r}{s}$ is a rational number. However, as is learned in geometry, the diagonal of a square and its side are incommensurable which means that there is no rational number that expresses the ratio of the lengths of the diagonal and the side of the same square.

We also cannot even take square roots of all natural numbers by merely using rational numbers. For example, if p is a prime then $\sqrt{p}$ is not a rational number. Indeed, if $\sqrt{p} = \frac{n}{k}$, where without loss of generality, we may assume that $\mathbf{GCD}(n, k) = 1$, then $p = \left(\frac{n}{k}\right)^2 = \frac{n^2}{k^2}$. It follows that $n^2 = pk^2$. Since $\mathbf{GCD}(n, k) = 1$, the integers n^2 and k^2 are relatively prime and by Proposition 9.1.15(i), p divides n^2. It follows that p divides n, so $n = mp$ for some integer m. Then $n^2 = (mp)^2 = m^2p^2$. We now have $m^2p^2 = pk^2$, or $m^2p = k^2$, and, as above, we prove that p divides k which now contradicts $\mathbf{GCD}(n, k) = 1$. Thus, $\sqrt{p}$ cannot be a rational number.

This proof was known in Ancient Greece and placed in Euclid's Elements. The existence of incommensurable numbers was known to Pythagoras. These problems naturally led to the creation of irrational numbers. The first axiom of real numbers was introduced by Archimedes, but ancient scientists were unable to develop a rigorous framework for irrational numbers.

In medieval times, people on the Indian subcontinent used irrational numbers without too much thought to rigor. In the seventeenth and eighteenth centuries, real numbers became one of the main subjects of investigation in calculus. During that time, real numbers were represented geometrically as points on a line, a plane or space. During the second part of the nineteenth century, Dedekind, Cantor, and Weierstrass constructed the real numbers in different ways using rigorous methods.

We start our construction using the Cartesian product $M = \prod_{n\in\mathbb{N}} A_n$, where $A_n = \mathbb{Q}$, for each $n \in \mathbb{N}$. As we saw in Section 1.1, this is the set of all sequences

$$\mathbf{a} = (a_1, \ldots, a_n, a_{n+1}, \ldots) = (a_n)_{n\in\mathbb{N}},$$

which we often abbreviate to (a_n), where $a_n \in \mathbb{Q}$ for each $n \in \mathbb{N}$.

Let

$$\mathbf{a} = (a_1, \ldots, a_n, a_{n+1}, \ldots) = (a_n) \quad \text{and}$$
$$\mathbf{b} = (b_1, \ldots, b_n, b_{n+1}, \ldots) = (b_n)$$

be elements of M and define addition and multiplication on M by

$$\mathbf{a} + \mathbf{b} = (a_1 + b_1, \ldots, a_n + b_n, a_{n+1} + b_{n+1}, \ldots) = (a_n + b_n) \text{ and}$$
$$\mathbf{ab} = (a_1 b_1, \ldots, a_n b_n, a_{n+1} b_{n+1}, \ldots) = (a_n b_n).$$

Clearly, the sequence $\mathbf{0} = (0, \ldots, 0, 0, \ldots)$ is the zero element for addition, the sequence $\mathbf{1} = (1, \ldots, 1, 1, \ldots)$ is the multiplicative identity, and the sequence $-\mathbf{a} = (-a_1, \ldots, -a_n, -a_{n+1}, \ldots)$ is the additive inverse of $\mathbf{a} = (a_n)_{n \in \mathbb{N}}$.

We define the difference of the sequences $\mathbf{a} = (a_n)$, $\mathbf{b} = (b_n)$ by

$$\mathbf{a} - \mathbf{b} = (a_1 - b_1, \ldots, a_n - b_n, a_{n+1} - b_{n+1}, \ldots) = (a_n - b_n)$$

and, as can be seen, addition and multiplication of sequences is reduced to addition and multiplication of the components. This makes it easy to observe that the commutative, associative, and distributive properties are valid.

We shall let $\mathbb{Q}_+$ (respectively $\mathbb{Q}_-$) denote the set of all nonnegative (respectively the set of all nonpositive) rational numbers.

10.4.1. Definition. *The sequence* $\mathbf{a} = (a_n) \in M$ *is called a Cauchy or fundamental sequence, if for every* $\varepsilon \in \mathbb{Q}_+$, *there exists a positive integer* $n(\varepsilon)$ *such that* $|a_k - a_j| < \varepsilon$, *whenever* $k, j \geq n(\varepsilon)$.

We observe the first important property of Cauchy sequences.

10.4.2. Lemma. *Let* $\mathbf{a} = (a_n)$ *be a Cauchy sequence. Then there exists* $r \in \mathbb{Q}_+$ *such that* $|a_n| < r$ *for each* $n \in \mathbb{N}$.

Proof. Let m be a positive integer such that $|a_k - a_j| < 1$ whenever $k, j \geq m$. The set $\{|a_1|, \ldots, |a_m|\}$ therefore has a largest element s. Let $r = s + 1$. If $1 \leq j \leq m$, then $|a_j| \leq s < s + 1 = r$. Let $j > m$. Then

$$|a_j| = |a_m + (a_j - a_m)| \leq |a_m| + |a_j - a_m| < s + 1 = r,$$

which proves the result.

10.4.3. Proposition. *Let* $\mathbf{a} = (a_n)$, $\mathbf{b} = (b_n)$ *be Cauchy sequences. Then* $\mathbf{a} + \mathbf{b}$, $\mathbf{a} - \mathbf{b}$, *and* $\mathbf{ab}$ *are Cauchy sequences.*

Proof. Let ε be a positive rational number. There are positive integers n_1 and n_2 such that $|a_k - a_j| < \frac{\varepsilon}{2}$ whenever $k, j \geq n_1$, and $|b_k - b_j| < \frac{\varepsilon}{2}$ whenever

$k, j \geq n_2$. Let $n = \mathbf{max}\{n_1, n_2\}$. Then for $k, j \geq n$, we have

$$\left|(a_k + b_k) - (a_j + b_j)\right| = \left|(a_k - a_j) + (b_k - b_j)\right| \leq$$
$$\left|a_k - a_j\right| + \left|b_k - b_j\right| < \frac{\varepsilon}{2} + \frac{\varepsilon}{2} = \varepsilon.$$

This shows that $\mathbf{a} + \mathbf{b}$ is a Cauchy sequence and the proof showing that $\mathbf{a} - \mathbf{b}$ is Cauchy is similar.

Next we prove that $\mathbf{ab}$ is a Cauchy sequence. By Lemma 10.4.2, there exists a positive rational r such that $|a_n|, |b_n| < r$ for each $n \in \mathbb{N}$. Let ε be a positive rational number. Since $\mathbf{a}, \mathbf{b}$ are Cauchy sequences, there are positive integers n_1 and n_2 such that $\left|a_k - a_j\right| < \frac{\varepsilon}{2r}$ whenever $k, j \geq n_1$, and $\left|b_k - b_j\right| < \frac{\varepsilon}{2r}$ whenever $k, j \geq n_2$. Let $n = \max\{n_1, n_2\}$. Then for $k, j \geq n$, we have

$$\begin{aligned}\left|a_k b_k - a_j b_j\right| &= \left|a_k b_k - a_k b_j + a_k b_j - a_j b_j\right| = \left|a_k(b_k - b_j) + b_j(a_k - a_j)\right| \\ &\leq \left|a_k(b_k - b_j)\right| + \left|b_j(a_k - a_j)\right| \\ &= |a_k| \left|b_k - b_j\right| + \left|b_j\right| \left|a_k - a_j\right| \leq r\frac{\varepsilon}{2r} + r\frac{\varepsilon}{2r} = \varepsilon.\end{aligned}$$

This shows that $\mathbf{ab}$ is a Cauchy sequence.

Let $\mathbb{F}$ denote the set of all Cauchy sequences. We say that a sequence $\mathbf{a} = (a_n)$ is a $\mathbf{0}$ sequence, if for every $\varepsilon \in \mathbb{Q}_+$, there exists a positive integer $n(\varepsilon)$ such that $|a_k| < \varepsilon$ whenever $k \geq n(\varepsilon)$.

10.4.4. Proposition.

(i) *Every* $\mathbf{0}$ *sequence is Cauchy.*

(ii) *Let* $\mathbf{a} = (a_n)$, $\mathbf{b} = (b_n)$ *be* $\mathbf{0}$ *sequences. Then* $\mathbf{a} + \mathbf{b}$, $\mathbf{a} - \mathbf{b}$ *are* $\mathbf{0}$ *sequences.*

(iii) *Let* $\mathbf{a} = (a_n)$ *be a* $\mathbf{0}$ *sequence and let* $\mathbf{b} = (b_n)$ *be a Cauchy sequence. Then* $\mathbf{ab}$ *is a* $\mathbf{0}$ *sequence.*

Proof.

(i) Let $\mathbf{a} = (a_n)$ be a $\mathbf{0}$ sequence and let ε be a positive rational number. There is a positive integer n such that $|a_k| < \frac{\varepsilon}{2}$ whenever $k \geq n$. Let $k, j \geq n$. Then

$$\left|a_k - a_j\right| = \left|a_k + (-a_j)\right| \leq |a_k| + \left|-a_j\right| = |a_k| + \left|a_j\right| \leq \frac{\varepsilon}{2} + \frac{\varepsilon}{2} = \varepsilon,$$

so $\mathbf{a}$ is Cauchy.

(ii) Let ε be a positive rational number. There are positive integers n_1 and n_2 such that $|a_k| < \frac{\varepsilon}{2}$ whenever $k \geq n_1$ and $|b_k| < \frac{\varepsilon}{2}$ whenever $k \geq n_2$. Let $n = \mathbf{max}\{n_1, n_2\}$. Then, for $k \geq n$, we have $|a_k + b_k| \leq |a_k| + |b_k| < \frac{\varepsilon}{2} + \frac{\varepsilon}{2} = \varepsilon$. This shows that $\mathbf{a} + \mathbf{b}$ is a $\mathbf{0}$ sequence and the proof that $\mathbf{a} - \mathbf{b}$ is a $\mathbf{0}$ sequence is similar.

(iii) By Lemma 10.4.2, there exists $r \in \mathbb{Q}_+$ such that $|b_n| < r$ for each $n \in \mathbb{N}$. Let ε be a positive rational number. Since a is a $\mathbf{0}$ sequence, there is a positive integer m such that $|a_k| < \frac{\varepsilon}{r}$ whenever $k \geq m$. For $k \geq m$, we have

$$|a_k b_k| = |a_k|\,|b_k| \leq r\left(\frac{\varepsilon}{r}\right) = \varepsilon.$$

This shows that **ab** is a **0** sequence.

We make the next definition very much in the spirit of this chapter.

10.4.5. Definition. *We say that the Cauchy sequences* **a** *and* **b** *are equivalent, if* $\mathbf{a} - \mathbf{b}$ *is a* **0** *sequence. Let*

$$\boldsymbol{\alpha} = \{\mathbf{b} \mid \mathbf{a} - \mathbf{b} \textit{ is a } \mathbf{0} \textit{ sequence}\}.$$

We shall write $\mathbf{a}R\mathbf{b}$ if **a** is equivalent to **b**, using this definition of equivalence. It is clear that the relation R is both reflexive and symmetric. Furthermore, Proposition 10.4.4 shows that if $\mathbf{a}R\mathbf{b}$ and $\mathbf{b}R\mathbf{c}$ then $\mathbf{a}R\mathbf{c}$ so that R is also transitive. Hence R is an equivalence relation and the equivalence class of **a** is precisely the set $\boldsymbol{\alpha} = \{\mathbf{b} \mid \mathbf{a} - \mathbf{b} \text{ is a } \mathbf{0} \text{ sequence}\}$. Hence the set of all subsets $\boldsymbol{\alpha}$ is a partition of $\mathbb{F}$.

10.4.6. Definition. *The set* $\mathbb{R}$ *is defined to be*

$$\mathbb{R} = \{\boldsymbol{\alpha} \mid \boldsymbol{\alpha} \textit{ is an equivalence class of Cauchy sequences of rationals}\}.$$

10.4.7. Definition. *Let* **a** *and* **b** *be two Cauchy sequences of rationals with corresponding equivalence classes* $\boldsymbol{\alpha}, \boldsymbol{\beta}$. *The operations of addition and multiplication are defined on* $\mathbb{R}$ *by*

$$\boldsymbol{\alpha} + \boldsymbol{\beta} = \boldsymbol{\gamma}, \textit{ where } \boldsymbol{\gamma} \textit{ is the equivalence class containing } \mathbf{a} + \mathbf{b},$$

$$\boldsymbol{\alpha\beta} = \boldsymbol{\delta}, \textit{ where } \boldsymbol{\delta} \textit{ is the equivalence class containing } \mathbf{ab}.$$

As in previous sections, we must be sure that these operations are well defined, which means that they are independent of the choice of elements of the sets $\boldsymbol{\alpha}, \boldsymbol{\beta}$. To see this, let $\mathbf{u} \in \boldsymbol{\alpha}$ and let $\mathbf{v} \in \boldsymbol{\beta}$. This means that $\mathbf{a} - \mathbf{u}$ and $\mathbf{b} - \mathbf{v}$ are **0** sequences. We have

$$(\mathbf{a} + \mathbf{b}) - (\mathbf{u} + \mathbf{v}) = (\mathbf{a} - \mathbf{u}) + (\mathbf{b} - \mathbf{v})$$

and, by Proposition 10.4.4, $(\mathbf{a} + \mathbf{b}) - (\mathbf{u} + \mathbf{v})$ is a **0** sequence. It follows that the sequences $(\mathbf{a} + \mathbf{b})$ and $(\mathbf{u} + \mathbf{v})$ are equivalent, so that the addition is well defined.

Furthermore,

$$\mathbf{ab} - \mathbf{uv} = \mathbf{ab} - \mathbf{ub} + \mathbf{ub} - \mathbf{uv} = \mathbf{b}(\mathbf{a} - \mathbf{u}) + \mathbf{u}(\mathbf{b} - \mathbf{v}).$$

By Proposition 10.4.4 again, $\mathbf{ab} - \mathbf{uv}$ is a $\mathbf{0}$ sequence, which implies that the sequences $\mathbf{ab}$ and $\mathbf{uv}$ are equivalent. Thus multiplication is also well defined.

Now we obtain some basic properties of these operations.

10.4.8. Theorem. *Let $\boldsymbol{\alpha}, \boldsymbol{\beta}, \boldsymbol{\gamma} \in \mathbb{R}$. The following properties hold:*

(i) $\boldsymbol{\alpha} + \boldsymbol{\beta} = \boldsymbol{\beta} + \boldsymbol{\alpha}$, *so addition is commutative;*

(ii) $(\boldsymbol{\alpha} + \boldsymbol{\beta}) + \boldsymbol{\gamma} = \boldsymbol{\alpha} + (\boldsymbol{\beta} + \boldsymbol{\gamma})$, *so addition is associative;*

(iii) *the subset* $\mathbf{0}$, *consisting of all* $\mathbf{0}$ *sequences, is the zero element for addition;*

(iv) *the subset* $-\boldsymbol{\alpha}$, *containing the sequence* $-\mathbf{a}$ *is the additive inverse of* $\boldsymbol{\alpha}$; *thus, every element of* $\mathbb{R}$ *has an additive inverse;*

(v) $\boldsymbol{\alpha}\boldsymbol{\beta} = \boldsymbol{\beta}\boldsymbol{\alpha}$, *so multiplication is commutative;*

(vi) $(\boldsymbol{\alpha}\boldsymbol{\beta})\boldsymbol{\gamma} = \boldsymbol{\alpha}(\boldsymbol{\beta}\boldsymbol{\gamma})$, *so multiplication is associative;*

(vii) *the subset* $\mathbf{1}$, *containing the sequence* $\mathbf{1} = (1, \ldots, 1, 1, \ldots)$, *is the multiplicative identity;*

(viii) *if* $\boldsymbol{\alpha} \neq \mathbf{0}$, *then there exists* $\boldsymbol{\alpha}^{-1} \in \mathbb{R}$ *such that* $\boldsymbol{\alpha}\boldsymbol{\alpha}^{-1} = \mathbf{1}$, *so every nonzero element of* $\mathbb{R}$ *has a multiplicative inverse or reciprocal;*

(ix) $\boldsymbol{\alpha}(\boldsymbol{\beta} + \boldsymbol{\gamma}) = \boldsymbol{\alpha}\boldsymbol{\beta} + \boldsymbol{\alpha}\boldsymbol{\gamma}$, *so multiplication is distributive over addition.*

Proof.

(i) We already noted that addition and multiplication of sequences are commutative and associative and connected via distributivity, so that (i), (ii), (v), (vi), (ix) hold.

(iii) Clearly, $\mathbf{0} = (0, \ldots, 0, 0, \ldots) \in \mathbf{0}$. Let $\mathbf{a} = (a_n) \in \boldsymbol{\alpha}$. Then $\boldsymbol{\alpha} + \mathbf{0}$ contains the sequence $\mathbf{a} + \mathbf{0} = \mathbf{a}$, so that $\boldsymbol{\alpha} + \mathbf{0} = \boldsymbol{\alpha}$.

(iv) Since $-\boldsymbol{\alpha} + \boldsymbol{\alpha}$ contains $-\mathbf{a} + \mathbf{a} = \mathbf{0}$ we have $-\boldsymbol{\alpha} + \boldsymbol{\alpha} = \mathbf{0}$.

(vii) The proof is similar to the proof of (iii).

(viii) Let $\mathbf{a} = (a_n)$ and $\mathbf{b} = (b_n)$ be two Cauchy sequences. If there is a positive integer k such that $a_n = b_n$ whenever $n \geq k$ then $a_n - b_n = 0$ whenever $n \geq k$ and it follows that $\mathbf{a} - \mathbf{b}$ is a $\mathbf{0}$ sequence, so $\boldsymbol{\alpha} = \boldsymbol{\beta}$.

Now suppose that $\mathbf{a} = (a_n)$ is not a $\mathbf{0}$ sequence. Then there exists $\varepsilon \in \mathbb{Q}_+$ such that, for each natural number n, there is natural number $m > n$ such that $|a_m| \geq \varepsilon$. Since $\mathbf{a}$ is a Cauchy sequence, there is a positive integer r such that $|a_k - a_j| < \frac{\varepsilon}{2}$ whenever $k, j \geq r$. If now $m > r$ is a positive integer such that $|a_m| \geq \varepsilon$, then for $j > m$, we have

$$|a_j| = |a_m + (a_j - a_m)| \geq \varepsilon - \frac{\varepsilon}{2} = \frac{\varepsilon}{2} = w > 0.$$

Hence there is a positive integer m such that $|a_j| \geq w$, whenever $j \geq m$. Now let $\mathbf{b} = (b_n)_{n \in \mathbb{N}}$ where $b_j = \begin{cases} w, & \text{if } j \leq m, \\ a_n, & \text{if } j > m. \end{cases}$

By construction, the sequences $\mathbf{a}$ and $\mathbf{b}$ are equivalent, so $\boldsymbol{\alpha} = \boldsymbol{\beta}$. By our choice of $\mathbf{b}$, $|b_j| \geq w$ for all j, in particular, $b_j \neq 0$ for all $j \in \mathbb{N}$. Now consider the sequence $\mathbf{c} = (c_n)_{n \in \mathbb{N}}$ where $c_n = \frac{1}{b_n}$, for $n \in \mathbb{N}$. Let ε be a positive rational number. Since $\mathbf{b}$ is a Cauchy sequence, there is a positive integer n such that $|b_k - b_j| < w^2\varepsilon$ whenever $k, j \geq n$. Then for $k, j \geq n$ we have

$$\left|\frac{1}{b_k} - \frac{1}{b_j}\right| = \frac{|b_j - b_k|}{|b_j|\,|b_k|} < \frac{|b_j - b_k|}{w^2} < \frac{w^2\varepsilon}{w^2} = \varepsilon.$$

It follows that $\mathbf{c}$ is a Cauchy sequence. Now

$$\mathbf{bc} = (b_n c_n)_{n \in \mathbb{N}} = (1, 1, \ldots, 1, \ldots) = \mathbf{1},$$

and this shows that $\boldsymbol{\alpha}^{-1} = \boldsymbol{\gamma}$.

The existence of additive inverses allows us to define the operation of subtraction. We define the difference of two elements $\boldsymbol{\alpha}$ and $\boldsymbol{\beta}$ by

$$\boldsymbol{\alpha} - \boldsymbol{\beta} = \boldsymbol{\alpha} + (-\boldsymbol{\beta}).$$

All properties of the difference of two rational numbers extend to real numbers.

The existence of reciprocals or multiplicative inverses allows us to define the operation of division by nonzero elements. We define the quotient of the elements $\boldsymbol{\alpha}$ and $\boldsymbol{\beta} \neq \mathbf{0}$ by

$$\frac{\alpha}{\beta} = \boldsymbol{\alpha}\boldsymbol{\beta}^{-1}.$$

Let $x \in \mathbb{Q}$. Clearly the sequence $\mathbf{x} = (x, x, \ldots, x, \ldots)$ is Cauchy. We consider the mapping $\iota : \mathbb{Q} \longrightarrow \mathbb{R}$, where $\iota(x)$ is the subset containing the sequence $\mathbf{x}$. We observe that if $x \neq y$ then $\mathbf{x} - \mathbf{y} = (x - y, x - y, \ldots, x - y, \ldots)$ and $x - y$ is a nonzero rational number. Therefore the sequence $\mathbf{x} - \mathbf{y}$ cannot be a $\mathbf{0}$ sequence, so that $\mathbf{x}$ and $\mathbf{y}$ cannot be equivalent. It follows that the mapping ι is injective. Furthermore, $\iota(x) + \iota(y)$ is the subset containing

$$\mathbf{x} + \mathbf{y} = (x + y, x + y, \ldots, x + y, \ldots),$$

and the latter is exactly $\iota(x + y)$.

Similarly, $\iota(x)\iota(y)$ is the subset containing

$$\mathbf{xy} = (xy, xy, \ldots, xy, \ldots),$$

and the latter is precisely $\iota(xy)$. Hence

$$\iota(x) + \iota(y) = \iota(x + y) \text{ and } \iota(x)\iota(y) = \iota(xy)$$

for every $x, y \in \mathbb{Q}$. It follows that ι induces a bijection between $\mathbb{Q}$ and $\mathbf{Im}\,\iota$, which respects the operations of addition and multiplication. Thus ι is a monomorphism.

Therefore, as we have done earlier, identifying a natural number m with its image $\mathbf{c}(m, 0)$ and an integer n with its image $\frac{nu}{u}$, we identify a rational number x with its image $\iota(x)$.

Thus, the set $\mathbb{R}$ is a field and, since it is an extension of $\mathbb{Q}$, it is of characteristic 0. The next natural step is to define an order on $\mathbb{R}$. However, we first need the following.

10.4.9. Lemma. *Let $\mathbf{a} = (a_n)$ be a Cauchy sequence. If $\mathbf{a}$ is not the $\mathbf{0}$ sequence, then there is a positive rational number w and a positive integer m such that either $a_j > w$ whenever $j \geq m$, or $a_j < -w$ whenever $j \geq m$.*

Proof. Since $\mathbf{a}$ is not the $\mathbf{0}$ sequence, there exists $\varepsilon \in \mathbb{Q}_+$ such that for each natural number n, there is a positive integer $s > n$ such that $|a_s| \geq \varepsilon$. Since $\mathbf{a}$ is a Cauchy sequence, there is a positive integer r such that $\left|a_k - a_j\right| < \frac{\varepsilon}{2}$, whenever $k, j \geq r$. If now $s > r$ is a positive integer such that $|a_s| \geq \varepsilon$, then for $j > s$, we have

$$\left|a_j\right| = \left|a_s + (a_j - a_s)\right| \geq \varepsilon - \frac{\varepsilon}{2} = \frac{\varepsilon}{2} = \nu > 0.$$

Hence there is a positive integer t such that $\left|a_j\right| \geq \nu$, whenever $j \geq t$. Since $\mathbf{a}$ is a Cauchy sequence, there is a positive integer r_1 such that $\left|a_k - a_j\right| < \frac{\nu}{2}$ whenever $k, j \geq r_1$. Now let $m > r$ be a positive integer such that $|a_m| > \nu$. If $a_m > 0$ then, for all $j \geq m$, we have

$$a_j = a_m + (a_j - a_m) > \nu - \frac{\nu}{2} = \frac{\nu}{2} = w.$$

If $a_m < 0$ then, as above, $-a_m > \nu$ and for all $j \geq m$, we have $-a_j > \frac{\nu}{2} = w$. It follows that $a_j < -w\varpi$.

10.4.10. Definition. *Let $\mathbf{a} = (a_n)$ be a Cauchy sequence. Then $\mathbf{a}$ is positive, if there is a positive rational number w and a positive integer m such that $a_j > w$ whenever $j \geq m$. Also $\mathbf{a}$ is negative, if there is a positive rational number w and a positive integer m such that $a_j < -w$ whenever $j \geq m$. Finally $\mathbf{a}$ is nonnegative (respectively nonpositive), if either $\mathbf{a}$ is positive or the $\mathbf{0}$ sequence (respectively either $\mathbf{a}$ is negative or the $\mathbf{0}$ sequence).*

The following result can be proved using similar arguments to those found earlier and we omit the details.

10.4.11. Lemma. *Let $\mathbf{a} = (a_n)$ and $\mathbf{b} = (b_n)$ be Cauchy sequences.*

(i) *If $\mathbf{a}, \mathbf{b}$ are positive then $\mathbf{ab}$ is positive.*
(ii) *If $\mathbf{a}, \mathbf{b}$ are negative then $\mathbf{ab}$ is positive.*
(iii) *If $\mathbf{a}$ is positive and $\mathbf{b}$ is negative then $\mathbf{ab}$ is negative.*
(iv) *If $\mathbf{ab}$ is a $\mathbf{0}$ sequence then at least one of $\mathbf{a}, \mathbf{b}$ is a $\mathbf{0}$ sequence.*

(v) *If* **a**, **b** *are nonnegative then* **ab** *is nonnegative.*

(vi) *If* **a**, **b** *are nonpositive then* **ab** *is nonnegative.*

(vii) *If* **a** *is nonpositive and* **b** *is nonnegative, then* **ab** *is nonpositive.*

Next, we extend these concepts to the set $\mathbb{R}$. For this, we need the following result.

10.4.12. Lemma. *Let* $\mathbf{a} = (a_n)$ *and let* $\mathbf{b} = (b_n)$ *be Cauchy sequences. Suppose that* **a** *and* **b** *are equivalent.*

(i) *If* **a** *is positive then* **b** *is also positive.*

(ii) *If* **a** *is negative then* **b** *is negative.*

Proof.

(i) By definition, there is a positive rational number w and a positive integer m such that $a_j > w$ whenever $j \geq m$. Since $\mathbf{b} - \mathbf{a}$ is a $\mathbf{0}$ sequence, there exists a positive integer t such that $|b_j - a_j| < \frac{w}{2}$ whenever $j \geq t$. Let $j > \mathbf{max}\{m, t\}$. Then

$$b_j = a_j + (b_j - a_j) > w - \frac{w}{2} = \frac{w}{2} > 0.$$

It follows that **b** is positive.

(ii) The proof is similar.

10.4.13. Definition. *An element* $\boldsymbol{\alpha} \in \mathbb{R}$ *is called positive if* $\boldsymbol{\alpha}$ *contains a positive sequence* **a**. *Also* $\boldsymbol{\alpha}$ *is called negative if* $\boldsymbol{\alpha}$ *contains a negative sequence* **a**. *We say that* $\boldsymbol{\alpha} \in \mathbb{R}$ *is nonnegative (respectively nonpositive) if either* $\boldsymbol{\alpha}$ *is positive or* $\mathbf{0}$ *(respectively either* $\boldsymbol{\alpha}$ *is negative or* $\mathbf{0}$*).*

Let $x \in \mathbb{Q}$. Then $\boldsymbol{\chi} = \iota(x)$ is a subset containing the sequence $\mathbf{x} = (x, x, \ldots, x, \ldots)$. If $x > 0$ then $\mathbf{x}$ is a positive sequence and, in this case, $\boldsymbol{\chi} > 0$. If $x < 0$ then $\mathbf{x}$ is a negative sequence and, in this case, $\boldsymbol{\chi} < 0$. Thus, if $x > 0$ (respectively, $x < 0$) then $\iota(x) > 0$ (respectively $\iota(x) < 0$), so the order that we defined on $\mathbb{R}$ respects the order defined on $\mathbb{Q}$.

Furthermore, we denote the set of all nonnegative (respectively the set of all nonpositive) real numbers by $\mathbb{R}_+$ (respectively $\mathbb{R}_-$).

10.4.14. Definition. *Let* $\boldsymbol{\chi}, \boldsymbol{\rho} \in \mathbb{R}$. *If* $\boldsymbol{\chi} - \boldsymbol{\rho}$ *is nonnegative then we say that* $\boldsymbol{\chi}$ *is greater than or equal to* $\boldsymbol{\rho}$ *or that* $\boldsymbol{\rho}$ *is less than or equal to* $\boldsymbol{\chi}$ *and denote these respectively by* $\boldsymbol{\chi} \geq \boldsymbol{\rho}$ *and* $\boldsymbol{\rho} \leq \boldsymbol{\chi}$. *If, in addition,* $\boldsymbol{\chi} \neq \boldsymbol{\rho}$, *we say that* $\boldsymbol{\chi}$ *is greater than* $\boldsymbol{\rho}$ *or* $\boldsymbol{\rho}$ *is less than* $\boldsymbol{\chi}$ *and denote these respectively by* $\boldsymbol{\chi} > \boldsymbol{\rho}$ *and* $\boldsymbol{\rho} < \boldsymbol{\chi}$.

The following results are similar to the corresponding results that we obtained already for the natural numbers, the integers, and the rational numbers. We assume that the reader has gained the necessary experience to prove them.

10.4.15. Theorem. *Let $\chi, \rho \in \mathbb{R}$. Then one and only one of $\chi = \rho$, $\chi < \rho$, and $\chi > \rho$ is valid.*

10.4.16. Proposition. *Let $\chi, \rho, \zeta \in \mathbb{R}$. Then the following properties hold:*

(i) $\chi \leq \chi$;
(ii) $\chi \leq \rho$ *and* $\rho \leq \zeta$ *imply* $\chi \leq \zeta$;
(iii) $\chi < \rho$ *and* $\rho \leq \zeta$ *imply* $\chi < \zeta$;
(iv) $\chi \leq \rho$ *and* $\rho < \zeta$ *imply* $\chi < \zeta$;
(v) $\chi < \rho$ *and* $\rho < \zeta$ *imply* $\chi < \zeta$;
(vi) $\chi \leq \rho$ *and* $\rho \leq \chi$ *imply* $\chi = \rho$.

10.4.17. Theorem. *Let $\chi, \rho, \zeta \in \mathbb{R}$. Then the following properties hold:*

(i) *if* $\chi \leq \rho$, *then* $\chi + \zeta \leq \rho + \zeta$;
(ii) *if* $\chi < \rho$, *then* $\chi + \zeta < \rho + \zeta$;
(iii) *if* $\chi + \zeta \leq \rho + \zeta$, *then* $\chi \leq \rho$;
(iv) *if* $\chi + \zeta < \rho + \zeta$, *then* $\chi < \rho$.

10.4.18. Theorem. *Let $\chi, \rho, \zeta \in \mathbb{R}$ and $\zeta \neq \mathbf{0}$. Then the following properties hold:*

(i) *if* χ, ρ *are nonzero then* $\chi\rho \neq \mathbf{0}$;
(ii) *if* $\chi\zeta = \rho\zeta$ *then* $\chi = \rho$;
(iii) *if* $\chi \leq \rho$ *and* $\zeta > \mathbf{0}$ *then* $\chi\zeta \leq \rho\zeta$;
(iv) *if* $\chi < \rho$ *and* $\zeta > \mathbf{0}$ *then* $\chi\zeta < \rho\zeta$;
(v) *if* $\chi \leq \rho$ *and* $\zeta < \mathbf{0}$ *then* $\chi\zeta \geq \rho\zeta$;
(vi) *if* $\chi < \rho$ *and* $\zeta < \mathbf{0}$ *then* $\chi\zeta > \rho\zeta$;
(vii) *if* $\chi\zeta < \rho\zeta$ *and* $\zeta > \mathbf{0}$ *then* $\chi < \rho$;
(viii) *if* $\chi\zeta \leq \rho\zeta$ *and* $\zeta > \mathbf{0}$ *then* $\chi \leq \rho$;
(ix) *if* $\chi\zeta < \rho\zeta$ *and* $\zeta < \mathbf{0}$ *then* $\chi > \rho$;
(x) *if* $\chi\zeta \leq \rho\zeta$ *and* $\zeta < \mathbf{0}$ *then* $\chi \geq \rho$;
(xi) *if* $\mathbf{0} < \chi \leq \rho$ *(respectively* $\chi \leq \rho < \mathbf{0}$*) then* $\chi^{-1} \geq \rho^{-1}$;
(xii) *if* $\mathbf{0} < \chi < \rho$ *(respectively* $\chi < \rho < \mathbf{0}$*) then* $\chi^{-1} > \rho^{-1}$.

The following result shows that the rational numbers are densely situated in the set of real numbers.

10.4.19. Theorem. *Let $\chi, \zeta \in \mathbb{R}$ and $\zeta > \chi$. Then there exists a rational number r such that $\chi < r < \zeta$.*

Proof. Let $\mathbf{x} = (x_n)$ be a sequence belonging to $\boldsymbol{\chi}$ and let $\mathbf{y} = (y_n)$ be a sequence belonging to $\boldsymbol{\zeta}$. Then $\mathbf{y} - \mathbf{x} > \mathbf{0}$ and, by Lemma 10.4.9, there is a positive rational number w and a positive integer m such that $y_j - x_j > w$, whenever $j \geq m$.

Since $\mathbf{x}$ is a Cauchy sequence, there is a positive integer n_1 such that $|x_k - x_j| < \frac{w}{3}$, whenever $k, j \geq n_1$. For the same reason, there is a positive integer n_2 such that $|y_k - y_j| < \frac{w}{3}$, whenever $k, j \geq n_2$. Let $t = \max\{m, n_1, n_2\}$ and set $r = \frac{(x_t + y_t)}{2}$. Then $r \in \mathbb{Q}$. Let $\mathbf{r} = (r, r, \ldots, r, \ldots)$ and suppose that $j \geq t$. Then

$$\begin{aligned} r - x_j &= \frac{(x_t + y_t)}{2} - x_j = \frac{x_t - x_j + y_t - x_j}{2} = \frac{(y_t - x_t) + 2(x_t - x_j)}{2} \\ &\geq \frac{(y_t - x_t) - 2(|x_t - x_j|)}{2} > \frac{w - \frac{2}{3}w}{2} = \frac{w}{6}. \end{aligned}$$

Similarly,

$$\begin{aligned} y_j - r &= \frac{y_j - x_t + y_j - y_t}{2} = \frac{(y_j - x_j) + (x_j - x_t) + (y_j - y_t)}{2} \\ &\geq \frac{(y_j - x_j) - (|x_j - x_t| - |y_j - y_t|)}{2} > \frac{w - \frac{2}{3}w}{2} = \frac{w}{6}. \end{aligned}$$

Now using Definitions 10.4.10, 10.4.13, and 10.4.14, we have $\boldsymbol{\chi} < r < \boldsymbol{\zeta}$.

As with rational numbers, we can define the absolute value (also sometimes called the modulus) of a real number. Let x be a real number and let

$$|x| = \begin{cases} x, & \text{if } x \geq 0, \\ -x, & \text{if } x \leq 0. \end{cases}$$

It is not hard to prove that for arbitrary $x, y \in \mathbb{R}$, the following assertions hold:

$|x| \geq 0$, and $|x| = 0$ if and only if $x = 0$;

$|xy| = |x|\,|y|$;

$|x + y| \leq |x| + |y|$.

We may now consider sequences consisting of real, rather than just rational numbers, which means we may now consider elements of the Cartesian product $\prod_{n \in \mathbb{N}} S_n$, where $S_n = \mathbb{R}$, for each $n \in \mathbb{N}$.

For these sequences, we make the following definition.

10.4.20. Definition. *We say that the sequence $A = (\boldsymbol{a}_n)_{n \in \mathbb{N}}$, where $\boldsymbol{a}_n \in \mathbb{R}$ converges to the real number $\boldsymbol{a}$ if for every $\varepsilon \in \mathbb{R}_+$, there exists a positive integer $n(\varepsilon)$ such that $|\boldsymbol{a}_k - \boldsymbol{a}| < \varepsilon$ whenever $k, j \geq n(\varepsilon)$.*

We can also define Cauchy sequences consisting of real numbers, in an analogous manner in the way we defined Cauchy sequences of rational numbers. Of course, every convergent sequence is Cauchy, but now the components are the real numbers.

We have already discussed some reasons for extending the set of rational numbers to the set of real numbers. One such reason for doing this is that not all convergent sequences of rational numbers converge to a rational number. It took a long time to understand this issue. Indeed, although Ancient Greece was the birthplace of irrational numbers, the place where such numbers were first conceptualized, it was not until the end of the eighteenth and the beginning of the nineteenth century that calculus was intensively developed. However, it was not until the second half of the nineteenth century that a full understanding of the real numbers was achieved. It turns out that the sequences of rational numbers that converge to a rational number are Cauchy sequences. Indeed, if some sequence (a_n), where $a_n \in \mathbb{Q}$ for each $n \in \mathbb{N}$, converges to some rational number r, then this sequence is Cauchy. To see this note that, for each positive rational number ε, there is a positive integer n such that $|r - a_j| < \frac{\varepsilon}{2}$ whenever $j \geq n$. Then for $k, j \geq n$, we have

$$\left|a_k - a_j\right| = \left|(r - a_j) + (a_k - r)\right| \leq \left|r - a_j\right| + |r - a_k| \leq \frac{\varepsilon}{2} + \frac{\varepsilon}{2} = \varepsilon.$$

Thus the Cauchy condition is necessary. However, in the set of rational numbers, it is not sufficient. There are Cauchy sequences of rationals that do not converge to rational numbers as can be seen using the example, 1, 1.4, 1.41, 1.414, ... of finite decimal approximations of the number $\sqrt{2}$.

Therefore, we are led to construct the "new" set $\mathbb{R}$ of numbers satisfying the following conditions:

(i) $\mathbb{R}$ contains $\mathbb{Q}$;
(ii) all Cauchy sequences of rational numbers converge in $\mathbb{R}$;
(iii) all algebraic operations in $\mathbb{Q}$ and the linear order in $\mathbb{Q}$ can be extended to $\mathbb{R}$.

Cantor suggested the construction based on condition (ii), which we have basically followed here. Here we need to take into account the fact that two different Cauchy sequences can converge to the same limit. This happens if the difference between the corresponding members of the sequences converge to zero, so the difference is a $\mathbf{0}$ sequence. This is why the new real numbers need to be defined as equivalence classes of Cauchy sequences.

Finally, we need to do one more step in order to justify the construction considered above. We need to consider sequences consisting of real numbers and ask whether they converge to a real number or whether the construction needs to be repeated extending the system of real numbers to some possibly larger set again. However, the following result shows that there is no need to do the construction again.

10.4.21. Theorem. *Let* $A = (\boldsymbol{a}_n)$, *be a Cauchy sequence, where* $\boldsymbol{a}_n \in \mathbb{R}$. *Then* A *converges to a real number* $\boldsymbol{a}$.

Proof. Let $\mathbf{a}_j = (a_{nj})_{n\in\mathbb{N}}$, where $a_{nj} \in \mathbb{R}$, be a Cauchy sequence, belonging to $\boldsymbol{\alpha}_j$, $j \in \mathbb{N}$. For every j, there is a positive integer $m(j)$ such that $\left|a_{mj} - a_{kj}\right| < \frac{1}{2^j}$ whenever $k \geq m(j)$. Put $c_j = a_{mj}$, where $j \in \mathbb{N}$. It is not difficult to check that the sequence $\mathbf{c} = (c_j)_{j\in\mathbb{N}}$ is a fundamental sequence converging to the real number $\boldsymbol{\alpha}$.

We would like to ask the reader to perform this check as an exercise.

ANSWERS TO SELECTED EXERCISES

CHAPTER 1

1.1.1. (i) Let A, B be arbitrary sets such that $A \notin B$; for example, $A = \{1\}, B = \{2, 3\}$. Put $C = \{A\}$; then $B \notin C$, but $A \in C$.

(ii) The same sets.

(iii) Since $A \neq B$, there exists an element $b \in B \backslash A$. The inclusion $B \subseteq C$ implies that $b \in C$. It follows that $C \nsubseteq A$.

(iv) Let $A = \{1\}, B = \{1, 2, 3\}$. Then if $A \subseteq B, A \neq B$. Put $C = \{\{A\}, \{B\}\}$; then $B \in C$ and $A \in C$.

1.1.5. Since $A \cap C = \emptyset$, $A \backslash C = A$, so we have $(A \cap B) \backslash C = (A \backslash C) \cap (B \backslash C) = A \cap (B \backslash C) = (A \cap B) \backslash C$. Since $A \cap B \subseteq A$ and $A \cap C = \emptyset$, $(A \cap B) \cap C = \emptyset$. It follows that $(A \cap B) \backslash C = A \cap B \neq \emptyset$.

1.1.6. Suppose that $(A \cap B) \cup C = A \cap (B \cup C)$. Let c be an arbitrary element of C. The inclusion $C \subseteq (A \cap B) \cup C$ implies that $c \in A \cap (B \cup C) = (A \cap B) \cup (A \cap C)$. Then either $c \in A \cap B$ or $c \in A \cap C$. Since $A \cap B \subseteq A$ and $A \cap C \subseteq A$, $c \in A$, it follows that $C \subseteq A$. Conversely, suppose that $C \subseteq A$. Then $(A \cap B) \cup C = (A \cup C) \cap (B \cup C) = A \cap (B \cup C)$.

1.1.12. Show that $X = B \cup (C \setminus A)$. Indeed, $A \cup X = A \cup B \cup (C \setminus A) = A \cup (C \setminus A) = C$, $A \cap X = A \cap (B \cup (C \setminus A)) = (A \cap B) \cup (A \cap (C \setminus A)) = B \cup \emptyset = B$.

1.2.2. No. For example, Φ does not contain the pairs whose first component is 1.

1.2.5. No. For example, Φ does not contain the pairs whose first component is 2.

1.2.7. Let a, b be arbitrary positive integers and suppose that $f(a) = f(b)$. Then $a/(a+1) = b/(b+1)$. It follows that $a(b+1) = b(a+1)$ or $ab + a = ba + b$ and $a = b$, so that f is injective. The mapping f is not surjective, because the rational number 1/3 has no preimage.

1.2.14. We have $f(0) = 2 = f(2)$, so that f is not injective. The mapping f is not surjective because, for example, the positive integer 9 has no preimage.

1.2.15. We have $f(6) = \{2, 3\} = f(18)$, so that f is not injective. The mapping f is not surjective because, for example, the singleton $\{4\}$ has no preimage.

1.2.17. Solution: Suppose that $\mathbf{Gr}(f_1) \cup \mathbf{Gr}(f_2) = \mathbf{Gr}(f_3)$ for some mapping $f_3 : A \longrightarrow B$. Let a be an arbitrary element of A. Then $(a, f_1(a))$, $(a, f_2(a)) \in \mathbf{Gr}(f_3)$. It follows by definition that $f_1(a) = f_2(a)$. Since this is true for all $a \in A$, we deduce that $f_1 = f_2$. The converse assertion is clear.

1.2.19. If A is finite and $f : A \longrightarrow B$ is a bijective mapping, then $|A| = |B|$. This shows that there is no bijective mapping from A to a proper subset of A.

Conversely, suppose that A is infinite. Then A contains a countable subset B. Let $B = \{b_n \mid n \in \mathbb{N}\}$, $B_1 = \{b_n \mid n > 1\}$. Define the mapping $f : A \longrightarrow B_1 \cup (A \setminus B)$ by $f(a) = a$ whenever $a \in A \setminus B$. If $a \in B$, then $a = b_n$ for some $n \in \mathbb{N}$. Put $f(b_n) = b_{n+1}$. It is not hard to prove that f is bijective.

1.3.3. Suppose that A is infinite. Then A contains a countable subset B. Let $B = \{b_n \mid n \in \mathbb{N}\}$. Define the mapping $\sigma_n : A \longrightarrow A$ by $\sigma_n(b_n) = b_{n+1}$, $\sigma_n(b_{n+1}) = b_n$, and $\sigma_n(a) = a$ whenever $a \notin \{b_n, b_{n+1}\}$. It is not hard to prove that σ_n is a permutation of A for every $n \in \mathbb{N}$ and $\sigma_n \neq \sigma_k$ whenever $n \neq k$. It follows that $S(A)$ is infinite.

Conversely, let $S(A)$ be infinite. If we suppose that A is finite and $|A| = n$, then as above, $|S(A)| = n!$. This contradiction shows that A must be infinite.

1.3.5. Define the mapping $f : A \times B \longrightarrow B \times A$ by $f(a, b) = (b, a)$ for each pair $(a, b) \in A \times B$. It is not hard to prove that f is bijection.

1.3.7. Let f be an arbitrary mapping from C to $A \times B$. Then for every $c \in C$, we have $f(c) = (a_c, b_c)$ where $a_c \in A$, $b_c \in B$. Define the mappings $f_A : C \longrightarrow A$ and $f_B : C \longrightarrow B$ by $f_A(c) = a_c$, $f_B(c) = b_c$. Finally, define the mapping $\Phi : (A \times B)^C \longrightarrow A^C \times B^C$ by $\Phi(f) = (f_A, f_B)$, $f \in (A \times B)^C$.

The mapping Φ is injective. Indeed, let $f, g : C \longrightarrow A \times B$ and suppose that $f \neq g$. Then, there exists an element $d \in C$ such that $f(d) \neq g(d)$. We have $f(d) = (a_d, b_d)$ where $a_d \in A, b_d \in B$, and $g(d) = (u_d, v_d)$ where $u_d \in A, v_d \in B$. In other words, $(a_d, b_d) \neq (u_d, v_d)$. If $a_d \neq u_d$, then $f_A(d) \neq g_A(d)$, so that $f_A \neq g_A$. If $a_d = u_d$, then $b_d \neq v_d$, so that $f_B(d) \neq g_B(d)$, and therefore $f_A \neq g_A$. In every case $\Phi(f) = (f_A, f_B) \neq (g_A, g_B) = \Phi(g)$.

The mapping Φ is injective. Indeed, let $(h_1, h_2) \in A^C \times B^C$. Then $h_1 : C \longrightarrow A$ and $h_2 : C \longrightarrow B$. Define the mapping $h : C \longrightarrow A \times B$ by the rule $h(c) = (h_1(c), h_2(c))$ for every $c \in C$. Clearly, $\Phi(h) = (h_1, h_2)$, so that Φ is surjective. Being surjective and injective, Φ is bijective.

1.3.9. Let k, n be arbitrary positive integers. Suppose that $k \neq n$. If k is even and n is odd, then $f(k) \geq 0$ and $f(n) < 0$; in particular, $f(n) \neq f(k)$. Suppose that k, n are even. Then $k = 2t$, $n = 2s$, and $t \neq s$. We have $f(k) = (n/2) - 1 = t - 1 \neq s - 1 = (n/2) - 1 = f(n)$. Suppose that k, n are odd. Then $k = 2t - 1, n = 2s - 1$, and $t \neq s$. We have $f(k) = -(k + 1/2) = -t \neq -s = -(n + 1/2) = f(n)$; hence, f is injective. Let q be an arbitrary integer. Suppose that $q \geq 0$, and put $m = 2(q + 1)$, then $f(m) = (m/2) - 1 = q + 1 - 1 = q$. Suppose that $q < 0$ and put $r = -1 - 2q$, then $f(r) = -(r + 1)/2 = q$. Hence f is surjective so f is bijective. Also

$$f^{-1}(n) = \left\{ \begin{array}{c} 2(n+1) \text{ whenever } n \geq 0; \\ -2(n+1) \text{ whenever } n < 0. \end{array} \right\}$$

1.3.13. Solution: Let (n, m) and (k, t) be arbitrary pairs of positive integers. Suppose that $f(n, m) = f(k, t)$, that is $2^{n-1}(2m - 1) = 2^{k-1}(2t - 1)$. Since the numbers $2m - 1$ and $2t - 1$ are odd, $2^{n-1} = 2^{k-1}$. It follows that $n - 1 = k - 1$ and $n = k$. Thus $2m - 1 = 2t - 1$, so that $m = t$. In other words, $(n, m) = (k, t)$; hence f is injective.

The mapping f is surjective. Indeed, if q is an arbitrary positive integer, then $q = 2^s r$ where r is odd and this representation is unique. We have $r = 2u - 1$. Then clearly, $q = f(s + 1, u)$; thus f is bijective. Also if $q = 2^s(2u - 1)$ is an arbitrary positive integer, then $f^{-1}(2^s(2u - 1)) = (s + 1, u)$.

1.3.15. The fact that f is a permutation of $\mathbb{Z}$ is clear. The mapping g is injective. Indeed, let n, m be arbitrary distinct integers. If n, m are even, then $g(n) = n \neq m = g(m)$. If n, m are odd, then $g(n) = n + 2 \neq m + 2 = g(m)$. If n is even and m is odd, then $g(n) = n$ is even and $g(m) = m + 2$ is odd, so that $g(n) \neq g(m)$.

The mapping g is surjective: let k be an arbitrary integer. If k is even, then $k = g(k)$; if k is odd, then $k = (k - 2) + 2 = g(k - 2)$. Similarly, we can prove that h is a permutation of $\mathbb{Z}$.

Finally, $(f \circ f)(x) = f(f(x)) = f(x + 1) = (x + 1) + 1 = x + 2$. Let x be even, then $(g \circ h)(x) = g(h(x)) = g(x + 2) = x + 2$. If x is odd, then $(g \circ h)(x) = g(h(x)) = g(x) = x + 2$. If again x is even, then $(h \circ g)(x) = h(g(x)) = h(x) = x + 2$. If x is odd, then $(h \circ g)(x) = h(g(x)) = h(x + 2) = x + 2$.

1.3.17. We have $(g \circ f)(x) = g(f(x)) = g(x^2 + 2) = ((x^2 + 2)/2) - 2 = (x^2/2) - 1$; $(f \circ g)(x) = (f(g(x)) = f((x/2) - 2) = ((x/2) - 2)2 + 2 = (x^2/4) - 2x + 6$; $((f \circ g) \circ f)(x) = (f \circ g)(f(x)) = (f \circ g)(x^2 + 2) = ((x^2 + 2)^2/4) - 2(x^2 + 2) + 6 = ((x^4 + 4x^2 + 4)/4) - 2x^2 + 2 = x^4/4 + x^2 + 1 - 2x^2 + 2 = x^4/4 - x^2 + 3$; $(f \circ (g \circ f))(x) = f((g \circ f)(x)) = f((x^2/2) - 1) = ((x^2/2) - 1)2 + 2 = x^4/4 - x^2 + 1 + 2 = x^4/4 - x^2 + 3$.

1.3.19. Define the mappings $f_j : \mathbb{Q} \longrightarrow \mathbb{Q}$ by $f_1(x) = 1 - x$, $f_2(x) = x(1/3)$, $f_3(x) = 1 + x$, $f_4(x) = x(1/5)$ for all $x \in \mathbb{Q}$. Then $(f_4 \circ f_3 \circ f_2 \circ f_1((x))) = f_4(f_3(f_2(f_1((x))))) = (f_4(f_3(f_2(1 - x)))) = f_4(f_3((1 - x)(1/3))) = f_4(1 + (1 - x)(1/3)) = (1 + (1 - x)(1/3))(1/5)$.

1.3.20. Suppose that f is surjective. Let y be an arbitrary element of B. Then, there exists an element $a \in A$ such that $b = f(a)$. We have $g(b) = g(f(a)) = (g \circ f)(a) = (h \circ f)(a) = h(f(a)) = h(b)$. Since domains and the value areas (ranges) of g and h coincide, $g = h$. Conversely, assume that f is not surjective. Then $\mathbf{Im} f \neq$ A. Let $u \in B \setminus \mathbf{Im} f$, $v \in \mathbf{Im} f$. Define the mapping $g : B \longrightarrow B$ by $g(u) = v$ and $g(b) = b$ for all $b \neq u$; then $g \neq \varepsilon_B$. We have $(g \circ f)(a) = g(f(a)) = f(a) = \varepsilon_B(f(a)) = (\varepsilon_B \circ f)(a)$. Since the domains and the ranges of $g \circ f$ and $\varepsilon_B \circ f$ coincide, $g \circ f = \varepsilon_B \circ f$. However, $g \neq \varepsilon_B$. This contradiction shows that f must be surjective.

1.4.1. Let $n = 2$, then $2^3 - 2 = 6 = 2 \times 3$. Suppose that $n > 2$ and we have already proved that 3 divides $k^3 - k$ for all $k < n$. Put $t = n - 1$, then $n^3 - n = (t + 1)^3 - t - 1 = t^3 + 3t^2 + 3t + 1 - t - 1 = (t^3 - t) + 3(t^2 + t)$. By induction hypothesis 3 divides $t^3 - t$, therefore 3 divides and $n^3 - n$.

This problem has an easier and more elegant solution: $n^3 - n = (n - 1)n(n + 1)$. Observe that a product of any three consecutive positive integers is divisible by 3.

1.4.3. Since n is odd, we can represent n as $2k + 1$, where k is a natural number. In other words, we must prove that 8 divides $(2k + 1)^2 - 1$ for each positive integer k. Let $k = 1$; then $3^2 - 1 = 8$. Suppose that $k > 1$ and we have already proved that 8 divides $(2m + 1)^2 - 1$ for all $m < k$. Put $t = k - 1$, then $(2k + 1)^2 - 1 = (2(t + 1) + 1)^2 - 1 = ((2t + 1) + 2)^2 - 1 = (2t + 1)^2 + 4(2t + 1) + 4 - 1 = (2t + 1)^2 - 1 + 8t + 4 + 4 = ((2t + 1)^2 - 1) + 8(t + 1)$. By induction hypothesis, 8 divides $(2t + 1)^2 - 1$, and therefore 8 divides $(2k + 1)^2 - 1$.

Here is another direct solution: Just observe that $(2k + 1)^2 - 1 = (2k)2(k + 1) = 4k(k + 1)$. One of the numbers k or $k + 1$ is even.

1.4.5. Let $n = 1$, then $13 = (1(1 + 1)/2)^2$. Suppose that $n > 1$ and we have already proved this equation for all $k < n$. Put $t = n - 1$; then by induction hypothesis $1 + 2^3 + 3^3 + \cdots + t^3 = (t(t + 1)/2)^2$. We have $1 + 2^3 + 3^3 + \cdots + n^3 = 1 + 2^3 + 3^3 + \cdots + t^3 + (t + 1)^3 = (t(t + 1)/2)^2 + (t + 1)^3 = (1/4)(t^2(t + 1)^2 + 4(t + 1)^3) = (1/4)((t+1)^2(t^2+4t+4)) = (1/4)(t + 1)^2(t^2 + 2)^2)) = ((t + 1)(t + 1 + 1)/2)^2 = (n(n + 1)/2)^2$.

1.4.7. Let $n = 0$; then $112 + 12 = 133$. Suppose that $n > 0$ and we have already proved that 133 divides $11^{m+2} + 12^{2m+1}$ for all $m < n$. Put $t = n - 1$; then

$11^{n+2} + 12^{2n+1} = 11^{t+1+2} + 12^{2t+2+1} = 11 \times 11^{t+2} + 12^2! \times 12^{2t+1} = (144 - 133)11^{t+2} + 144 \times 12^{2t+1} = 144 \times 11^{t+2} + 144 \times 12^{2t+1} - 133 \times 11^{t+2} = 144(11^{t+2} + 12^{2t+1}) - 133 \times 11^{t+2}$. By induction hypothesis, 133 divides $11^{t+2} + 12^{2t+1}$, so that 133 divides $144(11^{t+2} + 12^{2t+1}) - 133 \times 11^{t+2}$.

1.4.9. We have $x^2 + 2x - 3 = (x + 3)(x - 1)$. Since $x^2 + 2x - 3$ is a prime, one of the numbers $x + 3$ or $x - 1$ must be 1 or -1, and another must be prime. If $x + 3 = 1$, then $x = -2$ and $x - 1 = -3$. If $x + 3 = -1$, then $x = -4$ and $x - 1 = -5$. If $x - 1 = 1$, then $x = -2$ and $x + 3 = 1$. If $x - 1 = -1$, then $x = 0$ and $x + 3 = 3$. Hence the required values are $0, -2, -4$.

1.4.11. We have $n = x + 10y$. Furthermore, $n = 4(x + y) + 3$ and $n = 3xy + 5$. It follows that $x + 10y = 4x + 4y + 3$ and $x + 10y = 3xy + 5$. Then $6y = 3x + 3$, so that $x = 2y - 1$. Using the second equation, we see that $12y - 1 = 6y^2 - 3y + 5$ or $6y^2 - 15y + 6 = 0$. We have here only one integer root $y = 2$. It follows that $x = 3$ and $n = 23$.

1.4.13. Let $n = 3$; then $23 > 6 + 1$. Suppose that $n > 0$ and we have already proved that $2^k > 2k + 1$ for all $k < n$. Put $t = n - 1$; then $n = t + 1$. We have $2^n - 2n - 1 = 2^{t+1} - 2t - 2 - 1$. We observe that $-2t - 3 > -4t$, and therefore $2^{t+1} - 2t - 2 + 1 > 2^{t+1} - 4t - 2 = 2(2^t - 2t - 1) > 0$.

1.4.15. Suppose the contrary, let $n = pt$ where p is an odd prime. Then $2^n = 2^{tp} = (2^t)^p$. Put $a = 2^t$. We have $2^n + 1 = a^p + 1 = (a+1)(a^{p-1} - a^{p-2} + a^{p-3} - \cdots - a + 1)$. Since $a \geq 2$, $a + 1 \geq 3$. Suppose that $a^{p-1} - a^{p-2} + a^{p-3} - \cdots - a + 1 = 1$. Then $a^{p-1} - a^{p-2} + a^{p-3} - \cdots - a = 0$ which implies that either $a = 0$ or a is a root of the polynomial $X^{p-2} - X^{p-3} + X^{p-4} - \cdots - 1 = 0$. The first case is impossible. In the second case, a is a root of the polynomial $X^{p-1} - 1$. The last polynomial has only two real roots, 1 and -1, but this is impossible in the case we considered. Hence, $2^n + 1$ is not a prime. This contradiction shows that n has no odd divisors.

1.4.17. We have $1 + 2 + \cdots + k = \frac{k(k+1)}{2}$. It is not hard to check that the minimal positive integer k, for which $\frac{k(k+1)}{2}$ is a three digit number is 44. It follows that $k \leq 44$. it is given that $\frac{k(k+1)}{2} = a + 10a + 100a$ where $1 \leq a \leq 9$. It follows that $k(k+1) = 222a = 2 \times 3 \times 37 \times a$. Since 37 is a prime, it follows that either 37 divides k, or 37 divides $k + 1$. Since $k \leq 44$, it follows that either $k = 37$ or $k + 1 = 37$. In the first case, $\frac{k(k+1)}{2} = 703$, in the second case $\frac{k(k+1)}{2} = 666$; hence $k = 36$.

1.4.19. We have $(k!)^2 = (k(k-1)(k-2) \times \cdots \times 3 \times 2 \times 1)(1 \times 2 \times 3 \times \cdots \times (k-2)(k-1)k) = (k \times 1)((k-1)2)((k-2)3) \times \cdots \times ((k-t+1)t) \times \cdots \times (3(k-2))(2(k-1))(1 \times k)$. We show that $(k-t+1)t \geq k$ for all $t, 1 \leq t \leq k$. Indeed, $(k-t+1)t - k = kt - t^2 + t - k = (k-t)(t-1) \geq 0$. Hence $(k!)^2 \geq k^k$. $(k!)^2 \geq \underbrace{k \times k \times k \times \cdots \cdot \times k \times k \times k}_{k}$.

CHAPTER 2

2.1.2. Let $A = [a_{jk}]$, $B = [b_{jk}]$ and let $C = AB = [c_{jk}]$, $D = BA = [d_{jk}]$. Then $a_{jk} = 0$ whenever $j \neq k$. Now $c_{jk} = \sum_{1 \leq l \leq n} a_{jl} b_{lk} = a_{jj} b_{jk}$ and $d_{jk} = \sum_{1 \leq l \leq n} b_{jl} a_{lk} = b_{jk} a_{kk}$. Since $c_{jk} = d_{jk}$, we have $a_{jj} b_{jk} = b_{jk} a_{kk}$ or $b_{jk}(a_{jj} - a_{kk}) = 0$. For $j \neq k$, we know $a_{jj} \neq a_{kk}$ so $b_{jk} = 0$ and B is diagonal.

2.1.4. Let $A = [a_{jk}]$, $B = [b_{jk}]$, $AB = [u_{jk}]$. In A, interchange rows m and t. We obtain the matrix $C = [c_{jk}]$ defined by $c_{jk} = a_{jk}$ whenever $j \neq m, j \neq t$ and $c_{mk} = a_{tk}, c_{tk} = a_{mk}$, $1 \leq k \leq n$. Put $CB = [v_{jk}]$. Then $v_{jk} = \sum_{1 \leq s \leq n} c_{js} b_{sk} = \sum_{1 \leq s \leq n} a_{js} b_{sk}$, whenever $j \neq m, t$. Further, $v_{mk} = \sum_{1 \leq s \leq n} c_{ms} b_{sk} = \sum_{1 \leq s \leq n} a_{ts} b_{sk} = u_{tk}$; $v_{tk} = \sum_{1 \leq s \leq n} c_{ts} b_{sk} = \sum_{1 \leq s \leq n} a_{ms} b_{sk} = u_{mk}$.

Thus if we interchange rows m and t of AB, then we obtain the matrix CB.

2.1.6. Let $A = [a_{jk}]$, $B = [b_{jk}]$, $AB = [u_{jk}]$. Transform the matrix A multiplying row t by α and adding the result to row m. We obtain a matrix

$C = [c_{jk}]$ where $c_{jk} = a_{jk}$ whenever $j \neq m$, $c_{mk} = a_{mk} + \alpha a_{tk}$, $1 \leq k \leq n$. Put $CB = [v_{jk}]$. Then $v_{jk} = \sum_{1 \leq s \leq n} c_{js} b_{sk} = \sum_{1 \leq s \leq n} a_{js} b_{sk}$, whenever $j \neq m, t$. Further, $v_{mk} = \sum_{1 \leq s \leq n} c_{ms} b_{sk} = \sum_{1 \leq s \leq n} (a_{ms} + \alpha a_{ts}) b_{sk} = \sum_{1 \leq s \leq n} a_{ms} b_{sk} + \sum_{1 \leq s \leq n} \alpha a_{ts} b_{sk} = u_{mk} + \alpha \sum_{1 \leq s \leq n} a_{ts} b_{sk} = u_{mk} + \alpha u_{tk}$.

Thus if we transform the matrix AB multiplying row t by α and adding the result to row m, then we obtain the matrix CB.

2.1.7. $$\begin{pmatrix} 1 & 3 & 6 & 10 & \dots & n(n+1)/2 \\ 0 & 1 & 3 & 6 & \dots & n(n-1)/2 \\ 0 & 0 & 1 & 3 & \dots & (n-1)(n-2)/2 \\ 4 & 0 & 0 & 1 & \dots & 0 \\ \vdots & \vdots & \vdots & \vdots & \ddots & \vdots \\ 0 & 0 & 0 & 0 & \dots & 1 \end{pmatrix}.$$

2.1.9. $$\begin{pmatrix} 1 & 0 & 0 & 0 & \dots & 0 & 0 \\ 0 & 1 & 0 & 0 & \dots & 0 & 0 \\ 3 & 3 & 1 & 3 & \dots & 3 & 3 \\ 0 & 0 & 0 & 1 & \dots & 0 & 0 \\ \vdots & \vdots & \vdots & \vdots & \ddots & \vdots & \vdots \\ 0 & 0 & 0 & 0 & \dots & 1 & 0 \\ 0 & 0 & 0 & 0 & \dots & 0 & 1 \end{pmatrix}.$$

2.1.11. $$\begin{pmatrix} 1 & 8 & 28 & 56 & 70 & 56 & 28 & 8 & 1 \\ 0 & 1 & 8 & 28 & 56 & 70 & 56 & 28 & 8 \\ 0 & 0 & 1 & 8 & 28 & 56 & 70 & 28 & 8 \\ 0 & 0 & 0 & 1 & 8 & 28 & 56 & 70 & 56 \\ 0 & 0 & 0 & 0 & 1 & 8 & 28 & 56 & 700 \\ 0 & 0 & 0 & 0 & 0 & 1 & 8 & 28 & 56 \\ 0 & 0 & 0 & 0 & 0 & 0 & 1 & 8 & 280 \\ 0 & 0 & 0 & 0 & 0 & 0 & 0 & 1 & 8 \\ 0 & 0 & 0 & 0 & 0 & 0 & 0 & 0 & 1 \end{pmatrix}.$$

2.1.12. $A^7 = [b_{jk}]$, where $b_{1\,15} = 1, b_{2\,16} = 1, b_{3\,17} = 1, \cdots, b_{n-14\,n} = 1$, and all other elements are 0.

2.1.13. Put $A = [a_{jk}]$, $A^t = [b_{jk}]$, $AA^t = [c_{jk}]$ where $b_{jk} = a_{kj}$ for all $1 \leq j$, $k \leq n$. Then $c_{jk} = \sum_{1 \leq m \leq n} a_{jm} b_{mk} \sum_{1 \leq m \leq n} a_{jm} a_{km}$; $c_{kj} = \sum_{1 \leq m \leq n} a_{km} b_{mi} = \sum_{1 \leq m \leq n} a_{km} a_{jm}$. So $c_{jk} = c_{kj}$ for all $1 \leq j, k \leq n$.

2.1.15. We have $A^t = -A, B^t = -B$. Suppose that $AB = BA$, then $(AB)^t = B^t A^t = (-B)(-A) = BA = AB$; hence AB is a symmetric matrix. Conversely, suppose that AB is a symmetric matrix. Then $(AB)^t = AB$. On the other hand, $(AB)^t = B^t A^t = (-B)(-A) = BA$. It follows that $BA = AB$.

2.1.17. Using Theorem 2.1.10 we see $(ABAB\dots ABA)^t = A^tB^tA^tB^t\dots A^tB^tA^t = ABAB\dots ABA$.

2.1.19. Let $A = [a_{jk}]$, $A^2 = [b_{jk}]$. Then $a_{jm} = 0$ if $j \geq m$. We have $b_{j,j+1} = \sum_{1\leq m\leq n} a_{jm}a_{m,j+1} = a_{j1}a_{1,j+1} + a_{j2}a_{2,j+1} + \cdots + a_{jj}a_{j,j+1} + a_{j,j+1}a_{j+1,j+1} + a_{j,j+2}a_{j+2,j+1} + \cdots + a_{jn}a_{n,j+1} = 0$. Thus, all elements that are situated on the principal diagonal or on the diagonal just above this one are zero. Put $A^3 = [c_{jk}]$. Then $c_{j,j+1} = \sum_{1\leq m\leq n} b_{jm}a_{m,j+1} = b_{j1}a_{1,j+1} + b_{j2}a_{2,j+1} + \cdots + b_{jj}a_{j,j+1} + b_{j,j+1}a_{j+1,j+1} + b_{j,j+2}a_{j+2,j+1} + \cdots + b_{jn}a_{n,j+1} = 0$ and $c_{j,j+2} = \sum_{1\leq m\leq n} b_{jm}a_{m,j+2} = b_{j1}a_{1,j+2} + b_{j2}a_{2,j+2} + \cdots + b_{jj}a_{j,j+2} + b_{j,j+1}a_{j+1,j+2} + b_{j,j+2}a_{j+2,j+2} + b_{j,j+3}a_{j+3,j+2} + \cdots + b_{jn}a_{n,j+2} = 0$. Thus, all elements that are situated on the principal diagonal or on the two diagonals just immediately above this are zero; continuing this process we see that $A^n = O$.

2.2.5. Even.

2.2.6. Even.

2.2.12. $\mathbf{sign}\,\pi = (-1)^n$.

2.2.14. $\begin{pmatrix} 1 & 2 & 3 & 4 & 5 & 6 & \dots & 2n-3 & 2n-2 & 2n-1 & 2n \\ 3 & 4 & 5 & 6 & 7 & 8 & \dots & 2n-1 & 2n & 1 & 2 \end{pmatrix} =$

$(13)(46)(79)\dots(3n-2\,3n)$. It follows that $\mathbf{sign}\,\pi = (-1)^{3n}$.

2.2.15. $\begin{pmatrix} 1 & 2 & 3 & 4 & 5 & 6 & \dots & 3n-2 & 3n-1 & 3n \\ 2 & 3 & 1 & 5 & 6 & 4 & \dots & 3n-1 & 3n & 3n-2 \end{pmatrix} =$

$(123)(456)(789)\dots(3n-2\,3n-1\,3n)$. Since every cycle of length 3 is an even permutation, $\mathbf{sign}\,\pi = 1$.

2.2.17. We have

$$\begin{pmatrix} 1 & 2 & 3 & 4 & 5 & 6 \\ 2 & 3 & 5 & 1 & 6 & 4 \end{pmatrix}\begin{pmatrix} 1 & 2 & 3 & 4 & 5 & 6 \\ \pi(1) & \pi(2) & \pi(3) & \pi(4) & \pi(5) & \pi(6) \end{pmatrix}$$
$$= \begin{pmatrix} 1 & 2 & 3 & 4 & 5 & 6 \\ 2 & 3 & 5 & 1 & 6 & 4 \end{pmatrix}\begin{pmatrix} 1 & 2 & 3 & 4 & 5 & 6 \\ 2 & 4 & 1 & 6 & 3 & 5 \end{pmatrix}$$
$$= \begin{pmatrix} 1 & 2 & 3 & 4 & 5 & 6 \\ 6 & 1 & 3 & 2 & 5 & 6 \end{pmatrix}.$$

Then,

$$\begin{pmatrix} 1 & 2 & 3 & 4 & 5 & 6 \\ \pi(1) & \pi(2) & \pi(3) & \pi(4) & \pi(5) & \pi(6) \end{pmatrix}$$
$$= \begin{pmatrix} 1 & 2 & 3 & 4 & 5 & 6 \\ 3 & 1 & 2 & 6 & 4 & 5 \end{pmatrix}\begin{pmatrix} 1 & 2 & 3 & 4 & 5 & 6 \\ 6 & 1 & 3 & 2 & 5 & 6 \end{pmatrix}$$
$$= \begin{pmatrix} 1 & 2 & 3 & 4 & 5 & 6 \\ 3 & 6 & 1 & 6 & 2 & 5 \end{pmatrix}.$$

2.2.20. We have

$$\begin{pmatrix} 1 & 2 & 3 & 4 & 5 & 6 & 7 & 8 & 9 \\ 3 & 6 & 5 & 7 & 4 & 2 & 1 & 9 & 8 \end{pmatrix}^2 = \begin{pmatrix} 1 & 2 & 3 & 4 & 5 & 6 & 7 & 8 & 9 \\ 5 & 2 & 4 & 1 & 7 & 6 & 5 & 8 & 9 \end{pmatrix},$$

$$\begin{pmatrix} 1 & 2 & 3 & 4 & 5 & 6 & 7 & 8 & 9 \\ 3 & 6 & 5 & 7 & 4 & 2 & 1 & 9 & 8 \end{pmatrix}^3 = \begin{pmatrix} 1 & 2 & 3 & 4 & 5 & 6 & 7 & 8 & 9 \\ 4 & 6 & 7 & 3 & 1 & 2 & 5 & 8 & 9 \end{pmatrix},$$

$$\begin{pmatrix} 1 & 2 & 3 & 4 & 5 & 6 & 7 & 8 & 9 \\ 3 & 6 & 5 & 7 & 4 & 2 & 1 & 9 & 8 \end{pmatrix}^4 = \begin{pmatrix} 1 & 2 & 3 & 4 & 5 & 6 & 7 & 8 & 9 \\ 7 & 2 & 1 & 3 & 6 & 4 & 7 & 8 & 9 \end{pmatrix},$$

$$\begin{pmatrix} 1 & 2 & 3 & 4 & 5 & 6 & 7 & 8 & 9 \\ 3 & 6 & 5 & 7 & 4 & 2 & 1 & 9 & 8 \end{pmatrix}^5 = \begin{pmatrix} 1 & 2 & 3 & 4 & 5 & 6 & 7 & 8 & 9 \\ 1 & 6 & 3 & 4 & 5 & 2 & 7 & 8 & 9 \end{pmatrix}.$$

Therefore, it is easy to see that $\pi^{10} = (\pi^5)^2 = \varepsilon$. Then,

$$\pi^{97} = \pi^{90} \circ \pi^7 = (\pi^{10})^9 \circ \pi^5 \circ \pi^2 = \pi^5 \circ \pi^2 =$$

$$\begin{pmatrix} 1 & 2 & 3 & 4 & 5 & 6 & 7 & 8 & 9 \\ 5 & 2 & 4 & 1 & 7 & 6 & 5 & 8 & 9 \end{pmatrix} \begin{pmatrix} 1 & 2 & 3 & 4 & 5 & 6 & 7 & 8 & 9 \\ 1 & 6 & 3 & 4 & 5 & 2 & 7 & 8 & 9 \end{pmatrix}$$

$$= \begin{pmatrix} 1 & 2 & 3 & 4 & 5 & 6 & 7 & 8 & 9 \\ 5 & 6 & 4 & 1 & 7 & 2 & 3 & 8 & 9 \end{pmatrix}.$$

2.3.1. We have $\mathbf{det}(A) = \sum_{\pi \in \mathbf{S}_n} \mathbf{sign}\, \pi\, a_{1\pi(1)} a_{2\pi(2)} \ldots a_{n\pi(n)}$. Consider $d_\pi = \mathbf{sign}\, \pi\, a_{1\pi(1)} a_{2\pi(2)} \ldots a_{n\pi(n)}$. In the first row, we have the only one nonzero element a_{1n}. Therefore if $\pi(1) \neq n$, then $d_\pi = 0$. Suppose that $\pi(1) = n$. In the second row, only the elements a_{2n-1} and a_{2n} are nonzero. Since $\pi(1) = n$, $\pi(2) \neq n$. If $\pi(2) \neq n-1$, then $d_\pi = 0$. Suppose that $\pi(1) = n, \pi(2) \neq n-1$. Using similar arguments we can show that only the element d_π corresponding to the permutation

$$\pi = \begin{pmatrix} 1 & 2 & 3 & \ldots & n-1 & n \\ n & n-1 & n-2 & \ldots & 2 & 1 \end{pmatrix}$$

could be nonzero. The number of pairs forming inversions here is equal to $n - 1 + n - 2 + \cdots + 2 + 1 = n(n-1)/2$. Hence $\mathbf{det}(A) = (-1)^{\frac{n(n-1)}{2}} a_{1n} a_{2n-1} \ldots a_{n-12} a_{n1}$.

2.3.3. Add the second and the third rows of the matrix. By Corollary 2.3.9 we obtain the new matrix

$$\begin{pmatrix} a & b & c & 1 \\ b & c & a & 1 \\ b+c & c+a & a+b & 2 \\ \dfrac{b+c}{2} & \dfrac{a+c}{2} & \dfrac{b+a}{2} & 1 \end{pmatrix},$$

whose determinant is equal to the given one. Since the fourth row is a multiple of the third row, by Corollary 2.3.8 and Proposition 2.3.5, its determinant is 0.

2.3.5. We have $\mathbf{det}(A) = \mathbf{det}(A) = \sum_{\pi \in \mathbf{S}_5} \mathbf{sign}\,\pi\, a_{1\pi(1)}a_{2\pi(2)}a_{3\pi(3)}a_{4\pi(4)}a_{5\pi(5)}$. Consider $d_\pi = \mathbf{sign}\,\pi\, a_{1\pi(1)}a_{2\pi(2)}a_{3\pi(3)}a_{4\pi(4)}a_{5\pi(5)}$. Only the elements a_{11} and a_{13} in the first row are nonzero. Therefore if $\pi(1) \neq 1, 3$, then $d_\pi = 0$. Suppose that $\pi(1) \in \{1, 3\}$. In the second row, only the elements a_{21} and a_{23} are nonzero. If $\pi(1) = 1$, then $\pi(2) = 3$. If $\pi(1) = 3$, then $\pi(2) = 1$. In the third row, only the elements a_{31} and a_{33} are nonzero. If $\pi(1) = 1$ and $\pi(2) = 3$, then $\pi(3)$ is not equal to 1 or 3, and hence, $a_{3\pi(3)} = 0$. If If $\pi(1) = 3$ and $\pi(2) = 1$, then we have a similar case. In any case, $a_{3\pi(3)} = 0$, and therefore $\mathbf{det}(A) = 0$.

2.3.7. Exchange the positions of the first and the last rows, then the second and the $n-1$ rows, the third and the $n-2$ rows, and so on. Finally, we obtain the requested matrix. By Proposition 2.3.7, each of these transformations changes the determinant sign. The number of these transformations is $n/2$ if n is even, and $(n-1)/2$ if n is odd. Therefore, the determinant of the new matrix is $(-1)^{n/2}$ if n is even and $(-1)^{(n-1)/2}$ if n is odd.

2.3.9. The element $a_{1m}a_{23}a_{3j}a_{41}a_{5k}$ corresponds to the permutation

$$\begin{pmatrix} 1 & 2 & 3 & 4 & 5 \\ m & 3 & j & 1 & k \end{pmatrix}.$$

This permutation is odd if $m = 2$, $j = 4$, $k = 5$.

2.3.12. The element $a_{1n-1}a_{2n}a_{31}a_{42}\dots a_{nn-2}$ corresponds to the permutation

$$\begin{pmatrix} 1 & 2 & 3 & 4 & 5 & \dots & n \\ n-1 & n & 1 & 2 & 3 & \dots & n-2 \end{pmatrix}.$$

The numbers of pairs forming inversions relative to this permutation is $(n-2)+(n-2)$, which is even. Hence the element $a_{1n-1}a_{2n}a_{31}a_{42}\dots a_{nn-2}$ has the sign $+$ in the decomposition of the determinant.

2.3.13. After these transformations, we obtain the matrix

$$B = \begin{pmatrix} a_{nn} & a_{n-1n} & a_{n-2n} & \dots & a_{2n} & a_{1n} \\ a_{nn-1} & a_{n-1n-1} & a_{n-2n-1} & \dots & a_{2n-1} & a_{1n-1} \\ \vdots & \vdots & \vdots & \vdots & \vdots & \vdots \\ a_{n2} & a_{n-12} & a_{n-22} & \dots & a_{22} & a_{12} \\ a_{n1} & a_{n-11} & a_{n-21} & \dots & a_{21} & a_{11} \end{pmatrix}.$$

Permute the columns of the matrix A in reverse order. We have

$$A_1 = \begin{pmatrix} a_{1n} & a_{1n-1} & a_{1n-2} & \cdots & a_{12} & a_{11} \\ a_{2n} & a_{2n-1} & a_{2n-2} & \cdots & a_{22} & a_{21} \\ \vdots & \vdots & \vdots & \vdots & \vdots & \vdots \\ a_{n-1n} & a_{n-1n-1} & a_{n-1n-2} & \cdots & a_{n-12} & a_{n-11} \\ a_{nn} & a_{nn-1} & a_{nn-2} & \cdots & a_{n2} & a_{n1} \end{pmatrix}.$$

By exercise 2.3.7, $\mathbf{det}(A_1) = (-1)^{n/2}\mathbf{det}(A)$ if n is even and $\mathbf{det}(A_1) = (-1)^{(n-1)/2}\mathbf{det}(A)$ if n is odd. Transposing A_1, we obtain

$$A_2 = \begin{pmatrix} a_{1n} & a_{2n} & a_{3n} & \cdots & a_{n-1n} & a_{nn} \\ a_{1n-1} & a_{2n-1} & a_{3n-1} & \cdots & a_{n-1n-1} & a_{nn-1} \\ \vdots & \vdots & \vdots & \vdots & \vdots & \vdots \\ a_{12} & a_{22} & a_{32} & \cdots & a_{n-12} & a_{n2} \\ a_{11} & a_{2n} & a_{3n} & \cdots & a_{n-11} & a_{n1} \end{pmatrix}.$$

By Proposition 2.3.3, $\mathbf{det}(A_2) = \mathbf{det}(A_1)$. Permute the columns of the matrix A_2 in inverse order. We obtain the matrix B. By exercise 2.3.7, $\mathbf{det}(B) = (-1)^{n/2}\mathbf{det}(A_2) = (-1)^{n/2}\mathbf{det}(A_1) = (-1)^{n/2}(-1)^{n/2}$ $\mathbf{det}(A) = \mathbf{det}(A)$ if n is even, and $\mathbf{det}(B) = (-1)^{(n-1)/2}\mathbf{det}(A_2) = (-1)^{(n-1)/2}\mathbf{det}(A_1) = (-1)^{(n-1)/2}(-1)^{(n-1)/2}\mathbf{det}(A) = \mathbf{det}(A)$ if n is odd.

2.3.15. The matrix A is as follows:

$$A = \begin{pmatrix} 1 & 1 & 1 & 1 & 1 & 1 & 1 & 1 & 1 \\ 1 & 2 & 2 & 2 & 2 & 2 & 2 & 2 & 2 \\ 1 & 2 & 3 & 3 & 3 & 3 & 3 & 3 & 3 \\ 1 & 2 & 3 & 4 & 4 & 4 & 4 & 4 & 4 \\ 1 & 2 & 3 & 4 & 5 & 5 & 5 & 5 & 5 \\ 1 & 2 & 3 & 4 & 5 & 6 & 6 & 6 & 6 \\ 1 & 2 & 3 & 4 & 5 & 6 & 7 & 7 & 7 \\ 1 & 2 & 3 & 4 & 5 & 6 & 7 & 8 & 8 \\ 1 & 2 & 3 & 4 & 5 & 6 & 7 & 8 & 9 \end{pmatrix}.$$

Multiplying the first column by -1 and adding the result to all other columns, we obtain the matrix A_1, and then in the matrix A_1 multiplying the second column by -1 and adding the result to all following columns, we obtain the matrix A_2. Multiplying the third column by -1 A_2 and adding the result to all following columns, we obtain the matrix

$$A_3 = \begin{pmatrix} 1 & 0 & 0 & 0 & 0 & 0 & 0 & 0 & 0 \\ 1 & 1 & 0 & 0 & 0 & 0 & 0 & 0 & 0 \\ 1 & 1 & 1 & 0 & 0 & 0 & 0 & 0 & 0 \\ 1 & 1 & 1 & 1 & 0 & 0 & 0 & 0 & 0 \\ 1 & 1 & 1 & 1 & 1 & 0 & 0 & 0 & 0 \\ 1 & 1 & 1 & 1 & 1 & 1 & 0 & 0 & 0 \\ 1 & 1 & 1 & 1 & 1 & 1 & 1 & 0 & 0 \\ 1 & 1 & 1 & 1 & 1 & 1 & 1 & 1 & 0 \\ 1 & 1 & 1 & 1 & 1 & 1 & 1 & 1 & 1 \end{pmatrix}.$$

The determinant of this matrix is 1.

2.3.17. Multiply the last row of the determinant of A by -1 and add it to all other rows to obtain the matrix A_1 whose determinant, by Corollary 2.3.9, is equal to the given determinant. Multiply the first row of A_1 by $-1/x$ and add the result to the last row; then multiply the second row by $-1/y$ and add the result to the last row; then multiply the third row by $-1/z$ and add the result to the last row; then multiply the fourth row by $1/z$ and add the result to the last row; then multiply the fifth row by $1/y$ and add the result to the last row. Finally, we obtain the matrix

$$A_2 = \begin{pmatrix} x & 0 & 0 & 0 & 0 & x \\ 0 & y & 0 & 0 & 0 & x \\ 0 & 0 & z & 0 & 0 & x \\ 0 & 0 & 0 & -z & 0 & x \\ 0 & 0 & 0 & 0 & -y & x \\ 0 & 0 & 0 & 0 & 0 & x \end{pmatrix},$$

whose determinant by Corollary 2.3.9 is equal to $-(xyz)^2$.

2.3.19. Interchange the first and second rows to obtain the matrix A_1, whose determinant differs from the original only by sign (Proposition 2.3.7). Multiply the second row of A_1 by $-x$ and add it to all following rows. We will get the matrix A_2, whose determinant is equal to the determinant of the matrix A_1 (Corollary 2.3.9). Next, we add the third row to the first row; the fourth row and all following rows are added to the first row. We come to the matrix A_3, whose determinant is equal to the determinant of the matrix A_2 (Corollary 2.3.9). Now multiply the first row of A_2 by $\frac{-1}{n-1}$ and add it to the third and all following rows. We come to the matrix A_4, whose determinant is equal to the determinant of the matrix

A_2 (Corollary 2.3.9).

$$A_4 = \begin{pmatrix} n-1 & 0 & 0 & 0 & \dots & 0 & 0 \\ 0 & 1 & 1 & 1 & \dots & 1 & 1 \\ 0 & 0 & -x & 0 & \dots & 0 & 0 \\ 0 & 0 & 0 & -x & \dots & 0 & 0 \\ \dots & \dots & \dots & \dots & \dots & \dots & \dots \\ 0 & 0 & 0 & 0 & \dots & -x & 0 \\ 0 & 0 & 0 & 0 & \dots & 0 & -x \end{pmatrix}.$$

The determinant of the last matrix is equal to $(-1)^{n-2}(n-1)x^{n-2}$ (Proposition 2.3.11). This means that the given determinant is equal to $(-1)^{n-1}(n-1)x^{n-2}$.

2.4.4. 10.

2.4.5. -4.

2.4.6. 8.

2.4.7. 12.

2.4.8. $-(ayz + bxz + cxy)$.

2.4.9. $abc - x(bc + ca + ab)$.

2.4.10. -84.

2.4.11. -84.

2.4.12. 98.

2.4.13. 14.

2.5.2. It follows that $\mathbf{det}(A)^2 = 0$, so that $\mathbf{det}(A) = 0$. Hence

$$A = \begin{pmatrix} x & y \\ ax & ay \end{pmatrix}.$$

We have now

$$\mathrm{A}^2 = \begin{pmatrix} x^2 + axy & xy + ay^2 \\ ax^2 + a^2xy & axy + a^2y^2 \end{pmatrix}.$$

So we obtain the system

$$x^2 + axy = 0,$$
$$xy + ay^2 = 0,$$
$$ax^2 + a^2xy = 0,$$
$$axy + a^2y^2 = 0.$$

If $a = 0$, then $x = 0$ and

$$A = \begin{pmatrix} 0 & y \\ 0 & 0 \end{pmatrix}.$$

If $a \neq 0$, then we obtain the following system:

$$x^2 + axy = 0,$$
$$xy + ay^2 = 0.$$

If $x = 0$, then $y = 0$ and $A = O$. Suppose that $x \neq 0$, then we obtain the system

$$x + ay = 0,$$
$$xy + ay^2 = 0.$$

If $y = 0$, then $x = 0$ and $A = O$. Suppose that $y \neq 0$, then we obtain $x + ay = 0$. It follows that $y = \frac{-x}{a}$.

2.5.3. $A^{-1} = I - E_{12} + 2E_{13} - 2E_{23}$.

2.5.4. Yes.

2.5.6. $A^{-1} = \frac{1}{ad-bc}\begin{pmatrix} d & -b \\ -c & a \end{pmatrix}$.

2.5.7. $A^{-1} = \begin{pmatrix} \cos\alpha & \sin\alpha \\ -\sin\alpha & \cos\alpha \end{pmatrix}$.

2.5.8. $A^{-1} = \frac{1}{4}\begin{pmatrix} 1 & 1 & 1 & 1 \\ 1 & 1 & -1 & -1 \\ 1 & -1 & 1 & -1 \\ 1 & -1 & -1 & 1 \end{pmatrix}$.

2.5.9. $A^{-1} = \begin{pmatrix} 22 & -6 & -26 & 17 \\ -17 & 5 & 20 & -13 \\ -1 & 0 & 2 & -1 \\ 4 & -1 & -5 & 3 \end{pmatrix}$.

2.5.10. $A^{-1} = \begin{pmatrix} 1 & -1 & 1 & -1 & 1 & -1 & 1 & -1 & 1 \\ 0 & 1 & -1 & 1 & -1 & 1 & -1 & 1 & -1 \\ 0 & 0 & 1 & -1 & 1 & -1 & 1 & -1 & 1 \\ 0 & 0 & 0 & 1 & -1 & 1 & -1 & 1 & -1 \\ 0 & 0 & 0 & 0 & 1 & -1 & 1 & -1 & 1 \\ 0 & 0 & 0 & 0 & 0 & 1 & -1 & 1 & -1 \\ 0 & 0 & 0 & 0 & 0 & 0 & 1 & -1 & 1 \\ 0 & 0 & 0 & 0 & 0 & 0 & 0 & 1 & -1 \\ 0 & 0 & 0 & 0 & 0 & 0 & 0 & 0 & 1 \end{pmatrix}$.

2.5.11. $A^{-1} = \begin{pmatrix} 2-n & 1 & 1 & \dots & 1 \\ 1 & -1 & 0 & \dots & 1 \\ 1 & 0 & -1 & \dots & 1 \\ \vdots & \vdots & \vdots & \ddots & \vdots \\ 1 & 0 & 0 & \dots & -1 \end{pmatrix}.$

2.5.12. $X = \begin{pmatrix} -1 & -1 \\ 2 & 3 \end{pmatrix}.$

2.5.13. $X = \begin{pmatrix} 1 & 2 \\ 3 & 4 \end{pmatrix}.$

2.5.14. $X = \begin{pmatrix} 6 & 4 & 5 \\ 2 & 1 & 2 \\ 3 & 3 & 3 \end{pmatrix}.$

CHAPTER 3

3.1.1. We have $(m, a) * (n, b) = (m + an, ab), (n, b) * (m, a) = (n + bm, ba)$.

In particular, $(3, -1) * (2, 1) = (1, -1), (2, 1) * (3, -1) = (5, -1)$, so that this operation is not commutative. Further, $((m, a) * (n, b)) * (k, c) = (m + an, ab) * (k, c) = (m + an + abk, abc).(m, a) * ((n, b) * (k, c)) = (m, a) * (n + bk, bc) = (m + an + abk), abc)$, so that this operation is associative. The pair $(0, 1)$ is an identity element. Indeed, $(m, a) * (0, 1) = (m + a0, a1) = (m, a), (0, 1) * (m, a) = (0 + 1m, 1a) = (m, a)$. Furthermore, $(m, a) * (-am, a) = (m + a(-am), a2) = (m - m, 1) = (0, 1)$, so that every element has an inverse.

3.1.2.

	e	a	b	c
e	e	a	b	c
a	a	e	c	b
b	b	c	e	a
c	c	b	a	e

3.1.3. We have $a \perp b = a^2 + b^2 = b^2 + a^2 = b \perp a$, so that this operation is commutative. Further, $(a \perp b) \perp c = (a^2 + b^2) \perp c = (a^2 + b^2)^2 + c^2, a \perp (b \perp c) = a \perp (b^2 + c^2)^2 = a^2 + (b^2 + c^2)^2$. In particular, $(2 \perp 0) \perp 1 = 17$, $2 \perp (0 \perp 1) = 5$, $(a^2 + b^2) \perp c = (a^2 + b^2)^2 + c^2$, so that this operation is not associative. The identity element does not exist.

3.1.5. (i) $(a \bullet b) \bullet c = (a + b + ab) \bullet c = (a + b + ab) + c + (a + b + ab)c = a + b + c + ab + ac + bc + abc, a \bullet (b \bullet c) = a \bullet (b + c + bc) = a + (b + c + bc) + a(b + c + bc) = a + b + c + bc + ab + ac + abc$, so that this operation is associative.

(ii) We have $a \bullet b = a + b + ab = b + a + ba = b \bullet a$, so that this operation is commutative.

(iii) Suppose that $a \neq -1$. If $a \bullet b = a \bullet c$, then $a + b + ab = a + c + ac$. It follows that $b(1 + a) = c(1 + a)$ and hence $b = c$. If $b = c$, then clearly $a \bullet b = a \bullet c$.

(iv) The number 0 is an identity element relative to this operation: $a \bullet 0 = a + 0 + a0 = a$.

(v) Suppose that $0 = a \bullet b = a + b + ab$. It follows that $a + b(1 + a) = 0$ or $b = (-a)/(1 + a)$.

Hence if $a \neq -1$, then a has an inverse.

3.1.7. We have $((a, b) \bullet (c, d)) \bullet (u, v) = (ac - bd, bc + ad) \bullet (u, v) = ((ac - bd)u - (bc + ad)v, (bc + ad)u + (ac - bd)v) = (acu - bdu - bcv - adv, bcu + adu + acv - bdv)$; $(a, b) \bullet ((c, d) \bullet (u, v)) = (a, b) \bullet (cu - dv, du + cv) = (a(cu - dv) - b(du + cv), b(cu - dv) + a(du + cv)) = (acu - adv - bdu - bcv, bcu - bdv + adu + acv)$, so that this operation is associative. Further, $(a, b) \bullet (c, d) = (ac - bd, bc + ad)$, $(c, d) \bullet (a, b) = (ca - db, da + cb)$, so that this operation is commutative. The pair $(1, 0)$ is an identity element. Indeed, $(a, b) \bullet (1, 0) = (a, b)$.

3.1.8.

$$\begin{array}{ccccc} & e & a & b & c \\ e & e & a & b & c \\ a & a & a & c & c \\ b & b & c & b & c \\ c & c & c & c & c \end{array}.$$

3.1.9.

$$\begin{array}{ccccc} & e & a & b & c \\ e & e & a & b & c \\ a & a & b & c & c \\ b & b & c & e & a \\ c & c & e & a & b \end{array}.$$

3.1.11. By Theorem 1.1.10, S is a semigroup relative to the operations $\cap$ and $\cup$. Consider the mapping $\phi : S \longrightarrow S$, defined by: for every $A \subseteq M$ we put $\phi(A) = M \backslash A$. The mapping ϕ is injective: if $M \backslash A = \phi(A) = \phi(B) = M \backslash B$, then $A = M \backslash (M \backslash A) = M \backslash (M \backslash B) = B$. The mapping ϕ is surjective: for each subset C of M we have $C = M \backslash (M \backslash C) = \phi(M \backslash C)$. Furthermore, $\phi(A \cap B) = M \backslash (A \cap B) = (M \backslash A) \cup (M \backslash B) = \phi(A) \cup \phi(B)$, so that ϕ is an isomorphism.

3.1.13. We have $((a, b) \blacklozenge (c, d)) \blacklozenge (u, v) = (ac - 2bd, bc + ad) \blacklozenge (u, v) = ((ac - 2bd)u - 2(bc + ad)v, (bc + ad)u + (ac - 2bd)v) = (acu - 2bdu - 2bcv - 2adv, bcu + adu + acv - 2bdv)$; $(a, b) \blacklozenge ((c, d) \blacklozenge (u, v)) = (a, b) \blacklozenge (cu - 2dv, du + cv) = (a(cu - 2dv) - 2b(du + cv), b(cu - 2dv) + a(du + cv)) = (acu - 2adv - 2bdu - 2bcv, bcu - 2bdv + adu + acv)$, so that this operation is associative. Further, $(a, b) \blacklozenge (c, d) = (ac - 2bd, bc + ad)(c, d) \blacklozenge (a, b) =$

$(ca - 2db, da + cb)$, so that this operation is commutative. The pair $(1, 0)$ is an identity element. Indeed, $(a, b)\blacklozenge(1, 0) = (a, b)$.

Suppose that $(1, 0) = (a, b)\blacklozenge(x, y) = (ax - 2by, bx + ay)$. We obtain $ax - 2by = 1$, $bx + ay = 0$. If $(a, b) = (0, 0)$, then clearly such a pair (x, y) does not exists. Suppose that $(a, b)\blacklozenge(0, 0)$. Then $x = a/(a^2 + 2b^2)$, $y = (-b)/(a^2 + 2b^2)$.

3.1.15. We have $(a\blacktriangledown b)\blacktriangledown c = (pa + qb + r)\blacktriangledown c = p(pa + qb + r) + qc + r = p2a + pqb + pr + qc + r$; $a\blacktriangledown(b\blacktriangledown c) = a\blacktriangledown(pb + qc + r) = pa + q(pb + qc + r) + r = pa + qpb + q2c + qr + r$. Since $(a\blacktriangledown b)\blacktriangledown c = a\blacktriangledown(b\blacktriangledown c)$ for arbitrary a, b, c, then $p^2a + pqb + pr + qc + r = pa + qpb + q^2c + qr + r$ or $a(p^2 - p) + c(q - q^2) + (p - q)r = 0$. Put $a = c = 0$, we obtain $(p - q)r = 0$. It follows that either $r = 0$ or $p = q$. Put $a = 0, c = 1$, we obtain $q(q - 1) = 0$. It follows that either $q = 0$ or $q = 1$. Put $a = 1, c = 0$, we obtain $p(p - 1) = 0$. It follows that either $p = 0$ or $p = 1$. If $p = q = 0$, then $a\blacktriangledown b = r$ for all a, b. Clearly, in this case we obtain an associative operation. If $p = 1, q = 0$, then $r = 0$ and $a\blacktriangledown b = a$, if $p = 0, q = 1$, then again $r = 0$ and $a\blacktriangledown b = b$. In this case, we obtain an associative operation.

3.2.2. No solution.

3.2.3. $(-\frac{\sqrt{2}}{2}, \frac{\sqrt{2}}{2}), (\frac{\sqrt{2}}{2}, -\frac{\sqrt{2}}{2})$.

3.2.12. $\frac{1}{43}(5 + 9\alpha - \alpha^2)$.

3.3.1. $\mathbf{Re}\, z_1 \mathbf{Im}\, z_2 + \mathbf{Re}\, z_2\ \mathbf{Im}\, z_1 = 0$.

3.3.2. $\mathbf{Re}\, \alpha = -238$, $\mathbf{Im}\, \alpha = 0$.

3.3.3. $\mathbf{Re}\, \alpha = 1$, $\mathbf{Im}\, \alpha = 0$.

3.3.4. $-\frac{1}{4} + i$.

3.3.6. $\frac{-1 \pm \sqrt{5}}{2}$.

3.3.7. $-1 + i, -4 - i$.

3.3.8. $-1 - i$.

3.3.9. $\frac{1}{2\sqrt{2}}(\cos\frac{5\pi}{4} + i \sin\frac{5\pi}{4})$.

3.3.10. $\cos(\pi - \sigma) + i \sin(\pi - \sigma)$.

3.3.11. 0.

3.3.12. 1.

3.3.13. $\frac{1 - i\sqrt{3}}{2}, \frac{1 + i\sqrt{3}}{2}, \frac{-1 - i\sqrt{3}}{2}, \frac{-1 + i\sqrt{3}}{2}$.

3.3.14. $-1, -\alpha, -\alpha^2, -\alpha^3, -\alpha^4$ where $\alpha = \cos\frac{2\pi}{5} + i\sin\frac{2\pi}{5}$.

CHAPTER 4

4.1.1. No.

4.1.2. Yes, an action of M on A.

4.1.3. Yes, an action of M on A.

4.1.4. Yes, an action of M on A.

4.1.5. No.

4.1.6. No.

4.1.7. No.

4.1.8. Yes.

4.1.9. Yes.

4.1.10. Yes.

4.1.11. Yes.

4.1.12. Yes.

4.1.14. No.

4.1.17. No.

4.1.18. No.

4.1.19. No.

4.2.2. Yes.

4.2.3. For example, if $A = \{a_1, \ldots, a_n\}$, then $\{a_1\}, \{a_1, a_2\}, \{a_1, a_2, a_3\}, \ldots,$ $\{a_1, \ldots, a_{n-1}\}, \{a_1, \ldots, a_{n-1}, a_n\}$ is a basis of $\mathfrak{B}(A)$.

4.2.4. Yes, $\dim_{\mathbb{R}}(B) = 1$.

4.2.5. Yes, $\dim_{\mathbb{R}}(B) = 21$.

4.2.6. Yes, $\dim_{\mathbb{R}}(B) = 220$.

4.2.9. Yes, $\dim_{\mathbb{R}}(B) = 91$.

4.2.10. Yes, $\dim_{\mathbb{R}}(B) = 820$.

4.2.19. Yes, $\dim_{\mathbb{R}}(B) = 171$.

4.3.10. $\alpha = 0$, and the corresponding rank is 2.

4.4.3. $\dim_{\mathbb{R}}(A/B) = 1$.

4.4.4. $\dim_{\mathbb{R}}(A/B) = 55$.

CHAPTER 5

5.1.1. Yes.

5.1.2. No.

5.1.3. Yes.

5.1.4. Yes. $\mathbf{Im}\, f = \mathbb{R}^3$, $\mathbf{Ker}\, f = \{(0, 0, \gamma, -\gamma) \mid \gamma \in \mathbb{R}\}$.

5.1.5. Yes. $\mathbf{Im}\, f = \{(0, \alpha, \beta) \mid \alpha, \beta \in \mathbb{R}\}$, $\mathbf{Ker}\, f = \{(\alpha, -\alpha, \gamma, \lambda) \mid \alpha, \gamma, \lambda \in \mathbb{R}\}$.

5.1.10. Yes.

5.1.11. Yes.

5.2.1. $\begin{pmatrix} 1 & -1 & 0 \\ 1 & 0 & 1 \end{pmatrix}$.

5.2.2. $\begin{pmatrix} 1 & 0 \\ 0 & 0 \\ 1 & 1 \\ 0 & 0 \end{pmatrix}$.

5.2.5. 16.

5.2.6. 1089.

5.4.6. $\lambda_1 = \lambda_2 = \lambda_3 = -1$. The eigenvectors are $\alpha(1, 1, -1)$, where α is a nonzero real number.

5.4.7. $\lambda_1 = \lambda_2 = \lambda_3 = 2$. The eigenvectors are $\alpha(1, 2, 0) + \beta(0, 0, 1)$, where α and β are real numbers not simultaneously 0.

5.4.8. $\lambda_1 = 1, \lambda_2 = \lambda_3 = 0$. The eigenvectors for λ_1 are $\alpha(1, 1, 1)$ and for 0 are $\beta(1, 2, 3)$, where α and β are real numbers.

5.4.9. $\lambda_1 = \lambda_2 = \lambda_3 = 1$. The eigenvectors for λ_1 are $\alpha(3, 1, 1)$, where α is a nonzero real number.

5.4.10. Let M be the matrix of f relative to some basis. We have $\mathbf{det}(M - XI) = (\lambda_1 - X)\ldots(\lambda_n - X)$. Then $\mathbf{det}(M + XI) = (\lambda_1 + X)\ldots(\lambda_n + X)$, therefore $\mathbf{det}(M^2 - X^2 I) = \mathbf{det}((M - XI)(M + XI)) = \mathbf{det}(M - XI)\mathbf{det}(M + XI) = (\lambda_1 - X)\ldots(\lambda_n - X)(\lambda_1 + X)\ldots(\lambda_n + X) =$

$(\lambda_1^2 - X^2)\dots(\lambda_n^2 - X^2)$. Since X is a variable, we conclude that $\lambda_1^2, \dots, \lambda_n^2$ of the characteristic polynomial of f^2.

5.4.11. Let M (respectively R) be the matrix of f (respectively g) relative to some basis. We have $\mathbf{det}(RM - XI) = \mathbf{det}(M^{-1}(MR - XI)M) = \mathbf{det}(M^{-1})\mathbf{det}(MR - XI)\mathbf{det}(M) = \mathbf{det}(MR - XI)$.

CHAPTER 6

6.1.1. Yes.

6.1.2. Yes.

6.3.7. $y_1^2 - y_2^2 - y_3^2$.

6.3.8. $y_1^2 + y_2^2 - y_3^2$.

6.3.9. $y_1^2 - y_2^2 - y_3^2 - y_4^2$.

6.3.10. $y_1^2 - y_2^2$.

6.3.11. $y_1^2 + y_2^2 - y_3^2$.

6.4.10. $y = (1, -1, -1, 5), z = (3, 0, -2, -1)$.

6.4.11. $y = (3, 1, -1, -2), z = (2, 1, -1, 4)$.

6.4.12. $y = (5, -5, -2, -1), z = (2, 1, 1, 3)$.

CHAPTER 7

7.1.19. For each element $0_R \neq a \in R$, consider the mapping $\iota_a : R \to R$ defined by $\iota_a(x) = ax$, where $x \in R$. Suppose that x, y are elements of R such that $\iota_a(x) = \iota_a(y)$. Then $ax = ay$. Since R has no zero divisors, it follows that $x = y$. Hence the mapping ι_a is injective. In this case, $|R| = |\mathbf{Im}\, \iota_a|$, which implies that $R = \mathbf{Im}\, \iota_a$. Then, every element of R has a preimage relative to ι_a. In particular, there exists an element u such that $e = \iota_a(u) = au$. This means that $u = a^{-1}$. Hence, every nonzero element of R has a multiplicative inverse, so R is a field.

7.5.2. $a = -5, b = -1, c = 6$.

7.5.3. $a = 2, b = 5, c = 7$.

7.5.4. $a = 6, g(X) = \pm(X^2 + 3X + 1)$.

7.5.5. $a = 6, b = 2$.

7.5.6. Yes, if $a \in \mathbb{Z}$.

7.5.9. $a = -1$.

7.5.10. $a = -1, \frac{5}{3}$.

7.5.11. $b = -1 - a^2, c = a$.

7.5.15. $a = 1, b = -1$.

7.6.2. 63.

7.6.3. 728.

7.6.6. No.

7.6.7. Yes.

7.6.8. Yes.

7.6.18. (2,3), (3,2).

7.6.19. $(1, 2, -2), (1, -2, 2), (2, 1, -2), (2, -2, 1), (-2, 2, 1), (-2, 1, 2)$.

CHAPTER 8

8.3.4. Use the group $\mathbf{S}_3$.

8.3.10. Use the group $\mathbb{Z}_2 \times \mathbb{Z}_2$.

8.4.6. $\mathbf{S}_3$.

8.4.9. $\mathbf{S}_3/\mathbf{A}_3 \cong \mathbb{Z}_2$.

8.4.10. Using exercise 8.4.7., if H is subgroup of order 6 then H is normal in G. Also by exercise 8.4.3., we have, for example, $(1\,2\,3)^2 = (1\,3\,2) \in H$. We can show similarly that every cycle of length 3 is in H. However $\mathbf{A}_4$ has 8 cycles of length 3.

8.5.5. (a) Yes, isomorphism; (b) No; (c) No; (d) Yes, $\mathbf{ker}\, f = \{(a, -a)|a \in G\}$; (e) Yes $\mathbf{ker}\, f = N$.

8.5.12. $\mathbf{ker}\, f = \mathbb{Z}$ and $\mathbb{Q}/\mathbb{Z} \cong \{z \in \mathbb{C}|z^n = 1,\ \text{for some } n\}$.

CHAPTER 9

9.1.9. $5 + 9i = (1 + i)(7 + 21)$.

9.1.13. $-2 + 11i = (-2 + i)(3 - 4i)$.

9.1.20. For example, $20 = 2 \times 2 \times 5 = (3 + i\sqrt{11})(3 - i\sqrt{11})$.

9.2.1. Since $\mathbf{N}(-1+i) = 2, \mathbf{N}(2-i) = 5$, we have $\mathbf{GCD}(-1+i, 2-i) = 1, \mathbf{LCM}(-1+i, 2-i) = (-1+i)(2-i)$.

9.2.3. Since $\mathbf{N}(5-\varpi) = 25+5+1 = 31, \mathbf{N}(7+2\varpi) = 49-14+4 = 39$, we have $\mathbf{GCD}(5-\varpi, 7+2\varpi) = 1$ and $\mathbf{LCM}(5-\varpi, 7+2\varpi) = (5-\varpi)(7+2\varpi)$.

9.2.8. Since $\mathbf{GCD}(f(X), g(X)) = X - 2, \mathbf{LCM}(f(X), g(X)) = X^6 + 2X^5 + 4X^4 + 5X^3 + 2X^2 + 4X + 3$.

9.2.11. If $a = b$, then $f(X) = g(X)(X-a)^2 + f'(a)X + (f(a) - f'(a)a)$.

If $a \neq b$, then $f(X) = g(X)(X-a)(x-b) + ((f(a) - f(b))(a-b)^{-1})X + (f(b) - ((f(a) - f(b))(a-b)^{-1})b)$.

9.2.12. $\mathbf{GCD}(f(X), g(X)) = X + 1$.

9.2.14. $\mathbf{GCD}(f(X), g(X)) = X - 3$.

9.2.15. $\mathbf{GCD}(f(X), g(X)) = X^2 - X + 1$.

9.2.16. $\mathbf{GCD}(f(X), g(X)) = X^2 + (i+1)X + i$.

9.2.17. $\mathbf{LCM}(f(X), g(X)) = X^4 - 4X^3 + 4X^2 - 5X - 2$.

9.2.18. $\mathbf{LCM}(f(X), g(X)) = (2X^3 + 7X^2 + 4X - 3)(X - 1)$.

9.2.19. $\mathbf{LCM}(f(X), g(X)) = X^5 + 2iX^4 - 2X^3 - 2iX^2 + X$.

9.2.20. $u(X) = 1, v(X) = -X + 1$.

9.3.1. $3X^3 + 2X^2 + X + 1 = (X-2)(3X^2 + 3X + 2)$.

9.3.2. $X^4 + 2X^3 + 1 = (X-2)(X^3 + X^2 + 2X + 1)$.

9.3.9. $2X^5 - X^4 - 6X^3 + 3X^2 + 4X - 2 = (X-1)(X+1)(X-1)(X2-2)$.

9.3.13. $-1, -2, -3, 4$.

9.3.14. $\frac{1}{2}$.

9.3.15. -3.

9.3.16. -5.

9.3.17. 0.

9.3.18. $f(X) = (X-2)^2(X-1)$.

9.3.19. $f(X) = (X^2 + X + 1)^2(X+2)$.

9.3.20. $f(X) = (X + i)^3(X - 2i)^2$.

9.4.9. 22.

9.4.10. $1, 4, 1, 4$.

9.4.11. 21.

9.4.12. 6.

9.4.13. 2.

9.4.14. 01.

9.4.15. 61.

9.4.16. We have $42 = 2 \cdot 3 \cdot 7$, $a^7 - a = (a-1)a(a+1)(a^2 - a + 1)(a^2 + a + 1)$. It is obvious that the product $(a-1)a(a+1)$ is divisible by 2×3 for each $a \in \mathbb{N}$. If 7 divides a, then 7 divides $a^7 - a$, if $\mathbf{GCD}(a, 7) = 1$, then by Euler theorem $a^6 \equiv 1 \pmod{7}$ and $a^7 \equiv a \pmod{7}$. In other words, 7 divides $a^7 - a$ for each $a \in \mathbb{N}$. It follows that $42 = 2 \times 3 \times 7$ divides $a^7 - a$ for each $a \in \mathbb{N}$.

9.5.4. 5.

9.5.5. 74.

9.5.6. 14.

9.5.7. $x \equiv 18 \pmod{35}$.

9.5.8. $x \equiv 86 \pmod{315}$.

9.5.9. $a \equiv 1 \pmod{6}$.

9.5.10. 5,8,12.

9.5.12. No solution.

9.5.16. 2,9.

9.5.18. 6,7,10,11.

9.5.19. 9,12.

INDEX

List of Symbols